Fundamentals of
GENETICS

The following is a complete list of all HarperCollins Life Sciences texts available for faculty review and student purchase:

ALLIED HEALTH

Bastian
ILLUSTRATED REVIEW OF ANATOMY AND PHYSIOLOGY
  Series (1993)
  BASIC CONCEPTS OF CHEMISTRY, THE CELL, AND TISSUES
  THE MUSCULAR AND SKELETAL SYSTEMS
  THE NERVOUS SYSTEM
  THE ENDOCRINE SYSTEM
  THE CARDIOVASCULAR SYSTEM
  THE LYMPHATIC AND IMMUNE SYSTEMS
  THE RESPIRATORY SYSTEM
  THE DIGESTIVE SYSTEM
  THE URINARY SYSTEM
  THE REPRODUCTIVE SYSTEM

Kreier/Mortensen
PRINCIPLES OF INFECTION, RESISTANCE, AND IMMUNITY
  (1990)

Telford/Bridgman
INTRODUCTION TO FUNCTIONAL HISTOLOGY, Second Edition
  (1994)

Tortora
INTRODUCTION TO THE HUMAN BODY, Third Edition (1994)
PRINCIPLES OF HUMAN ANATOMY, Sixth Edition (1992)

Tortora/Grabowski
PRINCIPLES OF ANATOMY AND PHYSIOLOGY, Seventh Edition
  (1993)

Volk
BASIC MICROBIOLOGY, Seventh Edition (1992)

LIFE SCIENCES

Beck/Liem/Simpson
LIFE: AN INTRODUCTION TO BIOLOGY, Third Edition (1991)

Harris
CONCEPTS IN ZOOLOGY (1992)

Hermann
CELL BIOLOGY (1989)

Hill/Wyse
ANIMAL PHYSIOLOGY, Second Edition (1988)

Jenkins
HUMAN GENETICS, Second Edition (1989)

Kaufman
PLANTS: THEIR BIOLOGY AND IMPORTANCE (1990)

Kleinsmith/Kish
PRINCIPLES OF CELL BIOLOGY (1987)

Margulis/Schwartz/Dolan/Sagan
THE ILLUSTRATED FIVE KINGDOMS (1994)

Mix/Farber/King
BIOLOGY: THE NETWORK OF LIFE (1992)

Nickerson
GENETICS: A GUIDE TO BASIC CONCEPTS AND PROBLEM
  SOLVING (1989)

Nybakken
MARINE BIOLOGY: AN ECOLOGICAL APPROACH, Third Edition
  (1993)

Penchenik
A SHORT GUIDE TO WRITING ABOUT BIOLOGY, Second Edition
  (1993)

Rischer/Easton
FOCUS ON HUMAN BIOLOGY, Second Edition (1994)

Russell
GENETICS, Third Edition (1992)
FUNDAMENTALS OF GENETICS (1994)

Shostak
EMBRYOLOGY: AN INTRODUCTION TO DEVELOPMENTAL
  BIOLOGY (1991)

Wallace
BIOLOGY: THE WORLD OF LIFE, Sixth Edition (1992)

Wallace/Sanders/Ferl
BIOLOGY: THE SCIENCE OF LIFE, Third Edition (1991)

Webber/Thurman
MARINE BIOLOGY, Second Edition (1991)

ECOLOGY AND ENVIRONMENTAL STUDIES

Kaufman/Franz
BIOSPHERE 2000: PROTECTING OUR GLOBAL ENVIRONMENT
  (1993)

Krebs
ECOLOGY: THE EXPERIMENTAL ANALYSIS OF DISTRIBUTION
  AND ABUNDANCE, Fourth Edition (1994)
ECOLOGICAL METHODOLOGY (1988)
THE MESSAGE OF ECOLOGY (1987)

Pianka
EVOLUTIONARY ECOLOGY, Fifth Edition (1994)

Smith
ECOLOGY AND FIELD BIOLOGY, Fourth Edition (1991)
ELEMENTS OF ECOLOGY, Third Edition (1992)

Yodzis
INTRODUCTION TO THEORETICAL ECOLOGY (1989)

LABORATORY MANUALS

Beishir
MICROBIOLOGY IN PRACTICE: A SELF-INSTRUCTIONAL LAB-
  ORATORY COURSE, Fifth Edition (1991)

Donnelly
LABORATORY MANUAL FOR HUMAN ANATOMY: WITH CAT
  DISSECTIONS, Second Edition (1993)

Donnelly/Wistreich
LABORATORY MANUAL FOR ANATOMY AND PHYSIOLOGY:
  WITH CAT DISSECTIONS, Fourth Edition (1993)
LABORATORY MANUAL FOR ANATOMY AND PHYSIOLOGY:
  WITH FETAL PIG DISSECTIONS (1993)

Eroschenko
LABORATORY MANUAL FOR HUMAN ANATOMY WITH CADAV-
  ERS (1990)

Kleinelp
LABORATORY MANUAL FOR THE INTRODUCTION TO THE
  HUMAN BODY(1994)

Tietjen
THE HARPERCOLLINS BIOLOGY LABORATORY MANUAL (1991)
LABORATORY MANUAL TO ACCOMPANY BIOLOGY: THE NET-
  WORK OF LIFE (1992)

Tietjen/Harris
HARPERCOLLINS ZOOLOGY LABORATORY MANUAL (1992)

COLORING BOOKS

Alcamo/Elson
THE MICROBIOLOGY COLORING BOOK, Second Edition (1994)

Diamond/Scheibel/Elson
THE HUMAN BRAIN COLORING BOOK (1985)

Elson
THE ZOOLOGY COLORING BOOK (1982)

Griffin
THE BIOLOGY COLORING BOOK (1987)

Kapit/Elson
THE ANATOMY COLORING BOOK, Second Edition (1993)

Kapit/Macey/Meisami
THE PHYSIOLOGY COLORING BOOK (1987)

Niesen
THE MARINE BIOLOGY COLORING BOOK (1982)

Young
THE BOTANY COLORING BOOK (1982)

Fundamentals of
GENETICS

Peter J. Russell

Reed College

HarperCollinsCollegePublishers

Editor-in-Chief: Glyn Davies
Acquisitions Editor: Susan McLaughlin
Developmental Editor: Karen Trost
Project Editor: Steven Pisano
Design Supervisor: Jill Yutkowitz
Text Design: Claudia Durrell Design
Cover Design: Jill Yutkowitz
Cover Photos/Front Cover: A-DNA Top View ©Irving Geis, Photo Researchers, Inc.
   Back Cover: Cecropia Moth Larvae, Head Detail ©Rod Plank, Photo Researchers, Inc.
Photo Researcher: Lynn Mooney
Production Administrator: Jeffrey Taub
Compositor: Black Dot Graphics
Printer and Binder: R. R. Donnelly & Sons Company
Cover Printer: The Lehigh Press, Inc.

Fundamentals of Genetics

Library of Congress Cataloging-in-Publication Data

Russell, Peter J.
     Fundamentals of genetics / Peter J. Russell, Ben Pierce.
       p.   cm.
     Includes bibliographical references and index.
     ISBN 0-06-500640-2
     1. Genetics.   I. Pierce, Benjamin A.   II. Title.
QH430.R8685   1994
575.1—dc20
                                                         93-17903
                                                            CIP

94 95 96   9 8 7 6 5 4 3 2

Brief Contents

Detailed Contents

Preface

Genetics has long been one of the key or "core" areas in biology, and in the 40 years or so since the first unravelings of the DNA mysteries, we have seen an unparalleled explosion of knowledge and understanding of genetics and its related disciplines. Not only is knowledge of genetics accumulating rapidly, but its many applications are affecting our daily lives and bringing benefit to humanity. In writing *Fundamentals of Genetics,* my goal is to provide today's students with an accessible yet comprehensive treatment of this important field.

Fundamentals of Genetics is a shorter and less complex version of *Genetics,* now in its third edition. *Fundamentals of Genetics* has been carefully crafted to present full coverage of genetics while eliminating some of the details and simplifying some of the more complex explanations found in the larger text. This new text shares with the very successful *Genetics* text an "experimental" approach in which a solid treatment is given of many of the research experiments that contributed to our knowledge of the field. However, *Fundamentals of Genetics* is not simply a "cut-and-paste" version of *Genetics.* Rather, this new, updated text used the more comprehensive text as a starting point for the overall treatment of the subject, and much attention was given to reducing the complexity of the discussions and making the text easier to read, as well as selectively eliminating a few of the more complex topics that we determined, through reviewer feedback, were not essential to this type of text. Thus, *Fundamentals of Genetics* is a text ideally suited for courses whose students have a limited background in biology and chemistry, or for those in which time constraints prohibit the use of a more comprehensive text. The new text is approximately 25 percent shorter than *Genetics,* making it ideal not only for one-semester courses in genetics, but for one-quarter and summer courses as well.

The text retains the overall approach, logical progression of ideas, organization, and pedagogical features (such as "Principal Points," "Keynotes," and "Analytical Approaches for Solving Genetics Problems") that made *Genetics* readily accessible to students. As in *Genetics,* each chapter of *Fundamentals of Genetics* is self-contained so that you can use the chapters in a sequence that best accommodates your own teaching strategies.

ORGANIZATION AND COVERAGE

The three major areas of genetics—transmission genetics, molecular genetics, and population genetics—are covered in 21 chapters. The first six chapters deal with transmission of the genetic material. Chapter 1 reviews the structure of viruses and of prokaryotic and eukaryotic cells and the processes of asexual and sexual reproduction. Mitosis and meiosis are discussed in the context of both animal and plant life cycles. Chapter 2 focuses on Mendel's contributions to our understanding of the principles of heredity, and Chapters 2 and 3 both present the basic principles of genetics in relation to Mendel's laws. Chapter 3 covers experimental evidence for the relationship between genes and chromosomes, methods of sex determination, and Mendelian genetics in human beings. Mendelian genetics in humans is introduced in Chapter 2 with a focus on pedigree analysis and autosomal traits. The topic is continued in Chapter 3 with respect to sex-linked genes.

The exceptions to and extensions of Mendelian analysis (such as the existence of multiple alleles, the modification of dominance relationships, gene interactions and modified Mendelian ratios, essential genes and lethal alleles, and the relationship between genotype and phenotype) are described in Chapter 4. In Chapter 5, we describe how the order of and distance between the genes on eukaryotic chromosomes are determined in genetic experiments designed to quantify the crossovers that occur during meiosis. Specialized topics include the procedures for mapping genes to human chromosomes and tetrad analysis, primarily in fungal systems. In Chapter 6, we discuss the ways of mapping genes in bacteriophages and in bacteria, taking advantage of the processes of transformation, conjugation, and transduction. Fine structure analysis of bacteriophage genes concludes this chapter.

Chapters 7 through 13 comprise the "molecular core" of *Fundamentals of Genetics,* detailing the current level of our knowledge about the molecular aspects of genetics. In Chapter 7, we examine some aspects of gene function, such as the genetic control of the structure and function of proteins and enzymes, and the role of genes in directing and controlling biochemical pathways. A number of examples of human genetic diseases that result from enzyme deficiencies are described to reinforce the concepts. The discussion of gene function in Chapter 7 enables students to understand the important concept that genes specify proteins and enzymes, setting them up for the following chapters in which gene structure and expression are discussed.

In Chapter 8, we cover the structure of DNA, presenting the classical experiments that revealed DNA to be genetic material and that established the double helix model as the structure of DNA. The details of DNA structure and organization in prokaryotic chromosomes are set out in Chapter 9. We cover DNA replication in prokaryotes and eukaryotes in Chapter 10.

After thoroughly explaining the nature of the gene and its relationship to chromosome structure, we discuss in Chapter 11 the first step in the expression of a gene—transcription. First, we describe the general process of transcription and then present the currently understood details of the transcription of messenger RNA, transfer RNA, and ribosomal RNA genes, and the processing of the initial transcripts to the mature RNAs for both prokaryotes and eukaryotes. In Chapter 12, we describe the structure of proteins, the evidence for the nature of the genetic code, and a detailed expression of our current knowledge of translation in both prokaryotes and eukaryotes. In Chapter 13, we discuss recombinant DNA technology and other molecular techniques that are now essential tools of modern genetics. There are descriptions of using recombinant DNA technology to clone and characterize genes and to manipulate DNA, followed by a discussion of several of the many applications of recombinant DNA technology.

The next two chapters focus on regulation of gene expression in prokaryotes (Chapter 14) and eukaryotes (Chapter 15). In Chapter 14, we discuss the operon as a unit of gene regulation, the current molecular details in the regulation of gene expression in bacterial operons, and regulation of genes in bacteriophages. In Chapter 15, we explain how eukaryotic gene expression is regulated, stressing molecular changes that accompany gene regulation, short-term gene regulation in simple and complex eukaryotes, gene regulation in development and differentiation, and immunogenetics.

In the next three chapters, we describe the ways in which genetic material can change or be changed. Chapter 16 covers the processes of gene mutation, then the procedures that screen for potential mutagens and carcinogens (the Ames test) and that select for particular classes of mutants from a heterogeneous population. The various mechanisms cells use to repair damage to the DNA are also discussed. Chromosome aberrations—that

is, changes in normal chromosome structure or chromosome number—are described in Chapter 17. Chapter 18 presents the structures and movements of transposable genetic elements in prokaryotes and eukaryotes, as well as discussions of retroviruses, and oncogenes and their relationship to cancer.

In Chapter 19, we address the organization and genetics of extranuclear genomes of mitochondria and chloroplasts. We cover the classical genetic experiments which established that a gene is extranuclear. In this chapter we also discuss current molecular information about the organization of genes within the extranuclear genomes.

In Chapters 20 and 21, we describe quantitative genetics and the genetics of populations, respectively. In Chapter 20, "Quantitative Genetics," we develop the concept that some heritable traits—the quantitative traits—show continuous variation over a range of phenotypes. In this chapter we also discuss heritability: the relative extent to which a characteristic is determined by genes or by the environment. In Chapter 21, "Genetics of Populations," we present the basic principles in population genetics, extending our treatment of the gene from the individual organism to a population of organisms. This chapter includes an introduction to the developing area of conservation genetics, a topic new to the brief edition. Both Chapters 20 and 21 include discussions of the application of molecular tools to these areas of genetics. Chapter 21, for example, includes a section on measuring genetic variation with RFLPs and DNA sequencing and a discussion of molecular evolution.

PEDAGOGICAL FEATURES

Because the study of genetics is complex, making it potentially difficult, I have incorporated a number of special pedagogical features to enhance students' understanding and appreciation of genetic principles.

- With the exception of the introductory Chapter 1, all chapters have a section on "Analytical Approaches for Solving Genetics Problems." Genetics principles have always been taught best with a problem-solving approach. However, beginning students often do not acquire the necessary experience with basic concepts that would enable them to attack assigned problems methodically. In the "Analytical Approaches" sections (pioneered in *Genetics*), typical genetic problems are "talked through" in step-by-step detail to help students understand how to tackle a genetics problem by applying fundamental principles.
- Each chapter opens with an outline of its contents

and a section called "Principal Points." Principal Points are short summaries that alert students to the key concepts they will encounter in the material to come.

- In the problem sets that close the chapters there are approximately 400 questions and problems designed to give students further practice in solving genetics problems. The problems for each chapter represent a range of topics and difficulty. The answers to questions indicated by an asterisk (*) can be found at the back of the book, and answers to all questions are available in the *Problem-Solving Guide and Solutions Manual*.

- Throughout each chapter, strategically placed "Keynote" summaries emphasize important ideas and critical points.

- Chapter summaries close each chapter, further reinforcing the major points that have been discussed.

- Important terms and concepts—highlighted in boldface—are clearly defined where they are introduced in the text. For easy reference, they are also compiled in a Glossary at the back of the book. The Glossary includes the page numbers on which the term and concepts are introduced so students can look up the definitions and easily find the text location to read more about them. Special care has also been taken to provide the most useful Index—extensive, accurate, and well cross-referenced.

- Comprehensive and up-to-date suggested readings for each chapter are listed at the back of the book.

- Some chapters include boxes covering special topics related to chapter coverage, such as "What Organisms Are Suitable for Use in Genetic Experiments" (Chapter 2) and "Radioactive Labeling of DNA" (Chapter 13).

SUPPLEMENTS

A *Problem-Solving Guide and Solutions Manual* to accompany this text has been prepared by Gail Patt of Boston University and Peter Russell. In addition to detailed solutions for all the problems in the text, each chapter of the Guide contains the following features: a review of important terms and concepts a section called "Analytical Approaches for Solving Genetics Problems," which provides guidance and tips on solving problems and avoiding common pitfalls; and additional questions for practice and review.

A set of 125 full-color transparencies with increased label size for easier reading is available to adopters of the text.

ACKNOWLEDGMENTS

I would like to extend special thanks to Ben Pierce (Baylor University) for his rewriting and updating of Chapters 20 and 21, "Quantitative Genetics" and "Population Genetics" for this short text.

Also making an invaluable contribution to this text are John and Bette Woolsey of J/B Woolsey Associates, Inc., who developed and executed the full-color art program for the *Genetics* text that is adapted for this *Fundamentals* text.

I would also like to thank all of the reviewers involved in this edition: Michael A. Abruzzo, California State University, Chico; Wendall E. Allen, East Carolina University; Barry Bean, Lehigh University; John Belote, Syracuse University; John Boyer, Union College; Nathan M. Chu, Barnard College/Columbia University; Bruce Cochrane, University of South Florida; Peter S. Dawson, Oregon State University; Andrew A. Dewees, Sam Houston State University; Frank A. Einhellig, University of South Dakota; Gerald D. Elseth, Bradley University; DuWayne C. Englert, Southern Illinois University at Carbondale; Darrel S. English, Northern Arizona University; Paul Evans, Brigham Young University; Valerie R. Flechtner, John Carroll University; Chet Fornari, DePauw University; Michael A. Gates, Cleveland State University; Peter Gergen, State University of New York at Stony Brook; Ben Golden, Kennesaw State College; Michael A. Goldman, San Francisco State University; Elliott S. Goldstein, Arizona State University; Jeffrey C. Hall, Brandeis University; Richard B. Imberski, University of Maryland; Kenneth C. Jones, California State University, Northridge; Gene Kritsky, College of Mount St. Joseph; Michael L. Lockhart, Northeast Missouri State University; Paul F. Lurquin, Washington State University; Denson Kelly McLain, Georgia Southern University; Clint Magill, Texas A&M University; S.D. Michael, Binghamton University; William H. Nelson, Morgan State University; Harry Nickla, Creighton University; Chris Osgood, Old Dominion University; Thaddeus Osmolski, University of Massachusetts, Lowell; Ronald S. Ostrowski, University of North Carolina at Charlotte; Jack Parker, Southern Illinois University; Gail R. Patt, Boston University; Frederick J. Peabody, University of South Dakota; Dorene L. Petrosky, Delaware State College; Charles Rodell, St. John's University; Peter A. Rosenbaum, State University of New York—Oswego; Mark F. Sanders, University of California, Davis; Robert R. Schalles, Kansas State University; Ralph Seelke, University of Wisconsin, Superior; Joan Smith-Sonneborn, University of Wyoming; Martin L. Tracey, Jr., Florida International University; Monte E. Turner, University of Akron; Baldev K. Vig, University of Nevada; Dwight E. Wilson, Rensselaer Polytechnic Institute; Jong S. Yoon, Bowling Green State University; Robert M. Zarcaro, Providence College.

I am grateful to the Literary Executor of the late Sir Ronald A. Fisher, F.R.S., to Dr. Frank Yates, F.R.S., and to Longman Group Ltd. London, for permission to reprint Table IV from their book *Statistical Table for Biological, Agricultural, and Medical Research* (6th ed., 1974).

Finally, I wish to thank those who helped to make *Fundamentals of Genetics* a physical reality: Lynn Mooney, photo researcher; Barbara Littlewood, indexer; Clint Magill, accuracy checker; Jill Yutkowitz, design supervisor; Steve Pisano, project editor; Ellen Tweedy, copy editor; Jeffrey Taub, production administrator.

I especially thank Glyn Davies, editor-in-chief, and Karen Trost, developmental editor, for believing in this project and for their help and support in bringing it to fruition.

Peter J. Russell

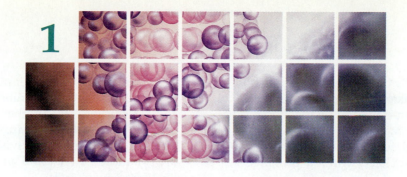

Cell Structure and Cellular Reproduction

- Prokaryotes are the simplest cellular organisms. All bacteria are prokaryotes. Prokaryotes lack a membrane-bound nucleus and divide by binary fission. The genetic material of prokaryotes is found in a single, circular, double-stranded DNA molecule that is associated with few proteins.

- Eukaryotes are organisms that have cells in which the genetic material is located in a membrane-bound nucleus. The genetic material is distributed among several linear chromosomes, each of which consists of a complex of double-stranded DNA and many proteins.

- Diploid eukaryotic cells have two sets of chromosomes, one set from each parent. The members of a pair of chromosomes are called homologous chromosomes. Haploid eukaryotic cells have only one set of chromosomes.

- Mitosis is the process of nuclear division in haploid or diploid eukaryotic cells. It results in the production of daughter nuclei that contain identical chromosome numbers and are genetically identical to one another and to the parent nucleus from which they arose. Typically cell division follows nuclear division so that two cells are produced. All haploid and diploid cells proliferate by mitosis.

- Meiosis occurs in certain specialized cells in all sexually reproducing eukaryotes. It is a process in which a diploid cell or nucleus is transformed through one round of DNA replication and two rounds of nuclear division into four haploid cells or four nuclei.

- Meiosis increases genetic variation through the various ways in which maternal and paternal chromosomes are combined in progeny nuclei and by crossing-over between members of a homologous pair of chromosomes.

WELCOME TO THE FASCINATING and exciting subject of genetics, the science of heredity. Genetics is of central importance to biology because genes are the principal determinants of all life processes from cell structure and function to reproduction of the organism. Learning what genes are, how genes are expressed, and how gene expression is regulated is the focus of this book. The subject of genetics is expanding so rapidly, though, that it is simply not possible to describe everything that is known about genetics here. The important principles are presented, with references at the end of the text for those students who want to go further.

Genetics has shown us that the differences between organisms are the result of differences in the genes they carry, differences which have resulted from the evolutionary processes of **mutation** (a change in the genetic material), **recombination** (exchange of genetic material between chromosomes), and **selection** (the favoring of particular combinations of genes in a given environment). The principles of heredity were recognized by Gregor Mendel in the 1860s, and the development of the subject began about 1900.

The study of genetics has examined three aspects of the genetic material of living organisms: (1) how physical traits are coded for and expressed in the organism; (2) how these traits are inherited (copied and passed on) from one generation to the next; and (3) how changes in the genetic material have led to past and present biological diversity. Presently we have a good understanding of the nature of genetic material, how it is replicated and passed on from generation to generation, how it is expressed in the cell, and how that expression is regulated, but ongoing extensive research in this area causes knowledge to change almost daily. Gene activity is of central importance to cell growth and function and to the development and differentiation of organisms. Thus the study of genetics is key to an understanding of molecular and cell biology, embryology, and developmental biology. And, with respect to the study of biological diversity, a knowledge of genetics is needed to appreciate fully specialties like ecology, evolutionary biology, and morphology.

Beyond the laboratory, genetics has made very important contributions to humankind. In agriculture genetics has significantly improved animals bred for food (such as reducing the amount of fat in beef and pork) and crop plants (such as increasing the amount of protein). In medicine a number of diseases have been shown to be caused by genetic defects, and great strides are being made in understanding the molecular bases of some of those diseases. And, in a health-conscious society such as ours, the public is becoming more educated about genetics as hopes for the eradication of genetic diseases are raised. Presentations in the media offer excellent opportunities for the student and the layperson to learn about genetics and its relevance to present and future society.

Public interest in genetics has also been heightened because of the development and application of particular molecular techniques known collectively as *recombinant DNA technology*. Research in this area has led to the development of industries dealing in what is called *biotechnology*, or *genetic engineering*. While most of this research has been undertaken to study fundamental genetic processes without regard to practical applications, there has been, and will continue to be, much applied work done that is of direct benefit to humankind. In the area of plant breeding genetic engineering is being applied to improve crop yields by developing plants resistant to pests and plant diseases. Genetic engineering is being employed in the beef, dairy cattle, and poultry industries to improve yields and to develop better strains. In medicine the results are equally impressive. For example, recombinant DNA technology is being used in the production of many antibiotics, hormones, and other medically important agents such as clotting factor, in the diagnosis and treatment of a number of human genetic diseases. DNA typing (DNA fingerprinting) is being used in forensics to obtain evidence in paternity cases and for solving crimes. In short, the science of genetics is currently in a dramatic and exciting growth phase. By understanding the basic principles of genetics presented in this text, you will have a greater capacity to comprehend, foster, and perhaps contribute to the new applications of this subject.

In this chapter some fundamental principles of biology relevant to genetics are reviewed. The first part of the chapter presents a brief discussion of the key organizational features of prokaryotic and eukaryotic cells. The remainder of the chapter concentrates on cellular reproduction in eukaryotes, with emphasis on mitosis and meiosis.

Cell Structure

Prokaryotes

Prokaryotes (meaning "prenuclear") are the simplest cellular organisms. Included in this group are all the bacteria.

Bacteria are spherical, rod-shaped, or spiral-shaped unicellular microorganisms lacking a defined nucleus and organelles. The most intensely studied bacterium in genetics is *Escherichia coli* (Figure 1.1), a rod-shaped bacterium common in human intestines. Studies of this bacterium have resulted in significant advances in our understanding of the regulation of gene expression and in the development of the whole field of molecular biology. Today *E. coli* is used extensively in recombinant DNA experiments.

Bacteria vary in size from about 100 nm (1 nm = 10^{-9} m) in diameter for a small spherical bacterium, to 10 μm (1 μm = 10^{-6} m) in diameter and 60 μm long for a large rod-shaped bacterium. The genetic material of bacteria is found in a single, circular, double-stranded DNA molecule that is associated with few proteins. In bacteria this genetic material is not separated from the rest of the cytoplasm by a nuclear membrane such as is found in eukaryotic ("true nuclear") organisms: This is a major distinguishing feature of prokaryotes. However, the genetic material is localized to a central area of the cell called the *nucleoid region*. The shape of the bacterium is maintained by a rigid cell wall located outside the cell membrane.

Bacteria reproduce by **binary fission**; that is, after their genetic material replicates, each cell simply divides in two (Figure 1.2).

Eukaryotes

Eukaryotes (meaning "true nucleus") are organisms that have cells in which the genetic material is located in the **nucleus**, a discrete organized struc-

■ *Figure 1.1* Scanning electron micrograph of *Escherichia coli*, a rod-shaped bacterium common in human intestines.

■ *Figure 1.2*   Binary fission of the bacterium, *E. coli.*

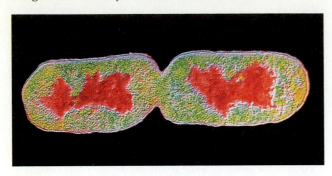

ture within the cell that is bounded by a nuclear membrane. Eukaryotes can be unicellular or multicellular and are often grouped by the terms *lower* and *higher* (lower really means less complex with respect to their DNA, while higher means more complex). Both groups of organisms are highly evolved for their particular niches. In this regard we must consider bacteria also to be highly evolved and highly adapted to their environments. Genetically, bacteria are even simpler than the simplest eukaryotes, having significantly less DNA. So we should consider prokaryotes and eukaryotes to be different, not more or less evolved. The same statement can be made for different eukaryotes.

Lower eukaryotes have relatively simple cellular organization and, while their DNA is significantly more complex than that of prokaryotes, it is less complex than that of the higher eukaryotes. Higher eukaryotes include genetically more complex multicellular organisms.

A number of eukaryotic organisms are commonly used in genetic research (Figure 1.3): for example, *Saccharomyces cerevisiae* (baker's yeast), *Neurospora crassa* (orange bread mold), *Chlamydomonas reinhardi* (a green alga), *Arabidopsis thaliana* (a small plant in the cabbage family), *Zea mays* (corn), *Drosophila melanogaster* (fruit fly), *Caenorhabditis elegans* (nematode), *Mus musculus* (mouse), and *Homo sapiens* (humans). Of these, *Chlamydomonas* and *Saccharomyces* are unicellular organisms, and the rest are multicellular. *Neurospora, Saccharomyces,* and *Chlamydomonas* are simple eukaryotes; the others are more complex eukaryotes.

Features of Eukaryotic Cells of Importance to Genetics   The structure of the eukaryotic cell is very different from that of a bacterial cell. In this section we will discuss some of the features of eukaryotic cells that are important to our study of genetics. Figure 1.4(a) is a diagram of a thin section through a generalized eukaryotic (animal) cell, and Figure 1.4(b) is a diagram of a thin section through a generalized higher plant cell. Surrounding the cyto-

plasm of both animal and plant cells is a membrane called the **plasma membrane**. Plant cells, but not animal cells, have a rigid cell wall outside of the plasma membrane.

Nucleus   The nucleus is separated from the rest of the cell, the cytoplasm and associated organelles, by the double membrane of the nuclear envelope. The membrane is selectively permeable and has pores (*nuclear pores*) about 20–80 nm in diameter. These two characteristics make it possible for materials to move between the nucleus and the cytoplasm.

The nucleus contains most of the genetic material—double-stranded DNA—of the cell. The genetic material is complexed with protein and is organized into a number of linear structures called **chromosomes**. Chromosome means "colored body" and is so named because the threadlike structures are visible under the light microscope only after they are stained with dyes and are visualized during the cell division process. However, in a nondividing cell the chromosomes are in an extended state and are relatively difficult to see even when stained. Also within the nucleus is the *nucleolus*, a structure within which the ribosomes are assembled.

Cytoplasmic Organelles   The **cytoplasm** contains a vast amount of different materials and organelles. Of special interest for geneticists are the *centrioles, endoplasmic reticulum* (ER), *ribosomes, Golgi apparatus, mitochondria,* and *chloroplasts.*

Centrioles (Figure 1.5, on page 7) (also called *basal bodies*) are found in nearly all animal cells, but not usually in plant cells (except those of lower plants). A centriole is a small, cylindrical organelle about 0.2 μm wide and 0.4 μm long. It consists of a ring of nine groups of three fused microtubules (a microtubule is a specialized protein filament). In animal cells a pair of centrioles is at the center of the *centrosome*, a region of undifferentiated cytoplasm which organizes the spindle fibers that function in mitosis and meiosis, processes that will be discussed later (pp. 9–19).

The **endoplasmic reticulum** (ER) is a double-membrane system that runs through the cell. Under the electron microscope we can see two types of ER: rough and smooth. The former has ribosomes attached to it, giving it a rough appearance; the latter does not. Ribosomes bound to the rough endoplasmic reticulum synthesize proteins to be secreted by the cell or to be localized in the cell membrane or particular vacuoles within the cell. These proteins are moved into the space between the two membranes of the ER (lumen) and are then

■ *Figure 1.3*   Eukaryotic organisms commonly used in genetics research: (a) *Saccha-romyces cerevisiae* (a budding yeast); (b) *Neurospora crassa* (orange bread mold); (c) *Chlamydomonas reinhardi* (a green alga); (d) *Zea mays* (corn); (e) *Arabidopsis thaliana* (a small plant of the cabbage family); (f) *Drosophila melanogaster* (fruit fly); (g) *Caenorhabditis elegans* (a nematode); (h) *Mus musculus* (mouse); (i) *Homo sapiens* (humans).

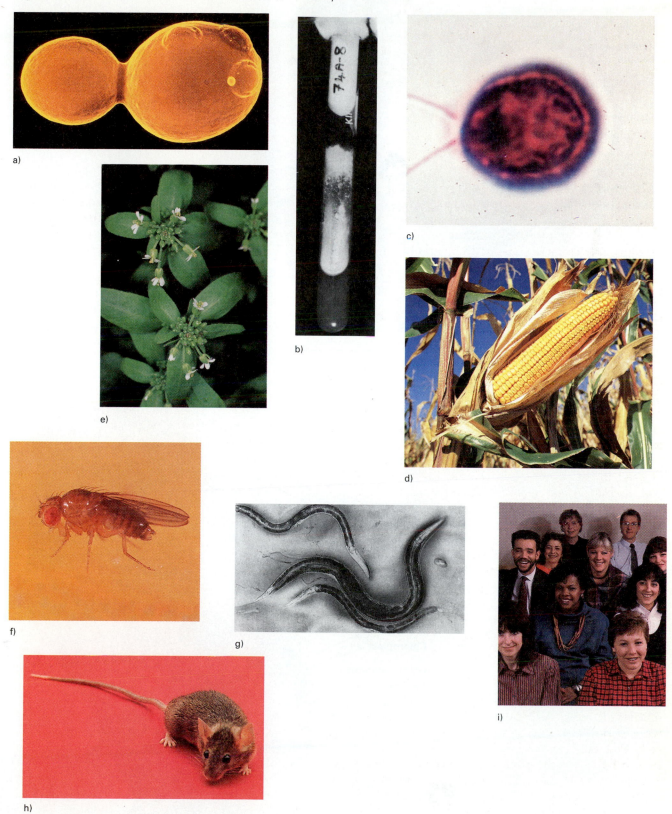

■ *Figure 1.4*   Eukaryotic cell: (a)   Drawing of a section through an animal cell, showing the main organizational features and the principal organelles;   (b) Drawing of a section through a plant cell.

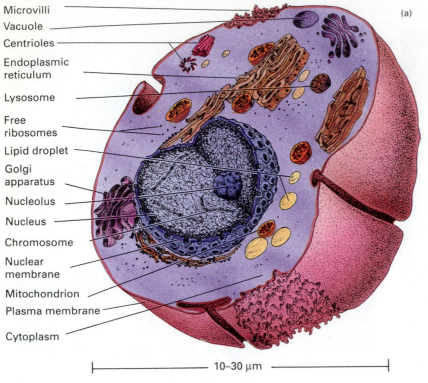

(a)

Microvilli
Vacuole
Centrioles
Endoplasmic reticulum
Lysosome
Free ribosomes
Lipid droplet
Golgi apparatus
Nucleolus
Nucleus
Chromosome
Nuclear membrane
Mitochondrion
Plasma membrane
Cytoplasm

|← 10–30 μm →|

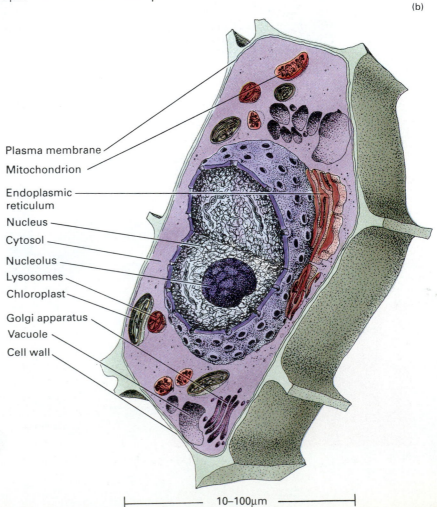

(b)

Plasma membrane
Mitochondrion
Endoplasmic reticulum
Nucleus
Cytosol
Nucleolus
Lysosomes
Chloroplast
Golgi apparatus
Vacuole
Cell wall

|← 10–100μm →|

■ *Figure 1.5*   Schematic of the centriole: (a) End view; (b) Side view.

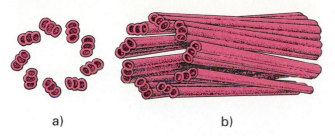

a)   b)

translocated to the **Golgi apparatus**, where they are packaged into vesicles. The vesicles are thought to carry material between the Golgi apparatus and different compartments of the cell. In secretory cells some of the vesicles migrate to the cell surface and, by fusion with the outer membrane, release their contents to the outside of the cell. The synthesis of proteins other than those that are distributed via the ER and Golgi apparatus, such as enzymes and cellular structural proteins, is performed by ribosomes that are free in the cytoplasm.

Eukaryotic cells contain mitochondria. A mitochondrion is illustrated in Figure 1.6. These large organelles are surrounded by a double membrane, the inner one of which is highly convoluted. Mitochondria play an extremely important role in energy processing for the cell. They also contain genetic material in the form of a circular, double-stranded molecule of DNA. As in bacteria, few structural proteins are associated with the genetic material of mitochondria. This DNA encodes some of the proteins that function in the mitochondrion, as well as some of the components of the mitochondrial protein synthesis machinery.

■ *Figure 1.6*   Cutaway diagram of a mitochondrion. Energy processing mechanisms involve interrelationships among the organelle's intermembranal space, the inner membrane, and the matrix.

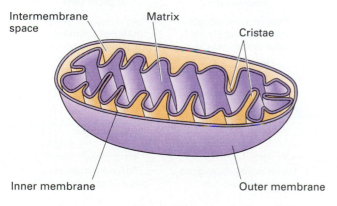

Intermembrane   Matrix
space
                                          Cristae

Inner membrane   Outer membrane

Finally, many plant cells contain chloroplasts, the large, double-membraned, chlorophyll-containing organelles involved in photosynthesis (Figure 1.7). Chloroplasts also contain genetic material and, as in mitochondria, its form is a circular, double-stranded DNA molecule. This DNA encodes some of the proteins that function in the chloroplast, as well as some of the components of the chloroplast protein synthesis machinery.

Cellular Reproduction in Eukaryotes

If all living things are composed of cells, and if all cells arise from other cells, then what processes give rise to accurate copies of a cell? To answer this question we now turn to the subject of cellular reproduction in eukaryotes.

Chromosome Complement of Eukaryotes

The genetic material of eukaryotes is distributed among multiple chromosomes; the number of chromosomes typically is characteristic of the species. For example, the nuclei of human cells have 46 chromosomes, those of the fruit fly *Drosophila melanogaster* have 8, and those of the bread mold *Neurospora crassa* have 7. Many eukaryotes have two copies of each type of chromosome in their nuclei, so their chromosome complement is said to be **diploid**, or 2N. Diploid eukaryotes are produced by the fusion of two **gametes**, one from the female parent and one from the male parent. The fusion produces a diploid **zygote**, which then divides. Each gamete has only one set of chromosomes and is said to be **haploid** (N). The complete haploid complement of genetic information is called the **genome** and the haploid DNA content of an organism is called the **C value**. Of the three organisms just mentioned, humans and fruit flies are diploid (2N) and the bread mold is haploid (N). Thus a human being has 23 *pairs* of chromosomes, *Drosophila melanogaster* has 4 *pairs* and *Neurospora crassa* has 7 *single* chromosomes.

In diploids the members of a chromosome pair are called **homologous chromosomes**; each individual member of a pair is called a **homolog**. Homologous chromosomes are identical with respect to the arrangement of genes they contain and to their visible structure. Chromosomes from different pairs are called **nonhomologous chromosomes**. Figure 1.8 illustrates the chromosomal organization of haploid and diploid organisms.

In animals and some plants there are differences in the chromosome complement of male and female cells. One sex typically has an identical pair of **sex chromosomes**, chromosomes related to the sex of

■ *Figure 1.7*   Diagram of a chloroplast. The organelle's energy-harvesting mechanisms involve interrelationships among the intermembrane space, the stroma, the thylakoids stacked in grana, and the area within each thylakoid.

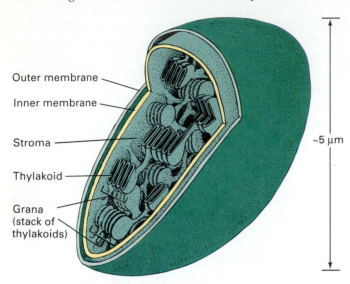

Outer membrane
Inner membrane
Stroma
Thylakoid
Grana (stack of thylakoids)

~5 μm

the organism; the other sex has two, nonidentical sex chromosomes. For instance, human females have two X chromosomes (XX), while human males have one X and one Y (XY). Chromosomes other than sex chromosomes are called **autosomes**.

Keynote

Diploid eukaryotic cells have a double set of chromosomes, one set coming from each parent. The members of a pair of chromosomes, one from each parent, are called homologous chromosomes. Haploid eukaryotic cells have only one set of chromosomes.

Under the microscope we see that chromosomes differ in their size and morphology within and between species. Each chromosome has a specialized region somewhere along its length that often looks under the microscope like a constriction. This constriction, called a **centromere** or **kinetochore**, is important in the activities of the chromosomes during cellular division and can be located in one of four general positions in the chromosome (Figure 1.9). A **metacentric chromosome** has the centromere in approximately the center of the chromosome so that it appears to have two approximately equal arms. **Submetacentric chromosomes** have one arm longer than the other; **acrocentric chromo-**

somes have one arm plus a stalk and satellite; and **telocentric chromosomes** have only one arm, since the centromere is so near the end that the small arm cannot be distinguished. Chromosomes also vary in relative size. Chromosomes of mice, for example, are all similar in length, whereas those of humans show a wide range of relative lengths.

Asexual and Sexual Reproduction

Eukaryotes can reproduce by asexual or sexual reproduction. In **asexual reproduction** a new individual develops from either a single cell or from a group of cells in the absence of any sexual process.

■ *Figure 1.8*   Chromosomal organization of haploid and diploid organisms.

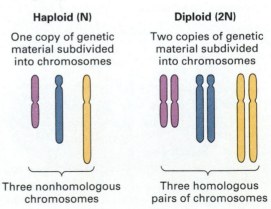

Haploid (N)

One copy of genetic material subdivided into chromosomes

Three nonhomologous chromosomes

Diploid (2N)

Two copies of genetic material subdivided into chromosomes

Three homologous pairs of chromosomes

■ *Figure 1.9*   General classification of eukaryotic chromosomes into metacentric, sub-metacentric, acrocentric, and telocentric types, based on the position of the centromere.

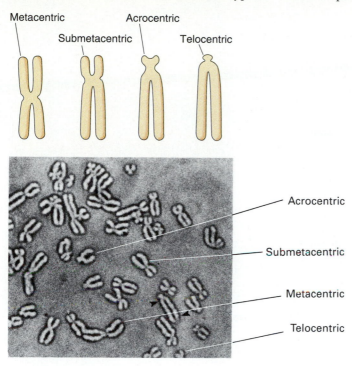

Metacentric   Acrocentric
Submetacentric   Telocentric

Acrocentric

Submetacentric

Metacentric

Telocentric

Asexual reproduction is found in both unicellular and multicellular eukaryotes. Single-celled eukaryotes (such as yeast) grow, double their genetic material, and generate two progeny cells. Each of these cells contains an exact copy of the genetic material found in the parental cell. This process repeats as long as there are sufficient nutrients in the growth medium. In multicellular organisms asexual reproduction is sometimes referred to as **vegetative reproduction**. Multicellular fungi, for example, can be propagated vegetatively by taking a small piece of the tissue and transferring it to a new medium. Many higher plants, such as roses and fruit trees, are commercially maintained by cuttings, which is an asexual means of propagation.

Sexual reproduction is the fusion of two haploid gametes (sex cells) to produce a single diploid zygote cell. From this zygote a new multicellular individual develops by mitotic division under programmed control from genetic material. Sexual reproduction involves the alternation of diploid and haploid phases. A key effect of sexual reproduction is that it achieves genetic recombination; that is, it generates gene combinations in the offspring that are distinct from those in the parents. With the exception of self-fertilizing organisms (such as many plants), the two gametes are from different parents, and genetic recombination takes place dur-

ing the production of the gametes. Figure 1.10 shows the cycle of growth and sexual reproduction in a higher eukaryote; here the haploid gametes are sperm and eggs. It also shows how the number of chromosomes is kept constant from generation to generation.

Sexually reproducing animals have two types of cells: somatic (body) cells and germ (sex) cells. Somatic cells are haploid or diploid, depending on the eukaryote. Lower eukaryotes such as yeast are often haploid, while all higher eukaryotes are diploid. All somatic cells reproduce by a process called mitosis. Germ cells (or gametes), which are always haploid, are only produced from specialized cells by a process called meiosis. Figure 1.11 illustrates the differences between asexual and sexual reproduction.

Mitosis: Nuclear Division

In both unicellular and multicellular eukaryotes, cellular reproduction is a cyclical process of growth: **mitosis** (*nuclear division* or *karyokinesis*) and (usually) **cell division** (**cytokinesis**). This process is called the **cell cycle** (Figure 1.12). The cell cycle consists of mitosis (M) during which the nucleus and (usually) the cell divide, and an interphase between divisions. Interphase consists of three stages: G_1 (gap 1),

■ *Figure 1.10*   Cycle of growth and reproduction in a sexually reproducing higher eukaryote. Sexual reproduction involves the alternation of diploid (somatic) and haploid (gametic) phases.

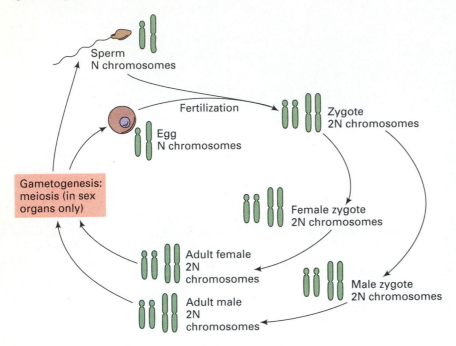

■ *Figure 1.11*   The differences between asexual and sexual reproduction.

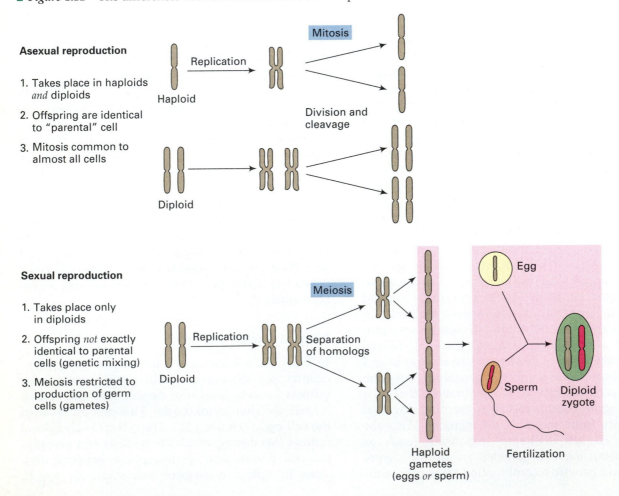

Asexual reproduction

1. Takes place in haploids *and* diploids

2. Offspring are identical to "parental" cell

3. Mitosis common to almost all cells

Sexual reproduction

1. Takes place only in diploids

2. Offspring *not* exactly identical to parental cells (genetic mixing)

3. Meiosis restricted to production of germ cells (gametes)

■ *Figure 1.12*   Eukaryotic cell cycle. This cycle assumes a period of 24 hours in culture, although there is great variation between cell types and organisms.

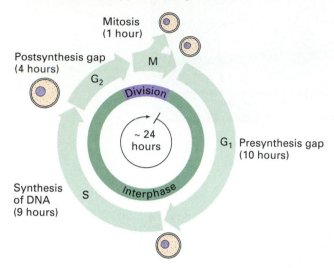

anaphase, and **telophase**. (The same four terms are also used for the similar stages in meiosis.) Figure 1.13 presents photographs showing the typical chromosome morphology in interphase and in the four stages of mitosis in plant (onion root tip) cells. Figure 1.14 shows simplified diagrams of the four stages of mitosis in an animal cell.

Prophase   At the beginning of **prophase** (see Figures 1.13[b] and 1.14) the chromatids are very elongated. In preparation for mitosis they begin to coil tightly so that they appear shorter and fatter under the microscope. By late prophase each chromosome, duplicated during the preceding S phase, now consists of two sister chromatids.

During prophase the mitotic spindle (spindle apparatus) assembles outside the nucleus. Each of the spindle fibers in the mitotic spindle consists of microtubules made of special proteins called *tubulins*. In most animal cells the focal points for spindle assembly are the centrioles (see Figure 1.5). (Higher-plant cells usually do not have centrioles, but they do have a mitotic spindle.) Prior to the S phase the cell's pair of centrioles have replicated and each new centriole pair becomes the focus for a radial array of microtubules called the *aster*. Early in prophase the two asters are adjacent to one another close to the nuclear membrane. By late prophase the two asters are far apart along the outside of the nucleus and are spanned by the microtubular spindle fibers.

Near the end of prophase the nuclear membrane breaks down and the nucleolus or nucleoli disappear. Specialized structures called *kinetochores* form on either face of the centromere of each chromosome and become attached to special microtubules called *kinetochore microtubules* (Figure 1.15). These microtubules radiate in opposite directions from each side of each chromosome and interact with the spindle microtubules.

Metaphase   During **metaphase** (see Figures 1.13[c] and 1.14) the chromosomes become arranged so that their centromeres are aligned in one plane halfway between the two spindle poles and with the long axes of the chromosomes at 90 degrees to the spindle axis. The kinetochore microtubules are responsible for this chromosome alignment event. The plane where the chromosomes become aligned is called the **metaphase plate**. Figure 1.16(a) shows an electron micrograph of a human chromosome at this stage of the cell cycle. Note the highly condensed state of the sister chromatids. The electron micrograph in Figure 1.16(b) shows a human chromosome from which much of the protein has been removed. In the center is a dense framework of pro-

S, and G_2 (gap 2). During G_1 (presynthesis stage) the cell grows and prepares for DNA and chromosome replication, which take place in the S stage. In G_2 (the postsynthesis stage) the cell prepares for the M phase. That is, mitosis follows the precise replication of chromosomes and results in the distribution of a complete chromosome set to each of the two progeny nuclei. In most, but not all, cases mitosis is followed by cytokinesis, the division of the cytoplasm to form two cells. Most of the cell cycle is spent in the G_1 stage, although the relative time spent in each of the four stages varies greatly among cell types.

During interphase of the cell cycle the individual chromosomes are extended and are impossible to see under the light microscope. The DNA of each chromosome replicates, and the DNA of the centromere also replicates, although *only one centromere structure is seen under the microscope*. The product of chromosome duplication is two exact copies, called **sister chromatids**. The sister chromatids are held together by the replicated but unseparated centromere. More precisely, a **chromatid** is one of the two visibly distinct, longitudinal subunits of all replicated chromosomes which become visible between early prophase and metaphase of mitosis (and between prophase I and the second metaphase of meiosis, discussed later). After mitotic anaphase, when the centromeres separate, they become known as **daughter chromosomes**.

Mitosis occurs in both haploid and diploid cells. It is a continuous process, but for purposes of discussion it is usually divided into four cytologically distinguishable stages called **prophase**, **metaphase**,

■ *Figure 1.13*   Stages of mitosis in onion root tip: (a) Interphase; (b) Prophase; (c) Metaphase; (d) Anaphase; (e) Telophase.

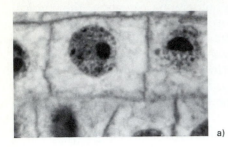
a)

b)

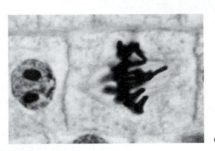
c)

d)

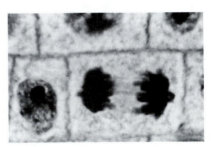
e)

tein called a *scaffold,* which retains the form of the chromosome. The scaffold is surrounded by a halo of DNA filaments that have uncoiled and spread outward.

Anaphase   Anaphase (see Figures 1.13[d] and 1.14) begins when sister chromatids break apart at the centromere, splitting the chromosome. Once the paired kinetochores on each chromosome separate, the sister chromatid pairs undergo **disjunction** (separation), and the daughter chromosomes (as each sister chromatid is now called) move toward the poles. In anaphase the two centromeres of the daughter chromosomes migrate toward the opposite poles of the cell as they are pulled by the contracting spindle fibers. Now the chromosomes assume characteristic shapes related to the location of the centromere along the chromosome's length. For example, a metacentric chromosome will appear as a V, with two roughly equal length chromosome arms trailing the centromere in its migration toward the pole. Similarly a submetacentric chromosome will appear as a J, with a long and a short arm. Cytokinesis (cell division) begins in the latter stages of anaphase.

Telophase   During **telophase** (see Figures 1.13[e] and 1.14) the migration of daughter chromosomes to the two poles is completed. The two sets of progeny chromosomes are assembled into two groups at opposite ends of the cell. The chromosomes begin to uncoil and assume the extended state characteristic of interphase. A nuclear membrane forms around each chromosome group, the spindle microtubules disappear, and the nucleolus or nucleoli reform. At this point nuclear division is complete: the cell has two nuclei. Cell division continues in telophase, giving rise to two cells, each with one nucleus.

Cytokinesis   Cytokinesis refers to division of the cytoplasm. Cytokinesis is completed by the end of telophase. Cytokinesis compartmentalizes the two new nuclei into separate daughter cells, completing the mitotic cell division process. In animal cells cytokinesis occurs by the formation of a constric-

■ *Figure 1.14*   Diagram of mitosis in an animal cell.

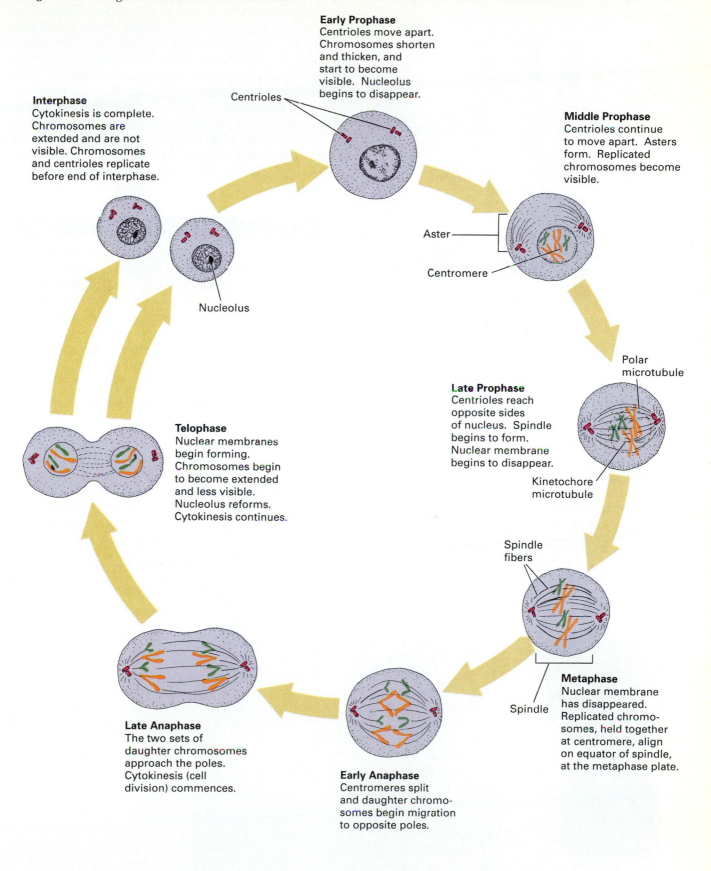

Early Prophase
Centrioles move apart.
Chromosomes shorten
and thicken, and
start to become
visible. Nucleolus
begins to disappear.

Centrioles

Interphase
Cytokinesis is complete.
Chromosomes are
extended and are not
visible. Chromosomes
and centrioles replicate
before end of interphase.

Middle Prophase
Centrioles continue
to move apart. Asters
form. Replicated
chromosomes become
visible.

Aster

Centromere

Nucleolus

Polar
microtubule

Late Prophase
Centrioles reach
opposite sides
of nucleus. Spindle
begins to form.
Nuclear membrane
begins to disappear.

Kinetochore
microtubule

Telophase
Nuclear membranes
begin forming.
Chromosomes begin
to become extended
and less visible.
Nucleolus reforms.
Cytokinesis continues.

Spindle
fibers

Metaphase
Nuclear membrane
has disappeared.
Replicated chromo-
somes, held together
at centromere, align
on equator of spindle,
at the metaphase plate.

Spindle

Late Anaphase
The two sets of
daughter chromosomes
approach the poles.
Cytokinesis (cell
division) commences.

Early Anaphase
Centromeres split
and daughter chromo-
somes begin migration
to opposite poles.

■ *Figure 1.15* Kinetochores and kinetochore microtubules.

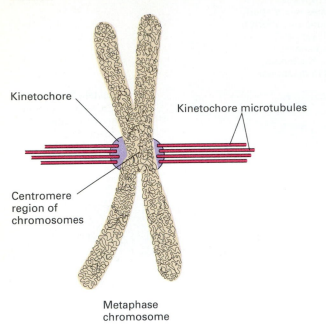

Kinetochore

Kinetochore microtubules

Centromere region of chromosomes

Metaphase chromosome

tion in the middle of the cell until two daughter cells are produced (Figure 1.17[a]). However, most plant cells do not divide by the formation of a constriction. Instead, a new cell membrane and cell wall are assembled between the two new nuclei to form a *cell plate* (Figure 1.17[b]). Cell wall material coats each side of the cell plate, and the result is two progeny cells.

Genetic Significance of Mitosis   Mitosis maintains the identical genetic content of a cell from generation to generation. It occurs in haploid or diploid cells after DNA and chromosome duplication have taken place. Mitosis is a highly ordered process in which a duplicated chromosome set is partitioned equally into the two daughter cells. Thus for a haploid (N) cell, chromosome duplication produces a cell with two sets of chromosomes. Mitosis then results in two progeny haploid cells, each with one set of chromosomes. For a diploid (2N) cell, which has two sets of chromosomes, chromosome duplication produces a cell in which each chromosome set has doubled its content. Mitosis then results in two progeny diploid cells, each with two sets of chromosomes.

Keynote

Mitosis is the process of nuclear division in eukaryotes. It is one part of the cell cycle (i.e., G_1, S, G_2 and M) and results in the production of daughter nuclei that contain identical chromosome numbers and are genetically identical to one another and to the parent nucleus from which they arose. Prior to mitosis the chromosomes duplicate. Mitosis is usually followed by cytokinesis. Both haploid and diploid cells can proliferate by mitosis.

Meiosis

Meiosis is the term applied to the two successive divisions of a diploid nucleus following only one DNA replication cycle. Meiosis occurs only at a special point in the organism's life cycle. Unlike mitosis, meiosis is not responsible for cell proliferation. Meiosis results in the formation of haploid gametes

■ *Figure 1.16*   Human metaphase chromosome: (a) Transmission electron micrograph of an intact chromosome; (b) Transmission electron micrograph of a chromosome from which much of the protein has been removed and the DNA filaments have uncoiled.

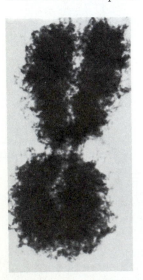

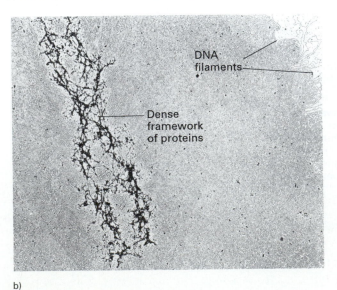

DNA filaments

Dense framework of proteins

a)

b)

■ *Figure 1.17*   Cytokinesis (cell division): (a) Cytokinesis in an animal cell; (b) Cytokinesis in a higher plant cell.

a) Animal cell   **b) Plant cell**

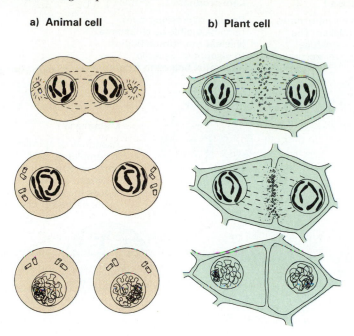

(**gametogenesis**) or of haploid meiospores (in *sporogenesis*). During meiosis homologous chromosomes replicate, then pair, then undergo two divisions. Consequently, each of the four cells resulting from the two meiotic divisions receives one chromosome of each chromosome set. The two nuclear divisions of a normal meiosis are called meiosis I and meiosis II. Diagrams of these divisions are presented in Figure 1.18. The first meiotic division results in reduction in the number of chromosomes from diploid to haploid (reductional division), and the second division results in separation of the chromatids (equational division). In most cases the divisions are accompanied by cytokinesis, so the result of the meiosis of a single diploid cell is four haploid cells.

Meiosis I: The First Meiotic Division   Meiosis I, in which the chromosome number is reduced from diploid to haploid, consists of four cytologically distinguishable stages: *prophase I, metaphase I, anaphase I,* and *telophase I.*

Prophase I.   As prophase I begins the chromosomes have already replicated. Then the chromosomes become shorter and thicker, crossing-over occurs, the spindle apparatus forms, and the nuclear membrane and nucleolus/nucleoli disappear (Figure 1.18). Except for the behavior of homologous pairs of chromosomes and crossing-over, prophase I of meiosis is very similar to the mitotic prophase.

In early prophase I the chromosomes have begun to coil and are now visible. A key event in this stage is pairing, that is, the alignment of homologous chromosomes. The threadlike chromosomes seen at this time are the sister chromatids still held together at the centromere. Once pairing has been completed, a most significant event begins: **crossing-over**, or the reciprocal exchange of chromosome segments at corresponding positions along pairs of homologous chromosomes. A chromosome that emerges from meiosis with a combination of genes that differs from the combination with which it started is called a **recombinant chromosome**. Therefore crossing-over is a mechanism that can give rise to **genetic recombination**, a concept we will examine more fully in later chapters (Chapters 5 and 6). If there are genetic differences between the homologs, crossing-over can produce new gene combinations in a chromatid. There is no loss or addition of genetic material to either chromosome, since crossing-over involves reciprocal exchanges.

In middle prophase I a key event is **synapsis**. Synapsis is the formation of an intimate association of homologous chromosomes brought about by the formation of a zipperlike structure along the length of the chromatids called a **synaptinemal complex** (Figure 1.19). The chromosomes are maximally condensed prior to the start of synaptinemal complex. Because of the replication that occurred earlier, each synapsed set of homologous chromosomes consists of four chromatids and is referred to as a **bivalent** or **tetrad**.

Next, in late prophase I, the chromosomes begin to move apart. The process of crossing-over becomes visible during this stage as a cross-shaped

■ *Figure 1.18*   Diagrams of the stages of meiosis in an animal cell.

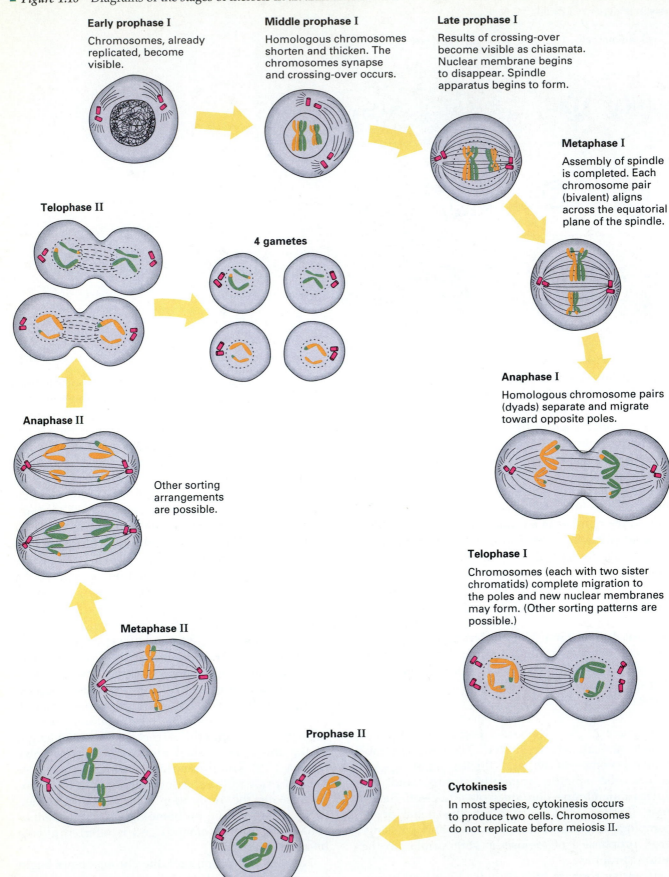

Early prophase I

Chromosomes, already replicated, become visible.

Middle prophase I

Homologous chromosomes shorten and thicken. The chromosomes synapse and crossing-over occurs.

Late prophase I

Results of crossing-over become visible as chiasmata. Nuclear membrane begins to disappear. Spindle apparatus begins to form.

Metaphase I

Assembly of spindle is completed. Each chromosome pair (bivalent) aligns across the equatorial plane of the spindle.

Telophase II

4 gametes

Anaphase II

Other sorting arrangements are possible.

Anaphase I

Homologous chromosome pairs (dyads) separate and migrate toward opposite poles.

Metaphase II

Prophase II

Telophase I

Chromosomes (each with two sister chromatids) complete migration to the poles and new nuclear membranes may form. (Other sorting patterns are possible.)

Cytokinesis

In most species, cytokinesis occurs to produce two cells. Chromosomes do not replicate before meiosis II.

■ *Figure 1.19* Electron micrograph of the synaptinemal complex separating two homologous chromatid pairs during the middle stage of prophase I.

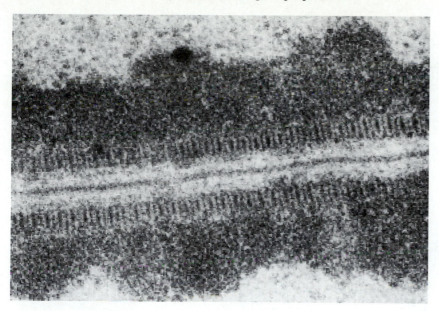

structure called a **chiasma** (plural: chiasmata) as shown in Figure 1.20(a). Since all four chromatids may be involved in crossing-over events along the length of the homologs, the chiasma pattern at this stage may be very complex. As prophase I is completed, the four chromatids of each bivalent (tetrad) become even more condensed, and the chiasmata often *terminalize*; that is, they move down the chromatids to the ends (Figure 1.20[b]). At the completion of prophase I the nucleoli disappear, and the nuclear membrane begins to break down. The chromosomes can be counted most easily at this stage of meiosis.

In most organisms prophase I is followed rapidly by the remaining stages of meiosis. However, in many animals the oocytes (egg cells) can remain in late prophase I for very long periods. For example, in the human female, germ cells go through meiosis

up to late prophase I by the seventh month of fetal development and then remain arrested in this stage—the primary oocyte—for many years. From the onset of puberty until menopause, one is released from the ovary each menstrual cycle, maturing into an ovum upon sperm penetration.

Metaphase I By the beginning of metaphase I (see Figure 1.19), the nuclear membrane has completely broken down, and the bivalents become aligned on the equatorial plane of the cell. The spindle apparatus is completely formed now, and the microtubules are attached to the centromeres of the homologs. Note particularly that it is the synapsed parts of homologs (the bivalents) that are found at the metaphase plate. In contrast, in mitosis of most organisms, replicated homologous chromosomes (sister chromatid pairs) align independently

■ *Figure 1.20* (a) Appearance of chiasmata, the visible evidence of crossing-over, in late prophase I;   (b) Terminalization of chiasmata as prophase I is completed.

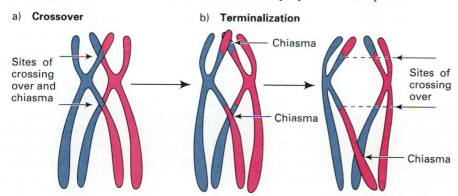

■ *Figure 1.21*   Comparison of mitosis and meiosis in a diploid cell.

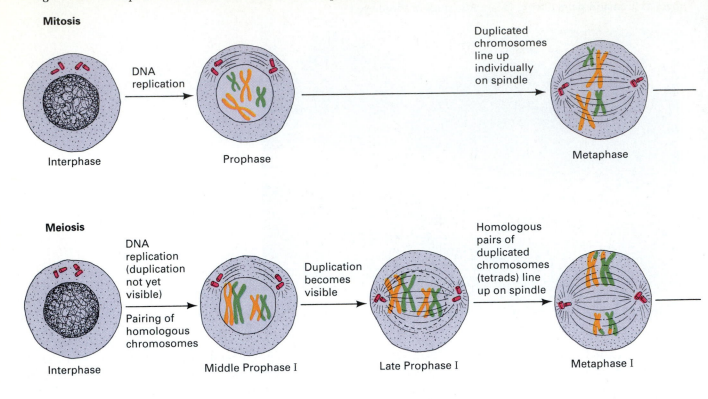

Mitosis

Interphase → DNA replication → Prophase → Duplicated chromosomes line up individually on spindle → Metaphase

Meiosis

Interphase → DNA replication (duplication not yet visible) / Pairing of homologous chromosomes → Middle Prophase I → Duplication becomes visible → Late Prophase I → Homologous pairs of duplicated chromosomes (tetrads) line up on spindle → Metaphase I

at the metaphase plate. *A key difference between mitosis and meiosis is that sister centromeres remain together in meiosis I, whereas in mitosis they are separate.*

Anaphase I   In anaphase I (see Figure 1.18) the chromosomes in each tetrad separate, so homologous pairs (the bivalents) disjoin and migrate toward opposite poles, the areas in which new nuclei will form. (At this stage each pair is called a *dyad*.) This migration occurs by contraction of the spindle fibers attached to the centromeres. The migration process gives rise to two things: (1) Maternally derived and paternally derived homologs are segregated randomly at each pole (except for the parts of chromosomes exchanged during the crossing-over process); and (2) There is a set of replicated chromosomes at each pole. The number of these chromosomes is equivalent to the haploid number of chromosomes for the organism. It is crucial to remember that *at this time the segregated sister chromatid pairs remain attached at their respective centromeres.*

Telophase I   In telophase I (see Figure 1.18) the dyads complete their migration to opposite poles of the cell, and new nuclear membranes form around each haploid grouping. In most species, cytokinesis follows, producing two haploid cells. Thus meiosis I,

which begins with a diploid cell that contains one maternally derived and one paternally derived set of replicated chromosomes, ends with two nuclei, each of which contains one mixed-parental set of replicated chromosomes. Each pair of replicated chromosomes is still joined at the centromere. After cytokinesis each of the two progeny cells has a nucleus with a haploid set of replicated chromosomes.

Meiosis II: The Second Meiotic Division   The second meiotic division is also quite similar to a mitotic division (see Figure 1.18). In **prophase II** chromosome contraction occurs and the already-replicated centromeres now divide. In **metaphase II** each of the two daughter cells organizes a spindle apparatus that attaches to the now-divided centromeres. The centromeres line up on the equator of the second-division spindles. During **anaphase II** the chromatids are pulled to the opposite poles of the spindle: One sister chromatid of each pair goes to one pole, while the other goes to the opposite pole. The separated chromatids are now referred to as chromosomes in their own right. In the last stage, **telophase II**, a nuclear membrane forms around each set of chromosomes, and cytokinesis takes place. After telophase II the chromosomes become more extended and again are invisible under the light microscope.

■ *Figure 1.21* Cont.

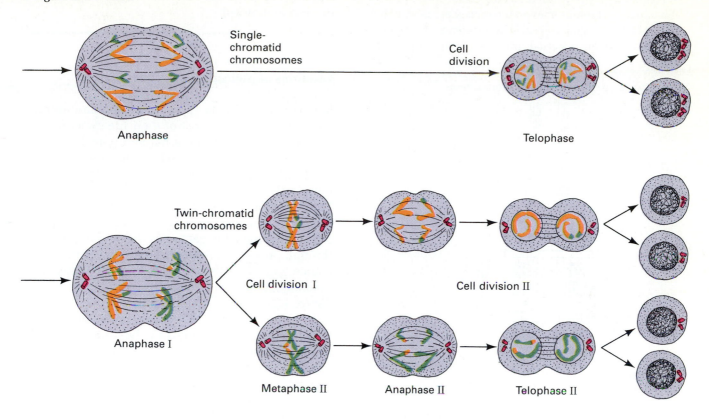

Since there is no chromosome duplication between meiosis I and meiosis II, the end products of the two meiotic divisions are four haploid cells from one original diploid cell. Each of the four progeny cells has one chromosome from each homologous pair of chromosomes. However, these chromosomes are not exact copies of the original chromosomes because of the crossing-over that occurs between chromosomes during prophase of meiosis I. Figure 1.21 compares mitosis and meiosis.

Genetic Significance of Meiosis Meiosis has three significant results.

1. Meiosis generates cells with half the number of chromosomes found in the diploid cell that entered the process because two division cycles follow only one cycle of DNA replication (S period). Fusion of the haploid nuclei (called fertilization or syngamy) restores the diploid number. Therefore, through a cycle of meiosis and fertilization, the chromosome number is maintained in sexually reproducing organisms.

2. In metaphase I of meiosis each maternally derived and paternally derived chromosome has an equal chance of aligning on one or the other side of the equatorial metaphase plate. Thus, each nucleus generated by meiosis will have a combination of maternal and paternal chromosomes.

The number of possible chromosome combinations is large, especially when the number of chromosomes in an organism is large. Consider a hypothetical organism with two pairs of chromosomes in a diploid cell entering meiosis. Figure 1.22 shows the two possible combinations of maternal and paternal chromosomes that can occur at the metaphase plate.

The general formula states that the number of possible chromosome arrangements is 2^{n-1}, where n is the number of chromosome pairs. In *Drosophila*, which has four pairs of chromosomes, the number of possible arrangements is 2^3, or 8; in humans, which have 23 chromosome pairs, over 4 million metaphase arrangements are possible. Therefore, since there are many gene differences between the maternally derived and paternally derived chromosomes, the nuclei produced by meiosis will be genetically quite different from the parental cell and from each other.

3. The crossing-over between maternal and paternal chromatid pairs during meiosis I results in

■ *Figure 1.22* Two possible arrangements of two pairs of homologous chromosomes on the metaphase plate of the first meiotic division. Paternal chromosomes are shown in yellow and green; maternal chromosomes in purple and tan.

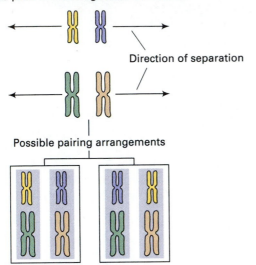

Two pairs of homologous chromosomes

Direction of separation

Possible pairing arrangements

still more variation in the final combinations. Crossing-over occurs during every meiosis, and because sites of crossing-over vary from one meiosis to another, the number of different kinds of progeny nuclei produced by the process is extremely large. Given the genetic significance of meiosis, this process is critical to understand the behavior of genes, as will be seen in the following chapters.

It is important to note that the events that occur in meiosis are the bases for the segregation of genes according to Mendel's Laws, to be discussed in Chapter 2.

Keynote

Meiosis occurs in all sexually reproducing eukaryotes. It is a process by which a specialized diploid (2N) cell or cell nucleus is transformed, through one round of chromosome replication and two rounds of nuclear division, into four haploid (N) cells or nuclei. In the first of two divisions, pairing (synapsis) of homologous chromosomes occurs. The meiotic process results in the conservation of the number of chromosomes from generation to generation. It also increases genetic variation through the various ways in which maternal and paternal chromosomes are combined in the progeny nuclei and by crossing-over (the physical

exchange of genes between maternally or paternally derived homologs).

Locations of Meiosis in the Life Cycle

Meiosis in Animals   Most multicellular animals are diploid through most of their life cycle. In such animals meiosis produces haploid gametes, which produce a diploid zygote when their nuclei fuse in the fertilization process. The zygote then divides mitotically to produce the new diploid organism. Thus the gametes are the only haploid stages of the life cycle. Gametes are only formed in specialized cells. In the male the gamete is the sperm, produced through a process called **spermatogenesis**. The female gamete is the egg, produced by **oogenesis**. Spermatogenesis and oogenesis are illustrated in Figure 1.23.

In male animals the **sperm cells** (also called **spermatozoa**) are produced by the testes. The testes contain the primordial germ cells (*primary spermatogonia*), which produce *secondary spermatogonia* through mitotic division. Spermatogonia transform into *primary spermatocytes* (*meiocytes*), each of which undergoes meiosis I and gives rise to two *secondary spermatocytes*. Each spermatocyte undergoes meiosis II. As a result of these two divisions, four haploid *spermatids* emerge, and eventually differentiate into the mature male gametes, the spermatozoa.

In female animals the ovary contains the primordial germ cells (*primary oogonia*), which, by mitosis, give rise to *secondary oogonia*. These cells transform into *oocytes*, which grow until the end of oogenesis. The diploid, primary oocyte goes through meiosis I and unequal cytokinesis to form two cells: a large cell called the **secondary oocyte**, and a very small one called the *first polar body*. In the second meiotic division the secondary oocyte produces two haploid cells. One is a very small cell called the *second polar body*; the other is a large cell called an *ootid* that rapidly matures into the mature egg cell, or **ovum**. The first polar body may or may not divide. Only the ovum is a viable gamete. Thus in the female animal only one mature gamete (the ovum) is produced by meiosis of a diploid cell.

Meiosis in Plants and Fungi   The role of meiosis in the life cycles of many plants and fungi is different from that in animals. The predominant vegetative cells in the life cycles of many single-celled plants and fungi are haploid. These cells are produced in the life cycle by meiosis and proliferate subsequently by mitosis. [For example, the life cycle

■ *Figure 1.23*   Spermatogenesis and oogenesis in an animal cell.

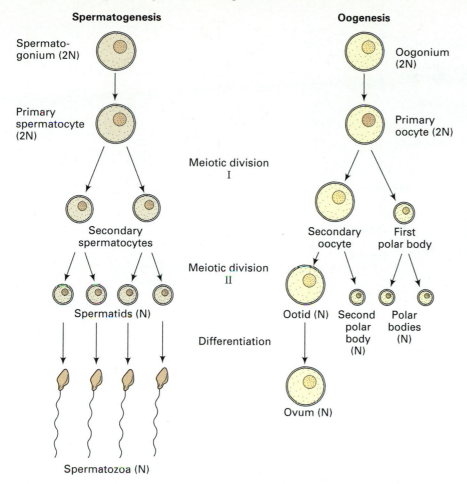

of the fungus *Neurospora crassa* is given in Chapter 5 (see Figure 5.17).]

Higher plants have two distinct reproductive phases called the **alternation of generations**. The diploid **sporophyte** phase produces haploid spores by meiosis. The spores develop into the haploid **gametophytes** which produce gametes capable of fertilization. Fusion of the gametes—fertilization—produces zygotes from which sporophytes develop, and so the cycle continues. Either the sporophyte or gametophyte phase may predominate in the life cycle, depending upon the plant species. Figure 1.24 shows the alternation of generations in a flowering plant.

Summary

In this chapter we reviewed the basic organizational features of cells that are relevant to genetics, and discussed cellular reproduction in eukaryotes, with particular emphasis on mitosis and meiosis.

As the simplest cellular organisms, prokaryotes lack a membrane-bound nucleus and divide by binary fission. The genetic material of prokaryotes is localized within the cell in an area called the nucleoid region. Prokaryotic genetic material consists of a single, circular, double-stranded DNA molecule that is associated with few proteins. Eukaryotic organisms may be single-celled or multicellular and have a membrane-bound nucleus. The genetic material of eukaryotes consists of double-stranded DNA distributed among several linear chromosomes. In these chromosomes the DNA is complexed with many proteins. Typically, eukaryotic cells contain either one set of chromosomes— haploid cells—or two sets of chromosomes— diploid cells. Proliferation of both haploid and diploid cells occurs by mitosis. Mitosis involves one round of DNA replication followed by one round of nuclear division, often accompanied by cell division. Thus mitosis results in the production of daughter nuclei that contain identical chromosome numbers and that are genetically identical to one another and to the parent nucleus from which they arose.

■ *Figure 1.24*   Alternation of gametophyte and sporophyte generations in flowering plants.

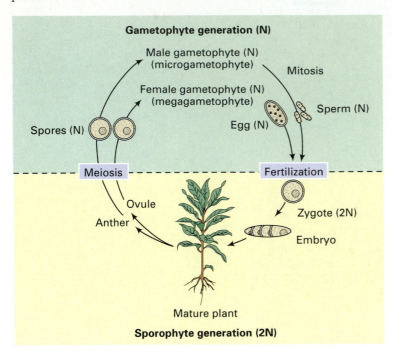

In all sexually reproducing organisms, meiosis occurs at a particular stage in the life cycle. Meiosis is the process by which a diploid cell (never a haploid cell) or cell nucleus undergoes one round of DNA replication and two rounds of nuclear division to produce four specialized haploid cells or nuclei. The products of meiosis are gametes or meiospores. Unlike mitosis, meiosis generates genetic variability two ways: (1) through the various means in which maternal and paternal chromosomes are combined in progeny nuclei; and (2) through crossing-over between maternally derived and paternally derived homologs to produce recombinant chromosomes with some maternal and some paternal genes.

Questions and Problems

*NOTE: Questions and Problems marked with an asterisk have solutions on pages 000–000.

For questions 1.1–1.3, select the correct answer.

*1.1   Interphase is a period corresponding to the cell cycle phases of
a. mitosis.
b. S.
c. $G_1 + S + G_2$.
d. $G_1 + S + G_2 + M$.

1.2   Chromatids joined together by a centromere are called
a. sister chromatids.
b. homologs.

c. alleles.
d. bivalents (tetrads).

*1.3   Mitosis and meiosis always differ in regard to the presence of
a. chromatids.
b. homologs.
c. bivalents.
d. centromeres.
e. spindles.

1.4   State whether each of the following statements is true or false. Explain your choice.
a. The chromosomes in a somatic cell of any organism are all morphologically alike.

b. During mitosis the chromosomes divide and the resulting sister chromatids separate at anaphase, ending up in two nuclei, each of which has the same number of chromosomes as the parental cell.

c. In prophase I any chromosome may synapse with any other chromosome in the same cell.

*1.5 Decide whether the answer to these statements is *yes* or *no*. Then explain the reasons for your decision.

a. Can meiosis occur in haploid species?

b. Can meiosis occur in a haploid individual?

1.6 The general life cycle of a eukaryotic organism has the sequence

a. 1N → meiosis → 2N → fertilization → 1N.

b. 2N → meiosis → 1N → fertilization → 2N.

c. 1N → mitosis → 2N → fertilization → 1N.

d. 2N → mitosis → 1N → fertilization → 2N.

*1.7 Give the names of the stages of mitosis or meiosis at which the following events occur:

a. Chromosomes are located in a plane at the center of the spindle.

b. The chromosomes move away from the spindle equator to the poles.

1.8 Given the diploid, meiotic mother cell in the following figure, diagram the chromosomes as they would appear

a. in middle prophase I;

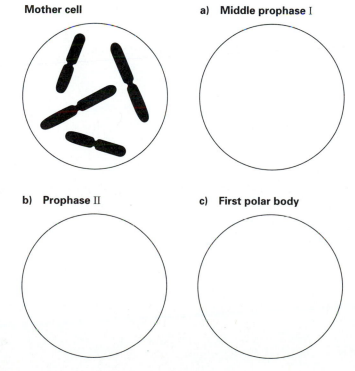

b. in a nucleus at prophase of the second meiotic division;

c. in the first polar body resulting from oogenesis in an animal.

1.9 The cells in the following figure were all taken from the same individual (a mammal). Identify the cell division events in each cell, and explain your reasoning. What is the sex of the individual? What is the diploid chromosome number?

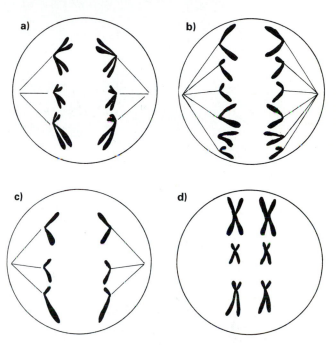

1.10 Does mitosis or meiosis have greater significance in genetics (i.e., the study of heredity)? Explain your choice.

*1.11 Consider a diploid organism that has three pairs of chromosomes. The organism receives chromosomes A, B, and C from the female parent and A′, B′, and C′ from the male parent. To answer the following questions, assume that no crossing-over occurs.

a. What proportion of the gametes of this organism would be expected to contain all the chromosomes of maternal origin?

b. What proportion of the gametes would be expected to contain some chromosomes from both maternal and paternal origins?

*1.12 Normal diploid cells of a theoretical mammal are examined cytologically at the mitotic metaphase stage for their chromosome complement. One short chromosome, two medium-length chromosomes, and three long chromosomes are present. Explain how the cells might have such a set of chromosomes.

*1.13 Is the following statement true or false? Explain your decision. "All of the sperm from one human male are genetically identical."

1.14 The horse has a diploid set of 64 chromosomes, and a donkey has a diploid set of 62 chromosomes. Mules (viable but usually sterile) are produced when a male donkey is mated to a female horse. How many chromosomes will a mule cell contain?

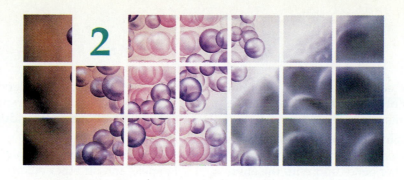

2

Mendelian Genetics

- The genotype is the genetic makeup of an organism, while the phenotype is the physical manifestation of genetic traits.

- The genes give the potential for the development of characteristics. This potential is affected by interactions with other genes and with the environment.

- Mendel's principle of segregation states that the two members of a gene pair (alleles) segregate from each other in the formation of gametes.

- A testcross is a cross of an individual of unknown genotype, usually expressing the dominant phenotype, with a homozygous recessive in order to determine the genotype of the unknown individual.

- Mendel's principle of independent assortment states that genes for different traits assort independently of one another during the production of gametes.

- Mendelian principles apply to humans as well as to peas and all other eukaryotes. The study of the inheritance of genetic traits in humans is complicated by the fact that no controlled crosses can be done. Instead, human geneticists analyze genetic traits by pedigree analysis; that is, by examining the occurrences of the trait in family trees of individuals who clearly exhibit the trait.

BY SIMPLE OBSERVATION it is evident that there is a lot of variation among individuals of a given species. Among dogs, for example, are many breeds, including Bernese mountain dogs, Dalmatians, pointers, dachshunds, Yorkshire terriers, and so on. These breeds are clearly distinguishable by their size, shape, and color, yet they are all representatives of the same species—*Canis familiaris*. Similarly, differences among individual humans include eye color, height, and hair color, even though they all belong to the species *Homo sapiens*. Within a species each generation perpetuates individual differences, yet the species remains clearly identifiable. The differences among individuals within and between species are the result of differences in the DNA sequences that constitute their genes. It is largely the genes that determine the structure, function, and development of the cell or organism. Thus the genetic information coded in DNA is responsible for species and individual variation.

The understanding of how genes are transmitted from parent to offspring began with the work of Gregor Johann Mendel (1822–1884), an Austrian monk. In this chapter we will learn the basic principles of transmission genetics by examining Mendel's work. Throughout this chapter we must remember that the segregation of genes is directly related to the behavior of chromosomes. However, although Mendel analyzed the patterns of segregation of hereditary traits that he called characters, he did not know that genes control the characters, or that genes are located in chromosomes.

Genotype and Phenotype

Before beginning our study of Mendel's work, we must distinguish between the nature of the genetic material each individual organism possesses and the physical characteristics expressed by this genetic material.

The characteristics of an individual that are transmitted from one generation to another are sometimes called **hereditary traits**, or **characters** as Mendel called them. These traits are under the control of DNA segments called **genes**. The genetic constitution of an organism is called its **genotype**. The physical manifestation of a genetic trait is called a **phenotype**.

The genes that every individual carries are essentially like a blueprint that can be interpreted in several ways. Genes give only the *potential* for the development of a particular phenotypic characteristic; the extent to which that potential is realized depends not only on interactions with other genes and their products but also on environmental influences (Figure 2.1). A person's height, for example, is controlled by many genes, the expression of

■ *Figure 2.1* Influences on the physical manifestation (phenotype) of the genetic blueprint (genotype): interactions with other genes and their products (such as hormones) and with the environment (such as nutrition).

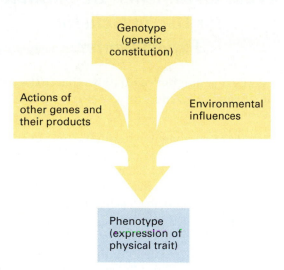

which, in turn, can be significantly affected by internal and external environmental influences. Notable among these influences are nutrition (an external environmental influence) and the effects of hormones during puberty (an internal environmental influence). While we can say that a child is "tall like her father," there is no simple genetic explanation for that statement. She may be tall because excellent nutritional habits are interacting with the genetic potential for height.

Because of the effects of environment, then, individuals may have identical genotypes (as in identical twins) but different phenotypes. Studies have shown that identical twins raised apart and exposed to different environmental effects tend to vary more in their appearance than identical twins raised together in the same environment. Similarly, individuals may have virtually identical phenotypes but very different genotypes. These examples show that (1) genes are a starting point for determining the structure and function of an organism, and that (2) the road to the mature phenotypic state is highly complex and involves a myriad of interacting biochemical pathways.

It is essential to understand for our future discussions, however, that while the phenotype is the product of the interaction between gene(s) and the environment, the contribution of the environment varies. In some cases the environmental influence is great; in others the environmental contribution is nonexistent. We will develop the relationship between genotype and phenotype in more detail as the text proceeds, but for our current task of exam-

ining Mendel's experiments, the aspects just mentioned are sufficient.

K e y n o t e

The genotype is the genetic constitution of an organism. The phenotype is the physical manifestation of the genetic traits. The genes give the potential for the development of characteristics and this potential is often affected by interactions with other genes and with the environment. Thus individuals with the same genotype can have different phenotypes, and individuals with the same phenotypes may have different genotypes.

Mendel's Experiments

Mendel's Experimental Design

The work of Gregor Johann Mendel (1822–1884) (Figure 2.2) is the foundation of modern genetics. In 1843 he was admitted to the Augustinian Monastery in Brno (now Brunn, Austria). In 1854 he began a series of breeding experiments with the garden pea *Pisum sativum* in an attempt to learn about the mechanisms of heredity. Likely as a result of his creativity, Mendel discovered some fundamental principles of genetics.

From the results of crossbreeding pea plants with different characteristics such as height, flower color, and seed shape, Mendel developed a simple theory

■ *Figure 2.2* Gregor Johann Mendel, the father of the science of genetics.

to explain the transmission of hereditary characteristics or traits from generation to generation. (Note that Mendel had no knowledge of mitosis and meiosis. Now, of course, we know that genes segregate according to chromosome behavior.) Although Mendel reported his conclusions in 1865, it wasn't until the late 1800s and early 1900s that their significance was fully realized. Today Mendel's methods of analysis continue to be used in genetic studies. In view of his significant contributions, Mendel is regarded as the father of genetics.

Mendel's experimental approach was effective because he made simple interpretations of the ratios of the progeny types he obtained from his crosses and then carried out direct and convincing experiments to test his hypotheses. In his initial breeding experiments, Mendel took the simplest approach of studying one trait at a time to see how it was inherited. (This is how you should work genetics problems.) He made carefully controlled matings (crosses) between strains of peas that showed differences in heritable traits and, most importantly, he kept precise records of the outcome of the crosses and the number of each type of pea produced. The numerical data he obtained enabled him to do rigorous analysis of the hereditary transmission of characteristics.

Genetic crosses are generally done as follows. Two diploid individuals differing in phenotype are allowed to produce haploid gametes by meiosis. Fusion of male and female gametes produces zygotes from which the diploid progeny individuals are generated. The phenotypes of the offspring are analyzed to provide clues to the heredity of those phenotypes.

Mendel performed all his significant genetic experiments with the garden pea. The pea plant was a good choice because it fits many of the criteria that make an organism suitable for use in genetic experiments (Box 2.1): It is easy to grow, bears flowers and fruit in the same year a seed is planted, and produces a large number of seeds. When the seeds are planted, they germinate to produce new pea plants.

Figure 2.3 presents a cross section of a flower of the garden pea, showing the stamens (the male reproductive organs) and the pistils (the female reproductive organs). The pea normally reproduces by **self-fertilization**; that is, pollen (the male [♂] gametophytes) produced from the stamens lands on the pistil (containing the female [♀] gametophytes) within the same flower and fertilizes the plant. This process is also called **selfing**. It is a relatively simple procedure to prevent self-fertilization of the pea by removing the stamens from a developing flower bud before they produce any mature

---

BOX 2.1

What Organisms Are Suitable for Use in Genetic Experiments?

There are a number of qualities that make an organism well suited for genetic experimentation.

1. To test a genetic hypothesis, the genetic history of the organism involved must be well known. Therefore the genetic background of the parents used in the experimental crosses must be known.

2. The organism must have a relatively short life cycle so that a large number of generations occur within a relatively short time. This allows data over many generations to be obtained rapidly.

3. A large number of offspring must be produced from a mating because much genetic information can be obtained only if there are numerous progeny to study.

4. The organism should be easy to handle. Hundreds of fruit flies can easily be kept in half-pint milk bottles for experimental purposes, but hundreds of elephants would certainly be much more difficult and expensive to maintain!

5. Most importantly, individuals in the population must differ in a number of ways. If there are no discernible differences among the individual organisms under study, it is impossible to study the inheritance of traits. The more marked the differences, the easier the genetic analysis.

---

pollen. The pistils of the emasculated flower can then be pollinated by taking pollen from the stamens of another flower and dusting it onto the pistil's stigma.

Cross-fertilization, or simply **cross**, is the term used for the fusion of male gametes (pollen) from one individual and female gametes (eggs) from another. Once cross-fertilization has occurred, the zygote develops into seeds (peas), which are then planted. The phenotypes of the plants that grow from the seeds are then analyzed.

■ *Figure 2.3*   Procedure for crossing pea plants.

Procedure for crossing pea plants.

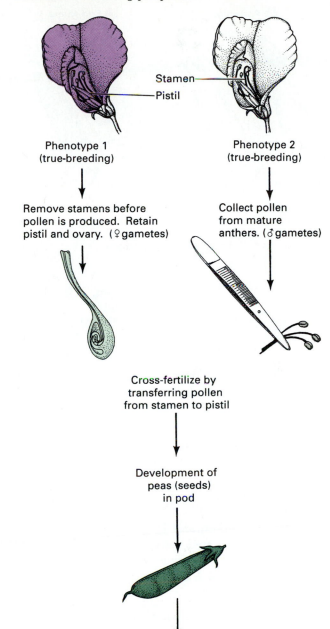

Stamen

Pistil

Phenotype 1
(true-breeding)

Phenotype 2
(true-breeding)

Remove stamens before
pollen is produced. Retain
pistil and ovary. (♀ gametes)

Collect pollen
from mature
anthers. (♂ gametes)

Cross-fertilize by
transferring pollen
from stamen to pistil

Development of
peas (seeds)
in pod

Plant seeds

Measure phenotypes of offspring

For his experiments Mendel obtained 34 strains
of pea plants that differed in a number of traits. He
allowed each strain to self-fertilize for many generations to ensure that the traits he wanted to study

were inherited and to remove from consideration
those strains that produced progeny with traits different from the parental type. This preliminary
work guaranteed that he worked only with pea
strains in which the trait under investigation
remained unchanged from parent to offspring for
many generations. Such strains are called **true-
breeding** or **pure-breeding strains**.

Next Mendel selected seven traits to study in
breeding experiments. Each trait had two easily distinguishable, alternative appearances (phenotypes),
as shown in Figure 2.4. These traits affect the
appearance of most parts of the pea plant, including: (1) flower and seed coat color (grey versus
white seed coats, and purple versus white flowers
[Note: A single gene controls these particular color
properties of both seed coats and flowers.]); (2)
seed color (yellow versus green); (3) seed shape
(smooth versus wrinkled); (4) pod color (green versus yellow); (5) pod shape (inflated versus
pinched); (6) stem height (tall versus short); and (7)
flower position (axial versus terminal).

And, finally, Mendel looked at only one or two
traits at a time, counted the number of each type of
progeny that resulted from the crosses, and *kept
careful records of his observations,* a critical responsibility with all experiments.

Monohybrid Crosses and Mendel's Principle of Segregation

Before discussing Mendel's experiment, we will
clarify the terminology encountered in breeding
experiments. The parental generation is called the **P
generation**. The progeny of the P mating is called
the **first filial generation** or **F₁**. The subsequent generation produced by breeding together the F₁ offspring is termed the **F₂ generation**. Interbreeding
the offspring of each generation results in F₃, F₄, F₅
generations, and so on.

To begin testing his hypothesis about the nature
of heredity, Mendel first performed crosses
between true-breeding strains of peas that differed
in a single trait. Such crosses are called **monohybrid crosses**. For example, he pollinated pea plants
that produced only smooth seeds with pollen from
a true-breeding variety that produced only wrinkled seeds.* As Figure 2.5 shows, the outcome of
this monohybrid cross was all smooth seeds. The

---

*Seeds are the diploid progeny of sexual reproduction. If a phe-
notype concerns the seed itself, the results of the cross can be
directly seen by inspection of the seeds. If a phenotype concerns a part of the mature plant, such as flower color, the seeds
must be germinated before that phenotype can be observed.

■ *Figure 2.4*   Seven character pairs in the garden pea that Mendel studied in his breeding experiments.

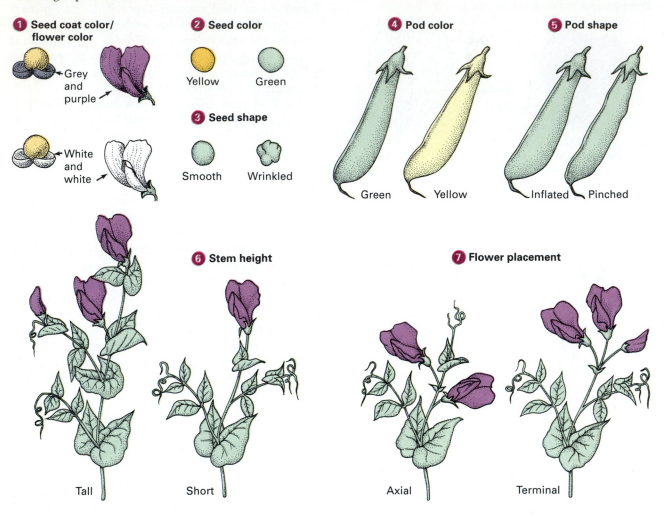

1 Seed coat color/ flower color

Grey and purple

White and white

2 Seed color

Yellow   Green

3 Seed shape

Smooth   Wrinkled

4 Pod color

Green   Yellow

5 Pod shape

Inflated   Pinched

6 Stem height

Tall   Short

7 Flower placement

Axial   Terminal

same result was obtained when the parental types were reversed; that is, when the pollen from a smooth-seeded plant was used to pollinate a pea plant that gave wrinkled seeds. In genetics matings that are done both ways, for instance, smooth female (♀) × wrinkled male (♂) and wrinkled female × smooth male, are called **reciprocal crosses**. (Conventionally the female is given first in crosses.) If the results of the reciprocal crosses are the same, the interpretation is that the trait is not dependent on the sex of the organism.

The significant point of this cross was that all the F_1 progeny seeds of the smooth × wrinkled reciprocal crosses were smooth; that is, they exactly resembled only one of the parents in this character rather than being a blend of both parental phenotypes.

Mendel then planted the seeds and allowed the F_1 plants to self-fertilize and produce the F_2 seed. Both smooth and wrinkled seeds appeared in the F_2 generation, and both seed phenotypes could be found within the same pod. Mendel found that

5474 were smooth and 1850 were wrinkled (Figure 2.6). The calculated ratio of smooth to wrinkled seeds was 2.96:1, which is very close to a 3:1 ratio.

Using the same quantitative approach, Mendel

■ *Figure 2.5*   Results of one of Mendel's breeding crosses. In the parental generation he crossed a true-breeding pea strain that produced smooth seeds with one that produced wrinkled seeds. All the F_1 progeny seeds were smooth.

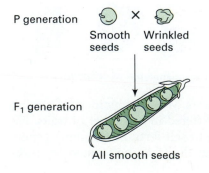

P generation

Smooth   Wrinkled
seeds   seeds

F_1 generation

All smooth seeds

■ *Figure 2.6*   The F₂ progeny of the cross shown in Figure 2.5. When the plants grown from the F₁ seeds were self-pollinated, both smooth and wrinkled F₂ progeny seeds were produced. Commonly, both seed types were found in the same pod. In Mendel's experiments he counted 5474 smooth and 1850 wrinkled F₂ progeny seeds, a ratio of 2.96: 1.

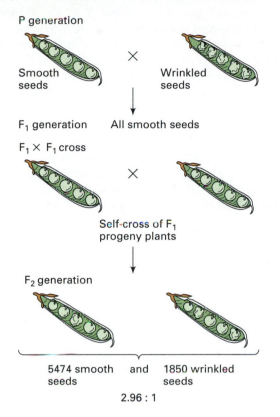

analyzed the behavior of the six other pairs of traits. Qualitatively and quantitatively, the same results were obtained (Table 2.1). From the seven sets of crosses he made the following general conclusions about his data:

1. The results of reciprocal crosses were always the same.
2. All F₁ progeny resembled one of the parental strains.
3. The parental trait that disappeared in the F₁ generation reappeared in the F₂ generation. Further, the trait seen in the F₁ was always found in the F₂ at about three times the frequency of the other trait.

How can a trait present in the P generation disappear in the F₁ and then reappear with full expression in the F₂? Mendel observed that while the F₁ progeny resembled only one of the parents in their phenotype, they did not breed true, a fact that distinguishes the F₁ from the parent they resembled. Moreover, the F₁ progeny could produce F₂ progeny with the parental phenotype that had disappeared in the F₁ generation. Mendel concluded that the alternative traits in the crosses—for example, smoothness or wrinkledness of the seeds—were determined by **particulate factors**. He reasoned that these factors, transmitted from parents to progeny through the gametes, carried hereditary information. We now know these factors by another name—*genes*.

Since Mendel was examining pairs of traits (e.g., wrinkled/smooth), each factor was considered to exist in alternative forms (which we now call **alleles**), each of which specified one of the traits.

Table 2.1

Mendel's Results in Crosses Between Plants Differing in One of Seven Characters

CHARACTER[a]	F₁	F₂ (NUMBER) DOMINANT	RECESSIVE	TOTAL	F₂ (PERCENT) DOMINANT	RECESSIVE
Seeds: smooth versus wrinkled	All smooth	5,474	1,850	7,324	74.7	25.3
Seeds: yellow versus green	All yellow	6,022	2,001	8,023	75.1	24.9
Seed coats[b]: grey versus white Flowers[b]: purple versus white	All grey All purple	705	224	929	75.9	24.1
Flowers: axial versus terminal	All axial	651	207	858	75.9	24.1
Pods: inflated versus pinched	All inflated	882	299	1,181	74.7	25.3
Pods: green versus yellow	All green	428	152	580	73.8	26.2
Stem: tall versus short	All tall	787	277	1,064	74.0	26.0
Total or average		14,949	5,010	19,959	74.9	25.1

[a] The dominant trait is always written first.
[b] A single gene controls both the seed coat and the flower color traits.

For the gene that controls the shape of the pea seed, for example, there is one form, or allele, that results in the production of a smooth seed and another allele that results in a wrinkled seed.

Mendel reasoned further that a true-breeding strain of peas must contain a pair of identical factors; the eggs and pollen of the plant contain only one factor. Since both traits were seen in the F_2, whereas only one appeared in the F_1, then each F_1 individual *must have contained both factors*, one for each of the alternative traits. In other words, crossing two true-breeding strains showing alternative traits brings together in the F_1 one factor from each strain. And since only one of the characters was seen in the F_1 progeny, the expression of the "missing" trait must somehow have been masked by the visible trait, a feature called *dominance*. In the smooth × wrinkled experiment the F_1 seeds were all smooth. Thus the allele for smoothness is **dominant** to the allele for wrinkledness. Conversely,

wrinkled is said to be **recessive** to smooth (Figure 2.7). Similar conclusions can be made for the other six pairs of traits. The dominant and the recessive forms for each pair of traits are indicated in Table 2.1.

A simple way to visualize the crosses is to use symbols for the alleles. For the smooth × wrinkled cross we can give the symbol S to the factor for smoothness and the symbol s to the factor for wrinkledness. The letter used is based on the dominant phenotype and the convention is that the dominant allele is given the uppercase (capital) letter and the recessive allele the lowercase (small) letter. Using these symbols we can diagram the cross as shown in Figure 2.7. The production of the F_1 generation is shown in Figure 2.7(a), the F_2 generation in Figure 2.7(b). (In the figure the genes are located on chromosomes. Keep in mind that gene segregation follows the behavior of chromosomes, although Mendel had no knowledge of the existence of chro-

■ *Figure 2.7*   The same cross as in Figures 2.5 and 2.6. Here, genetic symbols illustrate the principle of segregation of Mendelian factors. (a) Production of the F_1 generation; (b) Production of the F_2 generation.

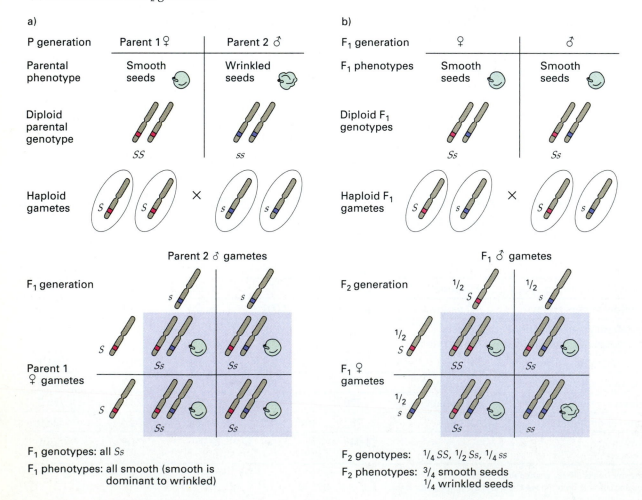

a)

| P generation | Parent 1 ♀ | Parent 2 ♂ |

| Parental phenotype | Smooth seeds | Wrinkled seeds |

Diploid parental genotype
SS ss

Haploid gametes
S S × s s

Parent 2 ♂ gametes

F_1 generation

Parent 1 ♀ gametes

	s	s
S	Ss	Ss
S	Ss	Ss

F_1 genotypes: all Ss
F_1 phenotypes: all smooth (smooth is dominant to wrinkled)

b)

| F_1 generation | ♀ | ♂ |

| F_1 phenotypes | Smooth seeds | Smooth seeds |

Diploid F_1 genotypes
Ss Ss

Haploid F_1 gametes
S s × S s

F_1 ♂ gametes

F_2 generation

F_1 ♀ gametes

	$\frac{1}{2}$ S	$\frac{1}{2}$ s
$\frac{1}{2}$ S	SS	Ss
$\frac{1}{2}$ s	Ss	ss

F_2 genotypes: $\frac{1}{4}$ SS, $\frac{1}{2}$ Ss, $\frac{1}{4}$ ss
F_2 phenotypes: $\frac{3}{4}$ smooth seeds
$\frac{1}{4}$ wrinkled seeds

mosomes.) Because each parent is true breeding, each must contain two copies of the same allele. Thus the genotype of the parental plant grown from the smooth seeds is *SS* and that of the wrinkled parent is *ss*. True-breeding individuals that contain only one specific allele are said to be **homozygous**.

When plants produce gametes by meiosis, each gamete contains only one copy of the gene (i.e., one allele); the plants from smooth seeds produce *S*-bearing gametes, and the plants from wrinkled seeds produce *s*-bearing gametes. When the gametes fuse during the fertilization process, the resulting zygote has one *S* and one *s* factor, a genotype of *Ss*. Plants that have two different alleles of a specific gene are said to be **heterozygous**. Because of the dominance of the smooth *S* allele, only smooth seeds develop from the F_1 zygotes.

The plants grown from the F_1 seeds differ from the smooth parent in that they produce two types of gametes in equal numbers: *S*-bearing and *s*-bearing. All the possible fusions of F_1 gametes are shown in the matrix in Figure 2.7, called a **Punnett square** after its originator, R. Punnett. These fusions give rise to the zygotes that produce the F_2 generation.

In the F_2 generation three types of genotypes are produced: *SS, Ss,* and *ss.* As a result of the random fusing of gametes, the relative proportion of these zygotes is 1:2:1, respectively. However, since the *S* factor is dominant to the *s* factor, both the *SS* and *Ss* seeds are smooth, and the F_2 generation seeds exhibit a phenotypic ratio of 3 smooth: 1 wrinkled. The results are the same for crosses involving the other six character pairs.

It is important to understand that the observed phenotypic ratios among progeny of a cross rarely match exactly the expected or predicted ratios even when the hypothesis on which the expected ratios are based is correct. For example, none of the F_2s of Mendel's monohybrid crosses showed an exact 3:1 ratio (see Table 2.1), although the observed ratios were very close. However, if the difference between the observed and predicted result is large enough, there may be cause to reject the hypothesis. A statistical test called the **chi-square** (χ^2) **test** can be used to decide if the observed data deviates from the expected to a degree sufficient to throw out the hypothesis. As used in the analysis of genetic data, this test is essentially a *goodness-of-fit test*. The chi-square test is discussed in Chapter 5 (pp. 101–103).

The Principle of Segregation

From the types of data we have discussed, Mendel proposed his **first law, the principle of segregation**, which states that *the two members of a gene pair (alle-les) segregate (separate) from each other in the formation of gametes.* As a result half the gametes carry one allele and the other half carry the other allele. In other words, each gamete carries only one allele of each gene. The progeny are produced by the random combination of gametes from the two parents.

In proposing the principle of segregation, Mendel had clearly differentiated between the factors (genes) that determined the traits (the genotype) and the traits themselves (the phenotype). From a modern perspective we know that genes are on chromosomes and the specific location of a gene on a chromosome is called a **locus** (or **gene locus**; plural: **loci**). Further, Mendel's first law means that at the gene level the members of a pair of alleles of a gene on a chromosome segregate during meiosis so that any offspring receives only one member of a pair from each parent. Thus **gene segregation** parallels the separation of homologous pairs of chromosomes at anaphase I in meiosis.

Box 2.2 presents a summary of the genetics concepts and terms discussed so far in this chapter. A thorough familiarity with these terms is essential to the study of genetics.

Keynote

Mendel's first law, the principle of segregation, states that the two members of a gene pair (alleles) segregate (separate) from each other in the formation of gametes; half the gametes carry one allele, and the other half carry the other allele.

Representing Crosses with a Branch Diagram

The use of a Punnett square to consider the pairing of all possible gamete types from the two parents in a monohybrid cross (shown in Figure 2.7) is a relatively simple way to learn how to predict the relative frequencies of genotypes and phenotypes in the next generation. In this section we will describe an alternative method—the branch diagram. (Box 2.3 discusses some elementary principles of probability that will help you understand this approach.) Figure 2.8 illustrates the application of the branch diagram analysis of the $F_1 \times F_1$ self of the smooth $\times$ wrinkled cross shown in Figure 2.7.

The F_1 seeds from the cross have the genotype *Ss.* In the meiotic process an equal number of *S* and *s* gametes are expected to be produced, so we can say that half the gametes are *S* and the other half are *s*. Thus 1/2 is the predicted frequency of each of these two types. But just as when you toss a coin many times you do not always get exactly heads half the

BOX 2.2

Genetic Terminology

Cross: mating between two individuals, leading to the fusion of gametes.

Zygote: the cell produced by the fusion of male and female gametes.

Gene (Mendelian particulate factor): the determinant of a characteristic of an organism.

Locus (gene locus): the specific place on a chromosome where a gene is located.

Alleles: alternative forms of a gene. For example, the *S* and *s* alleles represent the smoothness and wrinkledness of the pea seed.

Genotype: the genetic constitution of an organism. A diploid organism that has both alleles the same for a given gene locus is said to be *homozygous* for that allele. Homozygotes produce only one gametic type. For example, from our pea example, true-breeding smooth individuals have the genotype *SS*, and true-breeding wrinkled individuals have the genotype *ss*; both are homozygous. The smooth parent is *homozygous dominant*; the wrinkled parent is *homozygous recessive*.

Diploid organisms that have two different alleles at a specific gene locus are said to be *heterozygous*. So F$_1$ hybrid plants from the cross of *SS* and *ss* parents have one *S* allele and one *s* allele. Individuals heterozygous for two allelic forms of a gene produce two kinds of gametes (*S* and *s*).

Phenotype: the physical manifestation of a genetic trait that results from a specific genotype and its interaction with the environment. In our example the *S* allele was dominant to the *s* allele, so in the heterozygous condition the seed is smooth. Therefore both the homozygous dominant *SS* and the heterozygous *Ss* seeds have the same phenotype (smooth), even though they differ in genotype.

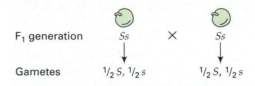

■ *Figure 2.8* Calculating the ratios of phenotypes in the F$_2$ generation of the Figure 2.7 cross by using the branch diagram approach.

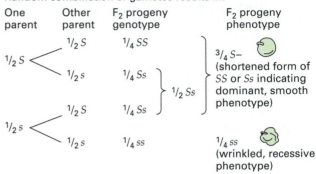

generation can be predicted. To produce an *SS* plant, an *S* egg must pair with an *S* pollen grain. The frequency of *S* eggs in the population of eggs is 1/2, and the frequency of *S* pollen grains in the pollen population is 1/2. Therefore the expected proportion of *SS* smooth plants in the F$_2$ is 1/2 × 1/2 = 1/4. Similarly, the expected proportion of *ss* wrinkled progeny in the F$_2$ is 1/2 × 1/2 = 1/4.

What about the *Ss* progeny? Again the frequency of *S* in one gametic type is 1/2, and the frequency of *s* in the other gametic type is also 1/2. However, there are two ways in which *Ss* progeny can be obtained. The first involves fusion of an *S* egg with *s* pollen; the second is a fusion of an *s* egg with *S* pollen. Using the product rule (Box 2.3), the probability of each of these events occurring is 1/2 × 1/2 = 1/4. Using the sum rule (also shown in Box 2.3), the probability of *one or the other occurring* is the sum of the individual probabilities, or 1/4 + 1/4 = 1/2.

Given the rules of probability, then, the prediction is that one-fourth of the F$_2$ progeny will be *SS*, half will be *Ss*, and one-fourth will be *ss*, exactly as was found with the method shown in Figure 2.7. Either method—the Punnett square or the branch diagram—may be used with any cross.

Confirming the Principle of Segregation: The Use of Testcrosses

When formulating his principle of segregation, Mendel performed a number of tests to ensure the validity of his results. He continued the self-fertil-

time and tails half the time, the two gametes may not be produced in an exactly 1:1 ratio. However, the more chances (e.g., tosses), the more likely you will approach the true frequency.

From the rules of probability, the expected frequencies of the three possible genotypes in the F$_2$

Elementary Principles of Probability

A **probability** is the ratio of the number of times a particular event occurs to the number of trials during which the event could have happened. For instance, the probability of picking a heart (from 13 hearts) from a deck of cards (52 cards) is p(heart) = 13/52 = 1/4. That is, on the average we would expect to pick a heart from a deck of cards once in every four trials.

Probabilities and the *laws of chance* are involved in the transmission of genes. As a simple example let's consider a couple and the chance that their child will be a boy or a girl. Assume that an exactly equal number of boys and girls are born (which is not actually true, but we can assume it to be so for the sake of discussion). The probability that the child will be a boy is 1/2, and the probability that the child will be a girl is 1/2.

A rule of probability, the **product rule**, *states that the probability of two independent events occurring together is the product of each of their individual probabilities.* Thus the probability that a family will have two children, both of which are girls is 1/4. That is, the probability of the first child being a girl is 1/2, the probability of the second being a girl is also 1/2, and by the product rule the probability of the first and second being girls is 1/2 × 1/2 = 1/4. The probability of having three boys a row is 1/2 × 1/2 × 1/2 = 1/8.

Another rule of probability is the **sum rule**, which states that *the probability that two mutually exclusive events will occur is the sum of their individual probabilities.* For example, if two dice are thrown, what is the probability of getting two sixes *or* two ones? The individual probabilities are calculated as follows: The probability of getting two sixes is found by using the product rule. The probability of getting one six, p(one six), is 1/6, since there are six faces to a die. Therefore the probability of getting two sixes, p(two sixes), when two dice are thrown is 1/6 × 1/6 = 1/36. Similarly, p(two ones) = 1/36. To roll two sixes *or* two ones involves mutually exclusive events, so the sum rule is used. The sum of the individual probabilities is 1/36 + 1/36 = 2/36 = 1/18. To return to our family example, the probability of having two boys *or* two girls is 1/4 + 1/4 = 1/2.

---

izations to the F_6 generation and found that in every generation both the dominant and recessive characters appeared. He concluded, therefore, that the principle of segregation was valid no matter how many generations were produced.

Another important test concerned the F_2 plants. As shown in Figure 2.7, a ratio of 1:2:1 occurs for the genotypes *SS*, *Ss*, and *ss* for the smooth × wrinkled example. Phenotypically, the ratio of smooth to wrinkled is 3:1. At the time of his experiments, the presence of segregating factors that were responsible for the smooth and wrinkled phenotypes was only an hypothesis. To test his factor hypothesis, Mendel allowed the F_2 plants to self-pollinate. As he expected, plants produced from wrinkled seeds bred true, supporting his conclusion that they were pure for the *s* factor (allele).

Selfing the plants derived from the F_2 smooth seeds produced two different types of progeny. One-third produced only smooth seeds, whereas the other two-thirds produced both smooth and wrinkled seeds in each pod (Figure 2.9). For the plants that produced both seed types in the progeny, the actual ratio of smooth: wrinkled seeds was 3:1; that is, the same ratio as seen for the F_2 progeny. These results completely support the principle of segregation of genes. The random combination

■ *Figure 2.9* Determining the genotypes of the F_2 smooth progeny of Figure 2.7 by selfing the plants grown from the smooth seeds.

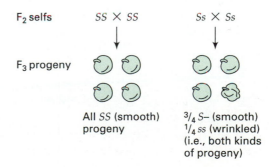

F_2 selfs $SS \times SS$ $Ss \times Ss$

F_3 progeny

All *SS* (smooth) progeny   3/4 *S–* (smooth)
1/4 *ss* (wrinkled) (i.e., both kinds of progeny)

of gametes that form the zygotes of the F$_2$ produces two genotypes that give rise to the smooth phenotype (see Figures 2.7 and 2.8); the relative proportion of the two genotypes *SS* and *Ss* is 1:2. The *SS* seeds give rise to true-breeding plants, whereas the *Ss* seeds give rise to plants that behave exactly like the F$_1$ plants when they are self-pollinated in that they produce a 3:1 ratio of smooth: wrinkled progeny. *Mendel explained these results by proposing that each plant had two factors, while each gamete had only one. He also proposed that the random combination of the gametes generated the progeny in the proportions he found. Mendel obtained the same results in all seven sets of crosses.*

The self-fertilization test of the F$_2$ plants is a use-ful way of confirming the genotype of a plant with a given phenotype. A more common test to determine the genotype of an organism is to perform a **testcross**, a cross of an individual of unknown genotype, usually expressing the dominant phenotype, with a homozygous recessive individual in order to determine the genotype of the individual.

Consider again the cross shown in Figure 2.7. We can predict the outcome of a testcross of the F$_2$ plants showing the dominant, smooth seed phenotype. If the F$_2$ plants are homozygous *SS*, the result of a testcross with an *ss* plant will be all smooth seeds. As Figure 2.10(a) shows, the Parent 1 smooth *SS* plants produce only *S* gametes. Parent 2 is homozygous recessive wrinkled, *ss*, so it produces

■ *Figure 2.10*   Determining the genotypes of the F$_2$ generation smooth seeds (Parent 1) of Figure 2.7 by testcrossing plants grown from the seed with a homozygous recessive wrinkled (*ss*) strain (Parent 2). (a) If Parent 1 is *SS*, all progeny seeds are smooth in phenotype; (b) If Parent 1 is *Ss*, 1/2 of the progeny seeds are smooth and 1/2 are wrinkled.

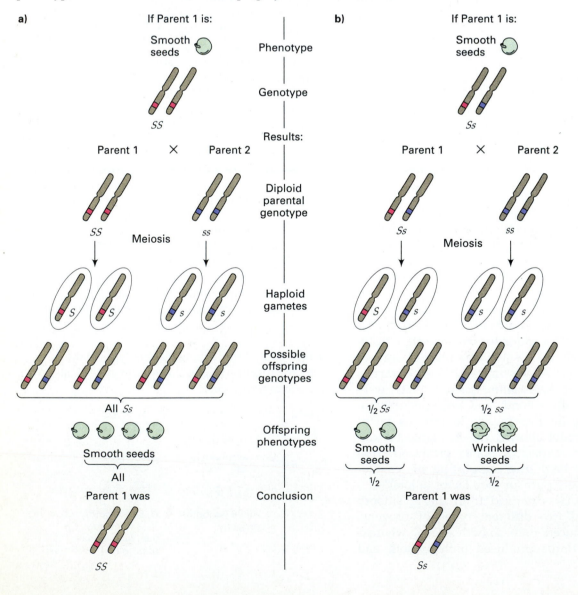

only s gametes. Thus, all zygotes are Ss, and all the resulting seeds are smooth in phenotype. In actual practice, therefore, only dominant phenotypes, the results indicate that the F_1 used in the testcross is homozygous for the dominant allele.

On the other hand, if heterozygous Ss F_2 plants are testcrossed with a homozygous ss plant, a 1:1 ratio of dominant to recessive phenotypes is expected. As Figure 2.10(b) shows, the Parent 1 smooth Ss produces both S and s gametes in equal proportion, while the homozygous ss Parent 2 produces only s gametes. As a result, half the progeny of the testcross are Ss heterozygotes and have a smooth phenotype because of the dominance of the S allele, and the other half are ss homozygotes and have a wrinkled phenotype. In actual practice, therefore, if one performs a testcross between a parent showing the dominant trait with a parent showing the recessive trait, and the progeny exhibit a 1:1 ratio of dominant to recessive phenotypes, the results indicate that the F_1 used in the testcross is heterozygous.

In summary, testcrosses of the F_2 plants from Mendel's crosses that showed the dominant phenotype indicated that there was a 1:2 ratio of homozygous dominant:heterozygous genotypes in the F_2 plants. That is, one third of the F_2 plants with the dominant phenotype gave only progeny with the dominant phenotype in crosses with the homozygous recessive. The remaining two-thirds of the F_2 plants with the dominant phenotype gave a 1:1 ratio of progeny with the dominant phenotype:progeny with the recessive phenotype.

Keynote

A testcross is a cross of an individual of unknown genotype, usually expressing the dominant phenotype, with a homozygous recessive in order to determine the genotype of the unknown individual. The phenotypes of the progeny of the testcross indicate the genotype of the individual tested. If the progeny all show the dominant phenotype, the individual was homozygous dominant. If there is an approximately 1:1 ratio of progeny with dominant and recessive phenotypes, the unknown individual was heterozygous.

Dihybrid Crosses and the Mendelian Principle of Independent Assortment

The Principle of Independent Assortment

The next logical extension of Mendel's experiments was to determine what happens when two pairs of traits are simultaneously involved in the cross. Mendel did a number of these crosses, and in each case he obtained the same results. From these experiments he proposed his **second law, the principle of independent assortment**, which states that *the factors (genes) for different traits assort independently of one another*. In modern terms this means that *genes on different chromosomes behave independently in the production of gametes*. There are also many examples of genes that are far enough apart on the same chromosome that they assort independently. This principle was unknown to Mendel and will be considered when linkage and crossing-over are discussed in Chapter 5.

Consider an example involving smooth (S)/wrinkled(s) and yellow (Y)/green(y) seed traits (yellow is dominant to green). Mendel made crosses between true-breeding smooth-yellow plants $(SS\ YY)$ and wrinkled-green plants $(ss\ yy)$, with the results shown in Figure 2.11. All the F_1 seeds from this cross were smooth and yellow, as predicted from the results of the monohybrid crosses. As Figure 2.11(a) shows, the smooth-yellow parent produces only $S\ Y$ gametes, which give rise to $Ss\ Yy$ zygotes upon fusion with the $s\ y$ gametes from the wrinkled-green parent. Because of dominance of the smooth and the yellow traits, all F_1 seeds are smooth and yellow.

The F_1 progeny are heterozygous for two pairs of alleles at two different loci. Such individuals are called dihybrids, and a cross between two dihybrids of the same type is called a **dihybrid cross**.

When Mendel self-pollinated the dihybrid F_1 plants to produce the F_2 generation (Figure 2.11[b]), he considered two possible outcomes. One was that the genes for the traits from the original parents would be transmitted together to the progeny. In this case a phenotypic ratio of 3:1 smooth-yellow:wrinkled-green would be predicted.

The other possibility was that the traits would be inherited independently of one another. The dihybrid F_1 progeny produce four types of gametes: $S\ Y$, $S\ y$, $s\ Y$, and $s\ y$. Because of the independence of the two pairs of genes, there is an equal frequency of each gametic type. In the $F_1 \times F_1$ selfing, the four types of gametes fuse randomly in all possible combinations to give rise to the zygotes and, hence, the progeny seeds. All the possible gametic fusions are represented in the Punnett square in Figure 2.11(b). In a dihybrid cross there are 16 possible gametic fusions. The result is 9 different genotypes but, because of dominance, only 4 phenotypes:

1 $SS\ YY$, 2 $Ss\ YY$, 2 $SS\ Yy$, 4 $Ss\ Yy$ = 9 smooth-yellow
1 $SS\ yy$, 2 $Ss\ yy$ = 3 smooth-green
1 $ss\ YY$, 2 $ss\ Yy$ = 3 wrinkled-yellow
1 $ss\ yy$ = 1 wrinkled-green

According to the rules of probability, if pairs of characters are inherited independently in a dihybrid cross, then the F_2 from an $F_1 \times F_1$ selfing will give a 9:3:3:1 ratio of the four possible phenotypic classes. This ratio is the result of the independent assortment of the two gene pairs into the gametes and of the random fusion of those gametes. The 9:3:3:1 ratio is the product of two 3:1 ratios; that is, $(3:1)^2 = 9:3:3:1$.

This prediction was met in all dihybrid crosses that Mendel performed. In every case the F_2 ratio was close to 9:3:3:1. For our example he counted 315 smooth-yellow, 108 smooth-green, 101 wrinkled-yellow, and 32 wrinkled-green seeds—very close to the predicted ratio. To Mendel this result meant that the factors (genes) determining the specific, different character pairs he was analyzing were transmitted independently.

Keynote

Mendel's second law, the principle of independent assortment, states that genes for different traits assort independently of one another in the production of gametes.

Branch Diagram of Dihybrid Crosses

As perhaps is already apparent, it is quite tedious to construct a Punnett square of gamete combinations and then to count up the numbers of each phenotypic class from all the genotypes produced. Although not too difficult for a dihybrid cross, for more than two gene pairs it becomes rather complex. *It is easier to calculate the expected ratios of phenotypic classes by using a branch diagram,* which is basically to consider the traits one at a time and use the laws of probability to determine the likelihood of a specific outcome.

Using the same example, in which the two gene pairs assort independently into the gametes, we will examine each gene pair in turn. Earlier we showed that an F_1 self of an Ss heterozygote gave rise to three-fourths smooth and one fourth wrinkled progeny. Genotypically the former class had at least one dominant S allele; that is, they were SS or Ss. A convenient way to designate this situation is to use a dash to indicate an allele that has no effect on the phenotype. Thus $S-$ means that phenotypically the seeds are smooth and genotypically they are either SS or Ss.

Now consider the F_2 produced from a selfing of Yy heterozygotes. Again, a 3:1 ratio is seen, with three-fourths of the seeds being yellow and one

■ *Figure 2.11* The principle of independent assortment in a dihybrid cross. This cross, actually done by Mendel, involves the smooth/wrinkled and yellow/green character pairs of the garden pea. (a) Production of the F_1 generation; (b) The F_2 genotypes and 9:3:3:1 phenotypic ratio of smooth-yellow:smooth-green: wrinkled-yellow: wrinkled-green are derived by using the Punnett square. (Note that compared with previous figures of this kind, only one box is shown in the F_1, instead of four. This is because only one class of gametes exists for Parent 2 and only one class for Parent 1. Previously we showed two gametes from each parent, even though those gametes were identical.)

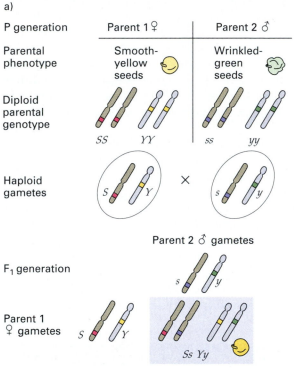

F$_1$ genotypes: all $Ss\ Yy$
F$_1$ phenotypes: all smooth-yellow seeds

fourth being green. Since this segregation occurs independently of the segregation of the smooth/wrinkled pair, we can consider all possible combinations of the phenotypic classes in the dihybrid cross. For example, the expected proportion of F_2 seeds that are smooth and yellow is the product of the probability that an F_2 seed will be smooth and the probability that it will be yellow, or $3/4 \times 3/4 = 9/16$. Similarly, the expected proportion of F_2s that are wrinkled and yellow is $3/4 \times 1/4 = 3/16$. Extending this calculation to all possible phenotypes, as shown in Figure 2.12, we obtain the ratio of 9 $S-Y-$ (smooth, yellow): 3 $S-yy$ (smooth, green): 3 $ss\ Y-$ (wrinkled, yellow): 1 $ss\ yy$ (wrinkled, green).

■ *Figure 2.11*   Cont.

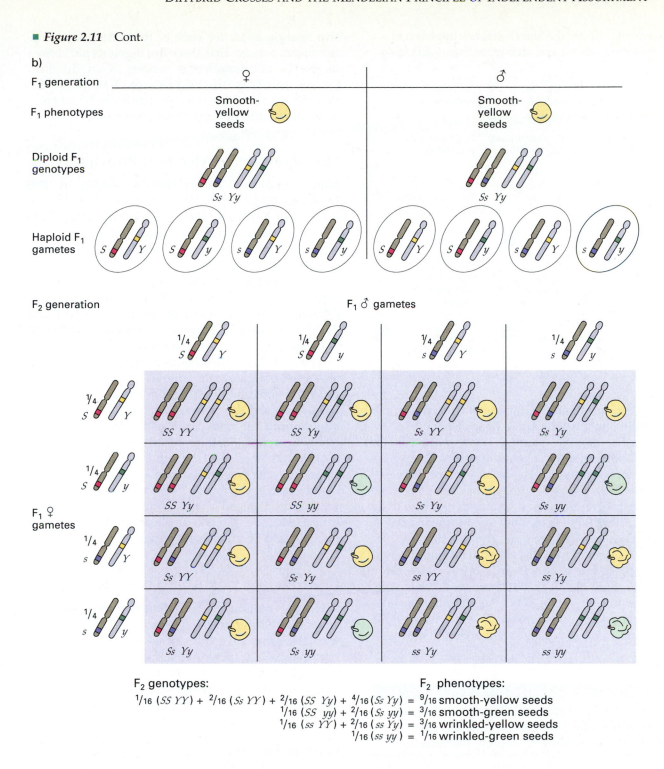

F₂ genotypes:

$$\frac{1}{16}(SS\ YY) + \frac{2}{16}(Ss\ YY) + \frac{2}{16}(SS\ Yy) + \frac{4}{16}(Ss\ Yy) = \frac{9}{16}\ \text{smooth-yellow seeds}$$

$$\frac{1}{16}(SS\ yy) + \frac{2}{16}(Ss\ yy) = \frac{3}{16}\ \text{smooth-green seeds}$$

$$\frac{1}{16}(ss\ YY) + \frac{2}{16}(ss\ Yy) = \frac{3}{16}\ \text{wrinkled-yellow seeds}$$

$$\frac{1}{16}(ss\ yy) = \frac{1}{16}\ \text{wrinkled-green seeds}$$

F₂ phenotypes:

Trihybrid Crosses

Mendel also confirmed his laws for three characters segregating in other garden pea crosses. Such crosses are called **trihybrid crosses**. Here the proportions of F₂ genotypes and phenotypes are predicted with precisely the same logic used before, considering each character pair independently. Figure 2.13 shows a branch diagram derivation of the F₂ phenotypic classes for a trihybrid cross. The independently assorting character pairs in the cross are smooth versus wrinkled seeds, yellow versus green seeds, and purple versus white flowers. There are 64 combinations of 8 maternal and 8 paternal gametes. Combination of these gametes gives rise to 27 dif-

■ *Figure 2.12*   Using the branch diagram approach to calculate the F_2 phenotypic ratio of the Figure 2.11 cross.

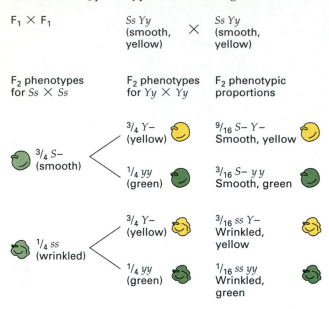

$F_1 \times F_1$

Ss Yy (smooth, yellow)	×	*Ss Yy* (smooth, yellow)

F_2 phenotypes for *Ss* × *Ss* F_2 phenotypes for *Yy* × *Yy* F_2 phenotypic proportions

3/4 *S–* (smooth)

3/4 *Y–* (yellow)   9/16 *S– Y–* Smooth, yellow

1/4 *yy* (green)   3/16 *S– y y* Smooth, green

1/4 *ss* (wrinkled)

3/4 *Y–* (yellow)   3/16 *ss Y–* Wrinkled, yellow

1/4 *yy* (green)   1/16 *ss yy* Wrinkled, green

ferent genotypes and 8 different phenotypes in the F_2 and the phenotypic ratio is 27:9:9:3:9:3:3:1.

Now that enough examples have been considered, we can make some generalizations about phenotypic and genotypic classes. In each of the cases discussed, the F_1 is heterozygous for each gene involved in the cross, and the F_2 is generated by selfing (where that is possible) or by allowing the F_1 individuals to interbreed. In monohybrid crosses there are 2 phenotypic classes in the F_2, in dihybrid crosses there are 4, and in trihybrid crosses there are 8. The general rule is that there are 2^n phenotypic classes in the F_2, where n is the number of independently assorting, heterozygous gene pairs (Table 2.2). (This rule holds *only* when a dominant-reces-

sive relation holds for each of the gene pairs.) Furthermore, we saw that there are 3 genotypic classes in the F_2 of monohybrid crosses, 9 in dihybrid crosses, and 27 in trihybrid crosses. A simple rule is that the number of genotypic classes is 3^n, where n is the number of heterozygous gene pairs.

"Rediscovery" of Mendel's Principles

Mendel published his treatise on heredity in 1865 in *Verhandlungen des Naturforschenden Vereines* in Brünn, but it received little attention from the scientific community at the time. At about the turn of the century three researchers working independently on breeding experiments came to the same conclusions as had Mendel. The three men were Carl Correns, Hugo de Vries, and Erich von Tschermak. Correns concentrated mostly on maize (corn) and peas, de Vries worked with a number of different plant species, and von Tschermak studied peas.

The first demonstration that Mendel's principles also applied to animals came in 1902 from the work of William Bateson, who experimented with fowl. Bateson coined the terms *genetics, zygote,* F_1, F_2, and **allelomorph** (literally, "alternative form," meaning one of an array of different forms of a gene). Allelomorph was shortened by other scientists to *allele*. The term *gene* was introduced by W. L. Johannsen in 1909 as a replacement for *Mendelian factor*. In the years after Bateson's work a number of investigators showed the general applicability of Mendelian principles to all eukaryotic organisms.

Gene Segregation Analysis in Humans

After the rediscovery of Mendel's laws in 1900, geneticists found that the inheritance of genes follows the same principles in all sexually reproducing eukaryotes, including humans. In this section we will introduce some of the methods used to determine the mechanism of hereditary transmission in humans.

Pedigree Analysis

The study of human genetics is complicated because controlled matings of humans cannot be made. This means that geneticists must study crosses that happen by chance rather than by control. The inheritance patterns of human traits are usually established by examining the way the trait occurs in the family trees of individuals who clearly exhibit the trait. The family tree investigation is called **pedigree analysis** and involves the careful compilation of phenotypic records of the family over sever-

---

Table 2.2

Number of Phenotypic and Genotypic Classes Expected from Self-crosses of Heterozygotes in Which All Genes Show Complete Dominance

NUMBER OF SEGREGATING GENE PAIRS	NUMBER OF PHENOTYPIC CLASSES	NUMBER OF GENOTYPIC CLASSES
1[a]	2	3
2	4	9
3	8	27
n	2^n	3^n

[a]For example, from *Aa* × *Aa*, two phenotypic classes are expected, with genotypic classes of *AA, Aa,* and *aa*.

■ *Figure 2.13*   Branch diagram derivation of the relative frequencies of the eight phenotypic classes in the F$_2$ of a trihybrid cross.

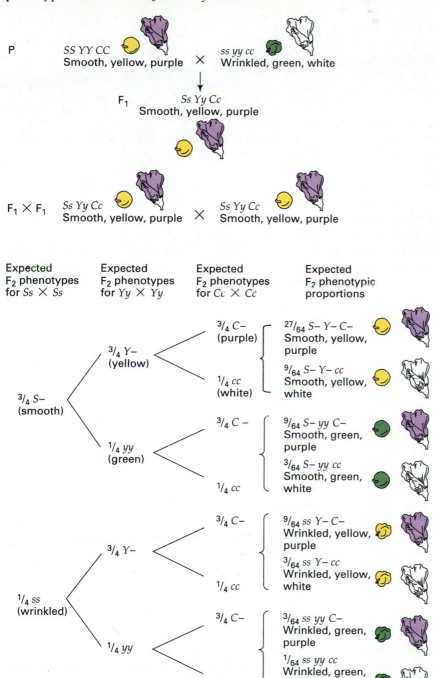

al generations. The individual through whom the pedigree is discovered is called the **propositus** (male) or **proposita** (female). The more information there is, the more likely the investigator will be able to make some conclusions about the mechanism of inheritance of the gene (or genes) responsible for the trait being studied.

One of the modern applications of pedigree analysis is called **genetic counseling**. A geneticist makes predictions about the probabilities of particular traits (deleterious or not) occurring among a couple's children. In most cases the couple comes to the counselor because there is some possibility of the existence of an undesirable heritable trait in one or both families. As we might expect, pedigree analysis is most useful for traits that are the result of a single gene difference.

Pedigree analysis has its own set of symbols, as shown in Figure 2.14. (The terms *autosomal* and *sex-linked*, used in the figure, are explained in Chapter 3). Figure 2.15 presents a hypothetical pedigree to show how the symbols are assigned to the family tree.

The trait presented in Figure 2.15 is determined by a recessive allele *a*. Generations are numbered with roman numerals, while individuals are numbered with Arabic numerals for ease of reference to particular people in the pedigree. The trait in this pedigree results from homozygosity for the rare allele, brought about by cousins marrying. Since cousins share a fair proportion of their genes, it can be expected that a number of genes will become homozygous, and in this case one of the genes

■ *Figure 2.15*   A human pedigree, illustrating the use of pedigree symbols.

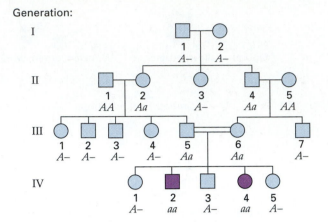

resulted in an identifiable genetic trait. Gene symbols are included in the pedigree to show the deductive reasoning possible with such pedigrees; normally such symbols would not be present. For example, the following reasoning could take place: The trait appears first in generation IV. Since neither parent (the two cousins) had the trait, while two children were produced with the trait (IV-2 and IV-4), the simplest hypothesis is that the trait is caused by a recessive allele. Thus IV-2 and IV-4 would both have the genotype *aa*, and their parents (III-5 and III-6) must both have the genotype *Aa*. All other individuals who did not have the trait must have at least one *A* allele; that is, they must be *A–*. Since III-5 and III-6 are both heterozygotes, then at least one of each of their parents must have carried an *a* allele. In Figure 2.15, II-2 and II-4 must both be *Aa* so that *Aa* III-5 and III-6 individuals can be produced to give rise to the *aa* offspring IV-2 and IV-4.

■ *Figure 2.14*   Symbols used in human pedigree analysis.

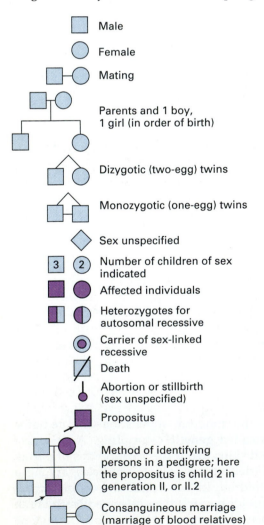

Examples of Human Genetic Traits

Recessive Traits   A large number of human traits are known to be caused by recessive genes. An individual expressing the recessive trait of albinism (deficient pigmentation) is shown in Figure 2.16(a) and a pedigree for this trait is shown in Figure 2.16(b).

For recessive traits to be expressed the allele must be homozygous (*aa*). Recessive mutant alleles are commonly associated with serious abnormalities or diseases. Individuals with albinism, for instance, do not produce the melanin pigment that protects the skin from harmful ultraviolet radiation. As a consequence, albinos have considerable skin and eye sensitivity to sunlight. Frequencies of recessive mutant alleles are usually higher than frequen-

■ *Figure 2.16* (a) Human albino (left); (b) A pedigree showing the transmission of the autosomal recessive trait of albinism.

b)   Pedigree

Generation:

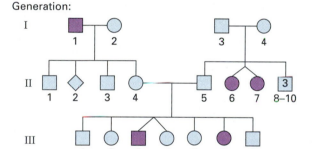

cies of dominant mutant alleles because heterozygotes for the recessive mutant allele are not at a significant selective disadvantage. Nonetheless, most recessive traits are rare. In the United States approximately 1 in 38,000 of the white population and 1 in 10,000 of the black population are albinos.

The following are some general characteristics of recessive inheritance for a relatively rare trait (refer to Figure 2.16[b]):

1. Affected individuals have two normal parents, both of whom are heterozygous. The trait appears in the F_1 since a quarter of the progeny are expected to be homozygous for the recessive allele. If the trait is rare or relatively rare, an individual expressing the trait is likely to mate with a homozygous normal individual; thus the next generation is represented by heterozygotes who do not express the trait. In other words recessive traits often "skip" generations. In the pedigree, for example, II-6 and II-7 must both

be *aa*, and this means both parents (I-3 and I-4) must be *Aa* heterozygotes. I-1 is also *aa* and so II-4 must be *Aa*. Since II-4 and II-5 produce some *aa* children, II-5 also must be *Aa*.

2. Matings between two normal heterozygotes should produce an approximately 3:1 ratio of normal progeny to progeny exhibiting the trait. However, it is difficult to obtain enough numbers to make the data statistically significant, especially if a biochemical test is necessary to confirm the presence of the trait. In such cases only the living members of a family can be surveyed. For recessive genes that have less deleterious effects, the allele can reach significant frequencies in the population. An example of such a recessive trait is attached ear lobes. There are significant numbers of heterozygotes and homozygous recessives for this trait in the population. As a result there is a strong possibility of $Aa \times aa$ matings, and half the progeny will have the trait.

3. When both parents are affected, all their progeny will exhibit the trait.

Dominant Traits There are many known dominant human traits. Figure 2.17(a) illustrates one such trait called woolly hair, in which an individual's hair is very tightly kinked, is very brittle, and breaks before it can grow very long. The best examples of pedigrees for this trait come from Norwegian families; one of these pedigrees is presented in Figure 2.17(b). Since it is a fairly rare trait and since not all children of an affected parent show the trait, it can be assumed that most woolly haired individuals are heterozygous for the dominant allele involved.

Dominant mutant alleles are expressed in a heterozygote when they are in combination with what, in experimental organisms, is usually called the **wild-type allele**—the allele that is designated as the standard ("normal") for the organism. Because many dominant mutant alleles that give rise to recognizable traits are rare, it is extremely unusual to find individuals homozygous for the dominant allele. Thus an affected person in a pedigree is likely to be a heterozygote, and most pairings that involve the mutant allele are between a heterozygote and a homozygous recessive (normal). Most dominant mutant genes that are clinically significant (that is, cause medical problems) fall into this category.

Most known dominant traits produce less severe clinical effects than do recessive traits. If the dominant trait has serious effects, natural selection tends

■ *Figure 2.17*   (a) Members of a Norwegian family, some exhibiting the trait of woolly hair; (b) Part of a pedigree showing the transmission of the autosomal dominant trait of woolly hair.

a)

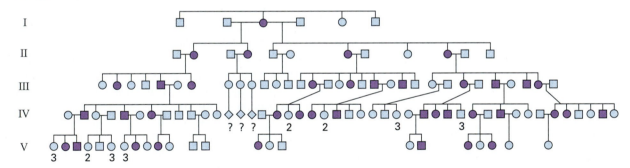

b)

to decrease the frequency of that allele or to remove that allele from the population. That is, mutant individuals may be unable to reproduce or may not survive.

Here are some general characteristics of a dominant trait (refer to Figure 2.17[b]):

1. Every affected person in the pedigree must have at least one affected parent.
2. Each generation of the pedigree must have individuals who express the mutant gene. In other words, the trait must not skip generations.
3. An affected heterozygous individual will, on the average, transmit the mutant gene to half of his or her progeny. Suppose the dominant

mutant allele is designated A, and its normal allele is a. Then most crosses will be $Aa \times aa$. According to basic Mendelian principles half the progeny will be aa (wild type) and the other half will be Aa and show the trait.

Keynote

Mendelian principles apply to humans as well as to peas and all other eukaryotes. The study of the human inheritance of genetic traits is complicated by the fact that no controlled crosses can be done. Instead, human geneticists analyze genetic traits by pedigree analysis, that is, by examining the occurrences of the trait in family trees of individuals who clearly exhibit the trait. Many recessively inherited and dominantly inherited genetic traits have been identified as a result of pedigree analysis.

Summary

In this chapter we discussed fundamental principles of gene function and gene segregation. Genes in this chapter are defined as DNA segments that control the biological characteristics that are transmitted from one generation to another; that is, the hereditary traits. An organism's genetic constitution is called its genotype, while the physical manifestation of a genetic trait is called the phenotype. An organism's genes only give the potential for the development of that organism's characteristics. That potential is influenced during development by interactions with other genes as well as with the environment. Thus individuals with the same genotype can have different phenotypes, and individuals with the same phenotype may have different genotypes.

The first person to begin to comprehend the principles of heredity, or the inheritance of certain traits, was Gregor Mendel. From his breeding experiments with garden peas, Mendel proposed two basic principles of genetics. In modern terms, the principle of segregation states that the two members of a single gene pair (the alleles) segregate from each other in the formation of gametes. For each gene with two alleles, half of the gametes carry one allele and the other half carry the other allele. The principle of independent assortment, proposed on the basis of experiments involving more than one gene, states that genes for different traits behave independently in the production of gametes. Both principles are recognized by characteristic phenotypic ratios—called Mendelian ratios—in particular crosses. For the principle of segregation, in a monohybrid cross between two true-breeding parents, one exhibiting a dominant phenotype and the other a recessive phenotype, the F_2 phenotypic ratio will be 3:1 for the dominant: recessive phenotypes. For the principle of independent assortment, in a dihybrid cross, the F_2 phenotypic ratio will be 9:3:3:1 for the four phenotypic classes. The gene segregation patterns can be studied more definitively by determining the genotype for each phenotypic class. This is done using a testcross, in which an individual of unknown genotype is crossed with a homozygous recessive to determine the genotype of the unknown individual. For example, in a 3:1 phenotypic ratio, the dominant class can be shown to consist of 1 homozygous dominant: 2 heterozygotes by using a testcross.

After the rediscovery of Mendel's laws in 1900, geneticists found that Mendelian principles of gene segregation apply to all other eukaryotes, including humans. The study of the inheritance of genetic traits in humans is complicated by the fact that no controlled crosses can be done. Instead, human geneticists analyze genetic traits by pedigree analysis, that is, by examining the occurrences of the trait in family trees of individuals who clearly exhibit the trait. Many recessively inherited and dominantly inherited genetic traits have been identified as a result of pedigree analysis.

Analytical Approaches for Solving Genetics Problems

The most practical way to reinforce Mendelian principles is to solve genetics problems. These problems present familiar and unfamiliar examples and are questions designed to make you think analytically. Here, and in all following chapters, we will discuss how to approach genetics problems by presenting examples of such problems and discussing their answers.

Q2.1 A purple-flowered pea plant is crossed with a white-flowered pea plant. All the F_1 plants produce purple flowers. When the F_1 plants are allowed to self-pollinate, 401 of the F_2 plants have purple flowers and 131 have white flowers. What are the geno-

types of the parental and F_1 generation plants?

A2.1 The ratio of plant phenotypes in the F_2 is very close to the 3:1 ratio expected of a monohybrid cross. More specifically, this ratio is expected to result from an $F_1 \times F_1$ cross in which both are identically heterozygous for a specific gene pair. In addition, since the two parents differed in phenotype and only one phenotypic class appeared in the F_1, it is likely that both parental plants were true breeding. Further, since the F_1 phenotype exactly resembled one of the parental phenotypes, we can say that purple is dominant to white flowers. Assigning the symbol P to the gene that determines purpleness of flowers and the symbol *p* to the alternative form of the gene that determines whiteness, we can write the genotypes:

P generation: *PP*, for the purple-flowered plant
pp, for the white-flowered plant
F₁ generation: *Pp*, which, because of dominance, is purple-flowered

We could further deduce that the F₂ plants have an approximately 1:2:1 ratio of *PP*:*Pp*:*pp*.

Q2.2 Consider three gene pairs—*Aa*, *Bb*, and *Cc*—each of which affects a different character. In each case the uppercase letter signifies the dominant allele and the lowercase letter the recessive allele. These three gene pairs assort independently of each other. Calculate the probability of obtaining:
a. an *Aa BB Cc* zygote from a cross of individuals that are *Aa Bb Cc* × *Aa Bb Cc*;
b. an *Aa BB cc* zygote from a cross of individuals that are *aa BB cc* × *AA bb CC*;
c. an *A B C* phenotype from a cross of individuals that are *Aa Bb CC* × *Aa Bb cc*;
d. an *a b c* phenotype from a cross of individuals that are *Aa Bb Cc* × *aa Bb cc*.

A2.2 We must again reduce the question into simple parts in order to apply basic Mendelian principles. The key is that the genes assort independently, so we must multiply the probabilities of the individual occurrences to obtain the answers.
a. First consider the *Aa* gene pair. The cross is *Aa* × *Aa*, so the probability of the zygote being *Aa* is 2/4 since the expected distribution of genotypes is 1 *AA*: 2 *Aa*: 1 *aa*, as we have discussed. Then the probability of *BB* from *Bb* × *Bb* is 1/4, and that of *Cc* from *Cc* × *Cc* is 2/4, following the same sort of logic. Using the product rule (see Box 2.3), the probability of an *Aa BB Cc* zygote is $1/2 \times 1/4 \times 1/2 = 1/16$.
b. Similar logic is needed here, although we must be sure of the genotypes of the parental types since they differ from one gene pair to another. For the *Aa* pair the probability of getting *Aa* from *AA* × *aa* has to be 1. And the probability of getting *BB* from *BB* × *bb* is 0, so on these grounds alone we cannot get the zygote asked for from the cross given.
c. This question and the next ask for the probability of getting a particular phenotype, so we must start thinking of dominance. First we break up the question and consider each character pair in turn. The probability of an *A* phenotype from *Aa* × *Aa* is 3/4, from basic Mendelian principles. The probability of a *B* phenotype from *Bb* × *Bb* is 3/4. Lastly, the probability of a *C* phenotype

from *CC* × *cc* is 1. Thus the probability of an *A B C* phenotype is $3/4 \times 3/4 \times 1 = 9/16$.
d. An *a b c* phenotype from *Aa Bb Cc* × *aa Bb cc* is $1/2 \times 1/4 \times 1/2 = 1/16$.

Q2.3 In chickens the white plumage of the leghorn breed is dominant over colored plumage, feathered shanks are dominant over clean shanks, and pea comb is dominant over single comb. Each of the gene pairs segregates independently. If a homozygous white, feathered, pea-combed chicken is crossed with a homozygous colored, clean, single-combed chicken, and the F₁s are allowed to interbreed, what proportion of the birds in the F₂ will produce only white, feathered, pea-combed progeny if mated to colored, clean-shanked, single-combed birds?

A2.3 This example is typical of a problem that presents the unfamiliar in an attempt to get at the familiar. The best approach to such questions is to reduce them to their simple parts and, wherever possible, to assign gene symbols for each character. We are told which character is dominant for each of the three gene pairs, so we can use *W* for white and *w* for colored, *F* for feathered and *f* for clean shanks, and *P* for pea comb and *p* for single comb. The cross involves true-breeding strains and can be written as follows:

P generation: *WW FF PP* × *ww ff pp*
F₁ generation: *Ww Ff Pp*

The question asks the proportion of birds in the F₂ that will produce only white, feathered, pea-combed progeny if mated to colored, clean-shanked, single-combed birds. The latter are homozygous recessive for all three genes; that is, *ww ff pp*, as in the parental generation. To produce the desired result, the F₂ birds must be white, feathered, pea-combed birds and they must be homozygous for the dominant alleles of the respective genes in order to produce progeny with the dominant phenotype. What we are seeking, then, is the proportion of the F₂ chickens that are *WW FF PP* in genotype. We know that each gene pair segregates independently, so the answer can be calculated by using simple probability rules. Consider each gene pair in turn. For the white/colored case the F₁ × F₁ is *Ww* × *Ww*, and from Mendelian principles the relative proportion of F₂ genotypes will be 1 *WW*: 2 *Ww*: 1 *ww*. Therefore the proportion of the F₂s that will be *WW* is 1/4. The same relationship holds for the other two pairs of genes. Since the segregation of the three gene pairs is independent, we must multiply the probabilities of each occurrence to calculate the probability for *WW FF PP* individuals. The answer is $1/4 \times 1/4 \times 1/4 = 1/64$.

Questions and Problems

*2.1 In tomatoes, red fruit color is dominant to yellow. Suppose a tomato plant homozygous for red is crossed with one homozygous for yellow. Determine the appearance of (a) the F_1; (b) the F_2; (c) the offspring of a cross of the F_1 back to the red parent; (d) the offspring of a cross of the F_1 back to the yellow parent.

2.2 A red-fruited tomato plant, when crossed with a yellow-fruited one, produces progeny about half of which are red-fruited and half are yellow-fruited. What are the genotypes of the parents?

2.3 In maize a dominant gene A is necessary for seed color as opposed to colorless (a). A recessive gene wx results in waxy starch as opposed to normal starch (Wx). The two genes segregate independently. Give phenotypes and relative frequencies for offspring resulting when a plant of genetic constitution $Aa\ WxWx$ is testcrossed.

*2.4 F_2 plants segregate 3/4 colored:1/4 colorless. If a colored plant is picked at random and selfed, what is the probability that more than one type will segregate among a large number of its progeny?

*2.5 In guinea pigs rough coat (R) is dominant over smooth coat (r). A rough-coated guinea pig is bred to a smooth one, giving 8 rough and 7 smooth progeny in the F_1.
a. What are the genotypes of the parents and their offspring?
b. If one of the rough F_1 animals is mated to its rough parent, what progeny would you expect?

2.6 In cattle the polled (hornless) condition (P) is dominant over the horned (p) phenotype. A particular polled bull is bred to three cows. With cow A, which is horned, a horned calf is produced; with a polled cow B a horned calf is produced; and with horned cow C a polled calf is produced. What are the genotypes of the bull and the three cows, and what phenotypic ratios do you expect in the offspring of these three matings?

*2.7 Purple flowers are dominant to white in the Jimsonweed. When a particular purple-flowered Jimsonweed is self-fertilized, there are 28 purple-flowered and 10 white-flowered progeny. What proportion of the purple-flowered progeny will breed true?

*2.8 Two black female mice (X and Y) are crossed with the same brown male. In a number of litters female X produced 9 blacks and 7 browns and female Y produced 14 blacks. What is the mechanism of inheritance of black and brown coat color in mice? What are the genotypes of the parents?

2.9 Bean plants may differ in their symptoms when infected with a virus. Some show local lesions that do not seriously harm the plant. Other plants show general systemic infection. The following genetic analysis was made:

> P local lesions × systemic lesions
> F_1 all local lesions
> F_2 785 local lesions: 269 systemic lesions

What is the probable genetic basis of this difference in beans? Assign gene symbols to all the genotypes occurring in this experiment. Design a testcross to verify your assumptions.

*2.10 Fur color in babbits, a furry little animal that is a popular pet, is determined by a pair of alleles, B and b. BB and Bb babbits are black; bb babbits are white. A farmer wants to breed babbits for sale. True-breeding white (bb) female babbits breed poorly. The farmer purchases a pair of black babbits, and these mate and produce 6 black and 2 white offspring. The farmer immediately sells his white babbits, and then comes to consult you for a breeding strategy to produce more white babbits.
a. If he performed random crosses between pairs of F_1 babbits, what proportion of the F_2 progeny would be white?
b. If he crossed an F_1 male to the parental female, what is the probability that this cross will produce white progeny?
c. What would be the farmer's best strategy to maximize the production of white babbits?

2.11 In Jimsonweed purple flower (P) is dominant to white (p), and spiny pods (S) are dominant to smooth (s). In a cross between a Jimsonweed homozygous for white flowers and spiny pods and one homozygous for purple flowers and smooth pods, determine the following:
a. the phenotype of the F_1;
b. the phenotype of the F_2;
c. the progeny of a cross of the F_1 back to the white, spiny parent;
d. the progeny of a cross of the F_1 back to the purple, smooth parent.

2.12 What progeny would you expect from the following Jimsonweed crosses? You are encouraged to use the branch diagram approach.

a. *PP ss* × *pp SS*
b. *Pp SS* × *pp ss*
c. *Pp Ss* × *Pp SS*
d. *Pp Ss* × *Pp Ss*
e. *Pp Ss* × *Pp ss*
f. *Pp Ss* × *pp ss*

*2.13   White fruit (*W*) in summer squash is dominant over yellow (*w*), and disk-shaped fruit (*D*) is dominant over sphere-shaped fruit (*d*). In the following problems the appearances of the parents and their progeny are given. Determine the genotypes of the parents in each case.

a. White, disk × yellow, sphere gives 1/2 white, disk and 1/2 white, sphere.
b. White, sphere × white, sphere gives 3/4 white, sphere and 1/4 yellow, sphere.
c. Yellow, disk × white, sphere gives all white, disk progeny.
d. White, disk × yellow, sphere gives 1/4 white, disk; 1/4 white, sphere, 1/4 yellow, disk; and 1/4 yellow, sphere.
e. White, disk × white, sphere gives 3/8 white, disk; 3/8 white, sphere, 1/8 yellow, disk; and 1/8 yellow, sphere.

*2.14   Genes *a*, *b*, and *c* assort independently and are recessive to their respective alleles *A, B,* and *C*. Two triply heterozygous (*Aa Bb Cc*) individuals are crossed.

a. What is the probability that a given offspring will be phenotypically *ABC*, that is, exhibit all three dominant traits?
b. What is the probability that a given offspring will be genotypically homozygous for all three dominant alleles?

2.15   In garden peas tall stem (*T*) is dominant over short stem (*t*), green pods (*G*) are dominant over yellow pods (*g*), and smooth seeds (*S*) are dominant over wrinkled seeds (*s*). Suppose a homozygous short, green, wrinkled pea plant is crossed with a homozygous tall, yellow, smooth one.

a. What will be the appearance of the F_1?
b. What will be the appearance of the F_2?
c. What will be the appearance of the offspring of a cross of the F_1 back to its short, green, wrinkled parent?
d. What will be the appearance of the offspring of a cross of the F_1 back to its tall, yellow, smooth parent?

2.16   Two homozygous strains of corn are hybridized. They are distinguished by six different pairs of genes, all of which assort independently and produce an independent phenotypic effect. The F_1 hybrid is selfed to give an F_2.

a. What is the number of possible genotypes in the F_2?
b. How many of these genotypes will be homozygous at all six gene loci?
c. If all gene pairs act in a dominant-recessive fashion, what proportion of the F_2 will be homozygous for all dominants?
d. What proportion of the F_2 will show all dominant phenotypes?

*2.17   The coat color of mice is controlled by several genes. The agouti pattern, characterized by a yellow band of pigment near the tip of the hairs, is produced by the dominant allele *A*; homozygous *aa* mice do not have the band and are nonagouti. The dominant allele *B* determines black hairs, and the recessive allele *b* determines brown. Homozygous $c^h c^h$ individuals allow pigments to be deposited only at the extremities (e.g., feet, nose, and ears) in a pattern called Himalayan. The genotype *C*– allows pigment to be distributed over the entire body.

a. If a true-breeding black mouse is crossed with a true-breeding brown, agouti, Himalayan mouse, what will be the phenotypes of the F_1 and F_2?
b. What proportion of the black agouti F_2 will be of genotype *Aa BB Cc^h*?
c. What proportion of the Himalayan mice in the F_2 are expected to show brown pigment?
d. What proportion of all agoutis in the F_2 are expected to show black pigment?

2.18   For pedigrees A and B shown in the figure, indicate whether the trait involved in each case could be (a) recessive or (b) dominant. Explain your answer.

Pedigree A

Generation:

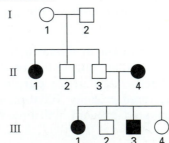

Pedigree B

Generation:

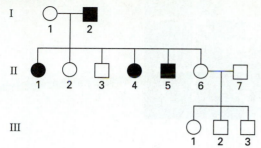

a. What is the genotype of the mother?
b. What is the genotype of the father?
c. What are the genotypes of the children?
d. Given the mechanism of inheritance involved, does the ratio of children with the trait to children without the trait match what would be expected?

*2.19   Consider the pedigree shown in the following figure. The allele responsible for the trait (*a*) is recessive to the normal allele (*A*).

Generation:

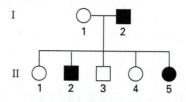

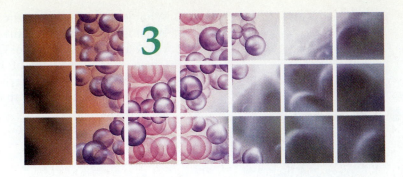

3

Chromosomal Basis of Inheritance, Sex Determination, and Sex Linkage

- The chromosome theory of heredity states that the chromosomes are the carriers of the genes.

- A sex chromosome is a chromosome in eukaryotic organisms that is represented differently in the two sexes. In most organisms with sex chromosomes, the female has two large X chromosomes (XX) and the male has one X and a smaller Y chromosome (XY). Though structurally different, the X and Y act as homologs in meiosis..

- Sex linkage is the association of genes with the sex determining chromosomes of eukaryotes. Such genes are referred to as sex-linked or X-linked genes.

- An unexpected inheritance pattern of an X-linked mutant gene in *Drosophila* correlated with a rare event during meiosis, called nondisjunction, in which members of a homologous pair of chromo-

somes do not segregate to the opposite poles. The correlation between gene segregation patterns and the patterns of chromosome behavior in meiosis supported the chromosome theory of heredity.

- In many eukaryotic organisms, sex determination is related to the sex chromosomes. In humans and other mammals, for example, the presence of a Y chromosome specifies maleness, while its absence results in femaleness. Several other sex-determination mechanisms are known in eukaryotes.

- In humans (and other eukaryotes with sex chromosomes), the gene responsible for a trait can be inherited several ways, including autosomal recessive, autosomal dominant, sex-linked recessive, sex-linked dominant, or Y-linked.

S OON AFTER THE BEGINNING of the twentieth century, scientists realized that Mendelian laws applied to a large variety of organisms and that the principles of Mendelian analysis could be used to predict the outcome of crosses in these organisms. On Mendel's foundation geneticists began to build genetic hypotheses that could be tested by appropriate crosses and to investigate the nature of Mendelian factors. We examine in this chapter the evidence showing that Mendelian factors are what we now know as genes and that genes are located in chromosomes. This is called the chromosome theory of heredity. We will learn about various mechanisms of sex determination, some of which involve special chromosomes (sex chromosomes) that are related to the sex of the organism. We will also discuss the genetic segregation patterns of genes located on the sex chromosomes and in the other chromosomes (the autosomes) of the cell.

Chromosome Theory of Heredity

Within a given organism the total number of chromosomes is constant in all cells, while the chromosome number varies considerably between species (Table 3.1). Based on observations that the transmission of chromosomes from one generation to the

next closely paralleled the pattern of transmission of genes from one generation to the next, the **chromosome theory of heredity** was proposed. This

Table 3.1
Chromosome Number in Various Organisms[a]

ORGANISM	TOTAL CHROMOSOME NUMBER
Human	46
Chimpanzee	48
Dog	78
Cat	72
Horse	64
Chicken	78
Bullfrog	26
Goldfish	94
Fruit fly	8
Nematode	♂ 11, ♀12
Neurospora (haploid)	7
Bread wheat	42
Baker's yeast	34

[a]Except as noted, all chromosome numbers are for diploid cells.

51

theory states that the chromosomes are the carriers of the genes. In this section we will consider some of the evidence to support this theory.

Sex Chromosomes

The chromosome theory of heredity was supported by experiments that related the hereditary behavior of particular genes to the transmission of the **sex chromosome**, the chromosome in eukaryotic organisms that is represented differently in the two sexes. For many animals the sex chromosome composition of the individual is directly related to the sex of the individual. The sex chromosomes typically are designated the **X chromosome** and the **Y chromosome**. In humans, for example, the female has two X chromosomes (i.e., she is XX with respect to the sex chromosomes), while the male has one X chromosome and one Y chromosome (i.e., he is XY). Similarly, in the fruit fly, *Drosophila melanogaster*, the female is XX and the male is XY. Figure 3.1 shows the appearance of male and female *Drosophila*, Figure 3.2 shows the chromosome complements of the two sexes, and Figure 3.3 shows the life cycle. Because the male produces two kinds of gametes with respect to sex chromosome content (X or Y),

■ *Figure 3.1*   Female (left) and male (right) *Drosophila melanogaster* (fruit fly), an organism used extensively in genetics experiments: (a) Adult flies; (b) Ventral abdominal surface showing differences in genitalia.

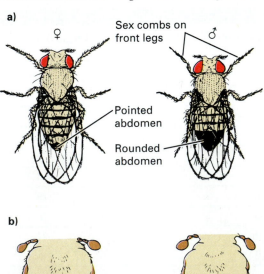

■ *Figure 3.2*   Chromosomes of *Drosophila melanogaster* diagrammed to show morphological differences. A female has four pairs of chromosomes in her somatic cells, including a pair of X chromosomes. The only difference in the male is an XY pair of sex chromosomes instead of two X chromosomes.

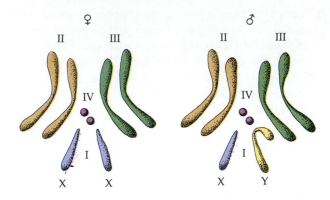

and because the female produces only one gametic type (X), the male is called the **heterogametic sex** and the female is called the **homogametic sex**. In *Drosophila* the X and Y chromosomes are similar in size but their shapes are different. (Note: In some organisms the male is homogametic and the female is heterogametic. This reversal will be discussed later in the chapter.)

As shown in Figure 3.4, the pattern of transmission of X and Y chromosomes from generation to generation is very straightforward. In this figure the X is represented by a straight structure much like a slash mark, and the Y by a similar structure topped by a hook to the right. The mother produces X-bearing gametes, while the male produces both X-bearing and Y-bearing gametes. Random fusion of these gametes produces an F_1 generation with half XX (female) and half XY (male) flies.

Keynote

A sex chromosome is a chromosome in eukaryotic organisms that is represented differently in the two sexes. In many of the organisms encountered in genetic studies, one sex possesses a pair of identical chromosomes (the X chromosomes). The opposite sex possesses a pair of visibly different chromosomes: one is an X chromosome, and the other, structurally and functionally different, is the Y chromosome. Commonly, the XX sex is female, and the XY sex is male. The XX and XY sexes are called the homogametic and heterogametic sexes, respectively.

■ *Figure 3.3*   Life cycle of *Drosophila melanogaster*.

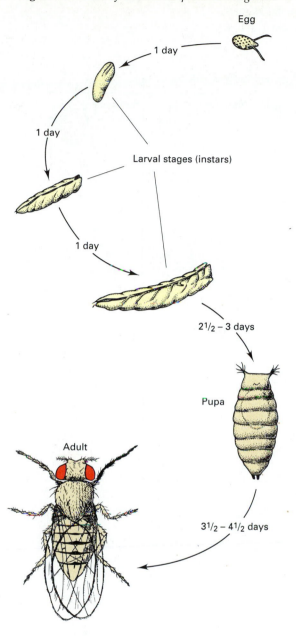

Egg

1 day

1 day

Larval stages (instars)

1 day

2½ – 3 days

Pupa

Adult

3½ – 4½ days

For example, a *Drosophila* strain with all wild-type genes will have bright red eyes. Variants of a wild-type strain arise from mutational changes of the wild-type genes that produce **mutant alleles**; the result is strains with mutant characteristics. Mutant alleles may be recessive or dominant to the wild-type allele; for example, the mutant allele that causes white eyes in *Drosophila* is recessive to the wild-type (red-eye) allele.

When Morgan crossed the white-eyed male with a red-eyed female from the same stock, he found that all the F_1 flies were red-eyed and concluded that the white-eyed trait was recessive. Next he allowed the F_1 progeny to interbreed and counted the phenotypic classes in the F_2 generation. There were 3470 red-eyed and 782 white-eyed flies. The number of individuals with the recessive phenotype was too small to fit the Mendelian 3:1 ratio. In addition, *Morgan noticed that all the white-eyed flies were male.* (Later he determined that there was another

■ *Figure 3.4*   Inheritance pattern of X and Y chromosomes in organisms such as *Drosophila melanogaster* where the female is XX and the male is XY.

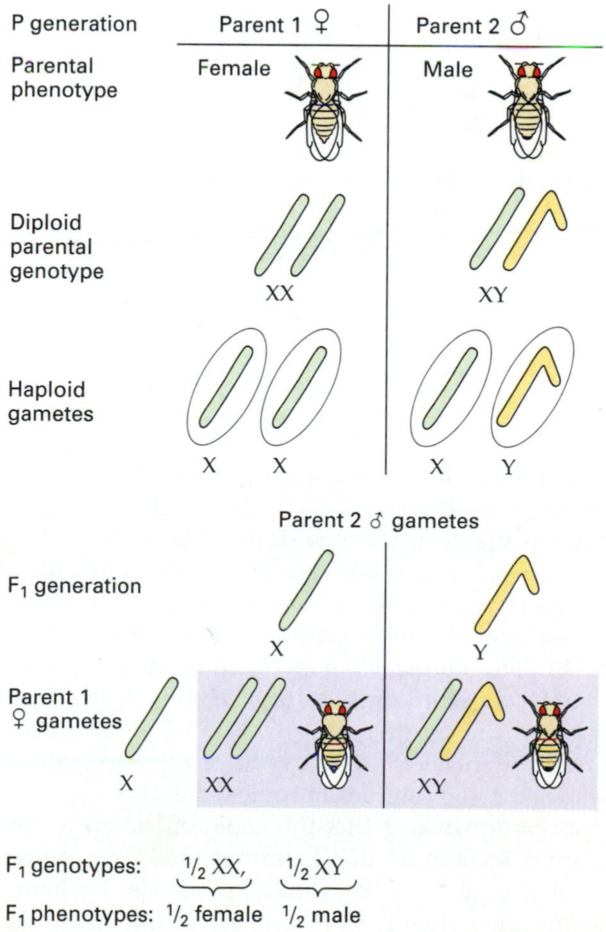

P generation	Parent 1 ♀	Parent 2 ♂
Parental phenotype	Female	Male
Diploid parental genotype	XX	XY
Haploid gametes	X X	X Y

Parent 2 ♂ gametes

F₁ generation	X	Y
Parent 1 ♀ gametes X	XX	XY

F_1 genotypes: ½ XX,   ½ XY

F_1 phenotypes: ½ female   ½ male

Sex Linkage

Further evidence to support the chromosome theory of heredity was obtained by using *Drosophila* (fruit flies) in genetic experiments. In 1910 Thomas Hunt Morgan reported his results of crossing experiments with the fruit fly.

Morgan found a male fly in one of his true-breeding stocks that had white eyes instead of the bright red eyes that are characteristic of the **wild type**. The term *wild type* refers to a strain, organism, or gene that is designated as the standard for the organism with respect to genotype and phenotype.

■ *Figure 3.5*   The X-linked inheritance of red eyes and white eyes in *Drosophila melanogaster*. The symbols *w* and w^+ indicate the white-eyed and red-eyed alleles, respectively. (a) A red-eyed female is crossed with a white-eyed male; (b) The F_1 flies are interbred to produce the F_2s.

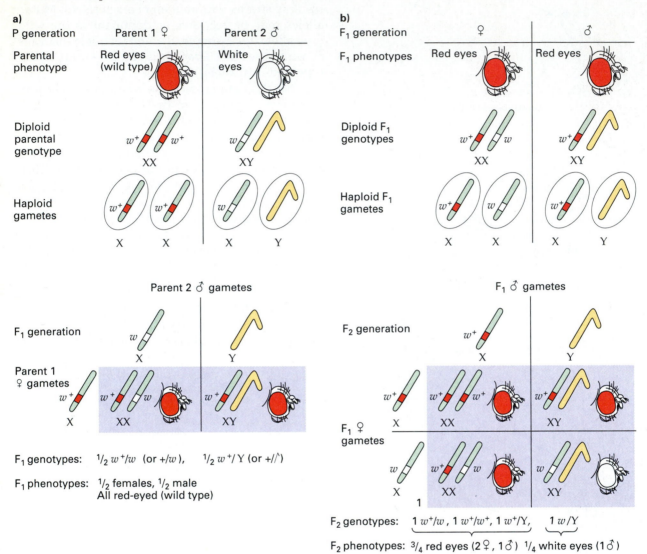

reason for the lower-than-expected number of flies with the recessive phenotype: flies with white eyes are not as viable as those with red eyes.)

Figure 3.5 diagrams this cross. The *Drosophila* gene symbolism used is different from the symbols we used for Mendel's crosses and is described in Box 3.1. The *Drosophila* gene symbolism should be understood before proceeding with this discussion. Note that the mother-son inheritance pattern presented in Figure 3.5 is the result of the segregation of genes located on a sex chromosome.

Morgan proposed that the gene for the eye color variant is located on the X chromosome. The condition of X-linked genes in males is said to be **hemizygous** since they have no corresponding allele on

the Y. For example, the white-eyed *Drosophila* males have an X chromosome with a white allele and no other allele of that gene in their genome: These males are said to be hemizygous for the white allele. Since the white allele of the gene is recessive, the original white-eyed male must have had the recessive allele for white eyes (designated *w*) on his X chromosome. The red-eyed female came from a true-breeding stock, so both of her X chromosomes must have carried the dominant allele for red eyes, w^+.

The F_1s are produced in the following way (see Figure 3.5[a]). The males receive their only X chromosome from the mother; hence they have the w^+ allele and are red-eyed. The females receive a domi-

Genetic Symbols Revisited

There is no one system of gene symbols used by geneticists. The gene symbols used for *Drosophila* are different from those used for peas in Chapter 2. The *Drosophila* symbolism is commonly, but not exclusively, used in genetics today. In this system the symbol + indicates a wild-type allele of a gene. A lowercase letter designates mutant alleles of a gene that are *recessive* to the wild-type allele, and an uppercase letter is used for alleles that are *dominant* to the wild-type allele. *The letters are chosen on the basis of the phenotype of the organism expressing the mutant allele.* For instance, a variant strain of *Drosophila* has bright orange eyes instead of the usual bright red. The mutant allele involved is recessive to the wild-type bright red allele, and its bright orange eye color is close to vermilion in tint. Therefore the allele is designated v and is called the vermilion allele. The wild-type allele of v is v^+, but when there is no chance of confusing it with other genes in the cross, it is often shortened to +. In the Mendelian terminology used up to now, the recessive mutant allele would be v, and its wild-type allele would be V.

A conventional way to represent the chromosomes (instead of the way we have been using in the figures) is to use the slash (/). Thus v^+/v or $+/v$ indicates that there are two homologous

chromosomes, one with the wild-type allele (v^+ or +) and the other with the recessive allele (v). The Y chromosome is usually symbolized as a Y or a bent slash (∧). Thus Morgan's cross of a true-breeding red-eyed female fly with a white-eyed male could be written $w^+/w^+ \times w/Y$ or $+/+ \times w//\wedge$.

The same rules apply when the alleles involved are dominant to the wild-type allele. Some *Drosophila* mutants, called *Curly*, have wings that curl up at the end rather than the typical straight wings. The symbol for this mutant allele is Cy, and the wild-type allele is Cy^+, or + in the shorthand version. Thus a heterozygote would be Cy/Cy^+ or $Cy/+$.

Both the A/a (Mendelian) and a^+/a (*Drosophila*) symbolisms will be used in the remaining chapters, so it is important that you be able to work with both. Since it is easier to verbalize the Mendelian symbols (e.g., big A, small a), many of our examples will follow that symbolism, even though the *Drosophila* symbolism is more informative in many ways. That is, with the *Drosophila* system, the wild-type and mutant alleles are readily apparent because the wild-type allele is indicated by a +. However, the Mendelian system is commonly used in animal and plant breeding.

nant w^+ allele from the mother and a recessive w allele from the father. As a result of this inheritance pattern, the F_1 females are also red-eyed.

To produce the F_2 flies Morgan interbred F_1 red-eyed females and red-eyed males (see Figure 3.5[b]). In the F_2 the males that received an X chromosome with the w allele from the mother are white-eyed; those that received an X chromosome with the w^+ allele are red-eyed. The gene transmission shown in this cross from a male parent to a female offspring ("child") to a male grandchild is called **crisscross inheritance**.

Morgan also crossed a red-eyed male with a true-breeding white-eyed female (Figure 3.6). (This cross is the *reciprocal cross* of Morgan's first cross performed—white male × red female—shown in Figure 3.5.) The parental male is hemizygous for

the w^+ allele, and the parental female is homozygous for the w allele. All the F_1 females receive a w^+-bearing X from their father and a w-bearing X from their mother. Consequently, they are heterozygous w^+/w and have red eyes. All the F_1 males receive a w-bearing X from their mother and a Y from their father, and so they have white eyes (Figure 3.6[a]). This result is distinct from that of the reciprocal cross in Figure 3.5 because of the sex linkage.

Interbreeding of the F_1 flies (Figure 3.6[b]) involves a $w//\wedge$ male and a w^+/w female giving approximately equal numbers of male and female red- and white-eyed flies in the F_2. This ratio differs from the results in the first cross, where an approximately 3:1 ratio of red-eyed:white-eyed flies was obtained and where none of the females and

■ *Figure 3.6*   Reciprocal cross of that shown in Figure 3.5. (a) A homozygous white-eyed female is crossed with a red-eyed (wild-type) male; (b) The F_1 flies are interbred to produce the F_2s. The results of this cross differ from those in Figure 3.5 because of the way sex chromosomes segregate in crosses.

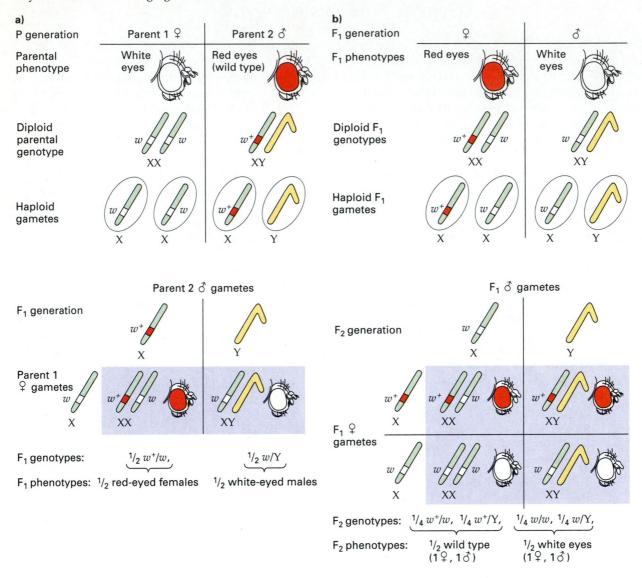

approximately half the males exhibited the white-eyed phenotype. The difference in phenotypic ratios in the two sets of crosses reflects the transmission patterns of sex chromosomes and the genes they contain.

Morgan's crosses of *Drosophila* involved eye color characteristics that we now know are coded for by genes found on the X chromosome. These characteristics, and the genes that give rise to them, are referred to as **sex-linked**, or, more correctly, **X-linked**. *Sex-linked inheritance* is the term used for the pattern of hereditary transmission of sex-linked genes. When the results of reciprocal crosses are

not the same, and different ratios are seen for the two sexes of the offspring, sex-linked characteristics may well be involved. By comparison, the results of reciprocal crosses are the same when they involve genes located on the **autosomes**, chromosomes other than the sex chromosomes. Morgan's results strongly supported the hypothesis that genes are located on chromosomes. Morgan found many other examples of genes on the X chromosome in *Drosophila* and in other organisms, thereby showing that his observations were not confined to a single species. Later in this chapter we will discuss the analysis of X-linked traits in humans.

Keynote

Sex linkage is the association of genes with the sex-determining chromosomes of eukaryotes. Such genes, as well as the phenotypic characteristics these genes control, are referred to as sex-linked. Morgan's pioneering work on the inheritance of sex-linked genes of *Drosophila* greatly supported, but did not prove, the chromosome theory of heredity.

Nondisjunction of X Chromosomes

Further support for the chromosome theory of inheritance came from the work of Morgan's student, Calvin Bridges. Morgan's work showed that from a cross of a white-eyed female (w/w ♀) with a red-eyed male (w^+/Y ♂), all the F_1 males should be white-eyed, and all the females should be red-eyed. Bridges found that there are rare exceptions to this result: About 1 in 2000 of the F_1 flies from such a cross are either white-eyed females or red-eyed males.

To explain these data Bridges hypothesized that a problem had occurred with chromosome segregation in meiosis. Normally homologous chromosomes (in meiosis I) or sister chromatids (in meiosis II or mitosis) move to opposite poles at anaphase (see p. 18). When this fails to take place, chromosome **nondisjunction** results. For the crosses being analyzed, the two X chromosomes occasionally failed to separate, so eggs were produced either with two X chromosomes or with no X chromosomes instead of the usual one. This particular example of nondisjunction is called **X chromosome nondisjunction** (Figure 3.7). Normal disjunction of the X chromosomes is illustrated in Figure 3.7(a), and nondisjunction of the X chromosomes in meiosis I and meiosis II is shown in Figures 3.7(b) and 3.7(c), respectively.

Nondisjunction of the X chromosomes can explain the exceptional flies in Bridges' first experimental cross. When primary nondisjunction occurs in the w/w female (Figure 3.8), two classes of exceptional eggs result with equal frequency: those with two X chromosomes and those with no X chromosomes. The XY male is w^+/Y and produces equal numbers of w^+-bearing and Y-bearing sperm. When these eggs are fertilized by the two types of sperm, the result is four types of zygotes. Two types usually do not survive: the YO class, which has a Y

■ *Figure 3.7* Nondisjunction during meiosis involving the X chromosome. Nondisjunction of autosomal chromosomes in meiosis occurs in the same way. (a) Normal X chromosome segregation; (b) Nondisjunction of X chromosomes in meiosis 1; (c) Nondisjunction of X chromosomes in meiosis II.

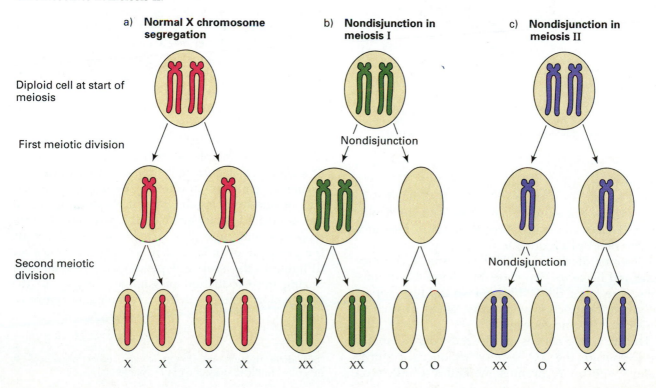

■ *Figure 3.8* Nondisjunction during meiosis in a white-eyed female *Drosophila melanogaster*, and results of a cross with a normal red-eyed male. Note the decreased viability of XXX and YO progeny. (This is a simplification of actual events to avoid confusion.)

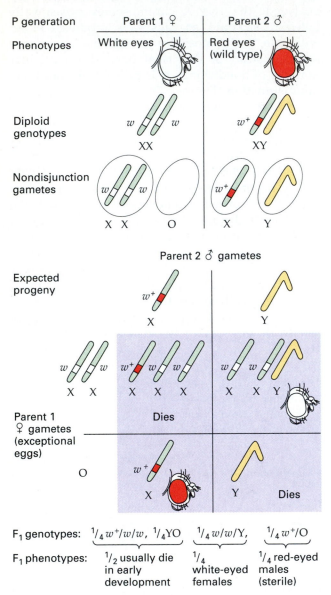

F₁ genotypes:	$\frac{1}{4}w^+/w/w$, $\frac{1}{4}$YO	$\frac{1}{4}w/w/$Y,	$\frac{1}{4}w^+/$O
F₁ phenotypes:	$\frac{1}{2}$ usually die in early development	$\frac{1}{4}$ white-eyed females	$\frac{1}{4}$ red-eyed males (sterile)

the X, and the white-eyed XXY females (in *Drosophila* the XXY pattern produces a fertile female), with a *w* allele on each X. The males are red-eyed because they received their X chromosome from their fathers. The females have white eyes because their two X chromosomes came from their mothers. This result is unusual because sons normally get their X from their mothers, and daughters get one X from each parent. Bridges' hypothesis was very plausible and was confirmed by examining the chromosome composition of the unexpected flies: the white-eyed females were XXY and the red-eyed males were XO. (**Aneuploidy** is the term used for the abnormal condition, as in the case here, in which one or more whole chromosomes of a normal set of chromosomes either are missing or are present in more than the usual number of copies.)

Bridges' experiments showed that the odd pattern of inheritance always went hand in hand with the specific aneuploid types (XO and XXY). Such a correlation could not be due to some accidental parallelism, proving without any doubt that a specific phenotype was associated with a specific chromosome.

In sum, gene segregation patterns follow the patterns of chromosome behavior in meiosis. In Figure 3.9 this parallel is illustrated for a diploid cell with a homologous pair of metacentric chromosomes and a homologous pair of acrocentric chromosomes. The cell is genotypically *Aa Bb*, with the *A/a* gene pair on the metacentric homologs and the *B/b* gene pair on the telocentric homologs. As the figure shows, the two homologous pairs of chromosomes align on the metaphase plate independently, giving rise to two different segregation patterns for the two gene pairs. Since each of the two alignments, and hence segregation patterns, is equally likely, the result of meiosis is cells that exhibit equal frequencies of the genotypes *AB*, *ab*, *Ab*, and *aB*. Genotypes *AB* and *ab* result from one chromosome alignment, and genotypes *Ab* and *aB* result from the other alignment. In terms of Mendel's laws, we can see how the principle of segregation (two members of a gene pair segregate from each other in the formation of gametes) applies to the segregation pattern of one homologous pair of chromosomes and the associated gene pair (e.g., the metacentric chromosomes and gene pair *A/a* in Figure 3.9), while the principle of independent assortment (genes for different traits assort independently of one another during the production of gametes) applies to the segregation pattern of both homologous pairs of

chromosome but no X chromosome, and the triplo-X (XXX) class, which has three X chromosomes and, as illustrated in Figure 3.8, the genotype *w/w/w⁺*. The former die because they lack the X chromosome and its genes that code for essential cell functions. The latter often die because the flies apparently cannot function with three doses of each of the X chromosome genes.

The surviving classes are the red-eyed XO males (in *Drosophila* the XO pattern produces a sterile male),* with no Y chromosome and a *w⁺* allele on

---

*Sex determination in *Drosophila* and other organisms is described in the next section of this chapter.

■ *Figure 3.9* Illustration of the parallel behavior between the segregation of Mendelian genes and chromosomes in meiosis. In the hypothetical *Aa Bb* diploid cell there is a homologous pair of metacentric chromosomes, which carry the *A/a* gene pair, and a homologous pair of acrocentric chromosomes, which carry the *B/b* gene pair. The independent alignment of the two homologous pairs of chromosomes at metaphase I results in equal frequencies of the four meiotic products, *AB*, *ab*, *Ab*, and *aB*, illustrating Mendel's principle of independent assortment.

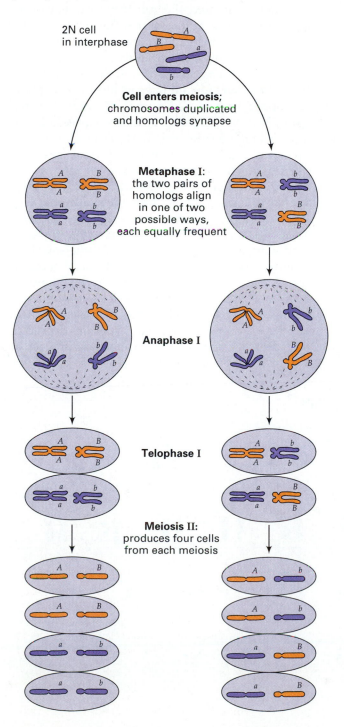

chromosomes and the associated two gene pairs in Figure 3.9.

K e y n o t e

Bridges observed an unexpected inheritance pattern of an X-linked mutant gene in *Drosophila*. He correlated this pattern directly with a rare event during meiosis, called nondisjunction, in which members of a homologous pair of chromosomes do not segregate to the opposite poles. The correlation between gene segregation patterns and the patterns of chromosome behavior in meiosis supported the chromosome theory of heredity.

Sex Determination and Sex Linkage in Eukaryotic Systems

In this section some of the mechanisms for sex determination will be discussed. In *genotypic sex determination systems*, sex is governed by the genotype of the zygote or spores. In *phenotypic (or environmental) sex determination systems*, sex is governed by internal and external environmental conditions.

Genotypic Sex Determination Systems

Genotypic sex determination, in which the sex chromosomes play a decisive role in the inheritance and determination of sex, may occur in one of two ways. In the **Y chromosome mechanism of sex determination** (seen in humans), the Y chromosome of the heterogametic sex is active in determining the sex of an individual. Individuals carrying the Y chromosome are genetically male, while individuals lacking the Y chromosome are genetically female. In the **X chromosome-autosome balance system** (seen in *Drosophila* and the nematode, *Caenorhabditis elegans*), the main factor in sex determination is the *ratio between the number of X chromosomes and the number of sets of autosomes*. In this system the Y chromosome has no effect on sex determination, but is required for male fertility.

Sex Determination in Mammals In humans and other mammals the presence of the Y chromosome determines maleness, and its absence determines femaleness. This mode of sex determination is called the *Y chromosome mechanism of sex determination*. If there is no Y chromosome present, the gonadal primordia develop into ovaries. If a Y chromosome is present, it regulates the production of a protein called *testis-determining factor* (encoded by the *TDF* gene), which causes the gonadal primordia to differentiate into testes instead of ovaries. This is the central event in mammalian sex determination. All other differences between the sexes are secondary effects resulting from hormone action or from the action of factors produced by the gonads. Therefore, sex determination is equivalent to testis determination, and the Y chromosomal gene responsible has been called *TDF* (testis-determining factor) in humans, and *Tdy* (testis-determining gene on the Y) in mice. Recently a gene called *SRY* in humans and *Sry* in mice has been identified and cloned (see Chapter 13). This gene has many of the genetic and biological properties expected of a testis-determining gene, and may in fact be *TDF* (*Tdy*). (For example, when introduced into XX mouse embryos, a DNA fragment carrying *Sry* is sufficient to induce testis differentiation and subsequent male development.)

The early evidence for the Y chromosome basis of sex determination came from studies of humans and other mammals in which meiotic nondisjunction produces an abnormal sex chromosome complement. While sex determination in these cases follows directly from the sex chromosomes present, those individuals who have unusual chromosome complements display many unusual characteristics.

Nondisjunction can produce, for example, exceptional XO individuals who have an X but no Y chromosome. In humans XO individuals with the normal two sets of autosomes are female and sterile, and exhibit **Turner syndrome**. A Turner syndrome individual is shown in Figure 3.10(a), and her set of metaphase chromosomes is shown in 3.10(b). (A complete set of metaphase chromosomes in a cell is called its *karyotype* and is described fully in Chapter 9.) Note that there is only one sex chromosome—an X chromosome. These aneuploid females have a genomic complement of 45,X, indicating that they have a total of 45 chromosomes (sex chromosomes + autosomes), in contrast to the normal 46, and that the sex chromosome complement consists of one X chromosome. Turner syndrome individuals occur with a frequency of 1 in 10,000 female births. It is estimated that up to 99 percent of all 45,X embryos die before birth. Surviving Turner syndrome individuals have few noticeable major defects until puberty, when they fail to develop secondary sexual characteristics. They tend to be shorter than average, and have weblike necks, poorly developed breasts, and immature internal sexual organs. They may have trouble with space orientation and are infertile. All of these defects in XO individuals indicate that two X chromosomes are needed in females for normal development.

Nondisjunction can also result in XXY individuals—males that exhibit **Klinefelter syndrome**. Figure 3.11(a) shows a Klinefelter syndrome individual, and Figure 3.11(b) shows a karyotype for such an individual. Note the presence of three sex chromosomes: two X chromosomes and one Y chromosome. About 1 in 1000 males born have Klinefelter syndrome. These 47,XXY males have underdeveloped testes, are often taller than the average male and have a lower than average IQ. Some degree of breast development is seen in about 50 percent of

■ *Figure 3.10*   Turner syndrome (XO): (a) Individual; (b) Karyotype.

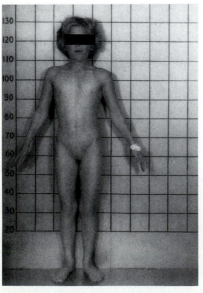

a)

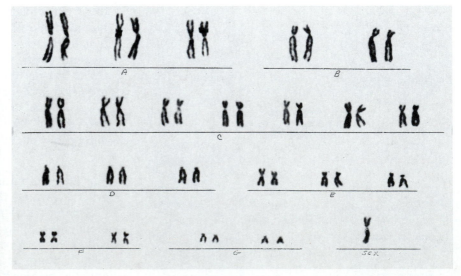

b)

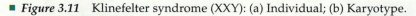

■ *Figure 3.11*   Klinefelter syndrome (XXY): (a) Individual; (b) Karyotype.

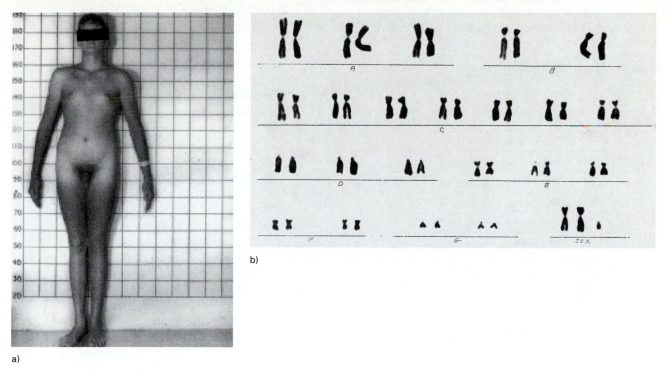

a)

b)

affected individuals. Individuals with similar phenotypes are also found with higher numbers of X and/or Y chromosomes (e.g., 48,XXXY, 49,XXXXY; and 48,XXYY). The defects in Klinefelter individuals indicate that one X and one Y chromosome are needed in males for normal development.

Some humans have one X and two Y chromosomes and have *XYY syndrome*. These 47,XYY individuals are male because of the presence of the Y chromosome. The XYY karyotype results from nondisjunction of the Y chromosome in meiosis. About 1 in 1000 males born have XYY syndrome. They tend to be taller than average and occasionally there are adverse effects on fertility.

About 1 in 1000 females born have three X chromosomes instead of the normal two. These 47,XXX (triplo-X) females are almost completely normal, although they are slightly less fertile and have slightly subnormal intelligence compared with the population as a whole.

Table 3.2 summarizes the consequences of exceptional X and Y chromosome complements in humans. In every case the normal two sets of autosomes are associated with the sex chromosomes. It is interesting that mammals can tolerate abnormalities in the complement of sex chromosomes quite well, whereas, with rare exceptions, they cannot tolerate any variation in the number of autosomes. Mammals with an unusual number of autosomes usually die because there is a **dosage compensation**

mechanism in mammals that compensates for X chromosomes in excess of the normal complement and no such mechanism exists for extra autosomes. In fact, the nuclei of normal XX females shows a highly condensed mass of chromatin not found in the nuclei of normal XY male cells. This mass has been named the **Barr body** after its discoverer, Murray Barr. The somatic cells of XX individuals have one Barr body, and the somatic cells of XY individuals have no Barr bodies (Figure 3.12). In 1961 this concept was expanded by Mary Lyon. In what is now called the *Lyon hypothesis*, Lyon proposed that the Barr body is a highly condensed and therefore inactive (or mostly inactive) X chromosome. The components of the Lyon hypothesis are as follows:

1. The Barr body is a genetically inactive X chromosome (it has become "lyonized"; the process is called **lyonization**).
2. The inactivation occurs at about the sixteenth day following fertilization.
3. The X chromosome to be inactivated is randomly chosen from the maternal and paternal X chromosomes in a process that is independent from cell to cell. (Once a maternal or paternal X chromosome is inactivated in a cell, all descendants of that cell inherit the inactivation pattern.)

When lyonization operates in cells with extra X chromosomes, all but one of the X chromosomes

Table 3.2

Consequences of Various Numbers of X and Y chromosome Abnormalities in Humans, Showing the Role of the Y in Sex Determination

CHROMOSOME CONSTITUTION[a]	DESIGNATION OF INDIVIDUAL	NUMBER OF BARR BODIES
46,XX	Normal ♀	1
46,XY	Normal ♂	0
45,X	Turner syndrome ♀	0
47,XXX	Triplo-X ♀	2
47,XXY	Klinefelter syndrome ♂	1
48,XXXY	Klinefelter syndrome ♂	2
48,XXYY	Klinefelter syndrome ♂	1
47,XYY	XYY syndrome ♂	0

[a]The first number indicates the total number of chromosomes in the nucleus, and the Xs and Ys indicate the sex chromosome complement.

typically becomes inactivated to produce Barr bodies. Thus a general formula for the number of Barr bodies is the number of X chromosomes minus one. This dosage compensation process therefore minimizes the effects of multiple copies of X-linked genes. In normal individuals lyonization results in males and females each with one active X chromosome per cell.

The number of Barr bodies associated with the normal and abnormal human X chromosome constitutions we have discussed are given in Table 3.2. Normal 46,XX females have one Barr body, while normal 46,XY males have no Barr bodies. Turner syndrome females (45,X) have no Barr bodies; 47,XXY Klinefelter syndrome males have one Barr body, triplo-X (47,XXX) females have two Barr bodies, and 47,XYY syndrome males have no Barr bodies.

Sex Determination in *Drosophila* *Drosophila melanogaster* has four pairs of chromosomes: one pair of sex chromosomes and three pairs of autosomes. In this organism the homogametic sex is the female (XX) and the heterogametic sex is the male (XY). However, the sex of the fly is not the consequence of the presence or absence of a Y chromosome: an XXY fly is female and an XO fly is male. The sex of the fly is determined by the ratio of the number of X chromosomes (X) to the number of *sets* of autosomes (A). Since *Drosophila* is diploid, there are two sets of autosomes in a wild-type fly, although abnormal numbers of sets can be produced as a result of nondisjunction. *Drosophila* is the prototype of the *X chromosome-autosome balance system of sex determination*.

Table 3.3 presents some chromosome complements and the sex of the resulting flies. In a normal female there are two Xs and two sets of autosomes; hence the X:A ratio is 1.00. A normal male has a ratio of 0.50. If the X:A ratio is greater than or equal to 1.00, the fly will be female; if the X:A ratio is less than or equal to 0.50 the fly will be male. If the ratio is between 0.50 and 1.00, the fly is neither male nor female; it is an intersex. Intersex flies are variable in appearance, and generally have complex mixtures of male and female attributes for the internal sex organs and external genitalia. Such flies are sterile.

Sex Chromosomes in Other Organisms Not all organisms have an X-Y sex chromosome makeup

■ *Figure 3.12* Barr bodies: (a) Nucleus of normal human female cells (XX), showing Barr bodies; (b) Nucleus of normal human male cells (XY), showing no Barr bodies.

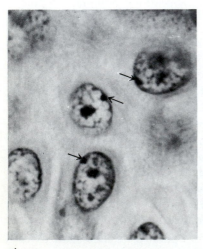
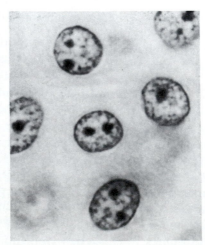

a)   b)

	Table 3.3		

Sex Balance Theory of Sex Determination in *Drosophila melanogaster*

SEX CHROMOSOME COMPLEMENT	AUTOSOME COMPLEMENT (A)	X:A RATIO[a]	SEX OF FLIES
XX	AA	1.00	♀
XY	AA	0.50	♂
XXX	AA	1.50	Metafemale (sterile)
XXY	AA	1.00	♀
XXX	AAAA	0.75	Intersex (sterile)
XX	AAA	0.67	Intersex (sterile)
X	AA	0.50	♂ (sterile)

[a]If the X chromosome:autosome ratio is greater than or equal to 1.00 (X:A ≥ 1.00), the fly will be a female. If the X chromosome:autosome ratio is less than or equal to 0.50 (X:A ≤ 0.50), the fly will be male. Between these two ratios, the fly will be an intersex.

like that found in mammals and in *Drosophila*. In birds, butterflies, moths, and some fish, the sex chromosome composition is the opposite of that in mammals: The male in these organisms is the homogametic sex and the female is the heterogametic sex. To prevent confusion with the X and Y chromosome convention, we designate the sex chromosomes in these organisms as Z and W. Thus males are ZZ and females are ZW. Genes on the Z chromosome behave just like X-linked genes in the earlier examples except that hemizygosity is found only in females and males can be homozygous or heterozygous. All the daughters of a male homozygous for a Z-linked recessive gene express the recessive trait, and so on.

Higher plants exhibit quite a variety of sexual situations. Some species (the ginkgo, for example) have plants of separate sexes, with male plants producing flowers that contain only stamens and female plants producing flowers that contain only pistils. These species are called **dioecious**. Other species have both male and female sex organs on the same plant. If the sex organs are in the same flower, the plant is said to be **hermaphroditic** (e.g., the rose and the buttercup), and the flower is said to be a *perfect flower*. If the sex organs are in different flowers on the same plant, it is said to be **monoecious** (e.g., corn), and the flower is said to be an *imperfect flower*.

Some dioecious plants have sex chromosomes that differ between the sexes, and a large proportion of these plants have an X-Y system. Such plants typically have an X chromosome-autosome balance system of sex determination like that in *Drosophila*. Many species, particularly eukaryotic microor-

ganisms, do not have sex chromosomes. Instead they rely on a *genic system* for the determination of sex, that is, a system in which the sexes are specified by simple allelic differences at one or a small number of gene loci. For example, the orange bread mold *Neurospora crassa* is a haploid fungus that has two "sexes" referred to as **mating types**. The mating types are morphologically indistinguishable and can act as either male or female, but crosses can occur only between individuals of opposite type. The mating types are determined by the *A* and *a* alleles of a single gene locus in chromosome I of this organism. Similarly, in the yeast *Saccharomyces cerevisiae*, there are two mating types: **a** and α. These mating types are controlled by the *MAT***a** and *MAT*α alleles, respectively, of a single gene.

Phenotypic Sex Determination Systems

The environment plays a major role in systems of **phenotypic sex determination**. These types of sex determination mechanisms are much rarer than the ones we have discussed thus far, and usually occur in organisms for which little genetic information is known. In the marine worm *Bonellia*, for example, the free-swimming larval forms are sexually undifferentiated. If an individual settles down alone, it becomes female. If a larva attaches to the body of an adult female, the larva will differentiate into a male. Thus sex differentiation in *Bonellia* is not determined at fertilization by some genetic component but is probably directed by environmental factors related either to the association or lack of association with other members of the species.

Keynote

Many eukaryotic organisms have sex chromosomes that are represented differentially in the two sexes; in humans and many other eukaryotes the male is XY and the female is XX. In many cases sex determination is related to the sex chromosomes. For humans and most other mammals, for instance, the presence of the Y chromosome confers maleness, and its absence results in femaleness. *Drosophila* has an X chromosome-autosome balance system of sex determination: The sex of the individual is related to the ratio of the number of X chromosomes to the number of sets of autosomes. Several other sex-determining systems are known in the eukaryotes, including genic systems, found particularly in the lower eukaryotes, and the rare phenotypic (environmental) systems.

Analysis of Sex-Linked Traits in Humans

In this section we will discuss examples of the analysis of X-linked and Y-linked traits in humans. Recall that we introduced the analysis of recessive and dominant traits in humans in Chapter 2. Those traits were not sex-linked but were the result of alleles carried on autosomes, chromosomes other than the sex chromosomes.

X-Linked Recessive Inheritance   A trait due to a recessive mutant gene carried on the X chromosome is called an **X-linked recessive trait**. At least 100 human traits are known for which inheritance has been traced to the X chromosome. Most of the traits involve X-linked recessive genes. The best-known X-linked recessive pedigree is that of hemophilia A in Queen Victoria's family (Figure 3.13). Hemophilia (the bleeder's disease) is a serious ailment in which the blood lacks a clotting factor. A cut or even a serious bruise can be fatal to a hemophiliac. In Queen Victoria's pedigree the first instance of hemophilia was in one of her sons, so she was either a carrier for this trait or a mutation had occurred in her germ cells.

In X-linked recessive traits the female must be homozygous for the recessive allele in order to express the mutant trait. The trait is expressed in the male who possesses one copy of the mutant allele on the X chromosome. Therefore affected males normally transmit the mutant gene to all their daughters but to none of their sons. The case of a father-to-son inheritance of a trait in a pedigree would tend to rule out X-linked recessive inheritance.

Other characteristics of X-linked recessive inheritance are as follows (refer to Figure 3.13):

1. For rare X-linked recessive genes many more males than females should exhibit the trait due to the different number of X chromosomes in the two sexes. Most females will be heterozygous under these conditions.

2. All sons of a homozygous mutant female should exhibit the trait, since males receive their only X chromosome from their mothers.

3. The sons of heterozygous (carrier) mothers should show an approximately 1:1 ratio of normal individuals to individuals expressing the trait. That is, $a^+/a \times a^+/Y$ gives half a^+/Y and half a/Y sons.

4. From a mating of a carrier female with a normal male all daughters will be normal, but half will be carriers. That is, $a^+/a \times a^+/Y$ gives half a^+/a^+ and half a^+/a females. In turn, half the sons of the F$_1$ carrier females will exhibit the trait.

5. A male expressing the trait, when mated with a homozygous normal female, will produce all normal children, but all the female progeny will be carriers. That is, $a^+/a^+ \times a/Y$ gives a^+/a females and a^+/Y (normal) males.

X-Linked Dominant Inheritance   A trait due to a dominant mutant gene carried on the X chromosome is called an **X-linked dominant trait**. Only a few dominant X-linked traits have been identified.

An example of an X-linked dominant trait that causes faulty tooth enamel and dental discoloration is shown in Figure 3.14(a) and a pedigree for this trait is shown in Figure 3.14(b). Note that all the daughters and none of the sons of an affected father (III.1) exhibit the trait, and that heterozygous mothers (IV.3) transmit the trait to half their sons and half their daughters. Webbing to the tips of the toes in a family in South Dakota (studied in the 1930s) is also an X-linked dominant mutant trait, as is a severe bleeding anomaly called constitutional thrombopathy. In the latter (also studied in the 1930s), bleeding is not due to the absence of a clotting factor, as in hemophilia, but to interference in the formation of blood platelets, which are needed for blood clotting.

The X-linked dominant traits follow the same sort of inheritance rules as the X-linked recessives, except that heterozygous females express the trait. In general, X-linked dominant traits tend to be milder in the female than in the male. Also, since females have twice the number of X chromosomes as males, X-linked dominant traits are more frequent in females than in males. For rare traits most females in a pedigree would be heterozygous rather than homozygous for the X-linked dominant allele. These heterozygous females should pass the trait

■ *Figure 3.13*   (a) Photograph of Queen Victoria as a young woman; (b) Pedigree of Queen Victoria (III.2) and her descendants, showing the inheritance of hemophilia. (Refer to Figure 2.15 for an explanation of symbols used in pedigrees. In the pedigree shown here, the marriage partner has been omitted in many cases as a way to save space. Those marriage partners were normal with respect to the trait.) Either Queen Victoria was heterozygous for the sex-linked recessive hemophilia allele, or the trait arose as a mutation in her germ cells (the cells that give rise to the gametes).

a)

b)

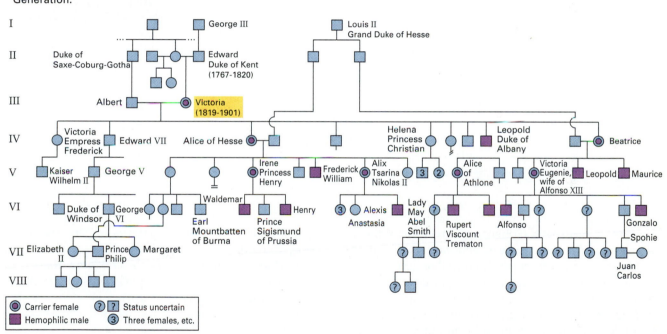

on to half their progeny, regardless of their sex. As in X-linked recessive traits, males who have an X-linked dominant trait transmit the allele to their daughters but not to their sons. As a result we see crisscross inheritance (father-to-daughter) of the trait.

Y-Linked Inheritance   A trait due to a mutant gene carried on the Y chromosome but with no counterpart on the X is called a **Y-linked**, or **holandric** ("wholly male"), **trait**. Such traits should be readily recognizable because every son of an affected male should have the trait and no females

should ever express it. Several traits with Y-linked inheritance have been suggested. In most cases the genetic evidence for such inheritance is poor or nonexistent. In fact there is no clear evidence for genetic loci on the Y chromosome other than those involved with sex determination in the male.

Keynote

Controlled matings cannot be made in humans, so analysis of the inheritance of genes in humans must rely on pedigree analysis. This technique involves

■ *Figure 3.14*   (a) A person with the X-linked dominant trait of faulty tooth enamel; (b) A pedigree showing the transmission of the faulty enamel trait. This pedigree uses the shorthand convention in which parents who do not exhibit the trait are omitted. Thus it is a given that the female in generation I paired with a male who did not exhibit the trait.

a)

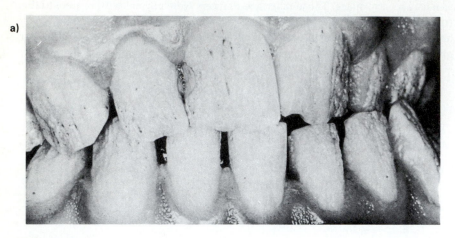

b)   Pedigree

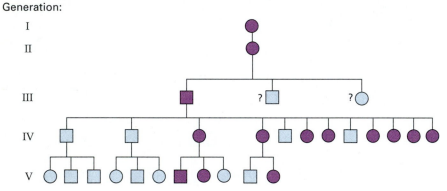

the careful study of the phenotypic records of the family extending over several generations. The data obtained from pedigree analysis enable geneticists to make judgments, with varying degrees of confidence, about whether a mutant gene is inherited as an autosomal recessive, an autosomal dominant, an X-linked recessive, an X-linked dominant, or a Y-linked allele.

Summary

In Chapter 2, genes were considered as abstract entities that control hereditary characteristics. What we now know as genes were called factors by Mendel. Cytologists working in the nineteenth century accumulated new information about cell structure and cell division. The fields of genetics and cytology came together in 1903 when it was hypothesized that genes are on chromosomes. The chromosome theory of heredity states that the chromosome transmission from one generation to the next closely parallels the patterns of gene transmission from one generation to the next.

Further support of the chromosome theory of heredity was proved by experiments that related the hereditary behavior of particular genes to the transmission of the sex chromosome. The sex chromosome in eukaryotic organisms is the chromosome that is represented differently in the two sexes. In most organisms with sex chromosomes the female has two X chromosomes, while the male has one X and one Y chromosome. The Y chromosome is structurally and genetically different from the X chromosome. The association of genes with the sex determining chromosomes of eukaryotes is called sex linkage. Such genes, and the phenotypes they control, are called sex-linked.

In this chapter a number of sex determination mechanisms were discussed. In many cases sex determination is related to the sex chromosomes. In humans, for example, the presence of a Y chromosome specifies maleness, while its absence results in femaleness. Several other sex determination systems are known in eukaryotes, including X chromosome-autosome balance systems (in which sex is determined by the ratio of the number of X chromosomes to the number of sets of autosomes), genic systems (in which a simple allelic difference determines the sex of an individual), and phenotypic systems (in which sex is determined by environmental cues).

The analysis of X-linked traits in humans was also examined. In Chapter 2 we saw that the inheritance patterns of traits in humans are usually studied by pedigree analysis. Here we considered examples of X- and Y-linked human traits to illustrate the features of those mechanisms of inheritance in pedigrees. Autosomal traits were discussed in Chapter 2. It is important to realize that collecting reliable human pedigree data is a difficult task. In many cases the accuracy of record keeping within the families involved is open to question. Also, particularly with small families, there may not be enough affected people to allow an unambiguous determination of the inheritance mechanism involved, especially in case of rare traits. Moreover, the degree to which a trait is expressed may vary, so that some individuals may be erroneously classified as normal. It is also possible for a mutant phenotype to be produced by alleles of different genes; therefore different pedigrees may, quite correctly, indicate different mechanisms of inheritance of the "same" trait.

Analytical Approaches for Solving Genetics Problems

The concepts introduced in this chapter may be reinforced by solving genetics problems. The types of problems are similar to those introduced in Chapter 2. When sex linkage is involved, remember that one sex has two kinds of sex chromosomes, whereas the other sex has only one; this feature alters the inheritance patterns slightly. Most of the problems presented focus on interpreting data and predicting the outcome of particular crosses.

Q3.1 A female from a pure-breeding strain of *Drosophila* with vermilion-colored eyes is crossed with a male from a pure-breeding wild-type, red-eyed strain. All the F_1 males have vermilion-colored eyes, and all the females have wild-type red eyes. What conclusions can you draw about the mode of inheritance of the vermilion trait, and how could you test them?

A3.1 This is a classic observation that suggests a sex-linked trait is involved. Since none of the F_1 daughters have the trait and all the F_1 males do, the trait is presumably X-linked recessive. The results fit this hypothesis because the F_1 males receive the X chromosome from their homozygous v/v mother, with the v gene on the X chromosome. Furthermore, the F_1 females are v^+/v because they receive a v^+-bearing X chromosome from the wild-type male parent and a v-bearing X chromosome from the female parent. If the trait were autosomal recessive, all the F_1 flies would have had wild-type eyes. If it were autosomal dominant, both the F_1 males and females would have had vermilion-colored eyes. If the trait were X-linked dominant, all the F_1 flies would have had vermilion eyes.

The easiest way to verify this hypothesis is to let the F_1 flies interbreed. This cross is v^+/v ♀ × v/Y ♂. The expectation is that there will be a 1:1 ratio of wild-type: vermilion eyes in both sexes in the F_2. That is, half the females are v^+/v and half are v/v; half the males are v^+/Y and half are v/Y. This ratio is certainly not the 3:1 ratio that would result from an $F_1 \times F_1$ cross for an autosomal gene.

Q3.2 In humans hemophilia A and B are each caused by a different X-linked recessive gene. A woman who is a nonbleeder had a father who was a hemophiliac. (Actually hemophiliac fathers are relatively rare, but we will consider one for this question.) She marries a nonbleeder, and they plan to have children. Calculate the probability of hemophilia in the female and male offspring.

A3.2 Since hemophilia is an X-linked trait, and since her father was a hemophiliac, the woman must be heterozygous for this recessive gene. If we assign the symbol h to this recessive mutation and h^+ to the wild-type (nonbleeder) allele, the women must be h^+/h. The man she marries is normal with regard to blood clotting and hence must be hemizygous for h^+, that is, h^+/Y. All their daughters receive an X chromosome from the father, and so each must have an h^+ gene. In fact half the daughters are h^+/h^+ and the other half are h^+/h. Since the wild-type allele is dominant, none of the daughters are hemophiliacs. However, all the sons of the marriage receive their X chromosome from the mother. Therefore they have a probability of 1/2 that they will receive the chromosome carrying the h allele and be hemophiliacs. Thus the probability of hemophilia among daughters of this marriage is 0, and among sons it is 1/2.

Q3.3 Tribbles are hypothetical animals that have an X-Y sex determination mechanism like that of humans. The trait bald (*b*) is X-linked and recessive to furry (*b*⁺), and the trait long leg (*l*) is autosomal and recessive to short leg (*l*⁺). You make reciprocal crosses between true-breeding bald, long-legged tribbles and true-breeding furry, short-legged tribbles. Do you expect a 9:3:3:1 ratio in the F₂ of either or both of these crosses? Explain your answer.

A3.3 This question focuses on the fundamentals of X chromosome and autosome segregation during a genetic cross and tests whether or not you have grasped the principles involved in gene segregation. We can discuss the answer by referring to the figure that diagrams the two crosses involved.

Let us first consider the cross of a wild-type female tribble (b^+/b^+, l^+/l^+) with a male double-mutant tribble (b/Y, l/l). Part (a) of the figure dia-

grams this cross. The F₁s are all normal, with furry bodies and short legs, because for the autosomal character both sexes are heterozygous, and for the X-linked character the female is heterozygous and the male is hemizygous for the b^+ allele donated by the normal mother. With the production of the F₂ progeny the best approach is to treat the X-linked and autosomal traits separately. For the X-linked trait, random combination of the gametes produced gives a 1:1:1:1 genotypic ratio of b^+/b^+ (furry female):b^+/b (furry female):b^+/Y (furry male):b/Y (bald male) tribbles. Collecting by phenotypes, we see that all the females are furry; half the males are furry and half are bald. For the autosomal leg trait the F₁ × F₁ is a cross of two heterozygotes, so we expect a 3:1 phenotypic ratio of short-legged:long-legged tribbles in the F₂. Since autosome segregation is independent of the inheritance of the X chromosome, we can multiply the probabilities of the occurrence of the X-linked and autosomal traits to calculate their relative frequencies. The calculations

---

a) Furry, short-legged (wild-type) ♀ × bald, long-legged ♂

P generation

$b^+/b^+ \; l^+/l^+$ ♀  × $b/Y \; l/l$ ♂
(furry, short)   (bald, long)

F₁ generation

$b^+/b \; l^+/l$ ♀  × $b^+/Y \; l^+/l$ ♂
(furry, short)   (furry, short)

F₂ generation

Sex-linked phenotypes and genotypes	Autosomal phenotypes and genotypes

Genotypic results:

$\frac{1}{2} \, b^+ \, (\frac{1}{2} \, b^+/b^+, \frac{1}{2} \, b^+/b;$ furry) ♀ ⎰ $\frac{3}{4} \, l^+ \, (l^+/l^+$ and $l^+/l;$ short) / $\frac{1}{4} \, l \, (l/l;$ long)

$\frac{1}{4} \, b^+ \, (b^+/Y;$ furry) ♂ ⎰ $\frac{3}{4} \, l^+$ (short) / $\frac{1}{4} \, l$ (long)

$\frac{1}{4} \, b \, (b^+/Y;$ bald) ♂ ⎰ $\frac{3}{4} \, l^+$ (short) / $\frac{1}{4} \, l$ (long)

Phenotypic ratios:

	Furry-short		Furry-long		Bald-short		Bald-long
	$b^+ l^+$		$b^+ l$		$b \, l^+$		$b \, l$
♀	6	:	2	:	0	:	0
♂	3	:	1	:	3	:	1
Total	9	:	3	:	3	:	1

---

b) Bald, long-legged ♂ × furry, short-legged (wild-type) ♀

P generation

$b/b \; l/l$ ♀ × $b^+/Y \; l^+/l^+$ ♂
(bald, long)   (furry, short)

F₁ generation

$b^+/b \; l^+/l$ ♀ 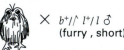 × $b/Y \; l^+/l$ ♂
(furry, short)   (bald, short)

F₂ generation

Sex-linked phenotypes and genotypes	Autosomal phenotypes and genotypes

Genotypic results:

$\frac{1}{4} \, b^+ \, (b^+/b^+;$ furry) ♀ ⎰ $\frac{3}{4} \, l^+ \, (l^+/l^+$ and $l^+/l;$ short) / $\frac{1}{4} \, l \, (l/l;$ long)

$\frac{1}{4} \, b \, (b/b;$ bald) ♀ ⎰ $\frac{3}{4} \, l^+$ (short) / $\frac{1}{4} \, l$ (long)

$\frac{1}{4} \, b^+ \, (b^+/Y;$ furry) ♂ ⎰ $\frac{3}{4} \, l^+$ (short) / $\frac{1}{4} \, l$ (long)

$\frac{1}{4} \, b \, (b/Y;$ bald) ♂ ⎰ $\frac{3}{4} \, l^+$ (short) / $\frac{1}{4} \, l$ (long)

Phenotypic ratios:

	Furry-short		Furry-long		Bald-short		Bald-long
	$b^+ l^+$		$b^+ l$		$b \, l^+$		$b \, l$
♀	3	:	1	:	3	:	1
♂	3	:	1	:	3	:	1
Total	6	:	2	:	6	:	2

are presented in part (a) of the figure, from which we see that the ratio of the four possible phenotypic classes differs in females and males.

The first cross, then, has a 9:3:3:1 ratio of the four possible phenotypes in the F_2. However, note that the ratio in each sex is not 9:3:3:1, owing to the inheritance pattern of the X chromosome. This result contrasts markedly with the pattern of two autosomal genes segregating independently, where the 9:3:3:1 ratio is found for both sexes.

The second cross (a reciprocal cross) is diagrammed in part (b) of the figure. Since the parental female in this cross is homozygous for the sex-linked trait, all the F_1 males are bald. Genotypically the F_1 males and females differ from those in the first cross with respect to the sex chromosome but are just the same with respect to the autosome. Again, considering the X chromosome first as we go to the F_2, we find a 1:1:1:1 genotypic ratio of furry females: bald females: furry males: bald males. In

this case, then, half of both males and females are furry and half of both are bald, in contrast to the results of the first cross in which no bald females were produced in the F_2 generation. For the autosomal trait we expect a 3:1 ratio of short: long in the F_2, as before. Putting the two traits together, we get the calculations presented in part (b) of the figure. (*Note*: We use the total 6:2:6:2 here rather than 3:1:3:1 because the numbers add up to 16, as does $9 + 3 + 3 + 1$.) So in this case we do not get a 9:3:3:1 ratio; moreover, the ratio is the same in both sexes.

This question has forced us to think through the segregation of two types of chromosomes and has shown that we must be careful about predicting the outcomes of crosses in which sex chromosomes are involved. Nonetheless, the basic principles for the analysis were no different from those used before: Reduce the questions to their basic parts and then put the puzzle together step by step.

Questions and Problems

3.1 In *Drosophila* white eyes are a sex-linked character. The mutant allele for white eyes (w) is recessive to the wild-type allele for brick-red eye color (w^+).

a. A white-eyed female is crossed with a red-eyed male. An F_1 female from this cross is mated with her father, and an F_1 male is mated with his mother. What will be the appearance of the offspring of these last two crosses with respect to eye color?

b. A white-eyed female is crossed with a red-eyed male, and the F_2 from this cross is interbred. What will be the eye color of the F_3?

*3.2 One form of color blindness (c) in humans is caused by a sex-linked recessive mutant gene. A woman with normal color vision (c^+) whose father was color-blind marries a man of normal vision whose father was also color-blind. What proportion of their offspring will be color-blind? (Give your answer separately for male and female children.)

*3.3 In humans red-green color blindness is recessive and X-linked, while albinism is recessive and autosomal. What types of children can be produced as the result of marriages between two homozygous parents who are a normal-visioned albino woman and a color-blind, normally pigmented man?

*3.4 In *Drosophila* vestigial (partially formed) wings (vg) are recessive to normal long wings (vg^+), and the gene for this trait is autosomal. The gene for the white eye trait is on the X chromosome. Suppose a homozygous white-eyed, long-winged female fly is crossed with a homozygous red-eyed, vestigial-winged male.

a. What will be the appearance of the F_1?

b. What will be the appearance of the F_2?

c. What will be the appearance of the offspring of a cross of the F_1 back to each parent?

3.5 In *Drosophila* two red-eyed, long-winged flies are bred together and produce the following offspring: females are 3/4 red, long and 1/4 red, vestigial; males are 3/8 red, long; 3/8 white, long; 1/8 red, vestigial; and 1/8 white, vestigial. What are the genotypes of the parents?

3.6 In poultry a dominant sex-linked gene (B) produces barred feathers, and the recessive allele (b), when homozygous, produces nonbarred feathers. Suppose a nonbarred cock is crossed with a barred hen.

a. What will be the appearance of the F_1 birds?

b. If an F_1 female is mated with her father, what will be the appearance of the offspring?

c. If an F_1 male is mated with his mother, what will be the appearance of the offspring?

*3.7 A man (A) suffering from defective tooth enamel, which results in brown-colored teeth, marries a normal woman. All their daughters have brown teeth, but the sons are normal. The sons of man A marry normal women, and all their children are normal. The daughters of man A marry normal men, and 50 percent of their children have brown teeth. Explain these facts.

3.8 Suppose gene A is on the X chromosome, and genes B, C, and D are on three different autosomes. Thus A– signifies the dominant phenotype in the male or female. An equivalent situation holds for B–, C–, and D–. The cross $AA\ BB\ CC\ DD$ females $\times$ $aY\ bb\ cc\ dd$ males is made.
  a. What is the probability of obtaining an A– individual in the F_1?
  b. What is the probability of obtaining an a male in the F_1?
  c. What is the probability of obtaining an A– B– C– D– female in the F_1?
  d. How many different F_2 genotypes will there be?
  e. What proportion of F_2s will be heterozygous for all four genes?
  f. Determine the probabilities of obtaining each of the following types in the F_2: (1) A– $bb\ CC\ dd$ (female); (2) $aY\ BB\ Cc\ Dd$ (male); (3) $AY\ bb\ CC\ dd$ (male); (4) $aa\ bb\ Cc\ Dd$ (female).

*3.9 As a famous mad scientist, you have cleverly devised a method to isolate *Drosophila* ova that have undergone primary nondisjunction of the sex chromosomes. In one experiment you used females homozygous for the sex-linked recessive mutation causing white eyes (w) as your source of nondisjunction ova. The ova were collected and fertilized with sperm from red-eyed males. The progeny of this "engineered" cross were then crossed separately to the two parental strains (this is called backcrossing). What classes of progeny (genotype and phenotype) would you expect to result from these backcrosses? (The genotype of the original parents may be denoted as ww for the females and w^+Y for the males.)

3.10 In *Drosophila* the bobbed gene (bb^+) is located on the X chromosome. Unlike most X-linked genes, however, the Y chromosome also carries a bobbed gene. The mutant allele bb is recessive to bb^+. For the following crosses, give the genotypes as well as the phenotypes and their frequencies in the offspring. Consider males and females separately in your answer.
  a. $X^{bb^+}X^{bb} \times X^{bb}Y^{bb^+}$
  b. $X^{bb^+}X^{bb} \times X^{bb^+}Y^{bb}$
  c. $X^{bb}X^{bb} \times X^{bb}Y^{bb^+}$

3.11 An XXY Klinefelter syndrome individual would be expected to have the following number of Barr bodies in the majority of cells:
  a. 0
  b. 1
  c. 2
  d. 3

*3.12 In human genetics the pedigree is used for analysis of inheritance patterns. The female is represented by a circle and the male by a square. The following figure presents three family pedigrees for a trait in human beings. Normal individuals are represented by unshaded symbols, and people with the trait by shaded symbols. For each pedigree (A, B, and C), state, by answering yes or no in the appropriate blank space, whether transmission of the trait can be accounted for on the basis of each of the listed simple mechanisms of inheritance.

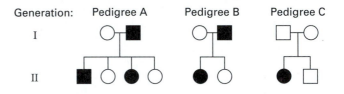

Generation: Pedigree A   Pedigree B   Pedigree C

I

II

	Pedigree A	*Pedigree B*	*Pedigree C*
Autosomal recessive	_____	_____	_____
Autosomal dominant	_____	_____	_____
X-linked recessive	_____	_____	_____
X-linked dominant	_____	_____	_____

3.13 Looking at the pedigree in the figure, in which shaded symbols represent a trait, which of the progeny (as designated by numbers) eliminate X-linked recessiveness as a mechanism of inheritance for the trait?

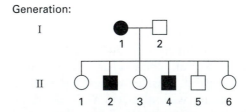

Generation:

I

II

*3.14 When constructing human pedigrees, geneticists often refer to particular persons by a number. The generations are labeled by roman numerals and the individuals in each generation by Arabic numerals. In the pedigree in the figure the female with the asterisk would be I.2. Use this system of labeling to designate specific individuals in the pedigree. Determine the probable mode for the trait shown in the affected individuals (the shaded symbols) by answering the following questions. Assume the condition is caused by a single gene.

Generation:

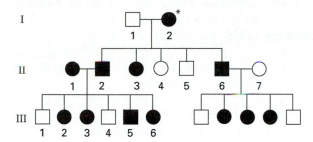

a. Y-linked inheritance can be excluded at a glance. What two other mechanisms of inheritance can be definitely excluded? Why can these be excluded?

b. Of the remaining mechanisms of inheritance, which is the most likely? Why?

3.15   For the more complex pedigrees shown in Figure 3A, indicate the probable mode of inheritance: autosomal recessive, autosomal dominant, X-linked recessive, X-linked dominant, Y-linked.

*3.16   If a rare genetic disease is inherited on the basis of an X-linked dominant gene, one would expect to find the following:

a. Affected fathers have 100% affected sons.
b. Affected mothers have 100% affected daughters.
c. Affected fathers have 100% affected daughters.
d. Affected mothers have 100% affected sons.

3.17   If a genetic disease is inherited on the basis of an autosomal dominant gene, one would expect to find the following:

a. Affected fathers have only affected children.
b. Affected mothers never have affected sons.
c. If both parents are affected, all of their offspring have the disease.
d. If a child has the disease, one of his or her grandparents also had the disease.

3.18   Which of the following statements is *not* true for a disease that is inherited as a rare X-linked dominant?

a. All daughters of an affected male will inherit the disease.
b. Sons will inherit the disease only if their mothers have the disease.
c. Both affected males and affected females will pass the trait to half their children.
d. Daughters will inherit the disease only if their father has the disease.

*3.19   Women who were known to be carriers of the X-linked, recessive, hemophilia gene were studied

■ *Figure 3A*

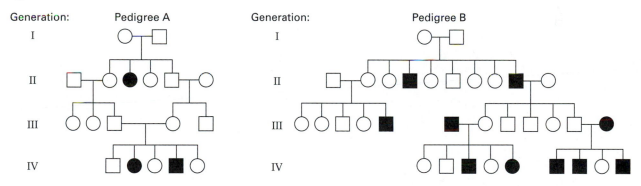

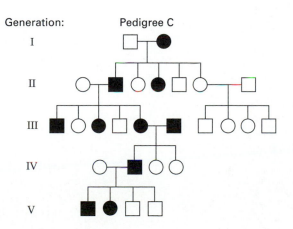

in order to determine the amount of time required for the blood clotting reaction. It was found that the time required for clotting was extremely variable from individual to individual. The values obtained ranged from normal clotting time at one extreme all the way to clinical hemophilia at the other extreme. What is the most probable explanation for these findings?

*3.20   In a species of amphibian sex can be reversed by placing hormones in the water where fertilized eggs are developing. Eggs exposed to estrogens develop as phenotypic females, whatever their genetic sex, while eggs exposed to androgens develop as phenotypic males.

In one experiment several phenotypic females raised in estrogen-containing water were mated to normal males. About half of these females pro-

duced only male offspring, while the other half produced males and females in the expected 1:1 ratio.

In another experiment several phenotypic males raised in androgen-containing water were mated to normal females. About half of these males produced the normal 1:1 ratio of daughters and sons, but the other half produced about three females for every male.

a. How is sex determined in these amphibians? Explain your reasoning.
b. If the offspring from the mating that yielded a ratio of 3 females to 1 male were crossed to normal males, how many kinds of matings would there be? What proportion of the two sexes would be seen in the offspring of each kind of mating?

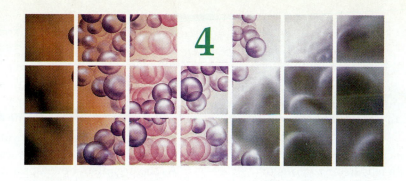

Extensions of Mendelian Genetic Analysis

- Many allelic forms of a gene can exist. This phenomenon is called multiple allelism. Any given diploid individual can possess only two different alleles.

- In complete dominance the same phenotype results whether an allele is heterozygous or homozygous. In incomplete dominance the phenotype of the heterozygote is intermediate between those of the two homozygotes, whereas in codominance the heterozygote exhibits the phenotypes of both homozygotes.

- In many cases nonallelic genes interact to determine phenotypic characteristics. In epistasis, for example, modified Mendelian ratios occur because of interactions of nonallelic genes: the phenotypic expression of one gene depends upon the genotype of another gene locus.

- Alleles of certain genes, when expressed, may be fatal to the individual. The existence of such lethal alleles of a gene indicates that the product usually produced by the gene is essential for the function of the organism.

- Penetrance is the frequency with which a dominant or homozygous recessive gene manifests itself in the phenotype of an individual. Expressivity is the kind or degree of phenotypic manifestation of a penetrant gene or genotype.

- The zygote's genetic constitution only specifies the organism's potential to develop and function. As the organism develops and differentiates, many things can influence gene expression. One influence is the organism's environment, both internal and external. For the former, examples may include age and sex of the individual. For the latter, factors include nutrition, light, chemicals, temperature, and infectious agents.

- Variation in most of the genetic traits considered in the discussion of Mendelian principles is determined predominantly by differences in genotype; that is, phenotypic differences resulted from genotypic differences. For many traits, however, the phenotypes are influenced by both genes and the environment. The debate over the relative contribution of genes and environment to the phenotype has been termed the nature versus nurture controversy.

I T WAS ORIGINALLY THOUGHT that Mendel's principles apply to all eukaryotic organisms and form the foundation for predicting the outcome of crosses in which segregation and independent assortment might be occurring. As more and more geneticists performed experiments, though, they found that there are exceptions and extensions to Mendel's principles. Several of these cases will be discussed in this chapter. We will examine examples of genes that have many different alleles rather than just two, cases in which one allele is not completely dominant to another allele at the gene locus, and situations in which products of different genes interact to produce modified Mendelian ratios. We will also look at lethal genes and at the effects of the environment on gene expression. This discussion will give us a broader knowledge of genetic analysis, particularly in terms of how genes relate to the phenotypes of an organism.

Multiple Alleles

So far in our genetic analyses we have considered only pairs of alleles controlling characters, such as smooth versus wrinkled seeds in peas, red versus white eyes in *Drosophila*, and unattached versus attached ear lobes in humans. The allele found in the standard laboratory strain of the organism is the **wild-type** allele, and the alternative allele is the variant or **mutant** allele. In a population of individuals, however, there can be many alleles of a given gene, not just two. Such genes are said to have **multiple alleles**, and the alleles are said to constitute a *multiple allelic series*. Although a gene may have multiple alleles in a given population of individuals, a *single diploid individual can have a maximum of two of these alleles, one on each of the two homologous chromosomes carrying the gene locus.*

Drosophila Eye Color

Alleles are alternative forms of a gene, each of which may affect a character differently. The w^+ and w alleles in *Drosophila*, for example, are involved in the development of eye color. The w^+ allele results in red eyes, and the w allele, when homozygous or hemizygous, results in white eyes.

Another eye color variant is eosin (reddish-orange). Like white, eosin is X-linked and is recessive to the wild-type color. However, when a female from an eosin-eyed strain is crossed with a male from a white-eyed strain, all F_1 females have eosin eyes. From these observations we can conclude that (1) red (wild-type) eye color is dominant to eosin and to white, and (2) eosin is recessive to wild type but dominant to white. These conclusions are based on an important assumption: eosin and white are both mutant alleles of a single gene. In other words, there are multiple alleles of the white gene.

Let us assign symbols to the alleles: w^+ is the wild-type allele of the white-eye gene, w is the recessive white allele, and w^e is the eosin allele. Figure 4.1(a) uses this notation to track the cross of an eosin-eyed female with a white-eyed male. The F_1 females are w^e/w* and have eosin eyes because w^e is dominant over w. When these F_1 females are crossed with red-eyed males (Figure 4.1[b]), all the female progeny are heterozygous and red-eyed since they contain the w^+ allele; they are either w^+/w^e or w^+/w. Half the male progeny are eosin-eyed (w^e/Y), and the other half are white (w/Y).

*As a reminder, the / represents the homologous chromosomes on which the alleles are found (refer back to Box 3.1).

■ *Figure 4.1* Results of crosses of *Drosophila melanogaster* involving two mutant alleles of the same locus, white (w) and white-eosin (w^e). (a) White-eosin-eyed (w^e/w^e) ♀ × white-eyed ($w//\wedge$) ♂; (b) F_1 (w^e/w) ♀ × red-eyed (wild-type) ($w^+//\wedge$) ♂.

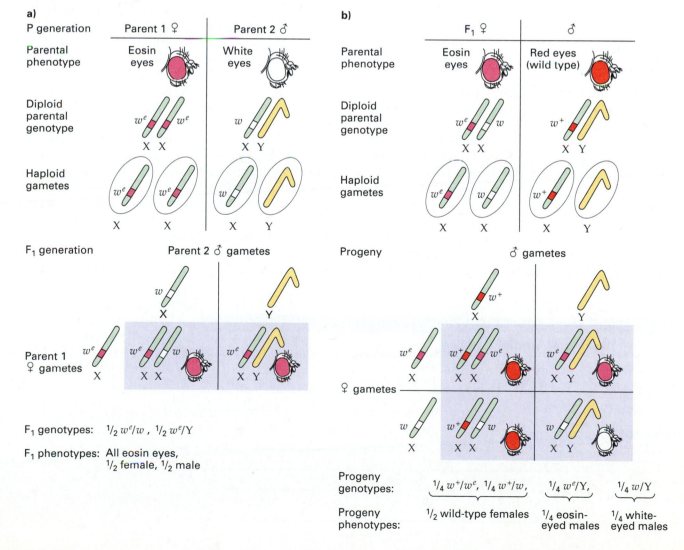

Multiple allelic series exist for all types of genes, not just X-linked ones. But let us expand on the example of the white-eyed gene of *Drosophila*. There are many known alleles of the white gene that are distinguishable because they produce different eye colors, ranging from white to near-wild type, when the alleles are homozygous or hemizygous. This is because the eye color phenotype is related to the amount of pigment deposited in the eye cells, and the amount of pigment deposited depends on the allele present. The original mutant allele *w* has the least amount of pigment.

The number of allelic forms of a gene is not restricted to three, and indeed many hundreds of alleles are known for some genes. The number of possible genotypes in a multiple allelic series depends on the number of alleles involved (Table 4.1). With one allele, only one genotype is possible. With two alleles, A^1 and A^2, three genotypes are possible: A^1/A^1 and A^2/A^2 homozygotes and the A^1/A^2 heterozygote. The formula for calculating the number of genotypes is as follows: When there are *n* alleles, $n(n + 1)/2$ genotypes are possible, of which *n* are homozygotes and $n(n - 1)/2$ are heterozygotes.

ABO Blood Groups

Another example of multiple alleles of a gene is found in the human ABO blood group series. Since certain ABO blood groups are incompatible, these alleles are of particular importance when blood transfusions are contemplated. (There are many blood group series other than ABO. They also can cause problems in blood transfusions and need to be checked through the process of cross-matching.)

Four blood group phenotypes occur in the ABO system: O, A, B, and AB. Table 4.2 gives their possible genotypes. The six genotypes that give rise to the four phenotypes represent various combina-

Table 4.2

ABO Blood Groups in Humans, Determined by the Alleles I^A, I^B, and I^O

PHENOTYPE (BLOOD GROUP)	GENOTYPE
O	I^O/I^O
A	I^A/I^A or I^A/I^O
B	I^B/I^B or I^B/I^O
AB	I^A/I^B

tions of three ABO blood group alleles, I^A, I^B, and I^O. People homozygous for the recessive I^O allele are of blood group O. Both I^A and I^B are dominant to I^O. Thus you will express blood group A if you are either I^A/I^A or I^A/I^O, and you will express blood group B if you are either I^B/I^B or I^B/I^O. Heterozygous I^A/I^B individuals express blood group AB.

The genetics of this system follows basic Mendelian principles. An individual who expresses blood group O, for example, must be I^O/I^O in genotype. The parents of this person could both be O ($I^O/I^O \times I^O/I^O$), they could both be A ($I^A/I^O \times I^A/I^O$, to produce one-fourth I^O/I^O progeny), they could both be B ($I^B/I^O \times I^B/I^O$), or one could be A and the other could be B ($I^A/I^O \times I^B/I^O$). Simply put, each parent would have to be either homozygous I^O or heterozygous, with I^O as one of the two alleles.

Blood typing (the determination of an individual's blood group) and the analysis of the blood group inheritance are sometimes used in legal medicine in cases of disputed paternity or maternity or in cases of an inadvertent baby switch in a hospital. In such instances genetic data cannot prove the identity of the parent. Genetic analysis on the basis of blood group can only be used to show that an individual is *not* the parent of a particular child. For

Table 4.1

Genotype Number of Multiple Alleles

NUMBER OF ALLELES	KINDS OF GENOTYPES	KINDS OF HOMOZYGOTES	KINDS OF HETEROZYGOTES
1	1	1	0
2	3	2	1
3	6	3	3
4	10	4	6
5	15	5	10
n	$n(n + 1)/2$	*n*	$n(n - 1)/2$

example, a child of phenotype AB (genotype I^A/I^B) could not be the child of a parent of phenotype O (genotype I^O/I^O). (Note: Blood typing alone is usually not sufficient for a legal decision according to the laws in most states. And more modern tests may be used, including DNA fingerprinting [see Chapter 13].)

When giving blood transfusions, the blood groups of donors and recipients must be carefully matched because the blood group alleles specify molecular groups, called *cellular antigens,* that attach to the outside of the red blood cells. The I^A allele specifies the A antigen, and in people with blood group A (I^A/I^A or I^A/I^O) the blood serum contains naturally occurring antibodies against the B antigen (called anti-B antibodies). Such antibodies will agglutinate, or clump, any red blood cells that have the B antigen on them. Since clumped cells cannot move through the fine capillaries, agglutination may lead to organ failure and possibly death. Conversely, people of the B blood group (I^B/I^B or I^B/I^O) have the B antigen on their red blood cells and have serum containing naturally occurring anti-A antibodies. For the AB blood type (I^A/I^B) both A and B antigens are on the blood cells, so neither anti-A nor anti-B antibodies occur in the serum of people with this blood type. In people with blood type O (I^O/I^O) the blood cells have neither A nor B antigen, and therefore the serum contains *both* anti-A and anti-B antibodies.

What transfusions are safe, then, between people with different blood groups in the ABO system?

- A individuals produce the A antigen so their blood can be transfused only into recipients who do *not* have the anti-A antibody; that is, people of blood groups A and AB.
- B individuals produce the B antigen so their blood can be transfused only into recipients who do *not* have the anti-B; that is, people of blood groups B and AB.
- AB individuals produce both the A and B antigens so their blood can be transfused only into recipients who do *not* have either the anti-A antibody or the anti-B antibody; that is, people of blood group AB.
- O individuals produce no antigens so their blood can be transfused into any recipient; that is, people of blood groups A, B, AB, and O.

You will note from this presentation that people of blood group AB can receive transfusions of blood from people of any of the four blood groups. Thus blood group AB individuals are called *universal recipients.* Similarly, blood group O blood can be used as donor blood for any recipient since it elicits

no reaction. Thus blood group O individuals are called *universal donors.*

K e y n o t e

Many allelic forms of a gene can exist in a population. When they do the gene is said to show multiple allelism, and the alleles involved constitute a multiple allelic series. Any given diploid individual can possess only two different alleles. Multiple alleles obey the same rule of transmission as alleles of which there are only two kinds, although the dominance relationships among multiple alleles vary from one group to another.

Modifications of Dominance Relationships

In the genetic examples discussed so far, one allele was dominant to the other, so that the phenotype of the heterozygote is the same as homozygous dominant. This phenomenon is called **complete dominance**. In **complete recessiveness** the allele is phenotypically expressed only when the organism is homozygous. Complete dominance and complete recessiveness are the two extremes of a range of dominance relationships. All the allelic pairs that Mendel studied showed such dominance/recessiveness relationships. Many allelic pairs, however, do not show this dominance relationship.

Incomplete Dominance

When one allele is not completely dominant to another allele, it is said to show **incomplete**, or **partial, dominance**. In incomplete dominance the heterozygote's phenotype is between that of individuals homozygous for either individual allele involved; no dominance is involved.

In the plant kingdom there are many examples of incomplete dominance; one is flower color in the snapdragon (Figure 4.2). There are two alleles, C^R and C^W. (Note that the symbols are designed to give equal weight to the two alleles since neither dominates the phenotype.) A cross of a red-flowered variety (C^RC^R which is homozygous for the Colored-Red allele) with a white-flowered variety (C^WC^W which is homozygous for the Colored-White allele) produces F_1 plants with pink flowers (Figure 4.2[a]). The F_1s are C^RC^W heterozygotes. Interbreeding the F_1s produces an F_2 generation with a 1:2:1 ratio of red:pink:white (Figure 4.2[b]). The red and white flowers are determined by homozygosity for the respective color alleles, the pink flowers by het-

■ *Figure 4.2* Incomplete dominance in snapdragons: (a) A cross of true-breeding red-flowered variety with a white-flowered variety produces F₁s with pink flowers. (b) A 1:2:1 ratio of red:pink:white is seen in the F₂.

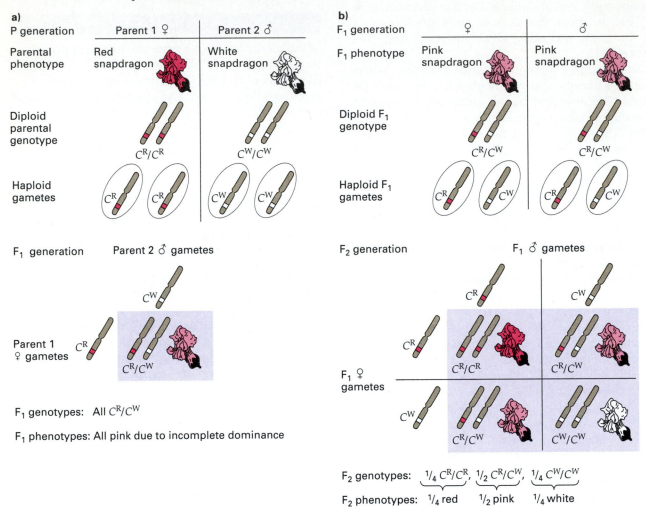

erozygosity for the two alleles. Thus plants with pink flowers do not breed true.

Another example of incomplete dominance is the Palomino horse, which has a golden-yellow body color and a mane and tail that are almost white (Figure 4.3[a]). Palominos do not breed true. When they are interbred one-quarter of the progeny, called cremellos, are extremely light in color (Figure 4.3[b]), one-half are Palomino, and one-quarter are light chestnuts (Figure 4.3[c]). The 1:2:1 ratio resulting from interbreeding is characteristic of incomplete dominance. That is, a Palomino is a light chestnut horse that has also inherited an incompletely dominant gene, *cremello* (C^{cr}). In homozygous C/C horses, in combination with other genes, a light chestnut (also called sorrel) color results. In heterozygous C/C^{cr} horses the normal reddish-brown of light chestnuts is diluted to yellow or cream to give the Palomino color. In homozygous

C^{cr}/C^{cr} horses an extreme dilution of the phenotype results in what is called a cremello color.

Codominance

Codominance is a modification of the dominance relationship that is related to incomplete dominance. In codominance the heterozygote exhibits the phenotypes of both homozygotes; again, no dominance is involved. It differs from incomplete dominance, in which the heterozygote exhibits a phenotype intermediate between the two homozygotes.

The ABO blood series discussed earlier in this chapter is a good example of codominance. Heterozygous I^A/I^B individuals are blood group AB because both the A antigen (product of the I^A allele) and the B antigen (product of the I^B allele) are produced. Thus the I^A and I^B alleles are codominant.

■ *Figure 4.3* Incomplete dominance in horses: (a) Palomino horse, genotype C/C^{cr}; (b) Cremello horse, genotype C^{cr}/C^{cr}; and (c) Light chestnut (sorrel) horse, genotype C/C.

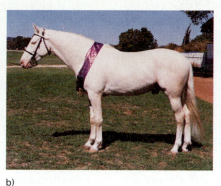

a)   b)   c)

There is no distinct dividing line between codominance and incomplete dominance. At the molecular level there may indeed be little difference with respect to the actions of the genes involved, with only more complex interactions of gene products making the phenotypic variations between the different genotypes.

Keynote

In complete dominance the same phenotype results whether an allele is heterozygous or homozygous. In complete recessiveness the allele is phenotypically expressed only when it is homozygous; the recessive allele has no effect on the phenotype of the heterozygote. Complete dominance and complete recessiveness are two extremes between which all transitional degrees of dominance are possible. In incomplete dominance, for example, the phenotype of the heterozygote is intermediate between those of the two homozygotes, while in codominance the heterozygote exhibits the phenotypes of both homozygotes (essentially, no dominance is involved).

Gene Interactions and Modified Mendelian Ratios

No gene acts by itself in determining an individual's phenotype; the phenotype is the result of highly complex and integrated patterns of molecular reactions that are under direct gene control. All the genetic examples we have discussed and will discuss have discrete biochemical bases, and in a number of cases the complex interactions between genes can be detected by genetic analysis. Some of these examples will be discussed in this section.

In Chapter 2 we discussed Mendel's principle of independent assortment, which states that genes on different chromosomes behave independently in the production of gametes. Let's assume that there are two independently assorting gene pairs, each with two alleles: *A* and *a*, and *B* and *b*. The outcome of a cross between individuals, each of whom is doubly heterozygous ($A/a\ B/b \times A/a\ B/b$), will be nine genotypes in the following proportions:

> 1/16 $A/A\ B/B$: 2/16 $A/A\ B/b$: 1/16 $A/A\ b/b$:
> 2/16 $A/a\ B/B$: 4/16 $A/a\ B/b$: 2/16 $A/a\ b/b$:
> 1/16 $a/a\ B/B$: 2/16 $a/a\ B/b$: 1/16 $a/a\ b/b$.

If the phenotypes determined by the two allelic pairs are distinct, such as smooth versus wrinkled peas, long versus short stems, then we get the familiar dihybrid phenotypic ratio of 9:3:3:1. That is, 9/16 of the progeny show both dominant phenotypes, 3/16 show one dominant phenotype and the other gene pair's recessive phenotype, 3/16 show the first gene pair's recessive phenotype and the other pair's dominant phenotype, and 1/16 show both recessive phenotypes. Any alteration in this standard 9:3:3:1 ratio indicates that the phenotype is the product of the interaction of two or more genes. As we discuss each modified Mendelian ratio we will refer to this distribution of genotypes.

For the 9:3:3:1 phenotypic ratio the genotypes can be represented in a shorthand way as, respectively, $A/-\ B/-$, $A/-\ b/b$, $a/a\ B/-$, $a/a\ b/b$. The dash indicates that the phenotype is the same, whether the gene is homozygous dominant or heterozygous (e.g., $A/-$ means A/A and/or A/a). This system cannot be used when incomplete dominance or codominance is involved.

The following sections discuss the main processes that result in modified Mendelian ratios. The first describes examples of interactions between nonallelic genes that control the same general phenotypic attribute. In the second section we will discuss cases of interactions of nonallelic genes in which the phenotypic expression of one gene depends on the genotype of another gene locus. This is called *epistasis*. In both cases the discussions are confined to dihybrid crosses in which the two pairs of alleles assort independently. In the "real world" there are many more complex examples of gene interactions, involving more than two pairs of alleles and/or genes that do not assort independently.

Gene Interactions That Produce New Phenotypes

If, in a dihybrid cross, the two allelic pairs affect the same phenotypic characteristic, there is a chance for gene product interaction to give novel phenotypes. The result can be modified phenotypic ratios, depending on the particular interaction between the products of the nonallelic genes.

A classic example of such gene interactions is comb shape in chickens. New comb shape phenotypes are a consequence of interactions between two allelic pairs. Figure 4.4 shows the four comb phenotypes that result from the interaction of the alleles of two gene loci. Each of these types can be true-breeding.

Crosses made between true-breeding rose-combed and single-combed varieties showed that

■ *Figure 4.4*   Four distinct comb shape phenotypes in chickens, resulting from all possible combinations of a dominant and a recessive allele at each of two gene loci: (a) Rose comb ($R/-\ p/p$); (b) Walnut comb ($R/-\ P/-$); (c) Pea comb ($r/r\ P/-$); (d) Single comb ($r/r\ p/p$).

a) Rose comb

b) Walnut comb

c) Pea comb

d) Single comb

rose was completely dominant over single. When the F_1 rose-combed birds were bred together, there was a clear segregation into 3 rose: 1 single in the F_2. Similarly, pea comb was found to be completely dominant over single, with a 3 pea: 1 single ratio in the F_2. When true-breeding rose and pea varieties were crossed, however, the result was new and interesting (Figure 4.5). Instead of showing either rose or pea combs, all birds in the F_1 showed a new comb form, different from the rose and the pea comb (Figure 4.5[a]). This new form was called walnut comb since it resembles half a walnut meat.

When the F_1 walnut-combed birds were bred together, another fascinating result was observed in the F_2 (Figure 4.5[b]): Not only did walnut-, rose-, and pea-combed birds appear, but so did single-combed birds. These four comb types occurred in a ratio of 9 walnut: 3 rose: 3 pea: 1 single. Such a ratio is characteristic in F_2 progeny from a cross of two parents each heterozygous for two genes. The doubly dominant class in the F_2 was walnut, while the proportion of the single-combed birds indicated that this class contained both recessive alleles.

The overall explanation of the results is as follows (see Figure 4.5[b]). The walnut comb depends on the presence of two dominant alleles, R and P, both located at two independently assorting gene loci. In $R/-\ p/p$ birds, a rose comb results; in $r/r\ P/-$ birds, a pea comb results; and in $r/r\ p/p$ birds, a single comb results.

Thus it is the interaction of two dominant alleles, each of which individually produces a different phenotype, that produces a new phenotype. No modification of typical Mendelian ratios is involved in this case. The biochemical basis for the four comb types is not known. At a very general level we can propose that the single-comb phenotype results from the activities of a number of genes other than the R and P genes. In other words, $r/r\ p/p$ birds do not produce any functional gene product that influences the comb phenotype beyond the basic single appearance. The dominant R allele might produce a gene product that interacts with the products of genes controlling the single-comb phenotype to produce a rose-shaped comb. Similarly, the dominant P allele might produce a gene product that interacts with the products of the single-comb genes to produce a pea-shaped comb. When the products of both the R and P alleles are present, they interact to produce another comb variation, the walnut comb.

Epistasis

Epistasis is a form of gene interaction in which one gene interferes with the phenotypic expression of

■ *Figure 4.5* Results of genetic crosses, showing the interaction of genes for comb shape in fowl: (a) The cross of a true-breeding rose-combed bird with a true-breeding pea-combed bird gives all walnut-combed offspring in the F_1; (b) When the F_1 birds are interbred a 9:3:3:1 ratio of walnut: rose: pea: single occurs in the F_2.

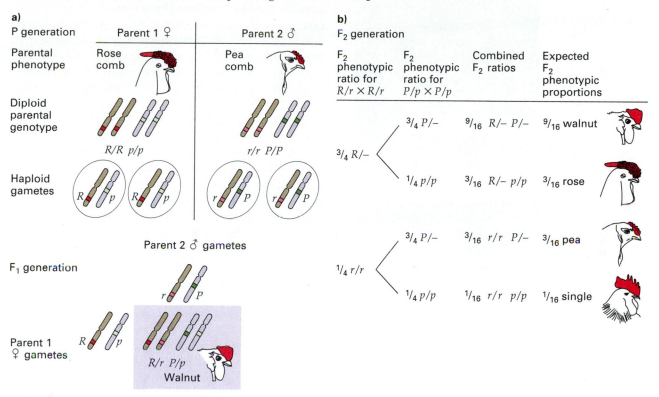

F₁ genotypes: All $R/r\ P/p$

F₁ phenotypes: All walnut comb

another, nonallelic gene so that the phenotype is governed by the former gene and not by the latter gene. Unlike the situation we have just discussed, no new phenotypes are produced by this type of gene interaction. If we consider the F_2 genotypes $A/-B/-$, $A/-b/b$, $a/a\,B/-$, and $a/a\,b/b$, epistasis may be caused by the presence of homozygous recessives of one gene pair, so that a/a masks the effect of the B allele. Or epistasis may result from the presence of one dominant allele in a gene pair. For example, the A allele might mask the effect of the B allele. Moreover, the epistatic effect need not be just in one direction, as in the examples we just mentioned. Epistasis can occur in both directions between two gene pairs. All these possibilities can produce quite a number of modifications of the 9:3:3:1 ratio.

Coat Color in Rodents (the 9:3:4 Ratio) In one type of epistasis $a/a\,B/-$ and $a/a\,b/b$ individuals have the same phenotype, so the overall phenotypic ratio in the F_2 is 9:3:4 rather than 9:3:3:1. An example is coat color in rodents. The ancestral coat color

of mice is the grayish color seen in ordinary wild mice, due to the presence of two pigments in the fur. Individual hairs are mostly black with narrow yellow bands near the tip (Figure 4.6). This is called the agouti pattern.

Several other coat colors are seen in domesticated rodents. The most familiar example is the albino, in which the complete absence of pigment in the fur and in the irises of the eyes causes a white coat and pink eyes. Albinos are true breeding, and this variation behaves as a complete recessive to any other color. Another variant has black coat color as the result of the absence of the yellow pigment found in the agouti pattern. Black also is recessive to agouti.

When true-breeding agouti mice are crossed with albinos, the F_1 progeny are all agouti. When these F_1 agoutis are interbred, the F_2 progeny consist of approximately 9/16 agouti, 3/16 black, and 4/16 albino as shown in Figure 4.7. This pattern occurs because the parents differ in a gene necessary for the development of any color, which the black mice have ($C/-$) but the albinos do not have (c/c), and in a gene for the agouti pattern ($A/-$ for

■ *Figure 4.6*   Pigment patterns in rodent fur: (a) Agouti; (b) Albino; and (c) Black.

a) Agouti
A/– C/–

b) Albino
(*A/– c/c* or *a/a c/c*)

c) Black
a/a C/–

agouti; *a/a* for nonagouti), which results in a banding of the black hairs with yellow. Genotypically *A/– C/–* are agouti, *a/a C/–* are black, and *A/– c/c* and *a/a c/c* are albino, giving a 9:3:4 phenotypic ratio of agouti:black:albino. Thus there is epistasis of *c/c* over *A/–* and white hairs are produced in *c/c* mice, no matter what the genotype is at the other locus.

Flower Color in Sweet Peas (the 9:7 Ratio)   In the sweet pea purple flower color is dominant to white and gives a typical 3:1 ratio in the F₂. White-flowered varieties of sweet peas breed true, and crosses between different white varieties usually produce white-flowered progeny. In some cases, however, crosses of two true-breeding white varieties give only purple-flowered F₁ plants. When these F₁ hybrids are self-fertilized, they produce an F₂ generation consisting of about 9/16 purple-flowered sweet peas and 7/16 white-flowered, as shown in Figure 4.8. All the F₂ white-flowered plants breed true when self-fertilized. One-ninth of the purple-flowered F₂ plants—the *C/C P/P* genotypes—breed true.

These results may be explained by considering the interaction of two nonallelic genes. The 9/16 purple-flowered F₂ plants suggests that colored flowers appear only when two independent dominant alleles are present together and that the color purple results from some interaction between them. White flower color would then be due to the absence of either or both of these dominant alleles. Thus gene pair *C/c* specifies whether or not the flower can be colored, and gene pair *P/p* specifies whether or not purple flower color will result. An interaction of two genes to give rise to a specific product is a form of epistasis called *complementary gene action*.

Figure 4.9 presents a hypothetical pathway for the production of purple pigment. In this pathway a colorless precursor compound is converted through several steps (via compounds 1, 2, and 3) to a purple end product. Each step is controlled by a functional gene product. To explain the F₂ ratio in the sweet pea example, we can propose an hypothesis that gene *C* controls the conversion of white compound 1 to white compound 2, and that gene *P* controls the conversion of compound 2 to compound 3. Therefore homozygosity for the recessive allele of either or both of the *C* and *P* genes will result in a block in the pathway, and only white pigment will accumulate. That is, *C/– p/p*, *c/c P/–*, and *c/c p/p* genotypes will *all* be white. The only plants that produce purple flowers will be those in which the two steps are completed so that the rest of the pathway leading to the colored pigment can be carried out. This situation occurs only in *C/– P/–* plants.

In sum, many types of dominance modifications are possible as a result of interactions between the products of different gene pairs. Geneticists detect such interactions when they observe deviations from the expected phenotypic ratios in crosses. We have discussed some examples in which two nonallelic genes assort independently and in which complete dominance is exhibited in each gene pair. The ratios we discussed would necessarily be modified further if the genes did not assort independently and/or if incomplete dominance or codominance

■ *Figure 4.7*   Generation of an F$_2$ 9:3:4 ratio for coat color in rodents.

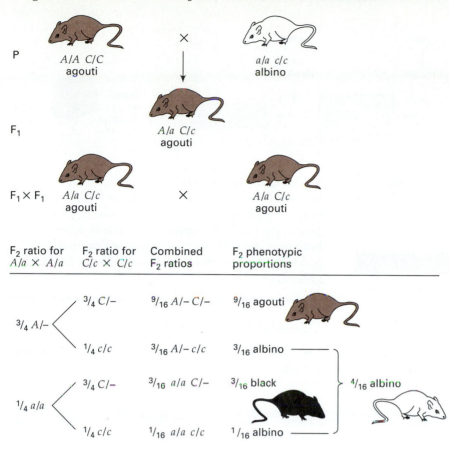

■ *Figure 4.8*   Generation of an F$_2$ 9:7 ratio for flower color in sweet peas.

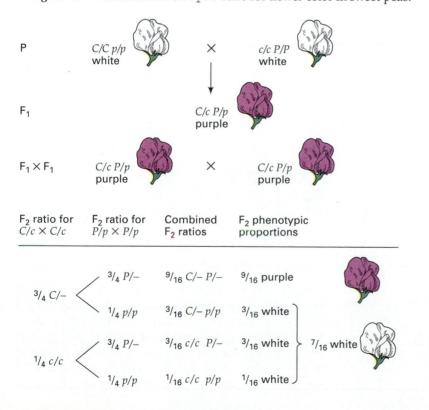

■ *Figure 4.9*   Hypothetical pathway for the production of purple pigment in sweet peas controlled by two genes, *C* and *P*.

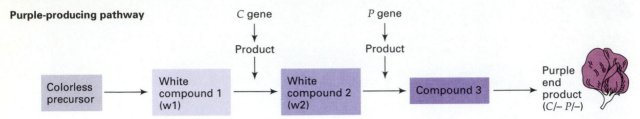

prevailed. Table 4.3 shows examples of epistatic F_2 phenotypic ratios from an $A/a\ B/b \times A/a\ B/b$ cross.

Keynote

In many instances nonallelic genes interact to determine phenotypic characteristics. In some cases interaction between gene products results in new phenotypes without modification of typical Mendelian ratios. In another type of gene interaction, called epistasis, interaction between gene

products causes modifications of Mendelian ratios because one gene product interferes with the phenotypic expression of another nonallelic gene (or genes). Epistasis is complicated further when one or both gene pairs involve incomplete dominance or codominance, or when gene pairs do not assort independently.

Essential Genes and Lethal Alleles

For a few years after the rediscovery of Mendelism, geneticists believed that mutations only changed

Table 4.3

Examples of Epistatic F_2 Phenotypic Ratios From an *A/a B/b* × *A/a B/b* Cross in Which Complete Dominance Is Shown for Each Gene Pair

		A/A B/B	A/A B/b	A/a B/B	A/a B/b	A/A b/b	A/a b/b	a/a B/B	a/a B/b	a/a b/b
Four phenotypic classes	A and B both completely dominant (classic ratio—no epistasis)	9				3		3		1
Fewer than four phenotypic classes	a/a epistatic to B and b; recessive epistasis	9				3		4		
	A epistatic to B and b; dominant epistasis	12						3		1
	a/a epistatic to B and b; b/b epistatic to A and a; duplicate recessive epistasis	9				7				
	A epistatic to B and b; B epistatic to A and a; duplicate dominant epistasis	15								1
	Duplicate interaction	9				6				1

the appearance of a living organism, but then they discovered that a mutant allele could cause the death of an organism. In a sense this mutation is still a change in phenotype, with the new phenotype being lethality. An allele that results in the death of an organism is called a **lethal allele**, and the gene involved is called an **essential gene**. Essential genes are genes which, when mutated, can result in a lethal phenotype. If the mutation is due to a **dominant lethal allele**, heterozygotes will show the lethal phenotype. If the mutation is due to a **recessive lethal allele**, only homozygotes for that allele will be lethal.

An example of an essential gene is the gene for yellow body color in mice. The yellow variety never breeds true. When yellows are crossed with nonyellows, the progeny shows an approximately 1:1 ratio of yellow:nonyellow mice. This ratio is what is expected from the mating of a heterozygote with a homozygous recessive, which suggests that yellow mice are heterozygous. When heterozygotes are interbred we would expect 1/4 to be pure yellow, 1/2 to be heterozygous yellow, and 1/4 to be nonyellow; instead, about 2/3 are yellow and 1/3 are nonyellow. (The nonyellow color depends on the other coat color genes carried by the mice.)

The yellow allele has a *dominant* effect with regard to coat color but, when homozygous, the yellow allele is lethal. In other words, the yellow allele is a recessive lethal allele. The yellow × yellow cross is shown in Figure 4.10. If we give the symbol Y to the yellow allele, and Y^+ to the normal, wild-type allele, the cross is $Y/Y^+ \times Y/Y^+$. We expect a genotypic ratio of 1 Y/Y: 2 Y/Y^+: 1 Y^+/Y^+ among the progeny. The 1/4 Y/Ys die before birth, leaving 2/4 Y/Y^+ and 1/4 Y^+/Y^+, giving a 2:1 phenotypic ratio of yellow Y/Y^+: nonyellow Y^+/Y^+ survivors. Characteristically when two heterozygotes are crossed, recessive lethal alleles are recognized by a 2:1 ratio of progeny types.

Numerous examples of essential genes are found in all diploid organisms. In humans there are many known recessive lethal alleles. One particular autosomal recessive mutant allele, when homozygous, causes *Tay-Sachs disease* (see Chapter 7, p. 163). Homozygotes appear normal at birth, but before about one year of age they begin to show symptoms of central nervous system deterioration. Progressive mental retardation, blindness, and loss of neuromuscular control follow. The afflicted children usually die at three to four years of age. The genetic defect in Tay-Sachs results in a deficiency in an enzyme called *hexosaminidase A* (hex A), a substance required for the metabolism of complex chemical structures called *sphingolipids,* necessary for proper nerve function.

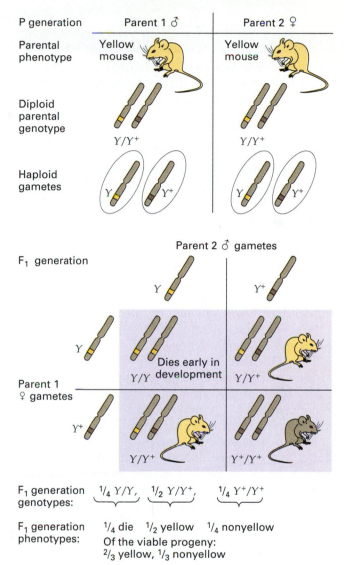

■ *Figure 4.10* Inheritance of a lethal gene Y in mice. A mating of yellow to yellow gives 1/4 nonyellow (black) mice, 1/4 yellow mice, and 1/4 dead embryos. The viable yellow mice are heterozygous Y/Y^+, and the dead individuals are homozygous Y/Y.

There are sex-linked lethal mutations as well as autosomal lethal mutations, and dominant lethal mutations as well as recessive lethals. By definition, dominant lethals exert their effect in heterozygotes, so, the organism typically dies at conception or quite young. Dominant lethals cannot be studied genetically unless death occurs after the organism has reached reproductive age. One example is the human autosomal dominant trait, Huntington's disease. This disease results in involuntary movements, progressive central nervous system degeneration, and eventually death. The onset of the symptoms may not occur until the early thirties, and death usually occurs when the afflicted persons are in their forties or fifties. As a result the individuals may have passed on the gene to their offspring before knowing that they are afflicted with the disease. The famous American folksinger Woody Guthrie died from Huntington's disease.

Keynote

An allele that is fatal to the individual is called a lethal allele. Recessive lethal and dominant lethal alleles exist, and they can be sex-linked or autosomal. The existence of recessive lethal alleles of a gene indicates that the gene's normal product is essential for the function of the organism.

The Relationship Between Genotype and Phenotype

The genetic constitution of the zygote only specifies the organism's potential to develop and function. Many factors can influence gene expression as the organism develops and differentiates; one such influence is the organism's environment.

A gene is a segment of DNA, and an organism's set of genes (its genome) is found in its chromosomes. The *development* of an organism from a zygote is a process of regulated *growth* and *differentiation* that results from the interactions of the genome with the internal cellular environment and the external environment. Development is a programmed series of phenotypic changes that are under temporal, spatial, and quantitative control. It is essentially irreversible under normal environmental conditions. Four major processes interact with one another to constitute the complex process of development: (1) replication of the genetic material, (2) growth, (3) differentiation of the various cell types, and (4) the aggregation of differentiated cells into defined tissues and organs.

We must think of development as a series of intertwined, complex, biochemical pathways whose steps are under gene control. Any of these pathways are susceptible to environmental influences if the products of the genes controlling the pathways are affected by the internal or external environment. Though genes influence development, no gene by itself totally determines a particular phenotype. In other words, a phenotype is the result of closely interwoven interactions among many factors during development. Therefore, even though individual organisms have the same genetic constitution, that does not necessarily mean the same phenotypes will result. This phenomenon is most readily studied in experimental organisms where the genotype is unequivocally known. The extent to which the gene manifests its effects under varying environmental conditions can then be seen. We will consider some examples in the following sections.

Penetrance and Expressivity

In some cases not all individuals who are known to have a particular genotype show the phenotype specified by that genotype. The frequency with which a dominant or homozygous recessive allele manifests itself in individuals in a population is called the **penetrance** of the gene. Penetrance depends on both the genotype (e.g., the presence of epistatic or other genes) and the environment. Figure 4.11 illustrates the concept of penetrance. Penetrance is complete (100 percent) when all the homozygous recessives show one phenotype, when all the homozygous dominants show another phenotype, and when all the heterozygotes are alike. If all individuals carrying a dominant mutant allele show the mutant phenotype, the allele is exhibiting complete penetrance. Many genes show complete penetrance: The seven gene pairs in Mendel's experiments and the alleles in the human ABO blood group system are two examples. If less than 100 percent of the carriers of a particular genotype exhibit the phenotype expected, penetrance is incomplete. For instance, an organism may be genotypically $A/-$ or a/a but may not display the phenotype typically associated with that genotype. If 80 percent, say, of the individuals carrying a particular gene show the corresponding phenotype, we say that there is 80 percent penetrance. Incomplete penetrance is seen in retinoblastoma (tumor of the eye) disease in humans, and polydactyly (extra fingers or toes) in humans, both of which are inherited as autosomal dominants.

We can also determine the degree to which a gene influences a phenotype. **Expressivity** refers to the degree to which a penetrant gene or genotype is phenotypically expressed. Expressivity may be described in either qualitative or quantitative terms (Figure 4.12); for example, expressivity may be referred to as severe, intermediate, or slight. Like penetrance, expressivity depends on both the genotype and the external environment, and it may be constant or variable. An example of variation in expression is found in the human condition *osteogenesis imperfecta*. The three main features of this disease are blue sclerae (the whites of the eyes), very fragile bones, and deafness. This condition is inherited as an autosomal dominant with almost 100 percent penetrance; that is, 100 percent of individuals with the genotype have the condition. However, the trait shows variable expressivity: A person with the

■ *Figure 4.11*   Complete and incomplete penetrance.

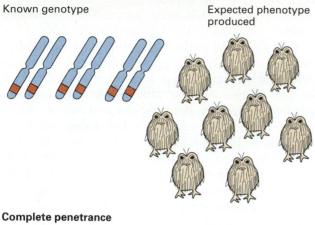

Known genotype

Expected phenotype
produced

Complete penetrance

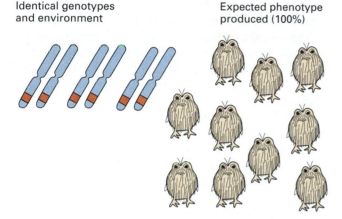

Identical genotypes
and environment

Expected phenotype
produced (100%)

Incomplete penetrance

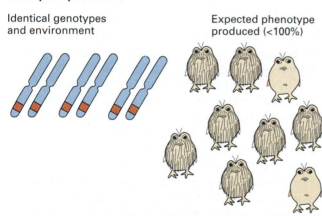

Identical genotypes
and environment

Expected phenotype
produced (<100%)

gene may have one or any combination of the three traits. Moreover, the fragility of the bones for those who exhibit this condition is also very variable. Therefore, we see that in medical genetics it is important to recognize that a gene may vary widely in its expression, a qualification that makes the task of genetic counseling that much more difficult.

Keynote

Penetrance is the frequency with which a dominant or homozygous recessive gene manifests itself in the phenotype of an individual. Expressivity is the type or degree of phenotypic manifestation of a penetrant gene or genotype.

Effects of the Internal Environment

Complex biochemical responses occur in cells and in the organism as a result of certain stimuli from the internal environment. These responses can have an effect on the phenotype.

Age   The age of the organism reflects internal environmental changes that can affect gene function. All genes do not continually function; instead, over time, there is programmed activation and deactivation of genes as the organism develops and functions. As discussed previously, the Hunting-

■ *Figure 4.12*   Expressivity of a gene: severe, intermediate, and slight.

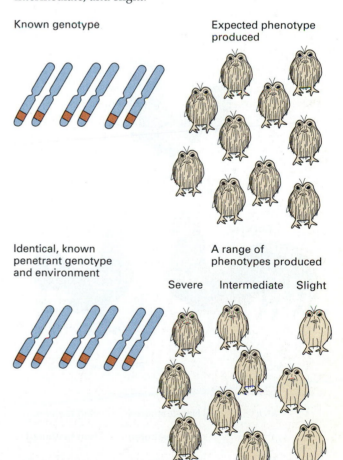

Known genotype

Expected phenotype
produced

Identical, known
penetrant genotype
and environment

A range of
phenotypes produced

Severe   Intermediate   Slight

ton's disease gene is one whose effects are not manifested until later in the organism's existence. Somehow, the gene responds to the age of the individual, and the devastating phenotypic effects commence. Numerous other age-dependent genetic diseases occur in humans—pattern baldness (i.e., the bald spot creeping forward from the crown of the head) appears in males between 20 and 30 years of age, and Duchenne severe muscular dystrophy appears in children between 2 and 5 years. In most cases the nature of the age dependency is not understood.

Sex   The expression of particular genes may be influenced by the sex of the individual. In the case of sex-linked genes, as mentioned earlier, differences in the phenotypes of the two sexes are related to different complements of genes on the sex chromosomes. However, in some cases genes that are not located on the sex chromosomes affect a particular character that appears in one sex but not the other. Traits of this kind are called **sex-limited traits**.

One such trait is the feathering pattern of chickens (Figure 4.13). A common difference between male and female fowls is in the structure of many of the feathers. Cocks often have long, narrow, pointed feathers on their hackles and saddle; pointed feathers on the cape back and wing; and long, curving, pointed sickle feathers on the tail. This pattern is called *cock-feathering*. Hens usually show the contrasting feather characteristics of short, broad, blunt, and straight feathers. This pattern is called *hen-feathering*. If heterozygous hen-feathered males and females are interbred, *all* progeny females are hen-feathered, whereas progeny males exhibit a 3 : 1 ratio of hen-feathered : cock-feathered.

This occurs because the feathering pattern trait is controlled by an autosomal gene, with hen feathering (h^+) dominant to cock feathering (h). In females h^+/h^+, h^+/h, and h/h genotypes all give rise to hen-feathering. In males the h^+/h^+ and h^+/h genotypes are hen-feathered while the h/h chickens are cock-feathered. So from a $h^+/h \times h^+/h$ cross, 1/4 of the male progeny will be h/h and exhibit cock-feathering, while the other 3/4 of the male progeny will exhibit hen-feathering. In other words, in the homozygous condition (h/h) the cock-feathered trait is male-limited, showing that sex as an internal environmental parameter can affect gene expression. Several experiments have indicated that biochemical differences in male and female sex hormones in association with the genetic constitution of the skin (in which the feathers originate) determine the state of feathering in the fowl.

A slightly different situation is found in **sex-influenced traits**. Such traits appear in both sexes, but either the frequency of occurrence in the two sexes is different or the relationship between genotype and phenotype is different. An example of a sex-influenced trait is *pattern baldness* in humans (Figure 4.14). One pair of alleles of an autosomal gene is involved in this trait, b^+ and b. The b/b genotype specifies pattern baldness in both males and females, and the b^+/b^+ genotype gives a nonbald phenotype. The difference lies in the heterozygote: In males it leads to the bald phenotype, and in females it leads to the nonbald phenotype. In other words, the b allele acts as a dominant in males but as a recessive in females; its expression is influenced by the sex hormones of the individual. This pattern of inheritance and gene expression explains why pattern baldness is far more frequent among men than among women. From a large sampling of the progeny of matings between two heterozygotes,

■ *Figure 4.13* Sex-limited inheritance of feathering in chickens. From a cross between heterozygous hen-feathered parents the male progeny show a 3:1 ratio of hen-feathered: cock-feathered birds, while all female progeny are hen-feathered. The gene controlling the phenotype is autosomal.

Hen-feathered cock (♂)

Cock (♂)   Hen (♀)

	Genotype	Phenotype ♀	Phenotype ♂
1	h^+/h^+	Hen-feathered	Hen-feathered
2	h^+/h	Hen-feathered	Hen-feathered
1	h/h	Hen-feathered	Cock-feathered

■ *Figure 4.14* Sex-influenced inheritance of pattern baldness in humans. The allele is recessive in one sex and dominant in the other, so a cross between two heterozygotes produces a 3:1 ratio of bald:nonbald in males and 1:3 ratio of bald:nonbald in females. (a) Production of F_1 generation; (b) Production of F_2 generation.

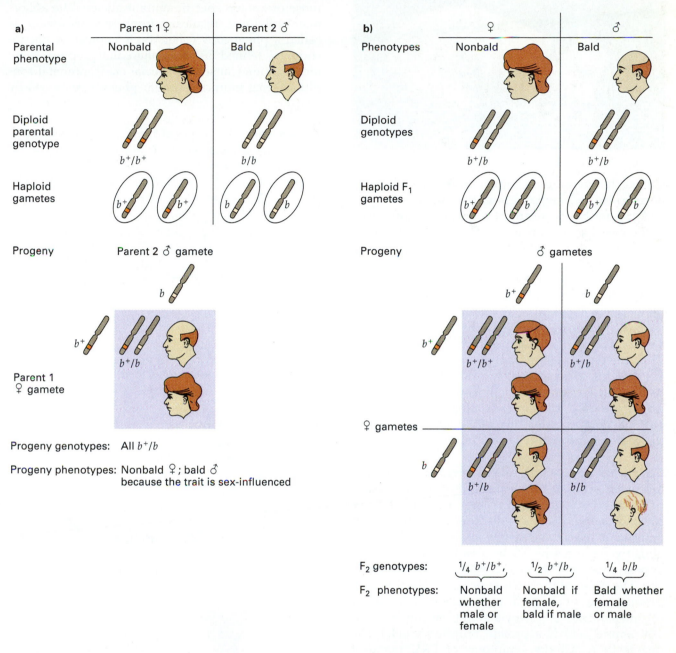

3/4 of the daughters are nonbald and 1/4 are bald, and 3/4 of the sons are bald and 1/4 are nonbald.

Effects of the External Environment

Although many factors in the external environment influence gene expression (temperature, nutrition, light, chemicals, infectious agents, and more), we will only discuss two factors here, temperature and chemicals.

Temperature The rate at which a chemical reaction occurs depends in large part on temperature. The same is also true, within limits, for biochemical reactions taking place within organisms. Therefore, it is reasonable to expect that temperature might affect gene expression, having significant effects on development. A good example is fur color in Himalayan rabbits (Figure 4.15). Certain genotypes of this white rabbit cause dark fur to develop at the extremities (ears, nose, and paws), where the local

■ *Figure 4.15*   Himalayan rabbit.

surface temperature is lower. Since all body cells develop from a single zygote, this distinct fur pattern cannot be the result of a genotypic difference of the cells in those areas. It must result from external environmental influences.

This hypothesis can be tested by rearing Himalayan rabbits under different temperature conditions (Figure 4.16). When a rabbit is reared at a temperature above 30°C, all its fur, including that of the ears, nose, and paws, is white (Figure 4.16[a]). If a rabbit is raised at a temperature of approximately 25°C, the typical Himalayan phenotype results (Figure 4.16[b]). When a rabbit is raised at 25°C while its left rear flank is artificially cooled to a temperature below 25°C, the rabbit develops the Himalayan coat phenotype, exhibiting additional dark fur on the cooled flank area (Figure 4.16[c]).

Chemicals   Ultimately, all the interactions we have discussed occur at the chemical level as gene products affect metabolism. The human disease *phenylketonuria* (PKU) is a disorder controlled by an autosomal recessive gene. In individuals homozygous for the recessive allele, a variety of symptoms appear, most notably mental retardation at an early age. The cause of the phenotypes is a defective gene product in a biochemical pathway that is used for the metabolism of phenylalanine, an amino acid. (An amino acid is a protein building block.) The degree to which the symptoms of PKU manifest themselves depends on diet. Problem foods include protein containing phenylalanine, such as the protein of mother's milk, which may be considered to be an external environment effect. PKU can be diagnosed soon after birth by a simple blood test, and then treated by restricting the amount of phenylalanine in the diet.

Changes in the chemical composition of the environment can also influence the expression of one or more genes. The most sensitive time period is early

development, since relatively small changes at that time can result in great changes later. When administered to a developing embryo, certain drugs, chemicals, and viruses produce effects that mimic those of known specific mutant alleles. The abnormal individual resulting from this treatment is called a **phenocopy** (phenotypic copies). A phenocopy is defined as a nonhereditary, phenotypic modification (caused by special environmental conditions) that mimics a similar phenotype caused by

■ *Figure 4.16*   Phenotypes of Himalayan rabbits raised under different temperature conditions: (a) White extremities, reared at ≥30°C; (b) Normal Himalayan pattern, reared at 25°C; (c) Himalayan pattern with dark patch on flank reared at 25°C, flank cooled to below 25°C.

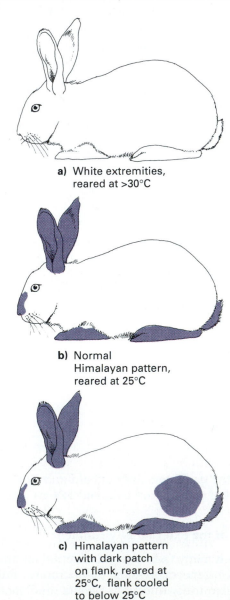

a) White extremities, reared at >30°C

b) Normal Himalayan pattern, reared at 25°C

c) Himalayan pattern with dark patch on flank, reared at 25°C, flank cooled to below 25°C

a gene mutation. In other words, although the individual expresses a mutant *phenotype*, the *genotype* is normal. The agent that produces a phenocopy is called a *phenocopying agent*. There are many examples of phenocopies, and, in some instances, studies of phenocopies have provided useful information about the actual molecular defects caused by the mutant counterpart. Remember that the abnormalities seen in these situations occur in a genotypically normal individual without any alteration of the genetic constitution.

Phenocopies occur in all sorts of organisms, from the simple to the complex. In humans, cataracts, deafness, and heart defects are sometimes produced when an individual is homozygous for rare recessive alleles; these disorders may also result if the mother is infected with rubella (German measles) virus during the first 12 weeks of pregnancy. Another human trait for which there is a phenocopy is *phocomelia*, a suppression of the development of the long bones of the limbs, which is caused by a rare dominant allele with variable expressivity. Between 1959 and 1961 similar phenotypes were produced by the sedative thalidomide when taken by expectant mothers during the 35th to 50th day of gestation.

Keynote

The phenotypic expression of a gene depends on several factors, including its dominance relations, the genetic constitution of the rest of the genome (e.g., the presence of epistatic or modifier genes), and the influences of the internal and external environments. In some cases special environmental conditions can cause a phenocopy, a nonhereditary and phenotypic modification that mimics a similar phenotype caused by a gene mutation.

Nature Versus Nurture

We are left with the nature versus nurture question; that is, what is the relative contribution of genes and environment to the phenotype? Up to this point, variation in most of the traits we have examined has been determined largely by differences in genotype; phenotypic differences have reflected genetic differences. Earlier in this section we learned that the phenotypes of many traits, however, are influenced by both genes and environment. Let us consider some other human examples.

Human height, or stature, is definitely influenced by genes. On the average tall parents tend to have tall offspring and short parents tend to have short offspring. There are also a number of genetic forms of dwarfism in humans. *Achondroplasia* is a type of dwarfism in which the bones of the arms and legs are shortened, but the trunk and head are of normal size; achondroplasia is due to the presence of a single dominant gene. But genes alone do not determine human height—environment also plays a role. For example, human height has increased about 1 inch per generation over the past 100 years. This increase is not the result of changes in the genotype, rather it is probably due to better diet and health care, both of which are known to affect human stature. Thus genes and the environment interact in determining human height.

For a trait like height *genes set certain limits or potential for the phenotype*. What phenotype an individual develops within these limits depends on the environment. The extent to which the phenotype varies according to the environment is termed the **norm of reaction**. For some genotypes, such as those producing the ABO blood types, the norm of reaction is small; the phenotype produced by a genotype is the same in different environments. For other genotypes the norm of reaction is large; the phenotype produced by the genotype varies greatly in different environments.

Many human behavioral traits are the result of interaction between genes and the external environment. Alcoholism is a major health problem in the United States. About 10 million Americans are problem drinkers and 6 million are severely addicted to alcohol. Numerous studies have shown that alcoholism is influenced by genes. For example, sons of alcoholic fathers who are separated from their biological parents at birth and adopted into foster homes are four times more likely to become alcoholic than sons adopted at birth whose biological fathers were not alcoholic. Thus genes affect alcoholism. However, there is no gene that forces a person to drink alcohol. That is, one cannot become an alcoholic unless one is exposed to an environment in which alcohol is available and drinking is encouraged. What genes do is make certain people more or less susceptible to alcohol abuse, increasing or decreasing the risk of developing alcoholism. How genes influence our susceptibility is not yet clear. They may affect the way we metabolize alcohol, which might affect how much we drink. Or genes may influence certain of our personality traits, which make us more or less likely to drink heavily. The important point is that a behavioral trait like alcoholism may be influenced by genes, but the genes alone do not produce the phenotype.

Nowhere has the role of genes and environment been more controversial than in the study of human intelligence. In the past there was a tendency to think of human intelligence as either genetically preprogrammed or produced entirely by the environment. The clash of these opposing views was termed the *nature–nurture* controversy. Today geneticists recognize that neither of these extreme views is correct; human intelligence is the product of both genes and environment. That genes influence human intelligence is clearly evidenced by genetic conditions that produce mental retardation. For example, PKU (see pp. 161–162) is an autosomal recessive disease that results in mental retardation if untreated; Down syndrome, caused by an extra copy of chromosome 21, produces mental retardation. Numerous studies also indicate that genes influence differences in IQ among non-retarded people. (IQ, or Intelligence Quotient, is a standardized measure of mental age compared to chronological age. It is a good predictor of scholastic achievement in school and is relatively stable over time.) Adoption studies show that the IQs of adopted children are closer to that of their biological parents than to the IQs of their foster parents. However, it is also clear that IQ is influenced by environment. Identical twins frequently differ in IQ, which can only be explained by environmental differences. Family size, diet, and culture are environmental factors that are known to affect IQ. Thus IQ results from the interaction of genes and environment. Consequently, if two people (other than identical twins) differ in IQ, it is impossible to attribute that difference to either genes or environment, because both interact in determining the phenotype. It is also essential to recognize that although we cannot change our genes, we can alter the environment and thus can affect a phenotypic trait like intelligence.

Keynote

Variation in most of the genetic traits considered in the discussion of Mendelian principles is determined predominantly by differences in genotype; that is, phenotypic differences resulted from genotypic differences. For many traits, however, the phenotypes are influenced by both genes and the environment. The debate over the relative contribution of genes and environment to the phenotype has been termed the *nature versus nurture controversy*.

Summary

A variety of exceptions and extensions to Mendel's principles were discussed in this chapter. These included:

1. Multiple alleles: A gene may have many allelic forms, and these alleles are called multiple alleles of a gene. Any given diploid individual can have only two different alleles of the multiple allelic series.

2. Modified dominance relationships: In complete dominance the same phenotype results whether an allele is heterozygous or homozygous. In incomplete dominance the phenotype of the heterozygote is intermediate between those of the two homozygotes. In codominance the heterozygote exhibits the phenotypes of both homozygotes.

3. Gene interactions and modified Mendelian ratios: In many cases nonallelic genes do not function independently in determining phenotypic characteristics. Modified Mendelian ratios occur in epistasis, because of interactions of nonallelic genes: The phenotypic expression of one gene depends on the genotype of another gene locus. In other interactions a new phenotype is produced.

4. Essential genes and lethal alleles: Alleles of certain genes, when expressed, are fatal. Such lethal alleles may be recessive or dominant. The existence of lethal alleles of a gene indicates that the normal product of the gene is essential for the organism to function.

5. It is not always the case that a gene's phenotypes are simply wild type and mutant. The phenomenon of penetrance describes the condition in which not all individuals who are known to have a particular gene show the phenotype specified by that gene. That is, penetrance is the frequency with which a dominant or homozygous recessive gene manifests itself in individuals in the population. The related phenomenon of expressivity describes the degree to which a penetrant gene or genotype is phenotypically expressed. Both penetrance and expressivity depend on both the genotype and the external environment.

6. An organism's potential to develop and function is specified by the zygote's genetic constitution. As an organism develops and differentiates, gene expression is influenced by a number of factors, including the aforementioned dominance relations, genetic constitution of the rest of the genome, and the influences of the internal and external environment. The phenotypes of many traits are influenced by both genes and environment. The debate over the relative contribution of genes and environment to the phenotype has been termed the nature versus nurture controversy.

Analytical Approaches for Solving Genetics Problems

Q4.1 In snapdragons red flower color is incompletely dominant to white, with the heterozygote being pink; normal flowers are completely dominant to peloric shaped ones; and tallness is completely dominant to dwarfness. The three gene pairs segregate independently. If a homozygous red, tall, normal-flowered plant is crossed with a homozygous white, dwarf, peloric-flowered one, what proportion of the F_2 will resemble the F_1 in appearance?

A4.1 Let us assign symbols: C^R = red and C^W = white; N = normal flowers and n = peloric; T = tall and t = dwarf. Then the initial cross is $C^R/C^R\ T/T\ N/N \times C^W/C^W\ t/t\ n/n$. From this cross we see that all the F_1 plants are triple heterozygotes with the genotype $C^R/C^W\ T/t\ N/n$ and the phenotype pink, tall, normal-flowered. Interbreeding the F_1 generation will produce 27 different genotypes in the F_2. This answer follows from the rule that the number of genotypes is 3^n, where n is the number of heterozygous gene pairs involved in the $F_1 \times F_1$ cross.

Here we are asked specifically for the proportion of F_2 progeny that resemble the F_1 in appearance. We can calculate this proportion directly, without displaying all the possible genotypes and collect those classes with the appropriate phenotype. First we calculate the frequency of pink-flowered plants in the F_2; then we determine the proportion of these plants that have the other two attributes. From an $C^R/C^W \times C^R/C^W$ cross we calculate that half of the progeny will be heterozygous C^R/C^W and therefore pink. Next we determine the proportion of F_2 plants that are phenotypically like the F_1 with respect to height (i.e., tall). Either T/T or T/t plants will be tall, and so 3/4 of the F_2 will be tall. Similarly, 3/4 of the F_2 plants will be normal-flowered like the F_1s. To obtain the probability of occurrence of all three of these phenotypes together (pink, tall, normal), we must multiply the individual probabilities since the gene pairs segregate independently. The answer is $1/2 \times 3/4 \times 3/4$, or 9/32.

Q4.2 a. An $F_1 \times F_1$ self gives a 9:7 phenotypic ratio in the F_2. What phenotypic ratio would you expect if you testcrossed the F_1?
   b. Answer the same question for an $F_1 \times F_1$ cross that gives a 9:3:4 ratio.
   c. Answer the same question for a 15:1 ratio.

A4.2 This question deals with epistatic effects. In answering the question we must consider the inter-

action between the different genotypes in order to proceed with the testcross. Let's set up the general genotypes that we will deal with throughout. The simplest are allelic pairs a^+ and a, and b^+ and b, where the wild-type alleles are completely dominant to the other member of the pair.

a. A 9:7 ratio in the F_2 implies that both members of the F_1 are double heterozygotes and that epistasis is involved. Essentially, any genotype with a homozygous recessive condition has the same phenotype, so the 3, 3, and 1 parts of a 9:3:3:1 ratio are phenotypically combined into one class. In terms of genotype 9/16 are $a^+/-\ b^+/-$ types, and the other 7/16 are $a^+/-\ b/b$, $a/a\ b^+/-$, and $a/a\ b/b$. (As always, the use of the $-$ after a wild-type allele signifies that the same phenotype results, whether the missing allele is a wild type or a mutant.) Now the testcross asked for is $a^+/a\ b^+/b \times a/a\ b/b$. Following the same logic used in questions like this one, we can predict a 1:1:1:1 ratio of $a^+/a\ b/b$: $a^+/a\ b/b$: $a/a\ b^+/b$: $a/a\ b/b$. The first genotype will have the same phenotype as the 9/16 class of the F_2, but because of epistasis the other three genotypes will have the same phenotype as the 7/16 class of the F_2. In sum, the answer is a phenotypic ratio of 1:3 in the progeny of a testcross of the F_1.

b. We are asked to answer the same question for a 9:3:4 ratio in the F_2. This question also involves a modified dihybrid ratio, where two classes of the 9:3:3:1 have the same phenotype. Complete dominance for each of the two gene pairs occurs here also, so the F_1 individuals are $a^+/a\ b^+/b$. Perhaps both the $a^+/-\ b^+/b$ and $a/a\ b/b$ classes in the F_2 will have the same phenotype, while the $a^+/-\ b^+/-$ and $a/a\ b^+/-$ classes will have phenotypes distinct from each other and from the interaction class. The genotypic ratio of a testcross of the F_1 is the same as in part a. Considering them in the same order as we did in part a, the second and fourth classes would have the same phenotype, owing to epistasis. So there are only three possible phenotypic classes instead of the four found in the testcross of a dihybrid F_1, where there is complete dominance and no interaction. The phenotypic ratio here is 1:1:2, where these phenotypes are listed in the same relative order as in the 9:3:4.

c. This question is yet another example of epistasis. Since $15 + 1$ is 16, this number gives the outcome of an F_1 self of a dihybrid where there is complete dominance for each gene pair and interaction between the dominant alleles. In this case the $a^+/-\ b^+/-$, $a^+/-\ b/b$, and $a/a\ b^+/-$ classes have one phenotype and include 15/16 of the F_2 progeny, and the $a/a\ b/b$ class has the other phenotype and 1/16 of the F_2. The genotypic results of a testcross of the F_1 are the same as in parts a and b;

that is, the F_2 progeny exhibit a 1:1:1:1 ratio of $a^+/a\ b^+/b : a^+/a\ b/b : a/a\ b^+/b : a/a\ b/b$. The first three classes have the same phenotype, which is the same as that of the 15/16 of the F_2s, and the last class has the other phenotype. The answer, then, is a 3:1 phenotypic ratio.

Questions and Problems

4.1 In rabbits C = agouti coat color, c^{ch} = chinchilla, C^h = Himalayan, and c = albino. The four alleles constitute a multiple allelic series. The agouti C is dominant to the three other alleles, c is recessive to all three other alleles, and chinchilla is dominant to Himalayan. Determine the phenotypes of progeny from the following crosses:
a. $C/C \times c/c$
b. $C/C^{ch} \times C/c$
c. $C/c \times C/c$
d. $C/c^h \times c^h/c$
e. $C/c^h \times c/c$
f. $c^{ch}/c^h \times c^h/c$
g. $c^h/c \times c/c$
h. $C/c^h \times c/c$
i. $C/c^h \times C/c^{ch}$

*4.2 If a given population of diploid organisms contains three, and only three, alleles of a particular gene (say w, $w1$, and $w2$), how many different diploid genotypes are possible in the populations? List all possible genotypes of diploids (considering *only* these three alleles).

4.3 In humans the three alleles I^A, I^B, and I^O constitute a multiple allelic series that determine the ABO blood group system, as we described in this chapter. For the following problems state whether the child mentioned can actually be produced from the marriage. Explain your answers.
a. An O child from the marriage of two A individuals.
b. An O child from the marriage of an A to a B.
c. An AB child from the marriage of an A to an O.
d. An O child from the marriage of AB to an A.
e. An A child from the marriage of an AB to a B.

*4.4 A woman of blood group AB marries a man of blood group A whose father was group O. They have two offspring. What is the probability that
a. their two children will both be group A?
b. one child will be group B and the other will be group O?

c. the first child will be a son of group AB and their second child a son of group B?

4.5 In snapdragons red flower color (C^R) is incompletely dominant to white (C^W) and the C^R/C^W heterozygotes are pink. A red-flowered snapdragon is crossed with a white-flowered one. Determine the flower color of the following:
a. the F_1;
b. the F_2;
c. the progeny of a cross of the F_1 to the red parent;
d. the progeny of a cross of the F_1 to the white parent.

*4.6 In peaches fuzzy skin (F) is completely dominant to smooth (nectarine) skin (f), and the heterozygous conditions of oval glands at the base of the leaves (G^O) and no glands (G^N) give round glands. A homozygous fuzzy, no-gland peach variety is bred to a smooth, oval-gland variety.
a. What will be the appearance of the F_1?
b. What will be the appearance of the F_2?
c. What will be the appearance of the offspring of a cross of the F_1 back to the smooth, oval-glanded parent?

4.7 In guinea pigs short hair (L) is dominant to long hair (l), and the heterozygous conditions of yellow coat (C^Y) and white coat (C^W) give cream coat. A short-haired, cream-colored guinea pig is bred to a long-haired, white guinea pig, and a long-haired, cream guinea pig is produced. When the baby matures it is bred back to the short-haired, cream parent. What phenotypic classes are expected among the offspring, and in what proportions?

4.8 In poultry the genes for rose comb (R) and pea comb (P), if both present, give walnut comb. The recessive alleles of each gene, when present together in a homozygous state, give single comb. What will be the comb characters of the offspring of the following crosses?

a. $R/R\,P/p \times r/r\,P/p$
b. $r/r\,P/P \times R/r\,P/p$
c. $R/r\,p/p \times r/r\,P/p$
d. $R/r\,P/p \times R/r\,P/p$
e. $R/r\,p/p \times R/r\,p/p$

4.9 For the following crosses involving the comb character in poultry, determine the genotypes of the two parents:

a. A walnut crossed with a single produces offspring 1/4 of which are walnut, 1/4 rose, 1/4 pea, and 1/4 single.

b. A rose crossed with a walnut produces offspring 3/8 of which are walnut, 3/8 rose, 1/8 pea, and 1/8 single.

c. A rose crossed with a pea produces 5 walnut and 6 rose offspring.

d. A walnut crossed with a walnut produces 1 rose, 2 walnut, and 1 single offspring.

4.10 In poultry, feathered shanks (*F*) are dominant to clean (*f*), and white plumage of white leghorns (*I*) is dominant to black (*i*).

a. A feathered-shanked, white, rose-combed bird crossed with a clean-shanked, white, walnut-combed bird produces these offspring: 2 feathered, white, rose; 4 clean, white, walnut; 3 feathered, black, pea; 1 clean, black, single; 1 feathered, white, single; 2 clean, white, rose. What are the genotypes of the parents?

b. A feathered-shanked, white, walnut-combed bird crossed with a clean-shanked, white, pea-combed bird produces a single offspring, which is clean-shanked, black, and single-combed. In further offspring from this cross, what proportion may be expected to resemble each parent, respectively?

*4.11 F_2 plants segregate 9/16 colored:7/16 colorless. If a colored plant from the F_2 is chosen at random and selfed, what is the probability that there will be no segregation of the two phenotypes among its progeny?

4.12 The gene *l* in *Drosophila* is a recessive, sex-linked gene, lethal when homozygous or hemizygous (the condition in the male). If a female of genotype *L/l* is crossed with a normal male, what is the probability that the first two surviving progeny to be observed will be males?

*4.13 A locus in mice is involved with pigment production. When parents heterozygous at this locus are mated, 3/4 of the progeny are colored and 1/4 are albino. Another phenotype concerns the coat color produced in the mice. When two yellow mice are mated, 2/3 of the progeny are yellow and 1/3 are agouti. The albino mice cannot express

whatever alleles they may have at the independently assorting agouti locus.

a. When yellow mice are crossed with albino, they produce an F_1 consisting of 1/2 albino, 1/3 yellow, and 1/6 agouti. What are the probable genotypes of the parents?

b. If yellow F_1 mice are interbred, what phenotypic ratio would you expect among the progeny? What proportion of the yellow progeny produced here would be expected to be true breeding?

4.14 In *Drosophila melanogaster* a recessive autosomal gene, ebony (*e*), produces black color when homozygous, and an independently assorting autosomal gene, black (*b*), also produces a black body color when homozygous. Flies with genotypes $e/e\,b^+/-$, $e^+/-\,b/b$, and $e/e\,b/b$ are phenotypically identical with respect to body color. If true-breeding $e/e\,b^+/b^+$ ebony flies are crossed with true-breeding $e^+/e^+\,b/b$ black flies, determine the following.

a. What will be the phenotype of the F_1s?

b. What phenotypes, and in what proportions, would occur in the F_2 generation?

c. What phenotypic ratios would you expect to find in the progeny of these backcrosses: (1) $F_1 \times$ true-breeding ebony and (2) $F_1 \times$ true-breeding black?

*4.15 Two genes, *Y* and *R*, affect flower color in four-o'clocks. Neither is completely dominant, and the two interact on each other to produce seven different flower colors:

$Y/Y\,R/R$ = crimson	$Y/y\,R/R$ = magenta
$Y/Y\,R/r$ = orange-red	$Y/y\,R/r$ = magenta-rose
$Y/Y\,r/r$ = yellow	$Y/y\,r/r$ = pale yellow
$y/y\,R/R$, $y/y\,R/r$, and $y/y\,r/r$ = white	

a. In a cross of a crimson-flowered plant with a white one ($y/y\,r/r$), what will be the appearances of the F_1, the F_2, and the offspring of the F_1 backcrossed to the crimson parent?

b. What will be the flower colors in the offspring of a cross of orange-red × pale yellow?

c. What will be the flower colors in the offspring of a cross of a yellow with a $y/y\,R/r$ white?

4.16 Two four-o'clock plants were crossed and gave the following offspring: 1/8 crimson, 1/8 orange-red, 1/4 magenta, 1/4 magenta-rose, and 1/4 white. Unfortunately the person who made the crosses was color-blind and could not record the flower colors of the parents. From the results of the cross, deduce the genotypes and flower colors of the two parents.

*4.17 Genes *A*, *B*, and *C* are independently assorting and control production of a black pigment.

a. Assume that A, B, and C act in a pathway as follows:

$$A \quad B \quad C$$
$$\text{colorless} \quad \rightarrow \quad \rightarrow \quad \rightarrow \quad \text{black}$$

The alternative alleles that give abnormal functioning of these genes are designated a, b, and c, respectively. A black $A/A\ B/B\ C/C$ is crossed with a colorless $a/a\ b/b\ c/c$ to give a black F_1. The F_1 is selfed. What proportion of the F_2 is colorless? (Assume that the products of each step except the last are colorless, so only colorless and black phenotypes are observed.)

b. Assume that C produces an inhibitor that prevents the formation of black by destroying the ability of B to carry out its function, as follows:

$$A \quad B$$
$$\text{colorless} \quad \rightarrow \quad \rightarrow \quad \text{black}$$
$$\uparrow$$
$$C \quad \text{(inhibitor)}$$

A colorless $A/A\ B/B\ C/C$ is crossed with a colorless $a/a\ b/b\ c/c$, giving a colorless F_1. The F_1 is selfed to give an F_2. What is the ratio of colorless to black in the F_2? (Only colorless and black phenotypes are observed, as in part a.)

4.18 In *Drosophila* a mutant strain has plum-colored eyes. A cross between a plum-eyed male and a plum-eyed female gives 2/3 plum-eyed and 1/3 red-eyed (wild-type) progeny flies. A second mutant strain of *Drosophila*, called stubble, has short bristles instead of the normal long bristles. A cross between a stubble female and a stubble male gives 2/3 stubble and 1/3 normal-bristled flies in the offspring. Assuming that the plum gene assorts independently from the stubble gene, what will be the phenotypes and their relative proportions in the progeny of a cross between two plum-eyed, stubble-bristled flies? (Both genes are autosomal.)

*4.19 In sheep white fleece (W) is dominant over black (w), and horned (H) is dominant over hornless (h) in males, but recessive in females. If a homozygous horned white ram is bred to a homozygous hornless black ewe, what will be the appearance of the F_1 and the F_2?

*4.20 A horned white ram is bred to four ewes and has one offspring by the first three and two by the fourth. Ewe A is hornless and black; the offspring is a horned white female. Ewe B is hornless and white; the offspring is a hornless black female. Ewe C is horned and black; the offspring is a horned white female. Ewe D is hornless and white; the offspring are one hornless black male and one horned white female. What are the genotypes of the five parents?

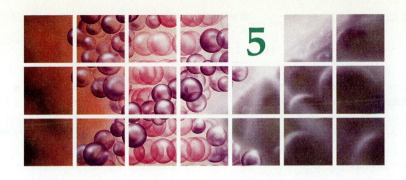

5

Linkage, Crossing-Over, and Gene Mapping in Eukaryotes

Principal Points

- The production of genetic recombinants results from physical exchanges between homologous chromosomes in meiosis. Crossing-over is the reciprocal exchange of chromosome parts at corresponding positions along homologous chromosomes by symmetrical breakage and crosswise rejoining. A chiasma is the site of crossing-over.

- Crossing-over is a reciprocal event that, in eukaryotes, occurs at the four-chromatid stage in prophase I of meiosis.

- The map distance between two genes is based on the frequency of recombination between those genes. The recombination frequency is an approximation of the frequency of crossovers between the two genes. As distance between genes increases, the incidence of multiple crossovers causes the recombination frequency to be an underestimate of the crossover frequency and hence of the map distance.

- The map distance between genes is calculated from the results of testcrosses between strains carrying appropriate genetic markers. The most accurate testcross is the three-point testcross using a diploid organism in which a triple heterozygote is crossed with a homozygous recessive for all three genes. The most accurate map distances are obtained when the three genes are reasonably close to one another. The unit of genetic distance is the map unit (mu).

- The occurrence of a chiasma between two chromatids may physically impede the occurrence of a second chiasma nearby, a phenomenon called chiasma interference.

- Mapping genes to human chromosomes can be accomplished using somatic cell hybridization techniques; for example, by fusing human and mouse cells and analyzing the descendants of the hybrid cells.

- Tetrad analysis is a mapping technique in which map distance is computed from the relative frequencies of tetrad types. This method can only be used to map the genes of certain haploid eukaryotic organisms in which the products of a single meiosis, the meiotic tetrad, are contained within a single structure.

I N THE EXAMPLES PRESENTED in Chapters 1 through 4, genes assorted independently during meiosis as a result of their location on nonhomologous chromosomes. In many instances, however, certain genes (and hence the phenotypic characters they control) are inherited together because they are located on the same chromosome. Genes on the same chromosome are called *linked genes* and are said to belong to a *linkage group* (Figure 5.1). The number of linkage groups in an organism equals the haploid number of chromosomes; for example, in humans there are 23 linkage groups. Through testcrosses we can determine which genes are linked to each other and can then construct a *linkage map*, or *genetic map*, of each chromosome.

Gene mapping and gene loci are fundamental to the entire concept of the gene. Genetic mapping provides information that is useful in many aspects of genetic analysis. Mapping can tell us whether genes that work together to produce a given phenotype are located together on the chromosome. Knowing the locations of genes on chromosomes has been useful in experiments directed toward determining the DNA sequences of genes. In this chapter we will learn the principles of genetic mapping in eukaryotes. We will then see how these principles are applied to mapping genes on human

- *Figure 5.1* Linked genes and linkage groups. Genes located on the same chromosome belong to the same linkage group.

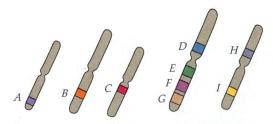

A, B, C:	Unlinked genes
D–E–F–G:	Genes on the same linkage group
H–I:	Genes on the same linkage group
[A], [B], [C], [D–E–F–G], [H–I]:	Five linkage groups

chromosomes and, in the form of tetrad analysis, to mapping genes in certain haploid organisms.

Genetic Linkage

In this section we will present some of the evidence that demonstrated that certain genes do not assort independently because they are on the same chromosome. Genes on the same chromosome are said to show **linkage** and are considered to be **linked genes.**

Morgan's Linkage Experiments with *Drosophila*

Our modern understanding of genetic linkage comes from T. H. Morgan's and colleagues' work with linkage in *Drosophila melanogaster*. By 1911 Morgan had identified a number of X-linked mutants. From crosses of the kind we discussed in Chapter 4, Morgan showed that two recessive genes, w (white eye) and m (miniature wing), were X-linked. Since both were on the X chromosome, the two genes were said to be *linked* to each other. Strains carrying these mutant genes were used in experiments to see whether or not genetic exchange occurred between the linked genes on the chromosome.

Morgan crossed a female white miniature ($w\ m/w\ m$) with a wild-type male ($w^+\ m^+//\wedge$) (Figure 5.2). (A special genetic symbolism is used for genes on the same chromosome. In the longhand form $\frac{w\ m}{w\ m}$ would indicate that genes w and m are on the same chromosome. The homologous pair of chromosomes is indicated by one or two lines. Sex-linked genes in a male are shown as $\xrightarrow{w\ m}$. The shorthand form was shown at the beginning of this paragraph. A slash signifies the pair of homologous chromosomes and indicates that the genes on either side of the slash are linked. Linked X-linked genes are shown in a similar way, with a slash indicating the X chromosome and a bent slash indicating a Y chromosome.) The F_1 males were white-eyed and had miniature wings (genotype $w\ m//\wedge$), while all females were wild type for both the eye color and wing size traits (genotype $w^+\ m^+/w\ m$). The F_1 flies were interbred and the F_2 flies were analyzed. (Note that in crosses of linked genes on the X chromosome set up as in Figure 5.2, the $F_1 \times F_1$ is equivalent to doing a testcross since the F_1 males produce X-bearing gametes with recessive alleles of both genes, and Y-bearing gametes that have no alleles for the genes being studied.) In the F_2 the most frequent phenotypic classes in both sexes were the grandparental phenotypes of mutant white eyes plus miniature wings or wild-type red eyes plus large wings. Conventionally we refer to the original genotypes of the two chromosomes as **parental genotypes, parental classes,** or, more simply, **parentals**. The term is also used to describe phenotypes, so the original white miniature females and wild-type males in these particular crosses are defined as the parentals.

Morgan observed a significant number of flies with the nonparental phenotypic combinations of white eyes and normal wings and red eyes and miniature wings. Such nonparental combinations are called **recombinants.** In all, 900 of 2441 F_2 flies, or 36.9 percent, had recombinant phenotypes. This percentage is significantly less than 50 percent, the figure that would have meant nonlinkage or independent assortment. To explain the recombinants Morgan proposed that in 36.9 percent of the meioses an exchange of genes had occurred between the two X chromosomes of the F_1 females. No such exchange needs to be postulated for sperm production since the males are hemizygous and no genetic exchange occurs between the nonhomologous X and Y chromosomes.

Morgan's group conducted a large number of other crosses of this type, and the conclusion was always the same. *In each case the parental phenotypic classes were the most frequent, while the recombinant classes occurred much less often.* Approximately equal numbers of each of the two parental classes were obtained, and similar results were obtained for the recombinant classes. Morgan's general conclusion was that *during segregation of alleles at meiosis, certain alleles tend to remain together because they lie near each other on the same chromosome.* To take this conclusion one step further, the closer two genes are on the chromosome, the more likely they are to remain together during meiosis. The recombinants are produced as a result of crossing-over between homologous chromosomes during meiosis.

The terminology relating to the physical exchange of homologous chromosome parts can be confusing. To clarify:

1. A chiasma (plural: chiasmata; see Chapter 1, Figure 1.21) is the place on a homologous pair of chromosomes at which a physical exchange is occurring. It is the site of crossing-over.

2. Crossing-over is the actual process of reciprocal exchange of chromosome segments at corresponding positions along homologous chromosomes. The process involves symmetrical breakage and crosswise rejoining.

3. Crossing-over is also defined as the events leading to genetic recombination between linked genes in eukaryotes.

■ *Figure 5.2*   Morgan's experimental crosses of white-eye and miniature-wing variants of *D. melanogaster*, showing evidence of linkage and recombination in the X chromosome.

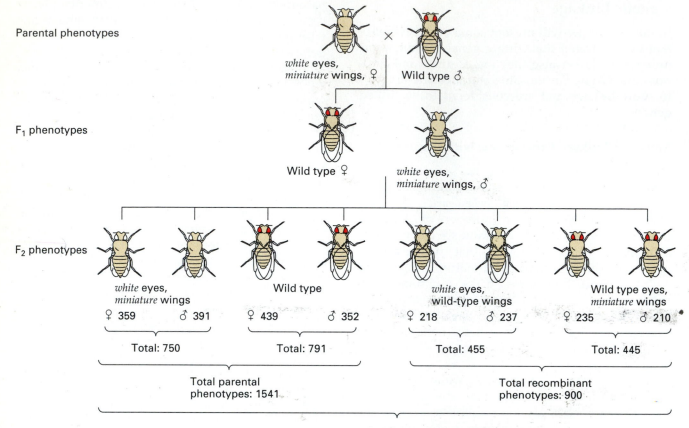

Parental phenotypes

white eyes,
miniature wings, ♀   Wild type ♂

F$_1$ phenotypes

Wild type ♀   *white* eyes,
miniature wings, ♂

F$_2$ phenotypes

white eyes,
miniature wings   Wild type   *white* eyes,
wild-type wings   Wild type eyes,
miniature wings

♀ 359   ♂ 391   ♀ 439   ♂ 352   ♀ 218   ♂ 237   ♀ 235   ♂ 210

Total: 750   Total: 791   Total: 455   Total: 445

Total parental
phenotypes: 1541   Total recombinant
phenotypes: 900

Total progeny: 1541 + 900 = 2441

Percent recombinants: $900/_{2441} \times 100 = 36.9$

Figure 5.3 presents a very simplified diagram of crossing-over. As we discussed in Chapter 1, crossing-over in eukaryotes occurs at the tetrad (four-chromatid) stage in prophase I of meiosis. Each crossover involves two of the four chromatids.

Crossing-over in eukaryotes takes place at the tetrad (four-chromatid) stage in prophase I of meiosis.

■ *Figure 5.3*   Mechanism of crossing-over. A highly simplified diagram of a crossover between two nonsister chromatids during meiotic prophase, giving rise to recombinant (nonparental) combinations of linked genes.

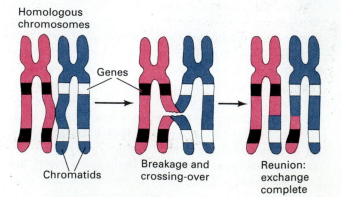

Homologous
chromosomes

Genes

Chromatids   Breakage and
crossing-over   Reunion:
exchange
complete

Keynote

The production of genetic recombinants results from physical exchanges between homologous chromosomes that have become tightly aligned during meiotic prophase I. A chiasma is the site of crossing-over. Crossing-over is the reciprocal exchange of chromosome parts at corresponding positions along homologous chromosomes by symmetrical breakage and crosswise rejoining. Crossing-over is also used to describe the events leading to genetic recombination between linked genes.

$\chi^2 = 1490.00$ with three degrees of freedom, the P value is much lower than 0.001; in fact it is not on the table. This is interpreted to mean that, if the hypothesis is correct, much lower than 1 time out of 1000, we could expect χ^2 values of this magnitude by chance alone. Thus we must consider the independent assortment hypothesis to be invalid, and we must think of an alternative hypothesis, the simplest of which is that the genes are linked.

Generally if the probability of obtaining the observed χ^2 values is greater than 5 in 100 ($P \geqslant 0.05$), the deviation is considered not statistically significant and is more likely to have occurred by chance alone. Thus the hypothesis being tested could apply to the data being analyzed. If $P \leqslant 0.05$ we consider the deviation from the expected values to be statistically significant and not due to chance alone, and the hypothesis may well be invalid. If $P \leqslant 0.01$ the deviation is highly significant and some nonchance factor is involved. The hypothesis is almost certainly invalid.

Let us consider that in another chi-square analysis of a different set of data, we obtained $\chi^2 = 0.4700$ with three degrees of freedom. By looking up the value in Table 5.2, we see that the P value is greater than 0.90 and less than 0.95 ($0.90 \leqslant P \leqslant 0.95$). Thus from 90 to 95 times out of 100 we could expect χ^2 values of this magnitude or greater. This P value, being greater than 0.05, indicates that the results are consistent with the hypothesis of independent assortment being tested.

The Concept of a Genetic Map   The data obtained by Morgan from *Drosophila* crosses indicated that the frequency of crossing-over (and hence of recombinants) for linked genes is characteristic of the gene pairs involved: For the X-linked genes for white (w) and miniature (m) the frequency of crossing-over is 36.9 percent, and for white (w) and yellow (y) genes it is 1.3 percent. It is important to realize that we are mapping the locations of the genes, and the allelic differences of the gene pairs give us phenotypes with which we can follow the recombination process. Thus the frequency of recombinants for two linked genes is approximately constant, regardless of how the alleles of the genes are arranged on the chromosomes with respect to each other. In an individual doubly heterozygous for the w and m alleles, for example, the alleles can be arranged two ways:

$$\frac{w^+ \; m^+}{w \;\; m} \quad \text{or} \quad \frac{w^+ \; m}{w \;\; m^+}$$

In the arrangement on the left, the two wild-type alleles are on one homolog and the two recessive mutant alleles are on the other homolog, an arrangement called **coupling** (or the *cis* configuration). Crossing-over between the two loci produces $w^+ \, m$ and $w \, m^+$ recombinants. (**Loci** is the plural of **locus,** the position of a gene on a genetic map.) In the arrangement on the right, each homolog carries the wild-type allele of one gene and the mutant allele of the other gene, an arrangement called **repulsion** (or the *trans* configuration). Crossing-over between the two genes produces $w^+ \, m^+$ and $w \, m$ recombinants. *While the actual phenotypes of the recombinant classes are different for the two arrangements, the percentage of recombinants among the total progeny will be approximately the same in each case.*

Morgan thought that the characteristic crossover frequencies for linked genes might be related to the physical distances separating the genes on the chromosome. In 1913 a student of Morgan's, Alfred Sturtevant, devised the testcross method to analyze the linkage relationships between genes. He suggested that the percentage of recombinants (produced by crossovers in gamete production) could be used as a quantitative measure of the genetic distance between two gene pairs on a genetic map. This distance is measured in **map units** (mu). A crossover frequency of 1 percent between two genes equals 1 map unit. That is, one map unit is the distance between gene pairs for which 1 product out of 100 is recombinant. The map unit is sometimes called a **centi-Morgan** (cM) in honor of T. H. Morgan.

On a genetic map the distances between a series of linked genes are additive. Thus the genes on a chromosome can be represented by a one-dimensional genetic map that shows in linear order the genes belonging to the chromosome. Crossover and recombination values give the linear order of the genes on a chromosome and provide information about the genetic distance between any two genes. Genetically speaking, the farther apart two genes are, the greater will be the crossover frequency. Thus in Figure 5.4, the probability of recombination occurring between genes A and B is much less than between genes B and C, because A and B are closer together than are B and C.

The first genetic map ever constructed was based on recombination frequencies from *Drosophila* crosses involving the sex-linked genes w, m, and y discussed earlier, where w gives white eyes, m gives miniature wings, and y gives yellow body. From these mapping experiments the recombination frequencies for the $w \times m$, $w \times y$, and $m \times y$ crosses were established as 32.6, 1.3, and 33.9 percent, respectively. The percentages are quantitative measures of the distance between the genes involved.

The logic used in constructing a genetic map of these three genes is as follows: The genes must be

■ *Figure 5.4* Recombination between linked genes and map distance. The farther apart two genes are, the greater the number of possible sites for recombination. Thus the probability of recombination occurring between genes *A* and *B*, is much less than that between genes *B* and *C*. The percentage of recombinants can provide information about the relative genetic distance between two linked genes.

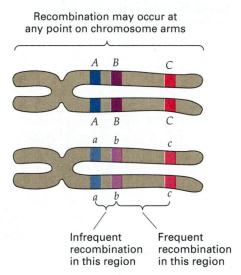

Recombination may occur at
any point on chromosome arms

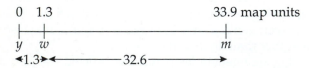

Infrequent
recombination
in this region

Frequent
recombination
in this region

arranged at different points on a line that best accommodates the data. The recombination frequencies show that *w* and *y* are closely linked, and that *m* is quite a distance away from the other two genes. Since the *w-m* genetic distance is less than the *y-m* distance (as shown by the smaller percentage of the recombinants in the *w* × *m* cross), the order of genes must be *y-w-m*. Thus the three genes are ordered and spaced as shown, with the *y* gene assigned an arbitrary value of 0 and the other map units indicating genetic distance from *y*:

```
0   1.3                        33.9 map units
├───┼──────────────────────────────┤
y    w                             m
◄1.3►◄───────────32.6──────────────►
```

Recombination frequencies are used to construct genetic maps in all gene mapping experiments, whether the organism is a eukaryote or a prokaryote.

Gene Mapping by Using Two-Point Testcrosses

We have seen that the percentage of recombinants resulting from crossing-over is used as a measurement of the genetic distance between two linked genes. By carrying out two-point testcrosses, such as those shown in Figure 5.5, we can determine the relative numbers of parental and recombinant classes in the progeny. For autosomal recessives a double heterozygote is crossed with a doubly homozygous recessive mutant strain (Figure 5.5). When the double heterozygous $a^+ b^+/a\ b$ F_1 progeny from a cross of $a^+ b^+/a^+ b^+$ with $a\ b/a\ b$ are backcrossed with $a\ b/a\ b$, four phenotypic classes are found among the F_2 progeny. Two of these classes have the parental phenotypes $a^+ b^+$ and $a\ b$ and derive from zygotes in which no crossover was detectable between the two genes. Since both classes were the

■ *Figure 5.5* Testcross to show that two genes are linked: Genes *a* and *b* are recessive mutant alleles linked on the same autosome. A homozygous $a^+ b^+/a^+ b^+$ individual is crossed with a homozygous recessive $a\ b/a\ b$ individual, and the doubly heterozygous F_1 progeny ($a^+ b^+/a\ b$) are testcrossed with homozygous $a\ b/a\ b$ individuals.

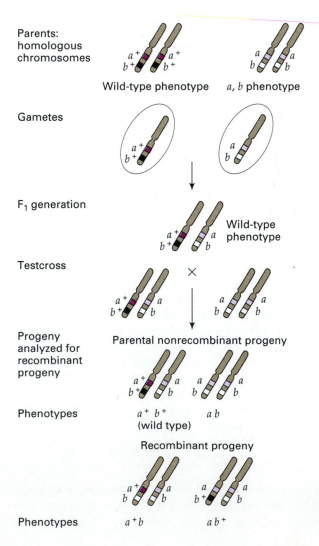

Parents:
homologous
chromosomes

Wild-type phenotype   *a, b* phenotype

Gametes

F_1 generation

Wild-type
phenotype

Testcross

Progeny
analyzed for
recombinant
progeny

Parental nonrecombinant progeny

Phenotypes $a^+ b^+$ $a\ b$
(wild type)

Recombinant progeny

Phenotypes $a^+ b$ $a\ b^+$

result of the same noncrossover event, approximately equal numbers of these two types are expected.

The other two F_2 phenotypic classes have recombinant phenotypes $a^+ b$ and $a b^+$, deriving from zygotes in which a single crossover occurred between the chromosomes. Again we expect approximately equal numbers of these two recombinant classes. Because a single crossover event occurs more rarely than no crossing over, an excess of parental phenotypes over recombinant phenotypes in the progeny of a testcross indicates linkage between the genes involved.

Two-point testcrosses are essentially the same when sex-linked genes are involved. For X-linked recessives a double heterozygous female is crossed with a hemizygous male carrying the recessive alleles

$$\frac{+\ \ +}{a\ \ b} \times \frac{a\ \ b}{\longrightarrow}$$

(Note that "+ +" here is shorthand for the linked wild-type alleles $a^+ b^+$; and $\longrightarrow$ symbolizes the Y chromosome. Refer to Box 3.1.) For X-linked dominants a doubly heterozygous female is crossed with a male carrying wild-type alleles on the X chromosome:

$$\frac{A\ \ B}{+\ \ +} \times \frac{+\ \ +}{\longrightarrow}$$

(Here since A and B are dominant mutations, + + is a shorthand for the linked wild-type alleles $A^+ B^+$.)

In all cases a two-point testcross should yield a pair of parental types that occur with about equal frequencies, and a pair of recombinant types that also occur with equal frequencies. As indicated previously, the actual phenotypes will, of course, depend on the relative arrangement of the two allelic pairs in the homologous chromosomes; that is, whether they are in coupling or repulsion. To count the representatives of each progeny class (to get the percentage of recombinant types), the following formula is used:

$$\frac{\text{number of recombinants}}{\text{total number of testcross progeny}} \times 100 = \text{percent recombinants}$$

For the

$$\frac{b^+\ \ vg}{b\ \ vg^+} \times \frac{b\ \ vg}{b\ \ vg}$$

testcross involving the black body and vestigial wing genes of *Drosophila* used in our discussion of the chi-square test, the number of recombinants

was $283 + 241 = 524$, and the total number of progeny flies was 3236. So the percent recombinants for b and vg is:

$$524/3236 \times 100 = 16.2\% \text{ recombinants}$$

The value for the percentage of recombinants is usually directly converted into map units, so for our example, 16.2 percent recombination between genes b and vg indicates that the two genes are 16.2 mu apart.

The two-point method of mapping is most accurate when the two genes examined are close together. When genes are far apart other factors can influence the production of recombinant progeny. (These will be discussed later on pp. 106–108.) When we are drawing a map of the genes on a chromosome, we must take these factors into consideration. Large numbers of progeny must also be counted (scored) to ensure a high degree of accuracy. We have already seen the logic behind constructing an actual linkage map from recombination frequencies involving all possible pairwise crosses of the genes under study. From mapping experiments carried out in all types of organisms, we know that genes are linearly arranged in linkage groups. There is a one-to-one correspondence of linkage groups and chromosomes; thus the sequence of genes on the linkage group reflects the sequence of genes on the chromosome.

Generating a Genetic Map

A genetic map is generated from estimating the number of times a crossover event occurred in a particular segment of the chromosome out of all meioses examined. In many cases the probability of crossing-over is not completely uniform along a chromosome; therefore, we must be cautious about how far we extrapolate the genetic map (which derives from data produced by genetic crosses) to the physical map of the chromosome (which derives from determinations of the locations of genes along the chromosome itself; e.g., from sequencing the DNA). Nonetheless, a simple working hypothesis is to consider crossovers as being randomly distributed along the chromosome.

The recombination frequencies observed between genes may also be used to predict the outcome of genetic crosses. For example, a recombination frequency of 20 percent between genes indicates that for a doubly heterozygous genotype (such as $a^+ b^+ / a b$), 20 percent of the gametes produced, on average, will be recombinants ($a^+ b$ and $a b^+$). Further, if we assume that crossing-over occurs randomly along a chromosome, more than one crossover may occur in a given region in a meiosis. The probabilities of these so-called **multiple crossovers** can readily be computed using the product rule (see Chapter 2, p. 35): the probability of two independent events occurring simultaneously is equal to the product of the individual probabilities of two single events. That is, the probability of two crossovers (called a **double crossover**) occurring between the genes in our example is $0.2 \times 0.2 = 0.04$. The probability of three crossovers (a triple crossover) is $0.2 \times 0.2 \times 0.2 = 0.008$, and so on.

For any testcross *the percentage of recombinants in the progeny cannot exceed 50 percent*. That is, if the genes are assorting independently, an equal number of recombinants and parentals is expected in the progeny, so the frequency of recombinants is 50 percent. With this value in hand, we state that the two genes are *unlinked*. Genes may be unlinked (show 50 percent recombination) in two ways. First, the genes may be on different chromosomes, a case we discussed before. Second, *the genes may be on the same chromosome but lie so far apart that at least one crossover is almost certain to occur between them*. Thus genes on the same chromosome may or may not be genetically linked. Genes that are known to be on the same chromosome, whether or not they are genetically linked, are said to be **syntenic.**

The case of unlinked genes on the same chromosome can be understood by referring to Figure 5.6, which shows the consequences of single and double crossovers on the production of parental and recombinant chromatids. The two loci are far apart on the same chromosome; as a result, in a given meiosis at least a single (and usually more) crossover always occurs between the two loci. For single crossovers occurring between any pair of nonsister chromatids, the result is two parental and two recombinant chromatids; that is, for two loci 50 percent of the products are recombinant (Figure 5.6[a]).

Double crossovers can involve two (Figure 5.6[b]), three (Figure 5.6[c]), or all four of the chromatids (Figure 5.6[d]). For double crossovers involving the same two nonsister chromatids (called a *two-strand double crossover*), all four resulting chromatids are parental. For *three-strand double crossovers* (double crossovers involving three of the four chromatids), two parental and two recombinant chromatids result. For a *four-strand double crossover*, all four resulting chromatids are recombinant. Considering all of the possible double crossover patterns together, 50 percent of the products are recombinant for the two loci. This is the reason for the recombination frequency limit of 50 percent exhibited by unlinked genes on the same chromosome.

When genes are more closely linked together, however, there will be no crossovers between the two loci in some meioses, resulting in four parental chromatid products. In mapping linked genes, therefore, the map distance depends on the ratio of the meioses with no detectable crossovers to meioses with any number of crossovers between the loci.

The point is if two genes are unlinked, and therefore show 50 percent recombination, it does not necessarily mean that they are on different chromosomes. More data would be needed to determine whether the genes are on the same chromosome or different chromosomes. One way to find out is to map a number of other genes in the linkage group. For example, if *a* and *m* show 50 percent recombination, perhaps we will find that *a* shows 27 percent recombination with *e*, and *e* shows 36 percent recombination with *m*. This result would indicate that *a* and *m* are in the same linkage group approximately 63 mu apart, as shown here:

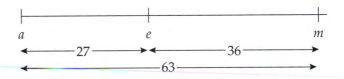

Double Crossovers

Data for the F_2 progeny of a cross between a yellow-bodied, white-eyed, miniature-winged female and a normal male *Drosophila* are shown in Table 5.3. We have already shown that the gene order is *y-w-m*. The data show 29 exchanges between the genes for body color and eye color (*y* and *w* loci) and 719 exchanges between genes for eye color and wing size (*w* and *m* loci), giving a total of 748. However, only 746 exchanges are apparent between the body color and wing size genes. The discrepancy is due to the single normal-bodied, white-eyed, normal-winged fly. In the production of the egg from which this fly developed, two exchanges (a double crossover) must have occurred between the X chromosomes, with the result that the body color and wing size genes (normal body color and normal-length wings) appeared together, as in one grandparent, although the intervening gene (for white eyes) was derived from the other grandparent. This example shows that large crossover frequencies give an inaccurate map. Unless intervening genes are available, there is no way of detecting double crossovers (or other even-numbered multiple crossovers).

To see the significance of the mechanics of double crossovers in genetic mapping experiments, consider a hypothetical case of two allelic pairs (a^+/a and b^+/b) in coupling (see p. 107) and separated by a great distance on the same chromosome. Figure 5.7(a) (page 109) shows that a single crossover results in recombination of the two allelic pairs.

■ *Figure 5.6*   Demonstration that the recombination frequency between two genes located far apart on the same chromosome cannot exceed 50 percent. (a) Single crossovers produce one-half parental and one-half recombinant chromatids; (b) Double crossovers (two-strand, three-strand, and four-strand) collectively produce one-half parental and one-half recombinant chromatids.

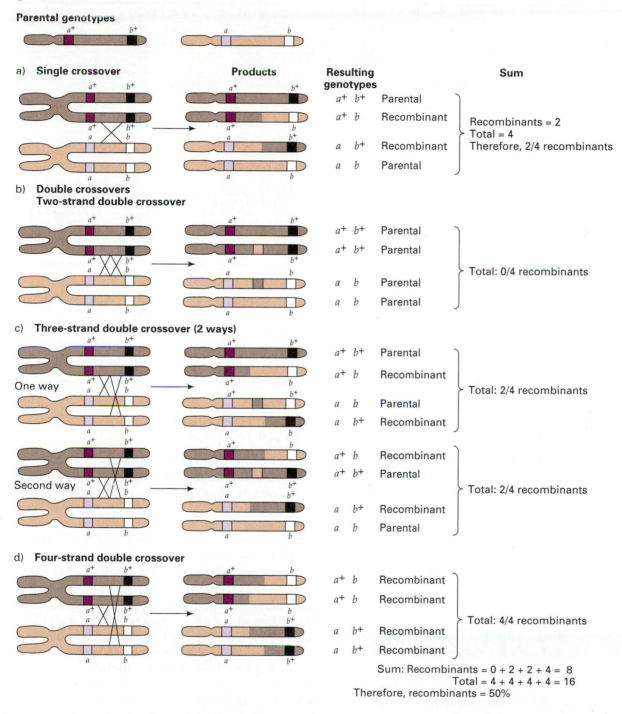

Figure 5.7(b) (top) shows that a two-strand crossover involving two of the four chromatids does not result in recombination of the allelic pairs, so only parental progeny result. Exactly the same would occur if there were no crossing-over between the two chromatids in this region. However, the percentage of crossing-over between genes is a measure of the distance between them. Therefore since the double crossover in Figure 5.7(b) (top) did not generate recombinants, two crossover events will be uncounted, and the estimate of map distance between genes *a* and *b* will be low, depending on

Table 5.3

Data for the F_2 Progeny of a Cross Between a Yellow-bodied, White-eyed, Miniature-winged ($y\ w\ m/y\ w\ m$) Mutant Female and a Wild-Type ($+\ +\ +/Y$) Male *Drosophila*

PHENOTYPES				CROSSOVERS		
BODY COLOR	EYE COLOR	WING SIZE	NUMBER	BODY COLOR AND EYE COLOR	EYE COLOR AND WING COLOR	BODY COLOR AND WING SIZE
normal	red (normal)	long (normal)	758	—	—	—
yellow	white	miniature	700	—	—	—
normal	red	miniature	401	—	401	401
yellow	white	long	317	—	317	317
normal	white	miniature	16	16	—	16
yellow	red	long	12	12	—	12
normal	white	long	1	1	1	—
yellow	red	miniature	0	0	0	—
		Total	2205	29	719	746
		Percentage	100	1.3	32.6	33.8

how many double crossovers occur per meiosis in the region being examined.

In sum, genetic map distance is derived from the average frequency of crossing-over occurring between linked genes, whereas recombination frequency is a measure of the crossovers that result in the detectable exchange of the gene markers. If no multiple crossovers occurred between genes, there would be a direct linear relationship between genetic map distance and recombination frequency. In practice such a relationship is seen only when genetic map distances are small; that is, when genes are closely linked. As genes become farther apart, the chances of multiple crossovers between them increases, and there is no longer a linear relationship between map distance and recombination frequency. Consequently, it is difficult to obtain an accurate measure of map distance when multiple crossovers are involved.

Keynote

The map distance between two genes is based on the frequency of recombination between the two genes (an approximation of the frequency of crossovers between the two genes). By using appropriate genetic markers in crosses, geneticists are able to compute the recombination frequency between genes in chromosomes as the percentage of progeny showing the reciprocal recombinant phenotypes. The more closely the recombination frequency parallels the crossover frequency, the closer

the genes are. But when the genetic distance increases, the incidence of multiple crossovers, and particularly double crossovers, causes the recombination frequency to be an underestimate of the crossover frequency and hence of the map distance.

Mapping Chromosomes by Using Three-Point Testcrosses

Double crossing-over occurs quite rarely within distances of 10 mu or less, and it does not always occur even in greater physical distances for certain chromosomal regions. Thus to get accurate map distances we can study closely linked genes. Another efficient way to obtain such data is to use a **three-point testcross**, which involves three genes within a relatively short segment of a chromosome. (The evidence for the closely linked nature of three genes would derive from separate two-point testcrosses and the drawing of a preliminary genetic map.)

The potential advantage of the three-point testcross can be seen in the theoretical case if we have a third allelic pair c^+/c between the a^+/a and b^+/b allelic pairs of Figure 5.7(a), as diagrammed in Figure 5.7(b) (bottom). A two-strand double crossover event between a and b might be detected by the recombination of the c^+/c allelic pair in relation to the other two allelic pairs. Thus the principle is that accurate mapping data can be obtained when the genes are relatively close together. In practice, then, three-point testcrosses are routinely used to map genes to determine their order in the chromosome

■ *Figure 5.7* Progeny of single and double crossovers. (In this figure, only the two chromatids involved in crossing over are shown.) (a) A single crossover between linked genes generates recombinant gametes; (b)(Top) A double crossover between linked genes gives parental gametes. Thus inaccurate map distances between genes result, since not all crossovers can be accounted for. (Bottom) A possible solution to the double crossover problem. The presence of a third allelic pair between the two allelic pairs of part a enables us to detect the double crossover event. In a double crossover the middle gene will change positions relative to the outside genes.

a) Single crossover

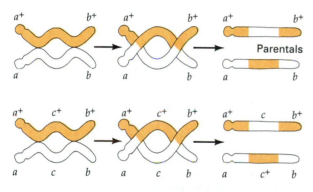

b) Double crossover

Parentals

Gametes are recombinant for the c^+/c gene pair relative to the a^+/a and b^+/b gene pairs

and to determine the distances between them, as we will now see.

In diploid organisms the three-point testcross is a cross of a triple heterozygote with a triply homozygous recessive. If the mutant genes in the cross are all recessive, a typical three-point testcross might be

$$\frac{+\ +\ +}{a\ \ b\ \ c} \times \frac{a\ \ b\ \ c}{a\ \ b\ \ c}$$

If any of the mutant genes are dominant, the triply homozygous recessive parent in the testcross will carry the wild-type allele of these genes. For example, a testcross of a strain heterozygous for two recessive mutations (*a* and *c*) and one dominant mutation (*B*) might be

$$\frac{+\ +\ +}{a\ \ B\ \ c} \times \frac{b\ +\ c}{a\ +\ c}$$

Let us consider a hypothetical plant in which there are three linked genes, all of which control fruit phenotypes. A recessive allele *p* of the first gene determines purple fruit color, in contrast to the white color of the wild type. A recessive allele *r* of the second gene results in a round fruit shape, as compared with an elongated fruit in the wild type. A recessive allele *j* of the third gene gives a juicy fruit instead of the wild-type dry fruit. The task before us is to determine the order of the genes on the chromosome and the map distance between the genes. To do so, we must carry out the appropriate testcross of a triple heterozygote (+ + +/*p r j*) with a triply homozygous recessive (*p r j/p r j*) and then count the different phenotypic classes in the progeny (Figure 5.8). For each gene in the cross, two different phenotypes occur in the progeny; therefore, for the three genes a total of $(2)^3 = 8$ phenotypic classes will appear in the progeny, representing all possible combinations of phenotypes.

Establishing the Order of Genes The first step in finding the genetic map distances of the three genes is to determine the order in which the genes are located on the chromosome; that is, which gene is in the middle? One parent carries the recessive alleles for all three genes; the other is heterozygous for all three genes. Therefore the phenotype of the resulting individual is determined by the alleles in the gamete from the triply heterozygous parent; the gamete from the other parent will carry only recessive alleles. We know from the genotypes of the original parents that all three genes are in coupling (recall that this means all wild-type alleles are on one homolog and all mutant alleles are on the other homolog). Since the heterozygous parent in the testcross was + + +/*p r j*, classes 1 and 2 in Figure 5.8 are parental progeny: Class 1 is produced by the fusion of a + + + gamete with a *p r j* gamete from the triply homozygous recessive parent. Class 2 is produced by the fusion of a *p r j* gamete from the triply heterozygous parent and a *p r j* gamete from the triply homozygous recessive parent. These classes are generated from meiosis in which no crossing-over occurs in the region of the chromosome in which the three genes are located.

The progeny classes deriving from a double crossover between the two loci can often be found by inspecting the numbers of each phenotypic class. Since the frequency of a double crossover in a region is expected to be lower than the frequency of a single crossover, *double crossover gametes are the least frequent to be produced.* Thus to identify the

■ *Figure 5.8*   Theoretical analysis of a three-point mapping, showing the testcross used and the resultant progeny.

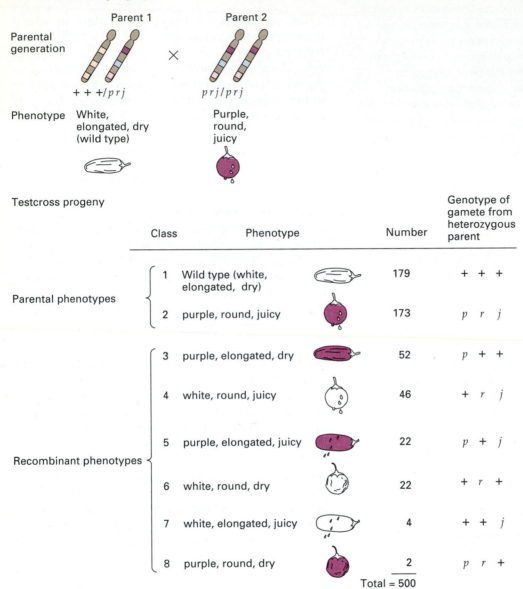

double crossover progeny, we can examine the progeny to find the *pair* of classes that have the lowest number of representatives. In Figure 5.8 classes 7 and 8 are such a pair. The genotypes of the gametes from the heterozygous parent that give rise to these phenotypes are + + *j* and *p r* +.

Referring now to Figure 5.7(b) (bottom), we see that a double crossover changes the orientation of the gene in the center of the sequence (here, c^+/c) with respect to the two flanking allelic pairs. Therefore genes *p*, *r*, and *j* must be arranged in such a way that the center gene switches to give classes 7 and 8 in Figure 5.8. To determine the arrangement, we must check the relative organization of the

genes in the parental heterozygote so that it is clear which genes are in coupling and which are in repulsion. In this example, the parental (noncrossover) gametes are + + + and *p r j*, so all are in coupling. The double crossover gametes are + + *j* and *p r* +, so the only possible gene order that is compatible with the data is *p j r*, with the genotype of the heterozygous parent being + + +/*p j r*. Figure 5.9 illustrates the generation of the double crossover gametes from that parent.

Calculating the Map Distances Between Genes
Now the data can be rewritten as shown in Figure 5.10 to reflect the newly determined gene order. For

■ *Figure 5.9*   Rearrangement of the three genes of Figure 5.8 to *p j r*. The evidence is that a double crossover generates the least frequent pair of recombinant phenotypes (in this case class 7 with four progeny and class 8 with two progeny).

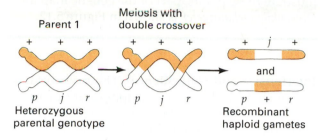

convenience in this analysis, the region between genes *p* and *j* is called region I, and that between genes *j* and *r* is called region II.

Map distances can now be calculated as described previously. The frequency of crossing-over (genetic recombination) is computed between two genes at a time. For the *p-j* distance all the crossovers that occurred in region I must be added together. Thus we must consider the recombinant progeny resulting from a single crossover in that

■ *Figure 5.10*   Rewritten form of the testcross and testcross progeny of Figure 5.8, based on the actual gene order *p j r*.

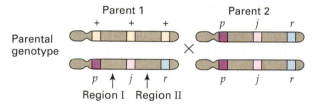

Testcross progeny

Class	Genotype of gamete from heterozygous parent	Number	Origin
1	+ + +	179	Parentals, no crossover
2	*p j r*	173	
3	*p* + +	52	Recombinants, single crossover region I
4	+ *j r*	46	
5	*p j* +	22	Recombinants, single crossover region II
6	+ + *r*	22	
7	+ *j* +	4	Recombinants, double crossover
8	*p* + *r*	2	

Total = 500

region (classes 3 and 4) *and* the recombinant progeny produced by a double crossover (classes 7 and 8). The latter must be included since each double crossover includes a crossover in region I and therefore is a recombination event between genes *p* and *j*. We see in Figure 5.10 that there are 98 recombinant progeny in classes 3 and 4, and 6 in classes 7 and 8, giving a total of 104 progeny that result from a recombination event in region I. Since there are a total of 500 progeny, the percentage of progeny generated by crossing-over in region I is 20.8 percent, determined as follows:

$$\frac{\text{single crossovers in region I } (p\text{-}j) + \text{double crossovers}}{\text{total progeny}} \times 100$$

$$= \frac{98 + 6}{500} \times 100$$

$$= \frac{104}{500} \times 100$$

$$= 20.8\%$$

So, the distance between genes *p* and *j* is 20.8 mu. This map distance, which is quite large, is chosen mainly for illustration. If this were an actual cross, we would have to be cautious about that map distance value because it is in the range where double crossovers might occur at a significant frequency and be undetected. In an actual cross the value of 20.8 mu would probably underestimate the true distance.

The same method is used to obtain the map distance between genes *j* and *r*. That is, we calculate the frequency of crossovers in the cross that gave rise to progeny recombinant for genes *j* and *r*, and directly relate that frequency to map distance. In this case all the crossovers that occurred in region II (see Figure 5.10) must be added (classes 5, 6, 7, and 8). The percentage of crossovers is calculated in the following manner:

$$\frac{\text{single crossovers in region II } (j\text{-}r) + \text{double crossovers}}{\text{total progeny}} \times 100$$

$$= \frac{44 + 6}{500} \times 100$$

$$= \frac{50}{500} \times 100$$

$$= 10.0\%$$

Thus the map distance between genes *j* and *r* is 10.0 map units.

extent of interference is as a **coefficient of coincidence**:

$$\text{coefficient of coincidence} = \frac{\text{observed double crossover frequency}}{\text{expected double crossover frequency}}$$

and

$$\text{interference} = 1 - \text{coefficient of coincidence}$$

For the portion of the map in our example the coefficient of coincidence is

$$0.012/0.0208 = 0.577$$

The coefficient of coincidence typically varies between zero and one and may be interpreted as follows: A coincidence of one means that in a given region all double crossovers occurred that were expected on the basis of two independent events; therefore there is no interference, so the interference value is zero. If the coefficient of coincidence is zero, none of the expected double crossovers occur. Here there is total interference, with one crossover completely preventing a second crossover in the region under examination. The interference value is one. We see that coincidence values and interference values are inversely related. In our example the coefficient of coincidence of 0.577 means that the interference value is 0.423. Only 57.7 percent of the expected double crossovers took place in the cross.

Keynote

The occurrence of a chiasma between two chromatids may physically impede the occurrence of a second chiasma nearby, a phenomenon called chiasma interference. The extent of interference is expressed by the coefficient of coincidence, which is calculated by dividing the number of observed double crossovers by the number of expected double crossovers. The coefficient of coincidence ranges from zero to one, and the extent of interference is measured as 1 − coefficient of coincidence.

Mapping Genes in Human Chromosomes

For practical reasons, with humans it is not possible to do genetic mapping experiments of the kind discussed for other organisms. Nonetheless, we have a strong interest in mapping genes in human chromosomes because there are so many known diseases and traits that have a genetic basis. In Chapters 2 and 3 we saw that pedigree analysis could be used to determine the mode by which a particular genetic trait is inherited. In this way many genes have been localized to the X chromosome. However, pedigree analysis cannot show on which chromosome a particular autosomal gene is located.

Although modern statistical and computer analysis has made it possible to obtain some linkage data for autosomal genes, many of the more than 1,200 known autosomal genes have not been located on one of the 23 autosomes. Moreover, probably between 10,000 and 50,000 genes have not even been identified yet, much less located on all the chromosomes. In this section we will discuss some traditional ways to map human genes. One method involves recombination analysis as described next. Another traditional method entails the fusion of human and rodent cultured cells in a process called **somatic cell hybridization.** Nowadays most mapping of human genes involves the application of recombinant DNA technology. A number of recombinant DNA techniques are described in Chapter 13.

Mapping Human Genes by Recombination Analysis

It is not possible to set up appropriate testcrosses for human genetic mapping by recombination analysis. In a very few cases certain pedigrees have included individuals with appropriate genotypes to permit analysis of linkage between autosomal genes. Recombination analysis in humans has been carried out more for X-linked genes, however, because the hemizygosity of the X chromosome in males has provided a rich source of beneficial genotypic pairings in pedigrees.

Consider the following theoretical example (Figure 5.13). A male with two rare X-linked recessive alleles a and b marries a woman who expresses neither of the traits involved. Since the traits are rare, it is likely that the woman is homozygous for the wild-type allele of each gene; that is, she is $a^+ b^+/a^+ b^+$. A female offspring from these parents would be doubly heterozygous $a^+ b^+/a b$. Recombinant gametes from this female ($a^+ b$ and $a b^+$) would be produced by crossing-over between the two genes at a frequency related to the genetic distance that separates them.

If the doubly heterozygous F_1 female pairs with a normal $a^+ b^+/Y$ male, all female progeny will be $a^+ b^+$ in phenotype because of the $a^+ b^+$ chromosome transmitted from the father. The male progeny, however, will express all four possible phenotypic classes because of the hemizygosity of the X chromosome; that is, the parental $a^+ b^+$ and $a b$, and the recombinant $a^+ b$ and $a b^+$. Thus analysis of the

Locating Genes on Chromosomes: Mapping Techniques

So far we have learned two significant concepts. The first is that genetic recombinants result from crossing-over between homologous chromosomes. More specifically, the number of genetic recombinants produced is characteristic of the two linked genes involved. The second concept is that crossing-over takes place at the tetrad stage in prophase I of meiosis, and each crossover involves only two of the four chromatids. We will now examine how genetic experiments can be used to determine the relative position of genes on chromosomes in eukaryotic organisms. This process is called **genetic mapping**.

Detecting Linkage Through Testcrosses

Before beginning any experiments to construct a **genetic map** (also called a **linkage map**), geneticists must show that the genes under consideration are linked. If they are not linked then the genes are assorting independently. Therefore, the simplest way to test for linkage is to analyze the results of crosses to determine whether the data deviate significantly from those expected by independent assortment.

The best cross to use to test for linkage is the testcross, a cross of an individual with a known or unknown genotype with an individual homozygous recessive for all genes involved. This is the case because the distribution of phenotypes is the result of segregation events in only one of the parents. The other parent contributes only recessive alleles to the progeny, and those alleles do not contribute to the phenotype of the progeny. We saw in Chapter 2 that a testcross between $a^+/a\ b^+/b$ and $a/a\ b/b$, where genes a and b are unlinked, gives a 1:1:1:1 ratio of the four possible phenotypic classes $a^+\ b^+$: $a^+\ b$: $a\ b^+$: $a\ b$. Any significant deviation from this ratio in the direction of too many parental types would suggest that the two genes are linked. Thus it is important to know how large a deviation must be in order to be considered "significant." The chi-square test can be used to make such a decision.

The Chi-Square Test The observed phenotypic ratios among progeny of a cross rarely match the expected or predicted ratios even when the hypothesis on which the expected ratios are based is correct. For example, none of the F_2s of Mendel's monohybrid crosses exhibited an exact 3:1 ratio, although the observed ratios were close enough so that he was not concerned by the discrepancies.

Observed results may not match the expected results for numerous reasons, such as inadequate sample size, sampling error, or decreased viability of some of the genotypes involved.

If the deviation between the observed and the expected result is large enough, the data may indicate that the hypothesis might be wrong. A common way to decide is to use a statistical test called the **chi-square (χ^2) test,** which, as used in the analysis of genetic data, is essentially a *goodness-of-fit test*.

To illustrate the use of the chi-square test, we will analyze the progeny data from a testcross involving fruit flies. In *Drosophila b* is a recessive autosomal mutation which, when homozygous, results in black body color, and *vg* is a recessive autosomal mutation which, when homozygous, results in flies with vestigial (short, crumpled) wings. Wild-type flies have grey bodies and long, uncrumpled (normal) wings. True-breeding black, normal ($b/b\ vg^+/vg^+$) flies were crossed with true-breeding grey, vestigial ($b^+/b^+\ vg/vg$) flies. The F_1 grey, normal ($b^+/b\ vg^+/vg$) flies were testcrossed to black, vestigial ($b/b\ vg/vg$) flies. The progeny data were:

	283 grey, normal
	1294 grey, vestigial
	1418 black, normal
	241 black, vestigial
Total	3236 flies

We hypothesize that the two genes are unlinked since that allows us to predict a defined ratio in the progeny and use the chi-square test to test the hypothesis, as shown in Table 5.1.

If the two genes are unlinked then a testcross should result in a 1:1:1:1 ratio of the four phenotypic classes. First we list the four phenotypes expected in the progeny of the cross in column 1. Then we list the observed (o) numbers for each phenotype, using actual numbers and not percentages or proportions (column 2). Next we calculate the expected (e) number for each phenotypic class (column 3), given the total number of progeny (3236) and the hypothesis under evaluation (in this case 1:1:1:1). Thus we list $1/4 \times 3236 = 809$, and so on. Now we subtract the expected number (e) from the observed number (o) for each class to find differences, called the deviation value (d). The sum of the d values is always zero (column 4).

In column 5 the deviation squared (d^2) is computed by multiplying each deviation value in column 4 by itself. In column 6 the deviation squared is then divided by the expected number (e). The chi-square value, χ^2 (item 7 in the table), is the total of

Table 5.1

Chi-Square Test of *Drosophila*

(1) PHENOTYPES	(2) OBSERVED NUMBER (o)	(3) EXPECTED NUMBER (e)	(4) DEVIATION VALUE d	(5) d^2	(6) d^2/e
grey, normal	283	809	-526	276,676	342.00
grey, vestigial	1,294	809	485	235,225	290.76
black, normal	1,418	809	609	370,881	458.44
black, vestigial	241	809	-568	322,624	398.79
Total	3,236	3,236			1,490.00

(7) $\chi^2 = 1,490.00$ (8) df 3

the four values in column 6. The more the observed data deviate from the expected data on the basis of the hypothesis being tested, the higher χ^2 will be. In our example χ^2 is 1490.00. The general formula is

$$\chi^2 = \sum d^2/e$$

The last value in the table, item 8, is the degrees of freedom (df) for the set of data. The degrees of freedom in a test involving n classes are usually equal to $n - 1$. That is, if the total number of progeny (3236 in our example) is divided among n classes (four phenotypic classes in the example), then once

the expected numbers have been computed for $n - 1$ classes (three in our example), the expected number of the last class is set. Thus in our example there are only three degrees of freedom in the analysis.

The χ^2 value and the degrees of freedom are next used to determine the probability (P) that the deviation of the observed values from expected values is due to chance. The P value for a set of data is obtained from tables of χ^2 values for various degrees of freedom. Table 5.2 presents part of a table of chi-square probabilities. For our example,

Table 5.2

Chi-Square Probabilities

df	PROBABILITIES									
	0.95	0.90	0.70	0.50	0.30	0.20	0.10	0.05	0.01	0.001
1	0.004	0.016	0.15	0.46	1.07	1.64	2.71	3.84	6.64	10.83
2	0.10	0.21	0.71	1.39	2.41	3.22	4.61	5.99	9.21	13.82
3	0.35	0.58	1.42	2.37	3.67	4.64	6.25	7.82	11.35	16.27
4	0.71	1.06	2.20	3.36	4.88	5.99	7.78	9.49	13.28	18.47
5	1.15	1.61	3.00	4.35	6.06	7.29	9.24	11.07	15.09	20.52
6	1.64	2.20	3.83	5.35	7.23	8.56	10.65	12.59	16.81	22.46
10	3.94	4.87	7.27	9.34	11.78	13.44	15.99	18.31	23.21	29.59
15	7.26	8.55	11.72	14.34	17.32	19.31	22.31	25.00	30.58	37.70
20	10.85	12.44	16.27	19.34	22.78	25.04	28.41	31.41	37.57	45.32
25	14.61	16.47	20.87	24.34	28.17	30.68	34.38	37.65	44.31	52.62
50	34.76	37.69	44.31	49.34	54.72	58.16	63.17	67.51	76.15	86.66

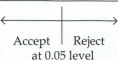

Accept | Reject
at 0.05 level

Source: Taken from Table IV of Fisher and Yates, *Statistical Tables for Biological, Agricultural and Medical Research*, published by Oliver & Boyd, Edinburgh, and by permission of the authors and publishers.

■ *Figure 5.13* Calculation of recombination frequency for two X-linked human genes by analyzing the male progeny of a woman doubly heterozygous for the two genes.

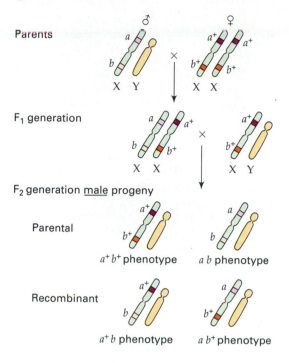

Parents

F$_1$ generation

F$_2$ generation <u>male</u> progeny

Parental

$a^+ b^+$ phenotype $a\ b$ phenotype

Recombinant

$a^+ b$ phenotype $a\ b^+$ phenotype

male progeny from pairings such as this ($a^+ b^+/a\ b$ × $a^+ b^+/Y$) in a large number of pedigrees will produce a value for the frequency of recombination between the two loci involved, and an estimate of genetic map distance can be obtained. Using this approach a number of genes have been mapped along the human X chromosome.

Mapping Human Genes by Somatic Cell Hybridization Techniques

Formation of Somatic Cell Hybrids Somatic cells can be isolated from a multicellular eukaryote, and under appropriate conditions can be cultured in vitro to give rise to what is known as a *cell culture line*. If the cell used to initiate the line is taken from mature somatic tissue, the cell line tends to have a finite lifetime, during which there is no change in the chromosomal constitution of progeny cells from that found in the parental cell. More useful are established cell lines. The cells for such lines have been selected to grow and divide essentially indefinitely in cell culture. Often the cell lines are derived from malignant tissues, and the chromosomal constitution frequently differs from that characteristic of the wild-type parental organism. For example, HeLa cells are a human cell line taken in 1951 from

a cervical cancer of Henrietta Lacks. (The name of the cell line comes from the first two letters of the woman's first and second names.) HeLa cells have been used in thousands of experiments since 1951, and are still widely used today.

Figure 5.14 diagrams the formation of a somatic cell hybrid of a human cell and an established cell

■ *Figure 5.14* Technique for producing a human-mouse hybrid somatic cell. Cell fusion results in cells that contain all the mouse chromosomes and a set of human chromosomes. (A fibroblast cell is a somatic cell from fibrous connective tissue.)

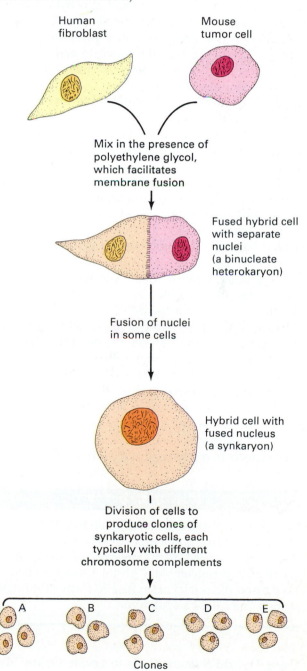

Human fibroblast

Mouse tumor cell

Mix in the presence of polyethylene glycol, which facilitates membrane fusion

Fused hybrid cell with separate nuclei (a binucleate heterokaryon)

Fusion of nuclei in some cells

Hybrid cell with fused nucleus (a synkaryon)

Division of cells to produce clones of synkaryotic cells, each typically with different chromosome complements

A   B   C   D   E

Clones

line from a mouse. To cause the cells to fuse, geneticists mix suspensions of the two cell types with polyethylene glycol (PEG), which causes the membranes to fuse. Once the cells have fused the nuclei of the now binucleate heterokaryon often fuse to form a single nucleus enclosing both sets of chromosomes. Each cell of this uninucleate cell line is called a **synkaryon.** The cell line reproduces by mitosis, so that each original synkaryotic cell gives rise to a colony of synkaryotic cells that can then be studied.

Using the HAT Technique to Locate Hybrid Cells
One of the problems in producing a hybrid cell is finding it among the parental cells. There is a convenient selective procedure that prevents parental cells from growing but allows hybrid cells to reproduce. This procedure is the *HAT technique* (hypoxanthine-aminopterin-thymidine). The HAT technique takes advantage of conditions under which only hybrid cells are able to synthesize DNA and thus divide and increase in number. The nonhybrid cells are unable to synthesize DNA and hence cannot divide.

Figure 5.15 illustrates the use of the HAT technique to produce a mouse-human hybrid cell. In this technique the two fused cells have different genetic defects. In the example, the mouse cell line is TK$^+$, HGPRT$^-$. It is defective in the activity of the enzyme hypoxanthine phosphoribosyl transferase, but it has the enzyme thymidine kinase (TK). The human cell line is TK$^-$ HGPRT$^+$. It has normal HGPRT activity but is deficient in TK activity. Because of the defective enzymes, neither cell line can grow in the presence of the chemical aminopterin. However, any hybrid cells formed by fusion of the two cell types will grow if aminopterin is present because the human chromosomes contribute a normal HGPRT gene and the mouse chromosomes contribute a normal TK gene. The HAT technique is an excellent example of using gene defects in a powerful selection scheme.

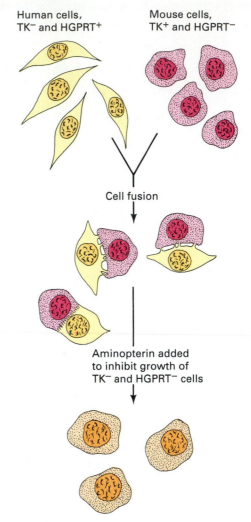

■ *Figure 5.15*   HAT technique for selecting fused, hybrid mouse-human cells.

Human cells, TK$^-$ and HGPRT$^+$

Mouse cells, TK$^+$ and HGPRT$^-$

Cell fusion

Aminopterin added to inhibit growth of TK$^-$ and HGPRT$^-$ cells

Only mouse-human hybrid cells (TK$^+$/TK$^-$ and HGPRT$^+$/HGPRT$^-$) proliferate.

K e y n o t e

Geneticists can make a somatic cell hybrid between a human cell line and a mouse cell line. The two cell types are fused by the presence of a suitable agent such as polyethylene glycol. Once the cells have fused the nuclei often fuse to form a uninucleate cell line called a synkaryon.

Using Hybrid Somatic Cells to Locate Genes   Two properties of human-rodent hybrid cells make them

particularly useful for somatic cell genetics. First, the human and rodent chromosomes are readily distinguishable under the microscope. Second, as the hybrid cell reproduces, the number of chromosomes decreases because some chromosomes are discarded; for unknown reasons human chromosomes are preferentially lost. On the selective medium with aminopterin just described, the eventual surviving, stable hybrid cell line is usually one that contains a complete set of mouse chromosomes plus a small number of human chromosomes, which vary in number and type from cell line to cell line.

To map human genes by using hybrid somatic cell lines, we start with a human cell that carries one or more genetic markers. Generally, we use markers that can readily be detected by analyzing tissue culture cells. Examples of markers include genes that control resistance to antibiotics and to other drugs, that code for enzymes, that determine nutritional requirements, and that specify cell surface antigens detectable by the use of fluorescence-labeled antibodies.

Next, various stable hybrid cell lines are analyzed for the presence or absence of the markers. These data are correlated for a number of somatic cell lines with the presence or absence of particular chromosomes. Given enough cell lines, we can show that a particular marker is present only when one particular chromosome is present, and the marker is absent when that chromosome is absent. For instance, if a drug-resistance phenotype is detected in several different cell lines that have human chromosome 14 in common, then we would conclude that the gene controlling resistance to the drug is on chromosome 14. In this way genes amenable to analysis can be localized to individual chromosomes. As we learned earlier, genes on the same chromosome, whether or not they are genetically linked, are said to be syntenic. This term was first coined as a result of somatic cell hybridization experiments. Figure 5.16 shows part of the genetic map of human chromosome X.

Keynote

Hybrid somatic cell lines help in localizing human genes to particular chromosomes. Within the nucleus the chromosomes of the two cell types are easily distinguished. Suitable genes for mapping are those that can be detected by analyzing tissue culture cells, such as genes that affect biochemical requirements or enzyme activities. As a somatic cell hybrid reproduces, a preferential and random loss of human chromosomes occurs. Eventually, stable hybrid lines are established with various subsets of the human chromosomes. With enough cell lines of this kind, a particular trait can be identified with a specific human chromosome.

Tetrad Analysis

Tetrad analysis is a mapping technique that can be used to map the genes of those eukaryotic organisms in which the products of a single meiosis, the meiotic tetrad, are contained within a single structure. The eukaryotic organisms in which this phenomenon occurs are either fungi or single-celled algae, all of which are haploid. Tetrad analysis is

■ *Figure 5.16*   Part of the genetic map of human chromosome X. The short and long arms of the chromosome are p and q, respectively.

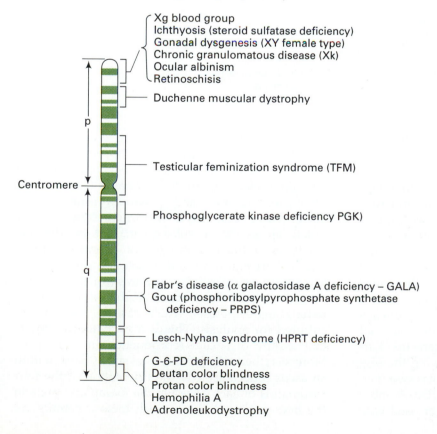

■ *Figure 5.17* Life cycle of the haploid, mycelial-form fungus *Neurospora crassa*. (Parts not to scale.)

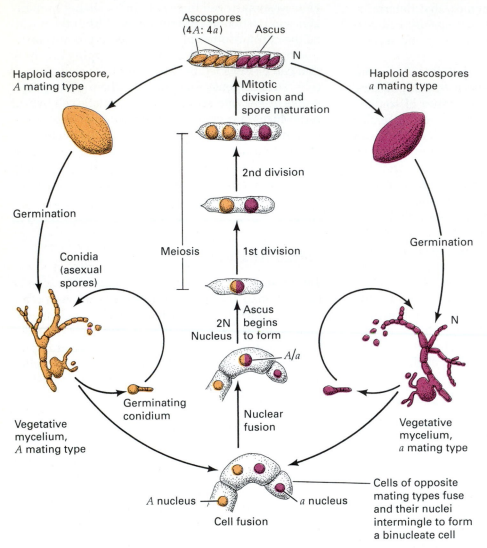

frequently used to map *Neurospora crassa* and yeast (both fungi) and *Chlamydomonas reinhardi* (a single-celled alga).

The life cycle of *Neurospora crassa* is shown in Figure 5.17. *Neurospora crassa* is a mycelial-form fungus, meaning that it spreads over its growth medium in a weblike pattern. Its designation as an orange bread mold comes from the color of the asexual spores (called conidia) it produces when it grows. *Neurospora* has several properties that make it useful for genetic and biochemical studies: It is a haploid organism, so there is only one copy of each gene. Therefore, the effects of mutations are not masked, as they can be in diploid organisms. Further, its short life cycle facilitates studying the segregation of genetic defects. *Neurospora* has two mating types (sexes, in a loose sense), called *A* and *a* (see Figure 5.17). When an *A* strain is crossed with

an *a* strain, cells of the two mating types fuse, and then two haploid nuclei fuse to produce an *A/a* diploid nucleus, which is the only diploid stage of the life cycle. The diploid nucleus immediately undergoes meioses and produces four haploid nuclei (two *A* and two *a*) within an elongating sac called an *ascus*. A subsequent mitotic division results in a linear arrangement of eight haploid nuclei (four pairs) around which spore walls form to produce eight sexual ascospores (four *A* and four *a*). Each ascus, then, contains all the products of the initial, single meiosis, which can be isolated and cultured for analysis. This is a particularly important feature of sexual reproduction in *Neurospora*. Moreover the order of the four spore pairs within an ascus reflects exactly the orientation of the four chromatids of each tetrad at the metaphase plate in the first meiotic division. Such meiotic tetrads are called **ordered tetrads.** By contrast, in yeast the four ascospores resulting from meiosis (there is no subsequent mitosis) are randomly located within a spherical ascus. Meiotic tetrads such as these are called **unordered tetrads**.

By analyzing the phenotypes of the meiotic tetrads, geneticists can directly infer the genotypes of each member of the tetrad. Haploid organisms exhibit no genetic dominance since there is only one copy of each gene; hence the genotype is expressed directly in the phenotype. Much valuable information about how and when genetic recombination (crossing-over) occurs and about the process of gene segregation has been discovered through tetrad analysis of various organisms. We will outline this technique in the following section.

Using Tetrad Analysis to Map Two Linked Genes

Figure 5.18 shows the origins of each tetrad type when two genes are linked on the same chromosome. If no crossing-over occurs between the genes (Figure 5.18[a]), then a **parental ditype** (PD) tetrad results. A parental ditype tetrad contains four nuclei, all of which are parental genotypes, with two of one parental type and two of the other parental type. In other words, there are two types of nuclei in a PD tetrad, and both are parentals. A single crossover (Figure 5.18[b]) produces two parental and two recombinant chromatids and hence a **tetratype** (T) tetrad. A tetratype tetrad contains two parental and two recombinant nuclei, one of each parental type and one of each recombinant type. In other words, there are four different types of nuclei in this tetrad.

As we learned earlier in this chapter, in double crossovers we must take into account the chromatid strands involved. In a two-strand double crossover (Figure 5.18[c]) the two crossover events take place between the same two chromatids. This crossover results in a PD tetrad, since no recombinant progeny are produced. Three-strand double crossovers (Figure 5.18[d]) include three of the four chromatids, and there are two possible ways in which this event can happen. In either case two recombinant and two parental progeny types are produced in each ascus, which is a T ascus. In four-strand double crossovers (Figure 5.18[e]) each crossover event involves two distinct chromatids, so all four chromatids of the tetrad are involved. This produces **nonparental ditype** (NPD) tetrads. An NPD tetrad contains four nuclei, all of which have recombinant (nonparental) genotypes. In other words, there are two types of nuclei in an NPD tetrad, and both types are recombinants.

Once we have data on the relative numbers of

each type of meiotic tetrad, the distance between the two genes can be computed by using the basic mapping formula:

$$\frac{\text{number of recombinants}}{\text{total number of progeny}} \times 100$$

In tetrad analysis, however, we analyze types of tetrads rather than individual progeny. By examining the nature of the four products in each tetrad, we see that a PD tetrad has all parental meiotic products, a T tetrad has two parental and two recombinant meiotic products, and an NPD tetrad has four recombinant meiotic products. To convert the basic mapping formula into tetrad terms, the recombination frequency between genes *a* and *b* becomes

$$\frac{1/2\,T + NPD}{\text{total tetrads}} \times 100$$

In this formula the 1/2 T and the NPD represent the only recombinants; the other 1/2 T and the PD represent the nonrecombinants (the parentals). For instance, if there are 200 tetrads with 162 PD, 36 T, and 2 NPD, the recombination frequency between the genes is

$$\frac{1/2(36) + 2}{200} \times 100 = 18\%$$

This conversion produces a formula for calculating the map distance between two linked genes by using the frequencies of the three possible types of tetrads rather than by analyzing individual progeny. If more than two genes are linked in a cross, the data may best be analyzed by considering two genes at a time and by classifying each tetrad into PD, NPD, and T for each pair.

Keynote

Tetrad analysis is a mapping technique in which map distance is computed from the relative frequencies of tetrad types. This method can only be used to map the genes of certain haploid eukaryotic organisms in which the products of a single meiosis, the meiotic tetrad, are contained within a single structure. In tetrad analysis the relative proportion of tetrad types is analyzed to compute the map distance between genes. The general formula when two linked genes are being mapped is

$$\frac{1/2\,T + NPD}{\text{total tetrads}} \times 100$$

Summary

Linkage, crossing-over, and gene mapping in

■ *Figure 5.18*   Origin of tetrad types for a cross *a b* × + + in which both genes are located on the same chromosome. The arrows indicate the sites at which crossovers occurred. (a) No crossover; (b) Single crossover; (c) 2-strand double crossover; (d) 3-strand double crossover; (e) 4-strand double crossover.

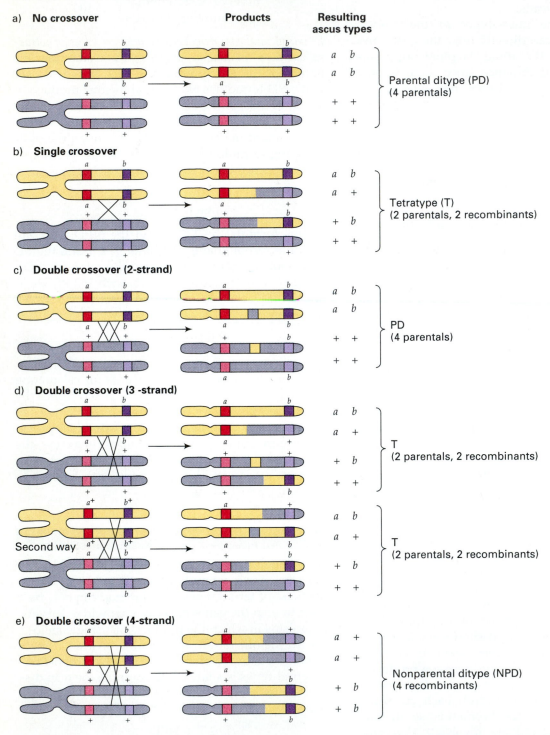

eukaryotes were discussed in this chapter. The production of genetic recombinants results from physical exchanges between homologous chromosomes in meiosis. The exchange of parts of chromatids is called crossing-over, and the site of crossing-over is called a chiasma. Proof that genetic recombination occurs when crossing-over takes place in meiosis came from experiments in which recombinants for genetic markers occurred only when exchanges of cytological markers occurred. Crossing-over is a reciprocal event that, in eukaryotes, occurs at the four-strand stage in prophase I of meiosis.

Genetic mapping is the process of locating the position of genes in relation to one another on the chromosome. To map genes it is first necessary to show that genes are linked (located on the same chromosome), a characteristic indicated by the fact that they do not assort independently in crosses. The map distance between two linked genes is then calculated based on the frequency of recombination between the two genes. The recombination frequency is an approximation of the frequency of crossovers between the two genes. The most accurate map distances are determined for genes that are closely linked because, as map distance increases, the incidence of multiple crossovers causes the recombination frequency to be an underestimate of the crossover frequency, and hence of the map distance.

Specialized mapping techniques were also presented in this chapter. Genes have been located on human chromosomes by using somatic cell hybrids of human and rodent cells; the hybrids preferentially and randomly discard human chromosomes. In certain microorganisms in which all four products of meiosis are kept together in a meiotic tetrad, genes can be mapped by analyzing the segregation patterns shown by the tetrads; this is called tetrad analysis.

Analytical Approaches for Solving Genetics Problems

Q5.1 In corn the gene for colored (C) seeds is completely dominant to the gene for colorless (c) seeds. For the character of the endosperm (the part of the seed that contains the food stored for the embryo), a single gene pair controls whether the endosperm is full or shrunken. Full (S) is dominant to shrunken (s). A true-breeding colored, full-seeded plant was crossed with a colorless, shrunken-seeded one. The F_1 colored, full plants (Cc Ss) were testcrossed to the doubly recessive type, that is, colorless and shrunken (cc ss). The result was as follows:

colored, full	4032
colored, shrunken	149
colorless, full	152
colorless, shrunken	4035
Total	8368

Is there evidence that the gene for color and the gene for endosperm shape are linked? If so, what is the map distance between the two loci?

A5.1 The best approach is to diagram the cross using gene symbols:

P: colored, full × colorless, shrunken
          CC SS   cc ss
                        ↓
F_1:   colored, full
                        Cc Ss
Testcross: colored, full × colorless, shrunken
              Cc Ss   cc ss

If the genes were unlinked, a 1:1:1:1 ratio of colored, full: colored, shrunken: colorless, full: colorless, shrunken would result in the progeny of this testcross. We can see that the actual progeny deviate a great deal from this ratio, showing a 27:1:1:27 ratio. If we did a chi-square test (using the actual numbers, not the percentages or ratios), we would see immediately that the hypothesis that the genes are unlinked is invalid, and we must consider the two genes to be linked in coupling. Specifically, the parental combinations (colored plus full and colorless plus shrunken) are more numerous than expected, while the recombinant types (colorless plus full, and colored plus shrunken) are less numerous than expected. This result comes directly from the inequality of the four gamete types produced by meiosis in the colored, full F_1 parent.

To calculate the map distance between the two genes, we need to compute the frequency of crossovers in that region of the chromosome during meiosis. We cannot do that directly, but we can compute the percentage of recombinant progeny that must have resulted from such crossovers:

Parental types:	colored, full	4032
	colorless, shrunken	4035
	Total	8067

Recombinant types:	colored, shrunken	149
	colorless, full	152
	Total	301

This calculation gives about 3.6 percent recombinant types ($\frac{301}{8368} \times 100$) and about 96.4 percent parental types ($\frac{8067}{8368} \times 100$). Since the recombination frequency can be used directly as an indication of map distance, especially when the distance is small, we can conclude that the distance between the two genes is 3.6 mu (3.6 cM).

We would get approximately the same result if the two genes were in repulsion rather than in coupling. That is, the crossovers are occurring between homologous chromosomes, regardless of whether or not there are genetic differences in the two homologs that we, as experimenters, use as markers in genetic crosses. This same cross in repulsion would be as follows:

P: colorless, full × colored, shrunken

$$\frac{c \quad S}{c \quad S} \qquad \frac{C \quad s}{C \quad s}$$

↓

F₁: colored, full

$$\frac{C \quad s}{c \quad S}$$

Data from an actual testcross of the F₁ with colorless, shrunken (*cc ss*) gave 638 colored, full (recombinant): 21,379 colored, shrunken (parental): 21,906 colorless, full (parental): 672 colorless, shrunken (recombinant) with a total of 44,595 progeny. Thus 2.94 percent were recombinants, for a map distance between the two genes of 2.94 mu, a figure reasonably close to the results of the cross made in coupling.

Q5.2 In the Chinese primrose, slate-colored flower (*s*) is recessive to blue flower (*S*); red stigma (*r*) is recessive to green stigma (*R*); and long style (*l*) is recessive to short style (*L*). All three genes involved are on the same chromosome. The F₁ of a cross between two true-breeding strains, when testcrossed, gave the following progeny:

PHENOTYPE	NUMBER OF PROGENY
slate flower, green stigma, short style	27
slate flower, red stigma, short style	85
blue flower, red stigma, short style	402
slate flower, red stigma, long style	977
slate flower, green stigma, long style	427
blue flower, green stigma, long style	95
blue flower, green stigma, short style	960
blue flower, red stigma, long style	27
Total	3000

a. What were the genotypes of the parents in the cross of the two true-breeding strains?

b. Make a map of these genes, showing gene order and distance between them.

c. Derive the coefficient of coincidence for interference between these genes.

A5.2 We analyze the data much as we did for the example in the chapter.

a. With three gene pairs, eight phenotypic classes are expected, and eight are observed. The recip-

rocal pairs of classes with the most representatives are those resulting from no crossovers, and these pairs can tell us the genotypes of the original parents. The two classes are slate, red, long and blue, green, short. Thus the F₁ triply heterozygous parent of this generation must have been *S R L/s r l*, so the true-breeding parents were *S R L/S R L* (blue, green, short) and *s r l/s r l* (slate, red, long).

b. The order of the genes can be determined by inspecting the reciprocal pairs of phenotypic classes that represent the results of double crossovers. These classes have the least numerous representatives, so the double crossover classes are slate plus green plus short (*s R L*) and blue plus red plus long (*S r l*). The gene pair that has changed its position relative to the other two pairs of alleles is the central gene, *S/s* in this case. Therefore the order of genes is *R S L* (or *L S R*). We can diagram the F₁ testcross as follows:

$$\frac{R \quad S \quad L}{r \quad s \quad l} \times \frac{r \quad s \quad l}{r \quad s \quad l}$$

A single crossover between the *R* and *S* genes gives the green plus slate plus long (*R s l*) and red plus blue plus short (*r S L*) classes, which have 427 and 402 members, respectively, for a total of 829. The double-crossover classes have already been defined, and they yield 54 progeny. The map distance between *R* and *S* is given by the crossover frequency in that region, which is the sum of the single crossovers and double crossovers divided by the total number of progeny, then multiplied by 100 percent. Thus

$$\frac{829 + 54}{3000} \times 100\% = \frac{883}{3000} \times 100\%$$
$$= 29.43\% \text{ or } 29.43 \text{ map units}$$

With similar logic, the distance between *S* and *L* is given by the crossover frequency in that region, which is the sum of the single-crossover and double-crossover progeny classes. The single-crossover progeny classes are green plus blue plus long (*R S l*) and red plus slate plus short (*r s L*), which have 95 and 85 members, respectively, for a total of 180. The map distance is given by

$$\frac{180 + 54}{3000} \times 100\% = \frac{234}{3000} \times 100\%$$
$$= 7.8\% \text{ or } 7.8 \text{ map units}$$

The data we have derived give us the following map:

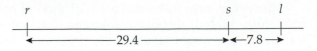

c. The coefficient of coincidence is given by

$$\frac{\text{frequency of observed double crossovers}}{\text{frequency of expected double crossovers}}.$$

The frequency of observed double crossovers is $(54/3000) \times 100\% = 1.8\%$. The expected frequency of double crossovers is the product of the map distances between r and s and between s and 1; that is, $29.4 \times 7.8\% = 2.3\%$. The coefficient of coincidence, therefore, is $1.8/2.3 = 0.78$. In other words, 78% of the expected double crossovers took place; there was only 22% interference.

Q5.3 A *Neurospora* strain that required both adenine (*ad*) and tryptophan (*trp*) for growth was mated to a wild-type strain (*ad⁺ trp⁺*). This cross produced the following ordered tetrads:

Spore pair 1:	*ad trp*	*ad +*	*ad trp*	*ad trp*
Spore pair 2:	*ad trp*	*ad +*	*ad +*	*+ trp*
Spore pair 3:	*+ +*	*+ trp*	*+ trp*	*ad +*
Spore pair 4:	*+ +*	*+ trp*	*+ +*	*+ +*
	(1) 63	(2) 3	(3) 15	(4) 9

Spore pair 1:	*ad trp*	*ad +*	*+ trp*
Spore pair 2:	*+ +*	*+ trp*	*+ +*
Spore pair 3:	*ad trp*	*ad +*	*ad trp*
Spore pair 4:	*+ +*	*+ trp*	*ad +*
	(5) 3	(6) 1	(7) 6

From the data given, calculate the map distance between the two genes.

A5.3 The linkage relationship between the genes can be determined by analyzing the relative number of parental ditype (PD), nonparental ditype (NPD), and tetratype (T) tetrads. Tetrads 1 and 5 are PD, 2 and 6 are NPD, and 3, 4, and 7 are T. The map distance between two genes is given by the general formula

$$\frac{1/2\,T + NPD}{total} \times 100$$

For this example the number of T tetrads is 30, and the number of NPD tetrads is 4. Thus the map distance between *ad* and *trp* is

$$\frac{(1/2 \times 30) + 4}{100} \times 100 = 19 \text{ mu}$$

and we have the following map:

ad ⊢————————————19.0————————————⊣ *trp*

Questions and Problems

5.1 A cross $a^+a^+\ b^+b^+ \times aa\ bb$ results in an F_1 of phenotype a^+b^+. The following numbers are obtained in the F_2 (phenotypes):

$a^+\ b^+$	110
$a^+\ b$	16
$a\ b^+$	19
$a\ b$	15
Total	160

Are genes at the a and b loci linked or independent? What F_2 numbers would otherwise be expected?

5.2 The F_1 from a cross of $A\ B/A\ B \times a\ b/a\ b$ is testcrossed, resulting in the following phenotypic ratios:

$A\ B$	308
$A\ b$	190
$a\ b$	292
$a\ B$	210

What is the frequency of recombination between genes a and b?

5.3 In *Drosophila* the mutant black (*b*) has a black body, and the wild type has a grey body; the mutant vestigial (*vg*) has wings that are much shorter and crumpled compared to the long wings of the wild type. In the following cross, the true-breeding parents are given together with the counts of offspring of F1 females × black and vestigial males:

P black and normal × gray and vestigial	
F_1 females × black and vestigial males	
grey, normal	283
grey, vestigial	1294
black, normal	1418
black, vestigial	241

From these data, calculate the map distance between the black and vestigial genes.

*5.4 A gene controlling wing size is located on chromosome 2 in *Drosophila*. The recessive allele *vg* results in vestigial wings when homozygous; the *vg⁺* allele determines long wings. A new eye mutation of the *m* gene is unknown, and you are asked to design an experiment to determine whether *m* is located on chromosome 2.

You cross true-breeding virgin maroon females to true-breeding *vg/vg* males and obtain all wild-type F_1 progeny. Then you allow the F_1 to interbreed. As soon as the F_2 start to hatch, you begin to classify the flies, and among the first six newly hatched flies, you find four wild type; one vestigial-winged, red-eyed fly; and one vestigial-winged, maroon-eyed fly. You immediately draw the conclusions that (1) maroonlike is not X-linked and (2) maroonlike is not linked to vestigial. Based on this small sample, how could you tell? On what chromosome is *m* located? (Hint: There is no crossing-over in *Drosophila* males.)

*5.5 Use the following two-point recombination data to map the genes concerned. Show the order and the length of the shortest intervals.

GENE LOCI	% RECOMBINATION	GENE LOCI	% RECOMBINATION
a,b	50	*b,d*	13
a,c	15	*b,e*	50
a,d	38	*c,d*	50
a,e	8	*c,e*	7
b,c	50	*d,c*	45

*5.6 Genes *a* and *b* are linked, with 10 percent recombination. What would be the phenotypes, and the probability of each, among progeny of the following cross?

$$\frac{a \quad b^+}{a^+ \quad b} \times \frac{a \quad b}{a \quad b}$$

*5.7 Genes *a* and *b* are sex-linked and are located 7 mu apart in the X chromosome of *Drosophila*. A female of genotype $a^+ b/a b^+$ is mated with a wild-type $(a^+ b^+)$ male.
a. What is the probability that one of her sons will be either $a^+ b^+$ or $a b^+$ in phenotype?
b. What is the probability that one of her daughters will be $a^+ b^+$ in phenotype?

*5.8 Assume that genes *a* and *b* are linked and show 20 percent crossing-over.
a. If a homozygous *A B/A B* individual is crossed with an *a b/a b* individual, what will be the genotype of the F_1? What gametes will the F_1 produce and in what proportions? If the F_1 is testcrossed with a doubly homozygous recessive individual, what will be the proportions and genotypes of the offspring?
b. If, instead, we have the original cross *A b/A b* × *a B/a B*, what will be the genotype of the F_1?

What gametes will the F_1 produce and in what proportions? If the F_1 is testcrossed with a doubly homozygous recessive, what will be the proportions and genotypes of the offspring?

5.9 In tomatoes tall vine is dominant over dwarf, and spherical fruit shape is dominant over pear shape. Vine height and fruit shape are linked, with a recombinant percentage of 20. A certain tall, spherical-fruited tomato plant is crossed with a dwarf, pear-fruited plant. The progeny are 81 tall, spherical; 79 dwarf, pear; 22 tall, pear; and 17 dwarf, spherical. Another tall and spherical plant crossed with a dwarf and pear plant produces 21 tall, pear; 18 dwarf, spherical; 5 tall, spherical; and 4 dwarf, pear. What are the genotypes of the two tall and spherical plants? If they were crossed, what would their offspring be?

*5.10 Genes *a* and *b* are in one chromosome, 20 mu apart; genes *c* and *d* are in another chromosome, 10 mu apart. Genes *e* and *f* are in yet another chromosome and are 30 mu apart. Cross a homozygous *A B C D E F* individual with an *a b c d e f* one, and cross the F_1 back to an *a b c d e f* individual. What are the chances of getting individuals of the following phenotypes in the progeny?
a. *A B C D E F*
b. *A B C d e f*
c. *A b c D E f*
d. *a B C d e f*
e. *a b c D e F*

*5.11 Genes *d* and *p* occupy loci 5 map units apart in the same autosomal linkage group. Gene *h* is a separate autosomal linkage group and therefore segregates independently of the other two. What types of offspring are expected, and what is the probability of each, when individuals of the following genotypes are testcrossed:

a. $\dfrac{D \quad P \quad h}{d \quad p \quad h}$

b. $\dfrac{D \quad P \quad h}{D \quad p \quad h}$

5.12 A hairy-winged (*h*) *Drosophila* female is mated with a yellow-bodied (*y*), white-eyed (*w*) male. The F_1 are all normal. The F_1 progeny are then crossed, and the following F_2 emerge:

Females:	wild type	757
	hairy	243
Males:	wild type	390
	hairy	130
	yellow	4
	white	3
	hairy, yellow	1

hairy, white	2
yellow, white	60
hairy, yellow, white	110

Give genotypes of the parents and the F_1, and note the linkage relations and distances, where appropriate.

5.13 Fill in the blanks. Continuous bars indicate linkage, and the order of linked genes is correct as shown. In the right two columns headed "Least frequent classes," show two gamete genotypes, unless all types are equally frequent, in which case write "none."

PARENT GENOTYPES	NUMBER OF DIFFERENT POSSIBLE GAMETES	LEAST FREQUENT CLASSES
$\dfrac{A\,b\,C}{a\,B\,c}$	_____	_____ _____
$\dfrac{A\,b\,C}{a\,B\,c}$	_____	_____ _____
$\dfrac{A\,b\,CD}{a\,B\,c\,d}$	_____	_____ _____
$\dfrac{A\,b\,CDE\,f}{a\,BCd\,e\,f}$	_____	_____ _____
$\dfrac{b\,D}{B\,d}$	_____	_____ _____

*5.14 Three of the many recessive mutations in *Drosophila melanogaster* that affect body color, wing shape, or bristle morphology are black (*b*) body versus grey in the wild type; dumpy (*dp*), obliquely truncated wings versus long wings in the wild type; and hooked (*hk*) bristles at the tip versus not hooked in the wild type. From a cross of a dumpy female with a black and hooked male, all the F_1 were wild type for all three characters. The testcross of an F_1 female with a dumpy, black, hooked male gave the following results:

wild type	169
black	19
black, hooked	301
dumpy, hooked	21
hooked	8
hooked, dumpy, black	172
dumpy, black	6
dumpy	305
Total	1000

a. Construct a genetic map of the linkage group (or groups) these genes occupy. If applicable, show the order and give the map distances between the genes.

b. Determine the coefficient of coincidence for the portion of the chromosome involved in the cross. How much interference is there?

5.15 The frequencies of gametes of different genotypes, determined by testcrossing a triple heterozygote, are as follows:

GAMETE GENOTYPE	%
+ + +	12.9
a b c	13.5
+ + c	6.9
a b +	6.5
+ b c	26.4
a + +	27.2
a + c	3.1
+ b +	3.5
Total	100.0

a. Which gametes are known to have been involved in double crossovers?

b. Which gamete types have not been involved in any exchanges?

c. The order shown is not necessarily correct. Which gene locus is in the middle?

*5.16 Genes *a*, *b*, and *c* are recessive. Females heterozygous at these three loci are crossed to phenotypically wild-type males. The progeny are phenotypically as follows:

Daughters:	All + + +	
Sons:	+ + +	23
	a b c	26
	+ + c	45
	a b +	54
	+ b c	427
	a + +	424
	a + c	1
	+ b +	0
	Total	1000

a. What is known of the genotype of the females' parents with respect to these three loci? Give gene order and the arrangement in the homologs.

b. What is known about the genotype of the male parents?

c. Map the three genes.

*5.17 The cross in *Drosophila* of

$$\frac{a^+\,b^+\,c\,\,d\,\,e}{a\,\,b\,\,c^+d^+e^+} \times \frac{a\,\,b\,\,c\,\,d\,\,e}{a\,\,b\,\,c\,\,d\,\,e}$$

gave 1000 progeny of the following 16 phenotypes:

GENOTYPE	NUMBER	GENOTYPE	NUMBER
(1) $a^+\,b^+\,c\,\,d\,\,e$	220	(9) $a\,\,b^+\,c^+d\,\,e^+$	14
(2) $a^+\,b^+\,c\,\,d\,\,e^+$	230	(10) $a\,\,b^+\,c^+d\,\,e$	13
(3) $a\,\,b\,\,c^+d^+e$	210	(11) $a^+\,b\,\,c\,\,d^+e^+$	8
(4) $a\,\,b\,\,c^+d^+e^+$	215	(12) $a^+\,b\,\,c\,\,d^+e$	8
(5) $a\,\,b^+c^+d^+e$	12	(13) $a^+\,b^+c^+d\,\,e^+$	7
(6) $a\,\,b^+c^+d^+e^+$	13	(14) $a^+\,b^+c^+d\,\,e$	7
(7) $a^+\,b\,\,c\,\,d\,\,e^+$	16	(15) $a\,\,b\,\,c\,\,d^+e^+$	6
(8) $a^+\,b\,\,c\,\,d\,\,e$	14	(16) $a\,\,b\,\,c\,\,d^+e$	7

a. Draw a genetic map of the chromosome, indicating the linkage of the five genes and the number of map units separating each. (Do not assume all the genes are linked.)

b. From the single crossover frequencies, what would be the expected frequency of $a^+\ b^+\ c^+\ d^+\ e^+$ flies?

5.18 The accompanying table shows the only human chromosomes present in stable human-mouse cell hybrid lines.

		HUMAN CHROMOSOMES			
		2	4	10	19
	A	−	+	+	−
HYBRID LINES	B	+	−	+	+
	C	−	+	+	+
	D	+	+	−	−

The presence of four enzymes, I, II, III, and IV, was investigated: I was present in *A*, *B*, and *C* but absent in *D*; II was in *B* and *D* but absent in *A* and *C*; III was in *A*, *C*, and *D* but not in *B*; and IV was in *B* and *C* but not in *A* and *D*. On what chromosomes are the genes for the four enzymes?

5.19 In *Saccharomyces*, *Neurospora*, and *Chlamydomonas*, what meiotic events give rise to PD, NPD, and T tetrads?

5.20 Double exchanges between two loci can be of several types, called two-strand, three-strand, and four-strand doubles.

a. Four recombination gametes would be produced from a tetrad in which the first of two exchanges was as depicted in the figure below. Draw in the second exchange.

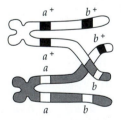

b. In the following figure, draw in the second exchange so that four nonrecombination gametes would result.

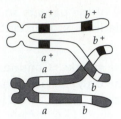

5.21 The following asci were obtained from the cross *leu* + × + *rib* in yeast. Draw the linkage map and determine the map distance.

110	45	6	39
leu +	*leu rib*	+ +	*leu* +
+ *rib*	*leu* +	*leu rib*	+ *rib*
leu +	+ +	*leu rib*	+ +
+ *rib*	+ *rib*	+ +	*leu rib*

***5.22** The genes *a*, *b*, and *c* are on the same chromosome in *Neurospora crassa*. The following asci were obtained from the cross *a b* + × + + *c*:

45	5	146	1
a b +	*a b* +	*a b* +	*a b* +
+ *b c*	*a* + +	*a b* +	+ + +
a + +	+ *b c*	+ + *c*	*a b c*
+ + *c*	+ + *c*	+ + *c*	+ + *c*

10	20	15	58
a b +	*a b* +	*a b* +	*a b* +
a + *c*	+ + *c*	*a b c*	+ *b* +
+ *b* +	*a b* +	+ + +	*a* + *c*
+ + *c*	+ + *c*	+ + *c*	+ + *c*

Determine the correct gene order and determine all gene–gene distances.

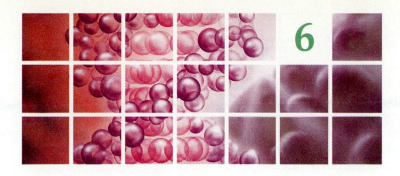

6

Genetic Analysis of Bacteria and Bacteriophages

- Transformation is the transfer of genetic material between organisms by means of extracellular pieces of DNA. By a genetic recombination event, part of the transforming DNA molecule can exchange with part of the recipient's chromosomal DNA. Transformation can be used to determine gene order and map distance between genes.

- Conjugation is a process in which there is a unidirectional transfer of genetic information through direct cellular contact between a donor and a recipient bacterial cell. The donor state is determined by the presence of a plasmid called the *F* factor. Conjugation results in the transfer of a copy of the *F* factor from donor to recipient.

- The *F* factor can integrate into the bacterial chromosome. Strains in which this has occurred—*Hfr* strains—can conjugate with recipient strains and transfer of the bacterial chromosome ensues. The sequential order of genes can be determined by the relative times at which donor genes enter the recipient during conjugation.

- Transduction is a process whereby bacteriophages (phages) mediate the transfer of bacterial DNA from one bacterium (the donor) to another (the recipient). Transduction can be used to map short chromosome segments.

- The same principles used to map eukaryotic genes are used to map phage genes. That is, genetic material is exchanged between strains differing in genetic markers and recombinants are detected and counted.

- The same general principles of recombinational mapping can be applied to mapping the distance between mutational sites in different genes (intergenic mapping) and to mapping mutational sites within the same gene (intragenic mapping).

- As a result of fine-structure analysis of a specific gene of bacteriophage T4 and other experiments, the unit of mutation and of recombination was determined to be the base pair in DNA.

- The number of units of function (genes) that cause a particular mutant phenotype is determined by the complementation, or *cis-trans*, test. If two mutants, each carrying a recessive mutation in a different gene, are combined, the mutations will complement and a wild-type function will result. If two mutants, each with a mutation in the same gene, are combined, the mutations will not complement and the mutant phenotype will still be seen.

I N THE PREVIOUS CHAPTER we considered the principles of genetic mapping in eukaryotic organisms. Just as we are interested in the locations and functions of genes in eukaryotes, so we are interested in the locations and functions of genes in bacteria and bacteriophages. To map genes in bacteria and bacteriophages, geneticists use essentially the same experimental strategies. Crosses are made between strains that differ in genetic markers, and recombinants—the products of the exchange of genetic material—are detected and counted. The analysis of data from such crosses is also the same; that is, the frequency with which exchanges occur between two sets of genes gives the map distance between the two loci. In this chapter we describe the classic genetic analysis of bacteria and bacteriophages. Currently the focus has changed from genetic mapping of genes to determining the DNA sequences of genes and, indeed, of entire chromosomes. At this writing

much of the sequence of the *Escherichia coli* genome has been determined, so we now have a very detailed understanding of the organization of genes in that bacterium.

In addition, we describe a series of classic genetic experiments which investigated the fine structure of the gene: that is, the detailed molecular organization of the gene as it relates to the mutational, recombinational, and functional events in which the gene is involved.

Mapping Genes in Bacteria

Escherichia coli (*E. coli*) is a bacterium used extensively for genetic and molecular analysis (Figure 6.1). Like other bacteria, *E. coli* can be grown both in liquid medium and on the surface of growth medium solidified with agar. Genetic analysis of bacteria typically is done by spreading ("plating") on agar-solidified media. Wherever a single bacterium

■ *Figure 6.1* Scanning electron micrograph of several *Escherichia coli* bacteria.

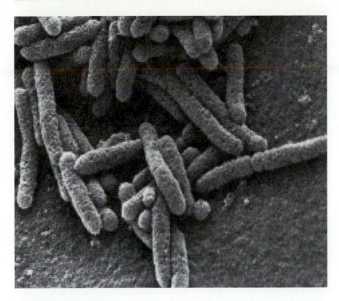

lands on the agar surface it will grow and divide repeatedly, resulting in the formation of a visible cluster of genetically identical cells called a *colony* (Figure 6.2). The concentration of bacterial cells in a culture (the *titer*) can be determined by spreading known volumes or dilutions of the culture on the agar surface, incubating the plates, and then counting the number of resulting colonies. By varying the nutrient composition of the media it is possible to detect and quantify the various genotypic classes in a genetic analysis.

■ *Figure 6.2* Bacterial colonies growing on a nutrient medium in a Petri dish.

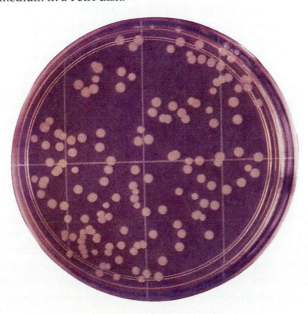

The genetic material in bacteria consists of a single circular, double-stranded DNA chromosome. Genetic material can be transferred between bacteria by three main processes: transformation, conjugation, and transduction. In each case: (1) transfer is unidirectional; (2) there is no true diploid zygote as in eukaryotes; and (3) only genes included in the circular chromosome will be stably inherited. It is possible to map bacterial genes by using any one of these methods. However, not all methods can be used for each species of bacterium, and the size of the region that can be mapped varies according to the method.

Bacterial Transformation

Transformation is a process in which there is a transfer of genetic information by means of extracellular pieces of DNA. The DNA fragments are derived from donor bacteria and may be taken up by another living bacterium, the recipient. If the donor and the recipient bacteria differ genetically, then stable genetic recombinants are produced by crossover events between the donor's DNA fragments and the recipient's chromosome. Recipient cells showing recombinant phenotypes are called **transformants.** Transformation is generally used to map genes in bacterial species in which conjugation (p. 131) or transduction (p. 137) do not occur.

There are two types of transformation: *natural transformation*, in which bacteria are naturally able to take up DNA and be genetically transformed; and *engineered transformation*, in which bacteria have been altered to enable them to take up DNA and be genetically transformed. Natural transformation occurs in, for example, *Bacillus subtilis*, and engineered transformation occurs in, for example, *E. coli*.

Once a donor DNA fragment has been taken up by a recipient bacterium with a different genetic constitution, we can detect the transformants by their recombinant phenotypes. Let us consider an example of natural transformation of *Bacillus subtilis* (Figure 6.3). (Other systems may differ in the details of the process.) The donor double-stranded DNA fragment is wild type (a^+) for a mutant allele *a* in the recipient (Figure 6.3[a]). During DNA uptake by the recipient, one of the two DNA strands is degraded so that only one DNA strand ultimately is found within the cell (Figure 6.3[b]). This DNA strand pairs with the homologous DNA of the recipient cell's chromosome, forming a triple-stranded structure over the length of the paired region (Figure 6.3[c]). Recombination (strand replacement) can then occur by a double crossover event involving the donor single-stranded DNA

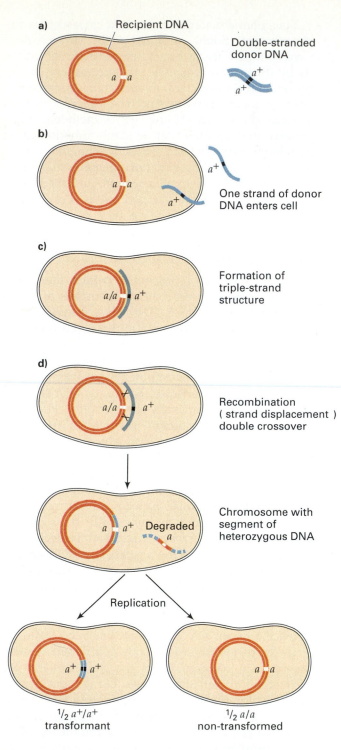

a) Recipient DNA

Double-stranded donor DNA

b) One strand of donor DNA enters cell

c) Formation of triple-strand structure

d) Recombination (strand displacement) double crossover

Chromosome with segment of heterozygous DNA   Degraded

Replication

½ a⁺/a⁺ transformant   ½ a/a non-transformed

■ *Figure 6.3* Natural transformation in *Bacillus subtilis*: (a) Linear donor double-stranded bacterial DNA fragment, which carries the a^+ allele and the recipient bacterium with the a allele; (b) One of the donor DNA strands enters the recipient; (c) The single, linear DNA strand pairs with the homologous region of the recipient's chromosome, forming a triple-stranded structure; (d) A double crossover produces a recombinant a^+/a recipient chromosome and a linear a DNA fragment. The linear fragment is degraded and, by replication (see Chapter 10), one-half of the progeny are a^+ recombinants and one-half are a nonrecombinants.

degraded.) Replication of the recipient chromosome then produces one progeny chromosome with donor genetic information on both DNA strands (an a^+ transformant) and another progeny chromosome with recipient genetic information on both DNA strands (an a nontransformant). Equal numbers of a^+ transformants and a nontransformants are expected. Transformation of most genes will occur at a frequency of about 10^{-3}, or in about 1 cell in every 10^3 cells.

Using transformation it is possible to determine gene linkage, gene order, and map distance using the same general principles we discussed for gene mapping of eukaryotes in Chapter 5. The principles for determining whether genes are linked and for determining gene order are as follows: If two genes, x^+ and y^+, are far apart on the donor chromosome, they will always be found on different DNA fragments because the average-sized DNA fragment is so much smaller than the entire chromosome. Thus given an $x^+ y^+$ donor and an $x y$ recipient, the probability of simultaneous transformation (*cotransformation*) of the recipient to $x^+ y^+$ (from the product rule) is the product of the probability of transformation with each gene alone. If transformation of x to x^+ or of y to y^+ occurred at a frequency of 1 in 10^3 (10^{-3}) cells per gene, according to the product rule, $x^+ y^+$ transformants would be expected to appear at a frequency of 1 in 10^6 recipient cells ($10^{-3} \times 10^{-3}$). If cotransformation occurs at a frequency that is substantially higher than the product of the two single-gene transformations, the two genes must be close together.

Gene order can also be determined from cotransformation data (Figure 6.4). For example, if genes p and q are often cotransformed, then these two genes must be relatively closely linked. Similarly, if genes q and o are often cotransformed, those two genes must be close to one another. To determine gene order we now need information about any cotransformation of genes p and o. If they are never cotransformed then they must not be closely linked.

and the recipient double-stranded DNA (Figure 6.3[d]). The result is a recombinant recipient chromosome: In the region between the two crossovers, one DNA strand has the donor a^+ DNA segment, and the other strand has the recipient a DNA segment. In other words, in that region *the two DNA strands are part donor–part recipient for the genetic information.* (The other product of the double crossover event is a single-stranded piece of DNA carrying an a DNA segment; that DNA fragment is

■ *Figure 6.4* Demonstration of determining gene order by cotransformation. DNA extracted from a population of $p^+ q^+ o^+$ donor bacteria were used to transform a population of $p\ q\ o$ recipient bacteria.

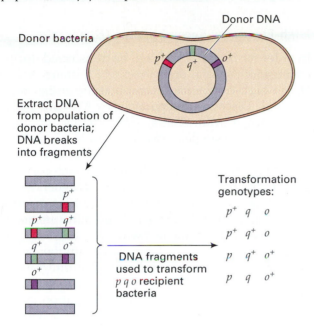

Since q is linked to p as well as to o, but p is not closely linked to o, the gene order logically must be p-q-o.

Keynote

Transformation is the transfer of genetic material between organisms by means of extracellular pieces of DNA. In transformation, DNA derived from a donor strain is added to recipient cells. A DNA fragment taken up by the recipient cell may associate with the homologous region of the recipient's chromosome. By a genetic recombination event, part of the transforming DNA molecule can exchange with part of the recipient's chromosomal DNA. Frequent cotransformation of donor genes indicates close physical linkage of those genes. Analysis of cotransformants can be used to determine gene order and map distance between genes. Transformation has been used to construct genetic maps for bacterial species for which other methods of genetic analysis are not possible.

Conjugation in Bacteria

Conjugation is a process in which there is a unidirectional transfer of genetic information through direct cellular contact between a donor bacterial cell and a recipient bacterial cell. The contact is followed by the formation of a bridge physically connecting both cells. Then a segment (rarely all) of the donor's chromosome may be transferred into the recipient and may undergo genetic recombination with a homologous chromosome segment of the recipient cell. The recipients receiving donor DNA are called **transconjugants.**

Conjugation was discovered in 1946 by Joshua Lederberg and Edward Tatum. They mixed together two *E. coli* strains that differed in their nutritional requirements. Strains that must have additives in the medium in order to grow are called **auxotrophs**. A strain that is wild type for all nutritional requirement genes and thus requires no supplements in its growth medium is called a **prototroph**. Conventionally, the gene symbol indicates the nutritional supplement in question. If the genotype has a superscript $^+$, then the gene is wild type and no nutritional requirement is involved. If no superscript is present, the gene is mutant and a nutritional requirement is involved. For example, a *met*$^+$ strain is wild type for a methionine supplement, and a *met* strain is mutant, requiring the amino acid methionine in order to grow.

In Lederberg and Tatum's experiment, strain A had the genotype *met bio thr*$^+$ *leu*$^+$ *thi*$^+$, and strain B had the genotype *met*$^+$ *bio*$^+$ *thr leu thi*. Strain A could only grow on a medium supplemented with the amino acid methionine (*met*) and the vitamin biotin (*bio*); it did not need the amino acids threonine (*thr*) or leucine (*leu*), or the vitamin thiamine (*thi*). Strain B could only grow on a medium supplemented with *thr*, *leu*, and *thi*; it did not require *met* or *bio*. Mixing strain A and strain B resulted in some prototrophic *met*$^+$ *bio*$^+$ *thr*$^+$ *leu*$^+$ *thi*$^+$ colonies, indicating that some type of genetic exchange had occurred between the two strains. The mixing, then, is a genetic cross from which recombinants can result. What mechanism was involved in the genetic exchange?

The Sex Factor *F*

In 1953 William Hayes discovered that in conjugation, one cell acts as a donor and the other cell acts as a recipient. The transfer of genetic material between the strains is mediated by a sex factor (also called a fertility factor) named *F* that the donor cell possesses (F^+) and the recipient cell lacks (F^-). The *F* factor found in F^+ *E. coli* cells is an example of a **plasmid,** a self-replicating, circular, double-stranded DNA molecule that is distinct from the main chromosome. The *F* factor contains a number of genes, including genes that specify hairlike cell surface components called **F-pili,** or sex-pili, which allow the physical union of F^+ and F^- cells to take place.

■ *Figure 6.5*   Electron micrograph of bacterial conjugation between an F^+ donor bacterium and F^- recipient *E. coli* bacterium. During conjugation between an F^+ donor and an F^- recipient, a bridge forms between the two cells and a copy of the F factor is transferred. Magnification 30,000x.

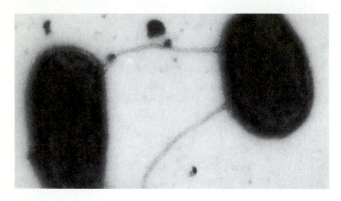

When F^+ and F^- cells are mixed, they conjugate as described above (Figure 6.5; Figure 6.6[a,1]). Genetic material is transferred during this process. One strand of the F factor is nicked at a region called the **origin**, and DNA replication proceeds from that point (Figure 6.6[a,2]). Beginning at the origin a single strand of DNA is transferred to the F^- cell as replication takes place (Figure 6.6[a,3]). Think of the process like a roll of paper towels unraveling. The origin is the first stretch of DNA unwound; as unwinding continues replication maintains the remaining circular F factor in a double-stranded form. Once the F factor DNA enters the F^- recipient, the complementary strand is synthesized (Figure 6.6[a,4]). In the transfer process, then, the origin is always transferred first, followed by the rest of the F factor. When the complete F factor has been transferred, the F^- cell becomes an F^+ cell as a result of the F factor genes (Figure 6.6[a,5]). In a population of cells, only some of which are F^+ to start with, the end result of the conjugation of F^+ and F^- cells and the subsequent transfer of the F factor is an epidemic conversion of the population to the F^+ state. In $F^+ \times F^-$ crosses none of the bacterial chromosome is transferred; only the F factor is transferred.

Keynote

Some *E. coli* bacteria possess a plasmid, called the F factor, that is required for mating. *E. coli* cells with the F factor are F^+ and those without it are F^-. The F^+ cells (donors) can mate with F^- cells (recipients) in a process called conjugation. This leads to the one-way transfer of a copy of the F factor from donor to recipient during replication of the F factor.

As a result, both donor and recipient are F^+. None of the bacterial chromosome is transferred during $F^+ \times F^-$ conjugation.

High-Frequency Recombination (*Hfr*) Strains

In order for bacterial genes to be transferred during conjugation, special derivatives of F^+ strains, called **Hfr** or **high-frequency recombination** strains, must be used. Discovered separately by William Hayes and Luca Cavalli-Sforza, *Hfr* strains originate through a rare single crossover event in which the F factor actually integrates into the bacterial chromosome (Figure 6.6[b,1–2]). When the F factor is integrated, it no longer replicates independently; instead, it is replicated as part of the host chromosome. Since the F factor genes are still functional, *Hfr* cells will conjugate with F^- cells (Figure 6.6[b,3]). When mating takes place, similar events occur as in the $F^+ \times F^-$ matings. The integrated F factor becomes nicked at the origin and replication begins (Figure 6.6[b,4]). During replication *part of the F factor first moves into the recipient cell* where the transferred strand is copied. With time, the donor bacterial chromosome begins to be transferred into the recipient. If there are genetic marker differences between the genes on the donor chromosome that has entered the recipient and the recipient chromosome, recombinants can be isolated (Figure 6.6[b,5]). This recombination process occurs by double crossovers between the donor DNA and the recipient chromosome, similar to that described for transformation. That is, a segment of donor DNA is exchanged for the homologous segment of recipient DNA.

In *Hfr* $\times$ F^- matings the F^- cell almost never acquires the F^+ phenotype. For the recipient cell to become F^+ it must receive a complete copy of the F factor. However, only part of the F factor is transferred at the beginning of conjugation; the rest of the F factor is at the end of the donor chromosome. All of the donor chromosome would have to be transferred for a complete functional F factor to be found in the recipient. This is an extremely rare event because while the bacteria are conjugating they are "jiggling" around, so it is very likely that the mating pair will break apart long before the second F factor part is transferred. As a result, transconjugants contain a complete copy of the circular recipient chromosome and a linear fragment of the donor chromosome. After recombination, the recombinant, recipient chromosome is passed on to progeny cells by the normal replication process, whereas the linear DNA fragment is degraded or lost.

■ *Figure 6.6*   Transfer of genetic material during conjugation in *E. coli*: (a) Transfer of the *F* factor from donor to recipient cell during $F^+ \times F^-$ matings; (b) Production of *Hfr* strain by integration of F factor and transfer of bacterial genes from donor to recipient cell during $Hfr \times F^-$ matings.

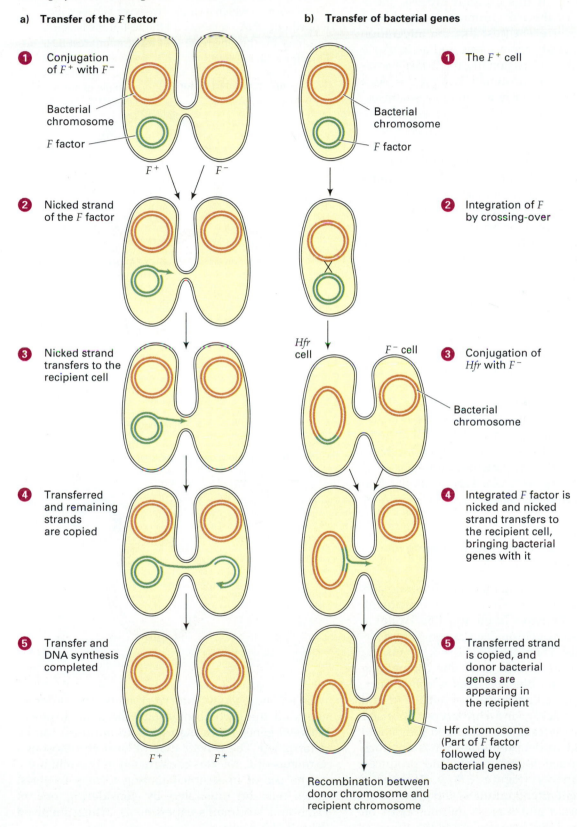

a)   Transfer of the F factor

① Conjugation of F^+ with F^-

Bacterial chromosome

F factor

F^+ F^-

② Nicked strand of the F factor

③ Nicked strand transfers to the recipient cell

④ Transferred and remaining strands are copied

⑤ Transfer and DNA synthesis completed

F^+ F^+

b)   Transfer of bacterial genes

① The F^+ cell

Bacterial chromosome

F factor

② Integration of F by crossing-over

Hfr cell F^- cell

③ Conjugation of *Hfr* with F^-

Bacterial chromosome

④ Integrated F factor is nicked and nicked strand transfers to the recipient cell, bringing bacterial genes with it

⑤ Transferred strand is copied, and donor bacterial genes are appearing in the recipient

Hfr chromosome (Part of F factor followed by bacterial genes)

Recombination between donor chromosome and recipient chromosome

F' Factors

Just as the *F* factor can integrate into the bacterial chromosome to produce an *Hfr* strain, the reverse process can occur. In this excision process, the *F* factor loops out of the *Hfr* chromosome, and by a single crossing-over event (just like the integration event), a circular host chromosome and a circular *F* factor are generated. Occasionally the *F* factor excision from the host chromosome is not precise. As a result, the excised *F* factor may contain a small section of the host chromosome that was adjacent to the *F* factor where it was integrated. Consider an *E. coli* strain in which the *F* factor has integrated next to the *lac*⁺ region, a set of genes required for the breakdown of lactose (Figure 6.7[a]). If the looping out is not precise, the adjacent *lac*⁺ host chromosomal genes may be included in the loop (Figure 6.7[b]). Then by a single crossover the looped-out DNA will be separated from the host chromosome (Figure 6.7[c]) to give an *F* factor that also contains the *lac*⁺ genes of the host chromosome. *F* factors containing bacterial genes are called *F'* (*F* prime) factors, and they are named for the genes they have picked up. For example, an *F'* with the *lac* region is called *F'* (*lac*). Cells with *F'* factors can conjugate with *F⁻* cells since all the *F* factor functions are present. As in regular conjugation, a copy of the *F'* factor is transferred to the *F⁻* cell, which then becomes *F'*. The recipient also receives a copy of the bacterial gene (*lac* in our example) attached to the *F* factor. Since the recipient has its own copy of that gene, the resulting cell will be partially diploid, having two copies of one or a few genes and only one copy of all the others. This particular type of conjugation is called **F-duction**, or *sexduction*.

Using Conjugation and Interrupted Mating to Map Bacterial Genes

In mapping genes by conjugation, an *Hfr* strain acts as the donor and an *F⁻* strain acts as the recipient. In the late 1950s Francois Jacob and Elie Wollman studied the transfer of chromosomal genes from *Hfr* strains to *F⁻* cells. They made an *Hfr* × *F⁻* mating and, at various times, broke apart the conjugating pairs using a kitchen blender. This is called *interrupted mating*. Since all pairings between an *Hfr* and *F⁻* formed very quickly, and chromosome transfer occurred at a steady rate, the separated cells could then be analyzed to determine the times at which donor genes entered the recipients and produced genetic recombinants (Figure 6.8[a]). If gene *a* enters the recipient after 9 minutes, and gene *b* after 15 minutes, genes *a* and *b* are 6 minutes apart on the genetic map. Thus by measuring the times at

■ *Figure 6.7*   Production of an *F'* factor: (a) Region of bacterial chromosome into which the *F* factor has integrated; (b) The *F* factor looping out incorrectly, so it includes a piece of bacterial chromosome, the *lac*⁺ genes; (c) Excision, in which a single crossover between the looped-out DNA segment and the rest of the bacterial chromosome (i.e., the reverse of integration) results in an *F'* factor called *F'* (*lac*).

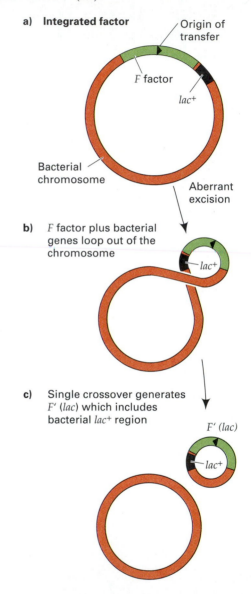

which donor genes enter recipients, the order of genes on the chromosome and the map distances between genes (with map units in minutes) can be determined. The genetic map of large chromosomal segments of *E. coli* was constructed in this manner.

The use of interrupted mating to map bacterial genes may be illustrated by considering one of Jacob and Wollman's experiments, which involved the following cross:

■ **Figure 6.8** Interrupted-mating experiment involving the cross *Hfr H thr⁺ leu⁺ azi^r ton^r lac⁺ gal⁺ str^s* × *F⁻ thr leu azi^s ton^s lac gal str^r*. The progressive transfer of donor genes with time is illustrated. Recombinants are generated by an exchange of a donor fragment with the homologous recipient fragment resulting from a double crossover event. (a) At various times after mating commences, the conjugating pairs are broken apart and the transconjugant cells are plated on selective media including streptomycin to eliminate *Hfr* donors in order to determine which genes have been transferred from the *Hfr* to the *F⁻*. (b) The graph shows the appearance of donor genetic markers in the *F⁻* cells as a function of time; that is, after the selected *thr⁺* and *leu⁺* genes have entered.

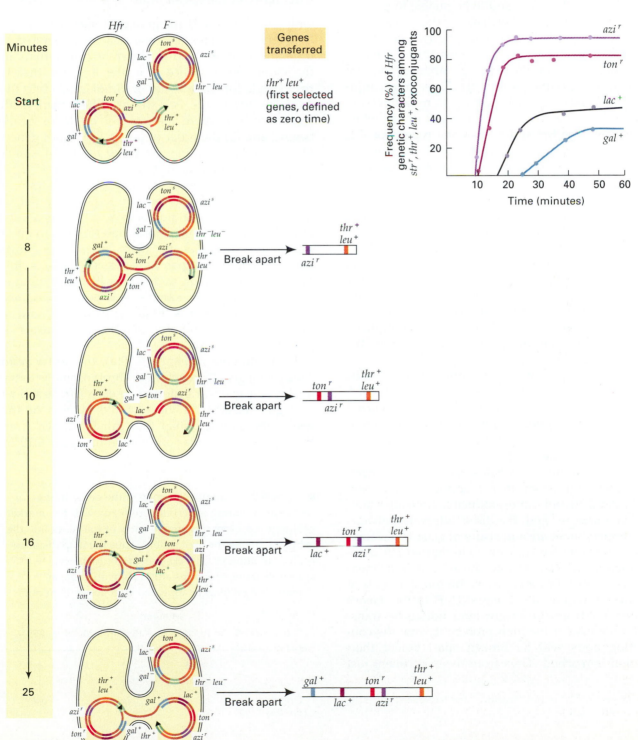

Donor:
Hfr H thr⁺ leu⁺ azi^r ton^r lac⁺ gal⁺ str^s *

Recipient:
F⁻ thr leu azi^s ton^s lac gal str^r

The *Hfr H* strain is prototrophic and is sensitive to the antibiotic streptomycin. The *F⁻* cell carries a streptomycin resistance gene and also a number of mutant genes. These genes cause the *F⁻* to be auxotrophic for threonine (*thr*) and leucine (*leu*), to be sensitive to sodium azide (*azi^s*) and to infection by bacteriophage T1 (*ton^s*), and to be unable to ferment lactose (*lac*) or galactose (*gal*) as a carbon source.

In a conjugation experiment the two cell types are mixed together and, at various times, samples are removed from the mating mixture and are blended to break the pairs apart. These **transconjugants** are plated on a selective medium designed to allow recombinant types to grow and divide while killing both the *Hfr* and *F⁻* parental types. For this particular cross the medium contains streptomycin to kill the *Hfr H* and lacks threonine and leucine so that the parental *F⁻* cannot grow. The threonine (*thr⁺*) and leucine (*leu⁺*) genes are the first donor chromosomal genes to be transferred to the *F⁻*, so recombinants formed by the exchange of those genes with the *thr leu* genes of the *F⁻* recipient are produced and will grow on the selective medium. Appropriate media can be used to test for the appearance of other donor genes (*azi^r ton^r lac⁺* and *gal⁺*) among the selected *thr⁺ leu⁺ str^r* transconjugants. For example, a medium with sodium azide added can test for the presence of *azi^r* from the donor, and so on.

Figure 6.8[b] shows an example of the results. The transconjugants here are *thr⁺*, *leu⁺*, and *str^r*. The first unselected marker gene to be transferred is *azi^r*, and recombinants for this gene are seen at about 9 minutes. The second gene transferred is *ton^r* at 10 minutes, followed by *lac⁺* at about 17 minutes, and *gal⁺* at about 25 minutes. In this experiment, then, each gene from an *Hfr* appears in recombinants at a different but reproducible time after mating commences, and the time intervals between gene appearances are a measure of genetic distance. From such data we can conclude that an *Hfr* chromosome is transferred into the *F⁻* cell in a linear way. As with *F* factor transfer, the transfer starts at the origin within the integrated *F* factor. Genes located far from the origin tend not to be transferred because of the high probability that the conjugating pairs will be broken apart before their location is reached. Thus from the experiment just

described (*Hfr H × F⁻*) and from the resulting time intervals, the map in Figure 6.9 may be constructed. The map units are in minutes: The entire *E. coli* chromosome is about 100 minutes. From transconjugants where two donor genes have entered, mapping can be done in terms of percentage recombination or map units. One minute in time is approximately equivalent to 20 percent recombination.

Circularity of the *E. coli* Map

Only one *F* factor is integrated in each *Hfr* strain. Different *Hfr* strains have the *F* factor integrated at different locations and in different orientations in the chromosome. Since it is the *F* factor that is responsible for the transfer of donor genes into the recipient strain, the *Hfr* strains differ, then, with respect to: (1) where the transfer of donor genes begins; and (2) the order of transfer of donor genes. Figure 6.10(a) diagrams the order of chromosomal gene transfer for four different *Hfr* strains H, 1, 2, and 3. In each experiment only one *Hfr* strain is used to cross with the recipient, and the order of gene transfer and the time between the appearance of each gene in the recipient is determined. In these experiments the genetic distance in time units between a particular pair of genes is constant, within experimental error, no matter which *Hfr* strain is used as donor; for example, the genetic distance between *thr* and *pro* is the same in H, 1, 2, and 3. This validates the use of time units as a measure of genetic distance in *E. coli*.

From the data in Figure 6.10(a), the best way to construct a genetic map of the chromosome is to try to align the genes transferred by each *Hfr*, as shown in Figure 6.10(b). In view of the overlap of the genes, the simplest map that can be drawn from these data is a circular one shown in Figure 6.10(c).

■ *Figure 6.9* Chromosome map of the genes in the interrupted-mating experiment of Figure 6.8. The marker positions represent the time of entry of the genes into the recipient during the experiment. Thus the map distance is given in minutes. The *azi* marker, for instance, appeared about 9 minutes after conjugation began, whereas the *gal* marker appeared after about 25 minutes.

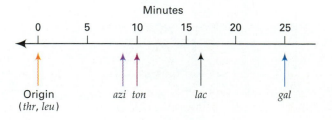

---

*The superscript "s" means "sensitive" and the superscript "r" means "resistant."

■ *Figure 6.10*   Interrupted-mating experiments with a variety of *Hfr* strains, showing that the *E. coli* linkage map is circular: (a) Orders of gene transfer for the *Hfr* strains *H*, *1*, *2*, and *3*; (b) Alignments of gene transfer for the *Hfr* strains; (c) Circular *E. coli* chromosome map derived from the *Hfr* gene transfer data. The map is a composite showing various locations of integrated *F* factors. A given *Hfr* strain has only one integrated *F* factor.

a)   Orders of gene transfer

Hfr strains:

H	origin – *thr* – *pro* – *lac* – *pur* – *gal* – *his* – *gly* – *thi*
1	origin – *thr* – *thi* – *gly* – *his* – *gal* – *pur* – *lac* – *pro*
2	origin – *pro* – *thr* – *thi* – *gly* – *his* – *gal* – *pur* – *lac*
3	origin – *pur* – *lac* – *pro* – *thr* – *thi* – *gly* – *his* – *gal*

b)   Alignments of gene transfer for the *Hfr* strains

H	*thr* – *pro* – *lac* – *pur* – *gal* – *his* – *gly* – *thi*
1	*pro* – *lac* – *pur* – *gal* – *his* – *gly* – *thi* – *thr*
2	*lac* – *pur* – *gal* – *his* – *gly* – *thi* – *thr* – *pro*
3	*gal* – *his* – *gly* – *thi* – *thr* – *pro* – *lac* – *pur*

c)   Circular *E. coli* chromosome map derived from *Hfr* gene transfer data

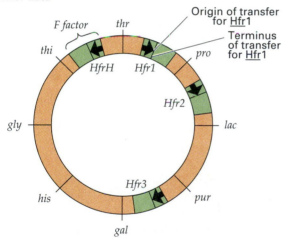

The map is a composite of the results of the individual matings. The circularity of the map was itself a truly significant finding, since all previous genetic maps—of eukaryotic chromosomes—were linear.

Keynote

The *F* factor can integrate into the bacterial chromosome by a single crossover event. Strains in which this integration has occurred are *Hfr* (high-frequency recombination) strains. *Hfr* strains can conjugate with *F⁻* strains, and transfer of the bacterial chromosome occurs from the *Hfr* to the *F⁻* beginning at a specific site called the origin. The farther a gene is from the origin, the later it is transferred to the *F⁻*,

which forms the basis for mapping genes by their times of entry into the *F⁻*. Conjugation allows mapping of large chromosome segments. This resulted in a circular map for the *E. coli* chromosome.

Transduction in Bacteria

Transduction is a process by which bacteriophages (bacterial viruses), or phages, for short, function as intermediaries in the transfer of bacterial genetic information from one bacterium (the donor) to another (the recipient). Since the amount of DNA a phage can carry is limited, the amount of genetic material that can be transferred is usually less than 1 percent of that in the bacterial chromosome. Once the donor genetic material has been introduced into the recipient, it may undergo genetic recombination with a homologous chromosome region. The recipients inheriting donor DNA in this way are called **transductants**.

Bacteriophages: An Introduction

Most bacterial strains have phages that are specific for them. For example, *E. coli* is susceptible to infection by DNA-containing phages such as T2, T4, T5, and λ (lambda), among others. Most of the genes located on the chromosomes of the phages commonly used in genetics have been identified and mapped.

The structure of all phages is simple. A phage contains its genetic material (either DNA or RNA) in a single chromosome surrounded by a coat of protein molecules. Variation in the number and organization of the proteins gives the phages their characteristic appearances. Figure 6.11 presents electron micrographs and diagrams of two phages of great genetic significance, T4 (Figure 6.11[a]) and λ (Figure 6.11[b]). Phage T4 is one of a series of similar phages collectively called T-even phages (T2, T4, and T6). It has a number of distinct structural components: a head (which contains the genetic material DNA), a core, a sheath, a base plate, and tail fibers (the latter two structures enable the phage to attach itself to a bacterium). The head and all other structural components consist of proteins.

The life cycles of the T4 and the λ phages are not identical. In the T4 life cycle, the phage particle first attaches to the surface of the bacterial cell, and the phage chromosome is then injected into the bacterium as the phage sheath contracts. Then, by the action of phage genes, the phage in essence takes over the bacterium, breaking down the bacterial chromosome and directing its metabolic processes to produce progeny phages. Finally, the progeny

■ *Figure 6.11*   Electron micrographs and diagrams of two bacteriophages (1 nm = 10^{-9} m): (a) T4 phage, which is representative of T-even phages; (b) λ phage.

a) T4 phage

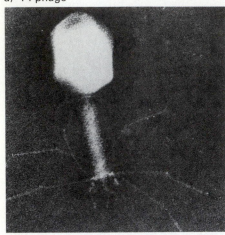

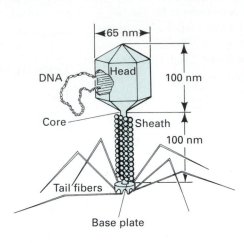

b) λ phage

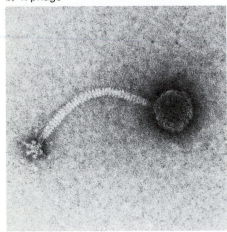

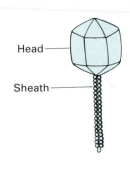

phages are released from the bacterium as the cell is broken open, or lysed. The suspension of released progeny phages is called a **phage lysate**. The T4 type of phage life cycle is called the **lytic cycle**, and phages that always follow that cycle when they infect bacteria are called **virulent phages.**

The λ life cycle, shown in Figure 6.12, is more complex than that of a T4 phage. Phage λ has a structure similar to that of T4. When its DNA is injected into *E. coli*, there are two alternative paths that the phage can follow. One is a lytic cycle, exactly like that of the T phages. The other is the **lysogenic pathway.** In the lysogenic pathway the λ chromosome does not replicate; instead, it inserts (integrates) itself physically into a specific region of the host cell's chromosome, much like *F* factor integration. In this integrated state the phage chromosome is called a **prophage.** Every time the host cell chromosome replicates, the integrated λ chromo-

some is also replicated. The bacterium that contains a phage in the prophage state is said to be **lysogenic** for that phage; the phenomenon of the insertion of a phage chromosome into a bacterial chromosome is called **lysogeny.** Phages that have a choice between lytic and lysogenic pathways are called **temperate phages.** Occasionally the mechanism that maintains the prophage state (which is a function of phage genes) breaks down. This can occur, for example, as a result of starvation for essential nutrients or by ultraviolet light irradiation of the cells. This *induction* event causes the lytic cycle to be initiated, and progeny λ phages are released.

Transduction Mapping of Bacterial Chromosomes

Transduction, the process in which a phage (called a **phage vector**) carries pieces of bacterial DNA between bacterial strains, is a useful mechanism for

■ *Figure 6.12*   Life cycle of a temperate phage such as λ. When a temperate phage infects a cell, the phage may go through either the lytic or lysogenic cycle.

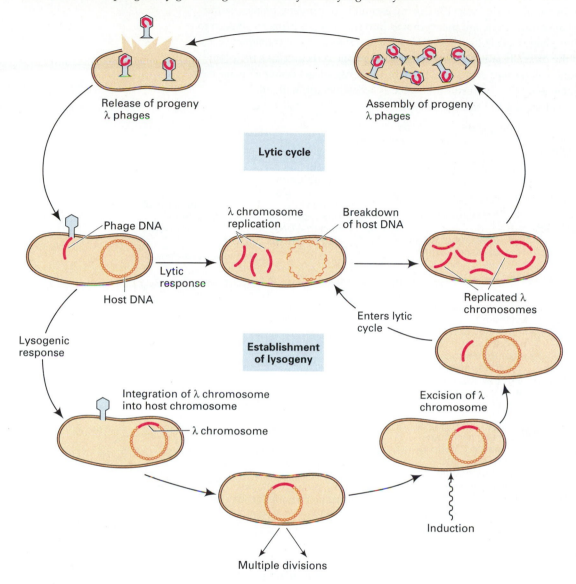

mapping bacterial genes. Examples are phages P1 and λ for *E. coli*, and phage P22 for *Salmonella typhimurium*. Two types of transduction occur: **generalized transduction**, in which any gene may be transferred between bacteria; **specialized transduction**, in which only specific genes are transferred. In the latter, an integrated phage genome excises imperfectly, picking up adjacent bacterial genes, while leaving some phage genes behind. We will discuss generalized transduction here.

Generalized Transduction. Generalized transduction was discovered by Joshua and Esther Lederberg and Norton Zinder in 1952. The mechanism for generalized transduction of *E. coli*

by temperate bacteriophage P1 is shown in Figure 6.13. When the P1 phage infects the cell (Figure 6.13[1]), it normally goes into the lysogenic pathway, existing in the prophage state in these cells. If the prophage state is not maintained, the phage goes through the lytic cycle and produces progeny phages (Figure 6.13[2]). During the lytic cycle the bacterial DNA is degraded and, rarely, a piece of bacterial DNA is packaged into the phage head instead of the phage genome. (Figure 6.13[3]). Since this event is rare only a very small proportion (about 1 in 10^5) of the progeny phages carry donor bacterial genes; the rest are normal phages. The phages containing donor DNA are called **transducing phages** since they are the vehicles by

■ *Figure 6.13*   Generalized transduction between strains of *E. coli*: (1) Wild-type (a^+ b^+ c^+) donor cell of *E. coli* infected with the temperate bacteriophage P1; (2) The host cell DNA is broken up during the lytic cycle; (3) During assembly of progeny phages, some pieces of the bacterial chromosome are incorporated into some of the progeny phages to produce transducing phages; (4) Following cell lysis, a low frequency of transducing phages is found in the phage lysate; (5) The transducing phage infecting an *a* recipient bacterium; (6) A double crossover event results in the exchange of the donor a^+ gene with the recipient mutant *a* gene; (7) The result is a stable a^+ transductant, with all descendants of that cell having the same genotype.

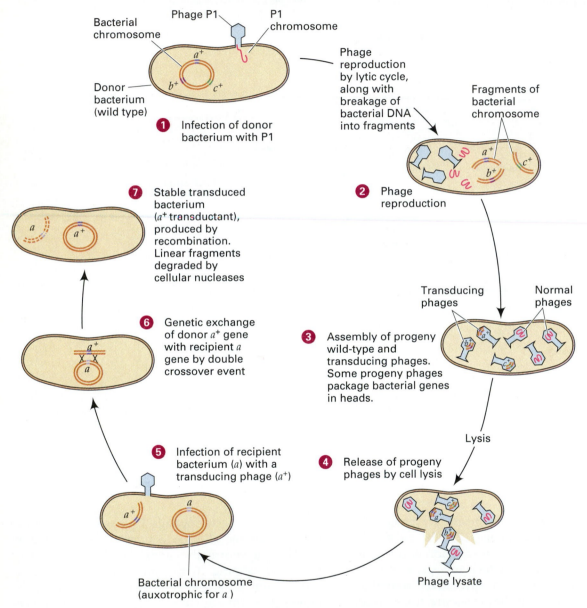

which genetic material is shuttled between donor and recipient bacteria. In the example shown in Figure 6.13, the transducing phages are those carrying the bacterial genes a^+, b^+, or c^+. The population of progeny phages (the phage lysate; Figure 6.13[4]), can now be used to infect a new population of bacteria (Figure 6.13[5]). The recipient bacteria are *a* in genotype. If the recipient receives a transducing phage carrying the a^+ gene, genetic exchange of the donor a^+ gene with the recipient *a* gene can occur by double crossing-over (Figure 6.13[6]). The result is stable transduced bacteria called *transductants*; in this case they are a^+ transductants (Figure 6.13[7]).

The process just described is called *generalized transduction* because the piece of bacterial DNA that the phage erroneously picks up and shuttles to a recipient bacterium is a random piece of the fragmented bacterial chromosome. Thus any genes can be transduced: All that are needed are a phage that packages random headfuls of DNA and bacterial strains carrying different genetic markers. Just as for transformation, generalized transduction may be used to determine gene order and to map the distance between genes. The logic is identical for the two processes. In a typical transduction experiment, transductants for one of two or more donor markers are selected and those transductants are then analyzed for the presence or absence of other donor markers.

Consider a donor bacterium with genotype $a^+ b^+$ and a recipient bacterium with genotype $a\ b$. If we do a transduction experiment and select for a^+ transductants, we can then test for whether they are also recombinant for b^+. Since the production of a transducing phage is a rare event, and since the amount of bacterial DNA the phage can carry is limited by the capacity of the phage head, only if the genes are close enough to be included in the transducing phage can an $a^+ b^+$ transductant—called a *cotransductant*—be produced. For illustration, Figure 6.14 shows the mechanisms by which different transductants are produced when donor $a^+ b^+$ DNA is introduced into an $a\ b$ recipient. The assumption for the figure is that genes a and b are closely linked such that both can be carried on the same transducing DNA fragment. Transductants of genotype $a^+ b$ are produced by a double crossover between the transducing DNA fragment and the recipient chromosome in which one exchange is to the left of the a locus and the other is between the a and b loci (Figure 6.14[a]). (Transductants of genotype $a^+ b$ can also result from a situation in which the transducing DNA fragment carries the a locus but not the b locus.) Transductants of genotype $a^+ b^+$ are produced by a double crossover in which one exchange is to the left of the a locus and the other is to the right of the b locus (Figure 6.14[b]). Clearly, if the a and b loci are far enough apart such that they cannot be included on the same transducing DNA fragment, $a^+ b^+$ transductants will never result. Thus, **cotransduction** of two or more genes is a good indication that the genes are closely linked.

Gene order and map distances between cotransduced genes may be determined by using generalized transduction, and it is by this procedure that fine structure (i.e., detailed) linkage maps of bacterial chromosomes have been constructed. For example, consider the mapping of some *E. coli* genes by using P1 transduction. The donor *E. coli* strain is $leu^+ thr^+ azi^r$ (able to grow on a minimal medium

■ *Figure 6.14* Mechanisms by which different transductants are produced when donor $a^+ b^+$ DNA is introduced into an $a\ b$ recipient and a^+ transductants are selected: (a) Production of an $a^+ b$ transductant involves a crossover to the left of a and a crossover between a and b; (b) Production of an $a^+ b^+$ transductant involves a crossover to the left of a and a crossover to the right of b.

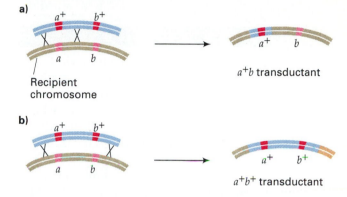

and resistant to the metabolic poison sodium azide). The recipient cell is *leu thr azi*ˢ (needs leucine and threonine as supplements in the medium and sensitive to azide). The P1 phages are grown on the bacterial donor cells, and the phage lysate is used to transduce the recipient bacterial cells. Transductants are selected for any one of the donor genes and then analyzed for the presence of the other two unselected donor genes. For instance, the donor gene *leu⁺* is selected by plating the cells on a medium lacking leucine: The resulting colonies are then plated on media lacking threonine, or containing the metabolic poison sodium azide to test for *thr⁺* and *azi*ʳ, respectively. Typical data from such an experiment are shown in Table 6.1.

Consider the *leu⁺* selected transductants. Of these transductants 50 percent also contain *azi*ʳ and 2 percent contain *thr⁺*. For the *thr⁺* transductants 3 percent are *leu⁺* and 0 percent are *azi*ʳ. The simplest interpretation is that the *leu* gene is closer to the *thr* gene than is the *azi* gene. So the order of genes is

Table 6.1

Transduction Data for Deducing Gene Order

SELECTED GENE	UNSELECTED GENES
leu⁺	50% = *azi*ʳ
	2% = *thr⁺*
thr⁺	3% = *leu⁺*
	0% = *azi*ʳ

Map distance can then be obtained for the genes in this transduction experiment by determining relative cotransduction frequencies similar to the way map distance was determined for *a* and *b* on page 141. By extension, similar analyses can be done with data from transduction experiments involving three or more genes.

Keynote

Transduction is the process by which bacteriophages mediate the transfer of genetic information from one bacterium (the donor) to another (the recipient). The capacity of the phage particle is limited, so the amount of DNA transferred is usually less than 1 percent of that in the bacterial chromosome. In generalized transduction any bacterial gene can be accidentally incorporated into the transducing phage during the phage life cycle and subsequently transferred to a recipient bacterium. Since only short segments of bacterial DNA can be packaged into the phages, transduction can be used only to map small regions of the genome at a time.

Mapping Genes in Bacteriophages

The same principles used to map eukaryotic genes are used to map phage genes. Crosses are made between phage strains that differ in genetic markers, and the proportion of recombinants among the total progeny is determined. Thus the basic procedure of mapping genes in two-, three-, or four-gene crosses involves the mixed infection of bacteria with phages of different genotypes.

In phages this basic experimental design must be adapted to the phage life cycle. First, we must be able to count phage types. We do so by plating a mixture of phages and bacteria on a solid medium. Each phage infects a bacterium and goes through the lytic cycle. The released progeny phages infect neighboring bacteria, and the lytic cycle is repeated. The result is a cleared patch in the lawn of bacteria. Each clearing is called a **plaque**, and each plaque derives from one of the original bacteriophages that was plated (Figure 6.15).

Second, we must be able to distinguish phage phenotypes easily. Since individual phages are visible only under the electron microscope, mutants that affect phage morphology cannot be used. However, several mutants affect the phage life cycle, giving rise to differences in the appearances of plaques on a bacterial lawn. For example, there are strains of T2 that differ in two phenotypes: plaque morphology and host range (i.e., which bac-

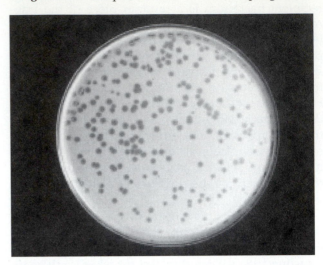

■ *Figure 6.15* Plaques of the *E. coli* bacteriophage T4.

terial strain the phage can lyse). Consider two phage strains, one with the genotype $h^+ r$, and the other with the genotype $h r^+$. The host range gene h^+ enables the phage to produce progeny only in the *B* strain of *E. coli*, while the allele *h* enables the phage to produce progeny in both the *B* and the *B/2* strains of *E. coli*. The plaque morphology gene r^+ produces small plaques with fuzzy borders, while its allele *r* produces large plaques with distinct borders.

To map these two genes, we must make a genetic cross. The cross is done by adding both types of phages ($h^+ r$ and $h r^+$) simultaneously (standardly, five copies of each phage per bacterial cell) to a culture of bacteria (Figure 6.16[a]). Each will replicate in the same bacterial cell (Figure 6.16[b]). If an $h^+ r$ and an $h r^+$ chromosome pair during that process, a crossover can occur between the two gene loci to produce $h^+ r^+$ and $h r$ recombinant chromosomes (Figure 6.16[c]), which are then assembled into progeny phages. When the bacterium lyses, the recombinant progeny are released into the medium, along with nonrecombinant (parental) phages (Figure 6.16[d]).

After the life cycle is completed, the progeny phages are plated onto a bacterial lawn containing a mixture of *E. coli* strains *B* and *B/2*. Four plaque phenotypes are found from this experiment (Figure 6.17): two parental types and two recombinant types. The parental type $h r^+$ gives a clear and small plaque; the other parental $h^+ r$ gives a fuzzy and large plaque. The reciprocal recombinant types give recombined phenotypes: The $h^+ r^+$ plaques are fuzzy and small, and the $h r$ plaques are clear and large. Once the progeny plaques are counted, we can calculate the recombination frequency for

■ *Figure 6.16* Schematic of the principles of performing a genetic cross with bacteriophages: (a) Bacteria are coinfected with the two parental bacteriophages, $h^+ r$ and $h r^+$; (b) Replication of both parental chromosomes; (c) There is pairing of some chromosomes of each parental type, and crossing-over takes place between the two gene loci to produce $h^+ r^+$ and $h r$ recombinants; (d) Progeny phages are assembled and are released into the medium when the bacteria lyse; both parental and recombinant phages are found among the progeny.

a) Coinfect bacteria with the two parental phages, $h^+ r$ and $h r^+$

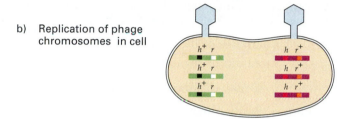

b) Replication of phage chromosomes in cell

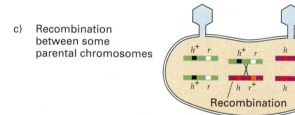

c) Recombination between some parental chromosomes

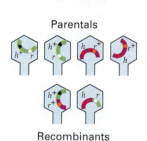

Recombination

d) Phage assembly, bacterial lysis, and release of progeny phages

Parentals

Recombinants

recombination events occurring between h and r. In this case the frequency is given by:

$$\frac{(h^+ r^+) + (h r) \text{ plaques}}{\text{total plaques}} \times 100$$

As with mapping eukaryotic genes, the recombination frequency gives the map distance (in map units) between the phage genes. Since many matings can occur where only one parent is present, the distances are relative, and comparisons can be made only if the same standard conditions are used.

Fine-Structure Analysis of a Bacteriophage Gene

All of the genetic mapping experiments described in this chapter for both bacteria and bacteriophages, and in Chapter 5 for eukaryotes, involved the use of mutant alleles of different genes. The recombinational mapping of the distance between genes is called *intergenic mapping* ("inter" means between) and is used to construct chromosome maps for organisms. In Chapter 4 we learned that genes have multiple alleles, each allele corresponding to a different mutation within the gene. It is possible to map the distance between alleles of the same gene using the general principles of recombi-

■ *Figure 6.17* Plaques produced by progeny of a cross of T2 strains $h r^+ \times h^+ r$. Four plaque phenotypes, representing both parental types and the two recombinants, may be discerned. The parental $h r^+$ type produces clear, small plaques, and the parental $h^+ r$ type produces fuzzy, large plaques. The recombinant $h r$ type produces a clear, large plaque, and the recombinant $h^+ r^+$ type produces a fuzzy, small plaque.

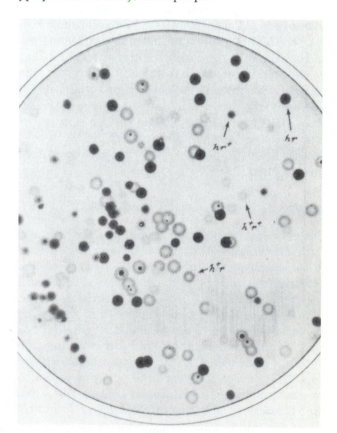

national mapping; this is called *intragenic mapping* ("intra" means within). Different alleles of one gene will obviously be very closely linked, genetically speaking. Constructing a detailed genetic map of alleles within a gene is called **fine-structure mapping**. Using particular *r* plaque morphology mutants of phage T4 that mapped to the *rII* locus, Seymour Benzer pioneered the fine-structure mapping of genes in the late 1950s and early 1960s. In this section we will see how Benzer's work helped define the *unit of recombination* (the smallest indivisible unit that could be distinguished by recombination) and the *unit of mutation* (the smallest indivisible unit that could be distinguished by mutation), concepts that were not clearly understood at the time of his work.

When mapping the distance between two alleles of the same gene, recombination frequencies will be very low. Benzer's experimental system was particularly suitable for this kind of analysis for two reasons. First, bacteriophages produce large numbers of progeny, making the measurement of very low recombination frequencies easy. Second, the *rII* mutants have a conditional lethal host range phenotype which, as we will see, enables extremely low recombination frequencies to be detected easily. That is, when cells of either *E. coli* strain *B* or strain *K12(λ)* are infected with *r*⁺ (wild-type) T4, small turbid plaques with fuzzy edges are produced. When cells of *E. coli* strain *B* are infected with *rII* mutants, large, clear plaques with distinct edges are produced. However, *rII* mutants are unable to grow in cells of *K12(λ)* so no plaques are produced. This is the conditional lethal phenotype because the phages can grow under one set of conditions but not another. Strain *B* is defined as the *permissive host* for *rII* mutants (allows propagation of *rII* mutant phages), while strain *K12(λ)* is the *nonpermissive host* (does not allow propagation of *rII* mutant phages).

Recombination Analysis of *rII* Mutants

Benzer's procedure for mapping the distance between *rII* alleles, that is, between different *rII* mutations, is shown in Figure 6.18. Two different *rII* mutants (*rII*1 and *rII*2 in the example) are crossed in the permissive host *E. coli B*, and progeny phages are collected. Four genetic classes of progeny were usually found: the two parental *rII* types (*rII*1 and *rII*2), and the two recombinant types resulting from a single crossover. One of the recombinant types was the wild type (*r*⁺), and the other was a double mutant carrying both *rII* alleles (*rII*1,2), which had an *r* phenotype. Map distance between *rII*1 and *rII*2

is calculated by determining the percentage of recombinants in the progeny:

$$\frac{\text{number of } r^+ \text{ progeny} + \text{number of } rII1,2 \text{ progeny}}{\text{total number of progeny}} \times 100\%$$

$$= \text{map distance (in map units)}$$

To count the total number of progeny, the progeny phages are plated on *E. coli B*, the permissive host, and the number of plaques counted. To count the *r*⁺ recombinants, the progeny phages are plated on the nonpermissive host *E. coli K12(λ)* because only *r*⁺ phages can grow on that strain. Since a single crossover between the two *rII* mutations gives *both* the *r*⁺ and double *rII* mutant recombinants in equal frequencies, the total number of recombinants from a cross between two *rII* mutants is *twice* the number of *r*⁺ plaques counted on plates of strain *K12(λ)*. The general formula for the map distance between two *rII* mutants then becomes:

$$\frac{2 \times \text{number of } r^+ \text{ recombinants}}{\text{total number of progeny}} \times 100\%$$

$$= \text{map distance (in map units)}$$

Benzer showed that the lowest frequency with which *r*⁺ recombinants were formed in any pairwise crosses of *rII* mutants was 0.01 percent. The minimum map distance figure of 0.01 percent can be used to make a rough calculation of the molecular distance—the distance in base pairs of DNA—involved in the recombination event. The T4 genetic map is about 1500 map units. If two *rII* mutants produce 0.01 percent *r*⁺ recombinants, the mutations are separated by 0.02 map units, or by about $0.02/1500 = 1.3 \times 10^{-5}$ of the total T4 genome. Since the total T4 genome contains about 2×10^5 base pairs, the minimum recombination distance is $(1.3 \times 10^{-5}) \times (2 \times 10^5)$, or about 3 base pairs. That means Benzer's data had shown that genetic recombination can occur in distances of 3 base pairs or less. Later experiments by other scientists showed conclusively that recombination can occur between mutations that affect adjacent base pairs in the DNA. In other words, genetic experiments had conclusively proved that the base pair is both the *unit of mutation* and the *unit of recombination*.

Deletion Mapping

In his initial series of crossing experiments, Benzer used *rII* **point mutants** for his fine-structure mapping. Point mutations are mutations that result from changes in individual base pairs in the DNA. To map the 3000 point mutations he had in the way we

■ *Figure 6.18*   Benzer's general procedure for determining the number of r^+ recombinants from a cross involving two *rII* mutants of T4.

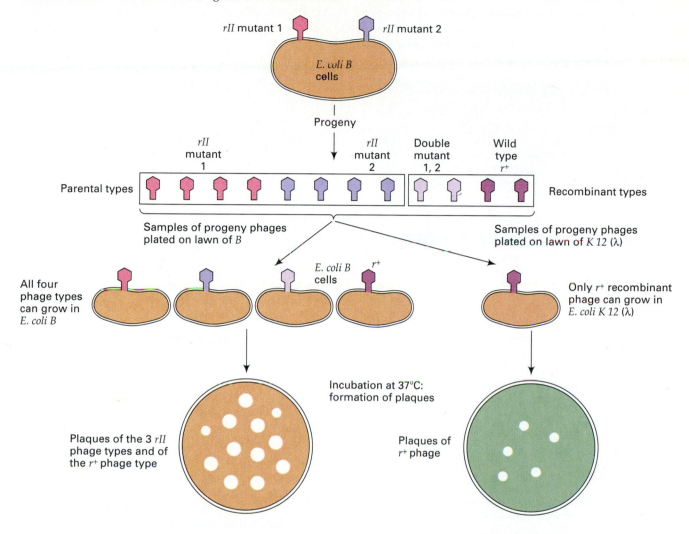

have just described would have been a monumental task. Therefore Benzer developed the technique of *deletion mapping* to localize unknown mutations. Deletion mapping involved *rII* mutants that had arisen, not by a change of one base pair to another, but by the loss of a segment of DNA; these mutants are called **deletion mutants**.

The principle of deletion mapping for localizing point mutations to segments of the *rII* region of phage T4 is illustrated in Figure 6.19. Wild type (r^+) has a complete *rII* region, whereas the *rII* deletion tester strains each have a different section of the *rII* region deleted (shown in brown in the figure). The deletion mutants overlap in the extent of the *rII* region lost. Thus deletion mutant *rG* has all of the *rII* region deleted, whereas deletion mutant *rH* has all but a small region of *rII* deleted; the region retained is designated *A1*. In all, the set of seven

deletion mutants in Figure 6.19 define seven segments of the *rII* region. Those segments are designated *A1* to *A6* and *B*. Point mutations can easily be localized to one of the seven segments as follows: Each point mutant is crossed with each of the seven deletion mutants by coinfecting bacteria, and progeny phages are tested for the presence of r^+ recombinants. The r^+ recombinants cannot be produced in crosses with deletions if the segment deleted in the deletion mutant includes the region of DNA containing the point mutation.

Consider the point mutants *1* through *7* in Figure 6.19. Point mutant *1* cannot give an r^+ recombinant when crossed with *rG* because the deleted segment of *rG* includes segment *A1* in which point mutant *1* is located. However, r^+ recombinants can be produced from crosses of point mutant *1* with each of the other six deletion mutants because the various

■ *Figure 6.19*   Principle of deletion mapping for localizing point mutations to segments of the *rII* region of phage T4. Deletion mutants divide the *rII* region into seven distinct segments. By crossing point mutants (bottom of figure) with each deletion mutant and examining the pattern of r^+ recombinants produced, each point mutation can be localized unambiguously to one of the segments (see text).

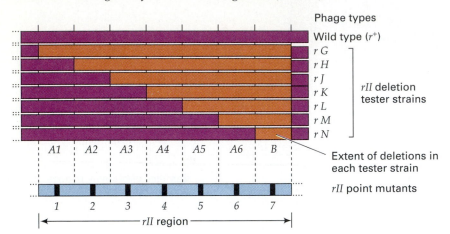

segments deleted in those strains do *not* overlap the segment in which point mutant *1* is located. Similarly, point mutant *2* will not give r^+ recombinants with either deletion *rG* or *rH*, but it will with the other five deletions. In actual practice the analysis is done in reverse: The pattern of r^+ recombinants is used to indicate the segment of the *rII* region in which the point mutation is located. Thus a point mutant that does not give r^+ recombinants with deletions *rG*, *rH*, *rJ*, *rK*, and *rL*, but does with *rM* and *rN*, indicates that the point mutation must be in the *A5* region. Using this deletion mapping approach, Benzer was able to localize any given *rII* point mutant to one of 47 regions defined by the deletions he had. Then all the point mutants within each segment could be crossed in all possible pairwise crosses to construct a detailed fine-structure genetic map. From a map of many hundreds of mutants, Benzer was able to conclude that the distribution of mutated sites was not random; some sites had no mutations while other sites, called *hot spots*, were represented by a large number of independently isolated point mutants. He concluded that hot spots were sites that were more likely to be mutated than the rest of the gene; that is, the sites were highly mutable.

Keynote

As a result of fine-structure analysis of the *rII* region of bacteriophage T4 and other experiments, it was determined that the unit of mutation and of recombination is the base pair in DNA.

Defining Genes by Complementation Tests

From the classical point of view the gene is a *unit of function*: each gene specifies one function. However, the same mutant phenotype can result if a number of different genes are mutated, as we saw in our discussion of epistasis in Chapter 4 (pp. 80–84). When a geneticist isolates mutants, he or she categorizes the mutants based on the mutant phenotypes observed. How can the number of genes associated that express the same mutant phenotypes be determined? Typically this is done by using the **complementation test.** We will describe this test in the context of Benzer's work to determine the number of genes in the *rII* region. To make it easier to understand the complementation test, it will help to know at the outset that the *rII* region actually consists of two genes, *rIIA* and *rIIB*. A mutation at any point in either gene will produce the mutant *rII* phenotypes. A functional product of the *rIIA* gene and a functional product of the *rIIB* gene are *both* needed for the phage to reproduce in *E. coli* K12(λ).

In Benzer's work with the *rII* mutants, the nonpermissive strain K12(λ) was infected with a pair of *rII* mutant phages to see whether the two mutants, each unable by itself to grow in strain K12(λ), are able to "work together" to produce progeny phages. If the phages do produce progeny, the two mutants are said to complement each other, meaning that the two mutations must be in different genes (units of function) that specify different functional products. The two products work together to allow progeny to be produced. If no progeny phages are produced, the mutants have not complemented,

■ *Figure 6.20*   Complementation tests for determining the units of function in the *rII* region of phage T4; the nonpermissive host *E. coli K12(λ)* is infected with two different *rII* mutants: (a) Complementation occurs; (b) Complementation does not occur.

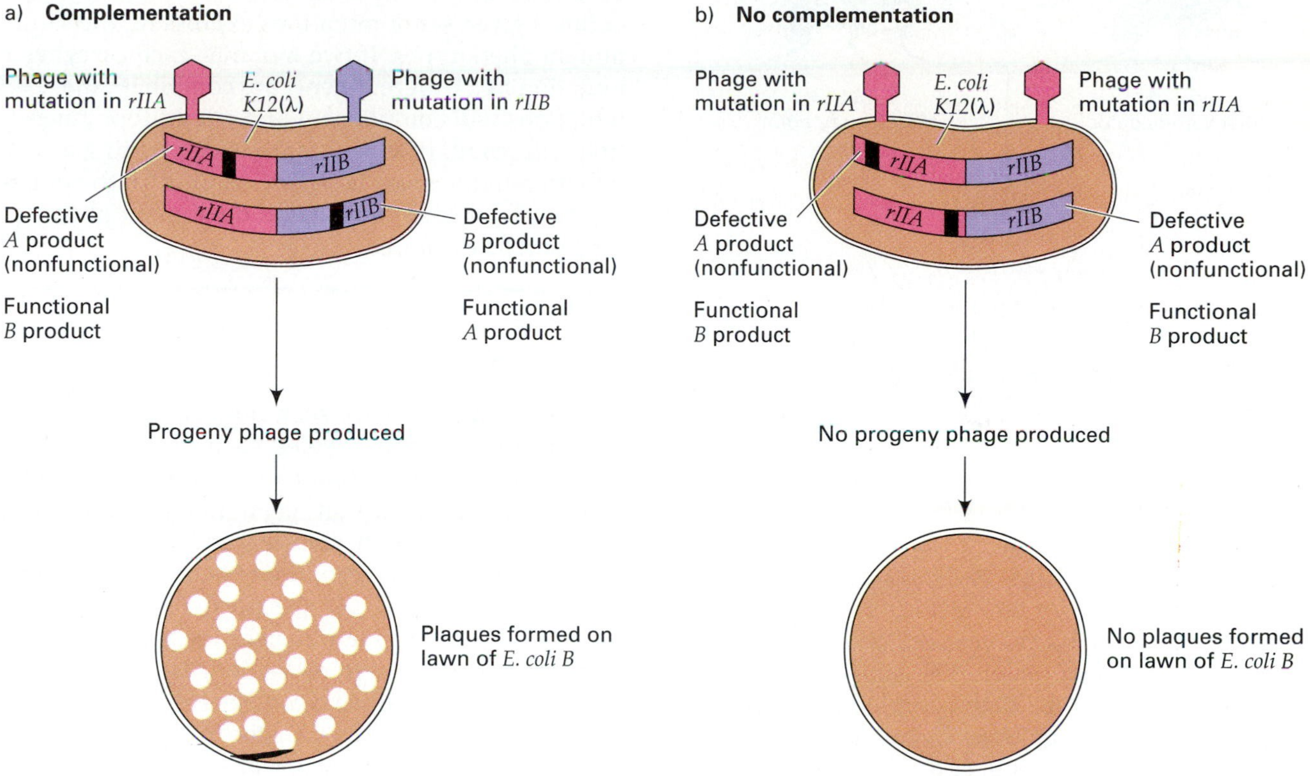

a) **Complementation**

Phage with mutation in *rIIA*
E. coli K12(λ)
Phage with mutation in *rIIB*

rIIA   *rIIB*

rIIA   *rIIB*

Defective A product (nonfunctional)

Functional B product

Defective B product (nonfunctional)

Functional A product

Progeny phage produced

Plaques formed on lawn of *E. coli B*

b) **No complementation**

Phage with mutation in *rIIA*
E. coli K12(λ)
Phage with mutation in *rIIA*

rIIA   *rIIB*

rIIA   *rIIB*

Defective A product (nonfunctional)

Functional B product

Defective A product (nonfunctional)

Functional B product

No progeny phage produced

No plaques formed on lawn of *E. coli B*

indicating that the mutations are in the same functional unit. Here both mutants produce the same defective product, so the phage life cycle cannot proceed, and no progeny phages result. (Note that genetic recombination is not necessary for complementation to occur.)

These two situations are diagrammed in Figure 6.20. In the first case the bacterium is infected with two phage genomes, one with a mutation in the *rIIA* gene and the other with a mutation in the *rIIB* gene (Figure 6.20a). Complementation occurs because the *rIIA* mutant still makes a functional *B* product, and the *rIIB* mutant makes a functional *A* product. As a result, two mutants work together to make both products necessary for phage growth in *E. coli K12(λ)*, and progeny phages are made. In the second case the bacterium is infected with two phage genomes, each with a different mutation in the same gene, *rIIA* (Figure 6.20[b]). Here, no complementation occurs because, while both produce a functional *rIIB* product, the two mutants do not both make functional *a* product, so the *A* function cannot take place. Consequently, phage reproduction in *E. coli K12(λ)* does not occur.

On the basis of the results of complementation tests, Benzer found that *rII* mutants fall into two

units of function, *rIIA* and *rIIB* (also called complementation groups, which in this case directly correspond to genes). All *rIIA* mutants complement all *rIIB* mutants, but *rIIA* mutants fail to complement other *rIIA* mutants, and *rIIB* mutants fail to complement other *rIIB* mutants.

Because complementation tests typically compare the effects of two mutations located on two different chromosomes (the *trans* configuration) with the effects of two mutations located on the same chromosome (the *cis* configuration), the test is also called the *cis-trans* test. (The complementation tests in Figure 6.20 involved mutations in the *trans* configuration; i.e., on two different chromosomes.) Because of this name for the test, Benzer called the genetic unit of function revealed by the cis-trans test the *cistron*. At the present time the term *gene* is more commonly used than *cistron*.

The complementation test is used to define the functional units (complementation groups or genes) for mutants with the same phenotype. The principles for performing a complementation test are always the same; only the practical details of performing the test are organism-specific. Let us consider an example of complementation in *Drosophila* (Figure 6.21; taken from Question 4.21). Two true-

■ *Figure 6.21*   Complementation between two black body mutations of *Drosophila melanogaster*.

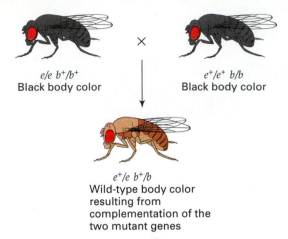

e/e b⁺/b⁺
Black body color

×

e⁺/e⁺ b/b
Black body color

e⁺/e b⁺/b
Wild-type body color resulting from complementation of the two mutant genes

breeding mutant strains of *Drosophila* have black body color instead of the wild-type grey-yellow. When the two strains are crossed, all of the F_1 flies have wild-type body color. The simplest explanation of these data is that complementation has occurred between mutations in two genes, each of which is involved in the body color phenotype. That is, a recessive autosomal gene, *ebony* (*e*), when homozygous, produces a black body color. On a different autosome a different recessive gene, *black* (*b*), also produces a black body color when homozygous. Since the two parents are homozygotes, they are genotypically *e/e b⁺/b⁺* and *e⁺/e⁺ b/b*, and each is phenotypically black. The F_1 genotype is *e⁺/e b⁺/b*, which is equivalent to the *trans* configuration of the *rII* cistron experiments. The F_1 have wild-type body color because complementation has occurred. Even without knowing in advance that two independently assorting genes were involved, we would conclude that two genes are involved. As for the phage experiments, no recombination is involved in the complementation of body color genes; the double heterozygote was set up simply by the fusion of gametes produced by the two true-breeding parents.

What if the F_1 from the cross between two independently isolated, true-breeding recessive black mutant strains were all phenotypically black? The interpretation would be that since the mutant phenotype was exhibited, the two mutations involved did not show complementation, and therefore the mutations are in the same complementation group; that is, they are different mutant alleles of the same gene.

Keynote

The complementation, or *cis-trans*, test is used to determine how many units of function (genes) define a given set of mutations expressing the same mutant phenotypes. If two mutants, each carrying a mutation in a different gene, are combined, the mutations will complement and a wild-type function will result. If two mutants, each carrying a mutation in the same gene, are combined, the mutations will not complement and the mutant phenotype will be exhibited.

Summary

In this chapter we have seen how genetic mapping can be done in prokaryotes such as bacteria and bacteriophages. Essentially the same experimental strategy is used for all gene mapping: Genetic material is exchanged between strains differing in genetic markers and recombinants are detected and counted. Depending on the particular bacterium, the mechanism of gene transfer may be transformation, conjugation, or transduction. In each process there is a donor strain and a recipient strain.

Transformation is the transfer of genetic material between organisms by means of extracellular pieces of DNA. Conjugation is a plasmid-mediated process in which there is a unidirectional transfer of genetic information through direct cellular contact between a donor and a recipient bacterial cell. Transduction is a process whereby bacteriophages mediate the transfer of bacterial DNA from the donor bacterium to the recipient. Bacteriophage chromosomes may be mapped by infecting bacteria simultaneously with two phage strains and analyzing the resulting phage progeny for parental and recombinant phenotypes. This method of mapping is the same as that used for mapping genes in eukaryotes. As a summary, the linkage map of *E. coli* is shown in Figure 6.22.

Insights into the relationships between mapping and gene structure were obtained from a fine-structure analysis of the bacteriophage T4 *rII* region. The mutational sites within a gene were mapped through intragenic mapping. The resulting map indicated that the unit of mutation and the unit of recombination is the base pair in DNA, replacing the classical definition that genes were indivisible by mutation and recombination.

The method of defining the number of units of function (genes) by complementation tests was discussed. Given a set of mutations expressing the same mutant phenotypes, two mutants are com-

■ *Figure 6.22*   Genetic map of *E. coli* based on conjugation experiments. Units are in minutes timed from an arbitrary origin at 12 o'clock.

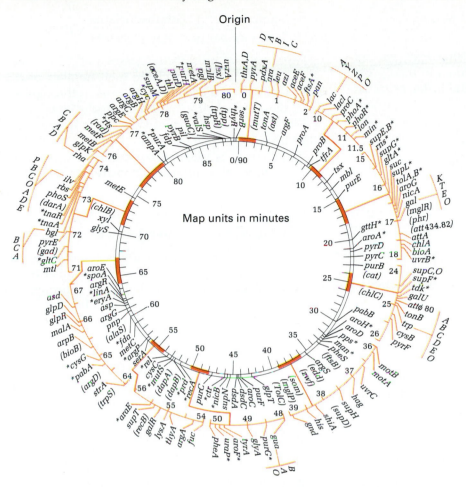

bined and the phenotype is determined. If the phenotype is wild type, the two mutations have complemented and must be in different units of function. If the phenotype is mutant, the two mutations have not complemented and must be in the same unit of function.

Analytical Approaches for Solving Genetics Problems

Q6.1   In *E. coli* the following *Hfr* strains donate the markers shown in the order given:

HFR STRAIN	ORDER OF GENE TRANSFER
1	G E B D N A
2	P Y L G E B
3	X T J F P Y
4	B E G L Y P

All the *Hfr* strains were derived from the same *F⁺*

strain. What is the order of genes in the original *F⁺* chromosome?

A6.1   This question is an exercise in piecing together various segments of the circumference of a circle. The best approach is to draw a circle and label it with the genes transferred from one *Hfr* and then to see which of the other *Hfr*'s transfers an overlapping set. For example, *Hfr 1* transfers *E*, then *B*, then *D*, and so on; and *Hfr 4* transfers *B*, then *E*, and so forth. Now we can juxtapose the two sets of genes transferred by the two *Hfr*'s and deduce that the polarities of transfer are opposite:

$$Hfr\ 1 \qquad G\ E\ B\ D\ N\ A$$
$$Hfr\ 4 \qquad P\ Y\ L\ G\ E\ B$$

Extending this reasoning to the other *Hfr*s, we can draw an unambiguous map as shown in the figure, with the arrowheads indicating the order of transfer.

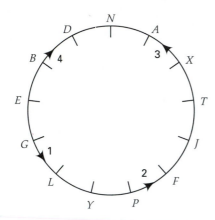

The same logic would be used if the question gave the relative time units of entry of each of the genes. In that case we would expect that the time "distance" between any two genes would be approximately the same regardless of the order of transfer or how far the genes were from the origin.

Q6.2   In a transduction experiment, the donor was $c^+\ d^+\ e^+$ and the recipient was $c\ d\ e$. Selection was for c^+. The four classes of transductants from this experiment were:

CLASS	GENETIC COMPOSITION	NUMBER OF INDIVIDUALS
1	$c^+\ d^+\ e^+$	57
2	$c^+\ d^+\ e$	76
3	$c^+\ d\ e$	365
4	$c^+\ d\ e^+$	2
	Total	500

a. Determine the cotransduction frequency for c^+ and d^+.

b. Determine the cotransduction frequency for c^+ and e^+.

c. Which of the cotransduction frequencies calculated in (a) and (b) represents the greater actual distance between genes? Why?

A6.2   **a.** The frequency with which d^+ is cotransduced with the c^+ gene is calculated using values for the total number of c^+ transductants, and the number of transductants for both c^+ and d^+. The

formula for the cotransduction frequency for c^+ and d^+ is:

$$\frac{\text{Number of } c^+\ d^+ \text{ cotransductants}}{\text{Total number of } c^+ \text{ transductants}} \times 100\%$$

From the data presented, Classes 1 and 2 are the $c^+\ d^+$ cotransductants, and the total number of c^+ transductants is the sum of Classes 1 through 4. Thus the number of c^+ and d^+ transductants is 57 + 76 = 133, and the cotransductant frequency is $133/500 \times 100 = 26.6$ percent.

b. The analysis is identical in approach to (a). The formula for the cotransduction frequency for c^+ and e^+ is:

$$\frac{\text{Number of } c^+\ e^+ \text{ cotransductants}}{\text{Total number of } c^+ \text{ transductants}} \times 100\%$$

From the data presented, Classes 1 and 4 are the $c^+\ e^+$ cotransductants, and the total number of c^+ transductants is the sum of Classes 1 through 4. Thus the number of c^+ and e^+ transductants is 57 + 2 = 59, and the cotransductant frequency is $59/500 \times 100 = 11.8$ percent.

c. The greater actual distance is for the c^+ and e^+ genes. The principle involved is that the closer two genes are on the chromosome, the greater the chance that they will be cotransduced. Thus as the distance between genes increases, concomitantly the cotransduction frequency decreases. Since the $c^+\ e^+$ cotransduction frequency is 11.8 percent and the $c^+\ d^+$ cotransduction frequency is 26.6 percent, genes c^+ and e^+ are farther apart than genes c^+ and d^+.

Q6.3   Five different *rII* deletion strains of phage T4 were tested for recombination by pairwise crossing in *E. coli* B. The following results were obtained, where $+ = r^+$ recombinants produced, and $0 =$ no r^+ recombinants produced:

	A	B	C	D	E
E	0	+	0	+	0
D	0	0	0	0	
C	0	0	0		
B	+	0			
A	0				

Draw a deletion map compatible with these data.

A6.3   The principle here is that if two deletion mutations overlap, then no r^+ recombinants can be produced. Conversely, if two deletion mutations do not overlap, then r^+ recombinants can be produced. To approach a question of this kind, we must draw overlapping and nonoverlapping lines from the given data.

Starting with A and B, these two deletions do not overlap since r^+ recombinants are produced. Therefore these two mutations can be represented as follows:

A   B

_____   _____

The next deletion, C, does not give r^+ recombinants with any of the other four deletions. We must conclude, therefore, that C is an extensive deletion that overlaps all of the other four, with endpoints that cannot be determined from the data given. One possibility is as follows:

C

_____

A   B

_____   _____

Deletion D does not give r^+ recombinants with A, B, or C, but it does with E. In turn, E gives r^+ recombinants with B and D, but not with A or C. Thus D must overlap both A and B but not E, and E must overlap A and C but not B. A compatible map for this situation follows. Other maps can be drawn in terms of the endpoints of the deletions.

C

_____

A   B

_____   _____

E   D

_____   _____

Q6.4 Seven different *rII* point mutants (*1* to *7*) of phage T4 were tested for recombination crosses in *E. coli* B with the five deletion strains described in question 6.3. The following results were obtained, where $+ = r^+$ recombinants produced, and $0 = $ no r^+ recombinants produced:

	A	B	C	D	E
1	0	+	0	+	+
2	+	0	0	+	+
3	0	+	0	+	0
4	+	+	0	+	0
5	+	0	0	0	+
6	0	+	0	0	+
7	+	+	0	0	+

In which regions of the map can you place the seven point mutations?

A6.4 If an r^+ recombinant is produced, the *rII* point mutation must be in the region covered by the deletion mutation with which it was crossed. Thus the matrix of results localizes the point mutations to the regions defined by the deletion mutants. The results potentially define the relative extent of deletion overlap. For example, point mutation 7 gives r^+ recombinants with A, B, and E but not with D. Logically, then, 7 is located in the region defined by the part of deletion D that is not involved in the overlap with A and B. Similarly, point mutation 4 gives r^+ recombinants with A, D, and B but not with E. Thus 4 must be in a region defined by a segment of deletion E that does not overlap deletion A. Furthermore, since 4 does not give r^+ recombinants with C, deletion C must overlap the site defined by point mutation 4. This result, then, refines the deletion map with regard to the E, C, and A endpoints. The map we can draw from the matrix of results is as follows:

C

_____

A   B

_____   _____

E   D

_____   _____

4   3   1   6   7   5   2
├───┼───┼───┼───┼───┼────────┼

Questions and Problems

***6.1** If an *E. coli* auxotroph *A* only grows on a medium containing thymine, and an auxotroph *B* only grows on one containing leucine, how would you test whether DNA from *A* could transform *B*?

***6.2** With the technique of interrupted mating, four *Hfr* strains were tested for the sequence in which they transmitted a number of different genes to an *F*⁻ strain. Each *Hfr* strain was found to transmit its genes in a unique sequence, as shown in the accompanying table (only the first six genes transmitted were scored for each strain).

| ORDER OF | Hfr STRAIN | | | |
TRANSMISSION	1	2	3	4
First	O	R	E	O
	F	H	M	G
	B	M	H	X
	A	E	R	C
	E	A	C	R
Last	M	B	X	H

What is the gene sequence in the original strain from which these *Hfr* strains derive? Indicate on your diagram the origin and polarity of each of the four *Hfr*'s.

6.3 At time zero an *Hfr* strain (strain 1) was mixed with an F^- strain, and at various times after mixing, samples were removed and agitated to separate conjugating cells. The cross may be written as:

$$Hfr\ 1:\ a^+\ b^+\ c^+\ d^+\ e^+\ f^+\ g^+\ h^+\ str^s$$

$$F^-:\ a\ \ b\ \ c\ \ d\ \ e\ \ f\ \ g\ \ h\ \ str^r$$

(No order is implied in listing the markers.)
The samples were then plated onto selective media to measure the frequency of $h^+\ str^r$ recombinants that had received certain genes from the *Hfr* cell. The graph of the number of recombinants against time is shown in the following figure:

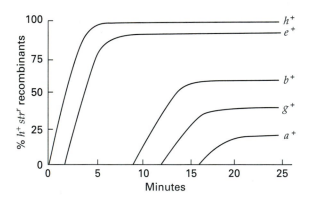

a. Indicate whether each of the following statements is true or false.
  i. All F^- cells which received a^+ from the *Hfr* in the chromosome transfer process must also have received b^+.
  ii. The order of gene transfer from *Hfr* to F^- was a^+ (first), then g^+, then b^+, then e^+, then h^+.
  iii. Most $e^+\ str^r$ recombinants are likely to be *Hfr* cells.
  iv. None of the $b^+\ str^r$ recombinants plated at 15 minutes are also a^+.
b. Draw a linear map of the *Hfr* chromosome indicating the
  i. point of insertion, or origin;

ii. the order of the genes a^+, b^+, e^+, g^+, and h^+;
iii. the shortest distance between consecutive genes on the chromosomes.

6.4 Consider the following transduction data:

DONOR	RECIPIENT	SELECTED MARKER	UNSELECTED MARKER	%
aceF⁺ dhl	*aceF dhl⁺*	*aceF⁺*	*dhl*	88
aceF⁺ leu	*aceF leu⁺*	*aceF⁺*	*leu*	34

Is *dhl* or *leu* closer to *aceF*?

*6.5 Consider the following P1 transduction data:

DONOR	RECIPIENT	SELECTED MARKER	UNSELECTED MARKER	%
cysB⁺ trpE	*cysB trpE⁺*	*cysB⁺*	*trpE*	37
cysB⁺ trpB	*cysB trpB⁺*	*cysB⁺*	*trpB*	53

Is *trpE* or *trpB* closer to *cysB*?

6.6 Order the mutants *trp*, *pyrF*, and *qts* on the basis of the following three-factor transduction cross:

Donor	*trp⁺ pyr⁺ qts*
Recipient	*trp pyr qts⁺*
Selected Marker	*trp⁺*

UNSELECTED MARKERS	NUMBER
pyr⁺ qts⁺	22
pyr⁺ qts	10
pyr qts⁺	68
pyr qts	0

*6.7 A stock of T4 phage is diluted by a factor of 10^{-8} and 0.1 mL of it is mixed with 0.1 mL of 10^8 *E. coli B*/mL and 2.5 mL melted agar, and poured on the surface of an agar Petri dish. The next day 20 plaques are visible. What is the concentration of T4 phages in the original T4 stock?

*6.8 Wild-type phage T4 grows on both *E. coli B* and *E. coli K12*(λ), producing turbid plaques. The *rII* mutants of T4 grow on *E. coli B*, producing clear plaques, but do not grow on *E. coli K12*(λ). This host range property permits the detection of a very low number of r^+ phages among a large number of *rII* phages. With this sensitive system it is possible to determine the genetic distance between two mutations within the same gene, in this case the *rII* locus. Suppose *E. coli B* is mixedly infected with *rIIx* and *rIIy*, two separate mutants in the *rII* locus. Suitable dilutions of progeny phages are plated on *E. coli B* and *E. coli K12*(λ). A 0.1-mL sample of a thousandfold dilution plated on *E. coli B* showed 672 plaques. A 0.2-mL sample of undiluted phage plated on *E. coli K12*(λ) showed 470 turbid plaques. What is the genetic distance between the two *rII* mutations?

6.9 Construct a map from the following two-factor phage cross data (show map distance):

CLASS	% RECOMBINATION
$r_1 \times r_2$	0.10
$r_1 \times r_3$	0.05
$r_1 \times r_4$	0.19
$r_2 \times r_3$	0.15
$r_2 \times r_4$	0.10
$r_3 \times r_4$	0.23

*6.10 The following two-factor crosses were made to analyze the genetic linkage between four genes in phage λ: *c, mi, s,* and *co*.

PARENTS	PROGENY
c + × + mi	1213 c +, 1205 + mi, 84 + +, 75 c mi
c + × + s	566 c +, 808 + s, 19 + +, 20 c s
co + × + mi	5162 co +, 6510 + mi, 311 + +, 341 co mi
mi + × + s	502 mi +, 647 + s, 65 + +, 56 mi s

Construct a genetic map of the four genes.

6.11 Three gene loci in T4 that affect plaque morphology in easily distinguishable ways are *r* (rapid lysis), *m* (minute), and *tu* (turbid). A culture of *E. coli* is mixedly infected with two types of phage *r m tu* and $r^+ m^+ tu^+$. Progeny phage are collected and the following genotype classes are found:

CLASS	NUMBER
$r^+ m^+ tu^+$	3,729
$r^+ m^+ tu$	965
$r^+ m\ tu^+$	520
$r\ m^+ tu^+$	172
$r^+ m\ tu$	162
$r\ m^+ tu$	474
$r\ m\ tu^+$	853
$r\ m\ tu$	3,467
TOTAL	10,342

Construct a map of the three genes. What is the coefficient of coincidence, and what does the value suggest?

*6.12 The *rII* mutants of bacteria T4 grow in *E. coli B* but not in *E. coli K12*(λ) . The *E. coli* strain *B* is doubly infected with two *rII* mutants. A 6×10^7 dilution of the lysate is plated on *E. coli B*. A 2×10^5 dilution is plated on *E. coli K12*(λ). Twelve plaques appeared on strain *K12*(λ), and 16 on strain *B*. Calculate the amount of recombination between these two mutants.

6.13 Choose the correct answer in each case:
a. If one wants to know if two different *rII* point mutants lie at exactly the same site (nucleotide pair), one should:
  i. Coinfect *E. coli K12*(λ) with both mutants. If phage are produced, they lie at the same site.
  ii. Coinfect *E. coli K12*(λ) with both mutants. If phage are not produced, they lie at the same site.
  iii. Coinfect *E. coli B* with both mutants and plate the progeny phage on both *E. coli B* and *E. coli K12*(λ). If plaques appear on *B* but not *K12*(λ), they lie at the same site.
  iv. Coinfect *E. coli K12*(λ) with both mutants and plate the progeny phage on both *E. coli B* and *E. coli K12*(λ). If plaques appear on *K12*(λ) but not *B* they lie at the same site.
b. If one wants to know if two different *rII* point mutants lie in the same cistron, one should:
  i. Coinfect *E. coli K12*(λ) with both mutants. If phage are produced, they lie at the same cistron.
  ii. Coinfect *E. coli K12*(λ) with both mutants. If phage are not produced, they lie in the same cistron.
  iii. Coinfect *E. coli B* with both mutants and plate the progeny phage on both *E. coli B* and *E. coli K12*(λ). If plaques appear on *B* but not *K12*(λ), they lie in the same cistron.
  iv. Coinfect *E. coli K12*(λ) with both mutants and plate the progeny phage on both *E. coli B* and *E. coli K12*(λ). If plaques appear on *K12*(λ) but not *B* they lie in the same cistron.

6.14 A set of seven different *rII* deletion mutants of bacteriophage T4, *1* through *7*, were mapped, with the following result:

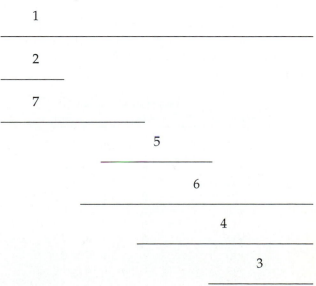

Five *rII* point mutants were crossed with each of the

deletions, with the following results, where + = r^+ recombinants were obtained, 0 = no r^+ recombinants were obtained:

POINT MUTANTS	DELETION MUTANTS						
	1	2	3	4	5	6	7
a	0	+	+	+	0	0	0
b	0	0	+	+	+	+	0
c	0	+	+	0	0	0	+
d	0	+	0	0	+	0	+
e	0	+	+	+	+	0	0

Map the locations of the point mutants.

*6.15 Given the following deletion map with deletions r31, r32, r33, r34, r35 and r36, place the point mutants r41, r42, and so, on the map. Be sure you show where they lie with respect to endpoints of the deletions.

POINT MUTANTS	DELETION MUTANTS (+ = r^+ RECOMBINANTS PRODUCED; 0 = No r^+ RECOMBINANTS PRODUCED)					
	r31	r32	r33	r34	r35	r36
r41	0	0	0	0	+	0
r42	0	0	0	+	0	+
r43	0	0	+	+	+	0
r44	0	0	0	0	+	+
r45	0	+	0	+	+	+
r46	0	0	+	0	+	0

Show the dividing line between the A cistron and the B cistron on your map above from the following data: [+ = growth on strain K12(λ), 0 = no growth on strain K12(λ)]:

MUTANT	COMPLEMENTATION WITH	
	rIIA	rIIB
r31	0	0
r32	0	0
r33	0	+
r34	0	0
r35	0	+
r36	0	0
r41	0	+
r42	0	+
r43	+	0
r44	0	+
r45	0	+
r46	0	+

6.16 Mutants in the ade2 gene of yeast require adenine and are pink because of the intracellular accumulation of a red pigment. Diploid strains were made by mating haploid mutant strains. The offspring exhibited the following phenotypes:

CROSS	DIPLOID PHENOTYPES
1 × 2	pink, adenine-requiring
1 × 3	white, prototrophic
1 × 4	white, prototrophic
3 × 4	pink, adenine-requiring

How many genes are defined by the four different mutants? Explain.

*6.17 In Drosophila mutants A, B, C, D, E, F, and G all have the same phenotype: the absence of red pigment in the eyes. In pairwise combinations in complementation tests, the following results were produced, where + = complementation and − = no complementation.

	A	B	C	D	E	F	G
G	+	−	+	+	+	+	−
F	−	+	+	−	+	−	
E	+	+	−	+	−		
D	−	+	+	−			
C	+	+	−				
B	+	−					
A	−						

a. How many genes are present?
b. Which mutants have defects in the same gene?

6.18 a. A homozygous white-eyed Martian fly (w_1/w_1) is crossed to a homozygous white-eyed fly from a different stock (w_2/w_2). It is well known that wild-type Martian flies have red eyes. This cross produces all white-eyed progeny. State whether the following is true or false. Explain your answer.
  i. w_1 and w_2 are allelic genes.
  ii. w_1 and w_2 are non-allelic.
  iii. w_1 and w_2 affect the same function.
  iv. The cross was a complementation test.
  v. The cross was a cis-trans test.
  vi. w_1 and w_2 are allelic by terms of the functional test.
b. The F_1 white-eyed flies were allowed to interbreed, and when you classified the F_2 you found 20,000 white-eyed flies and 10 red-eyed progeny. Concerned about contamination, you repeat the experiment and get exactly the same results. How can you best account for the presence of the red-eyed progeny? As part of your explanation give the genotypes of the F_1 and F_2 generation flies.

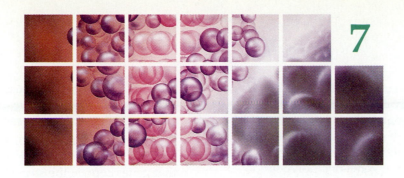

7

The Beginnings of Molecular Genetics: Gene Function

- There is a specific relationship between genes and enzymes, historically embodied in Beadle and Tatum's "one gene–one enzyme" hypothesis, which states that each gene determines the synthesis of a single functional enzyme. A more modern maxim is "one gene–one polypeptide."

- Many human genetic diseases are caused by deficiencies in enzyme activities. Most of these diseases are inherited as recessive traits.

- From the study of alterations in proteins other than enzymes, convincing evidence was obtained that genes control the structures of all proteins.

I N THE PRECEDING SIX CHAPTERS we have explored many aspects of genes. We learned that genes are defined by mutations, that genes segregate in genetic crosses according to how chromosomes segregate during meiosis, and that genes can be located or mapped relative to one another on chromosomes in both prokaryotes and eukaryotes by analyzing the results of appropriate genetic crosses. We also discovered that alleles of genes can have differing relationships, depending on the allele: most are recessive to the wild-type allele, some are dominant to the wild-type allele, and others are codominant or incompletely dominant. Throughout all of these discussions we considered the gene as an abstract entity, a factor that is located on a chromosome and that can give rise to a phenotype. We now know that a gene is a stretch of DNA in a chromosome that contains information for a specific product. This concept will be developed more fully in the following chapters. In fact the entire set of genes in the genome specifies a vast array of products that are responsible for all of a cell's and organism's structure and function. In the popular press the genetic information contained within an organism's genome is called the "blueprint for life." We now need to describe how a gene functions; that is, how the information in the DNA is used to specify a product and how that product is involved in the production of a phenotype.

In this chapter we will examine gene function: What do genes code for, and how do we know what they code for? We will present some of the classical evidence that genes code for enzymes and for other proteins. We examine the involvement of particular sets of genes in directing and controlling a particular biochemical pathway; that is, the series of enzyme-catalyzed steps required to break down or

synthesize a particular chemical compound. Instead of viewing the gene in isolation, as in the past few chapters, we will see that the gene must often work in cooperation with other genes in order for cells to function properly. The experiments we will discuss were the beginnings of molecular genetics, in that they were designed to understand a gene at the molecular level. In the following chapters the modern understanding of gene structure and function will be developed.

Gene Control of Enzyme Synthesis

Garrod's Hypothesis of Inborn Errors of Metabolism

In 1902 Archibald Garrod, an English physician, provided the first evidence of a specific relationship between genes and enzymes. Garrod studied *alkaptonuria*, a human disease characterized by urine that turns black upon exposure to the air and by the tendency to develop arthritis later in life.

In studying the occurrence of alkaptonuria in families of individuals with the disease, Garrod and his colleague William Bateson discovered two interesting facts that indicated that alkaptonuria is a genetically controlled trait: (1) Several members of the same families frequently had alkaptonuria, and (2) the disease was much more common among children of marriages involving first cousins than among children of marriages between unrelated partners. This finding was significant because first cousins have one-eighth of their genes in common and, therefore, the chance is greater for recessive alleles to become homozygous in children of first-cousin marriages.

Garrod found that people with alkaptonuria excrete in their urine all the homogentisic acid (HA) they produce, whereas normal people excrete none. Moreover, he showed that it is HA in the urine that turns black when exposed to air. This result indicated to Garrod that normal people are able to metabolize HA to its breakdown products but that people with alkaptonuria cannot. Thus alkaptonurics (individuals with alkaptonuria) lack the enzyme that metabolizes HA. Figure 7.1 shows part of the biochemical pathway in which HA is involved and the step that is blocked in alkaptonurics. From his data and the genetic evidence, Garrod concluded that alkaptonuria is a genetic disease caused by the absence of a particular enzyme required for the metabolism of HA. In Garrod's terms this disease is an example of an *inborn error of metabolism*. We now know that the mutation responsible for alkaptonuria is recessive, so only people homozygous for the mutant gene express the defect. The gene is on an autosome, but the specific autosome is still unknown.

An important aspect of Garrod's analysis of alkaptonuria was his understanding that the position of

■ *Figure 7.1* Part of the phenylalanine-tyrosine metabolic pathways. Each arrow represents an enzyme: (left) The biochemical steps that operate in normal individuals; (right) The metabolic block found in individuals with alkaptonuria.

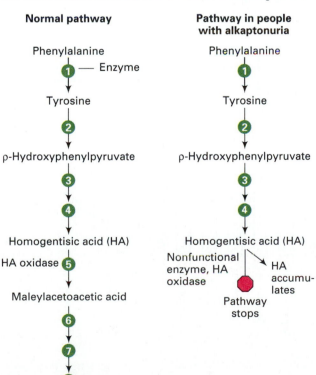

a block in a metabolic pathway can be determined by the accumulation of the chemical compound (HA in this case) that precedes the blocked step. It was another 40 years before the direct relationship between genes and enzymes was proved by Beadle and Tatum.

One Gene–One Enzyme Hypothesis

George Beadle and Edward Tatum in 1942 heralded the beginnings of biochemical genetics, a branch of genetics that combines genetics and biochemistry to elucidate the nature of metabolic pathways. Their results of studies involving the fungus *Neurospora crassa* showed that there was a direct relationship between genes and enzymes and resulted in the *one gene–one enzyme hypothesis*, a significant landmark in the history of genetics.

Isolation of Nutritional Mutants of *Neurospora* In Chapter 5 we discussed the life cycle of *Neurospora crassa* (see Figure 5.17 p. 118). This fungus has simple growth requirements. Wild-type *Neurospora*, by definition, is prototrophic: It can grow on a minimal medium that contains only inorganic salts, a carbon source (such as glucose or sucrose), and the vitamin biotin. Beadle and Tatum reasoned that *Neurospora* synthesized the materials it needed for growth (amino acids, nucleotides, vitamins, nucleic acids, proteins, etc.) from the various chemicals present in the minimal medium. They also realized that it should be possible to isolate nutritional mutants (auxotrophs) of *Neurospora*—mutant strains that required nutritional supplements in order to grow. These auxotrophic mutants could be isolated because they would not grow on the minimal medium.

Figure 7.2 shows how Beadle and Tatum isolated and characterized nutritional mutants. They treated asexual spores (conidia) with X rays to induce genetic mutants. Then they crossed the cultures derived from the surviving, mutated spores with a wild-type (prototrophic) strain. This cross was done because they wanted to study the genetics of biochemical events, and they had to identify those nutritional mutations that were heritable. By crossing the mutagenized spores with the wild type, they ensured that any nutritional mutant they isolated had segregated in a cross and therefore had a genetic basis rather than a nongenetic reason for requiring the nutrient.

One progeny spore per ascus from the crosses was allowed to germinate in a nutritionally complete medium that contained all necessary amino acids, purines, pyrimidines, and vitamins, in addition to the sucrose, salts, and biotin found in the minimal

■ *Figure 7.2*   Method devised by Beadle and Tatum to isolate auxotrophic mutations in *Neurospora.* Here the mutant strain isolated is a tryptophan auxotroph.

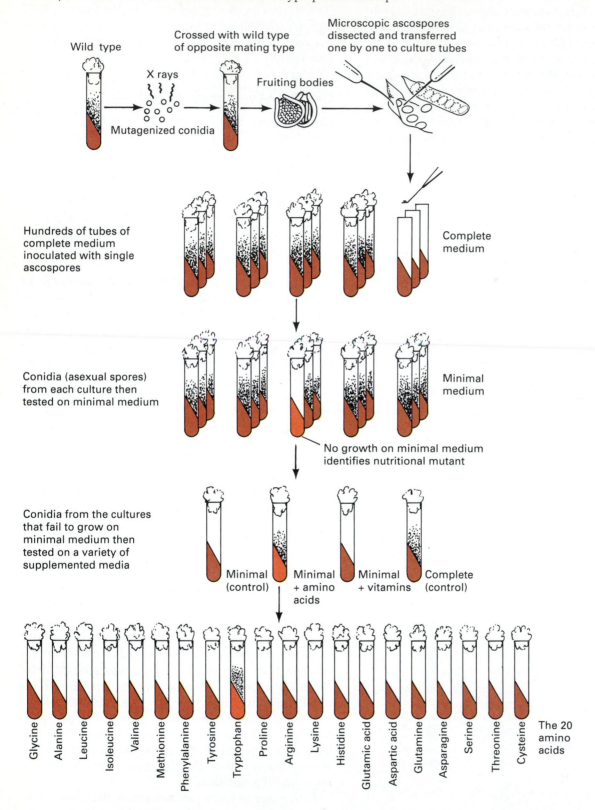

medium. Thus any strain that could not make one or more necessary compounds from the basic ingredients found in minimal medium could still grow by using the compounds supplied in the growth medium. Each culture grown from a progeny spore was then tested for growth on minimal medium. Those strains that did not grow were assumed to be auxotrophic mutants. These mutants were, in turn, individually tested for their abilities to grow on various supplemented minimal media. In this screening two media were used: minimal medium plus amino acids, and minimal medium plus vitamins. Theoretically an amino acid auxotroph—a mutant strain that requires a particular amino acid in order to grow—would grow on minimal medium plus amino acids but not on either minimal medium plus vitamins or minimal medium. Vitamin auxotrophs would only grow on minimal medium plus vitamins, and so on.

Having categorized the strains into nonauxotrophs, amino acid auxotrophs, and vitamin auxotrophs, Beadle and Tatum conducted a second round of screening to specify which chemical the strain needed for growth. Let us imagine an amino acid auxotroph was identified. To find out which of the 20 amino acids is required for a particular amino acid auxotroph to grow, the strain would be tested in 20 tubes, each containing minimal medium plus one of the 20 amino acids. In the example shown in Figure 7.2, a tryptophan auxotroph was identified because it grew only in the tube containing minimal medium plus tryptophan. Crosses between each mutant and wild type segregated in the 1:1 ratio expected for a single gene defect.

Genetic Dissection of a Biochemical Pathway
Once Beadle and Tatum had isolated and identified nutritional mutants, they set out to investigate the biochemical pathways affected by the mutations. They assumed that *Neurospora* cells, like all cells, function by the interaction of the products of a very large number of genes. Furthermore, they imagined that wild-type *Neurospora* converted the constituents of minimal medium into amino acids and

other required compounds by a series of small steps organized into biochemical pathways. In this way the synthesis of cellular components occurs by a series of small steps, each catalyzed by its own enzyme. But what was the relationship between genes and enzymes?

As the hypothetical pathway in Figure 7.3 shows, the product of each step is used as the substrate for the next enzyme. Beadle and Tatum called the proposed relationship between an organism's genes and the enzymes necessary for carrying out the steps in a biochemical pathway the **one gene–one enzyme hypothesis,** having concluded that a specific gene controls the production of a single active enzyme. Consequently, mutations that result in the loss of enzyme activity lead to the accumulation of precursors in the pathway, as well as to the absence of the end product of the pathway.

Using this important one gene–one enzyme hypothesis, we can make some predictions about the consequences of mutations affecting enzymes in the hypothetical pathway shown. If, for example, a mutation in gene *C* results in a nonfunctional enzyme C, then intermediate product B in the pathway cannot be converted to the end product C. The consequence to the organism is that end product C must be present in the growth medium in order for the organism to grow. Intermediate B also accumulates in the cells since it cannot be converted to end product C. Mutations in gene *A* or gene *B* will also result in auxotrophy for end product C.

However, the three mutants *A*, *B*, and *C* are distinguishable on other grounds, namely, their ability to grow or not to grow on various intermediates in the pathway. Gene *C* mutants can grow only if supplemented with end product C; gene *B* mutants can grow if supplemented with either B or C; and gene *A* mutants can grow on minimal medium plus either A, B, or C. A reciprocal pattern of pathway precursor accumulation occurs in these strains. Gene *C* mutants accumulate product B; gene *B* mutants accumulate product A; and gene *A* mutants accumulate the initial substrate for the pathway. In actual experiments where the pathway

■ *Figure 7.3* Hypothetical biochemical pathway for the conversion of a precursor substrate to an end product C in three enzyme-catalyzed steps. Each enzyme is coded for by one gene.

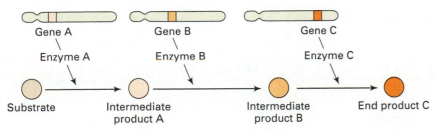

is unknown, the logic is carried out in the opposite direction. From the pattern of growth supplementation and precursor accumulation, the sequence of steps in a pathway can be deduced. In general, then, the later in a pathway a mutant strain is blocked, the fewer intermediate chemicals can enable the strain to grow. The earlier a mutant strain is blocked, the greater the number of intermediates can enable the strain to grow.

Figure 7.4 shows the biochemical pathway for the synthesis of the amino acid arginine in *Neurospora crassa*, along with the genes that code for the enzymes that catalyze each step. If we were to dissect the pathway genetically, we would start with a set of arginine auxotrophs. From genetic crosses and complementation tests (see Chapter 6) we would find that four genes are involved; a mutation in any one of them gives rise to a growth requirement for arginine. These four genes in a wild-type cell are designated $argE^+$, $argF^+$, $argG^+$, and $argH^+$.

Next, the sequence of biochemical steps in the pathway can be deduced by the growth pattern of the mutant strains on media supplemented with presumed arginine precursors. Table 7.1 shows the results obtained for this set of four mutants. By definition all four mutant strains can grow on arginine, and none can grow on unsupplemented minimal medium.

As shown in the table, the *argH* mutant strain can grow when supplemented with arginine but not when supplemented with any of the intermediates in the pathway. This result indicates that the *argH* gene codes for the enzyme that controls the last step in the pathway, which leads to the formation of arginine. The *argG* mutant strain grows on media supplemented with arginine or argininosuccinate, the *argF* mutant strain grows on media supplemented with arginine, argininosuccinate or citrulline, and the *argE* strain grows on arginine, argininosuccinate, citrulline, or ornithine. Using the principles

just described, we conclude that argininosuccinate must be the immediate precursor to arginine, since argininosuccinate permits all but the *argH* mutant to grow.

From this reasoning, the pathway can be hypothesized to be as shown in Figure 7.4. Gene $argF^+$ codes for the enzyme that converts ornithine to citrulline. An *argF* mutant strain can, therefore, grow on minimal medium plus citrulline, argininosuccinate, or arginine. As was discussed before, there is an accumulation of the intermediate chemical produced prior to the step blocked by the genetic mutation, and that information can be used to confirm any conclusions about the sequence of steps in a pathway. The *argF* mutant strains, for example, accumulate ornithine, and hence ornithine would be placed prior to citrulline, argininosuccinate, and arginine in the pathway.

With this sort of approach we can dissect a biochemical pathway genetically; that is, we can determine the sequence of steps in the pathway and relate each step to a specific gene or genes. We should not conclude, though, that there is necessarily only one gene for each step in a pathway, although that is usually true. In some cases an enzyme may consist of more than one polypeptide chain, each of which is coded for by a specific gene. In that event more than one gene would specify that enzyme. Therefore Beadle and Tatum's original name for their hypothesis has been modified to the *one gene–one polypeptide hypothesis*.

Keynote

A number of classical studies indicated the specific relationship between genes and enzymes, eventually embodied in Beadle and Tatum's one gene–one enzyme hypothesis, which states that each gene determines the synthesis of a single functional

■ *Figure 7.4*   Arginine biosynthetic pathway, showing the four genes in *Neurospora crassa* that code for the enzymes that catalyze each reaction.

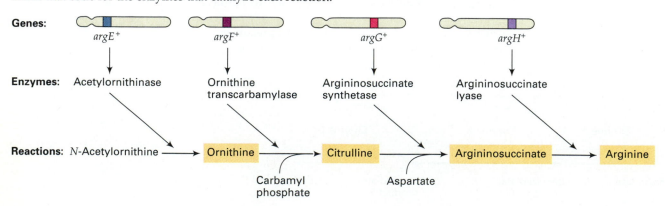

Table 70.1

Growth Responses of Arginine Auxotrophs

MUTANT STRAINS	GROWTH RESPONSE ON MINIMAL MEDIUM AND:				
	NOTHING	ORNITHINE	CITRULLINE	ARGININO-SUCCINATE	ARGININE
Wild type	+	+	+	+	+
argE	-	+	+	+	+
argF	-	-	+	+	+
argG	-	-	-	+	+
argH	-	-	-	-	+

enzyme. Since enzymes may consist of more than one polypeptide, a more modern title for this hypothesis is the one gene–one polypeptide hypothesis.

Human Genetic Diseases with Enzyme Deficiencies

Many human genetic diseases are caused by a single gene mutation that alters the function of an enzyme. In general, an enzyme deficiency caused by a mutation may have simple or pleiotropic (many) consequences. Table 7.2 presents a few of these diseases.

Phenylketonuria

Like alkaptonuria, *phenylketonuria (PKU)* is an inborn error of metabolism. An individual with PKU is called a *phenylketonuriac*. About 1 in 12,000 newborns has PKU. This autosomal recessive genetic disease is most commonly caused by a mutation in the gene for phenylalanine hydroxylase, which maps to chromosome 12. The absence of that enzyme activity prevents the conversion of the

Table 7.2

Selected Human Genetic Diseases with Demonstrated Enzyme Deficiencies

DISEASE	ENZYME DEFICIENCY
Albinism	Tyrosinase
Alkaptonuria	Homogentisic acid oxidase
G6PD deficiency (favism)	Glucose-6-phosphate dehydrogenase
Gout, primary	Hypoxanthine phosphoribosyltransferase
Hypoglycemia and acidosis	Fructose-1,6-diphosphatase
Intestinal lactase deficiency (adult)	Lactase
Lesch-Nyhan syndrome	Hypoxanthine phosphoribosyltransferase
Maple sugar urine disease	Keto acid decarboxylase
Phenylketonuria	Phenylalanine hydroxylase
Porphyria, acute	Uroporphyrinogen III synthetase
Pulmonary emphysema	α-l-Antitrypsin
Tay-Sachs disease	Hexosaminidase A
Tyrosinemia	para-Hydroxyphenylpyruvate oxidase
Xeroderma pigmentosum	Defective DNA repair enzyme

Information taken from V. A. McKusick, 1988. *Mendelian Inheritance in Man*, 8th ed. Baltimore: Johns Hopkins University Press.

■ *Figure 7.5*   Phenylalanine-tyrosine metabolic pathways: (a) Biochemical steps that operate in normal individuals; (b) Metabolic block in the pathway exhibited by individuals with phenylketonuria-phenylalanine cannot be metabolized to tyrosine and unusual metabolites of phenylalanine accumulate; (c) Metabolic block in the pathway exhibited by individuals with albinism cause no or very little melanin pigment to be produced.

a)   **Normal pathway**

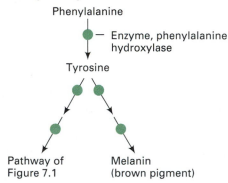

b)   **Pathway in people with phenylketonuria (PKU)**

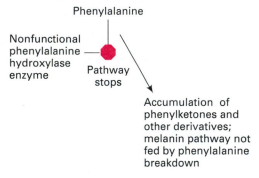

c)   **Pathway in people with albinism**

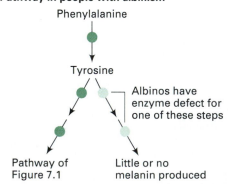

amino acid phenylalanine to the amino acid tyrosine (Figure 7.5[a] and [b]). Phenylalanine is one of the *essential amino acids*. It is an amino acid that must be included in the diet of humans because people are unable to synthesize it. Thus a serious problem for children born with PKU is that the absence of phenylalanine hydroxylase results in the accumulation of the phenylalanine they ingest in

their diets. (PKU children are unaffected prior to or at birth because excess phenylalanine that accumulates is metabolized by maternal enzymes.) However, unlike the accumulated precursor HA for alkaptonuria, which is excreted in the urine, the accumulated phenylalanine in phenylketonuriacs is converted by a secondary pathway to phenylpyruvic acid. This chemical drastically affects the cells of the central nervous system, producing serious symptoms: severe mental retardation, a slow growth rate, and early death.

A phenylketonuriac also cannot make tyrosine, an amino acid required for protein synthesis and for the production of the hormones thyroxine and adrenaline and the skin pigment melanin. This aspect of the phenotype is not critical because tyrosine can be obtained from food. Yet food does not normally have a lot of tyrosine. As a result, people with PKU make relatively little melanin, since their bodies can use only the limited amount of tyrosine found in food for melanin synthesis. Hence people with PKU tend to have very fair skin and blue eyes (even if they have brown-eye genes).

The symptoms of PKU depend on the amount of phenylalanine in the diet, so this genetic disease can be managed by controlling the dietary intake of that amino acid. A mixture of individual amino acids without phenylalanine is used as a protein substitute in the PKU diet. However, since it is impossible to devise a diet completely devoid of protein—and most proteins contain phenylalanine—the diet will contain some phenylalanine. The diet is expensive, costing more than $5,000 per year. It is continued until at least adolescence to ensure full intellectual development. Given the very serious consequences of allowing PKU to go untreated, all states require that newborns be screened for PKU. The screen—the Guthrie test—is conducted by placing a drop of blood on a filter containing a phenylalanine analog and bacteria that can grow only if excess phenylalanine is present.

You may have seen food and drink containing the artificial sweetener NutraSweet® carry the warning that people with PKU should not use them. NutraSweet® is aspartame, a chemical consisting of the amino acid aspartic acid attached to the amino acid phenylalanine. This combination signals to your taste receptors that the substance is sweet, yet it is not sugar and does not have the calories of sugar. Once ingested aspartame is broken down to the two amino acids, one of which has very serious consequences for a phenylketonuriac.

Albinism

Albinism is caused by an autosomal recessive mutation, and individuals must be homozygous for the mutation in order to exhibit the condition. About 1 in 33,000 Caucasians and 1 in 28,000 African Americans in the United States is albino. The mutation is in a gene for an enzyme used in the pathway from tyrosine to the brown pigment melanin (see Figure 7.5[c]). Since no melanin can be produced, albinos have white skin, white hair, and red eyes (owing to the lack of pigment in the iris). Melanin absorbs light in the UV range and is important in protecting the skin against harmful UV irradiation from the sun. Albinos are therefore very light-sensitive. No apparent problems result from the accumulation of precursors in the pathway prior to the block.

There are at least two kinds of albinism, since at least two biochemical steps can be blocked to prevent melanin formation. Therefore, if two albinos of different types mate, they can produce normal children as a result of complementation of the two non-allelic mutations (see Chapter 6, p. 146–147). This example serves also as a molecular explanation for epistasis (see Chapter 4, p. 80–86).

Tay-Sachs Disease

Lysosomes are organelles with a single membrane that contain 40 or more different digestive enzymes. These enzymes catalyze the breakdown of nucleic acids, proteins, polysaccharides, and lipids. A number of human diseases are caused by mutations in genes that code for lysosomal enzymes. These diseases, collectively called *lysosomal-storage diseases*, are generally caused by recessive mutations.

The best-known genetic disease of this type is *Tay-Sachs disease* (also called infantile amaurotic idiocy), which is caused by homozygosity for a rare autosomal recessive mutation that maps to chromosome 15. While Tay-Sachs disease is rare in the population as a whole, it has a relatively high incidence in Ashkenazi Jews of Central European origin, with about 1 in 3600 born with the disease.

The mutant gene codes for the enzyme hexosaminidase A, which functions in a pathway to convert a brain chemical (a particular ganglioside) to a different form. Because Tay-Sachs infants are deficient in that enzyme activity, the unprocessed ganglioside accumulates in the brain cells. Individuals with Tay-Sachs disease exhibit a number of clinical symptoms, illustrating the multiple effects of a single gene mutation. Typically, the symptom first recognized is an unusually enhanced reaction to sharp sounds. Early diagnosis is also made possible by the presence on the retina of a cherry-colored spot surrounded by a white halo. About a year after birth, there is rapid degeneration as the brain begins to lose control over normal function and activities. This degeneration involves generalized paralysis, blindness, a progressive loss of hearing,

and serious feeding problems. By two years of age the infants are essentially immobile, and death occurs at about three to four years of age, often from respiratory infections. There is no cure for Tay-Sachs disease.

Keynote

Many human genetic diseases can be attributed to deficiencies in enzyme activities. Most of these diseases are inherited as recessive traits.

Gene Control of Protein Structure

Earlier in the chapter evidence was presented that genes code for enzymes; this evidence was given because of the historical significance of those studies in advancing our understanding of gene function. While all enzymes are proteins, however, not all proteins are enzymes. To understand fully how genes function, we will look at the experimental evidence that genes are also responsible for the structure of nonenzymatic proteins such as hemoglobin. Nonenzymatic proteins are often easier to study than enzymes. Enzymes are usually present in small amounts, while nonenzymatic proteins occur in relatively large quantities in the cell, which makes them easier to isolate and purify.

Sickle-Cell Anemia: Symptoms and Causes

Sickle-cell anemia is a genetically based human disease that results from an amino acid change in the hemoglobin protein that transports oxygen in the blood. Sickle-cell anemia was first described in 1910 by J. Herrick. He found that red blood cells from individuals with the disease lose their characteristic disk shape in conditions of low oxygen tension and assume the shape of a sickle (Figure 7.6). The sickled red blood cells are more fragile than normal red blood cells and prone to break sooner. In addition,

■ *Figure 7.6* Scanning electron micrograph of (left) normal and (right) sickled red blood cells.

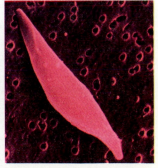

■ *Figure 7.7* The hemoglobin molecule, showing the two α polypeptides and two β polypeptides, each of which is associated with a heme group.

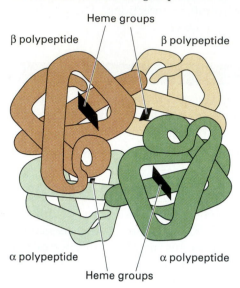

Heme groups

β polypeptide   β polypeptide

α polypeptide   α polypeptide

Heme groups

sickled cells are not as flexible as normal cells and therefore tend to jam up in the capillaries rather than squeeze through them. As a consequence, blood circulation is impaired, and tissues served by the capillaries become deprived of oxygen. Although oxygen deprivation occurs particularly at the extremities, the heart, lungs, brain, kidneys, gastrointestinal tract, muscles, and joints can also suffer from oxygen deprivation and its subsequent damage. The afflicted individual may suffer from a variety of health problems including heart failure, pneumonia, paralysis, kidney failure, abdominal pain, and rheumatism.

Sickle-cell anemia affects hemoglobin, the oxygen-transporting protein in red blood cells. As shown in Figure 7.7, hemoglobin consists of four polypeptide chains (two α polypeptides and two β polypeptides), each of which is associated with a heme group (involved in the binding of oxygen). Sickle-cell anemia is caused by homozygosity for a mutation (β^s) in the gene for the β polypeptide of hemoglobin. The sickle-cell mutation β^s is codominant with the wild-type allele β^A. Thus people who are heterozygous $\beta^A\beta^S$ make two types of hemoglobins. One, called Hb-A (hemoglobin A), is completely normal, with two normal α chains and two normal β chains specified by two wild-type α genes and one wild-type β gene (β^A). The other, called Hb-S, is the defective hemoglobin, with two normal α chains specified by wild-type α genes and two abnormal β chains specified by the mutant β gene β^S. $\beta^A\beta^S$ heterozygotes have *sickle-cell trait*, though under normal conditions they show few symptoms of the disease. However, after a sharp drop in oxy-

gen tension (e.g., in an unpressurized aircraft climbing into the atmosphere), sickling of red blood cells may occur, giving rise to symptoms similar to those of severe anemia.

The evidence that an abnormal hemoglobin molecule was present in individuals with sickle-cell anemia was obtained by Linus Pauling and his co-workers in the 1950s. They used the procedure of electrophoresis to separate the electrically charged protein molecules. In this technique proteins of the same molecular weight but different charge will migrate at different rates. Under the electrophoretic conditions Pauling used, the hemoglobin from normal people (Hb-A) migrated more slowly than the hemoglobin from people with sickle-cell anemia (Hb-S) (Figure 7.8). The hemoglobin from sickle-cell trait individuals had a 1:1 mixture of Hb-A and Hb-S. Pauling concluded that sickle-cell anemia results from a mutation that alters the chemical structure of the hemoglobin molecule. This experiment was one of the first rigorous proofs that protein structure is controlled by a gene.

The α chain is identical in hemoglobin types Hb-A and Hb-S. In the β chain, however, the amino acid glutamic acid at the sixth position from the N-terminal end is replaced in the mutant with the amino acid valine (Figure 7.9). Glutamic acid is a charged amino acid while valine is an uncharged amino acid. This particular substitution of amino acids causes the β polypeptide to fold up in a different way. This, in turn, leads to sickling of the red blood cells in individuals with sickle-cell anemia. The explanation for the amino acid change is that the β^S mutation involves a base pair change in the

■ *Figure 7.8* Electrophoresis of hemoglobin found in (left) normal $\beta^A\beta^A$ individuals, (center) $\beta^A\beta^S$ individuals with sickle-cell trait, and (right) $\beta^S\beta^S$ individuals who exhibit sickle-cell anemia. The two hemoglobins migrate to different positions in an electric field and hence must differ in electric charge.

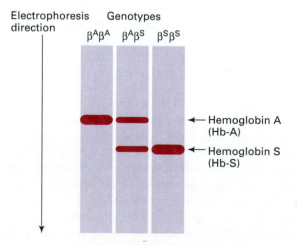

Electrophoresis direction

Genotypes

$\beta^A\beta^A$ $\beta^A\beta^S$ $\beta^S\beta^S$

← Hemoglobin A (Hb-A)

← Hemoglobin S (Hb-S)

■ *Figure 7.9*   The first seven N terminal amino acids in normal and sickled hemoglobin β polypeptides, showing the single amino acid change from glutamic acid to valine at the sixth position.

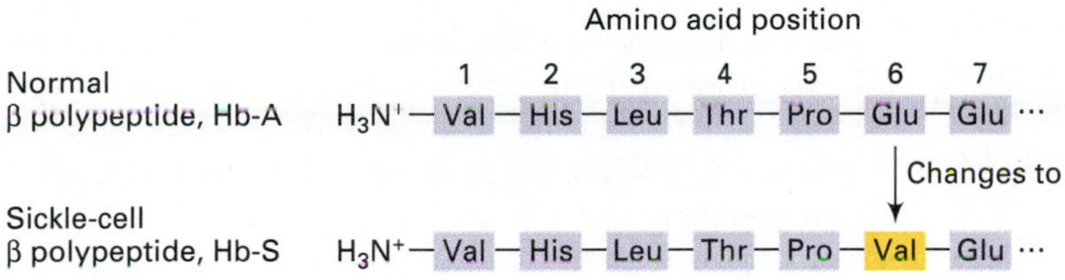

DNA so that valine is encoded instead of glutamic acid. This finding was strong evidence that genes specify the amino acid sequence of polypeptide chains. Furthermore, since the hemoglobin protein consists of two types of polypeptides, α and β, then according to the one-gene–one-polypeptide hypothesis, two genes encode hemoglobin.

Many other mutant hemoglobins have been detected in general screening programs in which hemoglobin is isolated from red blood cells and subjected to electrophoresis. Not many of the identified hemoglobin mutants have as drastic an effect as the sickle-cell anemia mutant because the consequences of the substitution depend both on the amino acids involved and on the location of the mutant amino acid in the chain. For example, in the Hb-C hemoglobin molecule there is a change in the β chain from a glutamic acid to a lysine. Like the Hb-S change, this substitution involves the sixth amino acid from the N terminal end of the polypeptide. However, since glutamic acid and lysine are both charged amino acids, the overall change in protein structure is much less than in Hb-S. Therefore people homozygous for the β^C mutation that results in the abnormal β chain of the Hb-C hemoglobin molecule experience only a mild form of anemia.

Biochemical Genetics of the Human ABO Blood Groups

The mechanism of inheritance of the human ABO blood groups was mentioned in Chapter 4. Recall that people of blood group A (genotypes I^A/I^A or I^A/I^O) carry the A antigen on their blood cells, people of blood group B (genotypes I^B/I^B or I^B/I^O) carry the B antigen, people with AB blood type (genotype I^A/I^B) carry both the A and B antigens, and people of blood group O (genotype I^O/I^O) have neither the A nor the B antigen.

The ABO gene locus is involved in the production of enzymes that are used in the biosynthesis of polysaccharides. The enzymes act to add sugar groups to a preexisting polysaccharide. The polysaccharides that are important here are those that attach to lipids to produce compounds called gly-

colipids. The glycolipids then associate with red blood cell membranes to form the blood group antigens. Most individuals produce a glycolipid called the H antigen. The I^A allele produces an enzyme that adds a particular sugar to the H antigen to produce the A antigen. So individuals with genotypes I^A/I^A or I^A/I^O have blood group A. The I^B allele produces a different enzyme, which also recognizes the H antigen but adds a different sugar to its polysaccharide to produce the B antigen. Consequently individuals with genotypes I^B/I^B or I^B/I^O have blood group B. (Note that this small difference in the structure of the A and B antigens is sufficient to induce an antibody response as discussed in Chapter 4.) In both cases some H antigen remains unconverted.

For the I^A/I^B heterozygote (individuals of blood group AB), both enzymes are produced, so some H antigen is converted to the A antigen and some to the B antigen. The blood cell has both antigens on the surface, so the individual is of blood type AB. It is the presence of both surface antigens that provides the molecular basis of the codominance of the I^A and I^B alleles (see Chapter 4, p. 78).

People who are homozygous for the I^O allele produce no enzymes to convert the polysaccharide component of the H antigen glycolipid. As a consequence, their red blood cells carry neither the A nor the B antigen, although they do carry the H antigen. The H antigen does not elicit an antibody response in people of different blood groups because its polysaccharide component is the basic component of the A and the B antigens as well, so it is not perceived as a foreign substance. Thus people with blood type O are universal donors (see Chapter 4, p. 77).

Keynote

From the study of alterations in proteins other than enzymes, such as the mutations responsible for sickle-cell anemia, convincing evidence was obtained that genes control the structures of all proteins.

Summary

In this chapter we began to think of the gene in molecular, rather than abstract, terms. Specifically, we discussed gene function—what genes code for, and what the evidence is for that coding. We considered a number of experiments showing that genes coded for enzymes and for nonenzymatic proteins. A number of these experiments were done prior to the unequivocal proof that DNA is genetic material and the elucidation of DNA structure.

As early as 1902 there was evidence for a specific relationship between genes and enzymes. This evidence was obtained by Archibald Garrod in his investigations of inborn errors of metabolism, human genetic diseases that resulted in enzyme deficiencies. The specific relationship between genes and enzymes is historically embodied in the one gene–one enzyme hypothesis proposed by Beadle and Tatum to explain their data on auxotrophic mutants of *Neurospora*. This hypothesis states that each gene determines the synthesis of a single enzyme. Since some enzymes and nonenzymatic proteins have more than one polypeptide (e.g., hemoglobin), and genes code for individual polypeptides, a more modern maxim is one gene–one polypeptide. Beadle and Tatum's methods for dissecting a biochemical pathway genetically, or determining the sequence of steps in the pathway and relating each step to a specific gene or genes, are landmarks in the development of molecular genetics.

In this chapter we also discussed a number of human examples of genetically based enzyme deficiencies that give rise to genetic diseases. Many of these are the result of homozygosity for recessive mutant genes. The severity of the disease depends on the effects of the loss of function of the particular enzyme involved. Thus albinism is a relatively mild genetic disease, while Tay-Sachs disease is lethal.

To make our understanding of gene function more complete, we examined experimental evidence that genes control the structure of nonenzymatic proteins as well as of enzymes.

Analytical Approaches for Solving Genetics Problems

Q7.1 k^+, l^+, and m^+ are independently assorting genes that control the production of a red pigment. These three genes act in a biochemical pathway as follows:

$$\text{colorless 1} \xrightarrow{k^+} \text{colorless 2} \xrightarrow{l^+} \text{orange} \xrightarrow{m^+} \text{red}$$

The mutant alleles that give abnormal functioning of these genes are k, l, and m; each is recessive to its wild-type counterpart. A red individual homozygous for all three wild-type alleles is crossed with a colorless individual that is homozygous for all three recessive mutant alleles. The F_1 is red. The F_1 is then selfed to produce the F_2 generation.
a. What proportion of the F_2 are colorless?
b. What proportion of the F_2 are orange?
c. What proportion of the F_2 are red?

A7.1 **a.** There are two ways to answer this question. One is to determine all of the genotypes that can specify the colorless phenotypes, and the other is to use subtractive logic, in which the proportions of orange and red are first calculated and then subtracted from one to give the proportion of colorless progeny in the F_2. We will begin with the second method.

To produce an orange phenotype three things are necessary: (1) at least one wild-type k^+ allele must be present so that the colorless 1 to colorless 2 step can occur; (2) at least one wild-type l^+ allele must be present so that the colorless 2 to orange step can occur; and (3) the individual must be m/m so that the orange to red step cannot proceed. Thus an orange phenotype results from the genotype $k^+/- \; l^+/- \; m/m$. From an $F_1 \times F_1$ self, the probability of getting an individual with that genotype is $3/4 \times 3/4 \times 1/4 = 9/64$.

To produce a red phenotype three things are needed: (1) at least one wild-type k^+ allele must be present so that the colorless 1 to colorless 2 step can occur; (2) at least one wild-type l^+ allele must be present so that the colorless 2 to orange step can occur; and (3) at least one m^+ allele must be present so that the orange to red step can proceed. Thus a red phenotype results from the genotype $k^+/- \; l^+/- \; m^+/-$. From an $F_1 \times F_1$ self, the probability of getting an individual with that genotype is $3/4 \times 3/4 \times 3/4 = 27/64$.

By default all other F_2 individuals are colorless. The proportion of F_2 individuals that are colorless is, therefore, $1 - 9/64 - 27/64 = 1 - 36/64 = 28/64$.

Using the first approach to calculate the proportion of F_2 colorless, we must determine all genotypes that will give a colorless phenotype and then add up the probabilities of each occurring. The genotypes and their probabilities are as follows:

GENOTYPES	PROBABILITIES	
$k/k\ l/l\ m/m$	1/64	(Cannot convert colorless 1)
$k/k\ l/l\ m^+/-$	3/64	(Cannot convert colorless 1)
$k/k\ l^+/-\ m/m$	3/64	(Cannot convert colorless 1)
$k/k\ l^+/-\ m^+/-$	9/64	(Cannot convert colorless 1)
$k^+/-\ l/l\ m/m$	3/64	(Cannot convert colorless 2)
$k^+/-\ l/l\ m^+/-$	9/64	(Cannot convert colorless 2)
TOTAL	28/64	

b. The proportion of the F_2 was calculated in (a): 9/64.

c. The proportion of the F_2 was calculated in (a): 27/64.

Q7.2 A number of auxotrophic mutant strains were isolated from wild-type, haploid yeast. These strains responded to the addition of certain nutritional supplements to minimal culture medium either by growth (+) or no growth (0). The following table gives the responses for single gene mutants:

MUTANT STRAINS	SUPPLEMENTS ADDED TO MINIMAL CULTURE MEDIUM				
	B	A	R	T	S
1	+	0	+	0	0
2	+	+	+	+	0
3	+	0	+	+	0
4	0	0	+	0	0

Diagram a biochemical pathway that is consistent with the data, indicating where in the pathway each mutant strain is blocked.

A7.2 The type of data to be analyzed are very similar to those discussed in the text for Beadle and Tatum's analysis of *Neurospora* auxotrophic mutants, from which they proposed the one gene–one enzyme hypothesis. Recall the important principle that the later in the pathway a mutant is blocked, the fewer nutritional supplements need be added to allow growth. In the data given we must assume that the nutritional supplements are not necessarily

listed in the order in which they appear in the pathway.

Analysis of the data indicates that all four strains will grow if given R, and that none will grow if given S. From this we can conclude that R is likely to be the end product of the pathway (all mutants should grow if given the end product), and that S is likely to be the first compound in the pathway (none of the mutants should grow given the first compound in the pathway). Thus the pathway as deduced so far is:

$$S \longrightarrow [B, A, T] \longrightarrow R$$

where the order of B, A, and T are as yet undetermined.

Now let us consider each of the mutant strains, and see how their growth phenotypes can help define the biochemical pathway:

Strain 1 will grow only if given B or R. Therefore the defective enzyme in strain 1 must act somewhere prior to the formation of B and R, and after the substances A, T, and S. Since we have deduced that R is the end product of the pathway, we can propose that B is the immediate precursor to R, and that strain 1 cannot make B. The pathway so far is:

$$S \longrightarrow [A, T] \overset{1}{\longrightarrow} B \longrightarrow R$$

Strain 2 will grow on all compounds except S, the first compound in the pathway. Thus the defective enzyme in strain 2 must act to convert S to the next compound in the pathway, which is either A or T. We do not know yet whether A or T follows S in the pathway, but the growth data at least allow us to conclude where strain 2 is blocked in the pathway, that is:

$$S \overset{2}{\longrightarrow} [A, T] \overset{1}{\longrightarrow} B \longrightarrow R$$

Strain 3 will grow on B, R, and T, but not on A or S. We know that R is the end product and S is the first compound in the pathway. This mutant strain allows us to determine the order of A and T in the pathway. Since strain 3 grows on T but not on A, T must be later in the pathway than A, and the defective enzyme in 3 must be blocked in the yeast's ability to convert A to T. The pathway now is:

$$S \overset{2}{\longrightarrow} A \overset{3}{\longrightarrow} T \overset{1}{\longrightarrow} B \longrightarrow R$$

Strain 4 will grow only if given the deduced end product R. Therefore the defective enzyme produced by the mutant gene in strain 4 must act before the formation of R, and after the formation of A, T, and B from the first compound S. The mutation in 4 must be blocked in the last step of the biochemical pathway in the conversion of B to R. The final deduced pathway, and the positions of the mutant blocks are:

$$S \xrightarrow{\text{2}} A \xrightarrow{\text{3}} T \xrightarrow{\text{1}} B \xrightarrow{\text{4}} R$$

Questions and Problems

7.1 Phenylketonuria (PKU) is a heritable metabolic disease of humans. One of its symptoms is mental deficiency. The gross phenotypic effect is due to:
a. an accumulation of phenylketones in the blood.
b. an accumulation of maple sugar in the blood.
c. a deficiency of phenylketones in the blood.
d. a deficiency of phenylketones in the diet.

*7.2 If a person were homozygous for both PKU (phenylketonuria) and AKU (alkaptonuria), what would you expect his or her phenotype to be? Refer to the pathway below.

Phenylalanine

↓ (blocked in PKU)

tyrosine → DOPA → melanin

↓

p-Hydroxyphenylpyruvic acid

↓

Homogentisic acid

↓ (blocked in AKU)

Maleylacetoacetic acid

7.3 Refer to the pathway shown in Question 7.2. What effect, if any, would you expect PKU (phenylketonuria) and AKU (alkaptonuria) to have on pigment formation?

*7.4 a^+, b^+, c^+, and d^+ are independently assorting Mendelian genes controlling the production of a black pigment. The alternate alleles that give abnormal functioning of these genes are a, b, c, and d. A black a^+/a^+ b^+/b^+ c^+/c^+ d^+/d^+ is crossed with a colorless a/a b/b c/c d/d to give a black F_1. The F_1 is then selfed. Assume that a^+, b^+, c^+, and d^+ act in a pathway as follows:

$$\text{colorless} \xrightarrow{a^+} \text{colorless} \xrightarrow{b^+} \text{colorless} \xrightarrow{c^+} \text{brown} \xrightarrow{d^+} \text{black}$$

a. What proportion of the F_2 are colorless?
b. What proportion of the F_2 are brown?

7.5 Using the genetic information given in Question 7.4, now assume that a^+, b^+, and c^+ act in a pathway as follows:

$$\text{colorless} \xrightarrow{a^+} \text{red} \searrow$$
$$\qquad\qquad\qquad\qquad \xrightarrow{c^+} \text{black}$$
$$\text{colorless} \xrightarrow{b^+} \text{red} \nearrow$$

Black can be produced only if both red pigments are present; that is, c^+ converts the two red pigments together into a black pigment.
a. What proportion of the F_2 are colorless?
b. What proportion of the F_2 are red?
c. What proportion of the F_2 are black?

7.6 a. Three genes on different chromosomes are responsible for three enzymes that catalyze the same reaction in corn:

$$\text{colorless compound} \xrightarrow{a^+, b^+, c^+} \text{red compound}$$

The normal functioning of any one of these genes is sufficient to convert the colorless compound to the red compound. The abnormal functioning of these genes is designated by a, b, and c, respectively. A red a^+/a^+ b^+/b^+ c^+/c^+ is crossed by a colorless a/a b/b c/c to give a red F_1, a^+/a b^+/b c^+/c. The F_1 is selfed. What proportion of the F_2 are colorless?
b. It turns out that another step is involved in the pathway. It is controlled by gene d^+, which assorts independently of a^+, b^+, and c^+:

$$\text{colorless compound 1} \xrightarrow{d^+} \text{colorless compound 2} \xrightarrow{a^+, b^+, c^+} \text{red compound}$$

The inability to convert colorless 1 to colorless 2 is designated d. A red a^+/a^+ b^+/b^+ c^+/c^+ d^+/d^+ is crossed by a colorless a/a b/b c/c d/d. The F_1 are all red. The red F_1s are now selfed. What proportion of the F_2 are colorless?

7.7 In hypothetical organisms called mongs, the recessive bw causes a brown eye and the (unlinked) recessive *st* causes a scarlet eye. Organisms homozygous for both recessives have white eyes. The genotypes and corresponding phenotypes, then, are as follows:

$bw^+/-$	$st^+/-$	red eye
bw/bw	$st^+/-$	brown
$bw^+/-$	st/st	scarlet
bw/bw	st/st	white

Outline a hypothetical biochemical pathway that would give this type of gene interaction. Demonstrate why each genotype shows its specific phenotype.

*7.8 The Black Riders of Mordor in *Lord of the Rings* ride steeds with eyes of fire. As a geneticist you are very interested in the inheritance of the fire-red eye color. You discover that the eyes contain two types of pigments, brown and red, that are usually bound to core granules in the eye. In wild-type steeds precursors are converted by these granules to the above pigments, but in steeds homozygous for the recessive X-linked gene *w* (white eye), the granules remain unconverted and a white eye results. The metabolic pathways for the synthesis of the two pigments are shown in the following figure.

Formylkynurenine $\xleftarrow{v^+}$ Tryptophan

Kynurenine $\xrightarrow{cn^+}$ 3-Hydroxykynurenine $\xrightarrow{st^+}$ Brown pigment

Brown pigment + Red pigment Combine to produce fire red eye color

$\xrightarrow{bw^+}$ Biopterin $\longrightarrow$ Sepiapterin (yellow pigment) $\xrightarrow{se^+}$ Red pigment

Each step of the pathway is controlled by a gene: Mutation *v* gives vermilion eyes; *cn* gives cinnabar eyes; *st* gives scarlet eyes; *bw* gives brown eyes; and *se* gives black eyes. All the mutations are recessive to their wild-type alleles and all are unlinked. For the following genotypes, show the phenotypes and proportions of steeds that would be obtained in the F_1 of the given matings.

a. $w/w\ bw^+/bw^+\ st/st \times w^+/Y\ bw/bw\ st/st$
b. $w^+/w^+\ se/se\ bw/bw \times w/Y\ se^+/se^+\ bw^+/bw^+$
c. $w^+/w^+\ v^+/v^+\ bw/bw \times w/Y\ v/v\ bw/bw$
d. $w^+/w^+\ bw^+/bw\ st^+/st \times w/Y\ bw/bw\ st/st$

*7.9 Upon infection of *E. coli* with bacteriophage T4, a series of biochemical pathways result in the formation of mature progeny phages. The phages are released following lysis of the bacterial host cells. Let us suppose that the following pathway exists:

$$A \xrightarrow{\text{enzyme}} B \xrightarrow{\text{enzyme}} \text{mature phage}$$

Let us also suppose that we have two temperature-sensitive mutants that involve the two enzymes catalyzing these sequential steps. One of the mutations is cold-sensitive (*cs*), in that no mature phages are produced at 17°C. The other is heat-sensitive (*hs*), in that no mature phages are produced at 42°C. Normal progeny phages are produced when a phage carrying either of the mutations infects bacteria at 30°C. However, let us assume that we do not know the sequence of the two mutations. Two models are therefore apparent:

$$(1)\quad A \xrightarrow{hs} B \xrightarrow{cs} \text{phage}$$
$$(2)\quad A \xrightarrow{cs} B \xrightarrow{hs} \text{phage}$$

Outline how you would experimentally determine which model is the correct model without artificially breaking phage-infected bacteria.

7.10 Four mutant strains of *E. coli* (*a*, *b*, *c*, and *d*) all require substance X in order to grow. Four plates were prepared, as shown in the following figure. In each case the medium was minimal, with just a trace amount of substance X, to allow a small amount of growth of mutant cells. On plate *a* cells of mutant strain *a* were spread over the agar, and grew to form a thin lawn. On plate *b* the lawn is composed of mutant *b* cells, and so on. On each plate cells of the four mutant types were inoculated

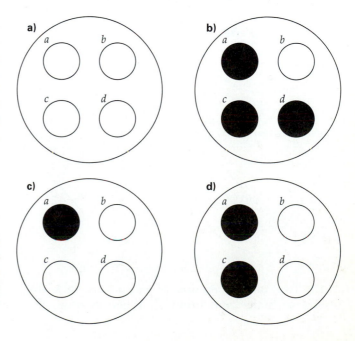

over the lawn, as indicated by the circles. Dark circles indicate luxuriant growth occurred. This experiment tests whether the bacterial strain spread on the plate can "feed" the four strains inoculated on the plate and allow them to grow. What do these results show about the relationship of the four mutants to the metabolic pathway leading to substance X?

*7.11 The following table indicates what enzyme is deficient in six different complementing mutants of *E. coli*, none of which can grow on minimal medium. All of them will grow if tryptophan (Trp) is added to the medium.

MUTANT	ENZYME MISSING
trpE	anthranilate synthetase
trpA	tryptophan synthetase
trpF	IGP synthetase
trpB	tryptophan synthetase
trpD	PPA transferase
trpC	PRA isomerase

Each of the plates in the following figure shows the results of streaking three of the mutants on minimal medium with just a trace of added tryptophan. Heavy shading shows regions of heavy growth, indicating where a strain can be fed by the strain streaked next to it on the plate. In what order do the enzymes listed above act in the tryptophan synthetic pathway?

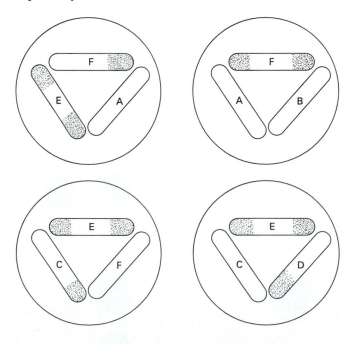

*7.12 Two mutant strains of *Neurospora* lack the ability to make compound Z. When crossed, the

strains usually yield asci of two types: (1) those with spores that are all mutant, and (2) those with four wild-type and four mutant spores. The two types occur in a 1:1 ratio.

a. Let *c* represent one mutant, and let *d* represent the other. What are the genotypes of the two mutant strains?

b. Are *c* and *d* linked?

c. Wild-type strains can make compound Z from the constituents of the minimal medium. Mutant *c* can make Z if supplied with X but not if supplied with Y, while mutant *d* can make Z from either X or Y. Construct the simplest linear pathway of the synthesis of Z from the precursors X and Y, and show where the pathway is blocked by mutations *c* and *d*.

7.13 The following growth responses (where + = growth, and 0 = no growth) of mutants 1 to 4 were seen on the related biosynthetic intermediates A, B, C, D, and E. Assume all intermediates are able to enter the cell, that each mutant carries only one mutation, and that all mutants affect steps after B in the pathway.

	GROWTH ON				
MUTANT	A	B	C	D	E
1	+	0	0	0	0
2	0	0	0	+	0
3	0	0	0	0	0
4	0	0	0	+	+

Which of the schemes in the following figure fits best with the data with regard to the biosynthetic pathway?

a)
$$B \longrightarrow A \longrightarrow D$$
with C and E branching from the point.

b)
$$B \longrightarrow A \longrightarrow D$$
with C and E branching from D.

c)
$$B \longrightarrow A$$
with E branching up and C → D branching down.

d)
$$B \longrightarrow A \longrightarrow E \longrightarrow D$$
with C branching down.

7.14 Four strains of *Neurospora*, all of which require arginine but have unknown genetic constitution, have the following nutrition and accumulation characteristics:

		GROWTH ON			
STRAIN	MINIMAL MEDIUM	ORNITHINE	CITRULLINE	ARGININE	ACCUMULATES
1	−	−	+	+	Ornithine
2	−	−	−	+	Citrulline
3	−	−	−	+	Citrulline
4	−	−	−	−	Ornithine

The pairwise complementation tests of the four strains gave the following results (+ = growth on minimal medium, and 0 = no growth on minimal medium):

	4	3	2	1
1	0	+	+	0
2	0	0	0	
3	0	0		
4	0			

Crosses among mutants yielded prototrophs in the following percentages:

1 × 2:   25 percent
1 × 3:   25 percent
1 × 4:   none detected among 1 million ascospores
2 × 3:   0.002 percent
2 × 4:   0.001 percent
3 × 4:   none detected among 1 million ascospores

Analyze the data and answer the following questions.

a. How many distinct mutational sites are represented among these four strains?

b. In this collection of strains, how many types of polypeptide chains (normally found in the wild type) are affected by mutations?

c. Give the genotypes of the four strains, using a consistent and informative set of symbols.

d. Give the map distances between all pairs of linked mutations.

e. Give the percentage of prototrophs that would be expected among ascospores of the following types: (1) strain *1* × wild type; (2) strain *2* × wild type; (3) strain *3* × wild type; (4) strain *4* × wild type.

7.15 In evaluating my teacher, my sincere opinion is that:

a. He (she) is a swell person whom I would be glad to have as a brother-in-law/sister-in-law.

b. He (she) is an excellent example of how tough it is when you don't have either genetics or environment going for you.

c. He (she) may have okay DNA to start with, but somehow all the important genes got turned off.

d. He (she) ought to be preserved in tissue culture for the benefit of other generations.

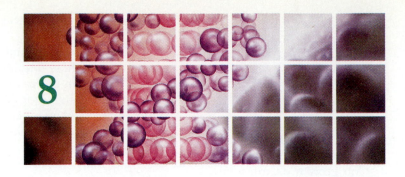

8

The Structure of Genetic Material

- Genetic material must contain all the information for the cell structure and function of an organism and must replicate accurately so that progeny cells have the same genetic information as the parental cell. In addition, genetic material must be capable of variation, one of the bases for evolutionary change. A series of experiments proved that the genetic material of organisms consists of one of two types of nucleic acids—DNA or RNA. Of the two, DNA is most common; only certain viruses have RNA as their genetic material.

- DNA and RNA are macromolecules composed of smaller building blocks called nucleotides. Each nucleotide consists of a 5-carbon sugar (deoxyribose in DNA, ribose in RNA) to which is attached one of four nitrogenous bases and a phosphate group.

- In DNA the four nitrogenous bases are adenine, guanine, cytosine, and thymine. In RNA the four bases are adenine, guanine, cytosine, and uracil.

- The DNA molecule consists of two polynucleotide chains joined by hydrogen bonds between pairs of bases (A and T, G and C) in a double helix. The diameter of the helix is 2 nm, and there are 10 base pairs in each complete turn (3.4 nm).

- Two important types of double-helical DNA have been identified by X-ray diffraction analysis: the right-handed B-DNA, and the left-handed Z-DNA. The common form found in cells is B-DNA. Z-DNA has been proved to be present in the *E. coli* chromosome, but its biological function, if any, is unknown.

I N THE NEXT SEVERAL CHAPTERS we will explore the molecular structure and function of genetic material—either DNA (deoxyribonucleic acid) or RNA (ribonucleic acid)—and will examine in detail the mechanisms by which genetic information is transmitted from generation to generation. We will see exactly what the genetic message is, and we will learn how DNA directs the manufacture of structural proteins and enzymes.

In this chapter we begin our study of the molecular aspects of genetics by discussing the properties of the genetic material, the evidence that DNA and RNA are genetic material, and the structure of DNA and RNA molecules.

The Nature of Genetic Material: DNA and RNA

Long before DNA and RNA were proved to carry genetic information, geneticists recognized that specific molecules must fulfill that function. They postulated that it would be necessary for the material responsible for heritable traits to have three principal characteristics:

1. It must contain, in a stable form, *the information* for an organism's cell structure, function, development, and reproduction.

2. It must *replicate accurately* so that progeny cells have the same genetic information as the parental cell.

3. It must be capable of *variation*. Without variation (such as through mutation and recombination), organisms would be incapable of change and adaptation, and evolution could not occur.

From experiments in the middle to late 1800s and early 1900s, scientists suspected that genetic material was composed of protein. Chromosomes were first seen under the microscope in the latter part of the nineteenth century. Chemical analysis during the first 40 years of this century revealed that the nucleus contained a unique molecular constituent, deoxyribonucleic acid (DNA). It was not understood, however, that DNA was the chemical constituent of genes. Rather, DNA was believed to be a molecular framework for a theoretical class of special protein that carried genetic information.

Evidence gathered from experiments in the middle of this century showed unequivocally that the genetic material of living organisms and many viruses consisted, not of protein, but of double-stranded DNA; in other viruses the genetic material may be single-stranded DNA, double-stranded RNA, or single-stranded RNA.

The Discovery of DNA as Genetic Material

One of the first studies that ultimately led to the identification of DNA as the genetic material involved the spherical bacterium *Streptococcus pneumoniae* (also called pneumococcus). Two strains of pneumococcus were used (Figure 8.1). The smooth (*S*) strain is infectious (virulent) and causes the death of the infected animal. An *S* type bacterium is surrounded by a polysaccharide coat that gives the strain its infectious properties and results in the smooth, shiny appearance of *S* colonies. The rough (*R*) strain is noninfectious (avirulent) because *R* cells lack the polysaccharide coat; colonies of this strain are not shiny.

In 1928 Frederick Griffith injected mice with *R* bacteria (Figure 8.2). Genetically *R* bacteria lacked the ability to make the polysaccharide coat, and they did not affect the mice. When Griffith injected mice with living *S* bacteria, the animals died, and living *S* bacteria could be isolated from their blood. If he first heat-killed *S* bacteria before injecting them, however, the mice survived. These two experiments showed that the bacteria had to be alive and had to be of the *S* type (have the polysaccharide coat) in order for them to be infectious.

Lastly, Griffith infected mice with a mixture of living *R* bacteria and heat-killed *S* bacteria. In this case the mice died, and living *S* bacteria were present in the blood. Griffith concluded that some *R* bacteria had somehow been transformed into smooth, infectious *S* cells by interaction with the dead *S* cells. The transformed *S* cells retained their infectious properties and polysaccharide coat in successive generations, indicating that the transformation was stable. Griffith believed that the unknown agent responsible for the change in genetic material was a protein. He referred to the agent as the *transforming principle.*

From a modern molecular perspective, the genetic material of the virulent *S* bacteria was released from the heat-killed cells and entered the living avirulent *R* cells where recombination occurred to generate a virulent *S* transformant (see the discussion of bacterial transformation in Chapter 6).

The Nature of the Transforming Principle   In 1944 Oswald T. Avery, Colin M. MacLeod, and Maclyn McCarty showed that the transforming principle was not a protein but was instead DNA. In their experiments they prepared an extract from *S* cells and showed that it was able to transform *R* bacteria to *S* in the test tube. Next, they progressively purified the extract until only proteins and the two types of *nucleic acids*, RNA and DNA, were left. Transformation still occurred. Protein was then ruled out as the genetic material because the transforming principle survived treatment of the extract with protein-degrading enzymes. Thus the nucleic acids (at this point not separated into DNA and RNA) were the only components of *S* cells that could transform the *R* cells into *S* cells.

Avery and his colleagues then used specific nucleases—enzymes that degrade nucleic acids—to determine whether DNA or RNA was the transforming principle (Figure 8.3). When they treated the nucleic acids with **ribonuclease (RNase)**, which degrades RNA but not DNA, the transforming activity was still present. However, when they used **deoxyribonuclease (DNase)**, which degrades only DNA, no transformation resulted. These results indicated that DNA was genetic material.

The Hershey and Chase Bacteriophage Experiments
Alfred D. Hershey and Martha Chase reported experimental results in 1953 that conclusively confirmed that DNA is genetic material. They were studying the replication of bacteriophage T2 to see which phage components were needed to complete the life cycle. At the time it was known that T2 infected *Escherichia coli* by first attaching itself with its tail fibers, then injecting into the bacterium some genetic material that somehow controlled the

■ *Figure 8.1*   Colonies of (left) smooth (*S*) and (right) rough (*R*) strains of *Streptococcus pneumoniae.* The smooth colonies "shine" under the lights.

■ *Figure 8.2*   Griffith's transformation experiment. Mice injected with type *S*
pneumococcus died, whereas mice injected with either type *R* or heat-killed type *S*
bacteria survived. When injected with a mixture of living type *R* and heat-killed type *S*
bacteria, however, the mice died.

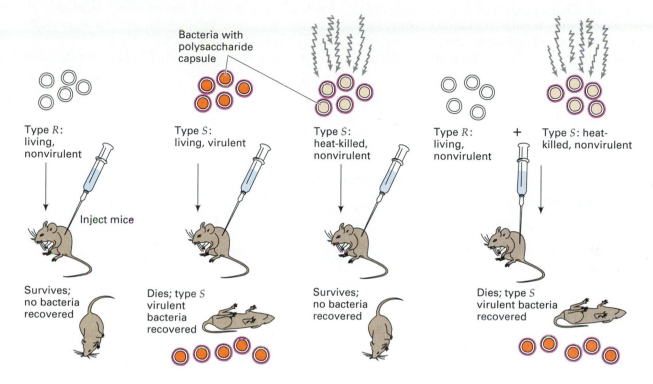

■ *Figure 8.3*   Experiment that showed DNA, and not RNA, was the transforming
principle. When the nucleic acid mixture of DNA and RNA was treated with
ribonuclease (RNase), transformants still resulted. However, when the DNA and RNA
mixture was treated with deoxyribonuclease (DNase), no transformants resulted.

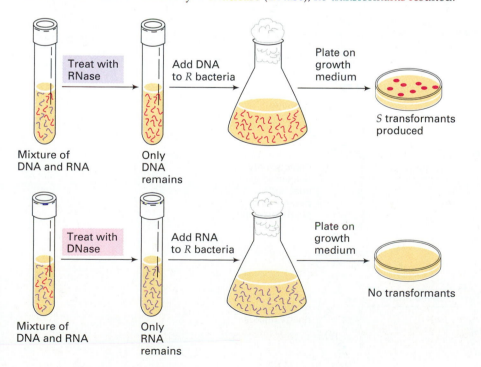

production of progeny T2 particles. Ultimately the bacterium lysed, releasing 100 to 200 phages that could infect other bacteria. However, the nature of T2's genetic material was not known.

Hershey and Chase used T2 to infect cultures of *E. coli* growing in media containing either a radioactive isotope of phosphorus (^{32}P) or a radioactive isotope of sulfur (^{35}S) (Figure 8.4[a]).

They used these isotopes because DNA contains phosphorus but no sulfur, and protein contains sulfur but no phosphorus. When they collected the progeny phages that were produced, they had one batch of T2 with DNA radioactively labeled with ^{32}P, and a second batch with the phage proteins radioactively labeled with ^{35}S.

Since they knew that each T2 phage consisted of

■ *Figure 8.4* Hershey and Chase experiment: (a) The production of T2 phages either with (1) ^{32}P-labeled DNA or with (2) ^{35}S-labeled protein; (b) The experimental evidence showing that DNA is the genetic material in T2; (1) The ^{32}P is found within the bacteria and appears in progeny phages, while (2) the ^{35}S is not found within the bacteria and is released with the phage ghosts.

a) **Preparation of radioactively labeled T2 bacteriophage**

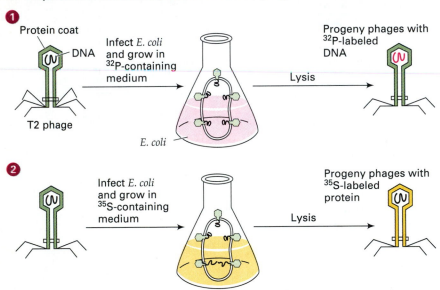

b) **Experiment that showed DNA to be the genetic material of T2**

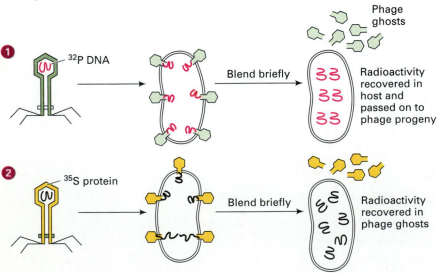

only DNA and protein, they realized that one of these two classes of molecules must be the genetic material. To determine which class it was, they infected unlabeled *E. coli* with the two types of radioactively labeled T2 with the results shown in Figure 8.4(b). When the infecting phage was ^{32}P-labeled, most of the radioactivity could be found within the bacteria soon after infection. Very little could be found in protein parts of the phage (the phage ghosts) released from the cell surface after mixing the cells in a kitchen blender. After lysis some of the ^{32}P was found in the progeny phages. In contrast, after infecting *E. coli* with ^{35}S-labeled T2, virtually none of the radioactivity appeared within the cell, and none was found in the progeny phage particles. Most of the radioactivity could be found in the phage ghosts released after treating the cultures in the kitchen blender.

Since genes serve as the blueprint for making the progeny virus particles, it was also presumed that the blueprint must get into the bacterial cell in order for new phage particles to be built. Therefore, since *it was DNA and not protein that entered the cell,* as evidenced by the presence of ^{32}P and the absence of ^{35}S, Hershey and Chase reasoned that *DNA must be the material responsible for the function and reproduction of phage T2.* The protein, they hypothesized, provided a structural framework to contain the DNA and the specialized structures required to inject the DNA into the bacterial cell.

Most of the organisms discussed in this book (such as humans, *Drosophila*, yeast, *E. coli*, and phage T2) have DNA as their genetic material. However, some bacterial viruses (such as Qβ), some animal viruses (such as poliovirus), and some plant viruses (such as tobacco mosaic virus) have RNA as their genetic material. No known prokaryotic or eukaryotic organism has RNA as its genetic material.

Keynote

Genetic material must contain all the information for the cell structure and function of an organism and must replicate accurately so that progeny cells have the same genetic information as the parental cell. In addition, genetic material must be capable of variation, one of the bases for evolutionary change. A series of experiments proved that the genetic material of organisms consists of one of two types of nucleic acids, DNA or RNA. Of the two, DNA is most common; only certain viruses have RNA as their genetic material.

The Chemical Composition of DNA and RNA

Both DNA and RNA are **macromolecules**, which means that they have a molecular weight of at least a few thousand daltons.* In this respect they differ markedly from many other molecules important to cell function, such as sugars and amino acids, which weigh from one hundred to a few hundred daltons.

Both DNA and RNA are polymeric molecules made up of four different monomeric units called **nucleotides**. Each nucleotide consists of three distinct parts: (1) a **pentose** (5-carbon) **sugar**; (2) a **nitrogenous** (nitrogen-containing) **base**; and (3) a **phosphate group**. Because they can be isolated from nuclei, and because they are acidic, these macromolecules are called *nucleic acids*.

For RNA the pentose sugar is **ribose**, and for DNA the sugar is **deoxyribose** (Figure 8.5). The two sugars differ by the chemical groups attached to the 2' carbon: a hydroxyl group (OH) in ribose, a hydrogen group (H) in deoxyribose.

The nitrogenous bases fall into two classes, the **purines** and the **pyrimidines**. In DNA the purines are **adenine** (A) and **guanine** (G), and the pyrimidines are **thymine** (T) and **cytosine** (C). The RNA molecule also contains adenine, guanine, and cytosine, but thymine is replaced by **uracil** (U). The chemical structures of these five bases are given in Figure 8.6. Note that thymine contains a methyl group (CH_3) not found in uracil.

In DNA and RNA bases are always attached to the 1' carbon of the pentose sugar by a covalent bond. The purine bases are bonded at the position 9 nitrogen, while the pyrimidines bond at the position 1 nitrogen (see Figure 8.6). The phosphate group (PO_4^{2-}) is attached to the 5' carbon of the sugar in both DNA and RNA. Examples of a DNA nucleotide (a **deoxyribonucleotide**) and an RNA nucleotide (a **ribonucleotide**) are shown in Figure

---

*One dalton is equivalent to a twelfth of the mass of the carbon 12 atom, or 1.67×10^{-24} g. One atomic mass unit (AMU), the approximate masses of protons and neutrons, = 1.67×10^{-24} g.

■ *Figure 8.5* Structures of ribose and deoxyribose, the pentose sugars of RNA and DNA, respectively. The difference between the two sugars is highlighted.

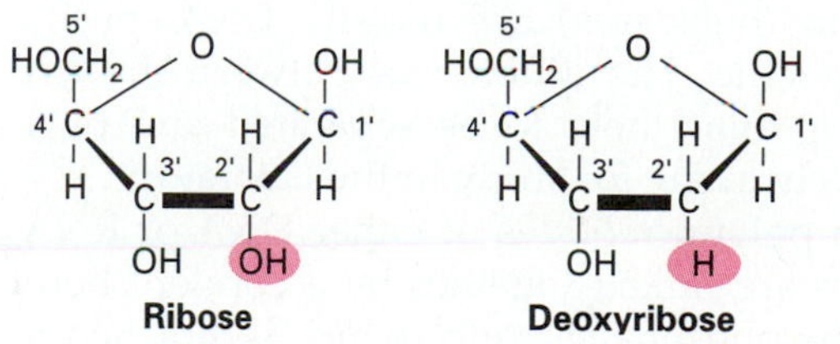

■ *Figure 8.6*   Structures of the nitrogenous bases in DNA and RNA. The parent compounds are purine (top), and pyrimidine (bottom). Differences between the bases are highlighted.

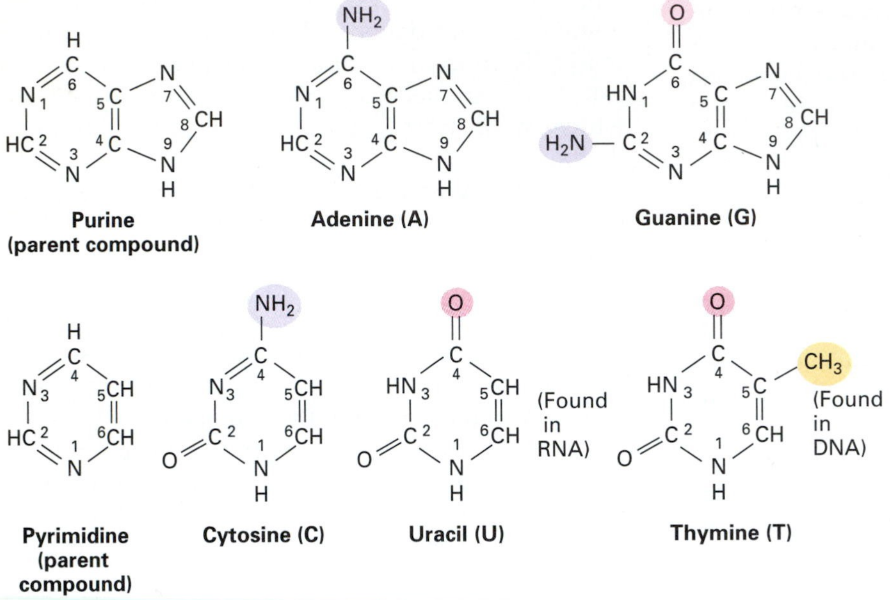

8.7(a). For future discussions, remember that a *nucleotide*, the basic building block of the DNA and RNA molecules, consists of the sugar, a base, and a phosphate group. The sugar plus base only is called a *nucleoside*, so a nucleotide is also called a **nucleoside phosphate.** A complete listing of the names for the bases, nucleosides, and nucleotides is presented in Table 8.1. The four deoxyribonucleotide subunits of DNA are deoxyadenosine 5'-monophosphate (dAMP), deoxyguanosine 5'-monophosphate (dGMP), deoxycytidine 5'-monophosphate (dCMP), and deoxythymidine 5'-monophosphate (dTMP). The four ribonucleotide subunits of RNA are adenosine 5'-monophosphate (AMP), guanosine 5'-monophosphate (GMP), cytidine 5'-monophosphate (CMP), and uridine 5'-monophosphate (UMP).

Because DNA contains the pentose sugar deoxyribose and the pyrimidine thymine, while RNA contains the pentose sugar ribose and the pyrimidine uracil, these two nucleic acids have different chemical and biological properties. For example, the cellular enzymes that catalyze nucleic acid synthesis (polymerases) and nucleic acid degradation (nucleases) are usually DNA-specific or RNA-specific. The differences between the two molecules permit them to be separated and purified relatively easily for study in the laboratory.

To form polynucleotides of either DNA or RNA, nucleotides are linked together by a covalent bond between the phosphate group (which is attached to

the 5' carbon of the sugar ring) of one nucleotide and the 3' carbon of the pentose sugar of another nucleotide. These 5'–3' phosphate linkages are called **phosphodiester bonds.** A short polynucleotide chain is diagrammed in Figure 8.7(b). The phosphodiester bonds are relatively strong, and as a consequence, the repeated sugar-phosphate-sugar-phosphate backbone of DNA and RNA is a stable structure.

To understand how a polynucleotide chain is synthesized (we will study this in a later chapter), we must be aware of one more feature of the chain: The two ends of the chain are not the same. The chain has a 5' carbon (with a phosphate group on it) at one end and a 3' carbon (with a hydroxyl group on it) at the other end, as shown in Figure 8.7(b). This asymmetry is referred to as *polarity* of the chain.

K e y n o t e

DNA and RNA occur in nature as macromolecules composed of smaller building blocks called nucleotides. Each nucleotide consists of a 5-carbon sugar (deoxyribose in DNA, ribose in RNA) to which is attached a phosphate group and one of four nitrogenous bases: adenine, guanine, cytosine, and thymine (in DNA), or adenine, guanine, cytosine, and uracil (in RNA).

■ *Figure 8.7*   Chemical structures of DNA and RNA: (a) Basic structures of DNA and RNA nucleosides (sugar plus base) and nucleotides (sugar plus base plus phosphate group), the basic building blocks of DNA and RNA molecules. Here the phosphate groups are orange, sugars are red, and bases are brown. (b) A segment of a polynucleotide chain, in this case a single strand of DNA. The deoxyribose sugars are linked by phosphodiester bonds between the 3' carbon of one sugar and the 5' carbon of the next sugar. The bases on this segment are adenine (A), guanine (G), and thymine (T).

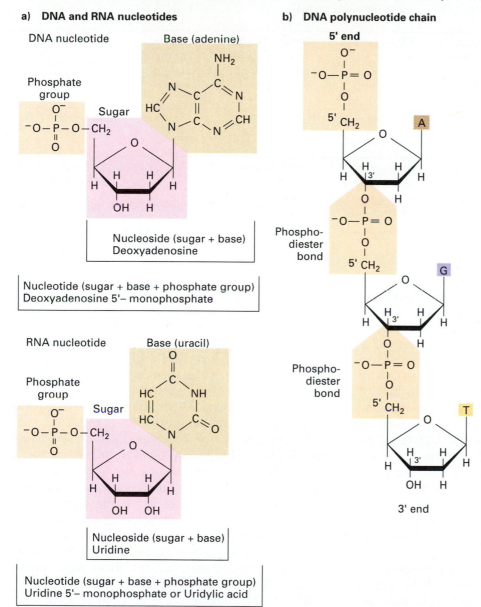

a) DNA and RNA nucleotides

b) DNA polynucleotide chain

The DNA Double Helix

In 1953 James D. Watson and Francis H. C. Crick published a paper in which they proposed a model for the physical and chemical structure of the DNA molecule. According to their model, most DNA consists of two polynucleotide chains wound around each other in a right-handed (clockwise)

helix. In generating their model, Watson and Crick used three main pieces of evidence:

1. The DNA molecule was known to be composed of bases, sugars, and phosphate groups linked together as a **polynucleotide** (deoxyribonucleotide) chain.

2. By chemical treatment Erwin Chargaff had

Table 8.1

Names of the Base, Nucleoside, and Nucleotide Components Found in DNA and RNA

		BASE: PURINES (PU)		BASE: PYRIMIDINES (PY)		
		ADENINE (A)	GUANINE (G)	CYTOSINE (C)	THYMINE (T) (DEOXYRIBOSE ONLY)	URACIL (U) (RIBOSE ONLY)
DNA	Nucleoside: deoxyribose + base	Deoxy-adenosine (dA)	Deoxy-guanosine (dG)	Deoxycytidine (dC)	Thymidine (dT)	
	Nucleotide: deoxyribose + base + phosphate group	Deoxyadenylic acid or deoxy-adenosine monophosphate (dAMP)	Deoxyguanylic acid or deoxy-guanosine monophosphate (dGMP)	Deoxycytidylic acid or deoxy-cytidine monophosphate (dCMP)	Thymidylic acid or thymidine monophosphate (TMP)	
RNA	Nucleoside: ribose + base	Adenosine (A)	Guanosine (G)	Cytidine (C)		Uridine (U)
	Nucleotide: ribose + base + phosphate group	Adenylic acid or adenosine monophosphate (AMP)	Guanylic acid or guanosine monophosphate (GMP)	Cytidylic acid or cytidine monophosphate (CMP)		Uridylic acid or uridine monophosphate (UMP)

hydrolyzed the DNA of a number of organisms and had quantified the purines and pyrimidines released. His studies showed that in all the DNAs the amount of the purines was equal to the amount of the pyrimidines. More important, the amount of adenine (A) was equal to that of thymine (T), and the amount of guanine (G) was equal to that of cytosine (C). These equivalencies have become known as Chargaff's Rules. In comparisons of DNAs from different organisms, the A/T and G/C ratios are always the same, although the (A + T)/(G + C) ratio (typically presented as %GC) varies.

3. Rosalind Franklin, working with Maurice H. F. Wilkins, studied isolated fibers of DNA by using the X-ray diffraction technique, a procedure in which a beam of parallel X rays is directed on a regular, repeating array of atoms. The beam is diffracted ("broken up") by the atoms in a pattern that is characteristic of the atomic weight and the atom's spatial arrangement. The diffracted X rays are recorded on a photographic plate. By analyzing the photograph, Franklin obtained information about the molecule's atomic structure. The analysis of X-ray diffraction patterns is extremely complicated. As a result, given diffraction patterns can usually be interpreted in more than one way, and models built of the analyzed molecules may not be accurate. Moreover, since the experiments usually use molecules in a crystalline or fiber formation, the structures deduced may not precisely reflect the form of the molecules in the cell.

The X-ray diffraction patterns of DNA indicated that the molecule is organized in a highly ordered, helical structure. An example of DNAs X-ray diffraction pattern and the method by which it was obtained are illustrated in Figure 8.8. Franklin interpreted these data to mean that DNA was a helical structure that had two distinctive regularities of 0.34 nm and 3.4 nm along the axis of the molecule (1 nanometer (nm) = $1/10^9$ meter).

Watson and Crick considered all the evidence just described and began to build three-dimensional models for the structure of DNA. The model they devised, which fit all the known data on the composition of the DNA molecule, is the now-famous double-helix model for DNA. Figure 8.9(a) shows a three-dimensional model of the DNA molecule, and Figure 8.9(b) is a diagram of the DNA molecule, showing the arrangement of the sugar-phosphate backbone and base pairs in a stylized way.

The double-helical model of DNA proposed by Watson and Crick has the following major features:

1. The DNA molecule consists of two polynucleotide chains wound around each other in a right-handed double helix; that is, viewed on end, the two strands wind around each other in a clockwise (right-handed) fashion.

2. The diameter of the helix is 2 nm.

3. The two chains are *antiparallel*, or show *opposite polarity*; that is, the two strands are oriented in opposite directions, with one strand oriented in the

■ *Figure 8.8*   X-ray diffraction analysis of DNA: The X-ray diffraction pattern of DNA that Watson and Crick used in developing their double-helix model. The dark areas that form an X shape in the center of the photograph indicate the helical nature of DNA. The dark crescents at the top and bottom of the photograph indicate the 0.34-nm distance between the base pairs.

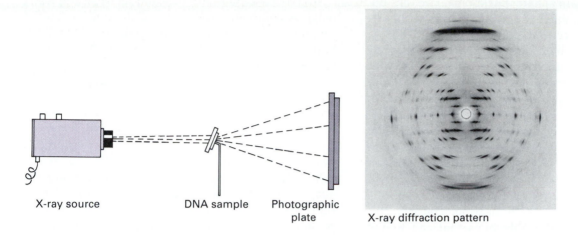

X-ray source   DNA sample   Photographic
                                          plate   X-ray diffraction pattern

■ *Figure 8.9*   Molecular structure of DNA: (a) Three-dimensional molecular model of DNA as prepared by Watson and Crick. H = hydrogen, O = oxygen, C = carbon, N = nitrogen, P = phosphorus; (b) Stylized representation of the DNA double helix showing the base pairs and the sugar-phosphate backbones; (c) Schematic showing the major and minor grooves in double-helical DNA.

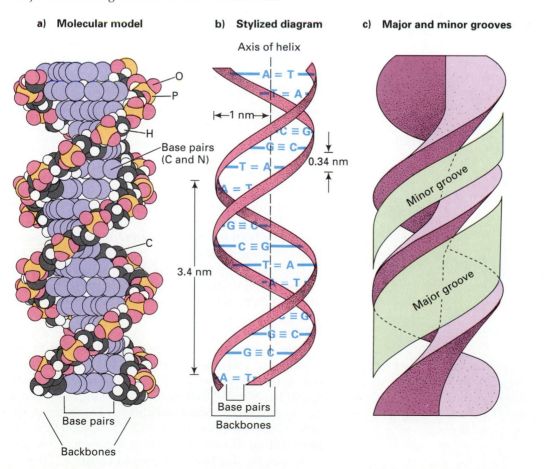

a) **Molecular model**   b) **Stylized diagram**   c) **Major and minor grooves**

5′ to 3′ direction and the other strand oriented 3′ to 5′. In simpler terms, if the 5′ end is the "head" of the chain and the 3′ end is the "tail" of the chain, antiparallel means that the head of one chain is against the tail of the other chain and *vice versa*.

4. The sugar-phosphate backbones are on the outsides of the double helix, while the bases are oriented toward the central axis (see Figure 8.9[a,b]). The bases of both chains are flat structures oriented perpendicularly to the long axis of the DNA; that is, the bases are stacked like pennies on top of one another (except for the "twist" of the helix).

5. The bases of the opposite strands are bonded together by relatively weak hydrogen bonds. The only specific pairings observed are A with T (two hydrogen bonds) and G with C (three hydrogen bonds) (Figure 8.10). The weak hydrogen bonds make it relatively easy to separate the two strands of the DNA, for example, by heating. The A-T and G-C base pairs are the only ones that can fit the physical dimensions of the helical model, and they are totally in accord with Chargaff's Rules. The specific A-T and G-C pairs are called *complementary*

base pairs, so the nucleotide sequence in one strand dictates the nucleotide sequence of the other. For instance, if one chain has the nucleotide sequence 5′-TATTCCGA-3′, the opposite, antiparallel chain must bear the sequence 3′-ATAAGGCT-5′.

6. The base pairs are 0.34 nm apart in the DNA helix. A complete (360°) turn of the helix takes 3.4 nm; therefore, there are 10 base pairs per turn. Each base pair, then, is twisted 36° clockwise with respect to the previous base pair.

7. The bonds that attach two complementary base pairs to their sugar rings are not directly opposite each other. Because of this, the two sugar-phosphate backbones of the double helix are not equally spaced along the helical axis, resulting in grooves of unequal size between the backbones called the *major groove* (the wider groove) and the *minor groove* (the narrower groove)(Figure 8.9[c]). Both of these grooves are large enough to allow protein molecules to make contact with the bases. The phenomenon of proteins "reading" specific base-pair sequences is common to many molecular processes as we will see in the following chapters.

■ *Figure 8.10*   Structures of the complementary base pairs found in DNA. In both cases a purine pairs with a pyrimidine: (a) The adenine-thymine bases, which pair through two hydrogen bonds; (b) The guanine-cytosine bases, which pair through three hydrogen bonds.

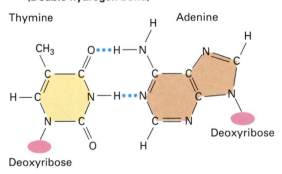

a) **Adenine-thymine base
(Double hydrogen bond)**

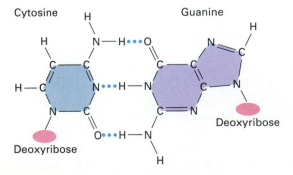

b) **Guanine-cytosine base
(Triple hydrogen bond)**

Keynote

The DNA molecule usually consists of two polynucleotide chains joined by hydrogen bonds between pairs of bases (A and T, G and C) in a double helix. The diameter of the helix is 2 nm, and there are 10 base pairs in each complete turn (3.4 nm). The double-helix model was proposed by Watson and Crick from chemical and physical analyses of DNA.

B-DNA and Z-DNA

A more thorough analysis of the early X-ray diffraction patterns of DNA fibers revealed that DNA can exist in different forms, depending on the conditions. The double-helical form deduced by Watson and Crick is in what is called the B form. A space-filling model is shown in Figure 8.11(a); Figure 8.12(a) shows a perspective drawing. B-DNA has a right-handed helix, so the helix axis passes through the base pairs, which are oriented perpendicular to that axis. There are 10.0 base pairs per 360° turn of the helix.

Other forms of DNA have been identified in the laboratory. The most intriguing is Z-DNA, which has a totally unexpected structure and is rarely found in cells. Z-DNA has a *left-handed* helix and a zigzag sugar-phosphate backbone. (The latter property gave this DNA form its "Z" designation.) A space-filling model is shown in Figure 8.11(b), and a perspective drawing is shown in Figure 8.12(b). Z-

■ *Figure 8.11* Space-filling models of: (a) B-DNA; (b) Z-DNA.

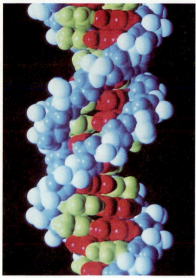
a)

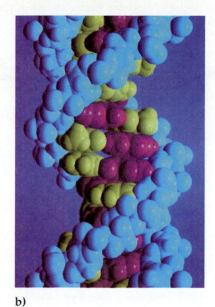
b)

DNA is found in stretches of DNA that are rich in guanine and cytosine. Compared with B-DNA, the Z-DNA double helix is elongated and thin, with no real major groove, but with a deep minor groove. Z-DNA has 12.0 base pairs per complete helical turn and the bases are inclined slightly away from the perpendicular.

While it is clear that B-DNA is the major form of DNA found in the cell, it has been difficult to prove the existence of Z-DNA in cells. In a number of cases where the presence of Z-DNA has been claimed, critics have argued that the observation is an artifact of the analytical procedures rather than being a natural feature of the chromosome. Recent experiments have shown, however, that Z-DNA *is* present in *E. coli*, but its biological function, if any, is unknown.

Keynote

X-ray diffraction analysis of single crystals of DNA oligomers of known sequence has produced detailed molecular models of two important DNA types: B-DNA and Z-DNA. B-DNA is a right-handed double helix; Z-DNA is a left-handed double helix. The number of base pairs per helical turn are 10.0 for B-DNA and 12.0 for Z-DNA. Compared with B-DNA, Z-DNA is elongated and thin. B-DNA is the form predominantly found in cells. Z-DNA has been proved to be present in the *E. coli* chromosome, but its biological function, if any, is not known.

Summary

In this chapter we have learned that DNA or RNA can be the genetic material. All prokaryotic and eukaryotic organisms and most viruses have DNA as their genetic material, while some viruses have RNA as their genetic material. The form of the genetic material varies among organisms and their viruses. In prokaryotic and eukaryotic organisms, the DNA is always double-stranded, whereas in viruses the genetic material may be double- or single-stranded DNA or RNA, depending on the virus.

Chemical analysis revealed that DNA and RNA are macromolecules composed of smaller building blocks called nucleotides. Each nucleotide consists of a 5-carbon sugar (deoxyribose in DNA, ribose in RNA) to which is attached one of four nitrogenous bases and a phosphate group. In DNA the four nitrogenous bases are adenine, guanine, cytosine, and thymine. In RNA the four bases are adenine, guanine, cytosine, and uracil. Thymine and uracil differ only by a methyl group that is present in thymine and absent in uracil. From chemical and physical analyses, it was determined that the DNA molecule consists of two polynucleotide chains joined by hydrogen bonds between pairs of bases (A and T, G and C) in a double helix. (The double-helix model was first proposed by Watson and Crick.) The diameter of the helix is 2 nm, and there are 10 base pairs in each complete turn of the helix (3.4 nm); each base pair is thus twisted 36° relative to the preceding base pair in the molecule.

A number of different types of double-helical DNA have been identified by X-ray diffraction

■ *Figure 8.12*   Comparison of (a) B-DNA; (b) Z-DNA. Short segments of each DNA are shown. The bases are red and the backbones are blue. Note the zigzag course of the sugar-phosphate backbone chain in the Z-DNA. (© Irving Geiss.)

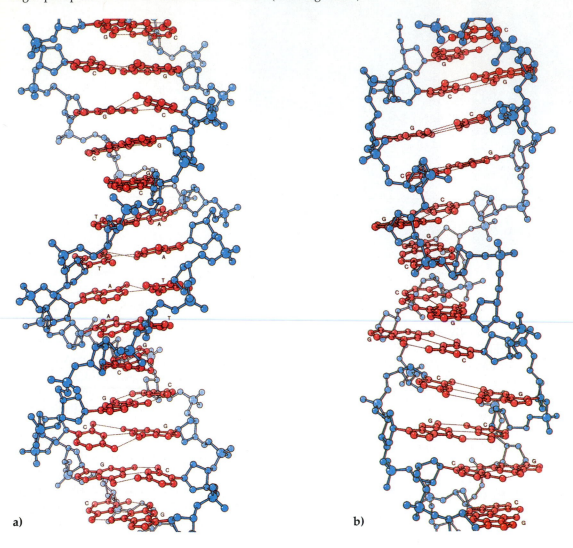

a)

b)

analysis. Two significant types are the right-handed B-DNA and the left-handed Z-DNA. The common form found in cells is B-DNA (the form analyzed by Watson and Crick). Z-DNA may exist in cells in stretches of DNA that are particularly rich in guanine and cytosine. The functional significance of Z-DNA is unknown.

Analytical Approaches for Solving Genetics Problems

Q8.1   The linear chromosome of phage T2 is 52 μm long. The chromosome consists of double-stranded DNA, with 0.34 nm between each base pair. The average weight of a base pair is 660 daltons. What is the molecular weight of the T2 molecule?

A8.1   This question involves the careful conversion of different units of measurement. The first step is to put the lengths in the same units: 52 μm is 52 millionths of a meter, or $52,000 \times 10^{-9}$ m, or 52,000 nm. One base occupies 0.34 nm in the double helix, so the number of base pairs in this chromosome is 52,000 divided by 0.34, or 152,941 base pairs. Each base pair, on the average, weighs 660 daltons; therefore the molecular weight of the chromosome is

152,941 × 660 = 1.01 × 10⁸ daltons, or 101 million daltons.

The average length of the double helix in a human chromosome is 3.8 cm, which is 3.8 hundredths of a meter or 38 million nm—substantially longer than the T2 chromosome! There are over 111.7 million base pairs in the average human chromosome.

Q8.2 The accompanying table lists the relative percentages of bases of nucleic acids isolated from different species. For each one, what type of nucleic acid is involved? Is it double-stranded or single-stranded? Explain your answer.

SPECIES	ADENINE	GUANINE	THYMINE	CYTOSINE	URACIL
(i)	21	29	21	29	0
(ii)	29	21	29	21	0
(iii)	21	21	29	29	0
(iv)	21	29	0	29	21
(v)	21	29	0	21	29

A8.2 This question focuses on the base-pairing rules and the difference between DNA and RNA. In analyzing the data, we should determine first whether the nucleic acid is RNA or DNA, and then whether it is double- or single-stranded. If the nucleic acid has thymine, it is DNA; if it has uracil, it is RNA. Thus species (i), (ii), and (iii) must have DNA as their genetic material, and species (iv) and (v) must have RNA as their genetic material.

Next, the data must be analyzed for strandedness. Double-stranded DNA must have equal percentages of A and T, and of G and C. Similarly, double-stranded RNA must have equal percentages of A and U, and of G and C. Hence species (i) and (ii) have double-stranded DNA, while species (iii) must have single-stranded DNA since the base pairing rules are violated, with A = G and T = C but A ≠ T and G ≠ C. As for the RNA-containing species, (iv) contains double-stranded RNA since A = U and G = C, and (v) must contain single-stranded RNA.

Q8.3 Here are four characteristics of one strand (the "original" strand) of a particular long, double-stranded DNA molecule: (1) 35% of the adenine-containing nucleotides (As) have guanine-containing nucleotides (Gs) on their 3′ sides; (2) 30% of the As have Ts as their 3′ neighbors; (3) 25% of the As have Cs; and (4) 10% of the As have As as their 3′ neighbors. Use this information to answer the following questions, explaining your reasoning in each case. (Some of the questions may be unanswerable without more information.)
a. In the complementary DNA strand, what will be the frequencies of the various bases on the 3′ side of A?
b. In the complementary strand, what will be the frequencies of the various bases on the 3′ side of T?
c. In the complementary strand, what will be the frequency of each kind of base on the 5′ side of T?
d. Why is the percentage of A not equal to the percentage of T (and the percentage of C not equal to the percentage of G) among the 3′ neighbors of A in the original DNA strand described?

A8.3 a. This cannot be answered without more information. Although we know that the As neighbored by Ts in the original strand will correspond to As neighbored by Ts in the complementary strand, there will be additional As in the complementary strand about whose neighbors we know nothing.
b. This also cannot be answered. All the As in the original strand correspond to Ts in the complementary strand, but we only know about the 5′ neighbors of these Ts, not the 3′ neighbors.
c. 10% will be T, 30% will be A, 25% will be G, and 35% will be C.
d. The A = T and G = C rule only applies when considering both strands of a double-stranded DNA. Here we are considering only the original single strand of DNA.

Questions and Problems

8.1 In the 1920s, while working with *Diplococcus pneumoniae*, the agent that causes pneumonia, Griffith discovered an interesting phenomenon. In the experiments mice were injected with different types of bacteria. For each of the following bacteria type(s) injected, indicate whether the mice lived or died:

a. type *R*;
b. type *S*;
c. heat-killed *S*;
d. type *R* and heat-killed *S*.

***8.2** By differentially labeling the coat protein and the DNA of phage T2, Hershey and Chase demon-

strated that (choose the correct answer):
a. only the protein enters the infected cell.
b. the entire virus enters the infected cell.
c. a metaphase chromosome is composed of two chromatids, each containing a single DNA molecule.
d. the phage genetic material is most probably DNA.
e. the phage coat protein directs synthesis of new progeny phage.

*8.3 In DNA and RNA, which carbon atoms of the sugar molecule are connected by a phosphodiester bond?

8.4 Which base is unique to DNA and which base is unique to RNA?

8.5 How do nucleosides and nucleotides differ?

8.6 What chemical group is found at the 5' end of a DNA chain? At the 3' end of a DNA chain?

*8.7 What is the base sequence of the DNA strand that would be complementary to the following single-stranded DNA molecules:
a. 5' AGTTACCTGATCGTA 3'
b. 5' TTCTCAAGAATTCCA 3'

*8.8 For double-stranded DNA, which of the following base ratios always equals 1?
a. $(A + T)/(G + C)$
b. $(A + G)/(C + T)$
c. C/G
d. $(G + T)/(A + C)$
e. A/G

8.9 If the ratio of $(A + T)$ to $(G + C)$ in a particular DNA is 1.00, does this result indicate that the DNA is most likely constituted of two complementary strands of DNA or a single strand of DNA, or is more information necessary?

8.10 Explain whether the $(A + T)/(G + C)$ ratio in double-stranded DNA is expected to be the same as the $(A + C)/(G + T)$ ratio.

*8.11 The percent cytosine in a double-stranded DNA is 17. What is the percent of adenine in that DNA?

8.12 Upon analysis, a DNA molecule was found to contain 32% thymine. What percent of this same molecule would be made up of cytosine?

8.13 A sample of double-stranded DNA has a percent GC content of 62. What is the percent of A in the DNA?

8.14 A double-stranded DNA polynucleotide contains 80 thymidylic acid and 110 guanylic acid residues. What is the total nucleotide number in this DNA fragment?

8.15 Analysis of DNA from a bacterial virus indicates that it contains 33% A, 26% T, 18% G, and 23% C. Interpret these data.

8.16 If a virus particle contains double-stranded DNA with 200,000 base pairs, how many complete 360° turns occur in this molecule?

*8.17 A double-stranded DNA molecule is 100,000 base pairs (100 kilobases long).
a. How many nucleotides does it contain?
b. How many complete turns are there in the molecule?
c. How long is the DNA molecule?

*8.18 If nucleotides were arranged at random in a single-stranded RNA 10^6 nucleotides long, and if the base composition of this RNA was 20% A, 25% C, 25% U, and 30% G, how many times would you expect the specific sequence (5')-GUUA-(3') to occur?

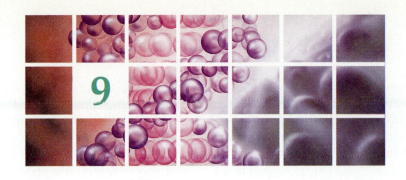

9

The Organization of DNA in Chromosomes

■ In bacteria the chromosome is a circular, double-stranded DNA molecule that is compacted by supercoiling of the DNA helix. In the *E. coli* chromosome about 100 independent looped domains of supercoiled DNA have been identified.

■ The complete set of metaphase chromosomes in a eukaryotic cell is called its karyotype. The karyotype is species-specific.

■ The amount of DNA in a prokaryotic chromosome or the haploid amount of DNA in a eukaryotic cell is called the C value. There is not a direct relationship between the C value and the structural or organizational complexity of an organism.

■ The nuclear chromosomes of eukaryotes are complexes of DNA and histone and nonhistone chromosomal proteins. Each chromosome consists of one linear, unbroken, double-stranded DNA molecule running throughout its length. The histones are constant from cell to cell within an organism, while the nonhistones vary significantly between cell types.

■ The large amount of DNA present in the eukaryotic chromosome is compacted by its association with histones in nucleosomes, and by higher levels of folding of the nucleosomes into chromatin fibers. Each chromosome contains a large number of looped domains of fibers attached to a protein scaffold.

■ The centromere region of each eukaryotic chromosome is responsible for the accurate segregation of the replicated chromosome to the daughter cells during mitosis and meiosis. The DNA sequences of centromeres are species-specific.

■ The ends of chromosomes, the telomeres, consist of simple, relatively short, tandemly repeated sequences that are species-specific.

■ Prokaryotic genomes consist mostly of unique-sequence DNA, with only a few repeated sequences and genes. Eukaryotes have both unique and repetitive sequences in the genome. The spectrum of complexity of repetitive DNA sequences among eukaryotes is extensive.

I
N THIS CHAPTER we will examine how DNA is organized into chromosomes. The chromosomes of eukaryotes are much more complex than the chromosomes of prokaryotes. Eukaryotic chromosomes consist of a highly ordered complex of DNA and proteins, with special regions—centromeres and telomeres—that are of particular importance for chromosome function. Ultimately, geneticists will need to have an even greater understanding of chromosome structure before the regulation of gene expression can be completely understood.

Bacterial Chromosomes

We will focus in this section on the chromosome of *E. coli*. The DNA of the bacterium *Escherichia coli* is located in a central region called the nucleoid (Figure 9.1). (Recall that *E. coli* cells do not have a nucleus.) If an *E. coli* cell is lysed gently, the DNA is released in a highly folded state. The double-stranded DNA is present as a single chromosome,

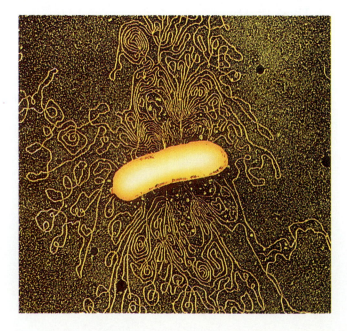

■ *Figure 9.1* Nucleoid region released from a lysed *E. coli* cell.

approximately 1100 μm long (4×10^3 kb [1 kb = 1 kilobase pairs = 1000 base pairs]). This amount of DNA is approximately 1000 times the length of the *E. coli* cell. How is all that DNA packaged into the nucleoid region of the cell? To answer this question we must investigate the properties of linear and circular forms of DNA.

Supercoiling of the Chromosome

Circular DNA can exist in a *relaxed* or a *supercoiled form*. Let us consider a 200-base-pair linear piece of DNA in the B-form (Figure 9.2[a]). Such a molecule has two free ends and, since there are 10 base pairs per helical turn of B-form DNA (see Chapter 8, p. 182), there are 20 helical turns in this molecule. If we now simply join the two ends, we have produced a circular DNA molecule (Figure 9.2[b]). This circular DNA molecule is said to be *relaxed*. Alternatively, if we first untwist one end of the linear DNA molecule by two turns (Figure 9.2[c]), and then join the two ends, the circular DNA molecule produced will have only 18 helical turns and a small unwound region (Figure 9.2[d]). Such a structure is not energetically favored, and will switch to a structure with 20 helical turns and 2 superhelical turns (Figure 9.2[e]). This structure is said to be *supercoiled*; that is, the DNA has supercoils introduced into it by having the double helix twisted in space about its own axis. It is important to remember that supercoiling can only occur in circular DNA, or in linear DNA in instances in which the

■ *Figure 9.2* Illustration of DNA supercoiling. (a) A linear B-form DNA with 20 helical turns (200 base pairs); (b) Relaxed circular DNA produced by joining the two ends of the linear molecule of (a); (c) The linear DNA molecule of (a) unwound from one end by two helical turns; (d) A possible circular DNA molecule produced by joining the two ends of the linear molecule of (c) showing 18 helical turns and a short unwound region; (e) The more energetically favored form of (d), a supercoiled DNA with 20 helical turns and two superhelical turns.

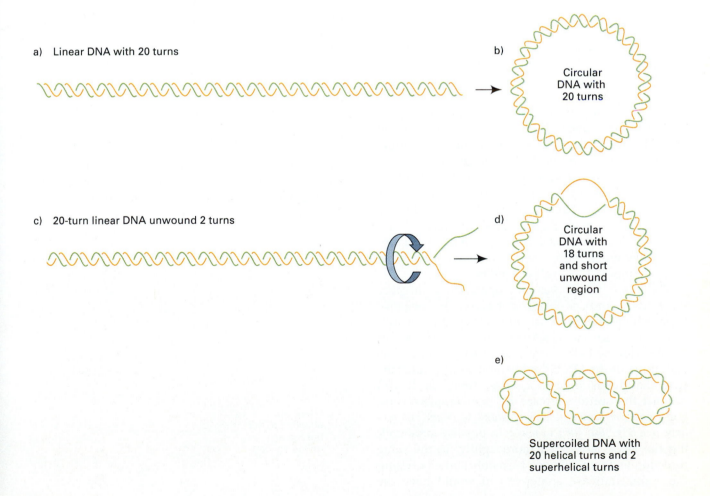

a) Linear DNA with 20 turns

b) Circular DNA with 20 turns

c) 20-turn linear DNA unwound 2 turns

d) Circular DNA with 18 turns and short unwound region

e) Supercoiled DNA with 20 helical turns and 2 superhelical turns

■ *Figure 9.3* Electron micrographs of plasmid DNA: (a) relaxed (nonsupercoiled) DNA: (b) supercoiled DNA. Both molecules are shown at the same magnification.

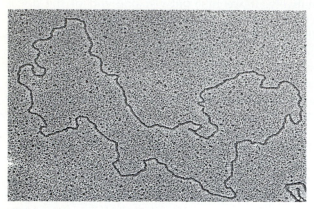

a)

b)

two ends are anchored in some way. To understand this more easily, think about a length of rope. If you twist the rope at one end without holding the other end, the rope just spins in the air and remains linear (i.e., relaxed). If you hold the other end while we twist it, the rope will eventually become knotted up (i.e., supercoiled). The very act of supercoiling produces tension in the DNA molecule. Therefore if a break (a nick) is introduced into the sugar-phosphate backbone of a supercoiled DNA molecule, spontaneous untwisting of the molecule produces a relaxed DNA circle. Figure 9.3 shows relaxed and supercoiled circular DNA of a plasmid to demonstrate how much more compact a supercoiled molecule is compared to a relaxed molecule.

There are two types of supercoiling—*negative supercoiling* and *positive supercoiling*. In negative supercoiling the DNA is *untwisted*, in the *opposite* direction from the clockwise turns of the right-handed double helix. To visualize a negative supercoil, think of the DNA double helix as a spiral staircase that turns in a clockwise direction. If you untwist the spiral staircase by one complete turn, you have *the same number of stairs to climb but you have one less 360° turn to make*. In positive supercoiling the DNA is twisted more tightly, in the same direction as the clockwise turns of the helix. Using the spiral staircase analogy, you would have the

same number of stairs to climb but now there is one more 360° turn to make.

The amount and type of DNA supercoiling is brought about by enzymes called **topoisomerases.** These enzymes convert one topological form of DNA into another. In *E. coli* topoisomerase II (also called *DNA gyrase*) untwists relaxed DNA to produce negatively supercoiled DNA. Topoisomerase I does the opposite: it converts negatively supercoiled DNA to the relaxed state. With both enzymes in the cell, then, DNA can be interchanged between the negatively supercoiled and relaxed states. By making extensively negatively supercoiled DNA, topoisomerase II enables the *E. coli* chromosome to pack into a very compact state.

Proteins Complexed to the Chromosome

Eukaryotic DNA is complexed with a number of discrete structural proteins called histones, which serve to compact the DNA into the chromosome structures characteristic of eukaryotic nuclei (see the discussion on, p. 000). Bacterial DNA also is associated with structural proteins, although not to the extent of eukaryotic DNA. In *E. coli* two structural proteins—HU and H—are bound to the DNA; they resemble two of the eukaryotic histone structural proteins. Based on analyses of the *E. coli* proteins and of the properties of nucleoid DNA, a model for the structure of the bacterial chromosome has been proposed (Figure 9.4). In the model the bacterial chromosome has about 100 independent *domains*. Each domain consists of a loop of

■ *Figure 9.4* Model for the structure of a bacterial chromosome. The chromosome is organized into looped domains, the bases of which are anchored in an unknown way.

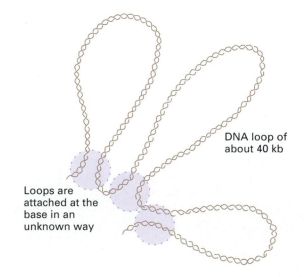

DNA loop of about 40 kb

Loops are attached at the base in an unknown way

about 40,000 base pairs (40 kilobase pairs [kb] = 13 μm) of negatively supercoiled DNA. The ends of each domain are held (by unknown means) such that the supercoiled DNA state in that domain is not affected by events that influence supercoiling of DNA in the other domains.

Keynote

In bacteria the chromosome consists of DNA with a number of associated proteins. The chromosome is a circular, double-stranded DNA molecule that is compacted by supercoiling of the DNA helix. In the *E. coli* chromosome about 100 independent looped domains of supercoiled DNA have been identified.

Bacteriophage λ Chromosome

Viruses are known that infect all living organisms. Viral genomes are known that consist of double-stranded DNA (e.g., bacteriophage T2 and bacteriophage λ), single-stranded DNA (e.g., bacteriophage ΦX174), double-stranded RNA (e.g., the plant virus, tobacco mosaic virus, and killer viruses of yeast), and single-stranded RNA (e.g., the animal virus, poliovirus). Most viruses have linear genomes (e.g., T2, λ, tobacco mosaic virus, poliovirus), and some have circular genomes (e.g., ΦX174). Given the scope of this textbook, it is impossible to describe the genome organizations of all viruses of genetic significance. We will discuss one example, the organization of the λ genome.

The structure and life cycle of the temperate bacteriophage λ was described in Chapter 6 (see Figure 6.12). This phage has been studied extensively for a number of years, so many of its gene functions are well understood. In Chapter 13 we will discuss how λ can be used to clone genes using recombinant DNA techniques, and in Chapter 14 we will discuss the gene regulation system that determines whether this bacteriophage will go through the lytic pathway or become lysogenic.

The phage λ chromosome is linear double-stranded DNA, and has no structural proteins associated with it. At each end there is a 12 nucleotide long, single-stranded DNA segment (Figure 9.5[a]). The base sequence of the two single-stranded ends are complementary, so they can pair together to form double-stranded DNA. This is important for the life cycle as we will now see.

Regardless of whether λ goes through the lytic or the lysogenic cycle, the first step after the λ DNA is injected into the host cell is the conversion of the linear molecule into a circular molecule (Figure 9.5[a]). The complementary ends ("sticky ends") pair and the single-stranded gaps are bonded in a reaction catalyzed by the enzyme DNA ligase (Figure 9.5[a]). The paired ends are called the *cos* sequence. In the lysogenic cycle the circular DNA finds a particular site in the host *E. coli* chromosome, and by a crossing-over event the DNA is integrated into the main chromosome (see Figure 6.12, p. 139).

In the lytic cycle the DNA replicates in the bacterium and produces a long, linear molecule consisting of many head-to-tail copies of the chromosome; this molecule is called a *concatamer*. Progeny phage λ chromosomes are generated from the concatamer as follows: The phage λ chromosome has a gene called *ter* (short for "terminus-generating activity") as shown in Figure 9.5(b). The *ter* gene codes for a DNA endonuclease (an enzyme that digests a nucleic acid chain somewhere along its length rather than at the termini). The endonuclease recognizes the *cos* sequence. Once *ter* is aligned on the DNA at the *cos* site, the endonuclease makes a staggered cut such that linear λ chromosomes with the correct complementary, 12-base-long, single-stranded ends are produced. These linear chromosomes are packaged in the assembled phage heads, complete progeny λ phage particles are assembled, then released from the cell when it lyses.

Eukaryotic Chromosomes

A fundamental difference between prokaryotes and eukaryotes is that prokaryotes have only one chromosome, while most eukaryotes have the diploid number of chromosomes in almost all somatic cells. A photograph of a stained set of metaphase chromosomes (see Chapter 1, p. 11) from a human male is shown in Figure 9.6. The number of chromosomes is 46. Recall that this is because humans are diploid (2N) organisms, possessing one haploid (N) set of chromosomes (23 chromosomes) from the egg, and another haploid set from the sperm.

The Karyotype

A complete set of all the metaphase chromosomes in a cell is called its **karyotype** (literally, "nucleus type"). For most organisms, all cells have the same karyotype. However, the karyotype is species-specific, so a wide range of number, size, and shape of metaphase chromosomes is seen among different eukaryotic organisms. Even closely related organisms may have quite different karyotypes. Table 9.1 gives the chromosome numbers of selected diploid eukaryotes.

■ *Figure 9.5*   Chromosome structure of phage λ varies at stages of lytic infection of
E. coli: (a) parts of the λ chromosome showing the nucleotide sequence of the two single-
stranded, complementary ("sticky") ends, and the chromosome circularizing after
infection by pairing of the ends, with the single-stranded gaps joined by DNA ligase to
produce a covalently closed circle: (b) Generation of the "sticky" ends of the λ DNA
during the lytic cycle. During replication of the λ chromosome, a giant concatameric
DNA molecule is produced; it contains tandem repeats of the λ genome. The diagram
shows the "join" between two adjacent λ chromosomes and the extent of the *cos*
sequence. The *cos* sequence is recognized by the *ter* gene product, an endonuclease that
makes two cuts at the sites shown by the arrows. These cuts produce one λ chromosome
from the concatamer.

a) Linear λ chromosome (~48,000 base pairs) forms circular λ chromosome

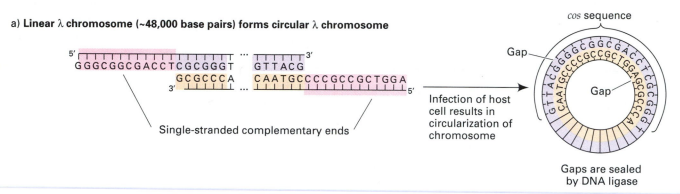

b) Production of progeny, linear λ chromosomes from concatamers (multiple copies linked end-to-end at complementary ends)

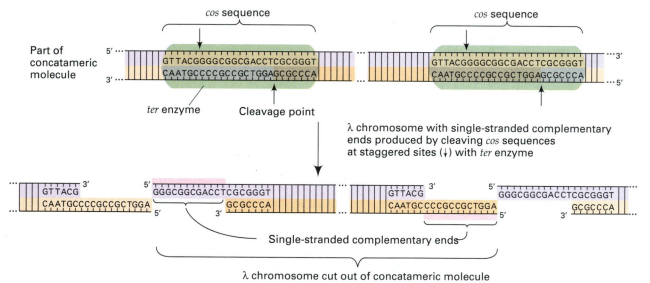

Figure 9.7 shows the karyotype for the cell of a
normal human male. It is customary, particularly
with human chromosomes, to arrange chromosomes in order according to size. This karyotype
shows 46 chromosomes: 2 pairs of each of the 22
autosomes and 1 of each of the X and Y sex chromosomes. In a human karyotype the chromosomes are
numbered for easy identification. Conventionally,
the largest pair of homologous chromosomes is
designated 1, the next largest 2, and so on. In
humans chromosomes 1 through 22 are called auto-

somes as distinguished from the pair of sex chromosomes. Although the chromosomes are numbered in order of size, for historical reasons, the
chromosome numbered 21 is actually smaller than
chromosome 22. Formally, the sex chromosomes
constitute pair 23 even though, in humans at least,
they do not fit properly in the size scale. As shown
in Figure 9.7, the X chromosome is a large metacentric chromosome, and the Y chromosome is a much
smaller chromosome.

A knowledge of the size, overall morphology,

■ *Figure 9.6* Mitotic metaphase chromosomes of a human male.

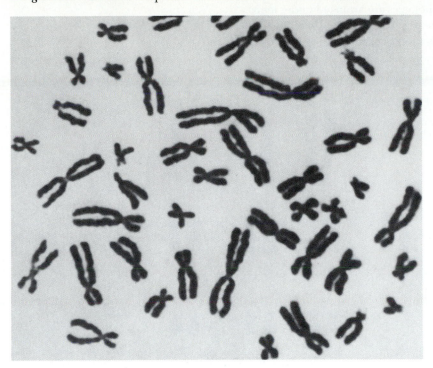

and banding patterns of chromosomes permits geneticists to identify certain chromosome mutations that correlate with congenital abnormalities or dysfunctions. For example, on rare occasions chromosomes undergo changes in morphology, such as by gaining or losing a portion of a chromosome or by exchanging pieces with a nonhomologous chromosome. In addition, changes in the number of chromosomes in the karyotype may occur because of an error in cell division, for instance, as a result

Table 9.1

Chromosome Numbers in Selected Diploid Eukaryotic Organisms

SCIENTIFIC NAME	COMMON NAME	DIPLOID NUMBER OF CHROMOSOMES
ANIMALS		
Homo sapiens	Human	46
Pan troglodytes	Chimpanzee	48
Equus caballus	Horse	64
Canis familiaris	Dog	78
Felis domesticus	Cat	38
Caenorhabditis elegans	Nematode	11♂/12♀
Musca domestica	Housefly	12
Drosophila melanogaster	Fruit fly	8
PLANTS		
Pisum sativum	Garden pea	14
Zea mays	Corn	20

Source: Data adapted from P. L. Altman and D. S. Dittmer (eds.), 1972. *Biology Data Book,* 2nd ed., vol. 1. Bethesda, MD: Federation of American Societies for Experimental Biology and from other sources.

■ *Figure 9.7*   Human male metaphase chromosomes arranged as a karyotype. Note that groups of chromosomes with similar morphologies are arranged under letter designations (A through G). This arrangement is based on the size of the condensed chromosomes. In the male all chromosomes except the X and Y sex chromosomes are present in pairs.

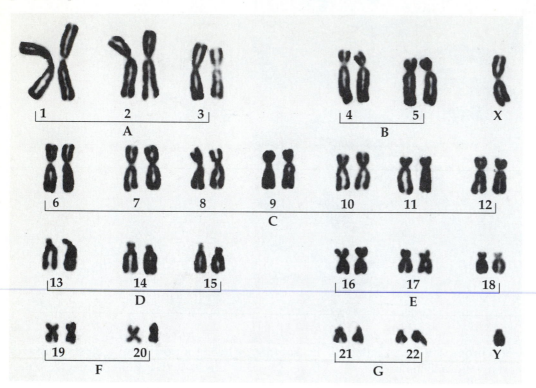

of chromosome nondisjunction (see Chapter 3, p. 57). Chromosome mutations will be discussed more fully in Chapter 17.

Chromosomal Banding Patterns

The human metaphase chromosomes shown in Figure 9.7 were seen after staining the chromosomes with Feulgen stain. This material stains the chromosomes uniformly, making it difficult to distinguish chromosomes that are similar in size and general morphology. For this reason, the human chromosomes in Figure 9.7 are grouped into subsets since the individual chromosomal cannot be clearly identified.

Fortunately a number of techniques have been developed that stain certain regions or *bands* of the chromosomes more intensely than other regions. Banding patterns are specific for each chromosome, enabling every chromosome in the karyotype to be distinguished clearly. This makes it easier to distinguish between chromosomes of similar sizes and shapes.

One of these staining techniques is called **G banding**. In this procedure metaphase chromosomes are first treated with mild heat or proteolytic enzymes (enzymes that digest proteins) and then stained with Giemsa stain (a permanent DNA dye) to produce a G banding pattern (Figure 9.8). G bands reflect regions of DNA that are rich in adenine and thymine. In humans approximately 2000 G bands can be identified. The G bands are stable and are even visible in scanning electron micrographs of chromosomes as constrictions at the banding regions.

Another staining technique is called **Q banding**. Metaphase chromosomes are stained with quinacrine mustard in this procedure, producing Q bands that can be visualized by their fluorescence under ultraviolet light. G bands and Q bands have the same locations on the chromosomes.

Cellular DNA Content and the Structural or Organizational Complexity of the Organism

At first glance it might be expected that the amount of DNA in an organism would be higher the more complex the organism is. The total amount of DNA in the haploid genome is characteristic of each living species, prokaryote or eukaryote, and is known as its **C value**. Table 9.2 gives the C values for a selection of prokaryotes and eukaryotes, as well as

■ *Figure 9.8*   Morphologies of human chromosomes in a female after G banding.

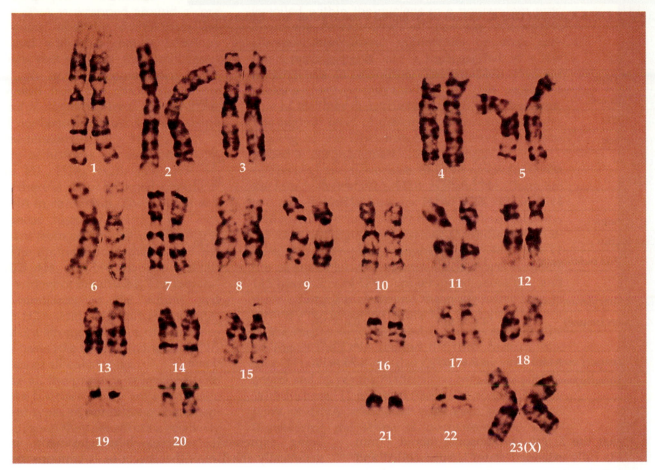

the length of the DNA as calculated from the number of base pairs.

The C value data show that the amount of DNA found among organisms varies considerably. Moreover, there is no direct correlation between DNA content and the complexity of the organism. For example, note that the lily has about 100 times the amount of DNA as humans. Further, there may or may not be significant variation in the amount of DNA among related organisms. Mammals, birds, and reptiles each show little variation, whereas amphibians, insects, and plants each vary over a wide range, often tenfold or so.

The lack of a direct relationship between the C value and the structural or organizational complexity of the organism is called the *C-value paradox*. We have no complete explanation of this paradox. At least one reason for variation in DNA amounts in related organisms may be the presence on chromosomes of different numbers of repeated segments of DNA that appear not to be expressed.

Keynote

In eukaryotes the complete set of metaphase chromosomes in a cell is called its karyotype. The karyotype is species-specific. The amount of DNA in a prokaryotic chromosome or the haploid (N) amount of DNA in a eukaryotic cell is called the C value. The amount of genetic material varies greatly among prokaryotes and eukaryotes. There is not a direct relationship between the C value and the structural or organizational complexity of the organism.

The Molecular Structure of the Eukaryotic Chromosome

In comparison with a prokaryotic cell, a eukaryotic cell contains a large amount of DNA in its nucleus. A human cell, for example, has more than a thousand times as much DNA as does *E. coli*. We saw

■ *Figure 9.10*   Model for the relationship between the nucleosome and histone H1.

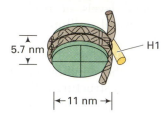

matin, there are about 200 base pairs of DNA per nucleosome; the DNA connecting adjacent nucleosomes is called *linker* DNA.

Under the electron microscope the DNA-protein complex is seen as 10-nm fibers (10-nm **nucleofilaments**); that is, fiber with a diameter of about 10 nm. Since DNA has a diameter of 2 nm, we must assume that in nucleofilaments the DNA is complexed with histone and nonhistone proteins. In its most unraveled state, the chromatin fiber of a nucleofilament has the appearance of "beads on a string," where the beads are the nucleosomes (Figure 9.11). The thinner "thread" connecting the "beads" is naked DNA.

In sum, the nucleosome is the fundamental structure of chromatin. The 10-nm fiber may be interpreted as a continuous string of nucleosomes, like tuna cans or hockey pucks stacked end to end or tilted slightly away from the long axis (Figure 9.12).

Higher-order Structures in Chromatin   In the living cell chromatin typically will not exist in a "beads-on-a-string" nucleofilament; rather, it is kept in a more highly compacted state. The details of these higher levels of folding, which are found predominantly in the chromosomes, are still not completely understood. Moreover, the details of the changes in chromatin folding that take place as a cell goes from interphase to mitosis (or meiosis) and back to interphase are not well understood.

There is general agreement, derived from microscopic observations, that the next level of packing above the nucleosome is the *30-nm chromatin fiber*. An electron micrograph of a 30-nm chromatin fiber is shown in Figure 9.13, and one of the models for that structure is shown in Figure 9.14. We can loosely think of the transition from the 10-nm nucleofilament to the 30-nm chromatin fiber like wrapping a thick rope around a thin stick from one end to the other. In the chromosome, histone H1 must play an important role in the formation of the 30-nm fiber, since H1-depleted chromatin can form 10-nm fibers, but not 30-nm fibers. However, the precise role of histone H1 in forming the 30-nm fiber is currently unknown.

The 30-nm fiber does not provide sufficient packing of the chromosomes to explain the degree to which chromosomes are condensed in the cell nucleus. A major step in modeling the structure of metaphase chromosomes came when data were obtained showing that metaphase chromosomes from which the histones have been removed still retain a residual folded structure with DNA *looped domains* of 30-nm fiber DNA extending from a condensed protein lattice. The central structure is called the *chromosome scaffold*. The scaffolding actually has the shape of the metaphase chromosome (Figure 9.15, p. 200), a shape that remains even when the DNA is digested away by nucleases.

Figure 9.16 (p. 200) presents a schematic of a series of looped domains of DNA in a chromosome. The looped domains extend at an angle from the main chromosome axis and are anchored to the protein scaffold. (Recall that bacterial chromosomal DNA is also organized into looped domains.) The amount of DNA in the loops ranges from 10,000 to 90,000 bp. An average human chromosome could contain approximately 2000 looped domains. There is evidence that the loops correspond to the units of replication of the chromosomes.

In highly purified metaphase chromosomes, only two nonhistone scaffold proteins are found. One of

■ *Figure 9.11*   Electron micrograph of unraveled chromatin showing the nucleosomes in a "beads on a string" morphology.

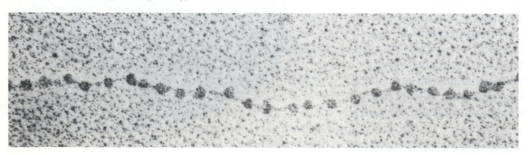

■ *Figure 9.12*   Model for the organization of nucleosomes in the 10-nm chromatin fiber. The nucleosomes are arranged in a linear array, either stacked end-to-end or tilted slightly away from the long axis of the fiber.

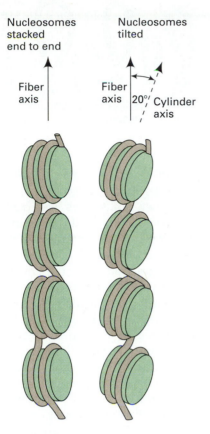

Nucleosomes stacked end to end

Nucleosomes tilted

Fiber axis

Fiber axis   20°   Cylinder axis

these proteins is eukaryotic topoisomerase II. This enzyme, like its bacterial counterpart, causes negative supercoiling of the DNA. Thus, as with bacterial chromosomes, the existence of looped domains permits localized supercoiling of the DNA without influencing the DNA in other looped domains.

Figure 9.17 (p. 201) shows the different orders of DNA packing that could give rise to the highly condensed metaphase chromosome. Interphase chromosomes would be packed less tightly than metaphase chromosomes. Very little is known about the mechanisms involved in the transitions between interphase and metaphase chromosome morphologies, or about the more subtle transitions in localized chromosome regions when genes are turned on or off.

K e y n o t e

The large amount of DNA present in the eukaryotic chromosome is compacted by its association with histones in nucleosomes and by higher levels of folding of the nucleosomes into chromatin fibers. Each chromosome contains a large number of looped domains of 30-nm chromatin fibers attached to a protein scaffold.

Centromeres and Telomeres

So far in this chapter we have described eukaryotic chromosomes as linear structures, each containing a single linear DNA molecule wrapped around histones and associated with nonhistone proteins. In terms of size, number, and morphology the chromosome complement of an organism is species-specific. Nonetheless, as described in Chapter 1, all chromosomes behave similarly at the time of cell division. You will recall that in mitosis, for example, the centromeres of all the replicated chromosomes become aligned at the metaphase plate, the sister chromatids separate at the centromeres, and one chromatid (now daughter chromosome) of each pair is distributed to each daughter cell. The behavior of chromosomes in mitosis and meiosis is a function of the centromeres, the sites at which chromosomes attach to the mitotic and meiotic spindles. In this section we discuss the structure of centromeres and of telomeres (the ends of the chromosomes).

■ *Figure 9.13*   Electron micrograph of a 30-nm chromatin fiber.

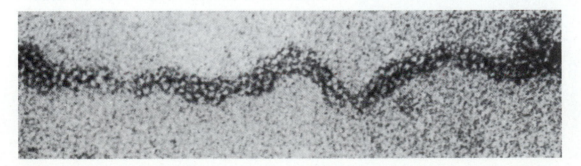

Centromeres The centromere region of each chromosome is responsible for the accurate segregation of the replicated chromosomes to the daughter cells during mitosis and meiosis. Nondisjunction of chromosomes (see Chapter 3) occurs when the centromere fails to function properly. And, if the centromere is absent, the chromosome loses its ability to attach to the spindle and thus will migrate through the cell randomly during the cell division process. Such acentric

■ *Figure 9.14* A model for the packaging of nucleosomes into the 30-nm chromatin fiber.

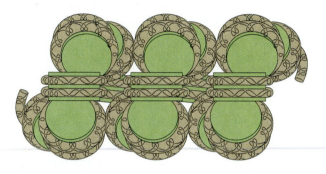

■ *Figure 9.15* Electron micrograph of a metaphase chromosome depleted of histones. The chromosome maintains its general shape by a nonhistone protein scaffolding from which loops of DNA protrude.

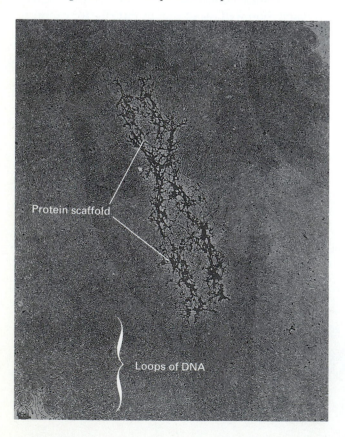

Protein scaffold

Loops of DNA

■ *Figure 9.16* Schematic model for the organization of 30-nm chromatin fiber into looped domains that are anchored to a nonhistone protein scaffold.

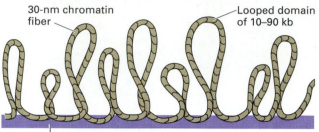

30-nm chromatin fiber

Looped domain of 10–90 kb

Nonhistone protein scaffold

chromosomes often are degraded so that they are "lost" to the segregation process.

In many higher eukaryotes the centromere is seen as a constricted region at one point along the chromosome. The region contains the kinetochore (see Chapter 1) to which the spindle fibers attach during cell division, and gives the chromosome its characteristic appearance at metaphase.

The DNA sequences (called *CEN* sequences, after the *cen*tromere) of at least 10 *S. cerevisiae* (yeast) centromeres have been determined. These centromeres are very similar, but not identical, to each another in nucleotide sequence and organization. They vary in length from 112 to 120 bp.

The yeast centromeres are the simplest known in terms of sequence. Centromeres of other eukaryotic organisms have much longer sequences and more complex sequence organization. This correlates with the fact that only one spindle fiber attaches to each yeast centromere, while several spindle fibers attach to each centromere of most other organisms.

It is clear that proteins must be involved in the attachment of the spindle microtubules to the centromere; some of these proteins presumably interact directly with centromeric DNA. Some progress is being made in identifying and characterizing centromere-binding proteins. One of the major problems yet to be solved is the mechanism of attachment of the centromere to the spindle.

Telomeres A telomere is the region of DNA at each end of a linear chromosome. It is required for replication and stability of that chromosome. Telomeres characteristically, but not necessarily, consist of heterochromatin. In most organisms that have been examined, the telomeres are positioned just inside the nuclear envelope, and are often found associated with each other as well as with the nuclear envelope.

The organization of telomeres is of considerable interest because without telomeres, "broken" chro-

■ *Figure 9.17*   Schematic drawing of the many different orders of chromatin packing that are thought to give rise to the highly condensed metaphase chromosome.

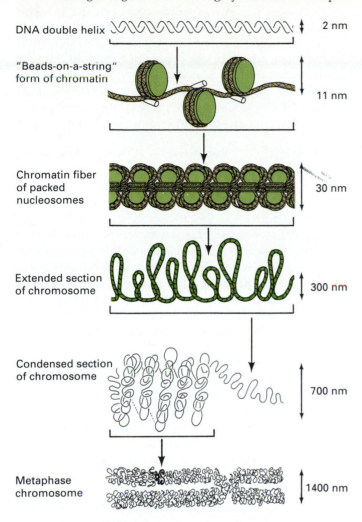

DNA double helix —————— 2 nm

"Beads-on-a-string" form of chromatin — 11 nm

Chromatin fiber of packed nucleosomes — 30 nm

Extended section of chromosome — 300 nm

Condensed section of chromosome — 700 nm

Metaphase chromosome — 1400 nm

mosomes stick readily to other chromosomes, giving broken chromosomes the potential to cause chromosomal mutations. Thus telomeres must function to stabilize chromosomes against such changes.

All telomeres in a given species share a common sequence. Telomeric sequences may be divided into two types:

1. **Simple telomeric sequences** are at the extreme ends of the chromosomal DNA molecules. These sequences are species-specific and consist of simple, tandemly repeated DNA sequences. Simple telomeric sequences are the essential functional components of telomeric regions in that they are sufficient to supply a chromosomal end with stability. In the ciliate *Tetrahymena*, for example, reading from the interior of the chromosome to its end, the repeated sequence

consists of elements that are 5'-TTGGGG-3' elements, and in the flagellate *Trypanosoma*, the repeated sequence is 5'-TTAGGG-3'. Figure 9.18 shows the telomere sequence at the end of a *Tetrahymena* chromosome. Note the unusual G–G base pairing. Much remains to be learned about these sequences in

■ *Figure 9.18*   Telomere sequence at the ends of *Tetrahymena* chromosomes.

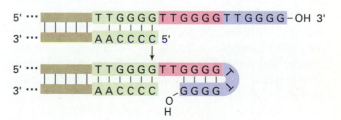

5' ⋯ T T G G G G T T G G G G T T G G G G –OH 3'
3' ⋯ A A C C C C 5'

5' ⋯ T T G G G G T T G G G G
3' ⋯ A A C C C C   G G G G
                        O
                        H

general and the properties of chromosomal ends.

2. **Telomere-associated sequences** are regions near, but not at the ends of chromosomes. These sequences often contain repeated, complex DNA sequences extending for many thousands of base pairs in from the chromosome end.

K e y · n o t e

The centromere region of each eukaryotic chromosome is responsible for the accurate segregation of the replicated chromosome to the daughter cells during both mitosis and meiosis. The DNA sequences of yeast centromeres (*CEN* sequences) are very similar in length and sequence.

The ends of chromosomes, the telomeres, often are associated with the nuclear membrane. Telomeres in a species share a common sequence. Characteristically, sequences at or very close to the extreme ends of the chromosomal DNA consist of simple, relatively short, tandemly repeated sequences. Repeated, often complex, DNA sequences, called telomere-associated sequences, are found farther in from the chromosome ends.

Sequence Complexity of Eukaryotic DNA

Now that we know about the basic structure of DNA and its organization in chromosomes, we can examine the distribution of certain sequences in the genomes of prokaryotes and eukaryotes.

Unique-Sequence and Repetitive-Sequence DNA

By using particular molecular procedures, the discussion of which are beyond the scope of this text, geneticists have determined the relative abundances of different types of sequences in DNA. For convenience, we speak of three different types of sequences: **unique** (or single-copy), **moderately repetitive,** and **highly repetitive.** In prokaryotes, with the exception of the ribosomal RNA genes and some transfer RNA genes, almost all of the genome is present as unique-sequence DNA. Eukaryotic genomes consist of a mixture of the three types of sequences, with the relative proportion of each type varying with the species. In mouse DNA, for example, the genome consists of about 70 percent unique-sequence DNA, about 20 percent moderately repetitive DNA, and about 10 percent highly repetitive DNA.

Unique-Sequence DNA   Unique sequences (sometimes called single-copy sequences) are defined as sequences present as single copies in the genome. (Thus there are two copies per diploid cell.) In current usage the term usually applies to sequences that have one to a few copies per genome. Most of the genes that we know about—those that code for proteins in the cell—are in the unique-sequence class of DNA. Conversely, not all unique-sequence material contains protein-coding sequences.

Moderately   Repetitive   DNA   Sequences
Depending on the sequence, moderately repetitive sequences are repeated from a few to as many as 10^3 to 10^5 times in the genome. While the vast majority of the moderately repetitive DNA sequences have no known function, some known genes are found in this class, notably the genes for ribosomal RNAs (rRNAs), transfer RNAs (tRNAs), and the histone genes. For example, there are about 450 copies of the major class of rRNA genes in the clawed toad (*Xenopus laevis*), 160–200 copies in humans, 260 copies in the sea urchin, and 3900 copies in the garden pea.

In most cases these repeated genes are found in one or more clusters in the genome, with each cluster consisting of an uninterrupted tandem array of genes. Sequences that are repeated one after another in a row are called **clustered repeated sequences.** Clusters also occur for some tRNA genes and, in some organisms, for the multiple copies of the histone genes.

Also found among the moderately repetitive sequences are *interspersed repeated sequences*, which are repeated sequences that are scattered, not clustered, among the unique sequences throughout the genome. The interspersed moderately repetitive sequences are characteristic of most eukaryotic DNAs, although no single unifying description of their arrangement can be applied to all the known examples. Two common patterns of interspersion have been found. In the first, the so-called *short-period pattern* (as found in *Xenopus* DNA) short, 100–300 base-pair repeated sequences are interspersed with longer unique sequences of about 1,000–2,000 base pairs in length. Between 50 and 80 percent of DNA typically shows this pattern. In the second, or *long-period pattern* (as found in *Drosophila* DNA), about 5,000 base-pair repeated sequences are dispersed among unique sequences that may be up to 35,000 base pairs or more in length. It is thought that interspersed repeated sequences may play a role in the regulation of gene expression, although the exact function of most of this repetitive DNA is unknown.

Highly Repetitive DNA Sequences   Highly repetitive sequences are repeated between 10^5 and 10^7 times in the genome. Among the highly repetitive class of DNA sequences are simple, clustered repeated sequences, some with repeated units no more than 6 base pairs long repeated as many as 10^6 to 10^7 times in the genome and others hundreds of base pairs long and repeated millions of times. No genes are found in highly repetitive DNA. The function of most highly repetitive DNA is unknown, although some of these sequences are localized around centromeres and at the telomeres.

Keynote

Prokaryotic genomes consist mostly of unique-sequence DNA, with only a few sequences and genes repeated. Eukaryotes have both unique and repetitive sequences in the genome. Particularly in higher eukaryotes, the spectrum of complexity of the repetitive DNA sequences is extensive. The highly repetitive sequences tend to be localized to heterochromatic regions around centromeres and chromosome ends, whereas the unique- and moderately repetitive sequences tend to be interspersed.

Summary

In this chapter we have learned some aspects of the molecular organization of prokaryotic, viral, and eukaryotic chromosomes. Such structural information contributes greatly to our understanding of gene regulation.

The chromosomes of a prokaryotic organism consist of a single, circular, double-stranded DNA molecule that is complexed with a number of proteins. The chromosome is compacted within the cell by the supercoiling of the DNA helix and the formation of looped domains of supercoiled DNA.

A distinguishing feature of eukaryotic chromosomes is that DNA is distributed among a number of chromosomes. The set of metaphase chromosomes in a cell is called its karyotype. We saw in Chapter 3 how karyotype analysis is useful for the diagnosis of certain human chromosomal mutations. This analysis is aided by the ability to stain human chromosomes so that each chromosome is distinguishable by its banding pattern.

The DNA in the genome specifies an organism's structure, function, and reproduction. The amount of DNA in a genome (made up of a prokaryotic chromosome or of the haploid set of chromosomes in eukaryotes) is called the C value. There is no direct relationship between the C value and the structural or organizational complexity of an organism.

The organization of DNA in eukaryotic chromosomes was detailed in this chapter. The nuclear chromosomes of eukaryotes are complexes of DNA with histone and nonhistone chromosomal proteins. Each chromosome consists of one linear, double-stranded DNA molecule that extends along its length. As a class the histones are constant from cell to cell within an organism and have very similar amino acid sequences among all eukaryotes. Nonhistone chromosomal proteins, on the other hand, vary significantly between cell types and among organisms. The DNA in a eukaryotic chromosome is highly compacted by its association with histones to form nucleosomes and by the several higher levels of folding of nucleosomes into chromatin fibers. The most highly compacted chromosome structure is seen in metaphase chromosomes; the least compacted structure is seen in interphase chromosomes. The factors controlling the transitions between different levels of chromosome folding as cells go through the cell division cycle are not known. Like prokaryotic chromosomes, eukaryotic chromosomes are organized into a large number of looped domains. These loops consist of 30-nm chromatin fibers and are attached to a protein scaffold.

Important elements for the chromosome segregation in mitosis and meiosis are centromeres and telomeres. Significant progress has been made in defining the sequences of centromeres and telomeres; for example, the DNA sequences of centromeres are relatively complex and are species-specific with some sequence variation within a species, and the DNA sequences of telomeres are simple, relatively short, tandemly repeated sequences that are also species-specific. Thus common chromosome activities in different eukaryotes are controlled by quite different DNA sequences.

From various molecular studies, we know that prokaryotic genomes consist mostly of unique-sequence DNA, with only a few repeated sequences. The genomes of eukaryotes contain both unique-sequence and repetitive-sequence DNA, the latter usually being classified into moderately repetitive and highly repetitive types. Unique-sequence DNA contains genes; moderately repetitive DNA includes repeated genes, such as rRNA and tRNA genes, and the histone genes; and highly repetitive DNA generally does not include genes. Some highly repetitive DNA is found at centromeres and telomeres. There is a wide spectrum of complexity of the repetitive DNA sequences among eukaryotes.

Analytical Approaches for Solving Genetics Problems

Q9.1 An organism has a haploid genome of 10^{10} nucleotide pairs, of which 70% is unique-sequence DNA with a copy number of one, 20% is moderately repetitive DNA with an average copy number of 1000, and 10% is highly repetitive DNA with an average copy number of 10^6. Assuming that an average DNA sequence is 10^3 nucleotide pairs, how many different sequences are in each of the three DNA classes?

A9.1 **a.** *Unique-sequence DNA:* the total DNA in this class is 70% of 10^{10} nucleotide pairs = 0.7×10^{10} = 7×10^9. Only one of each sequence exists so the number of different sequences = $(7 \times 10^9)/10^3$ = 7×10^6.

b. *Moderately repetitive DNA:* DNA in this class is 20% of 10^{10} bp = 0.2×10^{10} = 2×10^9. Given an average DNA sequence length of 10^3 bp, there are $(2 \times 10^9)/10^3 = 2 \times 10^6$ sequences in this class. Since the average copy number is 1000, the number of *different* sequences = $(2 \times 10^6)/10^3$ = 2×10^3 (2000).

c. *Highly repetitive DNA:* DNA in this class is 10% of 10^{10} bp = 0.1×10^{10} = 1×10^9. Given an average DNA sequence length of 10^3 bp, there are $(1 \times 10^9)/10^3 = 1 \times 10^6$ sequences in this class. Since the average copy number is 10^6, the number of different sequences = $(1 \times 10^6)/10^6 = 1$.

Questions and Problems

9.1 What are topoisomerases?

*9.2 In typical human fibroblasts in culture, the G1 period of the cell cycle lasts about ten hours, S lasts about 9 h, G2 takes 4 h, and M takes 1 h. Imagine you were to do an experiment in which you added radioactive (^{3}H) thymidine to the medium and left it there for 5 min (pulse), and then washed it out and put in ordinary medium (chase).
  a. What percent of cells would you expect to become labeled by incorporating the ^{3}H-thymidine into their DNA?
  b. How long would you have to wait after removing the ^{3}H medium before you would see labeled metaphase chromosomes?
  c. Would one or both chromatids be labeled?
  d. How long would you have to wait if you wanted to see metaphase chromosomes containing ^{3}H in the regions of the chromosomes that replicated at the beginning of the S period?

9.3 Assume you did the experiment in Question 9.2, but left the radioactive medium on the cells for 16 h instead of 5 min. How would your answers to the above questions change?

9.4 Karyotype analysis performed on cells cultured from an amniotic fluid sample reveals that the cells contain 47 chromosomes. The Feulgen stained chromosomes are classified into groups and the arrangement shows 6 chromosomes in A, 4 in B, 16 in C, 6 in D, 6 in E, 4 in F, and 5 in the G group. Based on the above information:
  a. What could be the genotype of the fetus? If more than one possibility exists, give all.
  b. How would you proceed to distinguish between the possibilities?

9.5 What is the relationship between cellular DNA content and the structural or organizational complexity of the organism?

*9.6 Match the DNA type with the chromatin type. (More than one DNA type may match a given chromatin type.)

DNA TYPE	CHROMATIN TYPE
Barr body (inactivated DNA)	Euchromatin
Centromere	Facultative heterochromatin
Telomere	Constitutive heterochromatin
Most expressed genes	

9.7 Eukaryotic chromosomes contain (choose the best answer):
  a. protein.
  b. DNA and protein.
  c. DNA, RNA, histone, and nonhistone protein.
  d. DNA, RNA, and histone.
  e. DNA and histone.

9.8 List four major features that distinguish eukaryotic chromosomes from prokaryotic chromosomes.

9.9 Discuss the structure and role of nucleosomes.

9.10 What are telomeres?

*9.11 Would you expect to find most protein-coding genes in unique-sequence DNA, in moderately repetitive DNA, or in highly repetitive DNA?

9.12 Would you expect to find tRNA genes in unique-sequence DNA, in moderately repetitive DNA, or in highly repetitive DNA?

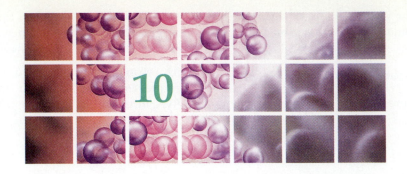

10

DNA Replication
and Recombination

- DNA replication in prokaryotes occurs by a semi-conservative mechanism in which the two strands of a DNA double helix separate and a new complementary strand of DNA is synthesized on each of the two parental template strands. This mechanism ensures the faithful copying of the genetic information at each cell division.

- The enzymes that catalyze the synthesis of DNA are called DNA polymerases. All DNA polymerases perform the same DNA synthesis reaction; new strands are made in the 5'-to-3' direction using deoxyribonucleoside 5'-triphosphate (dNTP) precursors.

- DNA polymerases are incapable of initiating the synthesis of a new DNA strand. All new DNA strands need a short primer of RNA, the synthesis of which is catalyzed by the enzyme DNA primase.

- DNA replication in *E. coli* requires two of the DNA polymerases and several other enzymes and proteins. Synthesis of DNA on one template strand is continuous while it is discontinuous on the other template strand, a process called semi-discontinuous replication.

- In eukaryotes DNA replication is biochemically and molecularly very similar to the replication process in prokaryotes. To enable long chromosomes to replicate efficiently, replication of DNA is initiated at a large number of sites along the chromosomes. When the DNA replicates new nucleosomes are assembled so the whole chromosome duplicates.

- In both prokaryotes and eukaryotes, genetic recombination involves the breakage and rejoining of homologous DNA double helices.

- Numerous models have been proposed to describe the molecular events involved with genetic recombination. Some of the enzymes involved are also involved in DNA replication.

- Mismatch repair of heteroduplex DNA that is an intermediate in genetic recombination can result in a non-Mendelian segregation of alleles, typically a 3:1 or 1:3 ratio rather than the expected 2:2 ratio. This is called gene conversion.

I N CHAPTER 8 WE LEARNED that one of the essential properties of genetic material is that it must replicate accurately so that progeny cells have the same genetic information as the parental cell. In this chapter we will learn about the mechanics of DNA replication and chromosome duplication in prokaryotes and eukaryotes and about some of the enzymes and other proteins required for replication. We will also learn some of the basic molecular details of DNA recombination.

DNA Replication in Prokaryotes

Semiconservative Replication of DNA

We presented Watson and Crick's double-helix model for DNA in Chapter 8. Watson and Crick reasoned that replication of the DNA would be straightforward if their model was correct. That is, by unwinding the DNA molecule and separating the two strands, each strand could be a template for the synthesis of a new, complementary strand of DNA. As the DNA double helix is progressively unwound from one end, the base sequence of the new strand would be determined by the base sequence of the template strand, following complementary base-pairing rules. When replication is completed, there would be two progeny DNA double helices, each consisting of one parental DNA strand and one new DNA strand. This model for DNA replication is known as the **semiconservative model** since each progeny molecule retains one of the parental strands (Figure 10.1[a]).

At the time, two other models for DNA replication were proposed—the **conservative** model (Figure 10.1[b]) and the **dispersive model** (Figure 10.1[c]). In the conservative model the two parental strands of DNA remain together or reanneal after replication and as a whole serve as a template for the synthesis of new progeny DNA double helices. Thus one of the two progeny DNA molecules is actually the parental double-stranded DNA molecule, and the other consists of totally new material. In the dispersive model the parental double helix is cleaved into double-stranded DNA segments that

■ *Figure 10.1*   Three models for the replication of DNA: (a) The semiconservative model
(the correct model); (b) The conservative model; (c) The dispersive model. The parental
strands are shown in grey and the newly synthesized strands are shown in red.

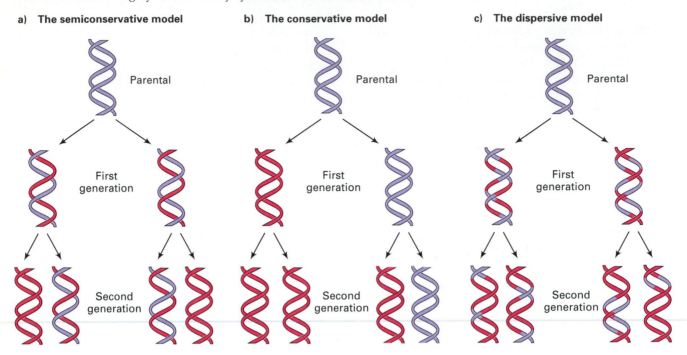

act as templates for the synthesis of new double-stranded DNA segments. Somehow the segments reassemble into complete DNA double helices, with parental and progeny DNA segments interspersed. Thus while the two progeny DNAs are identical with respect to base-pair sequence, the parental DNA has actually become dispersed throughout both progeny molecules.

Matthew Meselson and Frank Stahl demonstrated that the semiconservative replication model was the correct one. In their experiment, shown in Figure 10.2, *E. coli* was grown for several generations in a minimal medium in which the only nitrogen source was $^{15}NH_4Cl$ (ammonium chloride). In this compound the normal isotope of nitrogen, ^{14}N, is replaced with ^{15}N, the heavy isotope. (Note: Density is weight/volume so ^{15}N, with one extra neutron in its nucleus, is 1/14 more dense than ^{14}N.) As a result, all the bacteria's cellular nitrogen-containing compounds, including its DNA, contained ^{15}N instead of ^{14}N.

As the next step in Meselson and Stahl's experiment, the ^{15}N-labeled bacteria were transferred to a medium containing nitrogen in the normal ^{14}N form. The bacteria were allowed to replicate in the new conditions for several generations. Throughout the period of growth in the ^{14}N medium, samples of *E. coli* were taken and the densities of their DNA were analyzed (see Figure 10.2). Experimentally,

^{15}N DNA can be separated from ^{14}N DNA by using equilibrium density gradient centrifugation. Briefly, in this technique a concentrated solution of cesium chloride (CsCl) is centrifuged at high speed, causing the cesium chloride to form a density gradient (Figure 10.3). If DNA is present in the solution, it will band to a position where its buoyant density is the same as that of the surrounding cesium chloride. Thus if DNAs of more than one density are present during the centrifugation, they will band to different positions in the gradient according to their respective densities.

After one generation in ^{14}N medium, all the DNA had a density that was exactly intermediate between that of totally ^{15}N DNA and totally ^{14}N DNA. After two generations, half the DNA was of the intermediate density and half was of the density of DNA containing only ^{14}N. These observations (presented in Figure 10.2) and those for subsequent generations were exactly what the semiconservative model predicted.

If the conservative model of DNA replication had been correct, after one generation two bands of DNA would have been seen (see Figure 10.1[b]). One band would have been in the heavy-density position of the gradient, containing parental DNA molecules, and both strands would have consisted of ^{15}N-labeled DNA only. The other band would have been in the light-density position, containing

■ *Figure 10.2* The Meselson-Stahl experiment: The demonstration of semiconservative replication in *E. coli*. Cells were grown in ^{15}N-containing medium for several generations, and then transferred to ^{14}N-containing medium. At various times over several generations, samples were taken; the DNA was extracted and analyzed by CsCl equilibrium density gradient centrifugation. Shown in the figure are a schematic interpretation of the DNA composition at various generations, photographs of the DNA bands, and a densitometric scan of the bands.

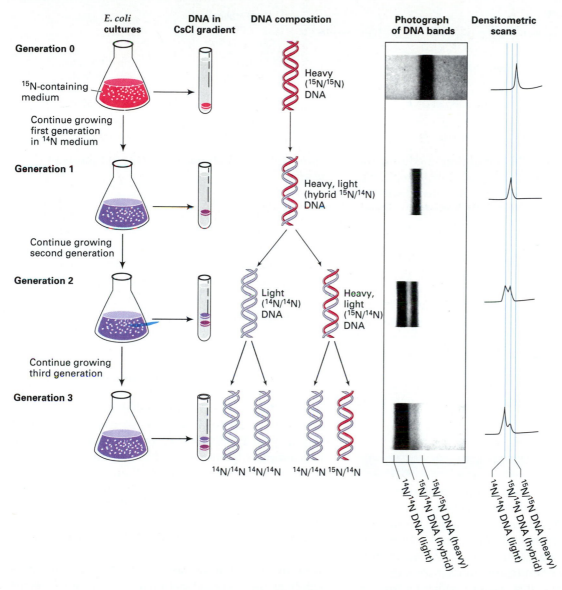

progeny DNA molecules with both strands totally ^{14}N-labeled. In subsequent generations the heavy parental DNA band would have been seen at each generation, and in the amount found at the start of the experiment. All new DNA molecules would have had both strands totally labeled with ^{14}N. Hence the amount of DNA in the light-density position would have increased with each generation. In the conservative model of DNA replication, the most significant prediction was that *at no time would*

any DNA of intermediate density have been found. The fact that intermediate-density DNA *was* found ruled out the conservative model.

In the dispersive model for DNA replication, the parental DNA is scattered in double-stranded segments throughout the progeny DNA molecules (see Figure 10.1[c]). According to this model, all DNA present after one generation in ^{14}N-containing medium would be of intermediate density, and this was seen in the Meselson-Stahl experiment. After a

■ *Figure 10.3*   Schematic diagram for separating DNAs of different buoyant densities by equilibrium centrifugation in a cesium chloride density gradient. Illustrated is the separation of DNA containing highly repetitive sequences (see Chapter 9, p.203) and main band DNA (the remainder of the DNA) from a eukaryote.

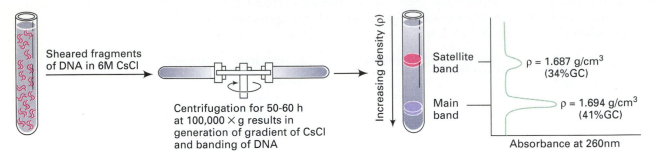

second generation in ^{14}N-containing medium, the dispersive model predicted that DNA segments from the first generation would be dispersed through the progeny DNA double helices produced. Thus the ^{15}N–^{15}N DNA segments dispersed among new ^{14}N–^{14}N DNA after one generation would then be distributed among twice as many DNA molecules after two generations. As a result, the DNA molecules would be found at one band located halfway between the intermediate-density and light-density band positions. With subsequent generations there would continue to be one band, and it would become lighter in density with each generation. Such a slow shift in DNA density was not seen in the results of the Meselson-Stahl experiment, and therefore the dispersive model was ruled out.

Keynote

DNA replication in *E. coli* and other prokaryotes occurs by a semiconservative mechanism in which the strands of a DNA double helix separate and a new complementary strand of DNA is synthesized on each of the two parental template strands. Semiconservative replication results in two doublestranded DNA molecules, each of which has one strand from the parent molecule and one newly synthesized strand. This mechanism ensures the faithful copying of the genetic information at each cell division.

DNA Synthesis Enzymes

Once the mechanism of DNA replication was figured out, attention turned to the biochemistry of DNA synthesis; that is, how is a new DNA chain synthesized in the cell. In the 1950s Arthur Kornberg and his colleagues demonstrated that four components were required for the synthesis of DNA *in vitro*:

1. four deoxyribonucleoside 5'-triphosphates (dATP, dGTP, dTTP, and dCTP) that we will abbreviate as dNTPs. (If any one dNTP is missing, no synthesis occurs.) These are the *precursors* for the nucleotide (phosphate-sugar-base) building blocks of DNA that we discussed in Chapter 8 (pp. 177–178);

2. magnesium ions (Mg^{2+});

3. a fragment of DNA to act as a template;

4. DNA polymerase.

If any one of the four components is omitted, DNA synthesis does not occur. Subsequent experiments showed that the new DNA that was made *in vitro* was a faithful base-pair for base-pair copy of the original DNA. That is, the original DNA acted as a template for the new DNA synthesis.

By definition, the enzymes that catalyze the synthesis of DNA are called **DNA polymerases.** Since Kornberg discovered the first DNA polymerase, it was named the *Kornberg enzyme* in his honor; it is now most commonly called **DNA polymerase I.**

Role of DNA Polymerases   All DNA polymerases catalyze the polymerization of nucleotide precursors (deoxyribonucleotides) into a DNA chain. The reaction that they catalyze (in the presence of magnesium ions) is:

$$\text{DNA (dNMP)}_n + \text{dNTP} \overset{\text{DNA polymerase}}{\rightleftharpoons} \text{DNA (dNMP)}_{n+1} + \text{PP}_i$$

Here the DNA is represented as a string of a number (n) of deoxyribonucleoside 5'-monophosphates (dNMP). The next nucleotide to be added is the precursor, dNTP. The DNA polymerase catalyzes a reaction in which one nucleotide is added to the existing DNA chain with the release of inorganic pyrophosphate (PP$_i$). The enzyme then repeats its action until the new DNA chain is complete.

The action of DNA polymerase in synthesizing a DNA chain is shown at the molecular level in Figure 10.4(a). The same reaction in shorthand notation is shown in Figure 10.4(b). The reaction has three main features:

1. At the growing end of the DNA chain, DNA polymerase catalyzes the formation of a **phosphodiester bond** between the 3'-OH group of the deoxyribose on the last nucleotide and the 5'-phosphate of the deoxyribonucleoside 5'-triphosphate (dNTP) precursor. The formation of the phosphodiester bond results in the release of two of three phosphates from the dNTP. The important concept here is that *the lengthening DNA chain acts as a primer in the reaction.* (A **primer** is a preexisting polynucleotide chain in DNA replication to which new nucleotides can be added.)

2. The addition of nucleotides to the chain is not random. Each deoxyribonucleotide is selected by the DNA polymerase, which is always bound to the DNA and which moves along the template strand as the polynucleotide chain is lengthened. The polymerase finds the precursor nucleotide (dNTP) that can form a complementary base pair with the nucleotide on the template strand of DNA. Since the DNA polymerase is bound to the template DNA, it ensures that the correct precursor has been chosen. This does not occur with 100 percent accuracy, but the error frequency is extremely low.

3. The direction of synthesis of the new DNA chain is only from 5' to 3' because of the properties of DNA polymerase.

All known DNA polymerases from both prokaryotes and eukaryotes carry out the same reaction and will make new DNA copies from any DNA added to the reaction mixture, provided all the necessary ingredients are present. Thus if given human DNA, *E. coli* DNA polymerase can replicate it faithfully, and vice versa.

One of the best-understood systems of DNA replication is that of *E. coli.* For several years after the discovery of the Kornberg enzyme (DNA polymerase I), scientists believed that this enzyme was the only DNA replication enzyme in *E. coli.* However, genetic studies proved that this was not so. In general, one way to study the action of a particular enzyme *in vivo* is to induce a mutation in the gene that codes for that particular enzyme. Then the phenotypic consequences of the mutation can be

compared with the wild-type phenotype. A mutation in the gene coding for an enzyme that is as essential to cell function as DNA polymerase, for instance, would be expected to be lethal. The first DNA polymerase I mutant, *polA1*, was isolated in 1969 by Peter DeLucia and John Cairns. Unexpectedly, *E. coli* cells carrying the *polA1* mutation grew and divided normally, showing that DNA polymerase I is not essential to cell function.

Subsequently, DNA polymerase I mutants that were temperature-sensitive were isolated. To study the consequences of mutations in genes coding for essential proteins and enzymes, geneticists find it easiest to work with **conditional mutants,** mutant organisms that are normal under one set of conditions but that become seriously impaired or die under other conditions (Figure 10.5). The most common types of conditional mutants are those that are temperature-sensitive—mutant organisms that function normally until the temperature is raised past some threshold level, at which time some temperature-sensitive defect is manifested. It was expected that the temperature-sensitive DNA polymerase I mutants would die at elevated temperatures. At *E. coli*'s normal growth temperature of 37°C, the temperature-sensitive *polA* mutant strains produce DNA polymerase I with normal catalytic activity. At 42°C, however, the mutant strains produce DNA polymerase I molecules that have only about 1 percent of the normal catalytic activity of the wild-type enzyme. Unexpectedly, the temperature-sensitive *polA* mutant cells multiplied at the same rate as the parental (wild-type) strain, even though the temperature was elevated, and DNA polymerase I activity was virtually nonexistent. These results indicated that there must be other DNA-polymerizing enzymes in the cell.

We know now that *E. coli* possesses three DNA polymerases, designated DNA polymerases I, II, and III. A comparison of the important properties of the three *E. coli* DNA polymerases is given in Table 10.1. Both DNA polymerase I and III are known to be involved in DNA replication, but the role of DNA polymerase II in the replication of the *E. coli* chromosome is unknown at this time.

As Table 10.1 shows, all three *E. coli* DNA polymerases have 3'→ 5' **exonuclease** activity; that is, they can catalyze the *removal* of nucleotides from the 3' end of a DNA chain. So in addition to selecting the correct nucleotide precursor to add to the primer DNA strand, the DNA polymerases also check the accuracy of the base pair linking the primer strand terminus and the template strand. If the wrong base pair has been generated in error, DNA synthesis cannot proceed. In such cases the 3'→ 5' exonuclease excises the erroneous nucleotide on the primer strand. The polymerase then cat-

■ *Figure 10.4*   DNA chain elongation catalyzed by DNA polymerase: (a) Mechanism at molecular level; (b)The same mechanism, using a shorthand method to represent DNA. In this method the vertical line symbolizes the sugar molecule; on the template strand (top in the diagram), the 5' carbon is at the top and the 1' is at the bottom of the line. The circled P is the phosphate group; the diagonal line symbolizes, then, the 3'-to-5' phosphodiester bond. The bases are shown in squares attached to the 1' carbon.

a)   **Mechanism of DNA elongation**

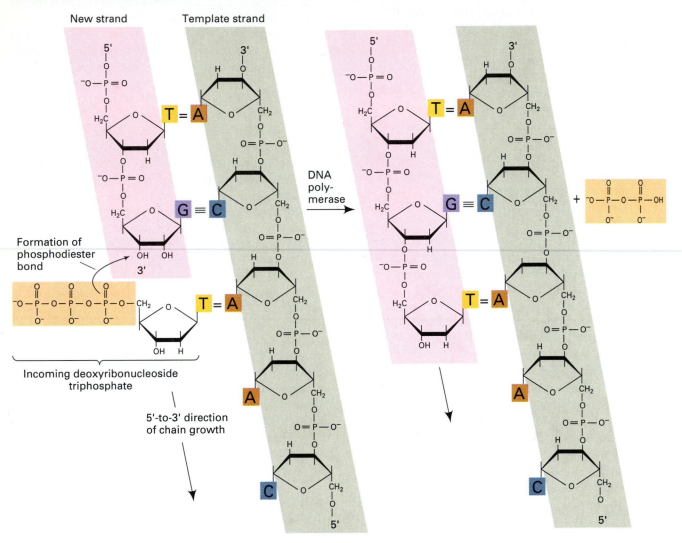

b)   **Shorthand notation**

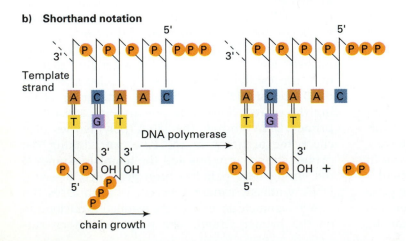

■ *Figure 10.5* The effect of a conditional mutation: (a) A normal gene, when transcribed and translated, produces a protein that functions normally at normal and high temperatures; (b) A gene with conditional mutation produces a slightly altered protein that functions normally at normal temperatures but at high temperatures changes its shape so that function is lost or reduced significantly.

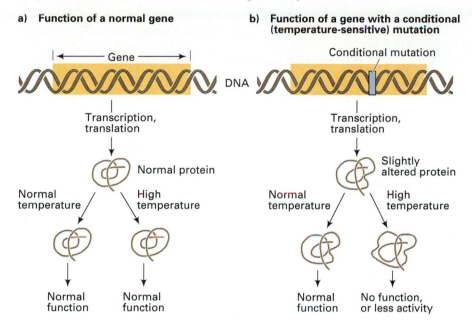

a) Function of a normal gene

b) Function of a gene with a conditional (temperature-sensitive) mutation

alyzes the formation of the correct base pair. Thus in DNA replication, the $3' \rightarrow 5'$ exonuclease that is part of the DNA polymerase enzyme is a **proofreading** mechanism that helps keep the frequency of DNA replication errors very low.

In addition, DNA polymerase I has $5' \rightarrow 3'$ exonuclease activity and can remove nucleotides from the 5' end of a DNA strand or of an RNA primer strand. This activity, also important in DNA replication, will be examined later in this chapter.

K e y n o t e

The enzymes that catalyze the synthesis of DNA are called DNA polymerases. Three DNA polymerases—I, II, and III—have been identified in *E. coli*; I and III are known to be involved in DNA replication. The three enzymes differ in a number of properties, including size, number of molecules per cell, and proofreading ability.

Molecular Details of DNA Replication

Initiation of DNA Replication. The initiation of DNA replication in prokaryotes requires the local separation of the two DNA strands (a process called *denaturation*) at a specific DNA sequence called an **origin of replication**.

Origin of Replication. Replication of prokaryotic and viral DNA usually starts at a specific site on the

Table 10.1

Comparison of the Biochemical Properties of the *E. coli* DNA Polymerases I, II, and III

DNA POLYMERASE	POLYMERIZATION $5' \rightarrow 3'$	EXONUCLEASE $3' \rightarrow 5'$	EXONUCLEASE: $5' \rightarrow 3'$	PROPERTIES MOLECULAR WEIGHT (DALTONS)	MOLECULES/CELL (APPROXIMATE)
I	yes	yes	yes	109,000	400
II	yes	yes	no	120,000	Not known
III	yes	yes	no	>250,000*	10–20

*(many different subunits)

chromosome, the **origin.** At that site the double helix separates into single strands, exposing the bases for the synthesis of new strands. In a circular chromosome, such as is found in *E. coli,* the local denaturing of the DNA produces what is called a **replication bubble.** The segments of untwisted single strands upon which the new strands are made (following complementary base pairing rules) are called the **template strands.** In *E. coli* there is a single origin of replication, called *oriC,* from which replication proceeds bidirectionally.

Initiation of DNA Synthesis. Figure 10.6 shows a model for the formation of a replication bubble at a replication origin in *E. coli,* and the initiation of the new DNA strand. An initiator protein binds to the parental DNA molecule at the origin of replication sequence (Figure 10.6[1]). The two DNA strands then separate in that region which involves the physical untwisting of the DNA. This untwisting is catalyzed by the enzyme **DNA helicase,** which first becomes bound to the initiator protein (Figure 10.6[2]), and then is loaded onto the DNA itself (Figure 10.6[3]). Next, another enzyme, **DNA primase,** binds to the helicase and the denatured DNA (Figure 10.6[4]). Primase is derived from **RNA polymerase** (the enzyme responsible for RNA synthesis in the cell) by the addition of at least two other proteins.

Primase is important because no known DNA polymerases can initiate the synthesis of a DNA strand; they can only catalyze the addition of deoxyribonucleotides (dNTPs) to a preexisting strand. The initiation of DNA synthesis involves the synthesis of a short **RNA primer,** catalyzed by primase (Figure 10.6[5]). The complex of the primase, helicase, and perhaps other proteins with the DNA is called the **primosome.** Only very short RNA primers, composed of from three to five nucleotides, are made. The primers function as preexisting polynucleotide chains to which new deoxyribonucleotides can be added by reactions catalyzed by DNA polymerase (Figure 10.6[5]). The RNA primers themselves do not remain as part of the new DNA chain; they are subsequently removed and replaced with DNA, as we shall discuss later in this chapter. Note that a primer is different from a template: A template strand is one upon which the new strand is made, following complementary base pairing rules.

Keynote

No DNA polymerase can initiate the synthesis of a new DNA chain. Instead, the initiation of DNA synthesis first involves the denaturation of double-

stranded DNA at an origin of replication, catalyzed by DNA helicase. Next, DNA primase binds to the helicase and the denatured DNA and synthesizes a short RNA primer. The RNA primer is extended by DNA polymerase as new DNA is made. The RNA primer is later removed.

Semidiscontinuous DNA Replication When a double-stranded DNA molecule unwinds to expose the two single-stranded template strands for DNA

■ *Figure 10.6* Schematic model for the formation of a replication bubble at a replication origin in *E. coli* and the initiation of the new DNA strand. From Alberts, 2nd ed., Figure 5.50, p. 235.

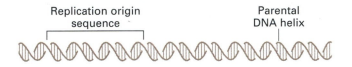

1 Initiator protein binds to replication origin

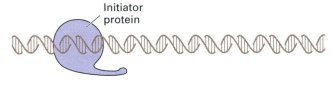

2 DNA helicase binds to initiator protein

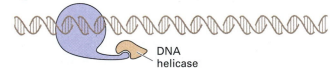

3 Helicase loads onto DNA

4 Helicase denatures helix and binds with DNA primase to form primosome

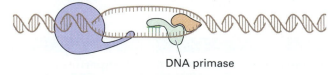

5 Primase synthesizes RNA primer which is extended as DNA chain by DNA polymerase

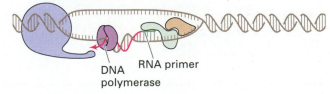

replication, a Y-shaped structure is formed, called a **replication fork.** A replication fork moves in one direction. This poses a problem since DNA polymerases can only make new DNA in the 5'-to-3' direction, yet the two DNA strands are of opposite polarity. As the DNA helix unwinds to provide templates for new synthesis, the new DNA cannot be polymerized continuously on the 3'-to-5' strand.

The work of Reiji and Tuneko Okazaki and their colleagues suggested a model to explain this process. DNA replication can involve the synthesis of short DNA segments, called **Okazaki fragments**. The Okazaki fragments are subsequently linked together by the action of DNA polymerase to remove the RNA primers, followed by the action of an enzyme called **DNA ligase**, which catalyzes formation of the final phosphodiester bond between Okazaki fragments to form a long polynucleotide chain. In this model DNA replication is **discontinuous.**

We can relate the discontinuous nature of DNA synthesis to the replication of the circular DNA chromosome in *E. coli*. One location on the chromosome serves as the point of origin for DNA replication. At this point the DNA denatures to expose the two template strands. Since denaturation occurs in the middle of a circular DNA molecule rather than

at the end, the result is the two Y-shaped structures linked head-to-head at the points of the Ys. These structures act as two replication forks, with DNA synthesis taking place in both directions (bidirectionally) away from the origin point.

The early stages of **bidirectional replication** are shown in Figure 10.7. This figure also shows how discontinuous replication occurs. As the replication fork migrates, synthesis of one new strand (the **leading strand**) is continuous since the 3'-to-5' template strand (the leading strand template) is being copied. Synthesis of the other new strand (the **lagging strand**) must be discontinuous because the helix must unwind to expose a new segment of template (the lagging strand template) so that the new strand can be made in the 5'-to-3' direction. Since one new DNA strand is synthesized continuously and the other discontinuously, DNA replication as a whole is considered to occur in a **semidiscontinuous** fashion.

DNA Replication Model   We will now consider a model for DNA replication which incorporates all of the facts we have just discussed (Figure 10.8). The model involves a single replication fork; however, the events described for a single replication fork also apply to the two replication forks formed

■ *Figure 10.7*   Bidirectional DNA replication. Synthesis of DNA is initiated at the origin and proceeds in the 5'-to-3' direction while the two replication forks are migrating in opposite directions. Replication is continuous on the leading strand and semidiscontinuous on the lagging strand. These events occur in the replication of all DNA in prokaryotes and eukaryotes.

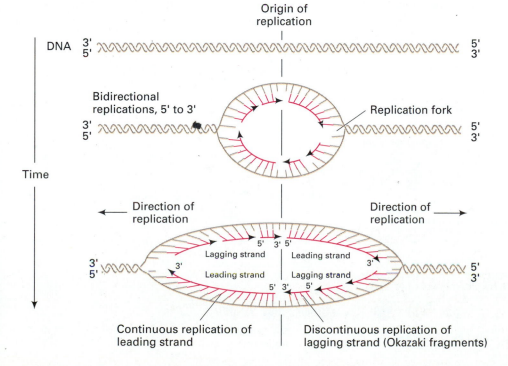

■ *Figure 10.8*   Model for the events occurring around the replication fork of the *E. coli* chromosome. (a) Untwisting; (b) Initiation; (c) Further untwisting and elongation of the new DNA strands; (d) Further untwisting and continued DNA synthesis; (e) Removal of the primer by DNA polymerase I; (f) joining of adjacent DNA fragments by the action of DNA ligase. (Adapted from James E. Watson, et al., *Molecular Biology of the Gene*, 4th ed. Copyright 1965, 1970, 1976, 1987 by the Benjamin/Cummings Publishing Company, Inc. Reprinted by permission.)

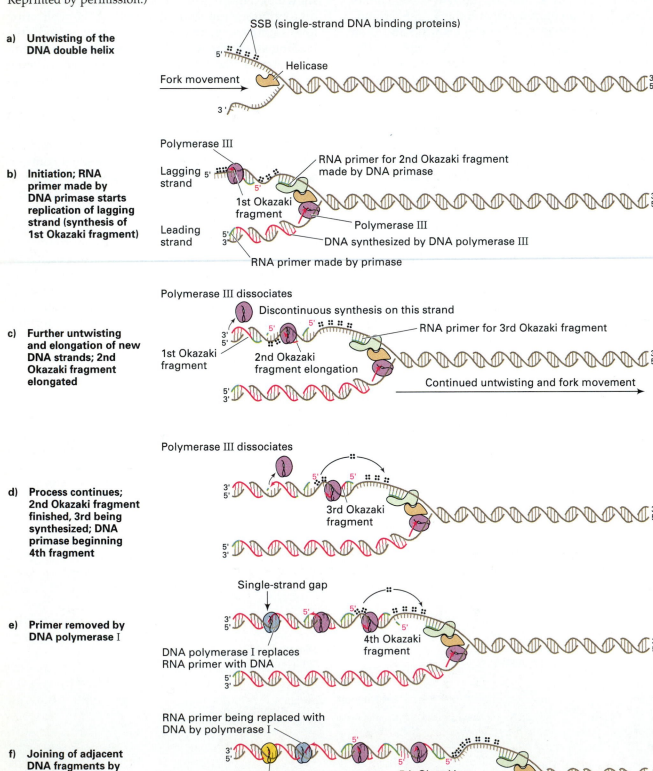

a) **Untwisting of the DNA double helix**

b) **Initiation; RNA primer made by DNA primase starts replication of lagging strand (synthesis of 1st Okazaki fragment)**

c) **Further untwisting and elongation of new DNA strands; 2nd Okazaki fragment elongated**

d) **Process continues; 2nd Okazaki fragment finished, 3rd being synthesized; DNA primase beginning 4th fragment**

e) **Primer removed by DNA polymerase I**

f) **Joining of adjacent DNA fragments by DNA ligase**

during bidirectional replication of circular bacterial chromosomes.

The key steps in DNA replication are as follows:

1. *Denaturation and untwisting of the double helix* (Figures 10.6 and 10.8[a]). DNA helicases denature and untwist the double helix. As the helicases move along the single strand, they untwist any double-stranded DNA they encounter.

2. *Stabilization of the single-stranded DNA in the replication fork.* In the single-stranded form, DNA is a flexible molecule. Therefore the single-stranded DNA produced after helicase untwists the double helix could potentially reform the original hydrogen bonds to reestablish a double-helical molecule. Such double-stranded regions would impede the path of DNA polymerase as it synthesizes a new strand on the exposed DNA template. This is avoided by the action of **single-strand DNA binding (SSB) proteins,** also called **helix-destabilizing proteins.** SSB proteins bind to the single-stranded DNA without covering the bases so that they are still readable by DNA polymerase (Figure 10.8). In binding to the single-stranded DNA, the SSB proteins help the DNA unwinding process by stabilizing the single-stranded DNA. Over 200 of the proteins bind to each replication fork.

3. *Initiation of synthesis of new DNA strands.* The primase-enzyme complex (primosome) binds to the single-stranded DNA and synthesizes the short RNA primer (Figures 10.6 and 10.8[b]). The RNA primers are lengthened by the action of DNA polymerase III, which synthesizes new DNA chains complementary to the template strands (Figure 10.8[b, c]). To maintain the 5'-to-3' polarity of DNA synthesis, and one overall direction of replication fork movement, the direction of DNA synthesis is different on the two template strands. That is, the *leading strand* is synthesized in the same direction as the direction of fork movement while the *lagging strand* is synthesized in the opposite direction. Each new piece of DNA synthesized on the lagging strand template is an Okazaki fragment.

4. *Elongation of the new DNA strands.* The DNA is untwisted further by the helicases (Figure 10.8[c]). On the leading strand template (bottom strand of Figure 10.8), the new leading strand is synthesized continuously as indicated above. Since DNA synthesis can only proceed in the 5'-to-3' direction, however, the DNA polymerase synthesis reaction on the lagging strand template (top strand of Figure 10.8) has gone as far as it can. For DNA replication to continue on that strand, a new initiation of DNA synthesis must occur on the single-stranded template that has been produced by the unwinding of the double helix. As before, an RNA primer is made, and this occurs close to the replication fork, catalyzed by the primase which is still bound to the helicase. The primer is lengthened by the DNA polymerase III, which displaces SSB proteins as it synthesizes the new Okazaki fragment.

In Figure 10.8(d) the process repeats itself: The DNA untwists, continuous DNA synthesis occurs on the leading strand template, and discontinuous DNA synthesis occurs on the lagging strand template.

5. *Joining of Okazaki fragments on the lagging strand to make a continuous strand.* Eventually, the unconnected Okazaki fragments on the lagging strand template are synthesized into a continuous DNA strand. That process requires two enzymes, DNA polymerase I and **DNA ligase.** Consider two adjacent Okazaki fragments. The 3' end of the newer DNA fragment is adjacent to, but not joined to, the primer at the 5' end of the previously made fragment. The DNA polymerase III dissociates from the DNA, and DNA polymerase I takes over. This enzyme continues the 5'-to-3' synthesis of the newer DNA fragment made by DNA polymerase III, simultaneously removing the primer section of the older fragment (Figure 10.8[e]). The removal of the RNA primer takes place nucleotide by nucleotide by the 5'→3' exonuclease activity of DNA polymerase I.

When DNA polymerase I has completed replacement of RNA primer nucleotides with DNA nucleotides, a single-stranded gap exists between adjacent nucleotides on the DNA strand between the two fragments. The two fragments are joined into one continuous DNA strand by the enzyme DNA ligase. The result is a longer DNA strand (Figure 10.8[f]). The catalytic reaction of DNA ligase is diagrammed in Figure 10.9. The whole process is repeated until all the DNA is replicated.

The complicated process of DNA replication in *E. coli* requires many different proteins, some of which function in other cellular processes such as the repair of the damaged DNA and genetic recombination. The roles of some of the essential genes for DNA replication are given in Table 10.2.

What has emerged from extensive studies of the proteins involved with replication is that the key proteins in the process are closely associated to form a **replication machine** or **replisome** (Figure 10.10). This close association is believed to increase the efficiency of replication significantly. Note that the lagging strand DNA is folded so that the DNA polymerase III on that strand is complexed with the DNA polymerase III on the leading strand. The folding also brings the 3' end of each completed Okazaki fragment near the site where the next

■ *Figure 10.9*   Action of DNA ligase in sealing the gap between adjacent DNA fragments (e.g., Okazaki fragments) to form a longer, covalently continuous chain. The DNA ligase catalyzes the formation of a phosphodiester bond between the 3'-OH and the 5'-phosphate groups on either side of a gap, sealing the gap.

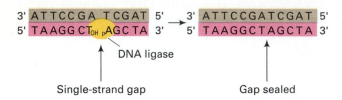

Okazaki fragment will start. The primase-helicase complex (the primosome) moves with the fork, synthesizing new RNA primers as it proceeds. Similarly, with the lagging strand DNA polymerase III complexed with the other replication proteins at the fork, that enzyme can be continually reused at the same replication fork, synthesizing a string of Okazaki fragments as it moves with the rest of the replication machine. In sum, the complex of replication proteins that forms at the replication fork moves as a unit along the DNA, and enables new DNA to be synthesized efficiently on both the leading strand template and lagging strand template.

Proofreading: Correcting Errors in DNA Replication   As indicated previously, occasionally an error is made in DNA replication and an incorrect nucleotide is incorporated into the DNA chain being synthesized. The mismatched base has a very high probability of being excised by the $3' \rightarrow 5'$ exonuclease activity of the DNA polymerase before the next base in the chain is added. This process, or proofreading ability, results in an extremely low error frequency in inserting the wrong base during DNA replication. The proofreading mechanism resembles that of a correcting typewriter, where a backspace is used to erase the incorrect character and the forward direction is resumed to insert the next correct character.

Replication of Circular DNA and the Twisting Problem   In *E. coli* the parental DNA strands remain in a circular form throughout the replication

---

Table 10.2

Functions of Some of the Genes and Sequences Involved in DNA Replication in *E. coli*

GENE PRODUCT AND/OR FUNCTION	GENE
DNA polymerase I	*polA*
DNA polymerase II	*polB*
DNA polymerase III	*polC* (*dnaE, N, Z*)
Initiation of chromosomal replication	*dnaA*
DNA replication	*dnaB*
DNA replication	*dnaC*
Primase—make primer for extension by DNA polymerase	*dnaG*
DNA replication	*dnaI*
DNA replication	*dnaJ*
DNA replication	*dnaK*
DNA replication	*dnaL*
Initiation of chromosomal replication	*dnaP*
Origin of chromosomal replication	*ori*
Helicase—unwinding activity to generate single-stranded arms of replication fork	*rep*
Single-stranded binding (SSB) protein—stabilize single-stranded arms of replication fork	*ssb*
DNA ligase—seal single-stranded gaps; join Okazaki fragments	*lig*
Terminus of chromosomal replication	*ter*

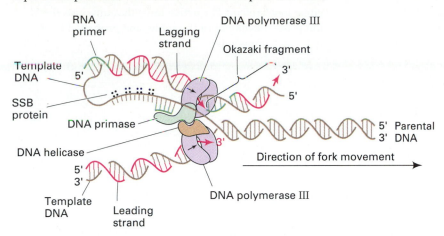

■ *Figure 10.10*   Model for the replication machine, or replisome, the complex of key replication proteins, with the DNA at the replication fork.

cycle. This is true of many, but not all, circular DNA molecules. During replication these circular DNA molecules exhibit a theta-like (θ) shape because of a replication bubble's initiation at the replication origin (Figure 10.11). In the case of the *E. coli* chromosome and a number of other circular chromosomes, replication proceeds bidirectionally from the origin. For other circular chromosomes— for example, certain plasmids—replication is unidirectional.

Since the two DNA strands in a circular chromosome must untwist for replication to occur, there is a problem that must be solved as replication proceeds. If both strands remain circular, then as a section of DNA double helix is untwisted to make the replication bubble, positive supercoils will form elsewhere in the molecule. By analogy, if you take a circular piece of double-stranded rope and try to separate the two strands at one point, the rope will become tightly supercoiled at the opposite side of the circle. In the DNA every 10 base pairs replicated at the fork represent one complete turn of the helix. For the replication fork to move, all of the chromosome ahead of the fork has to rotate. Given a rate of movement of the replication fork of 500 nucleotides per second, at 10 base pairs per turn, the helix ahead of the fork has to rotate at 50 revolutions per second, or 3000 rpm!

The twisting problem is solved by the action of topoisomerases (see Chapter 9), enzymes that introduce positive or negative supercoils into DNA. Topoisomerases play an important role in the replication process by preventing excessively twisted DNA from forming, thereby allowing both parental strands to remain intact during the replication cycle as the replication fork migrates. That is, the unreplicated part of the theta structure ahead of the replication fork repeatedly has negative supercoils intro-

duced into it by the action of topoisomerase II, relieving the positive supercoiling that occurs as DNA is untwisted during replication.

Rolling Circle Replication of DNA   The rolling circle model of DNA replication applies to the replication of several viral DNAs such as λ and to the replication of the *E. coli F* factor during conjugation and transfer of donor DNA to a recipient (see Figures 6.6 and 6.8 pp. 133 and 135). Recall that the λ chromosome enters the cell as a linear chromosome and immediately becomes circular. During a lytic cycle, copies of the circular chromosome are first made by replication mechanisms already described (via theta replication), and then multimeric head-to-tail λ chromosomes are made by rolling circle replication. The rolling circle model for DNA replication is shown in Figure 10.12. The first step is the generation of a specific cut (nick) in one of the two strands at the origin of replication. The 5' end of the cut strand is then displaced from the circular molecule. This creates a replication fork structure and leaves a single-stranded stretch of DNA that serves as a template for the addition of deoxyribonucleotides to the free 3' end by DNA polymerase III.

As the 5' cut end continues to be displaced from the circular molecule, new DNA is synthesized. This is the leading strand of the previous replication fork diagrams. As replication proceeds the 5' end of the cut DNA strand is rolled out as a free "tongue" of increasing length. (This is analogous to pulling out the end of a roll of paper towels.) This single-stranded DNA tongue becomes covered with SSB proteins.

As the circle continues to roll, the single-stranded tongue converts to a double-stranded form, as illustrated earlier in our discussion of lagging

■ *Figure 10.11*   Bidirectional replication of circular DNA molecules.

■ *Figure 10.12*   The replication process of double-stranded circular DNA molecules through the rolling circle mechanism. The active force that unwinds the 5′ tail is the movement of the replisome propelled by its helicase components.

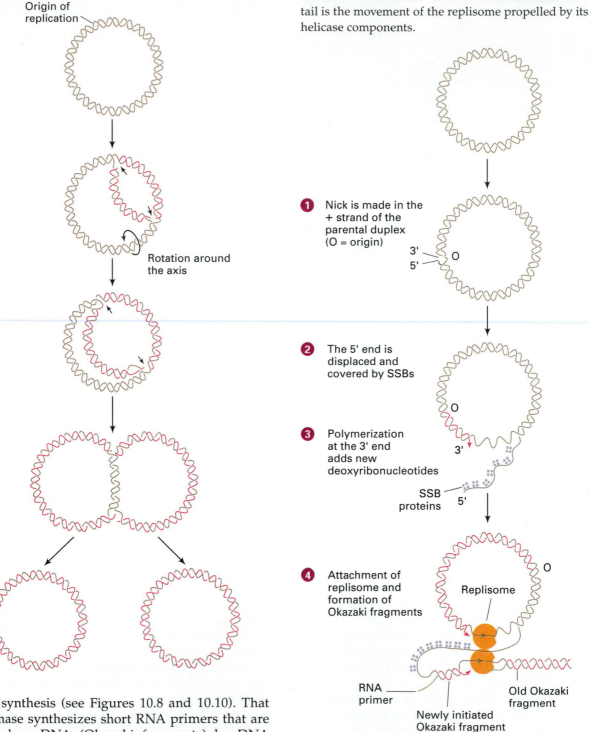

Origin of replication

Rotation around the axis

1 Nick is made in the + strand of the parental duplex (O = origin)

3′
5′
O

2 The 5′ end is displaced and covered by SSBs

O

3 Polymerization at the 3′ end adds new deoxyribonucleotides

3′

SSB proteins   5′

4 Attachment of replisome and formation of Okazaki fragments

O

Replisome

RNA primer

Newly initiated Okazaki fragment

Old Okazaki fragment

strand synthesis (see Figures 10.8 and 10.10). That is, primase synthesizes short RNA primers that are extended as DNA (Okazaki fragments) by DNA polymerase III. The RNA primers are ultimately removed and adjacent Okazaki fragments are joined through the action of DNA ligase. As the single-stranded DNA tongue rolls out, DNA synthesis continues on the circular DNA template.

Since the parental DNA circle can continue to roll, it is possible to generate a linear double-

stranded DNA molecule that is longer than the circumference of the circle. For example, as already mentioned, in the later stages of phage λ DNA replication, linear tongues that are many times the circumference of the original circle are produced by rolling circle replication. These molecules are cut into individual linear λ chromosomes by the *ter* enzyme (see Figure 9.5) and the unit-length molecules are then packaged into phage heads.

K e y n o t e

Replication of DNA in *E. coli* requires at least two of the DNA polymerases and several other enzymes and proteins. The DNA helix is untwisted by DNA helicase to provide templates for the synthesis of new DNA. Since new DNA is made in the 5′-to-3′ direction, chain growth is continuous on one strand and discontinuous (i.e., in segments that are later joined) on the other strand. This semidiscontinuous model is applicable to many other prokaryotic replication systems, each of which differs in the number and properties of the enzymes and proteins required. In the replication of circular DNA molecules, topoisomerases act to prevent the DNA from tangling ahead of the replication fork as the DNA untwists.

DNA Replication in Eukaryotes

The biochemistry and molecular biology of DNA replication are very similar in prokaryotes and eukaryotes. However, in eukaryotes there is an added complication because DNA is not found in only one chromosome but is distributed among many chromosomes, each of which is a complex aggregate of DNA and proteins. In every cell division cycle, each of these chromosomes must be faithfully duplicated and a copy of each distributed to each of the two progeny cells. This means that both the DNA and the histones must be doubled with each division cycle.

The cell cycle is qualitatively the same from eukaryote to eukaryote, although there are significant differences both in the relative amount of time spent in each phase of the cycle and in the total time spent in one cycle. Among higher eukaryotes, for example, some cells divide once every 3 hours and some cells divide once every 200 hours; human cells in culture divide approximately once every 24 hours.

Recall from Chapter 1 that the cell cycle in most somatic cells of higher eukaryotes is divided into four stages: gap 1 (G_1), synthesis (S), gap 2 (G_2), and mitosis (M). The DNA replicates and the chromosomes duplicate during the S phase, and the progeny chromosomes segregate into daughter cells during the M phase. Replication involves the coordination of the mechanisms for untwisting the DNA of each chromosome and producing two progeny DNA double helices, and for the duplication of the nucleosome organization of the eukaryotic chromosome. Although DNA is neither replicating nor being segregated during G_1 and G_2, the cell is metabolically active and growing. Events involved in the subsequent synthesis and mitotic activity also occur during G_1 and G_2.

Molecular Details of DNA Synthesis in Eukaryotes

The replication of the eukaryotic chromosome must involve the replication of the DNA and a doubling of the histones and nonhistones. As we saw earlier, many of the enzymes and proteins involved in prokaryote DNA replication have been identified. Less is known about the enzymes and proteins involved in eukaryotic DNA replication. It is clear, however, that the steps described for DNA synthesis in prokaryotes also occur for DNA synthesis in eukaryotes, namely, denaturation of the DNA double helix and the semiconservative, semidiscontinuous replication of the DNA. We will discuss features of DNA replication that are largely unique to eukaryotes and will not repeat the molecular details of DNA replication that are in common between prokaryotes and eukaryotes.

Initiation of DNA Replication   Each eukaryotic chromosome consists of one linear DNA double helix. If there was only one origin of replication per chromosome, the replication of each chromosome would take many, many hours. For example, there are 2.75×10^9 base pairs of DNA in the haploid human genome (23 chromosomes), and the average chromosome is roughly 10^8 base pairs long. With a replication rate of 2 kilobases (2000 bases) per minute in human cells, it would take approximately 830 hours to replicate one chromosome. If each cell cycle were at least that long for a developing human embryo, the gestation period would be many years instead of 9 months.

Actual measurements show that the chromosomes in eukaryotes replicate much faster than would be the case with only one origin of replication per chromosome. The diploid set of chromosomes in *Drosophila* embryos, for instance, replicates in 3 minutes. This is 6 times faster than the replication of the *E. coli* chromosome, even though there is about 100 times more DNA in *Drosophila* than there is in *E. coli*.

The rapidity of eukaryotic chromosome duplication is possible because DNA replication is initiated at many origins of replication throughout the genome. At each origin of replication, the DNA denatures (as in *E. coli*). Replication proceeds bidi-

■ *Figure 10.13*   Replicating DNA of *Drosophila melanogaster*: (a) Electron micrograph showing replication units (replicons); (b) Interpretive diagram of the electron micrograph shown in (a).

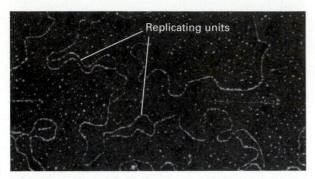

a)

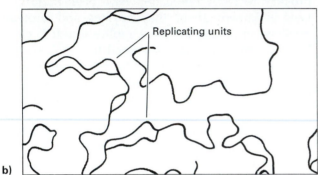

b)

rectionally, and the DNA double helix opens to expose single strands that act as templates for new DNA synthesis. Eventually, each replication fork will run into an adjacent replication fork, initiated at an adjacent origin of replication. In eukaryotes the stretch of DNA from the origin of replication to the two termini of replication (where adjacent replication forks fuse) on each side of the origin is called a **replicon** or a **replication unit.** Figure 10.13 presents an electron micrograph and an interpretive drawing showing a large number of replicons on a piece of *Drosophila* DNA. The large number of rela-

tively small replication units greatly decrease the overall time for duplication of the chromosomal DNA. Table 10.3 presents the number of replicons and their average size, and the rate of replication fork movement for a number of organisms. Note that the replicon size is much smaller and the rate of fork movement is much slower in eukaryotic organisms than in bacteria.

Replication of DNA does not occur simultaneously in all the replicons in an organism's genome. Instead, the initiation events occur during the replication phase with a particular timing. For many cell types this temporal ordering of initiation is characteristic of the cell type; that is, the same pattern occurs cell generation after cell generation. The molecular events that control which replicons replicate early and which replicate late are unknown. The diagram in Figure 10.14 gives a theoretical example of the temporal ordering of initiation phenomena at replication origins. The figure shows one segment of one chromosome in which there are three replicons that always begin replicating at distinct times. When the replication forks fuse at the margins of adjacent replicons, the chromosome has replicated into two sister chromatids. In general, replication of a segment of chromosomal DNA occurs following the synchronous activation of a cluster of origins.

Replication Enzymes and Proteins   Five different DNA polymerases have been identified in mammalian cells, designated α, β, δ, γ, and ε. Their properties are summarized in Table 10.4. The α, β, δ, and ε polymerases are located in the nucleus, while the γ polymerase is in the mitochondrion. DNA polymerases α and δ both replicate nuclear chromosomal DNA. They function essentially like the *E. coli* DNA polymerase III. DNA polymerase β is used in DNA repair, and DNA polymerase ε appears also to be used in that process. DNA polymerase γ replicates mitochondrial DNA. Both

	Table 10.3		
Comparison of Bacterial and Eukaryote Replicons			
ORGANISM	NO. OF REPLICONS	AVERAGE LENGTH	FORK MOVEMENT
Bacterium (*E. coli*)	1	4,200 kb	50,000 bp/min
Yeast (*Saccharomyces cerevisiae*)	500	40 kb	3,600 bp/min
Fruit fly (*Drosophila melanogaster*)	3,500	40 kb	2,600 bp/min
Toad (*Xenopus laevis*)	15,000	2,000 kb	500 bp/min
Mouse (*Mus musculus*)	25,000	150 kb	2,200 bp/min
Plant (*Vicia faba*)	35,000	300 kb	

■ *Figure 10.14*   Temporal ordering of DNA replication initiation events in replication units of eukaryotic chromosomes.

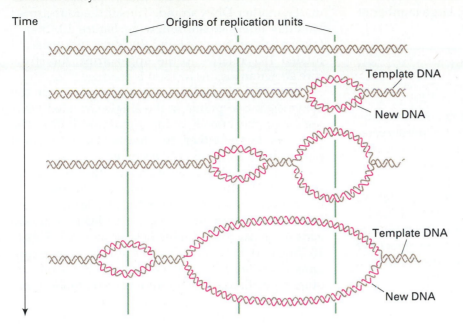

nuclear replication polymerases α and δ can use RNA primers for the initiation of DNA synthesis; δ can also use a DNA primer. A DNA primase activity has been shown to be associated only with the α polymerase. The mitochondrial polymerase γ uses a DNA primer for DNA initiation; that primer is synthesized by a separate primase enzyme, not by a DNA polymerase enzyme. All five enzymes are presumed to have 5′-to-3′ exonuclease activity, although there is no direct evidence for that. Only δ, ε, and γ have been shown to have proofreading (3′-to-5′ exonuclease) activity. Any proofreading done while the other enzymes replicate DNA must be performed by separate proteins with 3′-to-5′ exonuclease activity.

Keynote

In eukaryotes DNA replication is biochemically and molecularly very similar to the replication process in prokaryotic cells. Synthesis of DNA is initiated by RNA primers, occurs in the 5′-to-3′ direction, is catalyzed by DNA polymerases, requires a large

Table 10.4

Properties of DNA Polymerases in Mammalian Cells

	DNA POLYMERASE				
PROPERTY	α	β	δ	ε	γ
Location in cell:					
Nucleus	+	+	+	+	-
Mitochondrion	-	-	-	-	+
Polymerization activity:					
5′-to-3′	+	+	+	+	+
DNA primase associated					
with enzyme	+	-	-	-	-
Exonuclease activity:					
3′-to-5′ proofreading	-	-	+	+	+

number of other enzymes and proteins, and is a semiconservative and semidiscontinuous process. Replication of DNA is initiated at a large number of sites throughout the chromosomes.

DNA Recombination

In Chapters 5 and 6 we discussed genetic recombination and the mapping of genes in prokaryotes and eukaryotes. We learned that crossing-over involves the physical exchange of homologous chromosome parts as a result of breakage and rejoining of homologous DNA double helices. Now that we have an understanding of DNA structure (Chapter 8), the organization of DNA into chromosomes (Chapter 9), and DNA replication (this chapter), we return to a consideration of genetic recombination, this time concentrating on molecular aspects.

The Holliday Model for Recombination

In the mid-1960s Robin Holliday proposed a model for reciprocal recombination. Since then, the *Holliday model* has been refined by other geneticists, notably Matthew Meselson and Charles Radding.

The Holliday model is diagrammed in Figure 10.15 for genetically distinguishable homologous chromosomes, one with alleles a^+ and b^+ at opposite ends, and the other with alleles a and b. The two DNA double helices in the figure participate in the recombination event. The first stage of the recombination process is *recognition and alignment* (Figure 10.15[1]), in which two homologous DNA double helices align precisely. In the second stage one strand of each double helix breaks; each broken strand then invades the opposite double helix and base pairs with the complementary nucleotides of the invaded helix (Figure 10.15[2]). Each of these steps is catalyzed by specific cellular enzymes. The process leaves gaps that are sealed by DNA polymerase and DNA ligase, producing what is called a *Holliday intermediate*, with an internal branch point (Figures 10.15[3] and 10.16). The two DNA double helices in the Holliday intermediate can rotate, causing the branch point to move to the right or to the left. Figure 10.15(4) shows a branch migration event that has occurred to the right. The four-armed structure for the DNA strands is produced simply by pulling the four chromosome ends apart. The result of branch migration is the generation of complementary regions of hybrid DNA in both double helices (diagrammed as stretches of DNA helices with two different colors in Figure 10.15[5–8]).

The *cleavage and ligation* phase of recombination is best visualized if the Holliday intermediate is redrawn so that no DNA strand passes over or under another DNA strand. Thus if the four-armed Holliday intermediate following Figure 10.15(4) is taken as a starting point and the lower two arms are rotated 180° relative to the upper arms, the structure shown in Figure 10.15(5) is produced.

Next, the Holliday intermediate is cut by enzymes at two points in the single-stranded DNA region of the branch point (Figure 10.15[5]). The cuts can be in either the horizontal or vertical planes; both kinds of cuts occur with equal probability. Endonuclease cleavage in the horizontal plane (Figure 10.15[6, left]) produces the two double helices shown in Figure 10.15(7, left). In each helix is a single-stranded gap. DNA ligase seals the gaps to produce the double helices shown in Figure 10.15(8, left). Since each of the resulting helices contains a segment of single-stranded DNA from the other helix, flanked by nonrecombinant DNA, these double helices are called *patched duplexes*.

If the endonuclease cleavage in Figure 10.15(5) is in the vertical plane (Figure 10.15[6, right]), the gapped double helices of Figure 10.15(7, right) are produced. In this case there are segments of hybrid DNA in each duplex, but they are formed by what looks like a splicing together of two helices. The result is the double helices in Figure 10.15(8, right), which are called *spliced duplexes*.

In the example shown in Figure 10.15, the parental duplexes contained different genetic markers at the ends of the molecules. One parent was a^+ b^+ and the other was a b, as would be the case for a doubly heterozygous parent. However, the reciprocal recombination events shown in Figure 10.15 result in different products. In the patched duplexes (Figure 10.15[8, left]) the markers are $a^+ b^+$ and $a b$, which is the parental configuration. However, in the spliced duplexes (Figure 10.15[8, right]) the markers are recombinant, $a^+ b$ and $a b^+$. Since the enzymes that cut in the branch region during the final cleavage and ligation phase (Figure 10.15[6, left and right]) cut randomly with respect to the plane of the cut (horizontal or vertical), the Holliday model predicts that a physical exchange between two gene loci on homologous chromosomes should result in the genetic exchange of the outside chromosome markers about half of the time.

It is important to note that the Holliday model is only one model to explain the events of recombination. While its basic features are generally accepted, a number of other models have been proposed that attempt to explain recombination either at a more detailed level or recombination in special systems. Discussion of these models is beyond the scope of this book.

■ *Figure 10.15* Holliday model for reciprocal genetic recombination. Shown are two homologous DNA double helices that participate in the recombination process.

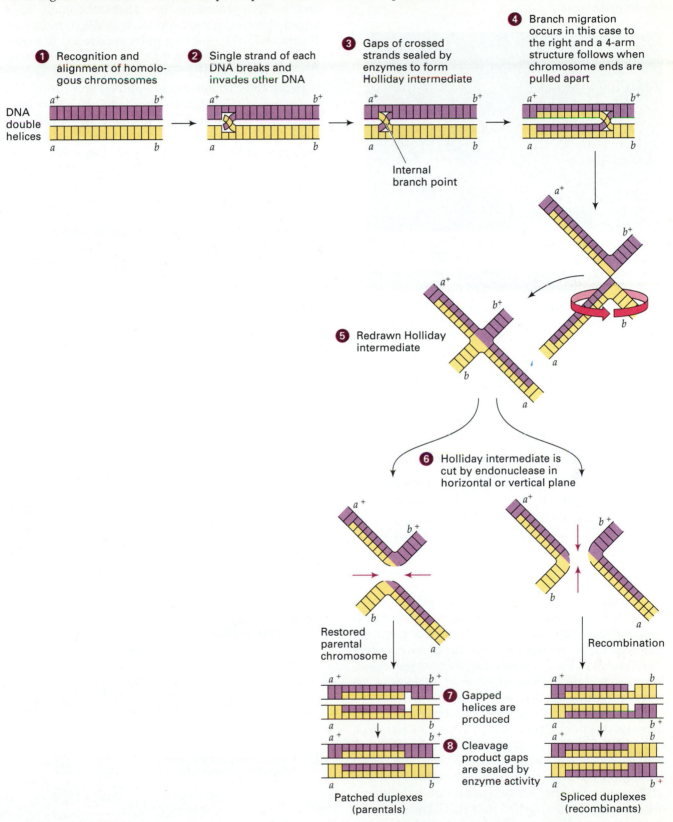

❶ Recognition and alignment of homologous chromosomes

❷ Single strand of each DNA breaks and invades other DNA

❸ Gaps of crossed strands sealed by enzymes to form Holliday intermediate

Internal branch point

❹ Branch migration occurs in this case to the right and a 4-arm structure follows when chromosome ends are pulled apart

DNA double helices

❺ Redrawn Holliday intermediate

❻ Holliday intermediate is cut by endonuclease in horizontal or vertical plane

Restored parental chromosome

Recombination

❼ Gapped helices are produced

❽ Cleavage product gaps are sealed by enzyme activity

Patched duplexes (parentals)

Spliced duplexes (recombinants)

■ *Figure 10.16* Electron micrograph of a Holliday intermediate with some single-stranded DNA in the branch point region.

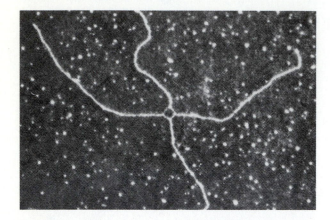

K e y n o t e

The Holliday model for genetic recombination involves a precise alignment of homologous DNA sequences of two parental double helices, endonucleolytic cleavage, invasion of the other helix by each broken end, ligation to produce the Holliday intermediate, branch migration, and finally, cleavage and ligation to resolve the Holliday intermediate into the recombinant double helices. Depending on the orientation of the cleavage events, the resulting double helices will be patched duplexes or spliced duplexes. If the recombination events occur between heterozygous loci, the patched duplexes produced are parental, whereas the spliced duplexes are recombinant for the two loci.

Mismatch Repair and Gene Conversion

In Chapter 5 we discussed tetrad analysis, in which all products of each meiotic event can be isolated and analyzed. Tetrad analysis is limited to relatively few organisms; among them are fungi such as *Saccharomyces cerevisiae* (yeast) and *Neurospora crassa*. Recall that from a cross of $a\ b \times a^+\ b^+$ in which a and b are linked on the same chromosome, but still reasonably distant, three types of tetrads can result: parental ditype (PD), $a\ b$, $a\ b$, $a^+\ b^+$, $a^+\ b^+$; tetratype (T), $a\ b^+$, $a\ b$, $a^+\ b^+$, $a^+\ b$; and the rare nonparental ditype (NPD), $a\ b^+$, $a\ b^+$, $a^+\ b$, $a^+\ b$. For each tetrad type, there is a 2:2 segregation of the alleles.

Occasionally, the Mendelian law of 2:2 segregation of alleles is not obeyed in that 3:1 and 1:3 ratios are seen. Tetrads showing these unusual ratios are proposed to derive from a process called *gene conversion*. Any pair of alleles is subject to gene conver-

sion. Consider, for example, the cross $a^+\ m\ b^+ \times a\ m^+\ b$, where gene m is located between genes a and b. The generation by mismatch repair of a tetrad showing gene conversion is diagrammed in Figure 10.17. The starting point is parental homologous chromosomes synapsed in meiotic prophase (Figure 10.17[1]). If a recombination event occurs between the two inner chromatids (Figure 10.17[2]), a patched duplex can be produced with two mismatches. That is, in these mismatches (called *heteroduplexes*) one of the two DNA strands has the m sequence from one parent and the other DNA strand in the helix has the m^+ sequence of the other

■ *Figure 10.17* Gene conversion by mismatch repair at two sites: (1) Parent homologs; (2) Recombination between inner two chromatids produces a patched duplex with two mismatches; (3) Both mismatches are repaired by excision and DNA synthesis; (4) Two meiotic divisions produce a tetrad with a 3:1 conversion for the m^+ allele.

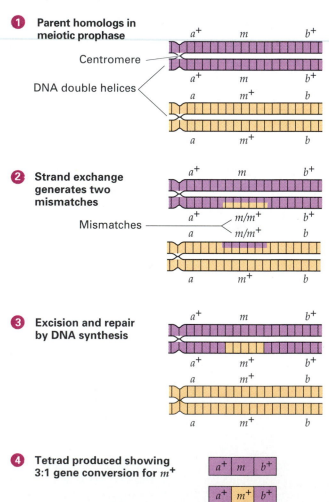

parent. These are labeled $\frac{m}{m^+}$ in the figure. By chance, both mismatches can be excised and repaired as shown in Figure 10.17(3). Following the subsequent two meiotic divisions, a tetrad is produced, showing a 3:1 gene conversion for the + allele of *m*, while the outside genetic markers a^+/a and b^+/b retain their parental configuration (Figure 10.17[4]).

If two markers are very close together, they undergo *co-conversion* and both will exhibit a 1:3 or 3:1 segregation in a tetrad, whereas outside markers will segregate in a normal 2:2 ratio.

Keynote

In organisms in which all four products of meiosis can be analyzed, mismatch repair of heteroduplex DNA can result in a non-Mendelian segregation of alleles, typically a 3:1 or 1:3 ratio rather than a 2:2 segregation. This phenomenon is known as gene conversion. Co-conversion of alleles can occur if the two alleles are very close together.

Summary

In this chapter we discussed DNA replication and DNA recombination. Many biochemical and molecular aspects of DNA replication are very similar in prokaryotes and eukaryotes—a semiconservative and semidiscontinuous replication mechanism, synthesis of new DNA in the 5'-to-3' direction, and the use of RNA primers to initiate DNA chains. The enzymes that catalyze the synthesis of DNA are the DNA polymerases. In *E. coli* there are three DNA polymerases, two of which are known to be involved in chromosomal DNA replication along with several other enzymes and proteins. The prokaryotic DNA polymerases have 3'-to-5' exonuclease activity, which permits proofreading to take place during DNA synthesis if an incorrect nucleotide is inserted opposite the template strand. In mammals four DNA polymerases are located in the nucleus and one in the mitochondrion. Two of the nuclear enzymes—α and δ—are responsible for chromosomal DNA replication; the other two nuclear enzymes—β and ε—are associated with DNA repair activities. The mitochondrial polymerase γ is responsible for replication of mitochondrial DNA. DNA polymerases δ, ε, and γ have been shown to have proofreading (3'-to-5' exonuclease) activity.

In prokaryotes DNA replication begins at specific chromosomal sites. A prokaryotic chromosome has one initiation site for DNA replication, while a eukaryotic chromosome has several initiation sites dividing the chromosome into replication units, or replicons. It is not clear whether the initiation sites used are the same sequences from cell generation to cell generation. The existence of replicons means that DNA replication of the entire set of chromosomes in a eukaryotic organism can proceed relatively quickly, in some cases faster than the single *E. coli* chromosome, despite the presence of orders of magnitude more DNA.

Some of the same enzymes used for DNA replication are also used in the genetic recombination process. A number of models have been proposed to describe the molecular events involved in the breakage and rejoining of DNA during crossing-over. One model, developed by Holliday, is described in this chapter. All molecular models for genetic recombination involve new DNA synthesis in small regions participating in crossing-over. In a later chapter (Chapter 16) we will see that DNA synthesis may also be involved in the repair of certain genetic damage. Thus three major cellular processes involve DNA synthesis: DNA replication, genetic recombination, and DNA repair.

Analytical Approaches for Solving Genetics Problems

Q10.1 a. Meselson and Stahl used ^{15}N-labeled DNA to prove that DNA replicates semiconservatively. The method of analysis was cesium chloride (CsCl) equilibrium density gradient centrifugation, in which bacterial DNA labeled in both strands with ^{15}N (the heavy isotope of nitrogen) bands to a different position in the gradient than DNA labeled in both strands with ^{14}N (the normal isotope of nitrogen).

Starting with a mixture of ^{15}N-containing and ^{14}N-containing DNA, then, two bands result after CsCl density gradient centrifugation. Now both DNAs are first heated to 100°C to denature the double helices into single strands of DNA and then cooled slowly. During cooling, any two single-stranded DNA molecules that are complementary will pair and form double-stranded DNA. After the cooling step the

DNAs were centrifuged and now the result is different from before. Two bands are seen in exactly the same positions as before, and a new, third band, is seen at a position halfway between the other two bands. From its position relative to the other two bands, the new band is interpreted to be intermediate in density between the other two bands. Explain the existence of the three bands in the gradient.

b. DNA from *E. coli* containing ^{15}N in both strands is mixed with DNA from another bacterial species, *Bacillus subtilis* containing ^{14}N in both strands. Two bands are seen after CsCl density gradient centrifugation. If the two DNAs are mixed, heated to 100°C, slowly cooled, and then centrifuged, only two bands result. These bands are in the same positions as in the unheated DNA experiment. Explain these results.

A10.1   a. When $^{15}N–^{15}N$ DNA and $^{14}N–^{14}N$ DNA from the same species are mixed, denatured by heat, and allowed to cool slowly, the single strands pair randomly during renaturation so that $^{15}N–^{15}N$, $^{14}N–^{14}N$, and $^{15}N–^{14}N$ double-stranded DNA is produced. The latter DNA will have a density intermediate between those of the two other DNA types, which accounts for the third band. Theoretically, if all DNA strands pair randomly, there should be a distribution of 1:2:1 of $^{15}N–^{15}N$, $^{15}N–^{14}N$, and $^{14}N–^{14}N$ DNAs, and this should be reflected in the relative intensities of the bands.

b. DNAs from different bacterial species have different sequences. In other words, DNA from one species typically is not complementary to DNA from another species. Therefore only two bands are seen because only the two *E. coli* DNA strands can renature to form $^{15}N–^{15}N$ DNA and only the two *B. subtilis* DNA strands can renature to form $^{14}N–^{14}N$ DNA. No $^{15}N–^{14}N$ hybrid DNA can form, so in this case there is no third band of intermediate density.

Q10.2 What would be the effect on chromosome replication in *E. coli* strains carrying deletions of the following genes:

a. *polC*
b. *polA*
c. *dnaG*
d. *lig*

e. *ssb*
f. *ori*
g. *rep*

A10.2 When genes are deleted, the function encoded by those genes is lost. All of the genes listed in the question are involved in DNA replication in *E. coli*, and their functions are briefly described in Table 10.2 and discussed in the text.

a. *polC* encodes DNA polymerase III, the principal DNA of DNA chains. A deletion of the *polC* gene would result in the loss of the ability to synthesize DNA strands from RNA primers; hence, synthesis of new DNA strands could not occur, and there would be no chromosome replication.

b. *polA* encodes DNA polymerase I, which is used in DNA synthesis to extend DNA chains made by DNA polymerase III while simultaneously excising the RNA primer by 5′-to-3′ exonuclease activity. Most DNA synthesis activity in *E. coli* is performed by DNA polymerase III, so chromosome replication would likely occur normally in an *E. coli* strain carrying a deletion of *polA*.

c. *dnaG* encodes DNA primase, the enzyme that synthesized the RNA primer on the DNA template. Without the synthesis of the short RNA primer, DNA polymerase III cannot initiate DNA synthesis and, therefore, chromosome replication will not take place.

d. *lig* encodes DNA ligase, the enzyme that catalyzes the ligation of Okazaki fragments together. In a strain carrying a deletion of *lig*, DNA synthesis would occur, but stable progeny chromosomes would not result because the Okazaki fragments could not be ligated together, so the lagging strand synthesized discontinuously on the lagging strand template would be in fragments.

e. *ssb* encodes the single-strand binding proteins that bind to and stabilize the single-stranded DNA regions produced as the DNA is unwound at the replication fork. In the absence of single-strand binding proteins, impeded or absent DNA replication would result because the replication bubble could not be kept open.

f. *ori* is the origin of replication region in *E. coli*, the location at which chromosome replication initiates. Without the origin the initiator protein cannot bind, no replication bubble can form, and therefore chromosome replication cannot take place.

g. *rep* encodes helicase, the enzyme that untwists the DNA at the replication fork. Without helicase the old chromosome cannot be untwisted and is therefore not accessible for replication; no chromosome replication will occur.

Questions and Problems

10.1 Compare and contrast the conservative and semiconservative models for DNA replication.

*10.2 In the Meselson and Stahl experiment ^{15}N-labeled cells were shifted to ^{14}N medium, at what we can designate as generation 0.
a. For the semiconservative model of replication, what proportion of $^{15}N–^{15}N$, $^{15}N–^{14}N$, and $^{14}N–^{14}N$ would you expect to find at generations 1, 2, 3, 4, 6, and 8?
b. Answer the above question in terms of the conservative model of DNA replication.

10.3 Suppose *E. coli* cells are grown on an ^{15}N medium for many generations. Then they are quickly shifted to an ^{14}N medium, and DNA is extracted from the samples taken after one, two, and three generations. The extracted DNA is subjected to equilibrium density gradient centrifugation in CsCl. In the figure below, using the reference positions of pure ^{15}N and pure ^{14}N DNA as guides, indicate where the bands of DNA would equilibrate if replication were semiconservative or conservative.

a) **Semiconservative model**

Pure ^{14}N DNA   Pure ^{15}N DNA

Generation:
1
2
3

b) **Conservative model**

Pure ^{14}N DNA   Pure ^{15}N DNA

Generation:
1
2
3

10.4 A spaceship lands on Earth with a sample of extraterrestrial bacteria. You are assigned the task of determining the mechanism of DNA replication in this organism.

You grow the bacteria in unlabeled medium for several generations, then grow it in the presence of ^{15}N exactly for one generation. You extract the DNA

and subject it to CsCl centrifugation. The banding pattern you find is shown in the figure below.

Control   **Experimental sample**

$^{15}N/^{15}N$ $^{14}N/^{14}N$

It appears to you that this is evidence that DNA replicates in the semiconservative manner, but this result does not prove that this is so. Why? What other experiment could you perform (using the same sample and technique of CsCl centrifugation) that would further distinguish between semiconservative and dispersive modes of replication?

10.5 Assume you have a DNA molecule with the base sequence T-A-T-C-A going from the 5′ to the 3′ end of one of the polynucleotide chains. The building blocks of the DNA are drawn as in the following figure. Use this shorthand system to diagram the completed double-stranded DNA molecule, as proposed by Watson and Crick.

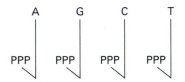

A   G   C   T

PPP   PPP   PPP   PPP

10.6 Base analogs are compounds that resemble the natural bases found in DNA and RNA, but are not normally found in those macromolecules. Base analogs can replace their normal counterparts in DNA during *in vitro* DNA synthesis. Four base analogs were studied for their effects on *in vitro* DNA synthesis using the *E. coli* DNA polymerase. The results were as follows, with the amounts of DNA synthesized expressed as percentages of that synthesized from normal bases only.

	NORMAL BASES SUBSTITUTED BY THE ANALOG			
ANALOG	A	T	C	G
A	0	0	0	25
B	0	54	0	0
C	0	0	100	0
D	0	97	0	0

Which bases are analogs of adenine? of thymine? of cytosine? of guanine?

*10.7 The length of the *E. coli* chromosome is about 1100 μm.

a. How many base pairs does the *E. coli* chromosome have?

b. How many complete turns of the helix does this chromosome have?

c. If this chromosome replicated unidirectionally and if it completed one round of replication in 60 minutes, how many revolutions per minute would the chromosome be turning during the replication process?

d. The *E. coli* chromosome, like many others, replicates bidirectionally. Draw a simple diagram of a replicating *E. coli* chromosome that is halfway through the round of replication. Be sure to distinguish new and old DNA strands.

10.8 The following events, steps, or reactions occur during *E. coli* DNA replication. For each entry in Column A, select the appropriate entry in Column B. Each entry in A may have more than one answer, and each entry in B can be used more than once.

COLUMN A	COLUMN B
___ a. Unwinds the double helix	A Polymerase I
	B Polymerase III
___ b. Prevents reassociation of complementary bases	C Helicase
	D Primase
___ c. Is an RNA polymerase	E Ligase
	F SSB protein
___ d. Is a DNA polymerase	G Gyrase
	H None of these
___ e. Is the "repair" enzyme	
___ f. Is the major elongation enzyme	
___ g. A 5'-3' polymerase	
___ h. A 3'-5' polymerase	
___ i. Has 5'-3' exonuclease function	
___ j. Has 3'-5' exonuclease function	
___ k. Bonds free 3'-OH end of a polynucleotide to a free 5' monophosphate end of polynucleotide	
___ l. Bonds 3'-OH end of a polynucleotide to a free 5' nucleotide triphosphate	
___ m. Separates daughter molecules and causes supercoiling	

*10.9 In *E. coli* distinguish between the activities of primase, single-stranded binding protein, helicase, DNA ligase, DNA polymerase I, and DNA polymerase III in DNA replication.

10.10 Chromosome replication in *E. coli* commences from a constant point, called the origin of replication. It is known that DNA replication is bidirectional. Devise a biochemical experiment to prove that the *E. coli* chromosome replicates bidirectionally. (Hint: Assume that the amount of gene product is directly proportional to the number of genes.)

10.11 A space probe returns from Jupiter and brings with it a new microorganism for study. It has double-stranded DNA as its genetic material. However, studies of replication of the alien DNA reveal that, while the process is semiconservative, DNA synthesis is continuous on both the leading strand and the lagging strand templates. What conclusion(s) can you make from that result?

10.12 Draw a eukaryotic chromosome as it would appear at each of the following cell cycle stages. Show both DNA strands, and use different line styles for old and newly synthesized DNA.

a. G_1

b. anaphase of mitosis

c. G_2

d. anaphase of meiosis I

e. anaphase of meiosis II

*10.13 Autoradiography is a technique that allows radioactive areas of chromosomes to be observed under the microscope. The slide is covered with a photographic emulsion, which is exposed by radioactive decay. In regions of exposure the emulsion forms silver grains upon being developed. The tiny silver grains can be seen on top of the (much larger) chromosomes. Devise a method for finding out which regions in the human karyotype replicate during the last 30 min of the S period. (Assume a cell cycle in which the cell spends 10 h in G_1, 9 h in S, 4 h in G_2, and 1 h in M.)

*10.14 How is mismatch repair related to recombination and repair of DNA?

*10.15 Crosses were made between strains, each of which carried one of three different alleles of the same gene, *a*, in yeast. For each cross, some unusual tetrads resulted at low frequencies. Explain the origin of each of these tetrads:

Cross:	$a1$ $a2^+$	$a1$ $a3^+$	$a2$ $a3^+$
	×	×	×
	$a1^+$ $a2$	$a1^+$ $a3$	$a2^+$ $a3$
Tetrads:	$a1^+$ $a2$	$a1^+$ $a3$	$a2^+$ $a3$
	$a1^+$ $a2^+$	$a1^+$ $a3$	$a2^+$ $a3^+$
	$a1$ $a2^+$	$a1^+$ $a3^+$	$a2$ $a3^+$
	$a1$ $a2^+$	$a1$ $a3^+$	$a2$ $a3^+$

10.16 From a cross of $y1$ $y2^+$ × $y1^+$ $y2$, where $y1$ and $y2$ are both alleles of the same gene in yeast, the following tetrad type occurs at very low frequencies:

$$y1^+ \quad y2$$
$$y1 \quad y2$$
$$y1 \quad y2$$
$$y1 \quad y2$$

Explain the origin of this tetrad at the molecular level.

10.17 In *Neurospora* the *a*, *b*, and *c* loci are situated in the same arm of a particular chromosome. *a* is near the centromere, *b* is near the middle, and *c* is near the telomere of the arm. Among the asci resulting from a cross of *ABC* × *abc*, the following ascus was found (the 8 spores are indicated in the order in which they were arranged in the ascus): *ABC, ABC, ABc, ABc, aBC, aBC, abc, abc*. How might this ascus have arisen?

10.18 In the population of asci produced in Question 10.17, an ascus was found containing, in this order, the spores *ABC, ABC, ABc, Abc, aBC, aBC, abc, abc*. How could this ascus have arisen?

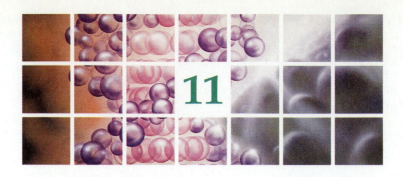

11

Transcription, RNA Molecules, and RNA Processing

- Transcription, the process of transcribing DNA base pair sequences into RNA base sequences is similar in prokaryotes and eukaryotes. The DNA unwinds in a short region next to a gene, and an RNA polymerase catalyzes the synthesis of an RNA molecule in the 5′-to-3′ direction. Only one strand of the double-stranded DNA is transcribed into an RNA molecule.

- In *E. coli* a single RNA polymerase synthesizes mRNA, tRNA, and rRNA. Eukaryotes have three distinct nuclear RNA polymerases, each of which transcribes different gene types: RNA polymerase I transcribes the genes for the large ribosomal RNAs; RNA polymerase II transcribes mRNA genes and some snRNA genes; and RNA polymerase III transcribes genes for the small 5S rRNAs, the tRNAs, and the remaining snRNAs.

- In *E. coli* the initiation of transcription of protein-coding genes requires the holoenzyme form of RNA polymerase (core enzyme + sigma factor) binding to the promoter. Once transcription has begun, the sigma factor dissociates and RNA synthesis is completed by the RNA polymerase core enzyme. Termination of transcription is signaled by specific sequences in the DNA. Two types of termination sequences are found and a particular gene will have one or the other. One type of terminator is recognized by the RNA polymerase alone, and the other type is recognized by the enzyme in association with the *rho* factor.

- The transcripts of protein-coding genes are linear precursor (pre-mRNAs). Prokaryotic mRNAs are modified little once they are transcribed, while eukaryotic mRNAs are modified by the addition of a 5′ cap and a 3′ poly(A) tail. Most eukaryotic pre-mRNAs contain sequences called introns, which do not code for amino acids. The introns are removed as the primary transcript is processed to produce the mature, functional mRNA molecule.

- Ribosomes are the cellular organelles on which protein synthesis takes place. In both prokaryotes and eukaryotes, ribosomes consist of two unequally sized subunits. Each subunit consists of a complex between one or more ribosomal RNA (rRNA) molecules and many ribosomal proteins. Eukaryotic ribosomes are larger and more complex than prokaryotic ribosomes.

- In eukaryotes three of the four rRNAs are encoded by tandem arrays of transcription units. Each transcription unit produces a single pre-rRNA molecule with the three rRNAs separated by spacer sequences. The individual rRNAs are generated by processing of the pre-rRNA to remove the spacers. The fourth rRNA is encoded by separate genes.

- In the precursor rRNAs of some organisms there are introns, the RNA sequences of which fold into a secondary structure that excises itself, a process called self-splicing. This process does not involve protein enzymes.

- Transfer RNA (tRNA) molecules bring amino acids to the ribosomes where the amino acids are polymerized into a protein chain. All tRNAs are about the same length, contain a number of modified bases, and have similar three-dimensional shapes. tRNAs are made as pre-tRNA molecules containing 5′-leader and 3′-trailer sequences, which are removed by enzymes.

T HE STRUCTURE, function, development, and reproduction of an organism depends on the properties of the proteins present in each cell and tissue. A protein consists of one or more chains of amino acids. Each chain of amino acids is called a polypeptide. The sequence of amino acids in a polypeptide chain is coded for by a gene, a specific base-pair sequence in DNA. (Recall the one gene–one polypeptide hypothesis discussed in Chapter 7.) When a protein is needed in the cell, the genetic code for that protein's amino acid sequence must be read from the DNA and processed into the finished protein. Two major steps occur in the process of protein synthesis: transcription and translation.

Transcription is the transfer of information from a double-stranded, template DNA molecule to a single-stranded RNA molecule. **Translation** (protein synthesis; see Chapter 12) is the conversion in the cell of the base sequence information in the messenger RNA (mRNA) into the amino acid sequence of a polypeptide. Unlike DNA replication, which typically occurs during only part of the cell cycle (at least in eukaryotes), transcription and translation generally occur throughout the cell cycle (although they are much reduced during the M phase of the cell cycle). In this chapter we will examine the transcription process, and the structures and properties of the different RNA classes: messenger RNA, transfer RNA, small nuclear RNA, and ribosomal RNA. We will see that the initial RNA transcripts (the **primary transcripts** or **precursor RNA molecules (pre-RNAs)**) of most genes must be modified and/or processed to produce the biologically active (mature) RNA molecules.

The Transcription Process

In 1956, three years after Watson and Crick proposed their double helix model for DNA, Crick gave the name **Central Dogma** to the two-step process of DNA → RNA → protein (transcription followed by translation) for the synthesis of proteins encoded by DNA. In this section we will discuss how an RNA chain is synthesized, and introduce the different classes of RNA and the genes that code for them.

RNA Synthesis

Only some of the DNA is transcribed at any one time. That is, the genome of an organism consists of specific sequences of base pairs distributed among a number of chromosomes. The sequences of base pairs that are transcribed are called *genes,* and thus the transcription process is referred to as *gene expression.* Associated with each gene are base-pair sequences called **gene regulatory elements,** which are involved in the regulation of gene expression (see Chapters 14 and 15).

In both prokaryotes and eukaryotes transcription occurs by a process that is catalyzed by an enzyme called **RNA polymerase** (Figure 11.1). The DNA double helix must unwind for a short region next to the gene before transcription can begin. In prokaryotes this occurs as part of the function of RNA polymerase, while in eukaryotes it is brought about by other proteins that bind to the DNA at the starting point for transcription.

Only one of the two DNA strands is transcribed into an RNA. During transcription RNA is synthesized in the 5'-to-3' direction. The 5'-to-3' DNA strand that has the *same* polarity of the resulting RNA strand is called the *coding strand.* Since the complementary 3'-to-5' DNA strand is the strand that is read to make the RNA strand, that strand is called the *template strand.*

In transcription the RNA precursors are the ribonucleoside triphosphates ATP, GTP, CTP, and UTP, collectively called NTPs. RNA synthesis occurs using polymerization reactions that are very similar to the polymerization reactions involved in DNA synthesis (Figure 11.2; DNA polymerization was shown in Figure 10.4). The next nucleotide to be added to the chain is selected by the RNA polymerase for its ability to pair with the exposed base on the DNA template strand. Unlike DNA polymerases, RNA polymerases are able to initiate new polynucleotide chains, and they have no proofreading abilities.

Recall that RNA chains contain nucleotides with the base uracil instead of thymine. Therefore where an A nucleotide occurs on the DNA template chain, a U nucleotide is placed in the RNA chain.

■ *Figure 11.1*   Transcription process. The DNA double helix is denatured by the action of RNA polymerase, which then catalyzes the synthesis of a single-stranded RNA chain, beginning at the "start of transcription" point. The RNA chain is made in the 5'-to-3' direction, using only one strand of the DNA as a template to determine the base sequence.

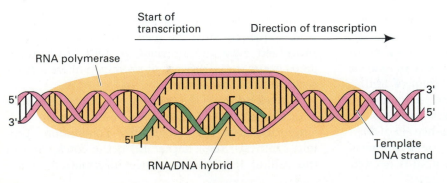

■ *Figure 11.2*   Chemical reaction involved in the RNA polymerase-catalyzed synthesis of RNA on a DNA template strand.

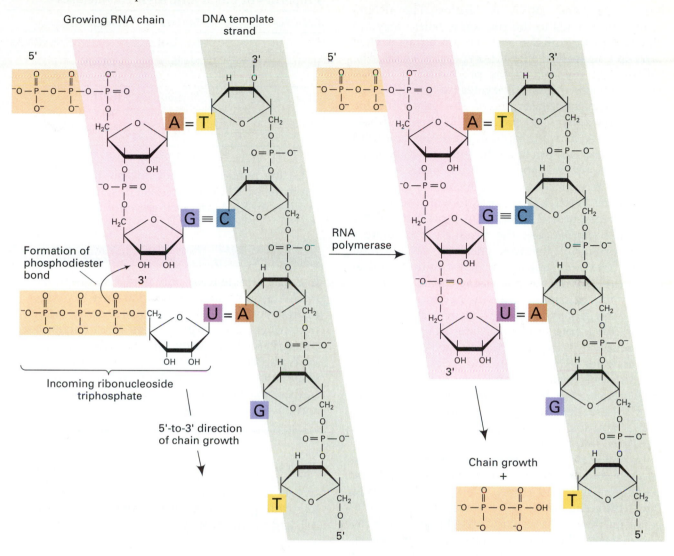

As an example, if the template DNA strand reads

3'-A T A C T G G A C-5'

then the RNA chain will be synthesized in the 5'-to-3' direction and will have the sequence

5'-U A U G A C C U G-3'

and the DNA sense strand is

5'-T A T G A C C T G-3'

K e y n o t e

Transcription, the process of transcribing DNA base-pair sequences into RNA base sequences, is similar in prokaryotes and eukaryotes. The DNA unwinds in a short region next to a gene, and an RNA polymerase catalyzes the synthesis of an RNA

molecule in the 5'-to-3' direction. Only one strand of the double-stranded DNA is transcribed into an RNA molecule.

Classes of RNA and the Genes that Code for Them

Four different major classes of RNA molecules or transcripts are produced by transcription: **messenger RNA (mRNA), transfer RNA (tRNA), ribosomal RNA (rRNA)**, and **small nuclear RNA (snRNA)**. The mRNAs, tRNAs, and rRNAs are found in both prokaryotes and eukaryotes, whereas snRNAs are found only in eukaryotes.

Only the mRNA molecule is translated to produce a polypeptide. Translation of mRNAs occurs on cellular organelles called **ribosomes**. (Translation is described in Chapter 12.) Ribosomes are composed of proteins and three (prokaryotes) or four (eukaryotes) rRNA molecules. The tRNAs bring amino acids to the ribosome, where they are matched to the mRNA and the amino acids they carry are assembled into a polypeptide chain. The snRNAs are involved in the processing of mRNA precursor molecules in eukaryotes (see pp. 247–248).

The primary transcripts of the genes for the RNAs generally are precursor-RNA (pre-RNA) molecules. After transcription, pre-RNA molecules are modified to produce the mature, functional RNAs. Two general types of modifications occur: 1) chemical modifications, in which the bases are altered; and 2) RNA processing, in which sequences of the precursor are removed in a precise, orderly way. In bacterial cells the production of mature RNAs primarily involves chemical modification events, while in eukaryotic cells both chemical modification and RNA processing events occur.

A gene that codes for an mRNA molecule, and hence for a protein, is called a **structural gene**, or a *protein-coding gene*. In this and later chapters we will be discussing primarily structural genes. The genes that code for tRNA, rRNA, and snRNA molecules are different from structural genes because their RNA transcripts (the products of transcription) are the final products of gene expression. That is, these RNAs function as RNA molecules; they are not translated into proteins.

In bacteria only one type of RNA polymerase is used to transcribe the structural genes and the genes for tRNA and rRNA, whereas in eukaryotes three different RNA polymerases transcribe the genes for the four types of RNAs (Table 11.1). **RNA polymerase I**, located exclusively in the nucleolus, catalyzes the synthesis of three of the RNAs found in ribosomes, the organelles responsible for protein synthesis: the two large ribosomal RNA (rRNA) molecules (18S and 28S rRNAs) and one of the two small rRNAs, the 5.8S rRNA. (In brief, the S value of a protein, RNA, or DNA molecule derives from the rate at which the molecules sediment in a gradient of sucrose in a centrifuge (see Box 11.1). The rate of sedimentation of a molecule is related both to the molecular weight of the molecule and to its three-dimensional configuration. Thus S values give a very rough comparison of molecular sizes.) **RNA polymerase II**, found only in the nucleoplasm of the nucleus, synthesizes messenger RNAs (mRNAs) and some snRNAs, some of which are involved in RNA processing events. **RNA polymerase III,** also found only in the nucleoplasm, synthesizes the following: (1) the transfer RNAs (tRNAs), which bring amino acids to the ribosome; (2) 5S rRNA, the other small rRNA molecule found in each ribosome; and

Table 11.1

Properties of Eukaryotic RNA Polymerases

RNA POLYMERASE	LOCATION	PRODUCTS
I	Nucleolus	28S, 18S, 5.8S rRNAs
II	Nucleus	mRNA some snRNAs
III	Nucleus	tRNA 5S rRNA some snRNAs

(3) the remaining small nuclear RNAs (snRNA) not made by RNA polymerase II, some of which are involved in RNA processing events.

Keynote

In *E. coli* a single RNA polymerase synthesizes mRNA, tRNA, and rRNA. Eukaryotes have three distinct nuclear RNA polymerases, each of which transcribes different gene types: RNA polymerase I transcribes the genes for the large ribosomal RNAs; RNA polymerase II transcribes mRNA genes and some snRNA genes; and RNA polymerase III transcribes genes for the small 5S rRNAs, the tRNAs, and the remaining snRNAs.

Transcription of Protein-coding Genes

In this section we discuss the transcription of protein-coding genes in prokaryotes, focusing on *E. coli*, and in eukaryotes.

Prokaryotes

A prokaryotic protein-coding gene may be divided into three sequences with respect to its transcription (Figure 11.3): (1) a sequence adjacent to the start of the gene that specifies where transcription will begin. This transcription start sequence is called the **promoter**; (2) the RNA coding sequence; that is, the sequence of base pairs transcribed by RNA polymerase into the single-stranded mRNA transcript; (3) a sequence adjacent to the end of the gene that specifies where transcription will stop. This sequence is called a **transcription terminator sequence,** or more simply, a **terminator**. Conventionally the promoter is considered to be "upstream" from the gene, and the terminator is "downstream" from the gene.

Initiation of Transcription In *E. coli* two DNA sequences in the promoter are critical for specifying

BOX 11.1

Sucrose Gradient Centrifugation

Sucrose gradient centrifugation is usually used to separate the components present in a mixture based on their relative rates of sedimentation. With this procedure we can measure the rates at which the components sediment in the gradient under centrifugal forces. In sucrose gradient centrifugation the sedimentation rates are converted to **Svedberg units,** using a formula the derivation of which is beyond the scope of this book. Svedberg units, or simply **S values,** are then used as a rough indication of relative sizes of the components being analyzed.

The rate of sedimentation of a component in a sucrose gradient is related both to the molecular weight of the component and to its three-dimensional configuration. Two components of the same molecular weight, for example, will have different S values if one is highly compact and thus sediments rapidly, while the other has a more extended shape and thus sediments slowly. Sucrose gradient centrifugation is typi-

cally used to separate, or estimate, the sizes of RNA molecules, ribosomes, ribosomal sub-units, proteins, various cellular organelles, and so on.

The sucrose gradient centrifugation method involves a tubeful of sucrose in which the concentration of sucrose increases toward the bottom of the tube. For the separation of different-sized RNA molecules, a continuous gradient of sucrose concentration, ranging from 10 percent at the top to 30 percent at the bottom, is prepared in a centrifuge tube (Box Figure 11.1[a]). Next, a small amount of the sample, in this case a solution containing the RNAs, is very carefully layered on top of the gradient (Box Figure 11.1[b]). The sucrose gradient is then centrifuged at high speeds for several hours, during which time the RNA molecules move through the gradient at different rates depending on their molecular weight and configuration (Box Figure 11.1[c]). At the completion of

■ *Box Figure 11.1*   Sucrose density gradient centrifugation technique for separating and isolating RNA molecules in a mixture: (a) 10% to 30% sucrose density gradient; (b) RNA sample added to top of gradient; (c) Three RNA types separate into bands after centrifugation; (d) Fractionation of gradient to collect the RNA types.

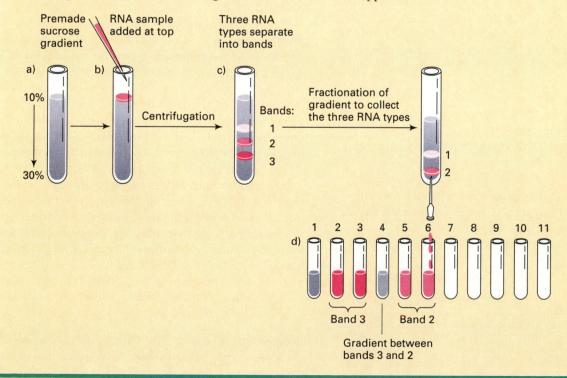

Continued ⟶

Continued

centrifugation, the RNAs of similar S values are located in discrete zones or bands in the gradient. These bands of RNA are not visible to the eye, but are determined after the analysis of the gradient.

To collect the different RNA types, the bottom of the tube is punctured with a needle, and the drops that run out are collected by using a fraction collector (Box Figure 11.1[d]). Fractions containing the RNAs are identified by measuring the degree to which each fraction absorbs ultraviolet light. Fractions with RNA will absorb ultraviolet light while those without

RNA will not. The relative positions the RNAs occupy in the gradient indicate the S values of the RNAs: Those with higher S values (larger and/or more compact) are found closer to the bottom of the gradient than those with lower S values.

Note that sucrose gradient centrifugation separates molecules on the basis of their relative rates of sedimentation. Hence it differs from cesium chloride equilibrium density gradient centrifugation (see Chapter 10, pp. 208–209), in which molecules are separated on the basis of buoyant density and not size.

---

the initiation of transcription. These sequences are at two distinct locations upstream from the first base pair to be transcribed into an RNA chain. These locations generally are found at −35 and −10; they are centered at 35 and 10 base pairs upstream from the base pair at which transcription starts. (Conventionally this initial base pair is designated as +1; base pairs upstream from the initial base pair (such as promoters) are given negative numbers; base pairs downstream from the initial base pair are given positive numbers.) From examination of the promoters of a large number of genes, the **consensus sequence** (the sequence showing the nucleotide found most frequently at each position for a number of sequences examined) for the −35 region is

5'-TTGACA-3'

The consensus sequence for the −10 region (also called the **Pribnow box** after the researcher who first discovered it, or a TATA box to reflect its sequence) is

■ *Figure 11.3*   Organization of a gene in terms of transcription into three regions: promoter, RNA coding sequence, and terminator. The promoter is considered to be "upstream" of the gene, and the terminator is considered to be "downstream" of the gene.

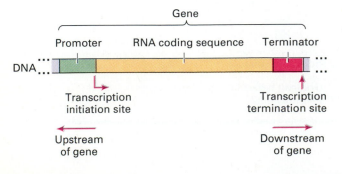

5'-TATAAT-3'

For transcription to begin, a form of RNA polymerase called the *holoenzyme* (or *complete enzyme*) must bind to the promoter. This holoenzyme consists of the **core enzyme** form of RNA polymerase (which has four polypeptides) bound to another polypeptide called the *sigma factor* (σ). The sigma factor is essential for recognition of a promoter sequence; if the sigma factor is not present, the core enzyme binds to DNA in various places but does not initiate transcription efficiently from any of them.

The RNA polymerase holoenzyme binds to a promoter in two distinct steps. First, it binds loosely to the −35 region of the promoter while the DNA is still in double-helical form (Figure 11.4[a]). Second, the enzyme untwists almost two complete helical turns of the DNA centered around the −10 region and becomes tightly bound at the −10 region (Figure 11.4[b]). At that point the RNA polymerase is correctly oriented to begin transcription at the correct nucleotide (Figure 11.4[b]).

Since promoters differ slightly in their actual sequence, the efficiency of RNA polymerase binding varies. As a result, the rate at which transcription is initiated varies from gene to gene, a fact that helps to explain why different genes have different rates of expression at the RNA level. For example, a −10 region sequence of 5'-GATACT-3' will have a lower rate of transcription initiation than the 5'-TATAAT-3'.

Elongation of the RNA Chain   RNA synthesis takes place in a region of the DNA that has denatured to form a transcription bubble. (This bubble is similar to a DNA replication bubble.) Once a few polymerizations have been completed, the sigma

■ *Figure 11.4*   Action of *E. coli* RNA polymerase in the initiation and elongation stages of transcription: (a) In initiation the RNA polymerase holoenzyme first binds loosely to the promoter at the -35 region; (b) As initiation continues RNA polymerase binds more tightly to the promoter at the -10 region, accompanied by a local untwisting of about 17 bp centered around the -10 region. At this point, the RNA polymerase is correctly oriented to begin transcription at +1; (c) After a few polymerizations have occurred, the sigma factor dissociates from the core enzyme; (d) As the RNA polymerase elongates the new RNA chain, the enzyme untwists the DNA ahead of it; as the double helix reforms behind the enzyme, the RNA is displaced away from the DNA.

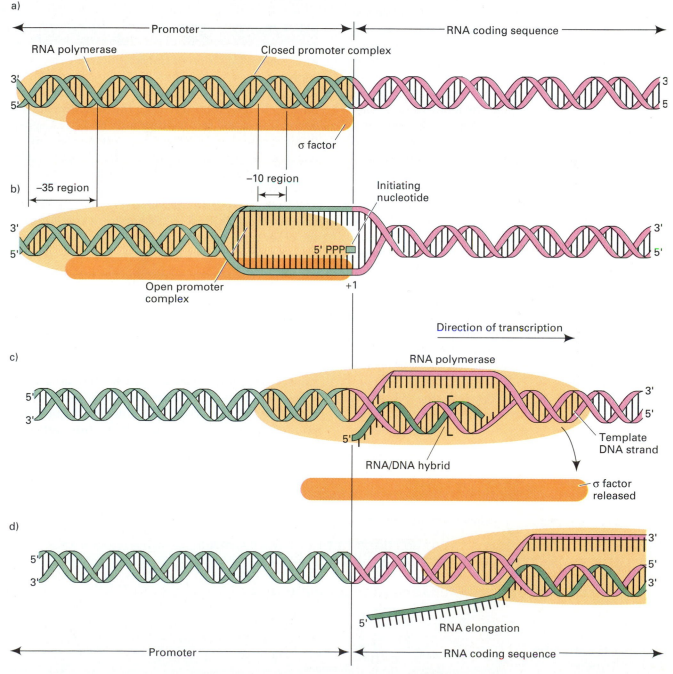

factor dissociates from the RNA polymerase core enzyme (Figure 11.4[c]) and can be used again in other transcription initiation reactions. The core enzyme completes the transcription of the gene.

As the core RNA polymerase moves along, it untwists the DNA double helix ahead of it. This produces torsional stress in the DNA much like unraveling a stretch of rope does, so that the DNA helix reforms behind the enzyme as the enzyme moves along the template (Figure 11.4[d]). Within the untwisted region, some bases of RNA are bonded to the DNA in a temporary RNA–DNA hybrid while the rest of the RNA is displaced away from the DNA (Figure 11.4[d]).

Termination of Transcription Termination of the transcription of a prokaryotic gene is signaled by controlling elements called *terminators.* One important protein involved in the termination of transcription of some genes in *E. coli* is *rho* (ρ). The terminators of such genes are called *rho-dependent terminators.* At many other terminators the core RNA polymerase itself can carry out the termination events. Those terminators are called *rho-independent terminators.*

Rho-independent terminators consist of sequences with twofold symmetry that are about 15 to 20 base pairs before the end of the gene, followed by a string of about six AT base pairs (Figure 11.5). A sequence with twofold symmetry is one that is approximately self-complementary about its center; that is, one half of the sequence is complementary to the other half. Thus the transcript of the region with twofold symmetry can form a hairpin loop as shown in Figure 11.5. The combination of the hairpin loop followed by the string of Us (transcribed from a string of AT base pairs) in the RNA leads to transcription termination. While it is not clear how the AT string and the hairpin loop bring about *rho*-independent termination, it is possible that the rapid formation of the hairpin loop destabilizes the RNA–DNA hybrid in the terminator region, which in turn causes the release of the RNA and transcription termination.

Rho-dependent terminators lack the AT string found in *rho*-independent terminators, and many cannot form hairpin structures. While the mechanism of *rho*-dependent termination is not clearly understood, it is known to require the interaction of *rho* and the growing RNA chain upstream of the terminator. Following this, the RNA transcript, RNA polymerase, and *rho* factor are released.

In sum, three key events occur at a terminator: (1) RNA synthesis stops, (2) the RNA chain is released from the DNA, and (3) RNA polymerase is released from the DNA.

Keynote

In *E. coli* the initiation of transcription of protein-coding genes requires the holoenzyme form of RNA polymerase (core enzyme + sigma factor) binding to the promoter. Once transcription has begun, the sigma factor dissociates and RNA synthesis is completed by the RNA polymerase core enzyme. Termination of transcription is signaled by specific sequences in the DNA. Two types of termination sequences are found and a particular gene will have one or the other. One type of terminator is recognized by the RNA polymerase alone, and the other type is recognized by the enzyme in association with the *rho* factor.

Eukaryotes

Transcription of Protein-coding Genes by RNA Polymerase II In eukaryotes RNA polymerase II transcribes protein-coding genes. The product of transcription is a **precursor-mRNA (pre-mRNA) molecule**, a transcript that must be modified and/or processed to produce the mature, function-

■ *Figure 11.5* Sequence of a ρ-independent terminator and structure of the terminated RNA.

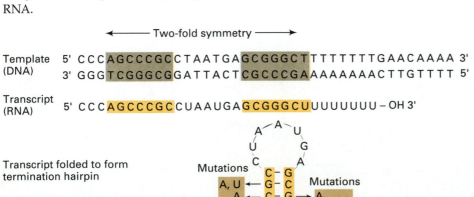

al mRNA molecule. RNA polymerase II also transcribes some snRNA genes.

Regulatory Elements   A protein-coding gene may have a large assortment of DNA sequences involved in the regulation of its transcription. These DNA sequences are called *regulatory elements*; they are located both upstream and downstream of the RNA initiation site for the gene. These regulatory elements can bind specific **transcription factors** (specific proteins required for the initiation of transcription by a eukaryotic RNA polymerase) and **regulatory factors** (proteins involved in the activation or repression of transcription of the gene). Generally, the regulatory elements are located within several hundred base pairs from the site of initiation of transcription, usually upstream from that point. However, some regulatory elements are much farther away. By definition the regulatory element adjacent to the transcription start site is called the *promoter.*

The promoter of a protein-coding gene is arranged as a series of **promoter elements**. Starting closest to the transcription initiation site, the promoter elements are the **TATA box** or **TATA element** (also called the **Goldberg-Hogness box** after its discoverers), the **CAAT element,** and the **GC element** (Figure 11.6). The elements are named for the general DNA base sequences they contain. Not all promoters have all three of the elements: some lack a TATA element, others lack a CAAT element or a GC element.

The TATA element has the consensus sequence 5'-TATAAA-3'. In higher eukaryotes, the TATA element is almost always located at position −30 (25 to 35 base pairs upstream from the first base pair transcribed into RNA). The CAAT element has the consensus sequence 5'-GGCCAATCT-3'. The CAAT element is found at approximately -80 in many genes, but it can also function at a number of other locations. The GC element has the consensus sequence 5'-GGGCGG-3'. Often there is more than one copy of a GC element in a promoter, and the

elements may function in either orientation—facing toward or away from the gene. The GC elements appear to help bind the RNA polymerase near the transcription start point.

While the promoter elements are crucial for determining whether transcription can occur, **enhancers,** or **enhancer elements,** are required for maximal transcription of the gene to occur. A gene's enhancer helps control transcription from the gene's promoter. Enhancers function in either orientation (facing toward or away from the gene) and at a large distance from the gene, often over 1000 base pairs from the promoter. In animal cells enhancers can activate genes when the enhancers are upstream or downstream from the RNA initiation site. In most cases, though, the enhancers are found upstream of the gene. In order for the enhancer to function, the DNA is folded to bring the enhancer sequence near the promoter. This folding is brought about by proteins binding at the enhancer and promoter sequences (see next section). Similar elements that have essentially the same properties as enhancer elements, except that they repress rather than activate gene transcription, are called **silencer elements**. Silencers are much less common than enhancers.

Transcription events   RNA polymerase II is unable to recognize the promoter on its own. Instead, specific transcription factors (TFs) are needed to bring about the initiation of transcription by RNA polymerase II. In general, all eukaryotic RNA polymerases require transcription factors for transcription initiation. The transcription factors are named for the RNA polymerase with which they work: TFI for RNA polymerase I, TFII for RNA polymerase II, and TFIII for RNA polymerase III. Because a number of different transcription factors are involved with transcription by each polymerase, the TFs are also lettered A, B, C, and so on.

For protein-coding genes, some transcription factors bind to specific DNA sequences of the promoter, while others appear to bind to the RNA poly-

■ *Figure 11.6*   Promoter elements (modules) for a eukaryotic protein-coding gene transcribed by RNA polymerase II. Each promoter element has a different function in transcription. The DNA sequences between the elements are not important for the transcription process. Transcription factors bind to the elements when the gene is transcribed.

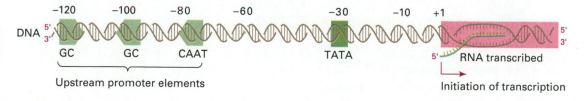

merase II when it initiates transcription. (Recall that the protein factor σ binds to RNA polymerase during the initiation of transcription in *E. coli*.) Figure 11.7 shows the events that may occur during the initiation and elongation stages of transcription, as catalyzed by RNA polymerase II. During initiation (Figure 11.7[a]) transcription factor TFIID (also called the **TATA factor**) binds to the TATA element of the promoter. TFIID is also bound to TFIIA and to an upstream DNA-binding protein (a regulatory factor) that is bound to the enhancer sequence. The binding between TFIID and the upstream DNA-binding protein results in the DNA forming a loop. When TFIID is bound to the TATA element, RNA polymerase II can bind and, along with transcription factors TFIIB, TFIIF, and TFIIE, forms a *transcription initiation complex*. Other transcription factors bind to the CAAT and GC promoter elements.

Once transcription has been initiated, and the RNA polymerase is moving down the DNA transcribing the gene (Figure 11.7[b]), TFIIB and TFIIE are released from the enzyme, leaving TFIIF bound. Another transcription factor, TFIIS, then binds to the polymerase during this phase of transcription. At the promoter TFIID remains bound to the TATA element and to TFIIA and the upstream DNA-bind-

ing protein. This means that the gene's promoter remains set up for another RNA polymerase II molecule to bind. In this way, the gene can be transcribed over and over until regulatory signals specify the repression of transcription.

The termination of transcription of eukaryotic protein-coding genes occurs in a different way from that of bacterial protein-coding genes. We will discuss that process in our discussion of eukaryotic mRNA molecules.

Keynote

Protein-coding genes in eukaryotes are transcribed by RNA polymerase II. The promoter for these genes consists of different combinations of promoter elements or modules, depending on the gene. These promoter elements are crucial for determining whether transcription can occur. They are the sites for the interaction of transcription factors and regulatory factors. Another important element associated with genes is the enhancer, which, through interaction with regulatory factors, functions to facilitate maximum transcription of the gene with which it is associated.

■ *Figure 11.7* Schematic of the events that may occur during: (a) the initiation; and (b) elongation stages of transcription catalyzed by RNA polymerase II. See text for detailed description of events.

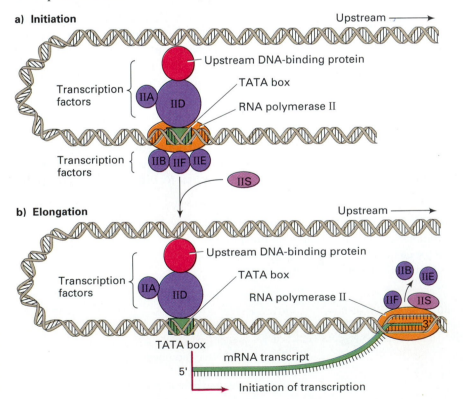

a) Initiation

Upstream →

Transcription factors { IIA   IID }

Upstream DNA-binding protein
TATA box
RNA polymerase II

Transcription factors { IIB   IIF   IIE }

IIS

b) Elongation

Upstream →

Transcription factors { IIA   IID }

Upstream DNA-binding protein
TATA box
RNA polymerase II

IIB   IIE
IIF   IIS

3'

TATA box

mRNA transcript

5'

Initiation of transcription

mRNA Molecules

Figure 11.8 shows the general structure of the mature, biologically active mRNA as it exists in both prokaryotic and eukaryotic cells. The mRNA molecule has three main parts. At the 5' end is a **leader sequence**, which varies in length from mRNA to mRNA. Within this leader sequence is the coded information that the ribosome reads to orient it correctly for beginning protein synthesis; none of the bases of the leader sequence are translated into amino acids. Following the 5' leader sequence is the actual **coding sequence** of the mRNA; this sequence determines the amino acid sequence of a protein during translation. The coding sequence varies in length, depending on the length of the protein for which it codes. Following the amino-acid-coding sequence, and constituting the rest of the mRNA at the 3' end of the molecule, is an untranslated **trailer sequence**.

The production of functioning mRNA is fundamentally different in prokaryotes and eukaryotes. In prokaryotes (Figure 11.9[a]) the RNA transcript functions directly as the mRNA molecule for translation, but in eukaryotes (Figure 11.9[b]), the RNA transcript must be modified in the nucleus by a series of events known as *RNA processing* in order to produce the mature mRNA. In addition, since prokaryotes lack a nucleus, an mRNA begins to be translated on ribosomes before it has been completely transcribed; this process is called *coupled transcription and translation* (Figure 11.10). In eukaryotes, however, the mRNA must migrate from the nucleus to the cytoplasm (where the ribosomes are located) before it can be translated. Thus a eukaryotic mRNA is always completely transcribed and processed before it is translated.

Production of Mature mRNA in Prokaryotes   In prokaryotes the initial transcript of a protein-coding gene *is* the mature mRNA molecule. That is, an exact point-by-point relationship exists between the order of base pairs in the gene and the order of the corresponding bases in the mature mRNA. This is referred to as *colinearity* between a gene and its primary transcript.

Production of Mature mRNA in Eukaryotes
Unlike prokaryotic mRNAs, eukaryotic mRNAs are usually modified at both the 5' and 3' ends. These *posttranscriptional modifications* are catalyzed by specific enzymes. In addition, most protein-coding genes have insertions of non-amino-acid coding sequences called **introns** between amino acid-coding sequences, or **exons** (also called **coding sequences**). Both exons and introns are copied into the primary mRNA transcript—the pre-mRNA—from which they are removed in the processing of pre-mRNA to the mature mRNA molecule. The synthesis and processing of pre-mRNA is discussed in this section.

5' Capping   The 5' end of the eukaryotic pre-mRNA is modified by the addition of a *cap* in a process called **5' capping.** 5' capping involves the addition of a methylated guanine nucleotide to the terminal 5' nucleotide by an unusual 5'-to-5' linkage (as opposed to the usual 5'-to-3' linkage) and the addition of two methyl groups (CH_3) to the first two nucleotides of the RNA chain (Figure 11.11) (see p. 245). 5' capping takes place as follows: RNA polymerase II starts transcription at the base to which the cap is added. That DNA site is called the *cap site*. There is no DNA template for the 5' cap. When the growing RNA transcript is about 20 to 30 nucleotides long, a methylated cap structure is added at the 5' end of the transcript by a *capping enzyme*. The 5' cap remains as the pre-mRNA is processed to the mature mRNA. The cap is essential for the ribosome to bind to the 5' end of the mRNA, an initial step of translation.

Addition of the 3' Poly(A) Tail   The posttranscriptional modification of the 3' ends of eukaryotic pre-mRNAs is the addition of a sequence of about 50 to 250 adenine nucleotides. This sequence is called a **poly(A) tail.** There is no DNA template for the poly(A) tails, and they are added only to pre-mRNA molecules. The poly(A) tail remains as the pre-mRNA is processed to the mature mRNA. Poly(A) tails are found on most of the mRNAs of all eukaryotic species. There are a few exceptions: for

■ *Figure 11.8*   General structure of mature prokaryotic and eukaryotic mRNA molecules.

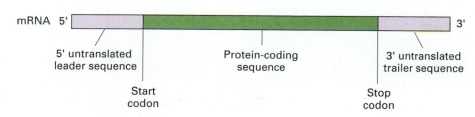

■ *Figure 11.9* Processes for synthesis of functional mRNA in prokaryotes and eukaryotes: (a) In prokaryotes the mRNA synthesized by RNA polymerase does not have to be processed before it can be translated by ribosomes. Also, since there is no nuclear membrane, translation of the mRNA can begin while transcription continues, resulting in a coupling of the transcriptional and translational processes. (b) In eukaryotes the primary RNA transcript is a precursor-mRNA (pre-mRNA) molecule, which is processed in the nucleus (addition of 5' cap and 3' poly(A) tail, and removal of introns to produce the mature, functioning mRNA molecule). Only when that mRNA is transported to the cytoplasm can translation occur.

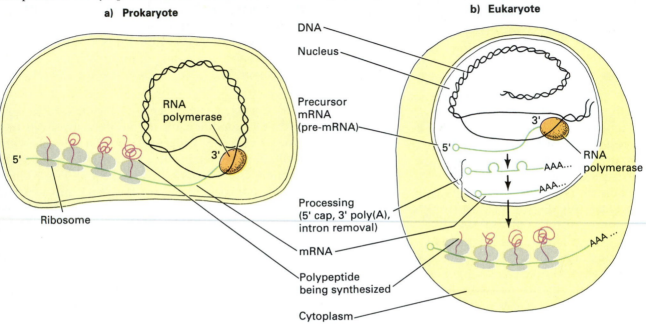

■ *Figure 11.10*   Electron micrograph of coupled mRNA transcription and translation in *E. coli*. From the faint DNA strand, a number of mRNA molecules are emerging. Since the length of the mRNA molecules increases from left to right, we assume that this strand is a single gene with a promoter to the left of the photograph. The globular structures on the mRNAs are ribosomes, which are translating the message into a protein before the synthesis of the mRNA is finished.

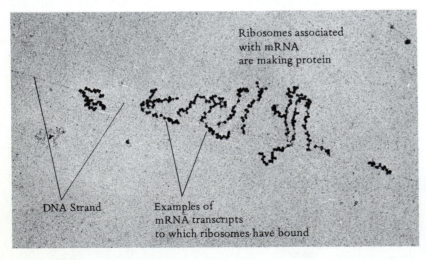

■ *Figure 11.11*   Cap structure at the 5′ end of a eukaryotic mRNA. The cap results from the addition of a guanine nucleotide and two methyl groups.

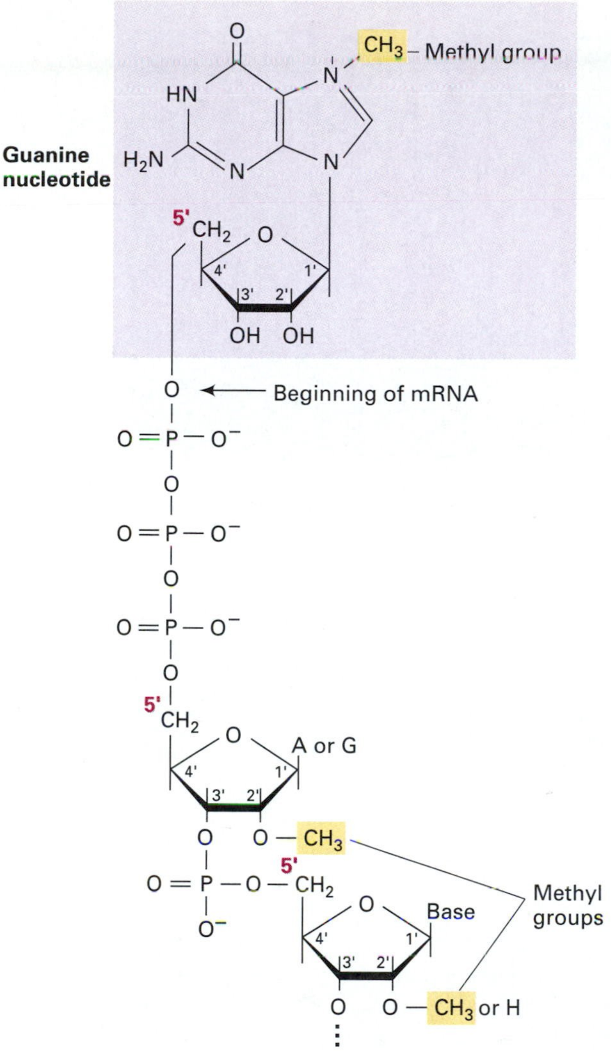

instance, histone mRNAs in mammalian cells have no poly(A) tails.

The addition of the poly(A) tail is part of the mechanism for signaling the 3′ end of the mRNA. While in prokaryotes specific transcription termination sequences specify the end of an mRNA molecule, in eukaryotes no such termination sequences are found in the DNA corresponding to the end of an mRNA molecule. Instead, mRNA transcription continues, in some cases for hundreds or thousands of nucleotides, past a site called the **poly(A) addition site.** The location of the poly(A) addition site is signaled in the transcript by an AAUAAA sequence 10 to 30 nucleotides upstream. At the poly(A) addition site the growing RNA is cleaved by an enzyme called an *RNA endonuclease.* The poly(A) tail is then added to the 3′ end of the RNA by repeated addi-

tion of A nucleotides in reactions catalyzed by the enzyme **poly(A) polymerase**.

The poly(A) tail is important for determining the stability of the mRNA. For example, mRNAs can be injected into frog oocytes where they will be translated. Globin mRNA, with its normal poly(A) tail, remains active in frog oocytes for a much longer time than globin mRNA from which the poly(A) tail has been removed because the non-poly(A) mRNA is more rapidly degraded. Thus the poly(A) tail probably protects mRNAs from degradation by ribonucleases that are present in the cytoplasm.

Introns   Pre-mRNAs often contain long insertions of non-amino-acid-coding sequences. These noncoding sequences are transcribed from the introns of the gene. Introns must be excised from each pre-mRNA in order to convert the transcript into a mature mRNA molecule that can be translated into a complete polypeptide. The mature mRNA, then, contains in a contiguous form the exon sequences that, in the gene, were separated by intron sequences.

Introns were first discovered in the genes for the rabbit and mouse β-globin polypeptide chains. (The β-globin polypeptide is part of a hemoglobin molecule.) Philip Leder's group studied the β-globin genes in mouse cells. They analyzed the β-globin mRNA by sucrose gradient centrifugation (see Box 11.1, p. 237). The data indicated that the β-globin mRNA has a size of about 10S, which is about the size expected if virtually all of the mRNA codes for the 147-amino-acid β-globin protein. Sequence analysis indicated the mature mRNA is 0.7-kb (700 nucleotides) long.

Next, they used **R-looping** experiments to study the organization of the gene encoding the mRNA. In R-looping RNA is hybridized with the double-stranded DNA that encoded the RNA. Under the conditions used, the RNA binds to its complementary sequence in the template DNA strand, displacing a loop of single-stranded (sense) DNA called an *R loop.* The DNA outside the region of the paired RNA and DNA remains double-stranded. The displaced R loops are visualized by electron microscopy.

Figure 11.12(a) shows the result of hybridizing the 0.7-kb β-globin mRNA with DNA containing the β-globin gene. Two R loops were seen to flank a loop of double-stranded DNA. This result indicated that there are two sequences in the β-globin gene that are complementary to the two ends of the mRNA. Between these two sequences, though, is a sequence that remains as double-stranded DNA

■ *Figure 11.12*   Demonstration that the mouse β-globin gene contains an intron: (a) R loops formed between 10S mature β-globin mRNA and the β-globin gene; (b) R loop formed between 15S β-globin pre-mRNA and the β-globin gene. Interpretative diagrams are alongside the micrographs.

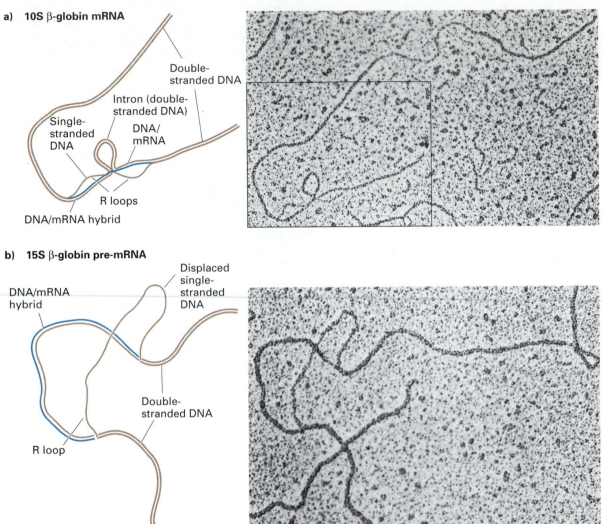

a)   10S β-globin mRNA

b)   15S β-globin pre-mRNA

because it is *not* complementary to any part of the mRNA. The conclusion was that the mouse β-globin gene and the β-globin mRNA were not colinear.

It was known at the time of discovering introns that the nucleus contains a large population of RNA molecules of various sizes. These RNA molecules are called **heterogeneous nuclear RNA**, or **hnRNA**, and it was thought that hnRNAs included pre-mRNA molecules. We now know that to be the case. Using procedures outside the scope of our discussion, a 15S (1.5-kb) RNA molecule was isolated from nuclear hnRNA that was the β-globin pre-mRNA. Like the mature mRNA, the pre-mRNA has a 5' cap and a 3' poly(A) tail. When the 1.5-kb pre-mRNA was hybridized to DNA containing the β-globin gene in an R-looping experiment, one continuous R loop was seen (Figure 11.12[b]). This

result indicated that the pre-mRNA was colinear with the gene that encoded it. The interpretation of this series of experiments was that the β-globin gene contains an intron of about 800 nucleotide pairs. Transcription of the gene results in a 1.5-kb pre-mRNA containing both exon and intron sequences. This RNA is found only in the nucleus. The intron sequence is excised by processing events, and the flanking exon sequences are spliced together to produce a mature mRNA. (Note: Subsequent research showed that the β-globin gene contains two introns; the second, smaller intron was not detected in the early research.)

Most, but not all, eukaryotic protein-coding genes contain introns. Typically, higher eukaryotes tend to have more genes with introns and longer introns than do lower eukaryotes. Interestingly, some bacteriophage genes have been shown to contain introns.

As an aside, we raise the question: What is a gene? Up until the discovery of introns, geneticists assumed that a gene was a contiguous stretch of DNA base pairs that was transcribed into a mature mRNA. That mRNA, in turn, was translated into an amino acid sequence. Prokaryotic genes fit this definition closely. However, the primary transcripts of most eukaryotic genes are not the molecules that are translated. Before the primary transcript can be translated, introns must be excised, and the adjacent segments—the exons—must be spliced together to produce the mature mRNA. If we define a gene as only the amino acid-coding regions of the DNA, then the presence of introns clearly means that eukaryotic genes are in pieces. On the other hand, if we define a gene as that region of DNA corresponding to the pre-mRNA, then the whole stretch of coding sequences (exons) plus introns constitute a gene. Both definitions may be encountered in your studies, and it is important to keep the distinctions between the two in mind.

Keynote

The transcripts of protein-coding genes are messenger RNAs. These molecules are linear and vary widely in length in correspondence to the variation in the size of the polypeptides they specify. Prokaryotic mRNAs are not modified once they are transcribed, whereas eukaryotic mRNAs are modified by the addition of a cap at the 5' end and a poly(A) tail at the 3' end. Many eukaryotic pre-mRNAs contain non-amino-acid-coding sequences called introns, which must be excised from the mRNA transcript to make a mature, functional mRNA molecule. The amino-acid-coding segments separated by introns are called exons.

Production of mRNA from pre-mRNA   A model for mRNA production from genes with introns is diagrammed in Figure 11.13. The sequence of steps is the same for genes without introns, except for the step involving intron removal: the transcription of the gene by RNA polymerase II, the addition of the methylated 5' cap and then the poly(A) tail to produce the pre-mRNA molecule, and finally the processing of the pre-mRNA in the nucleus to remove the introns and splice the exons together to produce the mature mRNA. 5' capping and 3' poly(A) addition were described earlier. Here we focus on intron removal.

Introns in pre-mRNAs are looped out with the aid of snRNAs (discussed later); the loop is then removed by nuclease cleavage. The adjacent exons are ligated together to generate a contiguous molecule. These events are called **mRNA splicing**. For mRNA splicing to take place, there must be some way for the machinery involved to determine what is an intron and what is an exon. In fact specific nucleotide sequences indicate where introns form junctions with exons. Introns typically begin with 5' GU and end with AG 3'. However, more than just those nucleotides are needed to specify a splicing junction between an intron and an exon: the 5' splice junction probably involves at least seven nucleotides and the 3' splice junction involves at least ten nucleotides of intron sequence.

Figure 11.14 (see p. 249) diagrams the sequence of events involved in splicing two exons (1 and 2) together with the elimination of an intron. The organization of the pre-mRNA is shown in Figure 11.14(a). The first step in splicing is a cleavage at the 5' splice junction that results in the separation of exon 1 from an RNA molecule that contains the intron and exon 2. The free 5' end of the intron loops and becomes joined to an A nucleotide that is part of a sequence called a **branch-point sequence**, which is located 18 to 38 nucleotides upstream of the 3' splice junction (Figure 11.14[b]). Because of its resemblance to the rope cowboys use, the looped back structure is called an *RNA lariat structure*. [For those unfamiliar with the word "lariat," think of the structure as a "p."] Next, the mature mRNA and a precisely excised lariat-shaped intron appear simultaneously as a result of a cleavage event at the 3' splice junction and ligation of the two coding sequences (Figure 11.14[c]).

The processing of pre-mRNA molecules occurs exclusively in the nucleus. The processing events occur in splicing complexes called **spliceosomes**, which consist of the pre-mRNA bound to **small nuclear ribonucleoprotein particles (snRNPs)**. These snRNPs are small nuclear RNAs (snRNAs) associated with proteins. There are six principal snRNAs (named U1–U6) in the nucleus, and they are associated with six to ten proteins each to form the snRNPs. U4 and U6 snRNAs are found within the same snRNP (U4/U6 snRNP), while the others are found within their own special snRNPs. The sequence and possible secondary structure of human U1 snRNA, which binds to the 5' splice junction, is shown in Figure 11.15 (see p. 250).

A model for the assembly of a spliceosome and its role in the removal of an intron is shown in Figure 11.16 (see p. 251). The steps are as follows:

1. U1 snRNP binds to the 5' splice site.
2. U2 snRNP binds to the branch-point region.

■ *Figure 11.13*   General sequence of steps in the formation of eukaryotic mRNA. Not all steps are necessary for all mRNAs.

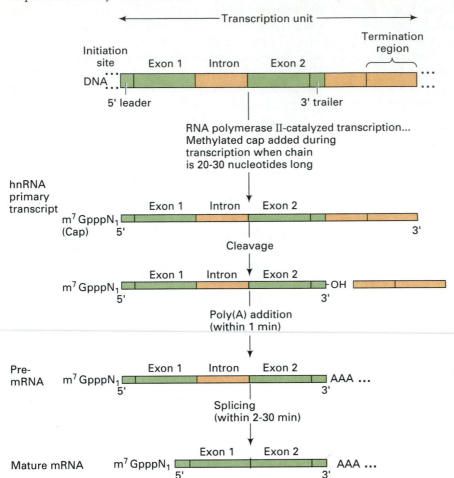

3. A preassembled U4/U6/U5 snRNP particle joins the complex.
4. U4 snRNP dissociates from the complex, resulting in the formation of the active spliceosome.

While the steps of spliceosome *assembly* are fairly well understood, it is less clear how the spliceosome *functions* in intron removal.

K e y n o t e

Introns are removed from pre-mRNAs in a series of well-defined steps. Introns typically begin with a 5′ GU and end with a 3′ AG. Intron removal begins with the cleavage of the pre-mRNA at the 5′ splice junction. The free 5′ end of the intron loops back and bonds to an A nucleotide in the branch-point consensus sequence, which is located upstream of the 3′ splice junction. Cleavage at the 3′ splice junction releases the intron, which is shaped like a lariat. The exons that flanked the intron are spliced

together once the intron is excised. The removal of introns from eukaryotic pre-mRNA occurs in the nucleus in complexes called spliceosomes. The spliceosome consists of several small nuclear ribonucleoprotein particles (snRNPs) bound specifically to each intron.

Transcription of Other Genes

In this section we discuss the transcription of non-protein-coding genes, focusing on genes for tRNA and rRNA, and the production of mature tRNA and rRNA molecules from their precursors.

Ribosomal RNA and Ribosomes

Ribosomes are the organelles within the cell on which protein synthesis takes place. Each cell contains thousands of ribosomes. Ribosomes bind to mRNA and facilitate the binding of the tRNA to the mRNA so that the polypeptide chain can be synthe-

■ *Figure 11.14*   Details of intron removal from a pre-mRNA molecule. At the 5' end of an intron is the sequence GU and at the 3' end is the sequence AG. Eighteen to 38 nucleotides upstream from the 3' end of the intron is an A nucleotide located within the branch-point sequence, which in mammals is YNCURAY, where Y = pyrimidine, N = any base, R = purine, and A = adenine. (a) Intron removal begins by a cleavage event at the first exon-intron junction. The G at the released 5' of the intron folds back and forms an unusual 2' → 5' bond with the A of the branch-point sequence; (b) This reaction produces a lariat-shaped intermediate; (c) Cleavage at the 3' exon-intron junction and ligation of the two exons completes the removal of the intron. Inset: Branch-point junction structure involving an unusual 2' → 5' phosphodiester bond.

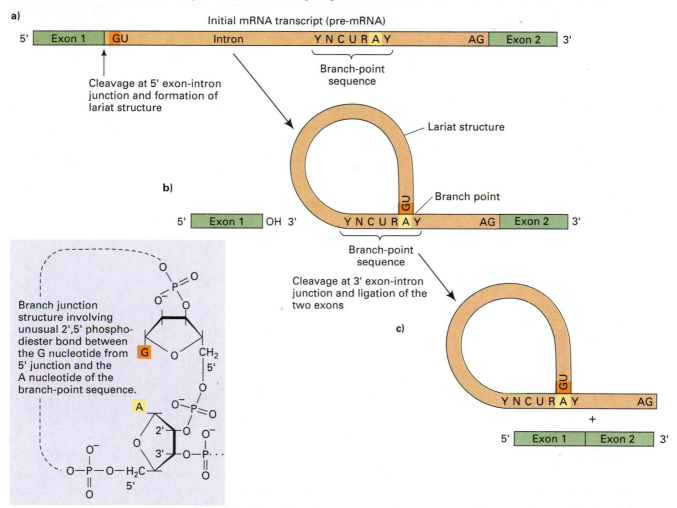

sized. In this section we discuss what ribosomes are, and how the rRNA molecules they contain are produced.

Ribosome Structure   Ribosomes are complex structures, whose function we do not yet fully understand. In both prokaryotes and eukaryotes the ribosomes consist of two unequally sized subunits (the large and small ribosomal subunits), each of which consists of a complex between RNA molecules and proteins. Each subunit contains at least one ribosomal RNA (rRNA) molecule and a large number of **ribosomal proteins.**

The Bacterial Ribosome   The *E. coli* ribosome is used as a model of a prokaryotic ribosome. The *E. coli* ribosome has a size of 70S, and the sizes of the two subunits are 50S (large subunit) and 30S (small subunit). (Recall from Box 11.1 that the S value is a measure of the rate of sedimentation of a component in a centrifuge, and that this is related both to the molecular weight and the three-dimensional shape of the component. Thus the reason that the 50S and 30S subunits together give a 70S ribosome is that when the two subunits are together, they actually have a three-dimensional shape that makes the ribosome sediment slower in the centrifuge

■ *Figure 11.15* Sequence and probable secondary structures for human U1 snRNA.

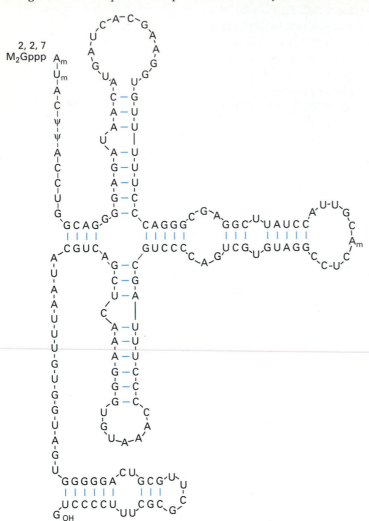

than the sum of the two parts would predict.) Figure 11.17 (see p. 252) presents models of ribosomal subunits and the whole ribosome of *E. coli*. Clearly the two subunits have distinct and recognizable three-dimensional shapes.

Approximately two-thirds of the *E. coli* ribosome consists of ribosomal RNA (rRNA), the rest consisting of ribosomal proteins. The large 50S subunit has 34 different proteins, a 23S rRNA, and a 5S rRNA. The small 30S ribosomal subunit has 20 different proteins and a 16S rRNA. Transcription of the ribosomal protein genes occurs by the mechanism we have already discussed for protein-coding genes.

The Eukaryotic Ribosome In general, eukaryotic ribosomes are larger and more complex than their prokaryotic counterparts. The size of the ribosome and the molecular weights of the rRNA molecules differ from organism to organism. Lower eukaryotes have the smallest ribosomes (although they are larger than ribosomes found in *E. coli*), while mammals have the largest ribosomes. Despite these differences, all eukaryotic ribosomes have many structural and chemical features in common. We will use the mammalian ribosome as a model for discussion.

Mammalian ribosomes have a size of 80S (versus 70S for the bacterial ribosomes), consisting of a large 60S subunit and a small 40S subunit (Figure 11.18) (see p. 252). The 80S mammalian ribosome consists of about equal weights of rRNA and ribosomal proteins. There are four rRNA types as shown in Figure 11.18: 18S, 28S (the large rRNAs),

■ *Figure 11.16* Model for spliceosome assembly and intron removal. (From M. R. Green, 1989. *Curr. Opinion Cell Biol.* 1:519–525.)

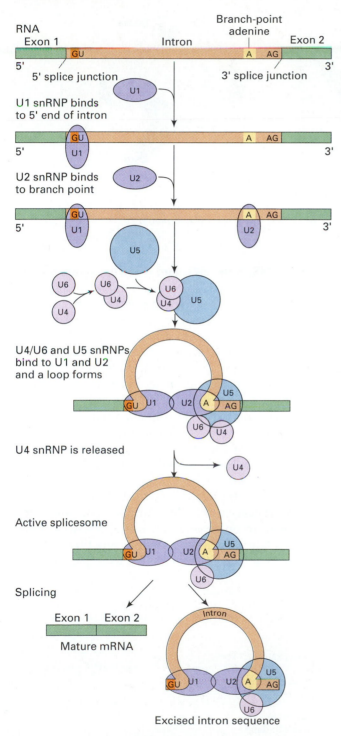

ribosomal proteins in the mammalian ribosome: about 35 are in the small subunit and about 50 are in the large subunit. Transcription of the ribosomal protein genes occurs by the mechanism we have already discussed for protein-coding genes.

K e y n o t e

Ribosomes, the organelles within the cell in which protein synthesis takes place, consist of two unequally sized subunits in both prokaryotes and eukaryotes. Each subunit contains ribosomal RNA and ribosomal proteins. Prokaryotic ribosomes contain three distinct rRNA molecules, whereas eukaryotic cytoplasmic ribosomes (larger and more complex) contain four.

Transcription of Prokaryotic rRNA Genes   In prokaryotes and eukaryotes the regions of DNA that contain the genes for rRNA are called **ribosomal DNA (rDNA).** In *E. coli* production of equal amounts of the three rRNAs is ensured by the transcription of the three adjacent genes for 16S, 23S, and 5S in rDNA into a *single* **precursor rRNA (pre-rRNA)** molecule. One rRNA transcription unit consists of one gene each for the three rRNAs. *E. coli* has seven such transcription units (*rrn* regions) scattered on the chromosome.

Figure 11.19(a) (see p. 253) shows the general organization of a transcription unit. In each transcription unit the three rRNA genes are arranged in the order 16S-23S-5S. In all seven transcription units one or two tRNA genes are also found in the internal spacer between the 16S and 23S rRNA coding sequences. Another one or two tRNA genes are found in the 3' spacer between the end of the 5S rRNA gene and the 3' end in three of the seven transcription units. During RNA synthesis from the transcription units, the tRNA genes are transcribed as part of the pre-rRNA molecule. The tRNA sequences are then removed from the precursor (see the discussion of tRNA synthesis on pp. 258–259).

Since there is only one RNA polymerase in *E. coli*, transcription of an rRNA transcription unit occurs in the same way as for protein-coding genes. Each transcription unit is transcribed into a 30S pre-mRNA (p30S), which contains a 5' leader sequence; the 16S, 23S, and 5S rRNA sequences (each separated by spacer sequences); and a 3' trailer sequence (Figure 11.19[b]) (see p. 253). The p30S molecule is cleaved by RNase III to produce the p16S, p23S, and p5S precursor molecules. Specific processing

5.8S, and 5S (the small rRNAs). The 40S subunit contains the 18S rRNA, and the large 60S subunit contains the 28S, 5.8S, and 5S rRNAs. The 5.8S rRNA is hydrogen-bonded to the 28S rRNA in the functional ribosome. There are about 85 different

■ *Figure 11.17*   Models of the *E. coli* small (30S) ribosomal subunit, large (50S) ribosomal subunit, and complete (70S) ribosome.

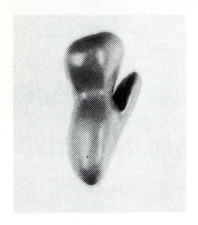

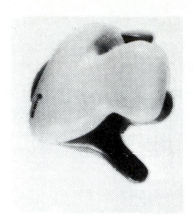

30S subunit   50S subunit   70S ribosome

enzymes then remove the spacers to generate the mature 16S, 23S, and 5S rRNAs (Figure 11.19[b]).

As the rRNA genes are being transcribed, the pre-rRNA transcript rapidly becomes associated with ribosomal proteins. The cleavage of the transcript takes place, then, within a complex formed between the rRNA transcript (as it is being transcribed) and ribosomal proteins. In this way, the functional ribosomal subunits are assembled.

Transcription of Eukaryotic rRNA Genes   Eukaryotic rRNA genes are organized in a special way. We discuss that first, before describing how they are transcribed.

rRNA Gene Organization   Most eukaryotes that have been examined contain a large number of copies of the genes for each of the four rRNA

species 18S, 5.8S, 28S, and 5S. The genes for 18S, 5.8S, and 28S rRNAs are usually found adjacent to one another in the order 18S-5.8S-28S, with each set of three genes repeated many times to form tandem arrays called **rDNA repeat units** (Figure 11.20[a]) (see p. 254). There are one or more clusters of rDNA repeat units in the genome, and around each cluster a nucleolus is formed. Within each nucleolus the rRNAs are synthesized; they associate with ribosomal proteins to produce the ribosomal subunits. The ribosomal subunits are transported to the cytoplasm where they function in protein synthesis; they are not functional in the nucleus.

There are typically 100 to 1000 copies of the rDNA repeat unit, the number varying from eukaryote to eukaryote. Thus the rRNA genes are representative of the moderately repetitive DNA in the genome (see Chapter 9). For example, yeast has

■ *Figure 11.18*   Composition of whole ribosomes and of ribosomal subunits in mammalian cells.

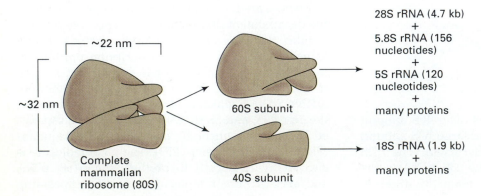

~22 nm

~32 nm

Complete mammalian ribosome (80S)

60S subunit

40S subunit

28S rRNA (4.7 kb)
+
5.8S rRNA (156 nucleotides)
+
5S rRNA (120 nucleotides)
+
many proteins

18S rRNA (1.9 kb)
+
many proteins

■ *Figure 11.19*   (a) General organization of an *E. coli* rRNA transcription unit (an *rrn* region); (b) Scheme for the synthesis and processing of a precursor rRNA (p30S) to the mature 16S, 23S, and 5S rRNAs of *E. coli*. (The tRNAs have been omitted from the processing scheme.)

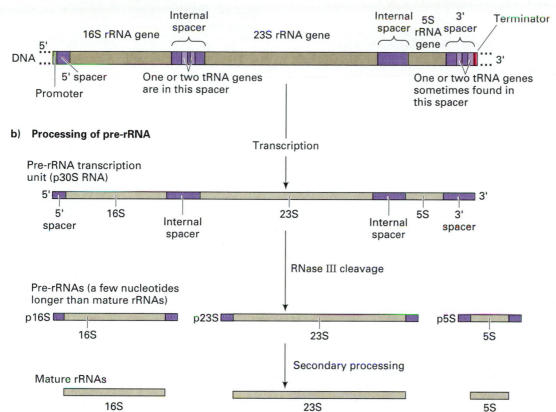

a) *E. coli* **rRNA transcription unit**

b)   Processing of pre-rRNA

140 rRNA gene sets and human cells have 1250. The 5S rRNA genes are also present in multiple copies and typically are located elsewhere in the genome from the other rRNA genes.

Introns have been found in the rRNA genes of only a few organisms (e.g., *Drosophila*, the slime mold *Physarum*, and the protozoan *Tetrahymena*). Thus they are certainly not as widespread as the introns in mRNA genes. The introns are excised from the pre-rRNA transcript as it is processed to produce the mature rRNAs. The splicing reactions involved are different from those involved in removing introns from pre-mRNA and pre-tRNA.

Transcription of rDNA Repeat Units by RNA Polymerase I   Each rDNA repeat unit is transcribed by RNA polymerase I to produce a large precursor-rRNA (pre-rRNA) molecule (45S pre-rRNA in human HeLa cells)(Figure 11.20[b]). This occurs in the nucleolus where ribosomes are assembled. An electron micrograph of pre-rRNA being transcribed from an rDNA repeat unit is shown in Figure 11.21.

The pre-rRNA contains the 18S, 5.8S, and 28S rRNA sequences as well as sequences between and flanking these three sequences. The latter are called **spacer sequences**. The external transcribed spacers—ETSs—are transcribed sequences that are located immediately upstream of the 5′ end of the 18S sequence and downstream of the 3′ end of the 28S sequence. The internal transcribed spacers—ITSs—are transcribed sequences that are located on either side of the 5.8S sequence; that is, between the 18S sequence and the 5.8S sequence, and between the 5.8S sequence and the 28S rRNA sequence. Between each rDNA repeat unit is the **nontranscribed spacer (NTS) sequence**, which is not transcribed (Figure 11.20[a]).

While RNA polymerases II and III each transcribe more than one type of gene, RNA polymerase I is unique in that it only transcribes the rDNA repeat units. The promoter for RNA polymerase I is upstream of the transcription initiation site in the NTS. Like other RNA polymerases, RNA polymerase I itself does not bind to the promoter;

■ *Figure 11.20*   (a) Generalized diagram of a eukaryotic, ribosomal DNA repeat unit. The coding sequences for 18S, 5.8S, and 28S rRNAs are indicated in light brown. NTS = nontranscribed spacer; ETS = external transcribed spacer; ITS = internal transcribed spacer. (b) The transcription and processing of 45S pre-rRNA in HeLa cells to produce the mature 18S, 5.8S, and 28S rRNAs.

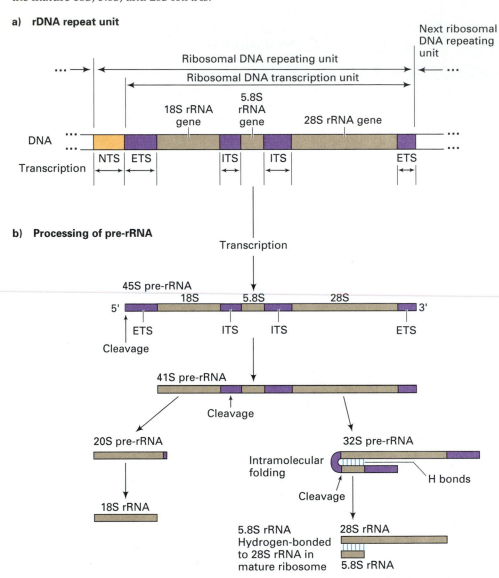

a)   **rDNA repeat unit**

b)   **Processing of pre-rRNA**

instead, specific transcription factors bind and form a complex to which RNA polymerase I binds and transcription begins. Termination of transcription of the pre-rRNA involves termination sites located downstream of the rDNA transcription unit in the next NTS.

organisms. Each transcription unit is separated from the next by a nontranscribed spacer (NTS) sequence. The promoter and terminator for each transcription unit are located within the NTS. As for the other eukaryotic RNA polymerases, specific transcription factors are required for the initiation of transcription by RNA polymerase I.

Keynote

RNA polymerase I transcribes the 18S, 5.8S, and 28S rRNA sequences into a single precursor molecule. A tandem array of 18S, 5.8S, and 28S rRNA transcription units is found in most eukaryotic

Processing of Pre-rRNA   To produce the mature 18S, 5.8S, and 28S rRNAs, the pre-rRNA is processed at specific sites by special ribonucleases to remove ITS and ETS sequences. As an example,

■ *Figure 11.21*   Electron micrograph of transcription of pre-rRNA molecules from rRNA genes in an oocyte from the spotted newt, *Triturus viridescens*. Transcription of the rRNA gene is from the tip to the base of each arrowhead shape.

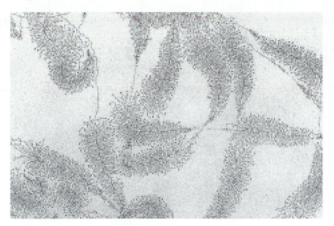

Figure 11.20(b) shows the pre-rRNA processing pathway in human HeLa cells. The first cleavage removes the 5′ ETS sequence and produces a precursor molecule containing all three rRNA sequences. A second cleavage produces the 20S precursor to 18S rRNA and the 32S precursor to 28S and 5.8S rRNAs. The 18S rRNA is produced from the 20S precursor by the removal of the ITS sequence. In the processing of the 32S precursor, the molecule folds so that the 5.8S sequence hydrogen-bonds to the 28S sequence. Then the ITS sequences are removed from the 5′ end and from between the 5.8S and 28S sequences, and the ETS is removed from the 3′ end.

All the pre-rRNA processing events take place in complexes formed between the pre-rRNA, 5S rRNA, and the ribosomal proteins. The 5S rRNA is produced by transcription of the 5S rRNA genes by RNA polymerase III, and the ribosomal proteins are produced by transcription of the ribosomal protein genes by RNA polymerase II and the subsequent translation of the mRNAs. As pre-rRNA processing proceeds, the complexes undergo shape changes resulting in the formation of the 60S and 40S ribosomal subunits. Once assembled, the two ribosomal subunits migrate out of the nucleolus, through the nucleus, and into the cytoplasm, where they associate with mRNAs and tRNAs and begin the process of protein synthesis.

Keynote

Three of the four rRNAs in eukaryotic ribosomes, the 18S, 5.8S, and 28S rRNAs, are transcribed from

the rDNA onto a single pre-rRNA molecule. In addition to the rRNA coding sequences, the pre-rRNA contains spacer sequences located at the ends of the molecule (external transcribed spacers) and internally between the rRNA sequences (internal transcribed spacers). The spacers are removed as the pre-rRNA is processed in the nucleolus to produce the mature rRNAs. The fourth rRNA, 5S rRNA, is transcribed separately from the other three rRNAs and is imported into the nucleolus where it is assembled with the mature rRNAs and the ribosomal proteins to produce the functional ribosomal subunits.

Self-splicing of Introns in Tetrahymena pre-rRNA   In some species of *Tetrahymena*, each 28S rRNA gene is interrupted by an intron. The removal of the intron unexpectedly was shown to occur by a *protein-independent reaction* in which the RNA intron becomes folded into a secondary structure that promotes its own excision. This process is called **self-splicing** and was discovered by Tom Cech and his research group. Other self-splicing introns have been found in rRNA genes and some mRNA genes in the mitochondria of yeast, *Neurospora*, and other fungi; and in rRNA genes and some mRNA and tRNA genes in bacteriophages.

Figure 11.22 diagrams the self-splicing reaction for the group I introns in *Tetrahymena* pre-rRNA. The steps are as follows:

1. The pre-rRNA is cleaved at the 5′ splice junction as guanosine is added to the 5′ end of the intron.
2. The intron is cleaved at the 3′ splice junction.
3. The two exons are spliced together.
4. The excised intron circularizes to produce a lariat molecule that is cleaved to produce a circular RNA and a short linear piece of RNA.

The self-splicing activity of the intron RNA sequence cannot be considered an enzyme activity because while it catalyzes the reaction, it is not regenerated in its original form at the end of the reaction, as is the case with protein enzymes.

Keynote

In some precursor rRNAs there are introns, the RNA sequences of which fold into a secondary structure that excises itself, a process called self-splicing. The self-splicing reaction requires guanine but does not involve any proteins.

■ *Figure 11.22*   Self-splicing reaction for the group I intron in *Tetrahymena* pre-rRNA.

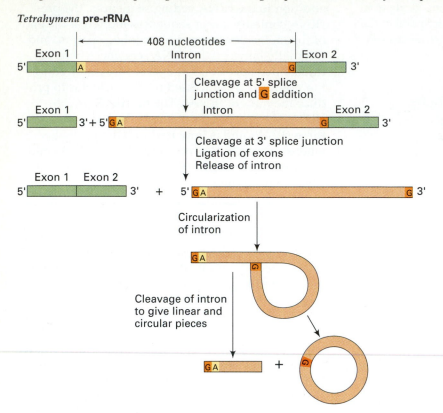

Transcription of 5S rRNA Genes   As we have just learned, the 5S rRNA genes (5S rDNA) are repeated genes located separately from the rDNA repeat units. The 5S rRNA genes encode the 120-nucleotide 5S rRNA; one 5S rRNA is found in each large ribosomal subunit. The 5S rRNA genes are transcribed by RNA polymerase III; this enzyme also transcribes tRNA genes and some snRNA genes. Unlike the arrangements we have discussed previously, the promoter for RNA polymerase III in 5S rRNA genes is *within the gene itself.* This internal promoter is called the **internal control region (ICR).** The tRNA genes (tDNA), which encode the 75 to 90 nucleotide tRNAs, also have internal promoters, although snRNA genes typically have promoters upstream of the genes.

The 5S rDNA and tDNA ICRs each have two functional domains, *box A* and *box C* for 5S rDNA and *box A* and *box B* for tDNA. The ICRs function through interaction with the transcription factors TFIIIA, TFIIIB, and TFIIIC. Figure 11.23 shows a model for the formation of a transcription initiation complex on a 5S rDNA ICR. TFIIIA first binds to *box C* of the ICR (Figure 11.23[1]), which allows

TFIIIC to bind to *box A* region (Figure 11.23[2]). TFIIIB then binds to the other TFs and not to DNA (Figure 11.23[3]). TFIIIB positions RNA polymerase III correctly on the gene; that is, TFIIIB functions as a *transcription initiation factor* (Figure 11.23[4]). RNA polymerase III then initiates transcription 50 base pairs *upstream* from the beginning of *box A,* at the beginning of the gene. Once the initiation complex is formed, it is stable, and repeated transcription of the whole gene by RNA polymerase III can take place.

Termination of transcription for both types of genes involves simple sequences at the 3' ends of the genes that signal the release of RNA polymerase III. For 5S rDNA, on the sense strand there is a cluster of four or more T nucleotides surrounded by GC-rich sequences. For tDNA, a cluster of T nucleotides is used to signal transcription termination.

Transcription of 5S rDNA directly produces the mature 5S rRNA; that is, there are no extra sequences that must be removed. Transcription of tDNA produces a **precursor-tRNA (pre-tRNA) molecule** that has extra sequences that must be

■ *Figure 11.23* Model for the formation of a transcription initiation complex on a 5S rDNA ICR. See text for detailed description of events.

1 TFIIIA binds to *Box C*

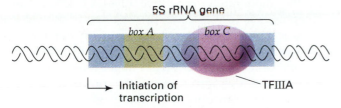

5S rRNA gene

box A   box C

→ Initiation of transcription   TFIIIA

2 This facilitates binding of TFIIIC to *box A*

TFIIIC

3 TFIIIB binds to other TFs, but not to DNA

TFIIIB

4 TFIIIB positions RNA polymerase III on gene

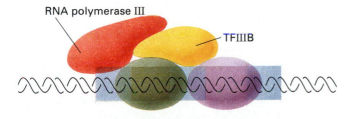

RNA polymerase III

TFIIIB

removed to produce the mature tRNA. This will be discussed later.

K e y n o t e

RNA polymerase III transcribes 5S rRNA genes (5S rDNA), tRNA genes (tDNA), and some snRNA genes. For 5S rDNA and tDNA, the promoter for RNA polymerase III is located within the gene itself. The internal promoter is called the internal control region and consists of different combinations of functional domains depending on the class of gene involved. The domains are binding sites for

transcription factors required for transcription by RNA polymerase III.

Transfer RNA

Transfer RNAs (tRNAs) bring amino acids to the ribosome-mRNA complex, where they are polymerized into protein chains in the translation (protein synthesis) process. In this section we discuss the structure and biosynthesis of tRNAs.

Structure of tRNA   Transfer RNA molecules have a size of 4S, and they consist of a single chain of 75 to 90 nucleotides. Each type of tRNA molecule has a different sequence, although all tRNAs have the sequence 5'-CCA-3' at their 3' ends.

The nucleotide sequences of all tRNAs can be arranged into what is called a *cloverleaf model of tRNA*. Figure 11.24 shows the general features of the cloverleaf model of yeast alanine tRNA. The cloverleaf shape results from complementary base pairing between different sections of the molecule which results in four base-paired "stems" separated by four loops—I, II, III, and IV. (Some tRNAs do not have loop III.) Loop II contains within it the three-nucleotide sequence called the **anticodon**, which pairs with a codon (three-nucleotide sequence) in mRNA by complementary base pairing during translation. This codon–anticodon pairing is crucial for adding the correct amino acid (as specified by the mRNA) to the growing polypeptide chain.

Because tRNA molecules are so small, they can be crystallized, and X-ray crystallography can be used to develop a three-dimensional model. Figure 11.25 (p. 259) shows the tertiary-structure model for yeast tRNA.Phe (this terminology indicates the amino acid specified by the anticodon of the tRNA, in this case phenylalanine). All other tRNAs that have been examined show similar three-dimensional structures.

tRNA Genes   A three-base sequence (codon) in an mRNA specifies each amino acid to be added to a polypeptide chain. While only 20 different amino acids can be used to make a protein, 61 different codons are actually used in an mRNA to specify the 20 amino acids. Three additional codons do not specify amino acids but instead are used as termination signals for protein synthesis. Since each

■ *Figure 11.24*   Cloverleaf structure of yeast alanine
tRNA. Py = pyrimidine. Modified bases are: I = inosine;
T = ribothymidine; Ψ = pseudothymidine;
D = dihydrouridine; GMe = methylguanosine;
GMe₂ = dimethylguanosine; IMe = methylinosine.

Yeast alanine tRNA

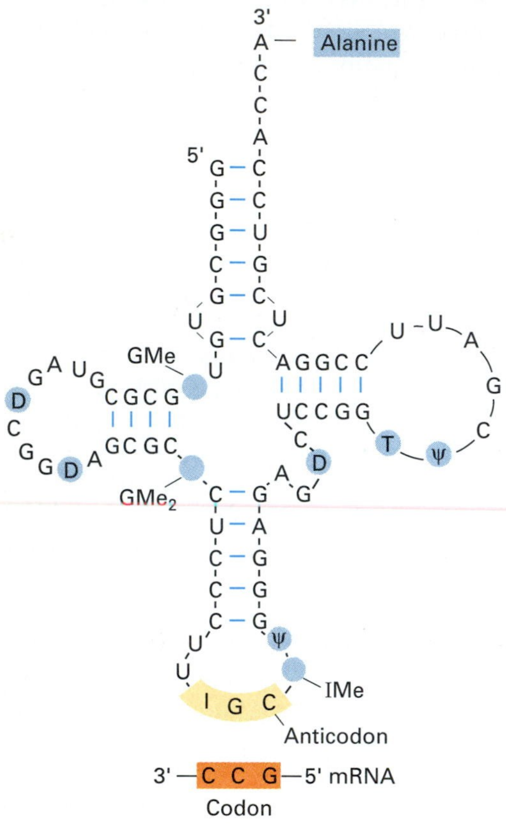

many more tRNA genes occur in eukaryotes than in
prokaryotes. In yeast, for example, about 400 tRNA
genes occur in the genome, while *Xenopus laevis* has
over 200 copies of *each* tRNA gene per genome.
As a class, then, eukaryotic tRNA genes are found
in the moderately repetitive class of DNA (see
Chapter 9).

At least some tRNA genes from many eukaryotic
organisms contain introns. For example, about 10
percent of the 400 tRNA genes in yeast have introns.
Removal of the intron from the precursor-tRNA
(pre-tRNA) transcripts of those genes involves
cleavage with a specific endonuclease. The two
RNA pieces generated by intron removal are then
spliced together by an enzyme called **RNA ligase** to
produce the mature tRNA molecule. These processing events are different than those for intron
removal from pre-mRNAs which, you will remember, involves the use of the spliceosome.

**Transcription of tRNA Genes and the Production
of tRNA**   In *E. coli* the tRNA genes are transcribed
by the same RNA polymerase that transcribes all
the genes; hence the promoter and terminator
sequences are the same as for other gene classes. In
eukaryotes the tRNA genes are transcribed by RNA
polymerase III. Recall from our discussion of 5S
rDNA transcription by RNA polymerase III that
eukaryotic tRNA genes have internal promoters
called internal control regions.

The primary transcripts of tRNA genes in both
prokaryotes and eukaryotes are the precursor
tRNAs (pre-tRNAs). The pre-tRNAs are extensively
modified and processed to produce the mature
tRNAs. Two types of modification occur: (1) addition of a 5'-CCA-3' sequence to the 3' end; and (2)
extensive chemical modification of a number of
nucleotides at many places within the chain. Examples of modified nucleotides produced are pseu-
douridine (ψ), inosine, and ribothymidine. The
specific modifications cause the particular two- and
three-dimensional configurations of tRNAs, which
in turn determine the functions of those molecules;
specifically, their ability to pick up a specific amino
acid and bind to ribosomes.

As is the case with mRNAs, pre-tRNAs are
longer than mature tRNAs and have a 5' leader
sequence and a 3' trailer sequence. Unlike mRNA,
in which the leader and trailer sequences remain,
these sequences are removed during the processing
of pre-tRNA into mature tRNA by specific
enzymes. In eukaryotes the pre-tRNA processing
occurs in the nucleus, so only mature tRNAs are
found in the cytoplasm.

amino-acid-specifying codon must be matched by
an appropriate anticodon on a tRNA molecule, any
cell could have 61 different tRNA types. Thus for
many amino acids there is more than one tRNA
with the appropriate codon (this will be described
in more detail in Chapter 14). And eukaryotes commonly have a number of different tRNA molecules
with the same anticodon. So that the instructions
for the manufacture of such large numbers of
tRNAs can be coded, the genome has many copies
of tRNA genes. tRNA genes are found scattered
around the genome in single copies and, in some
cases, grouped together in *gene clusters*.

On the *E. coli* chromosome specific tRNA genes
are present in one copy, while others are present in
two or more copies; this is called **gene redundancy**.
Recall also that some tRNA genes are found in the
rRNA transcription units (see p. 000). In general,

■ *Figure 11.25* (a) Schematic of the three-dimensional structure of yeast phenylalanine tRNA as determined by X-ray diffraction of tRNA crystals. Note the characteristic L-shaped structure. (b) Photograph of a space-filling molecular model of yeast phenylalanine tRNA. The CCA end of the molecule is at the upper right, and the anticodon loop is at the bottom.

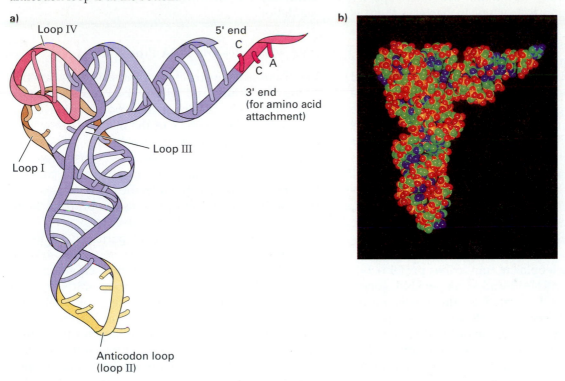

a)

Loop IV

5' end

C
C   A

3' end
(for amino acid
attachment)

Loop III

Loop I

Anticodon loop
(loop II)

b)

In prokaryotes, but not in eukaryotes, a cluster of tRNA genes may be transcribed to produce a single RNA transcript containing a number of tRNA sequences. The general organization of the multi-tRNA pre-tRNA molecules is

5'-leader–(tRNA-spacer)$_n$–tRNA–trailer-3'

where n is a number of tRNA spacers characteristic of a cluster. In such cases the leader, the trailer, and the spacer sequences are removed by specific enzymes.

Keynote

Molecules of transfer RNA bring amino acids to the ribosomes, where the amino acids are polymerized into a protein chain. All the tRNA molecules are between 75 and 90 nucleotides long, contain a number of modified bases, and have similar three-dimensional shapes. A CCA sequence is found at the 3' end of all tRNAs. tRNAs are made as pre-tRNA molecules containing 5'-leader and 3'-trailer

sequences, both of which are removed by enzyme activity. Some eukaryotic pre-tRNAs contain introns, which are removed in processing steps that are different from pre-mRNA processing steps.

Summary

Transcription

When a gene is expressed, the DNA base-pair sequence is transcribed into the base sequence of an RNA molecule. Four major classes of RNA transcripts are produced by transcription of four classes of genes: messenger RNA (mRNA), transfer RNA (tRNA), ribosomal RNA (rRNA), and small nuclear RNA (snRNA). snRNA is found only in eukaryotes, while the other three classes are found in both prokaryotes and eukaryotes. Only mRNA is subsequently translated to produce a protein molecule.

When a gene is transcribed, only one of the two DNA strands is copied. The direction of RNA synthesis is 5' to 3' and the reaction is catalyzed by RNA polymerase. In addition to the coding sequences, genes contain other sequences important for the regulation of transcription, including promoter sequences and terminator sequences. Promoter sequences specify where transcription of the gene is to begin, and terminator sequences specify where transcription is to stop. In bacteria there is only one type of RNA polymerase; hence all classes of RNA are transcribed by the same enzyme. Consequently, the promoters for all three classes of genes are very similar. The promoter is recognized by a complex between the RNA polymerase core enzyme and a protein factor called sigma. Once transcription is initiated correctly, the sigma factor dissociates from the enzyme and is reused in other transcription initiation events. Termination occurs in one of two ways.

In eukaryotes there are three different RNA polymerases located in the nucleus. RNA polymerase I, located exclusively in the nucleolus, transcribes the 18S, 5.8S, and 28S rRNA sequences. These rRNAs are part of ribosomes. RNA polymerase III, also located in the nucleoplasm, transcribes protein-coding genes (into mRNAs) and some snRNA genes. RNA polymerase II, located in the nucleoplasm, transcribes tRNA genes, 5S rRNA genes, and the remaining snRNA genes. None of the three eukaryotic RNA polymerases binds directly to promoters. Rather the promoter sequences for the genes they transcribe are first recognized by specific transcription factors. The transcription factors bind to the DNA and facilitate binding of the polymerase to the transcription factor-DNA complex for correct initiation of transcription.

The promoters for the three types of RNA polymerase differ. For genes transcribed by RNA polymerase II, the promoter consists of a number of sequence modules located within a relatively short distance upstream of the transcription initiation point. These modules serve as binding sites for transcription factors and for regulatory proteins that function to regulate transcription. The 18S, 5.8S, and 28S rRNA sequences transcribed by RNA polymerase I are organized into a single transcription unit so that transcription produces a precursor RNA molecule containing all three rRNAs plus extra RNA material. Uniquely for 5S rRNA genes and tRNA genes transcribed by RNA polymerase III, the promoter is located within the gene itself. For all three RNA polymerases, transcription factors first recognize the promoter; then the polymerase binds to initiate transcription.

In sum, while the transcription process is very similar in prokaryotes and eukaryotes, the molecular components of the transcription process itself differ considerably. Even within eukaryotes, three different promoters have evolved along with three distinct RNA polymerases. Much remains to be learned about the associated transcription factors and regulatory proteins before we have a complete understanding of transcription processes and their regulation.

RNA Molecules and RNA Processing

We discussed the structure, synthesis, and function of mRNA, tRNA, and rRNA. Each mRNA encodes the amino acid sequence of a polypeptide chain. The nucleotide sequence of the mRNA is translated into the amino acid sequence of the polypeptide chain on ribosomes. Ribosomes consist of two unequal-sized subunits, each of which contains both rRNA and protein molecules. The amino acids that are assembled into proteins are brought to the ribosome attached to tRNA molecules.

mRNAs have three main parts: a 5' leader sequence, the amino acid coding sequence, and the 3' trailer sequence. Since all three sequences vary from mRNA to mRNA, as a group mRNAs show extensive variability in length. In prokaryotes the primary gene transcript functions directly as the mRNA molecule, whereas in eukaryotes the primary RNA transcript must be modified in the nucleus by RNA modifying and processing events to produce the mature mRNA. These events are the addition of a 5' methylated cap, the addition of a 3' poly(A) tail, and the removal of any introns (internal sequences that do not code for amino acids) that are present. The latter is known as RNA splicing and involves specific interactions with snRNPs in molecular structures called spliceosomes. Only when all processing events have been completed is the mRNA functional; at that point it leaves the nucleus and can be translated in the cytoplasm.

Ribosomal RNAs are important structural components of ribosomes, the cellular organelles on which protein synthesis takes place. In both prokaryotes and eukaryotes, ribosomes consist of two unequal-sized subunits. The subunits consist of both rRNA and ribosomal proteins. In each case the smaller subunit contains one rRNA molecule—16S in prokaryotes and 18S in eukaryotes. The larger subunit in prokaryotes contains 23S and 5S rRNAs, while the larger subunit in eukaryotes contains 28S, 5.8S, and 5S rRNAs.

In prokaryotes the genes for the 16S, 23S, and 5S rRNAs are transcribed into a single pre-rRNA molecule. tRNA genes are found in the spacer regions of each transcription unit. The pre-rRNA molecules are processed in a number of enzymatically catalyzed steps to remove the noncoding sequences found at the ends of the molecules and between the rRNA sequences. At the same time the tRNAs are released. All of the processing events occur while the pre-rRNA is associating with the ribosomal proteins so that, when processing is complete, the functional 50S and 30S subunits have been assembled.

In eukaryotes there are many copies of the genes for each of the four rRNAs. The 18S, 5.8S, and 28S rRNA genes comprise a transcription unit, and many such transcription units are organized in a tandem array. The 5S rRNA genes are also present in many copies but they are located elsewhere in the genome. The 18S, 5.8S, and 28S sequences are transcribed into pre-rRNA molecules which, in addition to the rRNA sequences, contain spacer sequences at the ends and between the rRNA sequences. The spacers are removed by specific processing events that take place while the rRNA sequences are associated with ribosomal proteins. The 5S rRNA, which is transcribed elsewhere in the nucleus, becomes associated with the assembling 60S subunit so that, at the end of the processing steps, functional 60S and 40S ribosomal subunits have been produced. All of these events occur in the nucleolus. The mature ribosomal subunits then exit the nucleus and participate in protein synthesis in the cytoplasm.

In the pre-rRNA of *Tetrahymena* the 28S rRNA sequence is interrupted by an intron. The intron is removed during processing of the pre-rRNA in the nucleolus. The excision of this particular intron (and a few other examples in other systems) occurs by a protein-independent reaction in which the RNA sequence of the intron folds into a secondary structure that promotes its own excision. This process is called self-splicing.

All tRNAs carry out the same function of bringing amino acids to the ribosomes. Thus all tRNAs are very similar in length (75 to 90 nucleotides) and in secondary and tertiary structure. tRNAs are usually synthesized as precursor molecules that have extra nucleotides at both the 5' and 3' ends. The extra nucleotides are removed by specific processing events. In addition a number of the bases are modified during the maturation process and a CCA sequence is added to the 3' end of all tRNAs. Some tRNA genes in eukaryotes contain introns; these are removed from the transcripts of those genes by splicing events that are different from those used for intron removal from precursor mRNAs.

In sum, the functional RNA molecules of the cell are typically transcribed as precursor molecules that include extra nucleotides that are removed by specific processing events. In prokaryotes the processing events are confined to the tRNAs and rRNAs. In eukaryotes RNA processing is much more complex, particularly with mRNAs.

Analytical Approaches for Solving Genetics Problems

Q11.1 If two RNA molecules have complementary base sequences, they can hybridize to form a double-stranded structure just as DNA can do. Imagine that in a particular region of the genome of a certain bacterium one DNA strand is transcribed to give rise to the mRNA for protein A, while the other DNA strand is transcribed to give rise to the mRNA for protein B.
a. Would there be any problem in expressing these genes?
b. What would you see in protein B if a mutation occurred which affected the structure of protein A?

A11.1 **a.** mRNA A and mRNA B would have complementary sequences, so they might hybridize with each other and not be available for translation.
b. Every mutation in gene A would also be a mutation in gene B, so protein B might also be abnormal.

Q11.2 Compare and contrast the following two events in terms of what their consequences would be. Event #1: an incorrect nucleotide is inserted into the new DNA strand during replication, and not corrected by the proofreading or repair systems before the next replication. Event #2: an incorrect nucleotide is inserted into an mRNA during transcription.

A11.2 Event #1 would result in a mutation, assuming it occurred within a gene. The mistake would be inherited by future generations, and would affect the structure of all mRNA molecules transcribed from the region and therefore all molecules of the corresponding protein could be affected.

Event #2 would produce a single aberrant mRNA. This could produce a few aberrant protein molecules. Additional normal protein molecules would exist because other, normal, mRNAs would have been transcribed. The abnormal mRNA would soon be degraded. The mRNA mutation would not be hereditary.

Q11.3   You are given four different RNA samples. Sample I has a short lifetime; sample II has a homogeneous molecular weight; sample III is produced by processing of a larger precursor RNA; and sample IV has an additional sequence added onto the original transcript. For each sample, state whether the RNA could be rRNA, mRNA, or tRNA. If is not one of those, state what it might be. Note that for each sample more than one RNA could apply. Give reasons for your choices.

A11.3   Sample I: A short lifetime is characteristic of

mRNAs in prokaryotes, and many mRNAs in eukaryotes. Heterogeneous nuclear RNA of eukaryotes also has a short lifetime. Sample II: rRNA and tRNA species have homogeneous molecular weights since they carry out specific functions within the cell for which their length and three-dimensional configuration are important. Messenger RNA and heterogeneous nuclear RNA are heterogeneous in length. Sample III: rRNA, tRNA, and mRNA are all produced by the processing of a larger precursor RNA molecule. Sample IV: Both tRNA and eukaryotic mRNA have additional sequences added after they are transcribed. For tRNA this sequence is the CCA at the 3' end, and for mRNA this sequence is the poly(A) tail at the 3' end.

Questions and Problems

*11.1   Describe the differences between DNA and RNA.

11.2   Compare and contrast DNA polymerases and RNA polymerases.

11.3   All base pairs in the genome are replicated during the DNA synthesis phase of the cell cycle, but only *some* of the base pairs are transcribed into RNA. How is it determined *which* base pairs of the genome are transcribed into RNA?

11.4   Discuss the structure and function of the *E. coli* RNA polymerase. In your answer, be sure to distinguish between RNA core polymerase and RNA core polymerase-sigma factor complex.

*11.5   Discuss the similarities and differences between the *E. coli* RNA polymerase and eukaryotic RNA polymerases.

*11.6   Which classes of RNA do each of the three eukaryotic RNA polymerases synthesize? What are the functions of the different RNA types in the cell?

11.7   What is the Pribnow box? The Goldberg-Hogness box (TATA element)?

*11.8   What is an enhancer element?

11.9   A piece of mouse DNA was sequenced as follows (a space is inserted after every 10th base for ease in counting; "......." means a lot of unspecified bases):

```
AGAGGGCGGT   CCGTATCGGC   CAATCTGCTC   ACAGGGCGGA
TTCACACGTT   GTTATATAAA   TGACTGGGCG   TACCCCAGGG
TTCGAGTATT   CTATCGTATG   GTGCACCTGA   CT(.......)
GCTCACAAGT   ACCACTAAGC.......
```

What can you see in this sequence to indicate it might be all or part of a transcription unit?

11.10   Compare and contrast the structures of prokaryotic and eukaryotic mRNAs.

*11.11   Compare the structures of the three classes of RNA found in the cell.

11.12   Many eukaryotic mRNAs, but not prokaryotic mRNAs, contain introns. What is the evidence for the presence of introns in genes? Describe how these sequences are removed during the production of mature mRNA.

*11.13   Discuss the posttranscriptional modifications that take place on the primary transcripts of tRNA, rRNA, and protein-coding genes.

11.14   Distinguish between leader sequence, trailer sequence, coding sequence, intron, spacer sequence, nontranscribed spacer sequence, external transcribed spacer sequence, and internal transcribed sequence. Give examples of actual molecules in your answer.

11.15   Describe the organization of the ribosomal DNA repeating unit of a higher-eukaryotic cell.

*11.16   Which of the following kinds of mutations would be likely to be recessive lethals in humans? Explain your reasoning.
a. deletion of the U1 genes
b. deletion within intron 2 of β-globin
c. deletion of 4 bases at the end of intron 2 and 3 bases at the beginning of exon 3 in β-globin

11.17 The following figure shows the transcribed region of a typical eukaryotic protein-coding gene:

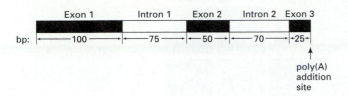

What is the size (in bases) of the fully processed, mature mRNA? Assume in your calculations a poly(A) tail of 200 As.

*11.18 Which of the following could occur in a single mutational event in a human? Explain.
a. deletion of 10 copies of the 5S ribosomal RNA genes only
b. deletion of 10 copies of the 18S rRNA genes only
c. simultaneous deletion of 10 copies of the 18S, 5.8S, and 28S rRNA genes only
d. simultaneous deletion of 10 copies each of the 18S, 5.8S, 28S, and 5S rRNA genes

11.19 During DNA replication in a mammalian cell a mistake occurs: 10 wrong nucleotides are inserted into a 28S rRNA gene. This mistake is not corrected. What will likely be the effect on the cell?

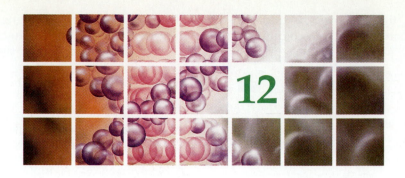

The Genetic Code and the
Translation of the Genetic Message

- A protein consists of one or more molecular subunits called polypeptides, which are themselves composed of smaller building blocks called amino acids. The amino acids are linked together in the polypeptide by peptide bonds.

- The primary amino acid sequence of a protein determines its secondary, tertiary, and quaternary structure, and hence its functional state.

- The genetic code is a triplet code in which each three-nucleotide codon in an mRNA specifies one amino acid. Some amino acids are represented by more than one codon. The code is almost universal, and it is read without gaps in successive, nonoverlapping codons.

- Translation of the mRNA into a protein chain occurs on ribosomes. Amino acids are brought to the ribosome on tRNA molecules. The correct amino acid sequence is achieved by the specific binding between the codon of the mRNA and the complementary anticodon of the tRNA, and by the specific binding of each amino acid to its specific tRNA.

- In prokaryotes and eukaryotes AUG (methionine) is the initiator codon for the start of translation.

- Elongation of the protein chain involves peptide bond formation between the amino acid attached to the tRNA in one site of the ribosome and the growing polypeptide attached to the tRNA in an adjacent site. Once the peptide bond has formed, the ribosome translocates one codon along the mRNA in preparation for the next tRNA with its bound amino acid to bind to the available codon.

- Translation continues until a chain termination codon (UAG, UAA, or UGA) is reached in the mRNA. These codons are read by one or more release factor proteins and then the polypeptide is released from the ribosome and the other components of the protein synthesis machinery dissociate.

- In eukaryotes proteins are found free in the cytoplasm, as well as in the various cell compartments such as the nucleus, mitochondria, chloroplasts, and secretory vesicles. Special mechanisms exist to sort proteins to their appropriate cell compartments. For example, proteins to be secreted have N-terminal signal sequences that facilitate their entry into the endoplasmic reticulum for later sorting in the Golgi complex and beyond.

T HE INFORMATION for making the proteins found in a cell is encoded in the structural genes of the cell's genome. Expression of a protein-coding gene occurs by transcription of the gene to produce an mRNA (discussed in Chapter 11), followed by translation of the mRNA; that is, the conversion of the mRNA base sequence information into an amino acid sequence of a polypeptide. The DNA base-pair information that specifies the amino acid sequence of a polypeptide is called the **genetic code.**

In this chapter we will study the translation of the genetic message and how the information for the amino acid sequence of proteins is encoded in the nucleotide sequence of messenger RNA. We will see how three classes of RNA we introduced in Chapter 11—messenger RNA, transfer RNA, and ribosomal RNA—are involved in the translation process.

Protein Structure

Chemical Structure of Proteins

A **protein** is one of a group of high molecular weight, nitrogen-containing organic compounds of complex shape and composition. Each cell type has a characteristic set of proteins that give that cell type its functional properties. A protein consists of one or more molecular subunits called **polypeptides,** which are themselves composed of smaller building blocks, the **amino acids,** linked together to form long chains. Each molecule of a given protein consists of the same number and kind of polypeptide chains, and each of these in turn is composed of the same number, kind, and sequence of amino acids. It is the sequence of amino acids in a polypeptide that gives the polypeptide its three-dimensional shape and its properties in the cell.

With the exception of proline, the amino acids have a common structure, which is shown in Figure 12.1. The structure consists of a central carbon atom (α-carbon) to which is bonded an amino group (NH_2), a carboxyl group (COOH), and a hydrogen atom. At the pH commonly found within cells, the NH_2 and COOH groups of the free amino acids are in a charged state; that is, $-NH_3^+$ and $-COO^-$, respectively. Bound to the α-carbon, each amino acid has an additional chemical group, called the *radical*, or *R group*. It is the R group that varies from one amino acid to another and gives each amino acid its distinctive properties. Since different proteins have different sequences and proportions of amino acids, the organization of the R groups gives a protein its structural and functional properties.

There are 20 naturally occurring amino acids; their names, three-letter abbreviations, and chemical structures are shown in Figure 12.2. The 20 amino acids are divided into subgroups, based on whether the R group is acidic (e.g., aspartic acid), basic (e.g., lysine), neutral-polar (e.g., serine), or neutral-nonpolar (e.g., leucine).

The amino acids of a polypeptide are held together by **peptide bonds,** covalent bonds that join the carboxyl group of one amino acid to the amino group of another amino acid (Figure 12.3). A polypeptide, then, is a linear, unbranched molecule that consists of many amino acids (usually 100 or more) joined by peptide bonds. Every polypeptide has a free α-amino group at one end (called the N terminus, or the N terminal end) and a free α-carboxyl group at the other end (called the C terminus, or the C terminal end). Polypeptides have polarity: By convention, and because the polypeptide is so constructed, the N terminal end is defined as the beginning of a polypeptide chain.

Molecular Structure of Proteins

The molecular structure of a protein is relatively complex. There are four levels of structural organization, as shown in Figure 12.4.

1. The primary structure of the polypeptide chain that constitutes a protein is the amino acid sequence (Figure 12.4[a]).
2. The secondary structure of a protein refers to the folding and twisting of a single polypeptide chain into a variety of shapes. A polypeptide's secondary structure is the result of weak bonds (e.g., electrostatic bonds or hydrogen bonds) that form between NH and CO groups of amino acids that are relatively near each other on the chain. One type of secondary structure found in regions of many polypeptides is the α-helix, which forms as a

■ *Figure 12.1*   General structural formula for an amino acid.

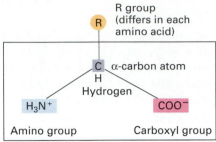

Structures common to all amino acids

result of hydrogen bonding between nearby amino acids (Figure 12.4[b]).

Another type of secondary structure is the β-pleated sheet (Figure 12.4[b]). In the β-pleated sheet, a polypeptide chain or chains is folded in a zigzag way with parallel regions or chains linked by hydrogen bonds. Most proteins contain a mixture of α-helical and β-pleated sheet regions.

3. A protein's tertiary structure (Figure 12.4[c]) is the three-dimensional structure (often called its *conformation*) into which the helices and other parts of a polypeptide chain are folded. Tertiary folding is a direct property of the amino acid sequence of the chain. Shown in Figure 12.4(c) is the tertiary structure of myoglobin, an oxygen-binding protein found in muscle.

4. An example of the quaternary structure of a protein is shown in Figure 12.4(d). Primary, secondary, and tertiary structures all refer to single polypeptide chains. However, proteins may consist of more than one polypeptide chain; such proteins are called multimeric ("many subunits") proteins. Quaternary structure refers to how the polypeptides of a multimeric protein are packaged into the whole protein molecule. Shown in Figure 12.4(d) is the quaternary structure of the oxygen-carrying protein hemoglobin, which consists of four polypeptide chains (two α polypeptides and two β polypeptides), each of which is associated with a heme group (involved in the binding of oxygen). The α and β polypeptides have different amino acid sequences. In the quaternary structure of hemoglobin, each α chain is in contact with each β chain; however, little interaction occurs between the two α chains or between the two β chains.

Keynote

A protein consists of one or more molecular subunits called polypeptides, which are themselves composed of smaller building blocks, the amino

■ *Figure 12.2*   Structures of the 20 naturally occurring amino acids.

Acidic

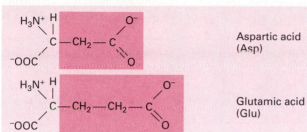

Basic

Neutral, nonpolar

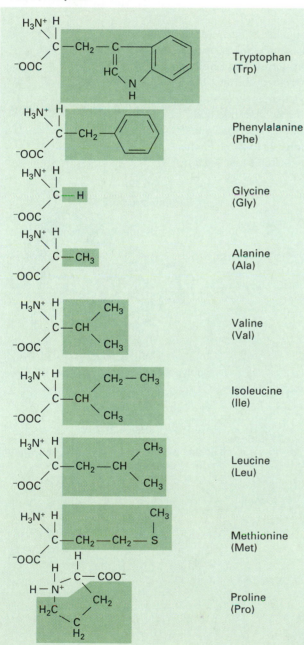

Neutral, polar

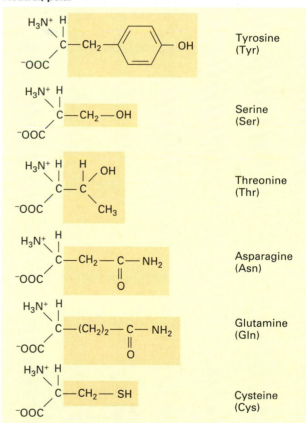

■ *Figure 12.3* Peptide bond between the carboxyl group of one amino acid and the amino group of another amino acid.

Two amino acids joined by a peptide bond

Amino (N-terminal) end H_3N^+—C—C—N—C—C   Carboxyl (C-terminal) end

Peptide bond

acids, linked together by peptide bonds to form long chains. The primary amino acid sequence of a protein determines its secondary, tertiary, and quaternary structure and hence its functional state.

The Nature of the Genetic Code

In Chapter 11 we learned that nucleotides in the mRNA molecule are read in groups of three (called **codons**) to specify the amino acid sequence in proteins. With four different nucleotides (A, C, G, U), a

■ *Figure 12.4* Four levels of protein structure: (a) Primary, the sequence of amino acids in a polypeptide chain. (b) Secondary, the folding and twisting of a single polypeptide chain into a variety of shapes. Shown are two types of secondary structure: the α-helix and β-pleated sheet. Both structures are stabilized by hydrogen bonds. (c) Tertiary, the specific three-dimensional folding of the polypeptide chain. Shown here is the main polypeptide chain of myoglobin, a 153-amino acid, heme-containing polypeptide that carries oxygen in muscles. (d) Quaternary, the specific aggregate of polypeptide chains. Shown here is hemoglobin, which carries oxygen in the blood; it consists of two α chains, two β chains, and four heme groups. The tertiary structures of the α-chains and β-chains closely resemble that of myoglobin.

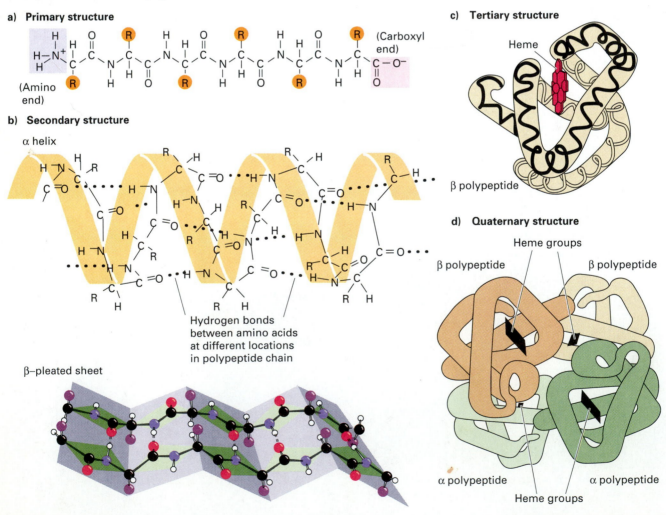

a) **Primary structure**

(Amino end)

(Carboxyl end)

b) **Secondary structure**

α helix

Hydrogen bonds between amino acids at different locations in polypeptide chain

β–pleated sheet

c) **Tertiary structure**

Heme

β polypeptide

d) **Quaternary structure**

Heme groups

β polypeptide   β polypeptide

α polypeptide   α polypeptide

Heme groups

three-letter code generates 64 possible codons, yet there are only 20 different amino acids. Thus we can deduce that at least some amino acids may be specified by more than one codon. In this section we discuss the evidence that showed the genetic code is a triplet code, how the code was deciphered, and the properties of the code.

The Genetic Code Is a Triplet Code

The evidence that the genetic code is a triplet code—that a set of three nucleotides (a codon) in mRNA code for one amino acid in a polypeptide chain—came from genetic experiments done by F. Crick, L. Barnett, S. Brenner, and R. Watts-Tobin in the early 1960s. The experiments used the bacteriophage T4. Recall from Chapter 6 (p. 144) that *rII* mutants of T4 produce clear plaques on *E. coli* B, whereas the wild type *r*$^+$ strain produces turbid plaques. Further, unlike the *r*$^+$ strain, *rII* mutants are unable to reproduce in *E. coli* K12(λ).

Crick and his colleagues began with an rII mutant strain that had been produced by treating the *r*$^+$ strain with a *mutagen*, a chemical that causes mutations (discussed in more detail in Chapter 16). The mutagen used was the chemical proflavin, which induces mutations by causing the addition or deletion of a base pair in the DNA. Such mutations are called *frameshift mutations* (see Chapter 16, p. 381). Crick et al. reasoned that, if the *rII* mutant phenotypes resulted from either an addition or a deletion, treatment of the *rII* mutant with proflavin could reverse the mutation to the wild-type *r*$^+$ state. The process of changing a mutant back to the wild-type state is called **reversion**, and the wild type produced is called a *revertant*. If the original mutation was an addition, it could be corrected by a deletion. They were able to isolate a number of *r*$^+$ revertant strains by plating a population of *rII* phages that had been treated with proflavin onto a lawn of *E. coli* K12(λ) in which only *r*$^+$ phages can grow. This made it very easy to isolate the low number of *r*$^+$ revertants produced by the proflavin treatment.

Some of the revertants resulted from an exact correction of the original mutation; that is, an addition corrected the deletion, or a deletion corrected the addition. A second type of revertant was much more useful for determining the nature of the genetic code. This revertant type resulted from a second mutation within the *rII* gene very close to, but distinct from, the original mutation site. If, for example, the first mutation was a deletion of a single base pair, the reversion of this type of mutation would involve an addition of a base pair nearby. Figure 12.5(a) shows hypothetical segment of DNA.

For the purposes of discussion, we have assumed that the code is a triplet code. Thus the mRNA transcript of the DNA would be read ACG ACG ACG, and so on, giving a polypeptide with a string of identical amino acids, each specified by ACG. If proflavin treatment causes a deletion of a second AT base pair, the mRNA will now read ACG CGA CGA and so on, giving a polypeptide starting with the amino acid specified by ACG, followed by a string of amino acids that are specified by CGA (Figure 12.5[b]). In essence, the reading frame of the message has been changed and the mutation is a frameshift mutation: the message is out of frame by one. Reversion of this mutation can occur by adding a base pair nearby (Figure 12.5[c]). For example, the insertion of a GC base pair after the GC in the third triplet results in a mRNA which reads ACG CGA

■ *Figure 12.5* Reversion of a deletion frameshift mutation by a nearby addition mutation: (a) Hypothetical segment of normal DNA, mRNA transcript, and polypeptide in the wild type; (b) Effect of a deletion mutation on the amino acid sequence of a polypeptide. The reading frame is disrupted; (c) Reversion of the deletion mutation by an addition mutation. The reading frame is restored, leaving a short segment of incorrect amino acids.

a) Wild type

DNA
5'···ACG ACG ACG ACG ACG···3'
3'···TGC TGC TGC TGC TGC···5'

mRNA
5'···ACG ACG ACG ACG ACG···3'

Polypeptide
····- 1 - 1 - 1 - 1 - 1 -····

b) Frameshift mutation by deletion

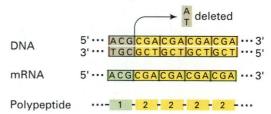

A
T deleted

DNA
5'···ACG CGA CGA CGA CGA···3'
3'···TGC GCT GCT GCT GCT···5'

mRNA
5'···ACG CGA CGA CGA CGA···3'

Polypeptide
····- 1 - 2 - 2 - 2 - 2 -····

c) Reversion of deletion mutation by addition

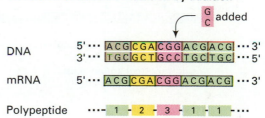

G
C added

DNA
5'···ACG CGA CGG ACG ACG···3'
3'···TGC GCT GCC TGC TGC···5'

mRNA
5'···ACG CGA CGG ACG ACG···3'

Polypeptide
····- 1 - 2 - 3 - 1 - 1 -····

CGG ACG ACG and so on. This gives a polypeptide consisting mostly of the amino acid specified by ACG but with two wrong amino acids—those specified by the CGA and CGG. Thus the second mutation has restored the reading frame and a nearly wild-type polypeptide is produced. As long as the incorrect amino acids in the short segment between the mutations do not significantly affect the function of the polypeptide, the double mutant will have a normal or at least near-normal phenotype.

In short, an addition mutation can be reverted by a nearby deletion mutation, and a deletion mutation can be reverted by a nearby addition mutation. We symbolize addition mutations as + mutations and deletion mutations as − mutations. The next step Crick and his colleagues took was to combine genetically distinct *rII* mutations of the same type (either all + or all − mutations*) in various numbers to see whether any combinations reverted the *rII* phenotypes. They found that the combination of either three nearby + mutations or three nearby − mutations gave *r*+ revertants. No other multiple combinations worked, except multiples of three. They concluded, therefore, that *the genetic code is a triplet code.* Figure 12.6 gives a hypothetical interpretation of the type of results they obtained, showing the effects of the mutations on the mRNA. The figure shows a 30-nucleotide segment of mRNA

---

*In actuality, Crick and his colleagues did not know whether an individual *rII* mutant resulted from a plus (+) or a minus (−) mutation. But they did know which of their single-mutant *rII* strains were of one sign and which were of the other sign. All mutants of one sign (e.g. +) could be reverted by nearby mutants of the other sign (i.e., −), and vice versa.

that codes for 10 different amino acids in the polypeptide. If we add three base pairs at nearby locations in the DNA coding for this mRNA segment, the results will be a 33-nucleotide segment that codes for 11 amino acids; that is, one more than the original. Note, though, that the amino acids between the first and third insertions are not the same as the wild-type mRNA. The reading frame is correct before the first insertion and again after the third insertion. The incorrect amino acids between those points may result in a near-wild-type phenotype for the revertant.

Deciphering the Genetic Code

With evidence in hand that the genetic code was a triplet code, the next step was to determine how each three-letter codon corresponded to an amino acid. The exact relationship of the 64 codons to the 20 amino acids was determined by experiments done mostly in the laboratories of Marshall Nirenberg and Ghobind Khorana. Essential to these experiments was the use of **cell-free, protein-synthesizing systems**, which contained ribosomes, tRNAs with amino acids attached, and all the necessary protein factors for polypeptide synthesis. When mRNA is added to such a system, proteins are made.

Synthetic mRNAs played an important role in working out the genetic code. Various synthetic mRNAs were made and added to cell-free, protein-synthesizing systems. The polypeptides made in these systems were then analyzed to determine what the synthetic mRNAs coded for. Synthetic poly(U) mRNA, for example, directed the synthesis

■ *Figure 12.6*   Hypothetical example showing how three nearby + (addition) mutations restore the reading frame, giving normal or near-normal function. The mutations are shown here at the level of the mRNA.

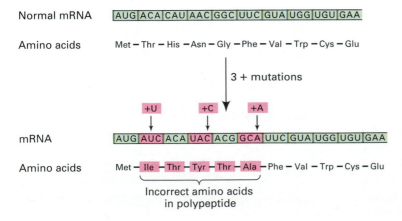

of a polypeptide consisting entirely of phenylala-nines. Since the genetic code is a triplet code, this result indicated that UUU is a codon for phenylala-nine. Similarly, a synthetic poly(A) mRNA directed the synthesis of a polypeptide consisting entirely of lysine, indicating that AAA is a codon for lysine. Poly(CCC) mRNA directed the synthesis of a pro-line chain, indicating that CCC is a codon for pro-line. Poly(G) did not make any polypeptide because it folds up in a way that prevents ribosomes from reading it.

Other synthetic mRNAs were made from a mix-ture of nucleotides and analyzed in the same way. These types of mRNAs are called *copolymers*. In one type of synthesis, the nucleotides were incorporat-ed randomly into the synthetic mRNA to produce *random copolymers*. For example, poly(AC) mole-cules made from a random mixture of As and Cs would contain eight different codons (CCC, CCA, CAC, ACC, CAA, ACA, AAC, and AAA). The pro-portions of each of the codons would depend on the relative amounts of As and Cs used to make the polymer. In the cell-free, protein-synthesizing sys-tem, poly(AC) synthetic mRNAs caused the incor-poration of asparagine, glutamine, histidine, and threonine into polypeptides, in addition to the lysine expected from AAA codons and the proline expected from CCC codons. The proportions of asparagine, glutamine, histidine, and threonine incorporated into the polypeptides produced depended on the A/C ratio used to make the mRNA, and these observations were used to deduce information about the codons that specify the amino acids. For example, since a poly(AC) mRNA containing much more A than C resulted in the incorporation of many more asparagines than histidines, researchers concluded that asparagine is coded by two As and one C and histidine by two Cs and one A. From experiments of this kind, the base composition (*but not the base sequence*) of the codons for a number of amino acids was determined.

In another type of synthesis, synthetic mRNAs were made with more than one nucleotide, with the nucleotides assembled in a known sequence rather than a random one. For example, a repeating copolymer of U and C gives a synthetic mRNA of UCUCUCUC, which results in a polypeptide with a repeating amino acid pattern of leucine-ser-ine-leucine-serine . . . . This meant that UCU and CUC specify leucine and serine, although which coded for which could not be determined from the results.

A different approach used a *ribosome-binding assay*, developed in 1964 by Nirenberg and Philip Leder. In this assay the 64 different codons were synthesized in the laboratory. When a synthetic codon is added to ribosomes, it will bind to them. Then, if a tRNA carrying an amino acid is present that can recognize the codon, it will bind to the codon-ribosome complex. So for each codon, a ribo-some-binding reaction was set up involving 20 dif-ferent tubes. In each tube the ribosome and codon were the same. Each tube also contained tRNAs and enzymes to attach the amino acids to the tRNAs and all 20 amino acids; each tube differed in the amino acid that was radioactively labeled. For example, when synthetic UUU is in the reaction, only the tRNA with phenylalanine attached and with the correct anticodon (AAA, for the UUU codon) will bind to the UUU codon-ribosome com-plex. In the tube in which phenylalanine is radioac-tively labeled, the tRNA-codon-ribosome complex will be radioactive. By passing the mixture through a filter, this complex will be trapped on the filter which can be tested to see if it is radioactive. In the other 19 reactions, a different amino acid is labeled, so the tRNA with amino acid attached will not bind to the codon-ribosome complex, and will pass through the filter leaving the filter nonradioactive. Note that in this particular approach the *specific nucleotide sequence of the codon is determined*, not merely the nucleotide composition. Using the ribo-some-binding assay, many of the ambiguities that had arisen from other approaches were resolved. For example, UCU was found to code for serine, and CUC to code for leucine.

In sum, no single approach produced an unam-biguous set of codon assignments, but information obtained through all the approaches enabled all codon assignments to be deduced with high degrees of certainty.

Nature and Characteristics of the Genetic Code

From the types of experiments just described, all 64 codons were assigned as shown in Figure 12.7. Each codon is written as it appears in mRNA and reads in a 5'-to-3' direction. The characteristics of the genetic code are as follows:

1. *The code is a triplet code.* Each mRNA codon that specifies an amino acid in a polypeptide chain consists of three nucleotides.

2. *The code is comma-free, or continuous.* The mRNA is read continuously, three nucleotides (one codon) at a time, without skipping any nucleotides of the message.

3. *The code is nonoverlapping.* The mRNA is read in successive groups of three nucleotides. A mes-sage of AAGAAGAAG . . . in the cell would be read as lysine-lysine-lysine . . . which is what the AAG specifies. Theoretically, three readings are possible

■ *Figure 12.7* The genetic code. Of the 64 codons, 61 specify 20 amino acids. One of those 61 codons, AUG, is used in the initiation of protein synthesis. The other 3 codons are chain termination codons and do not specify any amino acid.

Second letter

	U	C	A	G	
U	UUU Phe UUC UUA Leu UUG	UCU UCC Ser UCA UCG	UAU Tyr UAC **UAA Stop** **UAG Stop**	UGU Cys UGC **UGA Stop** UGG Trp	U C A G
C	CUU CUC Leu CUA CUG	CCU CCC Pro CCA CCG	CAU His CAC CAA Gln CAG	CGU CGC Arg CGA CGG	U C A G
A	AUU AUC Ile AUA AUG Met	ACU ACC Thr ACA ACG	AAU Asn AAC AAA Lys AAG	AGU Ser AGC AGA Arg AGG	U C A G
G	GUU GUC Val GUA GUG	GCU GCC Ala GCA GCG	GAU Asp GAC GAA Glu GAG	GGU GGC Gly GGA GGG	U C A G

First letter (left axis) — Third letter (right axis)

■ = Chain termination codon (stop)
■ = Initiation codon

from this message, depending on where the reading is begun (the *reading frame* in which the message is read); namely, the repeating AAG, the repeating AGA, and the repeating GAA. Later in this chapter we will examine the mechanisms in the cell that ensure that the translation of the genetic code in an mRNA begins at the correct point.

4. *The code is almost universal.* All organisms share the same genetic language. Thus, for example, lysine is coded for by AAA or AAG in the mRNA of all organisms, arginine by CGU, CGC, CGA, CGG, AGA, and AGG, and so on. Hence we can isolate an mRNA from one organism, translate it by using the machinery isolated from another organism, and produce the protein as if it had been translated in the original organism. The code, however, is not completely universal. For example, the mitochondria of some organisms, such as mammals, have minor changes in the code, as does the nuclear genome of the protozoan, *Tetrahymena* (discussed in Chapter 19).

5. *The code is degenerate.* With two exceptions (AUG, which codes for methionine, and UGG, which codes for tryptophan), more than one codon occurs for each amino acid. This multiple coding is

called the **degeneracy** of the code. A close examination of Figure 12.7 reveals particular patterns in this degeneracy. When the first two nucleotides in a codon are identical and the third letter is U or C, the codon always codes for the same amino acid. For example, UUU and UUC both specify phenylalanine; similarly, CAU and CAC specify histidine. Also, when the first two nucleotides in a codon are identical and the third letter is A or G, the same amino acid is often specified. For instance, UUA and UUG specify leucine, and AAA and AAG specify lysine. In a few cases, when the first two nucleotides in a codon are identical and the base in the third position is U, C, A, or G, the same amino acid is often specified. For example, CUU, CUC, CUA, and CUG all code for leucine.

6. *The code has start and stop signals.* Specific start and stop signals for protein synthesis are contained in the code. In both eukaryotes and prokaryotes AUG (which codes for methionine) is usually the start codon for protein synthesis, although in rare cases GUG is used.

As Figure 12.7 shows, only 61 of the 64 codons specify amino acids; these codons are called the **sense codons**. The other three codons—UAG, UAA, and UGA—do not specify an amino acid, and no tRNAs in normal cells carry the appropriate anticodons. These three codons are the **stop codons;** also called **nonsense codons** or **chain-terminating codons.** They specify the end of the translation process of a polypeptide chain. Thus when we read a particular mRNA sequence, we look for the presence of a stop codon *in the same reading frame* as the AUG start codon to determine where the amino-acid-coding sequence for the polypeptide ends.

7. *Wobble occurs in the anticodon.* Since 61 sense codons specify amino acids in mRNA, a total of 61 tRNA molecules could have the appropriate anticodons. Theoretically, though, the complete set of 61 sense codons can be read by fewer than 61 distinct tRNAs because of wobble in the anticodon. The **wobble hypothesis,** proposed by Francis Crick, is shown in Table 12.1. Structural analysis had shown that the base at the 5′ end of the anticodon (complementary to the base at the 3′ end of the codon; i.e., the third letter) is freer in space to pair with more than one base at the 3′ end of the codon; it can wobble. As Table 12.1 shows, no single tRNA molecule can recognize four different codons. But if the tRNA molecule contains the modified nucleoside inosine at the 5′ end of the anticodon, then three different codons can be read by that one tRNA. Figure 12.8(a) gives an example of how a single leucine tRNA can read two different leucine codons by wobble pairing, and Figure 12.8(b) shows how a *single* glycine tRNA can read *three* different glycine codons by wobble pairing.

Table 12.1
Wobble in the Genetic Code

NUCLEOTIDE AT 5' END OF ANTICODON		NUCLEOTIDE AT 3' END OF CODON
G	can pair with	U or C
C	can pair with	G
A	can pair with	U
U	can pair with	A or G
I (inosine)	can pair with	A, U, or C

Keynote

The genetic code is a triplet code in which each codon (a contiguous set of three bases) in an mRNA specifies one amino acid. Since 64 code words are possible and only 20 amino acids exist, some amino acids are specified by more than one codon. The genetic code in mRNA is read without gaps and in successive, nonoverlapping codons. The code is universal: The same codons specify the same amino acids in most systems. Protein synthesis is typically initiated by codon AUG (methionine), and it is terminated by three codons singly or in combination: UAG, UAA, UGA (these codons do not normally code for any amino acid).

Translation of the Genetic Message

In brief, protein synthesis takes place on ribosomes, where the genetic message encoded in mRNA is translated. As we learned in Chapter 11, ribosomes are composed of ribosomal RNA (rRNA) and proteins. The mRNA molecule is translated in the 5'-to-3' direction (the same direction in which it is made), and the polypeptide is made in the N-terminal to C-terminal direction. Amino acids are brought to the ribosome bound to tRNA molecules. The correct amino acid sequence is achieved as a result of (1) the specific binding between the codon of the mRNA and the complementary anticodon in the tRNA, and (2) the specific binding of each amino acid to its own specific tRNA.

The three basic stages of protein synthesis—*initiation*, *elongation*, and *termination*—are similar in prokaryotes and eukaryotes. In the following sections we will discuss each of these stages in turn, concentrating, as before, on the processes in *E. coli*. In the discussions we will note where significant differences in translation occur in prokaryotes and eukaryotes.

Aminoacyl-tRNA Molecules

The important function of tRNA molecules in protein synthesis is that they bring specific amino acids to the mRNA-ribosomal complex so that the correct polypeptide chain can be assembled. In this section we will see how an amino acid becomes attached to its appropriate tRNA molecule to produce an aminoacyl-tRNA molecule.

Attachment of Amino Acids to tRNA   The correct amino acid is attached to the tRNA by an enzyme called an **aminoacyl-tRNA synthetase** in a process often referred to as amino acid activation. Since 20 different amino acids exist, 20 types of aminoacyl-tRNA synthetases also exist. Further, since degeneracy is common in the genetic code (i.e., a single amino acid may be specified by more than one codon), all the tRNAs that are specific for a particular amino acid have that amino acid added by the same aminoacyl-tRNA synthetase.

Figure 12.9 shows how an amino acid is attached to a tRNA molecule to produce an **aminoacyl-tRNA,**

■ *Figure 12.8*   Example of base-pairing wobble: (a) Two different leucine codons (CUC, CUU) can be read by the same leucine tRNA molecule, contrary to regular base-pairing rules; (b) Three different glycine codons (GGU, GGC, GGA) can be read by the same glycine tRNA molecule by base-pairing wobble involving inosine in the anticodon.

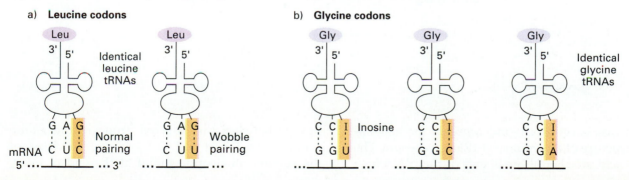

■ *Figure 12.9*   Molecular details of the attachment of an amino acid to a tRNA molecule: (a) In the first step the amino acid (serine here) reacts with ATP to produce an aminoacyl-AMP complex. This reaction is catalyzed by an aminoacyl-tRNA synthetase (seryl-tRNA synthetase). (b) In the second step (also catalyzed by aminoacyl-tRNA synthetase) the aminoacyl-AMP complex then reacts with the appropriate tRNA molecule (tRNA.Ser) to produce an aminoacyl-tRNA (Ser-tRNA.Ser). (c) In the general molecular structure of an aminoacyl-tRNA molecule (charged tRNA), the carboxyl group of the amino acid is attached to the 3′-OH or 2′-OH group of the 3′ terminal adenine nucleotide of the tRNA.

a) **First step**

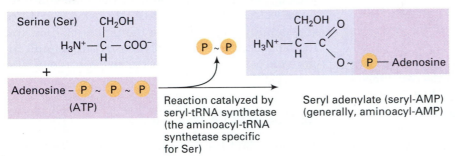

b) **Second step**

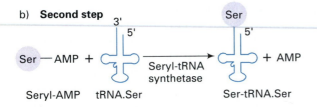

c) **General structure**

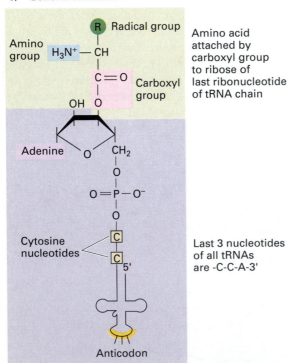

in this case seryl-tRNA, and presents the generalized structure of an aminoacyl-tRNA molecule. The amino acid attaches at the 3′ end of the tRNA by a linkage between the carboxyl group of the amino acid and the 3′-OH or 2′-OH group of the ribose of the adenine nucleotide found at the end of every tRNA (the amino acid is attached to the 3'-OH in the figure). The act of adding the amino acid to the tRNA is called **charging,** and the product is commonly referred to as a **charged tRNA** or *aminoacylated-tRNA*.

K e y n o t e

Protein synthesis occurs on ribosomes, where the genetic message encoded in mRNA is translated. Amino acids are brought to the ribosome on charged tRNA molecules. The correct amino acid sequence is achieved as a result of (1) the specific binding between the codon of the mRNA and the complementary anticodon in the tRNA and (2) the specific binding of each amino acid to its own specific tRNA.

Initiation of Translation

Initiator Codon and Initiator tRNA   In both prokaryotes and eukaryotes translation begins at the AUG initiator codon in the mRNA, which specifies methionine. As a result, newly made proteins in both types of organisms begin with methionine; in some cases this methionine is subsequently removed.

Prokaryotes   In prokaryotes the initiator methionine is a modified form of methionine, called **formylmethionine** (abbreviated fMet), in which a formyl group has been added to the methionine's amino group. The fMet is brought to the ribosome attached to a special tRNA, called tRNA.fMet, which has the anticodon 5'-CAU-3' that is complementary to the initiator codon 5'-AUG-3' in the mRNA. The charged tRNA is designated fMet-tRNA.fMet (this terminology indicates that the tRNA is specific for the attachment of fMet and that, in fact, fMet is attached to it).

The tRNA.fMet molecule is a special tRNA that is *only used in the initiation of translation*. When an AUG codon in an mRNA molecule is encountered at a position other than at the start of the amino acid–coding sequence, a different species of tRNA is used to insert methionine at that point in the polypeptide chain. This tRNA is called tRNA.Met, and it is charged by the same aminoacyl-tRNA synthetases as is tRNA.fMet to produce Met-tRNA.Met. The tRNA.Met and tRNA.fMet molecules are coded for by different genes and have different sequences.

Eukaryotes   In eukaryotes protein synthesis initiation occurs in much the same way. AUG is the usual initiation codon, but the N-terminal methionine has no formyl group. Nonetheless, a special methionine tRNA is used for initiation of translation in eukaryotes, and a different species of tRNA.Met is used to read AUG codons elsewhere in an mRNA molecule.

Formation of the Initiation Complex   The initiation of translation is very similar in prokaryotes and eukaryotes. In both types of organisms the small ribosomal subunit first binds to the mRNA with the aid of special proteins, and then the large ribosomal subunit binds to produce an *initiation complex* that has all the components to begin protein synthesis.

Prokaryotes   In addition to the AUG initiation codon, the initiation of protein synthesis in prokaryotes requires a sequence in the mRNA molecule upstream (to the 5' side) of the initiation codon. This sequence, called the **Shine-Dalgarno sequence,** aligns the ribosome on the message in the proper reading frame so that polypeptide synthesis can proceed correctly. Figure 12.10 shows the Shine-Dalgarno sequence 5'-UAAGGAGG-3' and how it can form complementary base pairs with a region (which always contains the sequence 5'-CCUCC-3') at the 3' end of 16S rRNA of the small ribosomal subunit. Apparently the formation of base pairs between the two RNAs allows the ribosome to locate the true initiator region of the message.

Figure 12.11 shows the formation of the initiation complex for starting translation in *E. coli*. At the beginning of this process, three protein **initiation factors,** IF1, IF2, and IF3, bind to the 30S ribosomal subunit along with a molecule of GTP (guanosine triphosphate). The fMet-tRNA.fMet and the mRNA then attach to the 30S–IF–GTP complex to form the

■ *Figure 12.10*   Binding of ribosomes to the mRNA in the initiation of protein synthesis in prokaryotes: Example of how the 3' end of 16S rRNA base-pairs with the Shine-Dalgarno sequence 5' upstream from the AUG initiation codon.

a)   **Sequence at 3' end of 16S rRNA**

3'...AUUCCUCCAUAG...5'

b)   **Example of mRNA leader and 16S rRNA pairing**

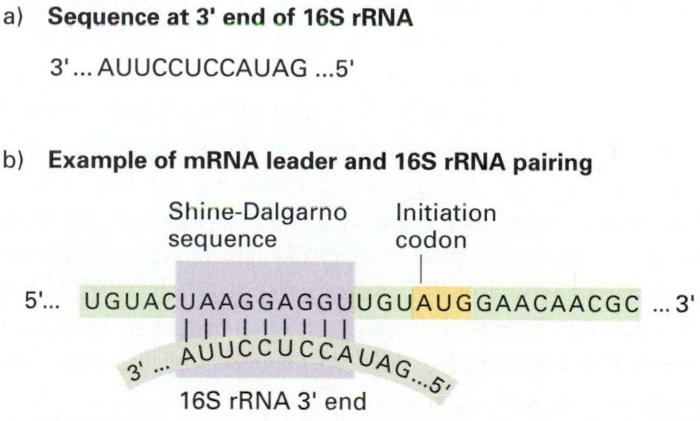

■ *Figure 12.11*　Initiation of protein synthesis in prokaryotes. A 30S ribosomal subunit, complexed with initiation factors and GTP, binds to mRNA and fMet-tRNA.fMet to form a 30S initiation complex. Then the 50S ribosomal subunit binds, forming a 70S initiation complex. During the latter event, the initiation factors are released and GTP is hydrolyzed.

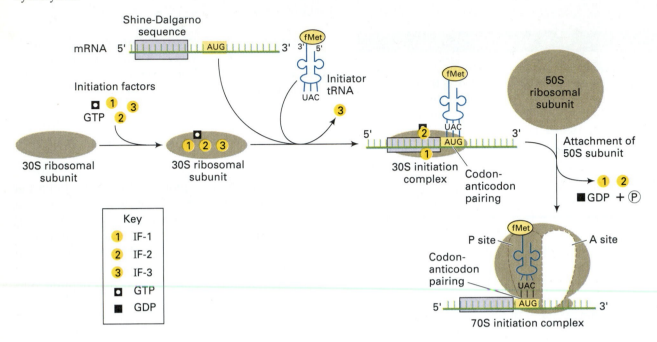

30S initiation complex. IF3 is released as a result of this process. Next, the 50S subunit binds, causing GTP to be hydrolyzed and IF1 and IF2 to be released. This gives the final complex which has the large and small ribosomal subunits bound to the mRNA and fMet-tRNA.fMet bound to the AUG codon. The final complex is called the *70S initiation complex*. At this point the ribosome is oriented in the correct reading frame for the initiation of protein synthesis. The 70S ribosome has two sites to allow the aminoacyl-tRNAs to bind to the mRNA—the peptidyl (P) and aminoacyl (A). The fMet-tRNA.fMet is bound to the mRNA in the P site.

Eukaryotes Nothing like a Shine-Dalgarno sequence is found in eukaryotic mRNAs. Instead, the eukaryotic ribosome uses another way to initiate protein synthesis on mRNA. First, a special **eukaryotic initiator factor**, eIF4A, recognizes and binds to the cap at the 5′ end of the mRNA (see Chapter 11). Then, a complex of the 40S ribosomal subunit with the initiator Met-tRNA.Met, several eIFs, and GTP binds, along with other eIFs, and migrates along the mRNA, scanning for the initiator AUG codon. This AUG codon is embedded in a short sequence that indicates it is the initiator codon. Once the 40S subunit finds this AUG, it binds to it. The 60S ribosomal subunit then binds to form the eukaryotic *80S initiation complex*. Like its prokaryotic counterpart, the eukaryotic 80S ribosome has a P site and an A site and the initiator Met-tRNA.fMet is bound to the mRNA in the P site.

Elongation of the Polypeptide Chain

After initiation the elongation phase of translation begins. This phase has three steps:

1. The binding of aminoacyl-tRNA to the ribosome;
2. The formation of a peptide bond; and
3. The movement (translocation) of the ribosome along the mRNA, one codon at a time. Figure 12.12 diagrams these events.

Binding of Aminoacyl′-tRNA　At the start of elongation, the fMet-tRNA.fMet is hydrogen-bonded to the AUG initiation codon in the P site of the 70S initiation complex. The next codon in the mRNA is exposed in the A site (Figure 12.12[1]). In Figure 12.12 this codon (UCC) specifies the amino acid serine (Ser).

Next, the appropriate aminoacyl-tRNA (in this case Ser-tRNA.Ser) binds to the exposed mRNA codon in the A site (Figure 12.12[2]). This aminoacyl-tRNA is brought to the ribosome complexed with the protein **elongation factor** EF-Tu and a molecule of GTP. When the aminoacyl-tRNA binds to

■ *Figure 12.12*   Elongation stage of polypeptide synthesis in prokaryotes.

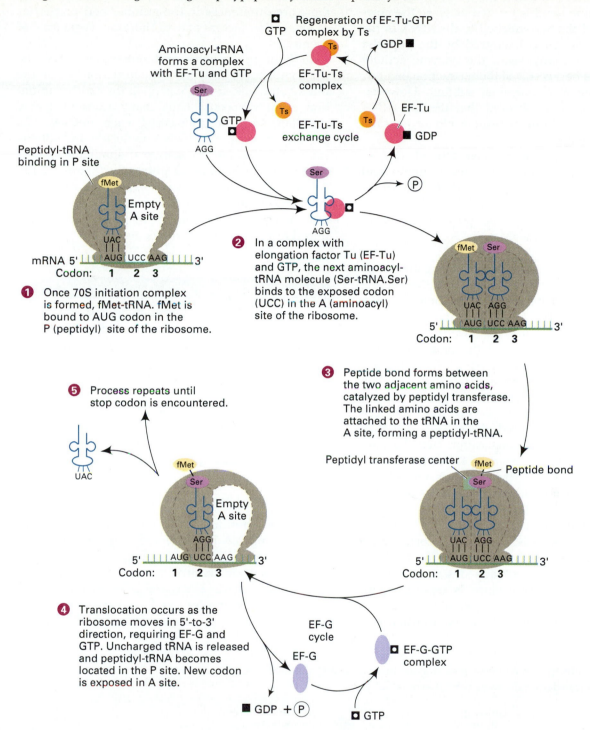

the codon in the A site, the GTP is hydrolyzed, and EF-Tu is released bound to the GDP produced.

As shown in Figure 12.12(2), EF-Tu is recycled. First, elongation factor EF-Ts binds to EF-Tu and displaces the GDP. Next, GTP binds to the EF-Tu–EF-Ts complex to produce an EF-Tu–GTP complex simultaneously with the release of EF-Ts. The aminoacyl-tRNA binds to the EF-Tu–GTP, and that

complex can then bind to the A site in the ribosome when the appropriate codon is exposed.

Peptide Bond Formation   The ribosome maintains the two aminoacyl-tRNAs in the correct positions so that a peptide bond can form between the two amino acids (Figure 12.12[3]). This occurs by breaking the bond between the carboxyl group of the amino acid (here, fMet) and the tRNA in the P site. Then the peptide bond is formed between the now-freed fMet and the Ser attached to the tRNA in the A site. This reaction is catalyzed by the **peptidyl transferase**. For many years the enzyme activity was thought to be a result of the interaction of a few proteins of the 50S ribosomal subunit. However, Harry Noller in 1992 showed that the 23S rRNA molecule of that subunit is responsible for peptidyl transferase activity.

Once the peptide bond has formed (Figure 12.12[3]), a tRNA without an attached amino acid (an uncharged tRNA) is left in the P site. The tRNA in the A site—now called peptidyl-tRNA—has the first two amino acids of the polypeptide chain attached to it, in this case fMet-Ser.

Translocation   The last step in the elongation cycle is **translocation** (Figure 12.12[4]). Once the peptide bond is formed and the growing polypeptide chain is on the tRNA in the A site, the ribosome moves one codon along the mRNA toward the 3' end. Translocation requires the activity of another protein elongation factor, EF-G. An EF-G–GTP complex binds to the ribosome, and translocation then takes place along with ejection of the uncharged tRNA from the P site. The EF-G is then released in a reaction requiring GTP hydrolysis; EF-G can then be reused. During the translocation step the peptidyl-tRNA remains attached to its codon on the mRNA. And, since the ribosome has moved, the peptidyl-tRNA is now located in the P site (hence the name, the *peptidyl* site).

After translocation is completed the A site is vacant. An aminoacyl-tRNA with the correct anticodon binds to the newly exposed codon in the A site, using the process already described (Figure 12.12[5]). The whole procedure is repeated until translation terminates at a stop codon.

In eukaryotes the elongation and translocation steps are similar to those in prokaryotes, although there are differences in the number and properties of elongation factors and in the exact sequences of events.

In both prokaryotes and eukaryotes once the ribosome moves away from the initiation site on the mRNA, the initiation site is open for another initiation event to occur. Thus many ribosomes may simultaneously be translating each mRNA. The complex between an mRNA molecule and all the ribosomes that are translating it simultaneously is called a **polyribosome** or **polysome** (Figure 12.13). An average mRNA may have eight to ten ribosomes synthesizing **protein** from it. Simultaneous translation enables a large amount of protein to be produced from each mRNA molecule.

Keynote

The AUG (methionine) initiator codon signals the start of translation in prokaryotes and eukaryotes. Elongation proceeds when a peptide bond forms between the amino acid attached to the tRNA in the A site of the ribosome and the growing polypeptide attached to the tRNA in the P site. Translocation occurs when the now-uncharged tRNA in the P site is released from the ribosome and the ribosome moves one codon down the mRNA.

Termination of Translation

Elongation continues until the polypeptide coded for in the mRNA is completed. The termination of translation is diagrammed in Figure 12.14. The end of a polypeptide chain is signaled by one of three stop codons—UAG, UAA, and UGA—which are the same in prokaryotes and eukaryotes (Figure 12.14[1]). The stop codons do not code for any

■ *Figure 12.13*   Electron micrograph and diagram of a polysome, a number of ribosomes each translating the same mRNA sequentially.

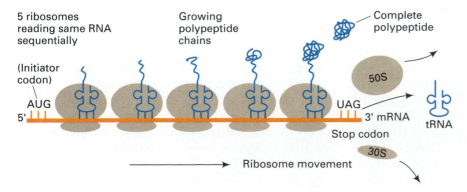

■ *Figure 12.14*   Termination of translation. The ribosome recognizes a chain termination codon (UAG) with the aid of release factors. The release factor reads the stop codon, and this initiates a series of specific termination events leading to the release of the completed polypeptide.

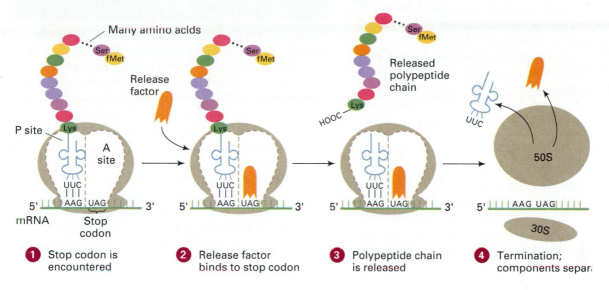

1 Stop codon is encountered

2 Release factor binds to stop codon

3 Polypeptide chain is released

4 Termination; components separ

amino acid, and so no tRNAs in the cell have anti-codons for them. The ribosome recognizes a chain termination codon only with the help of proteins called **termination factors**, or **release factors** (RF), which read the chain termination codons (Figure 12.14[2]), and then initiate a series of specific termination events. *E. coli* has three RFs—RF1, RF2, and RF3. RF1 recognizes UAA and UAG, while RF2 recognizes UAA and UGA. RF3, which does not recognize any of the stop codons, stimulates the termination events. In eukaryotes only one release factor, called eukaryotic release factor (eRF), is needed. eRF recognizes all three stop codons. No stimulatory factor analogous to RF3 has been found in eukaryotes.

The specific termination events triggered by the release factors are: (1) release of the polypeptide from the tRNA in the P site of the ribosome in a reaction catalyzed by peptidyl transferase (Figure 12.14[3]), and then (2) release of the tRNA from the ribosome and dissociation of the two ribosomal subunits from the mRNA (Figure 12.14[4]).

K e y n o t e

Protein synthesis continues until a chain-terminating codon is located in the A site of the ribosome. These codons are read by one or more proteins. Then the polypeptide and its tRNA are released from the ribosome, and the ribosome disengages from the reading frame of the mRNA.

Protein Sorting in the Cell

In eukaryotes some proteins may be secreted, depending on cell type, while other proteins need to be located in different cell compartments in order to function. That is, each cell compartment, such as the nucleus, mitochondria, chloroplasts, and lysosomes, contains a specific set of proteins. The sorting of proteins to their appropriate compartments is under genetic control. Similarly, in prokaryotes certain proteins become localized in the membrane and others are secreted. A complete discussion of protein localization mechanisms in eukaryotes and prokaryotes is beyond the scope of this text. Here, we will describe briefly protein transport into the ER of eukaryotes.

As we discussed in Chapter 1, rough ER has ribosomes associated with it. Proteins synthesized on ribosomes of the rough ER are moved into the space between the two membranes of the ER (the cisternal space) and are then transferred to the Golgi complex. Proteins to be secreted are then packaged into secretory vesicles. From these vesicles the proteins are secreted to the outside of the cell by the fusion of the vesicles with the cell membrane. Significantly, a common mechanism translocates: (1) proteins destined to be secreted from the cell; (2) proteins that will become embedded in the plasma membrane; and (3) proteins that are packaged into lysosomes. That is, initially the proteins are moved into the ER cisternal space. Thereafter, sorting to their final destinations occurs primarily at the Golgi complex.

Our current understanding of the transport of proteins into the ER is as follows (Figure 12.15). The growing proteins (those being synthesized on ribosomes) that are to be sorted by the Golgi complex all have an N-terminal stretch of about 15 to 30 amino acids, called the **signal sequence**. The signal sequence is hydrophobic (water hating). Only proteins that have signal sequences will be transported into the ER. When a protein destined for the ER exposes its signal sequence at the ribosome surface as it is being synthesized, the **signal recognition particle** (SRP, a complex of the small RNA molecule 7SL RNA, with six proteins) recognizes the signal sequence, binds to it, and blocks further translation of the mRNA. Translation stops until the growing protein-SRP-ribosome-mRNA complex reaches and binds to the ER. The SRP binds to a membrane protein of the ER called the **docking protein** (also called the *signal recognition protein receptor*). This enables the protein's signal sequence and the ribosome from which it emerged to attach to the ER. Translation resumes, and the SRP is released. The growing protein (with its signal sequence) now is translocated through the membrane into the cisternal space of the ER. The transport of the protein molecule into the ER simultaneously with its synthesis is an example of **cotranslational transport.**

Once the signal sequence is fully into the cisternal space of the ER, it is removed from the polypeptide by the action of an enzyme called **signal peptidase.** When the complete protein is entirely within the ER cisternal space, it is typically modified further by the addition of specific carbohydrate groups to produce *glycoproteins*. The glycoproteins are then sorted for their final destinations in the Golgi complex. Discussion of this sorting is beyond the scope of this text.

K e y n o t e

In eukaryotes proteins are found free in the cytoplasm, as well as in the various cell compartments such as the nucleus, mitochondria, chloroplasts, and secretory vesicles. Special mechanisms exist to sort proteins to their appropriate cell compartments. Proteins that enter the ER have signal sequences at their N-terminal ends. Characteristically, the signal sequences contain a significant number of hydrophobic amino acids. Proteins destined for the ER are translocated into the cisternal space of the ER, where the signal sequence is removed by signal peptidase. They are then sorted to their final destinations by the Golgi complex.

Summary

In this chapter we discussed the features of the structure of proteins, the genetic code, and the process of translation. From a number of experiments using synthetic mRNAs, the genetic code was found to have the following features: it is a triplet code, it is comma-free, it is nonoverlapping, it is universal, and, with two exceptions, it is degenerate. Of the 64 codons, 61 specify amino acids, while the other three codons are chain termination codons. One codon, AUG (for methionine), is typically used to specify the first amino acid in a protein chain.

Protein synthesis occurs on ribosomes, where the

■ *Figure 12.15*   Model for the translocation of proteins into the endoplasmic reticulum in eukaryotes.

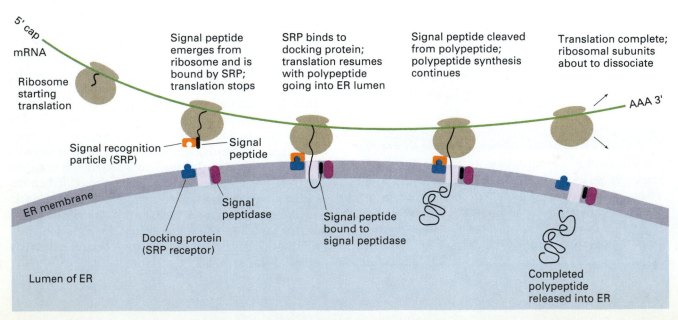

genetic message encoded in mRNA is translated. Amino acids are brought to the ribosome on charged tRNA molecules. Each tRNA has an anticodon, which binds specifically to a codon in the mRNA. As a result, the correct amino acid sequence is achieved by (1) the specific binding between the codon of the mRNA and the complementary anticodon in the tRNA and (2) the specific binding of each amino acid to its own specific tRNA.

In both prokaryotes and eukaryotes an AUG codon is the initiator codon for the start of protein synthesis. In prokaryotes the initiation of protein synthesis requires an upstream sequence of the AUG codon, to which the small ribosomal subunit binds. This sequence is the Shine-Dalgarno sequence, and it binds specifically to the 3' end of the 16S rRNA of the small ribosomal subunit, thereby associating the small subunit with the mRNA. No functionally equivalent sequence occurs in eukaryotic mRNAs; instead, the ribosomes load onto the mRNA at the 5' end of the mRNA and scan until the initiator AUG codon is found.

In both prokaryotes and eukaryotes the initiation of protein synthesis requires protein factors called initiation factors. Initiation factors are bound to the ribosome-mRNA complex during the initiation phase and dissociate once the polypeptide chain has been initiated. During the elongation phase, in which the polypeptide chain is elongated one amino acid at a time concomitantly with the movement of the ribosome toward the 3' end of the mRNA one codon at a time, protein factors called elongation factors play important catalytic roles. The signal for polypeptide chain growth to stop is the presence of a chain termination codon (UAG, UAA, or UGA) in the mRNA. No naturally occurring tRNA has an anticodon that can read a chain termination codon. Instead, specific protein factors called release factors read the stop codon and initiate the events characteristic of protein synthesis termination; namely, the release of the completed polypeptide from the ribosome, the release of the tRNA from the ribosome, and the dissociation of the two ribosomal subunits from the mRNA.

Eukaryotic proteins that enter the endoplasmic reticulum, the mitochondrion, the chloroplast, or the nucleus are different from proteins that remain free in the cytoplasm in that they possess specific targeting sequences—sequences that are required for localizing them into the appropriate cellular compartment. Proteins destined for the ER, for example, have specific hydrophobic signal sequences at their amino ends. When a protein being synthesized exposes its signal sequence, a signal recognition particle (SRP) binds to it and blocks further translation of the mRNA. Translation is blocked until the nascent polypeptide-SRP-ribosome-mRNA complex reaches and binds to the ER by interaction between the SRP and an integral membrane protein of the ER called the docking protein. Resumption of translation then results in the growing polypeptide being translocated into the cisternal space of the ER where the signal sequence is removed. The protein is then sorted for its final location by the Golgi complex.

In sum, protein synthesis is a complex process involving the interaction between three major classes of RNA (mRNA, tRNA, and rRNA) and a large number of accessory protein factors that act catalytically in the process. With very few exceptions the genetic code that specifies the amino acids for each mRNA codon is the same in every organism, prokaryotic and eukaryotic. By repeated translation of an mRNA molecule by a string of ribosomes (producing a polysome), a large number of protein molecules can be produced. Thus from a single gene, large quantities of a protein can be produced by two amplification steps: (1) the production of multiple mRNAs from the gene; and (2) the production of many protein molecules by repeated translation of each mRNA.

Analytical Approaches for Solving Genetics Problems

Q12.1 a. How many of the 64 codon permutations can be made from the three nucleotides A, U, and G?
b. How many of the 64 codon permutations can be made from the four nucleotides A, U, G, and C, with one or more Cs in each codon?

A12.1 a. This question involves probability. There are four bases, so the probability of a cytosine at the first position in a codon is 1/4. Conversely, the probability of a base other than cytosine in the first position is $(1 - 1/4) = 3/4$. These same probabilities apply to the other two positions in the codon. Therefore the probability of a codon without a cytosine is $(3/4)^3 = 27/64$, and thus the number of codons that can be made from A, U, and G is 27.

b. This question involves the relative frequency of codons that have one or more cytosines. We have already calculated the probability of a codon not having a cytosine, so all the remaining codons have one or more cytosines. The answer to this question, therefore, is $(1 - 27/64) = 37/64$.

Q12.2 Random copolymers were used in some of the experiments directed toward deciphering the genetic code. For a ribonucleotide mixture of 2U:1C, give the expected codons and their frequencies, and give the expected proportions of the amino acids that would be found in a polypeptide directed by the copolymer in a cell-free, protein-synthesizing system.

A12.2 The probability of a U at any position in a codon is 2/3, and the probability of a C at any position in a codon is 1/3. Thus the codons, their relative frequencies, and the amino acids for which they code are:

UUU = (2/3)(2/3)(2/3) = 8/27 = 0.296 = 29.6% Phe
UUC = (2/3)(2/3)(1/3) = 4/27 = 0.148 = 14.8% Phe
UCC = (2/3)(1/3)(1/3) = 2/27 = 0.0743 = 7.43% Ser
UCU = (2/3)(1/3)(2/3) = 4/27 = 0.148 = 14.8% Ser
CUU = (1/3)(2/3)(2/3) = 4/27 = 0.14 = 14.8% Leu
CUC = (1/3)(2/3)(1/3) = 2/27 = 0.0743 = 7.43% Leu
CCU = (1/3)(1/3)(2/3) = 2/27 = 0.0743 = 7.43% Pro
CCC = (1/3)(1/3)(1/3) = 1/27 = 0.037 = 3.7% Pro

In sum, 44.4% Phe, 22.23% Ser, 22.23% Leu, 11.13% Pro. (Does not quite add up to 100% because of rounding-off errors.)

Questions and Problems

*12.1 The form of genetic information used directly in protein synthesis is (choose the correct answer):
a. DNA
b. mRNA
c. rRNA
d. Ribosomes

12.2 The process in which ribosomes engage is (choose the correct answer):
a. replication
b. transcription
c. translation
d. disjunction
e. cell division

12.3 What are the characteristics of the genetic code?

12.4 Base-pairing wobble occurs in the interaction between the anticodon of the tRNAs and the codons. On the theoretical level, determine the minimum number of tRNAs needed to read the 61 sense codons.

12.5 Antibiotics have been very useful in elucidating the steps of protein synthesis. If you have an artificial messenger of the sequence of AUGUUUUUUUUUUUUU . . . , it will produce the following polypeptide in a cell-free, protein-synthesizing system: fMet-Phe-Phe-Phe. . . . In your search for new antibiotics you find one called putyermycin, which blocks protein synthesis. When you try it with your artificial mRNA in a cell-free

system, the product is fMet-Phe. What step in protein synthesis does putyermycin affect? Why?

12.6 Describe the reactions involved in the aminoacylation (charging) of a tRNA molecule.

12.7 Compare and contrast the following in prokaryotes and eukaryotes: (a) protein synthesis initiation; (b) protein synthesis elongation; (c) protein synthesis termination.

12.8 Discuss the two species of methionine tRNA, and describe how they differ in structure and function. In your answer include a discussion of how each of these tRNAs binds to the ribosome.

*12.9 Random copolymers were used in some of the experiments that revealed the characteristics of the genetic code. For each of the following ribonucleotide mixtures, give the expected codons and their frequencies, and give the expected proportions of the amino acids that would be found in a polypeptide directed by the copolymer in a cell-free, protein-synthesizing system:
a. 4G:1C
b. 1A:3U:1C

*12.10 Other features of the reading of mRNA into proteins being the same as they are now (i.e., codons must exist for 20 different amino acids), what would the minimum WORD (CODON) SIZE be if the number of different bases in the mRNA were, instead of four:
a. two

b. three

c. five

12.11 Suppose that at stage A in the evolution of the genetic code only the first two nucleotides in the coding triplets led to unique differences, and that any nucleotide could occupy the third position. Then, suppose there was a stage B in which differences in meaning arose depending on whether a purine (A or G) or pyrimidine (C or T) was present at the third position. Without reference to the number of amino acids or multiplicity of tRNA molecules, how many triplets of different meaning can be constructed out of the code at stage A? At stage B?

*12.12 A gene makes a polypeptide 30 amino acids long containing an alternating sequence of phenylalanine and tyrosine. What are the sequences of nucleotides corresponding to this sequence in the following:

a. The DNA strand which is read to produce the mRNA, assuming Phe = UUU and Tyr = UAU in mRNA

b. The DNA strand which is not read

c. tRNA

12.13 A segment of a polypeptide chain is Arg-Gly-Ser-Phe-Val-Asp-Arg. It is encoded by the following segment of DNA:

Which strand is the template strand? Label each strand with its correct polarity (5′ and 3′).

*12.14 Two populations of RNAs are made by the random combination of nucleotides. In population A the RNAs contain only A and G nucleotides (3A:1G), while in population B the RNAs contain only A and U nucleotides (3A:1U). In what ways *other than amino acid content* will the proteins produced by translating the population A RNAs differ from those produced by translating the population B RNAs?

12.15 In *E. coli* a particular tRNA normally has the anticodon 5′-GGG-3′, but because of a mutation in the tRNA gene, the mutant tRNA has the anticodon 5′-GGA-3′.

a. What amino acid would this tRNA carry?

b. What codon would the normal tRNA recognize?

c. What codon would the mutant tRNA recognize?

d. What would be the effect of the mutation on the proteins in the cell?

*12.16 A particular protein found in *E. coli* normally has the N-terminal sequence Met-Val-Ser-Ser-Pro-Met-Gly-Ala-Ala-Met-Ser. . . . In a particular cell a mutation alters the anticodon of a particular tRNA from 5′-GAU-3′ to 5′-CAU-3′. What would be the N-terminal amino acid sequence of this protein in the mutant cell? Explain your reasoning.

12.17 The gene encoding an *E. coli* tRNA containing the anticodon 5′-GUA-3′ mutates so that the anticodon now is 5′-UUA-3′. What will be the effect of this mutation? Explain your reasoning.

*12.18 The normal sequence of the coding region of a particular mRNA is shown in the figure, along with several mutant versions of the same mRNA. Indicate what protein would be formed in each case. (. . . equals many [a multiple of 3] unspecified bases.)

Normal: AUGUUCUCUAAUUAC(. . .)AUGGGGUGGGUGUAG
Mutant *a*: AUGUUCUCUAAUUAG(. . .)UGGGGUGGGUGUAG
Mutant *b*: AGGUUCUCUAAUUAC(. . .)AUGGGGUGGGUGUAG
Mutant *c*: AUGUUCUCGAAUUAC(. . .)AUGGGGUGGGUGUAG
Mutant *d*: AUGUUCUCUAAAUAC(. . .)AUGGGGUGGGUGUAG
Mutant *e*: AUGUUCUCUAAUUUC(. . .)AUGGGGUGGGUGUAG
Mutant *f*: AUGUUCUCUAAUUAC(. . .)AUGGGGUGGGUGUGG

12.19 The normal sequence of a particular protein is given below, along with several mutant versions of it. For each mutant, explain what mutation occurred in the coding sequence of the gene.

Normal: Met-Gly-Glu-Thr-Lys-Val-Val-. . .- Pro
Mutant 1: Met-Gly
Mutant 2: Met-Gly-Glu-Asp
Mutant 3: Met-Gly-Arg-Leu-Lys
Mutant 4: Met-Arg-Glu-Thr-Lys-Val-Val-. . .-Pro

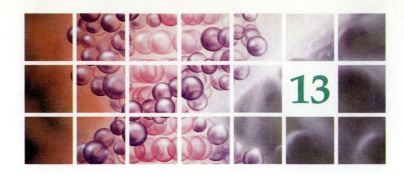

13

Gene Cloning and Recombinant DNA Technology

- Genes are cloned by splicing DNA from an organism into a cloning vector to make a recombinant DNA molecule, and then introducing that molecule into a host cell in which it will replicate. Essential to cloning are restriction enzymes (restriction endonucleases). These enzymes are useful for cloning because they recognize specific nucleotide pair sequences in DNA (restriction sites) and cleave at a specific point within the sequence.

- Different kinds of cloning vectors (DNA molecules capable of replication in a host organism) have been developed to construct and clone recombinant DNA molecules. All cloning vectors must: (1) replicate within their host organism; (2) have one or more restriction sites into which foreign DNA can be inserted; and (3) have one or more dominant selectable markers to detect those cells which contain the vectors. The main classes of vectors are plasmids, bacteriophages, and cosmids. Shuttle vectors have the above properties but, in addition, are able to replicate in more than one type of host.

- Genomic libraries are collections of clones that contain one or more copies of every DNA sequence in an organism's genome. Genomic libraries are useful for isolating specific genes and for studying the organization of the genome.

- Individual chromosomes can be isolated and chromosome-specific libraries made from them. If a gene has been localized to a specific chromosome by genetic means, the existence of chromosome-specific libraries makes it easier to isolate a clone of the gene.

- DNA copies, called complementary DNAs (cDNAs), can be made of mRNA molecules isolated from the cell. The cDNAs can be cloned using appropriate cloning vectors.

- Specific sequences in genomic libraries and cDNA libraries can be identified using a number of approaches, including the use of specific DNA or cDNA probes, heterologous probes, complementation or specific antibodies.

- Genes and cloned DNA sequences can be analyzed to determine the arrangement and specific locations of restriction sites, a process called restriction mapping. Gene transcripts can be analyzed qualitatively and quantitatively using recombinant DNA procedures.

- Rapid methods have been developed for determining the sequence of a cloned piece of DNA. One method, the Maxam-Gilbert procedure, uses specific chemicals to modify and cleave the DNA chain at specific nucleotides. The other method, the dideoxy or Sanger procedure, uses enzymatic synthesis of a new DNA chain on a template DNA strand. With this procedure, synthesis of new strands is stopped by the incorporation of a dideoxy analog of the normal deoxyribonucleotide. Using four different dideoxy analogs, the new strands stop at all possible nucleotide positions, thereby allowing the complete DNA sequence to be determined.

- The polymerase chain reaction (PCR) uses synthetic oligonucleotides to amplify a specific segment of DNA many thousandfold in an automated procedure. A major benefit of PCR is that small amounts of DNA are needed. Thus DNA from a single cell, or a single intact DNA molecule, can be amplified using PCR. PCR is finding increasing applications both in research and in the commercial arena, including cloning, sequencing, amplifying DNA to detect the presence of specific genetic defects, and forensics.

- There are many applications for recombinant DNA technology and related procedures. For example, with appropriate probes, detection of specific genetic diseases is possible, and many products in the clinical, veterinary, and agricultural areas have been developed. A project is underway to determine the complete sequence of the human genome. The knowledge obtained will expand greatly our understanding of human genetics. With continued advances in genetic engineering of plants, it is expected that many types of improved crops (increased yields, disease resistance) will be forthcoming.

EXPERIMENTAL PROCEDURES have been developed that enable researchers to construct **recombinant DNA molecules** in test tubes. In this process genetic material from two different sources is combined into a single DNA molecule. The technology has opened the way for new and exciting research possibilities and affirms the plausibility of **genetic engineering,** the alteration of the genetic constitution of cells or individuals by directed and selective modification, insertion, or deletion of an individual gene or genes. In this chapter we discuss gene cloning and discuss the manipulation of DNA using recombinant DNA techniques. We also present some examples of how **recombinant DNA technology** is furthering our knowledge of the structure and function of the prokaryotic and eukaryotic genomes, how cloned DNA sequences are being used for genetic diagnosis, and the ability to manipulate DNA sequences has opened new ways for producing commercial products.

Gene Cloning

In brief, genes are cloned by taking a piece of DNA from an organism and splicing it into a **cloning vector** to make a recombinant DNA molecule. A cloning vector is an artificially constructed DNA molecule capable of replication in a host organism, such as a bacterium, and into which a piece of DNA to be studied can be specifically inserted at known positions. The recombinant DNA molecule is introduced into a host such as *E. coli*, yeast, animal cell, or plant cell. Reproduction of the host cell results in the replication of the recombinant DNA molecule (**molecular cloning**), thereby producing many identical copies. This is useful for many reasons. For example, suppose we want to study the gene for a particular human protein in order to determine its DNA sequence, and how its expression is regulated. Each human cell contains only two copies of that gene, making it an almost impossible task to isolate enough copies of the gene for analysis. By contrast, an essentially unlimited number of copies of the gene can be produced by cloning. Similarly, if we can clone a gene selectively, we can design experiments for manipulating the gene (e.g., changing its DNA sequence), or for the synthesis of large amounts of the gene's products. For example, human insulin (called *humulin*), produced from a recombinant DNA molecule, is now substituted for insulin isolated from pig pancreas in the treatment

of diabetes. In this section we will describe how DNA sequences can be cloned.

Restriction Enzymes

Recombinant DNA molecules could not be constructed without the use of **restriction enzymes** (or **restriction endonucleases**). The important feature of a restriction enzyme is that it is able to cleave double-stranded DNA molecules at a specific nucleotide pair sequence called a *restriction site*. Restriction enzymes are used to produce a pool of cut DNA molecules to be cloned. Restriction enzymes are also used to analyze the positioning of restriction sites in a piece of cloned DNA or in a segment of DNA in the genome (see later in this chapter, pp. 298–299).

Restriction enzymes are found naturally only in bacteria. Their natural function in bacteria is to protect the organism against invading viruses. That is, the bacterium modifies its own restriction sites so that the restriction enzyme it makes cannot cut the DNA. Then, when a virus injects its DNA, the restriction enzyme is able to cleave the viral DNA without affecting the bacterial DNA, thereby protecting the bacterium against the viral infection.

Over 150 different restriction enzymes have been isolated. Restriction enzymes are named for the bacterium from which they are isolated. A three-letter system is used, italicized or underlined, followed by a roman numeral. Additional letters are sometimes added to signify a particular strain of the bacterial species from which the enzymes were obtained. For example, *Bgl*II is from *Bacillus globigi*; *Eco*RI is from *E. coli* strain RY13, and *Hin*dIII is from *Haemophilus influenzae* strain R_d. The names are pronounced in particular ways but those pronunciations follow no set pattern. For instance, *Bam*HI is "bam-H-one," *Bgl*II is "bagel-two," *Eco*RI is "echo-R-one," *Hin*dIII is "hin-D-three," *Hha*I is "ha-ha-one," and *Hpa*II is "hepa-two."

The restriction enzymes most commonly used in recombinant DNA experiments recognize a specific nucleotide pair sequence in DNA and cleave the DNA within that sequence. In each case the restriction enzyme makes its cut by cleaving the DNA backbone between the oxygen attached to the 3' carbon of the deoxyribose sugar and the phosphate on the subsequent deoxyribose in the chain. Thus all DNA fragments produced by cutting with a restriction enzyme have a phosphate on their 5' ends and a hydroxyl group on their 3' ends. Since all DNA fragments generated by cleavage with a particular restriction enzyme have been cut at the same sequence, these enzymes are very valuable for constructing recombinant DNA molecules, as we will see.

The restriction sites for most restriction enzymes are symmetrical: the base sequence from 5′ to 3′ on one DNA strand is the same as the base sequence from 5′ to 3′ on the complementary DNA strand. For example, Figure 13.1 shows the restriction site for the enzyme *Eco*RI; that is, $\frac{5'\text{-GAATTC-}3'}{3'\text{-CTTAAG-}5'}$. These symmetrical sequences are also called *palindromes*, as are words and phrases in the English language that read the same backwards as forwards (e.g., "rotator" or "nurses run").

Since each of the restriction enzymes cuts DNA at a restriction site specific for the enzyme, the number of cuts the enzyme makes in a particular DNA molecule depends on the number of times the particular restriction site is present in the DNA. In some cases the same restriction site is recognized and cleaved by restriction enzymes called *isoschizomers* (iso = same; schizo = to split) that are isolated from different bacteria.

The cleavage sites for a number of restriction enzymes are shown in Table 13.1. Some restriction enzymes recognize and cut within restriction sites consisting of four nucleotide pairs (e.g., *Hae*III ["hay-three"], *Hha*I) or of six nucleotide pairs (e.g., *Bam*HI, *Eco*RI). Some other enzymes cut within a five-nucleotide-pair restriction site (e.g., *Rsa*I ["rosa-one"]), while a very few enzymes cut within eight-nucleotide-pair restriction sites (e.g., *Not*I ["not-one"]).

Figure 13.1 Restriction enzyme recognition sequence in DNA, showing symmetry of the sequence about the center point. The sequence is a palindrome, reading the same from left to right (5′-to-3′) on the top strand (GAATTC), as it does from right to left (5′-to-3′) on the bottom strand. Shown here is the recognition sequence for the restriction enzyme *Eco*RI.

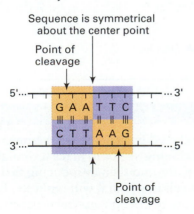

Sequence is symmetrical about the center point

Point of cleavage

5′ ··· G A A T T C ··· 3′
3′ ··· C T T A A G ··· 5′

Point of cleavage

For DNA with a random distribution of nucleotide pairs (rare among organisms), there is a clear relationship between the number of nucleotide pairs in the restriction site and the frequency of cutting the DNA. That is, an enzyme that recognizes a four-nucleotide-pair sequence will cut more frequently than one that recognizes a five-nucleotide-pair sequence, and so on. More specifically, an enzyme that cuts a four-nucleotide-pair restriction site will cut every 256 bp on the average, and an enzyme that cuts a six-nucleotide-pair restriction site will cut every 4096 bp on the average.*

Note, though, that DNA from living organisms does not have a random distribution of base pairs. Rather, some DNA is GC-rich and other DNA is AT-rich. The latter, for example, would have few restriction sites for *Sma*I ("sma-one"), which recognizes $\frac{5'\text{-CCCGGG-}3'}{3'\text{-GGGCCC-}5'}$, so this enzyme would cut AT-rich DNA less frequently than the average value of once every 4096 bp.

As Table 13.1 indicates, some enzymes, such as *Hae*III or *Sma*I, cut both strands of DNA between the same two nucleotide pairs to produce *blunt ends* (Figure 13.2[a]), while others, such as *Eco*RI, *Bam*HI, and *Hind*III, make staggered cuts in the symmetrical nucleotide pair sequence to produce *sticky* or *staggered ends*. The staggered ends may either have an overhanging 5′ end, as in the case of cleavage with *Bam*HI or *Eco*RI (Figure 13.2[b]), or an overhanging 3′ end, as in the case of cleavage with *Pst*I ("P-S-T-one") (Figure 13.2[c]).

Restriction enzymes that produce staggered ends are of particular value in cloning DNA fragments because every DNA fragment generated by cutting a piece of DNA with a particular restriction enzyme (such as *Eco*RI) has the same base sequence at the two staggered ends (Figure 13.3). Therefore, the single-stranded sticky ends of two such pieces of DNA can pair by complementary base pairing; the two DNAs are said to *anneal* (Figure 13.3). DNA ligase can then be used to seal the gaps in the sugar-phosphate backbones to produce a circular DNA molecule consisting of the two original DNA molecules joined together. At each of the two junctions of the two molecules is a reconstituted restriction

*For example, for DNA with a random distribution of nucleotide pairs, there is an equal chance of finding one of the four possible nucleotide pairs $\frac{G}{C}, \frac{C}{G}, \frac{A}{T},$ or $\frac{T}{A}$ at any one position. Therefore, the probability of a particular six-nucleotide-pair restriction site sequence occurring in the DNA is found by multiplying the probabilities of the nucleotide pairs found at each position in the sequence; i.e., $1/4 \times 1/4 \times 1/4 \times 1/4 \times 1/4 \times 1/4 = 1/4096$. Thus a restriction site for an enzyme which recognizes a six-nucleotide-pair sequence will occur on the average every 4096 nucleotide pairs along DNA which has a random distribution of nucleotide pairs.

Table 13.1

Characteristics of Some Restriction Enzymes

	ENZYME NAME	PRONUNCIATION	ORGANISM IN WHICH ENZYME IS FOUND	RECOGNITION SEQUENCE AND POSITION OF CUT[A]
Enzymes with 6-bp Recognition Sequences	*Bam*HI	"bam-H-one"	*Bacillus amyloliquefaciens H*	↓ 5'GGATCC3' 3'CCTAGG5' ↑
	*Eco*RI	"echo-R-one"	*E. coli RY13*	↓ GAATTC CTTAAG ↑
	*Hind*III	"hin-D-three"	*Haemophilus influenzae* R_d	↓ AAGCTT TTCGAA ↑
	*Pst*I	"P-S-T-one"	*Providencia stuartii*	↓ CTGCAG GACGTC ↑
	*Sma*I	"sma-one"	*Serratia marcescens*	↓ CCCGGG GGGCCC ↑
Enzymes with 4-bp Recognition Sequences	*Hae*III	"hay-three"	*Haemophilus egyptius*	↓ GGCC CCGG ↑
	*Hpa*II	"hepa-two"	*Haemophilus parainfluenzae*	↓ CCGG GGCC ↑
	*Sau*3A	"sow-three-A"	*Staphylococcus aureus 3A*	↓ GATC CTAG ↑
Enzymes with 8-bp Recognition Sequences	*Not*I	"not-one"	*Nocardia otitidis-caviarum*	↓ GCGGCCGC CGCCGGCG ↑

[a]In this column the two strands of DNa are shown with the sites of cleavage indicated by arrows. Since the sequence is symmetrical about a center point in each recognition sequence, the DNA molecules resulting from the cleavage are symmetrical.

Key: R = purine; Y = pyrimidine.

Adapted from R. J. Roberts, 1976. *CRC Critical Reviews in Biochemistry* 4:123. Copyright © 1976 CRC Press, Inc., Boca Raton, Fl. Reprinted with permission.

---

enzyme site. Even DNA fragments with blunt ends can be ligated together with DNA ligase at high concentrations of the enzyme. This ligation of two DNA fragments with identical sticky ends or of two DNA fragments with blunt ends by DNA ligase is the principle behind the formation of recombinant DNA molecules, as we will see in the next section.

Keynote

Genes are cloned by splicing DNA from an organism into a cloning vector to make a recombinant DNA molecule, and then introducing that molecule into a host cell in which it will replicate. Essential to cloning are restriction enzymes, or restriction endonucleases. These enzymes are useful for cloning because they recognize specific nucleotide pair sequences in DNA (restriction sites) and cleave at a specific point within the sequence. Cleavage of the DNA with a restriction enzyme can be staggered, producing DNA fragments with single-stranded, sticky ends, or blunt.

■ *Figure 13.2* Examples of how restriction enzymes cleave DNA: (a) *Sma*I results in blunt ends; (b) *Bam*HI results in overhanging (sticky) 5' ends; (c) *Pst*I results in overhanging (sticky) 3' ends.

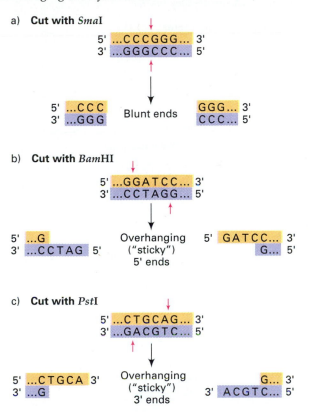

Cloning Vectors and the Cloning of DNA

Three major types of vectors are commonly used for cloning DNA sequences in bacterial cells: plasmids, bacteriophages, and cosmids. Each differs in the way it must be manipulated to clone pieces of DNA and in the maximum amount of DNA that can be cloned using it. Each type of vector has been specially constructed in the laboratory to possess the features necessary to make it an efficient cloning vector.

Plasmid Cloning Vectors. Plasmids (see Chapter 6 and Chapter 18, p. 428) are extrachromosomal genetic elements that replicate autonomously within bacterial cells. Their DNA is circular and double-stranded. The particular plasmids used for cloning experiments are all derivatives of naturally occurring plasmids that have been "engineered" to put together DNA sequences to give the plasmids properties that are useful for gene cloning. Because they are most commonly used, we focus here on features of *E. coli* plasmid cloning vectors.

An *E. coli* plasmid cloning vector must have three features:

1. An *ori* sequence, which allows the plasmid to replicate in *E. coli* because it provides a DNA

■ *Figure 13.3* Cleavage of DNA by the restriction enzyme *Eco*RI. *Eco*RI makes staggered, symmetrical cuts in DNA, leaving sticky ends. A DNA fragment with a sticky end produced by *Eco*RI digestion can bind by complementary base pairing (anneal) to any other DNA fragment with a sticky end produced by *Eco*RI cleavage. The gaps may then be sealed by DNA ligase.

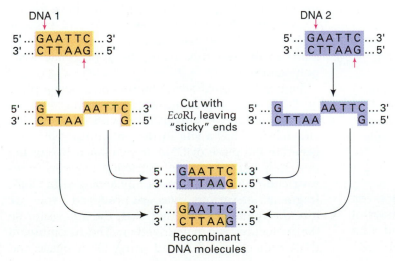

sequence that is recognized by the replication enzymes in the cell.

2. A *selectable marker*, which enables *E. coli* cells that carry the plasmid to be distinguished easily from cells that lack the plasmid. A common selectable marker is the *amp*^R gene for ampicillin resistance. When plasmids carrying the *amp*^R gene are introduced by transformation (see Chapter 6) into a plasmid-free and therefore ampicillin-sensitive (*amp*^s) *E. coli* cell, the bacterium will become ampicillin resistant and be detected easily since it will grow in medium to which ampicillin has been added.

3. *Unique restriction enzyme cleavage sites* for the insertion of the DNA sequences that are to be cloned. Cloning involves cutting the plasmid at one of the unique sites with the appropriate restriction enzyme, and splicing into that site a piece of DNA that has been cut with the same enzyme.

As an example, Figure 13.4 diagrams the plasmid vector pUC19 ("puck-19"). This 2686 bp vector has the following features that make it useful for cloning DNA:

1. It replicates efficiently in the cell, producing up to 500 copies per cell.

2. It carries the *amp*^R selectable marker.

3. It contains a number of unique restriction sites engineered to cluster in one region. Such a region of clustered unique restriction sites useful for inserting DNA is called a **polylinker** or **multiple cloning site** (MCS).

4. The polylinker is located in the *E. coli lacZ*^+ gene, which encodes the enzyme β-galactosidase. When pUC19 is introduced into an *E. coli* cell which carries a *lacZ*^- mutation, functional β-galactosidase is produced from the plasmid. However, when DNA is inserted into the polylinker, the β-galactosidase sequence is disrupted so functional β-galactosidase cannot be produced in *E. coli*. A simple color test is used to distinguish colonies of *E. coli* containing pUC19 with no inserted DNA (blue) from colonies containing pUC19 with inserted DNA (white). This test provides a rapid, visual way to identify cells containing potentially interesting recombinant DNA molecules.

Similar vectors are available with variants on these features, for example with different arrays of unique restriction sites in the polylinker, and with phage promoters flanking the polylinker-*lacZ*^+

■ *Figure 13.4* Restriction map of the plasmid cloning vector pUC19. This plasmid has an origin of replication (*ori*), an *amp*^R selectable marker, and a polylinker located within the β-galactosidase gene, *lacZ*^+. In a bacterial cell pUC19 without a DNA insert can make β-galactosidase, while pUC19 with a DNA insert cannot make β-galactosidase. This makes it possible to use a color selection system for distinguishing vectors with and without DNA inserts (see text).

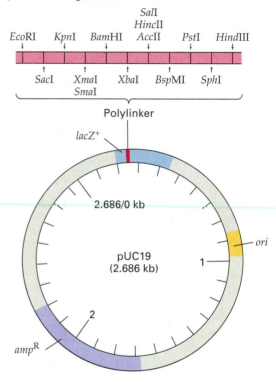

pUC19 cloning vector

ori = Origin of replication sequence
amp^R = Ampicillin resistance gene
lacZ^+ = β-galactosidase gene

region. The phage promoters, such as those for T7, T3, or SP6 RNA polymerases, are useful because RNA copies of the inserted DNA sequences can be made in the test tube using the appropriate RNA polymerase.

Figure 13.5 illustrates how we can insert a piece of DNA into a plasmid cloning vector such as pUC19. First, we cut pUC19 with a restriction enzyme that has a site in the polylinker. Next, we generate the piece of DNA to be cloned by cutting the DNA with the same restriction enzyme. Since restriction sites are nonrandomly arranged in DNA, fragments of various sizes are produced. Now, we mix the fragments together with the cut vector so the molecules can pair together. The recombinant DNA molecules are sealed using DNA ligase and

■ *Figure 13.5*   Insertion of a piece of DNA into the plasmid cloning vector pUC19 to produce a recombinant DNA molecule. pUC19 contains several unique restriction enzyme sites localized in a polylinker. Insertion of a DNA fragment into the multiple cloning site disrupts the β-galactosidase gene, making it possible to distinguish colonies containing pUC19 that lack an insert (blue colonies) from colonies containing pUC19 that have an insert (white colonies).

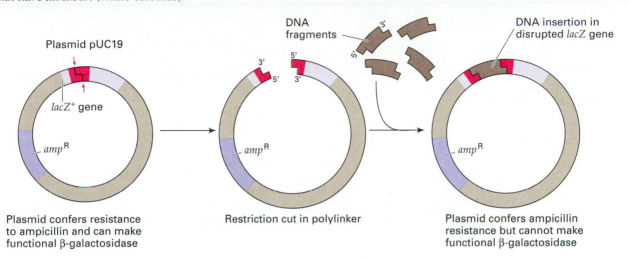

Plasmid confers resistance to ampicillin and can make functional β-galactosidase

Restriction cut in polylinker

Plasmid confers ampicillin resistance but cannot make functional β-galactosidase

introduced into *E. coli* by transformation. Resulting ampicillin-resistant colonies indicate cells that were transformed by plasmids, and white colonies (versus blue) on color selection plates indicate the recombinant DNA clones (versus plasmids that were resealed without including a DNA fragment).

Plasmid vectors have been developed to introduce recombinant DNA molecules into organisms other than *E. coli*. For example, there are vectors that can be used to transform mammalian cells in culture, vectors to transform other animal cells, and vectors to transform plant cells. Other plasmid vectors—**shuttle vectors**—can introduce the same recombinant DNA into two or more host organisms. Shuttle vectors have sequences (e.g., *ori* for *E. coli*) and selectable markers to permit replication and selection in each host organism, as well as multiple cloning sites. An *E. coli*-yeast shuttle vector makes it possible to produce large quantities of DNA in *E. coli* for restriction mapping purposes or for DNA sequencing, and to introduce the DNA into yeast to study or for manipulating the yeast genome.

Phage Cloning Vectors. Commonly used phage cloning vectors are derivatives of bacteriophage λ (see Chapter 6) which have been engineered so that the lytic cycle is possible, but lysogeny is not possible. The λ cloning vectors possess restriction enzyme sites that allow them to be used to clone DNA fragments (Figure 13.6). Such vectors have chromosomes in which there is a "left" arm and a "right" arm that collectively contain all the essential

genes for the lytic cycle. Between the two arms is a segment of DNA that is "disposable" since it does not contain any genes needed for phage propagation. The junctions between the disposable central segment and the two arms each have unique cleavage sites for one or more restriction enzymes; here, *Eco*RI (Figure 13.6).

The following steps are used to clone a DNA fragment in a λ vector: (1) We cut DNA with a restriction enzyme appropriate for the vector (*Eco*RI in Figure 13.6). This produces restriction fragments of different sizes; (2) We cut the vector with the same restriction enzyme and then separate the left and right arms from the "disposable" segment; (3) We mix the cut DNA fragments with the λ arms and ligate the pieces together with DNA ligase; and (4) We package the recombinant DNA molecules into λ particles which then inject the recombinant DNA molecules into *E. coli*.

The only splicing events that produce λ chromosomes capable of replication in *E. coli* are those in which the foreign DNA is inserted between the left arm and right arm and in which the total length of the recombinant DNA molecule is approximately 45 to 50 kb. The latter is the size of DNA required for packaging into lambda particles. Therefore, only restriction fragments of about 15 kb can be inserted into a λ cloning vector.

The result of replication in *E. coli* is the production of 100 to 200 phage particles per cell. The phages lyse the cell and the cloned recombinant DNA molecules may be extracted from the released progeny phage particles.

■ *Figure 13.6* Scheme for using phage λ DNA as a cloning vector. DNA fragments of approximately 15 kb in length can be cloned in λ vectors.

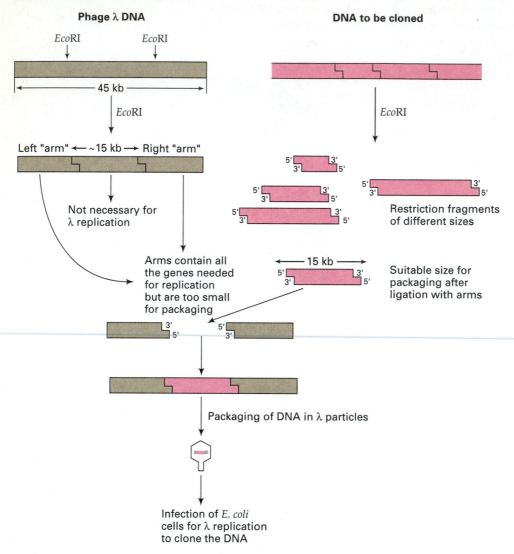

Cosmid Cloning Vectors.

Cosmid cloning vectors allow DNA fragments of about 40 to 45 kb to be cloned; these fragments are much larger than those that can be cloned in plasmid and λ vectors. Cosmids have features of both plasmid and λ vectors. They have a plasmid *ori* sequence for replication in *E. coli*, a selectable marker such as *amp*ᴿ, and unique restriction sites for the insertion and cloning of DNA fragments. In addition, cosmids have a *cos* site that is derived from phage λ. Recall from Chapter 9 (p. 000) that the *cos* site is the site at which multiple, tandem copies of the λ genome are cleaved into 34-kb to 45-kb pieces for packaging into phage particles. Recall also that the cut at the *cos* site is staggered and results in a linear DNA molecule with sticky (single-stranded), complementary ends. In fact, all that is needed for packaging DNA into a phage particle are *cos* sites that are a specified dis-

tance apart. Thus the *cos* site on a cosmid permits packaging of DNA into a λ phage particle. Since λ can hold about 50 kb of DNA, a 5 kb cosmid vector permits a 45-kb DNA fragment to be packaged into a phage particle. The phage then injects the recombinant cosmid into the *E. coli* host cell, where it replicates like a plasmid.

K e y n o t e

Different kinds of vectors are used to construct and clone recombinant DNA molecules. All vectors must (1) be able to replicate within their host organism; (2) have one or more restriction enzyme cleavage sites at which foreign DNA fragments can be inserted; and (3) have one or more dominant selectable markers. Three major types of cloning vectors are used to clone recombinant DNA: plasmids, which are introduced by transformation into *E. coli* or other hosts, where they replicate; and bacteriophage λ and cosmids, both of which are packaged into λ phage particles, which, in turn, inject the DNA into the *E. coli* host. In λ clones the injected λ DNA replicates and progeny phages are produced, whereas in cosmid clones the cosmids replicate like plasmids.

Construction of Recombinant DNA Libraries

Researchers are interested in cloning individual genes so that those genes can be studied in detail. It is relatively easy to isolate DNA from cells of an organism and to cleave that DNA into specific fragments with restriction enzymes. The restriction fragments collectively represent the entire genome of the organism, and they may be cloned in a cloning vector. The collection of clones that contains at least one copy of every DNA sequence in the genome is called a **genomic library.** Clearly, a genomic library is an extremely valuable repository of genetic information. A genomic library can be searched to identify and single out for study a recombinant DNA molecule that contains a particular gene or DNA sequence of interest. This will be described a little later in the chapter. Genomic libraries have been made for many organisms of genetic interest, including humans (see discussion on p. 312.)

The genome size of many eukaryotic organisms, including humans, is so large that many thousands of clones are needed to represent the entire genome. For example, cloning the 3×10^9-bp human genome into cosmids, which can accommodate 45,000 bp of DNA, would give 66,667 different clones. This makes searching a genomic library for a gene of interest very time consuming. One approach that has been used to reduce the screening time of large genomes is to make *chromosome-specific libraries*, libraries of the individual chromosomes in the genome. In humans this gives 24 different libraries (one for each of the 22 autosomes, and one each for the X and Y). Then, if a gene has been localized to a chromosome by genetic means, researchers can restrict their attention to the library of that chromosome when they search for its DNA sequence. Individual chromosomes of an organism can be separated by morphology, size, or unique DNA sequence, as is the case for human chromosomes. The most efficient procedure currently used to isolate chromosomes individually is *flow cytometry*. In this procedure chromosomes from cells in the mitotic phase of the cell cycle are stained with a fluorescent dye. Chromosomes released from cells are passed through a laser beam connected to a light detector. This system sorts and fractionates the chromosomes based on their differences in dye binding and resulting light scattering.

Keynote

A genomic library is a collection of clones that contains at least one copy of every DNA sequence in an organism's genome. Like regular book libraries, genomic libraries are great resources of information; in this case the information is about the genome. Genomic libraries are used for isolating specific genes and for studying the organization of the genome, among many other things. In organisms where the chromosomes are different enough in size and morphology or DNA sequence, the individual chromosomes can be isolated and chromosome-specific libraries made from them. If a gene has been localized to a specific chromosome by genetic means, the existence of chromosome-specific libraries makes it easier to isolate a clone of the gene.

In a different approach DNA copies, called **complementary DNA (cDNA)**, can be made from mRNA molecules isolated from cells (as described below). These cDNA molecules can then be cloned. Thus if a specific mRNA molecule can be isolated, the corresponding cDNA can be made and cloned. The analysis of that cloned cDNA molecule can then provide information about the gene that encoded the mRNA. More typically, the entire mRNA population of a cell is isolated and a corresponding set of cDNA molecules is made and inserted into a cloning vector to produce a **cDNA library.** Since a cDNA library reflects the gene activity of the cell type at the time the mRNAs are isolated, the construction and analysis of cDNA libraries is useful for comparing gene activities in different cell types of the same organism, because there would be similarities and differences in the clones represented in the cDNA libraries of each cell type.

Recognize that the clones in the cDNA library represent the *mature mRNAs* found in the cell. In eukaryotes mature mRNAs are processed molecules, so the sequences obtained are *not* equivalent to gene clones. In particular, intron sequences are typically present in gene clones but *not* in cDNA clones. For any mRNA, cDNA clones can be useful for subsequently isolating the gene that codes for

■ *Figure 13.7* The synthesis of double-stranded complementary DNA (cDNA) from a polyadenylated mRNA, using reverse transcriptase RNase H, DNA polymerase I, and DNA ligase.

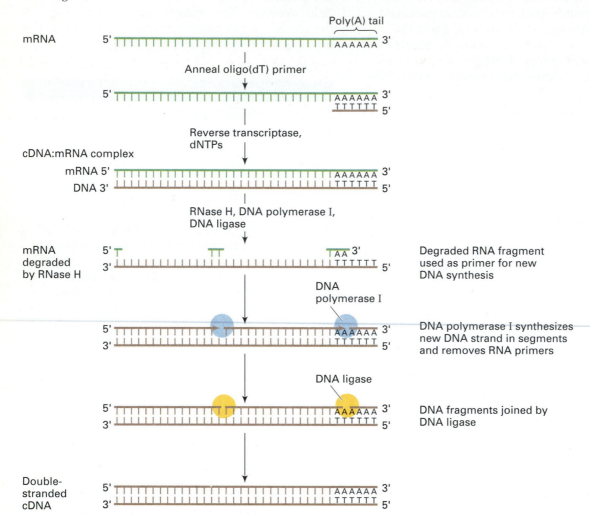

that mRNA. The gene clone can provide more information than can the cDNA clone, for example, on the presence and arrangement of introns, and on the regulatory sequences associated with the gene.

cDNA libraries are mostly made from eukaryotic mRNAs. That is, uniquely among the RNAs found in eukaryotes, only mRNAs contain a poly(A) tail (see Chapter 11, p. 243–245). These poly(A)⁺ mRNAs can be purified from a mixture of RNAs in a cell extract by passing the RNA molecules over a column to which short chains of deoxythymidylic acid (called *oligo(dT) chains*) have been attached. As the RNA molecules pass through an oligo(dT) column, the poly(A) tails on the mRNA molecules form complementary base pairs with the oligo(dT) chains. As a result the mRNAs are captured on the column while the other RNAs pass through. The captured mRNAs are subsequently released by heating and collected from the bottom of the column.

Once a population of mRNA molecules has been isolated, double-stranded complementary DNA (cDNA) copies are made in the test tube, using the enzyme reverse transcriptase (Figure 13.7).* First, a short oligo(dT) primer (a short, single-stranded DNA chain containing only T nucleotides) is hybridized to the poly (A) tail at the 3' end of each mRNA strand. The oligo(dT) acts as a primer for reverse transcriptase (RNA-dependent DNA polymerase), which makes a complementary DNA copy of the mRNA strand. Next, RNase H, DNA polymerase I, and DNA ligase are used to synthesize the second DNA strand. RNase H degrades the RNA

*The enzyme's name comes from the fact that it catalyzes a reaction that is the reverse of transcription: RNA → DNA. In nature, reverse transcriptase is found in so-called retroviruses that replicate by making a DNA copy of their RNA genome that integrates into the host cell's genome from which new RNA genomes are transcribed. Human Immunodeficiency Virus-1 [HIV-1], the causative agent of Acquired Immunodeficiency Syndrome [AIDS], is an example of such a virus (See Chapter 18).

strand in the hybrid DNA-mRNA; DNA polymerase I makes new DNA fragments using the partially degraded RNA fragments as primers; and finally DNA ligase ligates the new DNA fragments together to make a complete chain. The result is a double-stranded cDNA molecule, the sequence of which is derived from the original poly(A)-mRNA molecule.

Once cDNA molecules have been synthesized, they must be cloned. Figure 13.8 illustrates the cloning of cDNA using a **restriction site linker,** or **linker,** which is a relatively short, double-stranded piece of DNA (oligodeoxyribonucleotide) about 8 to 12 nucleotide pairs long. The linker contains a restriction site; for example, the linker shown in Figure 13.8 contains the *Bam*HI restriction site. Both the cDNA molecules and the linkers have blunt ends, and they can be ligated together at high concentrations of T4 DNA ligase. Sticky ends are produced in the cDNA molecule by cleaving the cDNA (with linkers now at each end) with *Bam*HI. The resulting DNA is precipitated, leaving the tiny end fragments in solution. The precipitated DNA is

resuspended and inserted into a cloning vector that has also been cleaved with *Bam*HI, and the recombinant DNA molecule produced is transformed into an *E. coli* host cell for cloning.

K e y n o t e

It is possible to make DNA copies of mRNA molecules. First, the enzyme reverse transcriptase makes a single-stranded DNA copy of the mRNA; then RNase H, DNA polymerase I, and DNA ligase are used to make a double-stranded DNA copy called complementary DNA (cDNA). This cDNA can be spliced into cloning vectors using restriction site linkers.

Identifying Specific Clones in Libraries

Recombinant DNA libraries are large collections of cloned sequences. Scientists are usually interested in studying only one or a few genes, so it is usually necessary to search through a library to "pull out" the gene or genes of interest. In this section some of the ways to identify specific clones in libraries are described.

Using DNA Probes to Identify Specific Clones in a Library

In molecular genetics a **probe** is any molecule (usually labeled radioactively or in a nonradioactive way that permits easy detection) that is used to identify or isolate a gene, gene product, or a protein. For example, DNA or RNA probes are used to identify clones in a library, or to study RNA transcripts produced by a cell. Such DNA or RNA probes are derived from cloned DNA molecules. Given the existence of a probe, such as a labeled cloned cDNA molecule, it is possible to identify in a genomic library the cloned gene that codes for the mRNA molecule from which the cDNA was made, and then to isolate it for characterization.

The process of screening for specific DNA sequences in a plasmid or cosmid library is shown in Figure 13.9. The library consists of a population of recombinant DNA molecules (Figure 13.9[1]). First, *E. coli* cells are transformed with the recombinant DNA clones (Figure 13.9[2]) and the cells are plated onto growth medium containing an antibiotic appropriate for the marker on the vector (the selective medium, Figure 13.9[3]). The resulting colonies are transferred to the wells of microtiter dishes containing selective medium for growth and storage (Figure 13.9[4]). A large number of

■ *Figure 13.8* The cloning of cDNA by using *Bam*HI linkers.

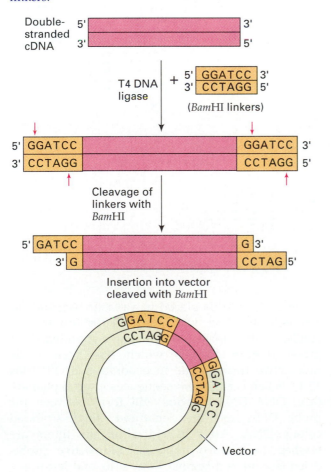

■ *Figure 13.9*   Using DNA probes to screen plasmid and cosmid libraries for specific DNA sequences.

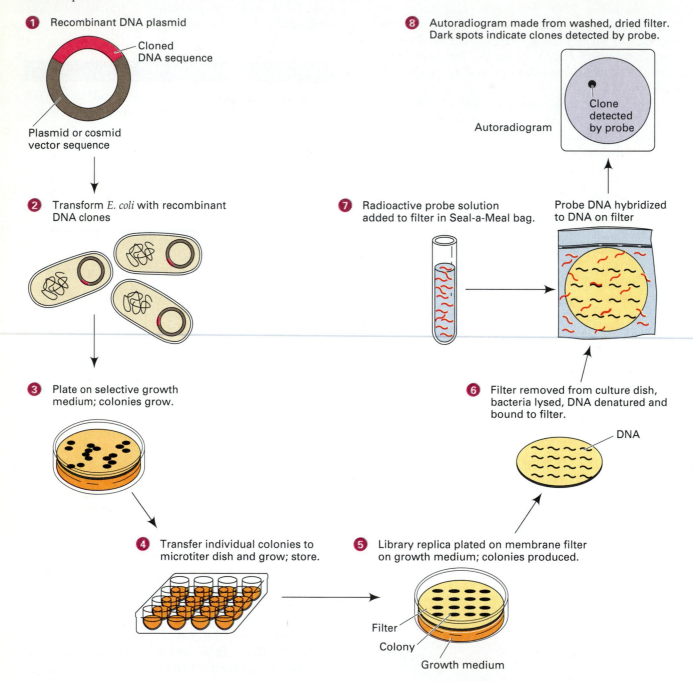

1 Recombinant DNA plasmid

Cloned DNA sequence

Plasmid or cosmid vector sequence

2 Transform *E. coli* with recombinant DNA clones

3 Plate on selective growth medium; colonies grow.

4 Transfer individual colonies to microtiter dish and grow; store.

5 Library replica plated on membrane filter on growth medium; colonies produced.

Filter
Colony
Growth medium

6 Filter removed from culture dish, bacteria lysed, DNA denatured and bound to filter.

DNA

7 Radioactive probe solution added to filter in Seal-a-Meal bag.

Probe DNA hybridized to DNA on filter

8 Autoradiogram made from washed, dried filter. Dark spots indicate clones detected by probe.

Autoradiogram

Clone detected by probe

microtiter dishes is needed to hold a genomic library. To screen the library to find a specific clone, replicas of the clones stored in each microtiter dish are placed (printed) onto a membrane filter that has been placed on a Petri dish of selective growth medium (Figure 13.9[5]). Incubation of the plate produces colonies growing on the filter in the same pattern as the clones in the microtiter dish. The filter is peeled from the dish and treated to lyse the bacteria, denature the released DNA to single strands, and then bind the single-stranded DNA to the filter (Figure 13.9[6]).

Next, the filter is placed in a plastic bag, such as a Seal-A-Meal bag, which is used for boiling vegetables, and incubated with a single-stranded DNA probe (Figure 13.9[7]) which has been made radioactive through the incorporation of ^{32}P (Box 13.1). Wherever the two sequences are complementary, DNA–DNA hybrids will form between the probe DNA and the denatured, single-stranded colony DNA. After hybridization the filters are washed to remove unbound radioactive probe, dried, and placed against X-ray film, and left in the dark for a period of time to produce what is called

BOX 13.1

Radioactive Labeling of DNA

The radioactive labeling of the cDNA probe may be done by the process of **nick translation** (Box Figure 13.1). The DNA to be labeled is incubated in a buffer containing the enzymes DNase I and DNA polymerase I, and the four deoxyribonucleoside triphosphate DNA precursors (dATP, dGTP, dCTP, and dTTP), one or more of which has ^{32}P in its phosphate group. (In Box Figure 13.1 the dCTP is ^{32}P-labeled.) In the reaction mixture the DNase I enzyme makes "nicks," that is, single-stranded gaps in the backbones of the DNA molecules. DNA polymerase I recognizes the nicks and proceeds to extend the DNA chain from the free 3' end, simultaneously removing DNA ahead of it by its 5'-to-3' exonuclease activity. When the dCTP is ^{32}P-labeled, wherever a G nucleotide is on the intact template strand, the new DNA strand will have a ^{32}P-labeled C nucleotide incorporated. In this way the DNA becomes radioactively labeled with ^{32}P. To be used as a probe, the radioactive DNA is denatured to single strands by heating.

■ *Box Figure 13.1*   Radioactive labeling of DNA fragments by nick translation. Here the reaction is shown occurring on one strand; in actuality both strands become labeled by nick translation.

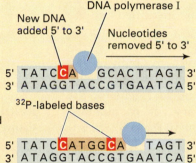

DNase I randomly nicks one strand of DNA

DNase I nick

5' T A T C C A T G G C A C T T A G T 3'
3' A T A G G T A C C G T G A A T C A 5'

New DNA added 5' to 3'

DNA polymerase I

Nucleotides removed 5' to 3'

Polymerizing activity of DNA polymerase I makes new DNA in 5' to 3' direction, starting from nick, simultaneously removing nucleotides ahead of it by 5'-to-3' exonuclease activity. The reaction mixture contains ^{32}P-labeled dCTP and unlabeled dATP, dTTP, and dGTP.

5' T A T C C A   G C A C T T A G T 3'
3' A T A G G T A C C G T G A A T C A 5'

^{32}P-labeled bases

5' T A T C C A T G G C A   T A G T 3'
3' A T A G G T A C C G T G A A T C A 5'

---

an *autoradiogram* (Figure 13.9[8]). The process is called *autoradiography*. When the film is developed dark spots are seen wherever the radioactive probe bound to the target DNA sequence on the filter. (The dark spots are the result of the decay of the radioactive atoms changing the silver grains of the film.) From the position(s) of the spot(s) on the film, the locations of the original bacterial culture(s) in the microtiter dish(es) can be determined and the clone(s) of interest isolated for further study.

A variety of types of probes can be used in screening for specific DNA sequences in libraries. As already mentioned, a cloned cDNA probe can be used. For example, if the DNA probe is a cDNA derived from the mRNA for β-globin, that cDNA probe will hybridize with DNA bound to the filter that encodes the β-globin mRNA thereby identifying a clone containing the β-globin gene. Another common type of probe used is the *heterologous probe*. That is, it is possible to screen libraries made from the DNA of one organism for a particular gene by using a clone of that gene isolated from another organism. Since the gene used as a probe will not have exactly the same DNA sequence as the gene to be cloned, the probe is called a heterologous probe. The effectiveness of such probes depends on a good degree of DNA sequence homology between the probes and the genes. For that reason, the greatest success with this approach has come either with genes that have very similar sequences (genes that are *highly conserved* in the evolutionary sense) or with probes from a species closely related to the organism from which a particular gene is to be isolated. And, reduced hybridization temperature permits more mismatched bases when two sequences anneal.

Identifying Specific cDNA Clones in a Library

A specific cDNA clone can be identified in a cDNA library by selecting the cDNA clone that codes for a specific protein. This approach requires that the cDNAs be cloned in a special kind of plasmid vector called an *expression vector*. In the expression vector the cDNA is placed next to a promoter sequence and to translation start signals. Downstream is a transcription terminator sequence. In the *E. coli* host an mRNA will be transcribed corresponding to the cDNA and the translation of the mRNA will produce the encoded protein.

cDNA libraries are screened in a very similar way to that described for genomic libraries (see Figure 13.9). Replicas of cDNA clones in microtiter dishes are printed onto nitrocellulose filters placed on selective medium; colonies will grow on the filter. Each filter is peeled from the plate and treated to lyse the cells. The proteins within the cell, including those expressed from the cDNA, become stuck to the filter. The filter is then incubated with a radioactive antibody that is specific for the protein of interest. The antibody will bind to the filter wherever the protein it recognizes is located and that site can be visualized by autoradiography (see Figure 13.9[8]). In this way, the clone or clones that synthesized the target protein are identified and the corresponding cDNA clone(s) can be isolated from the original microtiter dishes. That clone can be used as a probe, for example, to analyze the genome of the same or other organisms for homologous sequences, to isolate the nuclear gene for the mRNA from a genomic library, or to quantify mRNA production from the gene in the cell.

K e y n o t e

Specific sequences in genomic libraries and cDNA libraries can be identified using a number of approaches, including the use of specific DNA or cDNA probes, heterologous probes, or complementation.

Analysis of Genes and Gene Transcripts

Cloned DNA sequences are resources for experiments designed to answer many kinds of biological questions. The following experimental techniques will be described in this section:

1. *The cloned DNA may be mapped with respect to the number and arrangement of restriction sites.* The resulting map, analogous to the arrangement of genes on a linkage map, is called a **restriction map**.

This is commonly done for both cloned genes (contain introns) and cloned cDNAs (correspond to mRNAs, so no introns) to produce a restriction map of the cloned DNA. Such information might be useful for making clones of subsections of the gene or the cDNA, or for comparing the gene and the cDNA.

2. *A cloned cDNA or a cloned gene may be used to analyze the transcription of the corresponding gene in the cell.* For example, we can study the size of the initial transcript of the gene, the processing steps it goes through (if any) to produce a mature RNA, the amount of the gene transcript, the time of expression in the cell cycle, and the time and amount of expression in different tissues and/or during development.

3. *The complete sequence of the cloned DNA may be determined.* In the case of a cloned gene, the sequence information can be useful in studies of how the expression of the gene is regulated. DNA sequences may also be compared with other DNA sequences in a computer database to determine the extent of similarity between related genes. If the cloned DNA sequence is not a known gene, a computer search might provide insights into what the sequence might be. Further, the DNA sequence of a protein-coding gene can be "translated" by computer to provide information about the properties of the protein for which it codes. Such information can be helpful for an investigator who wishes to isolate and study an unknown protein product of a gene for which a clone is available.

Restriction Enzyme Analysis of Cloned DNA Sequences

Cloned DNA sequences are often analyzed to determine the arrangement and specific locations of restriction sites. Because cloned DNA sequences represent a homogeneous population of DNA molecules, restriction enzyme cleavage of cloned DNA sequences produces a relatively small number of discretely sized DNA fragments. These DNA fragments can easily be seen following agarose gel electrophoresis* and ethidium bromide staining, permitting restriction maps to be constructed without the need for hybridization with a radioactive probe and autoradiography.

---

*Agarose gel electrophoresis of DNA involves the movement of DNA in an electric field through the pores of a rectangular slab of agarose (a firm, gelatinous material). Since DNA is negatively charged due to its phosphates, the DNA migrates toward the positive pole. Small DNA fragments can "squirm" more readily through the small pores in the gel, so the small fragments move through the gel more rapidly than large DNA fragments. Thus the smallest fragments migrate the farthest distance while the largest fragments migrate the least.

Example of Restriction Mapping. Let us consider the construction of a relatively simple restriction map (Figure 13.10). We have cloned a 5.0-kb piece of DNA and wish to construct a restriction map of it (Figure 13.10[1]). One sample of the DNA is digested with *Eco*RI, a second sample is digested with *Bam*HI, and a third sample is digested with both *Eco*RI and *Bam*HI. The results of each reaction are analyzed on an agarose gel alongside a control of the same DNA not cut with any enzyme, and size markers so that the sizes of the DNA fragments can be computed (Figure 13.10[2]). After electrophoresis the gel is stained with ethidium bromide and photographed under ultraviolet light. The distance each DNA band migrated can be measured from the photograph. For the marker, the molecular size of each DNA band is known, so a calibration curve can be drawn of DNA size (in log kb) versus migration distance (in mm) (Figure 13.10[3]). The migration distances for the DNA bands from the uncut and cut DNA are then used with the calibration curve to determine the molecular sizes of the DNA fragments in the bands (Figure 13.10[4]). For our theoretical example the results are shown in Figure 13.10(5).

The results are analyzed as follows (Figure 13.10[6–8]):

1. When the 5.0-kb DNA is cut with *Eco*RI, 4.5-kb and 0.5-kb DNA fragments result. This indicates that there is only one restriction site for *Eco*RI in the DNA, and that this site is 0.5 kb from one end of the molecule.

2. Using similar logic, there is one restriction site for *Bam*HI located 2 kb from one end of the molecule.

3. At this point we know there is one restriction site for each enzyme, but we don't know the relationship between the two. We can, however, make two models (see Figure 13.10[7]). In model A the *Eco*RI site is 0.5 kb from one end and the *Bam*HI site is 2.0 kb from that same end. In model B the *Eco*RI site is 0.5 kb from one end and the *Bam*HI site is 3.0 kb from that end (i.e., 2.0 kb from the other end). Model A predicts that cutting with *Eco*RI and *Bam*HI will produce three fragments of 0.5, 1.5, and 3.0 kb (going from left to right along the DNA). In model B cutting with both enzymes will produce three fragments of 0.5, 2.5, and 2.0 kb. The actual data show three fragments with sizes 2.5, 2.0, and 0.5 kb, thereby validating model B.

Restriction mapping typically involves more complicated data than in our example. There are more restriction enzymes and many sites for each enzyme. Analysis may be done with a complete plasmid (which is circular), with just the cloned sequence, or with part of the cloned sequence cut out of the plasmid and purified.

Restriction Enzyme Analysis of Genes

As part of the analysis of genes in the genome, the arrangement and specific locations of restriction sites is often determined. This information is useful, for example, *for comparing homologous genes in different species, for analyzing intron organization, or for planning experiments to clone parts of a gene into a vector.* The arrangement of restriction sites in a gene can be analyzed without actually cloning the gene by using a cDNA probe or a closely related heterologous gene probe. The process of analysis is as follows:

1. DNA is cut with different restriction enzymes (Figure 13.11[1,2]), each of which will produce DNA fragments of different lengths (depending on the locations of the restriction sites on the DNA molecules).

2. The DNA restriction fragments are separated according to their molecular size by agarose gel electrophoresis (Figure 13.11[3]). After electrophoresis the DNA is stained with ethidium bromide so that it can be seen under ultraviolet light illumination. When total cellular DNA is digested with a restriction enzyme, the result is usually a continuous smear of fluorescence down the length of the gel lane because the enzyme produces many fragments of all sizes.

3. The DNA fragments are transferred to a membrane filter so that they are in exactly the same relative position on the filter as they were on the gel (Figure 13.11[4]). The transfer to the membrane filter is done by the **Southern blot technique** (named after its inventor, Edward Southern). In brief, the gel is treated to denature the double-stranded DNA into single strands. The gel is then placed on a piece of blotting paper that spans a glass plate. The ends of the paper are in a container of buffer and act as wicks. A piece of membrane filter is laid down so that it just covers the gel. Additional sheets of blotting paper and a weight are stacked on top of the membrane filter. The buffer solution in the bottom tray is wicked up by the blotting paper, passes through the gel, through the membrane filter, and finally into the stack of blotting paper. During this process the DNA

■ *Figure 13.10*   Construction of a restriction map for *Eco*RI and *Bam*HI in a DNA fragment.

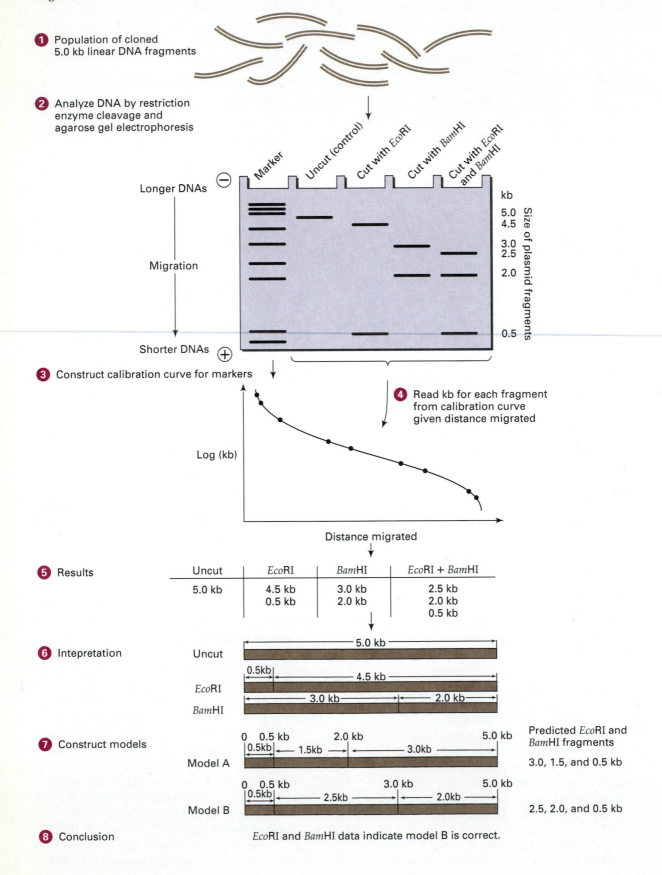

■ *Figure 13.11* Procedure for the analysis of cellular DNA for the presence of sequences complementary to a radioactive probe, such as a cDNA molecule made from an isolated mRNA molecule. The DNA fragments with homology to the cDNA probe, shown as three bands in this theoretical example, are visualized by autoradiography.

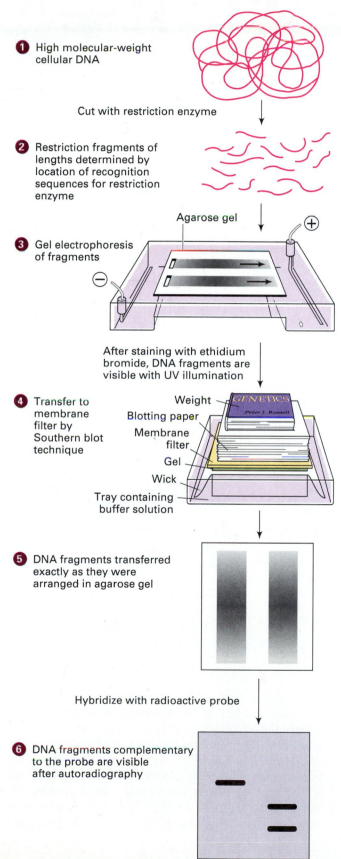

1 High molecular-weight cellular DNA

Cut with restriction enzyme

2 Restriction fragments of lengths determined by location of recognition sequences for restriction enzyme

Agarose gel

3 Gel electrophoresis of fragments

After staining with ethidium bromide, DNA fragments are visible with UV illumination

4 Transfer to membrane filter by Southern blot technique

Weight
Blotting paper
Membrane filter
Gel
Wick
Tray containing buffer solution

5 DNA fragments transferred exactly as they were arranged in agarose gel

Hybridize with radioactive probe

6 DNA fragments complementary to the probe are visible after autoradiography

fragments are transferred from the gel to the membrane filter, to which they stick. The fragments on the filter are arranged in exactly the same way as they were in the gel (Figure 13.11[5]).

4. Once the Southern blot technique is completed, the filter is placed in a bag, a radioactive DNA probe is added and the mixture is incubated 3–24 hours at a selected temperature (usually about 60°C). The probe will hybridize to any complementary DNA fragments that are bound to the filter.

5. Next, the filter is washed to remove radioactive probes that have not hybridized. The filter is dried, and an autoradiogram is prepared to determine the position(s) of the hybrids (Figure 13.11[6]). If DNA fragment size markers are separated in a different lane in the agarose gel electrophoresis process, the sizes of the genomic restriction fragments that hybridized with the probe can be calculated. From the fragment sizes obtained, a restriction map can be generated to show the relative positions of the restriction sites. The construction of restriction maps was described previously.

Analysis of Gene Transcripts

A related blotting technique to the Southern blot technique has been developed to analyze RNA rather than DNA. This technique is called **northern blot analysis** (the name is derived, not from a person, but to indicate that it is a technique related to the Southern technique). In northern blotting RNA extracted from a cell is separated by size using gel electrophoresis, and the RNA molecules are transferred and bound to a filter in a procedure that is essentially identical to Southern blotting in which DNA is transferred. After hybridization with a radioactive probe, an autoradiogram is prepared of the bands corresponding to the RNA species that were complementary to the probe. Given appropriate RNA size markers, the sizes of the RNA species identified with the probe can be determined.

Northern blotting is useful in several kinds of experiments. For example, northern blotting can reveal the size(s) of the specific mRNA encoded by a gene. In some cases a number of different mRNA species encoded by the same gene have been identified in this way, suggesting that either different promoter sites can be used, different terminator sites can be used, or alternative mRNA processing can occur. Northern blotting can also be used to investigate whether or not an mRNA species is present in a cell type or tissue, and how much of it is present. This type of experiment is useful for determining levels of gene activity during development, in different cell types of an organism, or in cells before and after they are subjected to various physiological stimuli.

Keynote

Genes and cloned DNA sequences are often analyzed to determine the arrangement and specific locations of restriction sites. The result is a restriction map.

Gene transcripts can be analyzed by separating mRNA species by gel electrophoresis, transferring the mRNAs to a filter by the northern blot technique, and hybridizing with a specific radioactive probe.

DNA Sequence Analysis

Cloned DNA fragments may be analyzed to determine the nucleotide pair sequence of the DNA. This information is helpful, for example, for identifying gene sequences and controlling sequences within the fragment, and for comparing the sequences of homologous genes from different organisms.

Two techniques for the *rapid sequencing of DNA molecules* were developed in the 1970s. One method, developed by Allan Maxam and Walter Gilbert, uses chemicals to cleave DNA at different sites and is called **Maxam-Gilbert sequencing**, after its developers. The other method, called **dideoxy sequencing** or **Sanger sequencing,** was developed by Fred Sanger and involves enzymatic extension of a short primer. The dideoxy method is more commonly used today.

Dideoxy DNA Sequencing. Cloned DNA molecules are denatured to single strands. Next, a short, radioactive, oligonucleotide primer is annealed to one of the two DNA strands (Figure 13.12). The oligonucleotide is designed so that its 3' end will be next to the DNA sequence of interest. The oligonucleotide acts as a *primer* for DNA synthesis; the DNA made is a *complementary copy of the DNA sequence of interest* (Figure 13.12).

■ *Figure 13.12*   Dideoxy DNA sequencing of a theoretical DNA fragment.

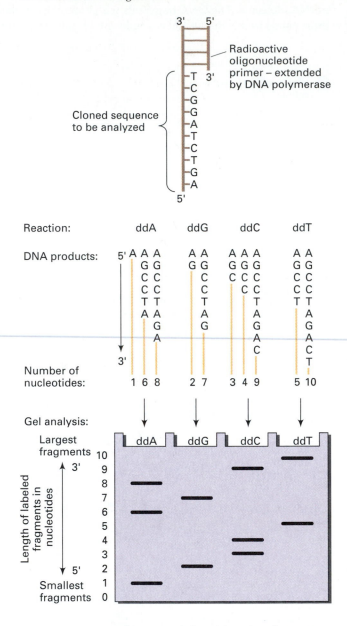

Sequence deduced from banding pattern of autoradiogram made from gel:

5' A-G-C-C-T-A-G-A-C-T 3'

For each sequencing experiment four separate reactions are set up with single-stranded DNA to which the radioactive primer has been annealed. Each reaction contains the four normal precursors of DNA (dATP, dTTP, dCTP, and dGTP), and DNA polymerase. The reactions differ in the presence of a different modified nucleotide called a **dideoxy nucleotide** (Figure 13.13). The only difference between a dideoxy nucleotide and the deoxynu-

■ *Figure 13.13*   A dideoxy nucleotide DNA precursor.

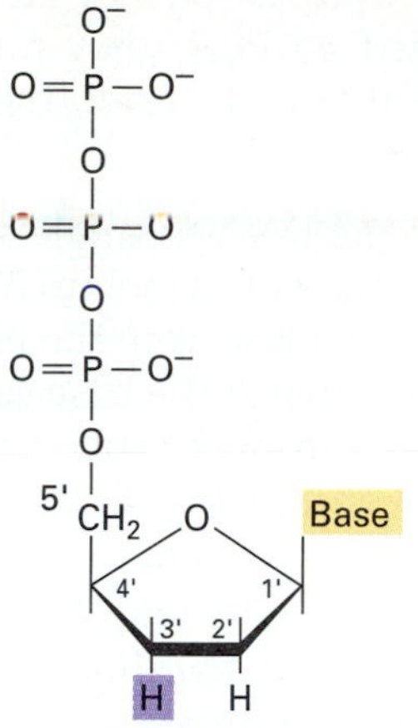

Dideoxynucleoside
triphosphate

(Normal DNA precursor
has OH at 3' position)

tion, and so on (see Figure 13.12). It is important to realize that each DNA chain synthesized started from the same fixed point and ended at a particular base, the latter determined by the dideoxy nucleotide incorporated. The DNA chains in each reaction mixture are separated by polyacrylamide gel electrophoresis, and the locations of the DNA bands are revealed by autoradiography. The DNA sequence of the newly synthesized strand is determined from the autoradiogram by reading from the bottom to the top to give the sequence in a 5'-to-3' orientation. In the example, the band that moved the farthest ended with ddA, the band that moved the second farthest ended with ddG, and so on. The complete sequence determined is 5'-AGCCTA-GACT-3'; this is complementary to the sequence on the template strand (refer to Figure 13.12).

An example of a dideoxy sequencing gel result is shown in Figure 13.14. The various DNA chains that are synthesized in each of the dideoxy reac-

cleotides normally used in DNA synthesis is a dideoxy nucleotide has 3'-H on the deoxyribose sugar rather than 3'-OH. If a dideoxy nucleoside triphosphate (ddNTP) is used in a DNA synthesis reaction, the dideoxy nucleotide can be incorporated into the growing DNA chain. However, once that happens no further DNA synthesis can then occur because the *absence of a 3'-OH prevents the formation of a phosphodiester bond with an incoming DNA precursor.*

The four sequencing reactions, then, are ddA, ddG, ddC, or ddT; that is, they have *one* of ddATP, ddGTP, ddCTP, and ddTTP, in addition to the four normal precursors dATP, dGTP, dCTP, and dTTP. So that some DNA synthesis occurs in the dideoxy sequencing reactions, only a small proportion of the nucleotides are dideoxy nucleotides. The primer is extended by DNA polymerase, and when a particular nucleotide is specified by the template strand, there is a small chance that the dideoxy nucleotide will be incorporated instead of the normal nucleotide in the appropriate reaction mixture. For example, if an A is specified by the template strand, a ddA could be incorporated rather than dA. Once a dideoxy nucleotide is incorporated, elongation of the chain stops. In a population of molecules in the same DNA synthesis reaction, then, new DNA chains will stop at all possible positions where a particular nucleotide is required because of the incorporation of the dideoxy nucleotide at that position. In the ddA reaction the many different chains that are produced all end with ddA, all chains end with ddG in the ddG reac-

■ *Figure 13.14*   Autoradiogram of a dideoxy sequencing gel.

A C G T

tions are given to illustrate the principle involved in dideoxy sequencing.

Keynote

Rapid methods have been developed for determining the sequence of a cloned piece of DNA. One method, the Maxam-Gilbert procedure, uses specific chemicals to modify and cleave the DNA chain at specific nucleotides. Another method, the dideoxy or Sanger procedure, uses enzymatic synthesis of a new DNA chain on a cloned template DNA strand. With this procedure synthesis of new strands is stopped by the incorporation of a dideoxy analog of the normal deoxyribonucleotide. Using four different dideoxy analogs, the new strands stop at all possible nucleotide positions, thereby allowing the complete DNA sequence to be determined.

Methods for sequencing DNA are continually being improved and automated procedures are now available which enable DNA sequencing to proceed much more rapidly than is the case with the manual dideoxy method. Such procedures are of great utility as research teams determine the complete sequences of various genomes, including that of humans.

Polymerase Chain Reaction (PCR)

The generation of large numbers of identical copies of DNA by the construction and cloning of recombinant DNA molecules was made possible in the 1970s. Indeed, recombinant DNA techniques revolutionized molecular genetics by making it possible to analyze genes and their functions in new ways. However, the cloning of DNA is time consuming, involving the insertion of DNA into cloning vectors and the screening of libraries to detect specific DNA sequences. In the mid-1980s the **polymerase chain reaction (PCR)** was developed, which resulted in yet another revolution in the way genes can be analyzed. PCR is a method for producing an extremely large number of copies of a *specific* DNA sequence from a DNA mixture without having to clone it, a process called *amplification*. PCR permits the selective amplification of DNA sequences.

The starting point for PCR is the DNA mixture containing the DNA sequence to be amplified, and a pair of oligonucleotide primers that flank that DNA sequence of interest (Figure 13.15). The primers are made synthetically, so it is essential that some DNA sequence information be available

about the DNA sequence of interest so that it can be amplified. For that reason not all DNA sequences can be amplified by PCR; they must be cloned by the conventional way. In brief, the PCR procedure is as follows:

1. Denature the DNA to single strands by incubating at 94°C. Cool (to 37–55°C, depending on how well the base sequence of the primers match the base sequence of the DNA) and anneal the specific pair of primers (primer A and primer B) that flank the targeted DNA sequence (Figure 13.15[1]).

2. Extend the primers with DNA polymerase (Figure 13.15[2]). For this a special heat-resistant DNA polymerase from a hot-springs bacterium called *Taq* ("tack") *polymerase* is used.

3. Repeat the denaturation of DNA to single strands and anneal new primers (Figure 13.15[3]). (The further amplification of the original strands is omitted in the remainder of the figure.)

4. Repeat the primer extension with *Taq* DNA polymerase (Figure 13.15[4]). In each of the two double-stranded molecules produced in the figure, one strand is of unit length—the length of DNA between the 5' end of primer A and the 5' end of primer B or the length of the target DNA. The other strand in both molecules is longer than unit length.

5. Repeat the denaturation of DNA and anneal new primers (Figure 13.15[5]). (For simplification, the further amplification of the longer-than-unit-length strands continues with linear increase only and is omitted in the rest of the figure.)

6. Repeat the primer extension with *Taq* DNA polymerase (Figure 13.15[6]). This produces unit-length, double-stranded DNA. Note that it took three cycles to produce the two molecules of target DNA. Repeated denaturation, annealing, and extension cycles results in the geometric increase of the unit-length DNA. Amplification of the longer-than-unit-length DNA also occurs simultaneously, but only in a linear fashion.

Using PCR the amount of new DNA generated increases geometrically. Starting with 1 molecule of DNA, 1 cycle of PCR produces 2 molecules, 2 cycles produces 4 molecules, and 3 cycles produces 8 molecules, 2 of which are the target DNA. A further 10 cycles produces 1,024 copies (2^{10}) of the target DNA and in 20 cycles there will be 1,048,576 copies (2^{20}) of the target DNA! The procedure is rapid, each cycle taking only a few minutes using a *thermal*

■ *Figure 13.15*   The polymerase chain reaction (PCR) for selective amplification of DNA sequences.

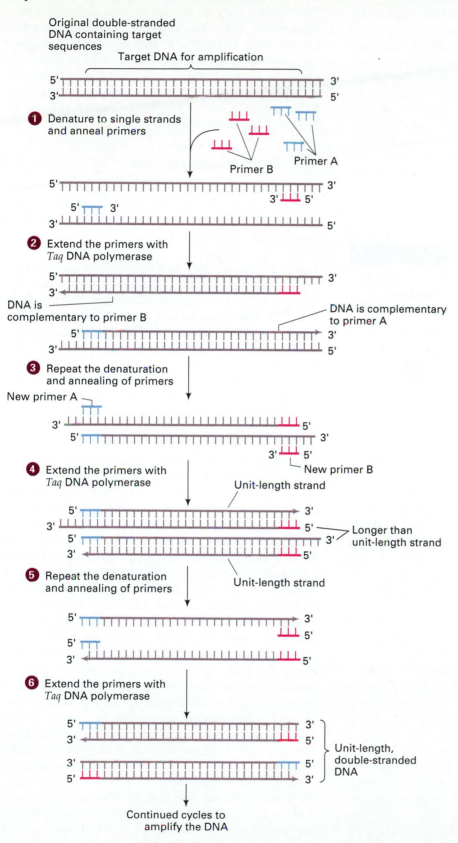

Original double-stranded DNA containing target sequences

Target DNA for amplification

① Denature to single strands and anneal primers

Primer B

Primer A

② Extend the primers with *Taq* DNA polymerase

DNA is complementary to primer B

DNA is complementary to primer A

③ Repeat the denaturation and annealing of primers

New primer A

New primer B

④ Extend the primers with *Taq* DNA polymerase

Unit-length strand

Longer than unit-length strand

Unit-length strand

⑤ Repeat the denaturation and annealing of primers

⑥ Extend the primers with *Taq* DNA polymerase

Unit-length, double-stranded DNA

Continued cycles to amplify the DNA

cycler, a machine that automatically cycles through the temperature changes in a programmed way.

There are many applications for PCR, including amplifying DNA for cloning, sequencing without cloning, mapping DNA segments, disease diagnosis, sex determination of embryos, forensics, and studies of molecular evolution. In each case some DNA sequence information must be available so that appropriate pairs of primers can be synthesized. In disease diagnosis, for example, PCR can be used to detect viral pathogens such as HIV-1. Because PCR makes it easier to detect single-copy sequences in total DNA isolated from cells, it can also be used in genetic disease diagnosis, which is discussed in the next section.

Keynote

The polymerase chain reaction (PCR) uses specific oligonucleotides to amplify a specific segment of DNA many thousandfold in an automated procedure. PCR is finding increasing applications both in research and in the commercial arena, including generating specific DNA segments for cloning or for sequencing, and for amplifying DNA to detect the presence of specific genetic defects.

Applications of Recombinant DNA Technology

Recombinant DNA and PCR technologies have many applications, including the diagnosis of human and animal genetic diseases; the synthesis of commercially important products such as human insulin and human growth hormone; *in vitro* modifications of genes; and genetic engineering of plants. Some new applications will be briefly outlined in this section.

Analysis of Biological Processes

One of the fundamental and widespread applications of recombinant DNA and PCR methodologies is in basic research to explore biological functions. Questions in most areas of biology are being addressed with these techniques. In genetics researchers are investigating such areas as the functional organization of genes and the regulation of gene expression. In developmental biology key regulatory genes and target genes responsible for developmental events are being discovered and analyzed, and gene changes associated with aging and cancer are being investigated. In evolutionary biology DNA sequence analysis is adding new information about the evolutionary relationships between organisms. PCR is being used to amplify DNA from ancient tissues to compare the DNA sequences with modern-day descendants. For instance, the DNA of a 17-million-year-old magnolia was shown to be little different from magnolia DNA of today. Indeed, it is rare today to find a research scientist who is not aware of advances being made in his or her field through the use of recombinant DNA and PCR techniques.

Diagnosis of Genetic Diseases

We have encountered many examples of human genetic diseases throughout this text. In this section we will discuss some of the tools available to diagnose genetic diseases in humans.

Genetic Counseling.   In Chapter 7 we learned that many human genetic diseases are caused by enzyme or protein defects; those defects ultimately are the result of mutations at the DNA level. Many other genetic diseases arise from chromosome defects. It is now possible to assay for many enzyme or protein deficiencies, and for many of the DNA changes associated with genetic diseases and, thereby, to determine whether or not an individual has a genetic disease. Further, it is possible to determine whether individuals have any chromosomal defects. **Genetic counseling** is the analysis of the probability that individuals have a genetic defect, or of the risk that prospective parents may produce a child with a genetic defect. In the latter case genetic counseling involves presenting the available options for avoiding or minimizing those possible risks. Thus genetic counseling provides people with an understanding of the genetic problems that are or may be inherent in their families or prospective families. Counseling utilizes a wide range of information on human heredity. In many instances the risk of having a child with a genetic defect may be stated in terms of rather precise probabilities; in others, where the role of heredity is not sufficiently clear, the risk may be estimated only in general terms. It is the responsibility of the genetic counselors to supply their clients with clear, unemotional, and nonprescriptive statements based on the family history and on their knowledge of all relevant scientific information and of the probable risks of giving birth to a child with a genetic defect. In sum, genetic counseling is a lot more than the simple presentation of risk facts and figures to patients; it is the prevention of disease, the relief of pain, and the maintenance of health, all of which are goals of the medical profession in general.

Genetic counseling usually begins with pedigree

analysis of both families to determine the likelihood that a genetic disease is present in either group. (Pedigree analysis was described in Chapters 2 and 3.) Once evidence is found of a genetic disease in a family or families, prospective parents need to be informed of the probability that they will produce a child with that disease. Early detection of a genetic disease occurs at one or both of two levels. One is the detection of heterozygotes (carriers) of recessive mutations, and the other is the determination of whether or not the developing fetus shows the biochemical defect. For tests that measure enzyme activities or protein amounts, these two types of procedures are limited to those genetic diseases in which the biochemical defect is expressed in the parents and/or the developing fetus. For tests that measure actual changes in the DNA, these procedures do not depend on expression of the gene in the parents or fetus.

Although we can identify carriers of many genetic diseases and can determine if fetuses have a genetic disease, in most cases there is no cure for the diseases. Carrier detection and fetal analysis serve mainly to inform parents of the risks and probabilities of having a child with the defect.

Carrier Detection. *Carrier detection* is the detection of individuals who are heterozygous for a recessive gene mutation. The heterozygous carrier of a mutant gene usually is normal in phenotype. Nonetheless, the individual carries a recessive mutation that can be passed on to the progeny. In the case of a recessive mutation that has serious deleterious consequences to an individual homozygous for that mutation, there is great value in determining whether two people who are contemplating having a child are both carriers, because in that situation one-quarter of the children would be born with the trait. Carrier detection can be used in those cases in which a gene product (protein or enzyme) can be assayed. In such cases the heterozygote (carrier) is expected to have approximately half of the enzyme activity or protein amount as homozygous normal individuals. We will see later how carriers can be detected by DNA tests.

Fetal Analysis. A second important aspect of genetic counseling is finding out whether a fetus is normal. **Amniocentesis** is one way in which this can be done (Figure 13.16). As a fetus develops in the amniotic sac, it is surrounded by amniotic fluid, which serves to cushion it against shock. In amniocentesis a sample of the amniotic fluid is taken by carefully inserting a syringe needle through the mother's uterine wall and into the amniotic sac. The amniotic fluid contains cells that have sloughed

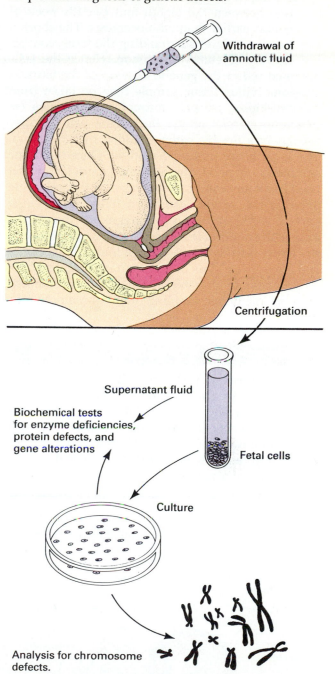

■ *Figure 13.16* Steps in amniocentesis, a procedure used for prenatal diagnosis of genetic defects.

Withdrawal of amniotic fluid

Centrifugation

Supernatant fluid

Biochemical tests for enzyme deficiencies, protein defects, and gene alterations

Fetal cells

Culture

Analysis for chromosome defects.

off the fetus's skin; these cells can be cultured in the laboratory. Once cultured they may be examined for protein or enzyme alterations or deficiencies, DNA changes, and chromosomal abnormalities. Amniocentesis is possible at any stage of pregnancy, but the quantity of amniotic fluid available and the increased risk to the fetus makes it impractical to perform the procedure before the twelfth week of pregnancy. Because amniocentesis is complicated and costly, it is primarily used in high-risk cases.

Another method for fetal analysis is **chorionic villus sampling** (Figure 13.17). The procedure can be done between the eighth and twelfth week of pregnancy, earlier than amniocentesis. The chorion is a membrane layer surrounding the fetus, consisting entirely of embryonic tissue. Hence the cells obtained reflect the genetic makeup of the fetus. A chorionic villus tissue sample may be taken from the developing placenta through the abdomen (as in amniocentesis) or via the vagina using biopsy forceps or a flexible catheter, aided by ultrasound. The latter is the preferred method. Once the tissue sample is obtained, the analysis is similar to that used in amniocentesis. Advantages of the technique are that the parents can learn if the fetus has a genetic defect earlier in the pregnancy than is the case with amniocentesis, and that it is not necessary to culture cells to obtain enough to do the biochemical assays. Fetal loss and inaccurate diagnoses due to the presence of maternal cells are more common in chorionic villus sampling than in amniocentesis, however.

K e y n o t e

Genetic counseling is the analysis of the risk that prospective parents may produce a child with a genetic defect, and the presentation to appropriate family members of the available options for avoiding or minimizing those possible risks. Early detection of a genetic disease is done by carrier detection and by fetal analysis.

DNA Analysis. For more and more genetic diseases we can screen individuals for the actual DNA mutation, rather than for the resulting biochemical change. For example, we can use DNA probes and other molecular techniques to diagnose genetic diseases such as Duchenne muscular dystrophy, Huntington's chorea, hemophilia, cystic fibrosis, Tay-Sachs disease, and sickle-cell anemia. Unlike analysis for enzyme or protein defects, these techniques are limited not by whether a gene product is expressed in the developing fetus, but by whether or not there is a known DNA difference that can be used to distinguish the wild-type from the mutant gene.

Recombinant DNA and/or PCR approaches for detecting genetic disease require cellular DNA as the starting point. Such DNA can be isolated from fetal cells obtained by amniocentesis or chorionic villus sampling, and from blood samples of children and adults. The DNA is digested with a restriction enzyme, producing restriction fragments of lengths determined by the locations of the restriction sites along the DNA molecules. The restriction fragments are analyzed as described earlier: The fragments are separated according to size by agarose gel electrophoresis, then Southern blotted to a nitrocellulose filter for hybridization with a specific ^{32}P-labeled DNA probe. (If insufficient DNA is available for the diagnostic procedures, and some sequence information is known for the target DNA sequence, PCR can be used to produce enough DNA for analysis.)

These analytical procedures are most useful

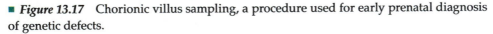

■ *Figure 13.17*   Chorionic villus sampling, a procedure used for early prenatal diagnosis of genetic defects.

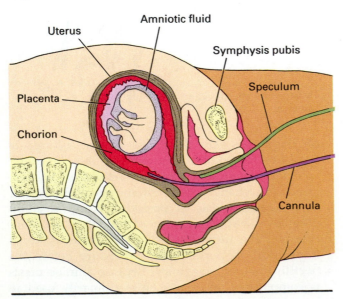

when the genetic mutation that causes a disease is associated with a change in the number or distribution of restriction sites, either within the gene or in a flanking region. That is, in the human genome, and the genomes of other eukaryotes, different restriction maps may be found among individuals for the same region of a chromosome (detected by the same probe). The different restriction maps result from different patterns of distribution of restriction sites and are called **restriction fragment length polymorphisms,** or **RFLPs,** because they are detected by the presence of restriction fragments of different lengths on gels. ("Polymorphism" means the existence of many different forms; here a region of DNA that can have different lengths on different homologs and/or in different individuals.) RFLPs arise, for example, by the addition or deletion of DNA between **restriction sites,** or by base-pair changes that create or abolish a restriction site sequence. A restriction map is independent of gene function, so an RFLP is detected whether the DNA sequence change responsible affects a detectable phenotype or not. An RFLP can be used as a genetic marker in the same way as the "conventional" genetic markers we have discussed previously. In this case we assay the DNA—the genotype—directly in the form of a restriction map. Moreover, because we are looking directly at DNA, both parental types are seen in heterozygotes, so carriers can easily be identified.

RFLPs are useful both for mapping chromosome regions as well as in human disease diagnosis. For the latter there are many cases of an RFLP being associated with a gene known to cause a disease, as the following example illustrates. The genetic disease *sickle-cell anemia* (discussed in more detail in Chapter 7) results from a single base-pair change in the gene for the hemoglobin's β-globin polypeptide, resulting in an abnormal form of hemoglobin, Hb-S, instead of the normal Hb-A form. This change from $\frac{A}{T}$ to $\frac{T}{A}$ changes the codon from CAT to CTT, which results in the substitution of a valine for a glutamic acid in the sixth amino acid of the polypeptide, which, in turn, produces abnormal associations of hemoglobin molecules, sickling of the red blood cells, tissue damage, and sometimes death.

Using a cDNA probe for human β-globin, an RFLP has been shown for the restriction enzyme *Dde*I ("d-d-e-one"). That is, the base-pair change for the sickle-cell mutation generates a new *Dde*I restriction site that is not present in DNA of normal individuals. In normal individuals there are two relevant *Dde*I sites: one is upstream of the start of the β-globin gene, and the other is within the coding

■ *Figure 13.18*   Detection of a sickle-cell gene by the *Dde*I restriction fragment length polymorphism: (a) Diagrams of DNA segments showing the *Dde*I restriction sites; (b) Schematic drawing of the results of analysis of DNA cut with *Dde*I, subjected to gel electrophoresis, blotted, and probed with a β-globin probe.

a) *Dde* I restriction sites

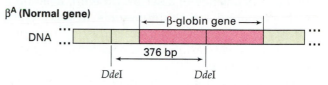

b)   Results of *Dde* I cuts with beginning of β-globin gene

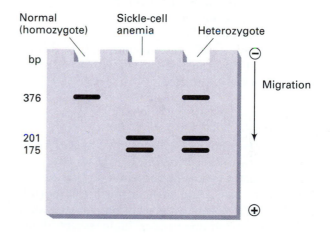

sequence itself (Figure 13.18[a]). When DNA from normal individuals is cut with *Dde*I, and the fragments separated by gel electrophoresis, blotted, and probed with the first part of the β-globin gene, a 376-base pair fragment is seen (Figure 13.17[b]). DNA from individuals with sickle cell anemia analyzed in the same way gives two fragments of 175 bp and 201 bp because of the additional *Dde*I site. Heterozygotes can be detected by the presence of three bands of 376 bp, 201 bp, and 175 bp.

Not all RFLPs result from changes in restriction sites directly related to the gene mutations. Many result from changes to the DNA flanking the gene, sometimes a fair distance away. In these cases detection of the genetic disease relies on the flanking RFLP segregating most of the time with the gene mutation. Recombination occurring between the RFLP and the gene of interest can occur, of course, and this can cause some difficulty in interpreting the results.

Other examples of human genetic diseases for which recombinant DNA technology can or will soon provide early diagnosis include PKU (see Chapter 7), four types of thalassemia (hemoglobin diseases resulting in anemia), α-antitrypsin deficiency (a deficiency of a serum protein), hemophilia A, hemophilia B, Huntington's disease, cystic fibrosis, and Duchenne muscular dystrophy (a progressive disease resulting in muscle atrophy and muscle dysfunction).

The beauty of the recombinant DNA approach is that it directly assays for a DNA genotype (RFLP), so detection does not depend on the expression of the gene (phenotype). Thus this approach is not limited to detecting genetic diseases in which the gene involved is active in the parent or the fetus at the time of analysis. For example, phenylalanine hydroxylase, the enzyme defective in individuals with PKU, is found in the liver but is not found either in blood serum or in fibroblast cells, the cells usually cultured following amniocentesis. The DNA of fibroblast cells can be analyzed for RFLPs, however, and both an individual with the disease *and a carrier* can be detected.

DNA Fingerprinting

Everyone is familiar with the use of fingerprints in forensic science. The principle is that no two individuals have the same fingerprints so that fingerprints left at the scene of a crime are important identification evidence in a criminal investigation. Similarly, no two human individuals (except identical twins) have exactly the same genome, base pair for base pair; this has led to the development of DNA techniques for use in forensic science and in paternity and maternity testing. Such a use might seem unfounded at first, since it is clear that the similarities in the DNA of different individuals greatly exceed the differences, as we would expect of a genome which specifies the human species. In fact, estimates indicate that about 1 base pair in 1000 base pairs is the site of polymorphisms among humans. For **DNA fingerprinting,** or the use of DNA analysis to identify an individual, scientists have identified *highly polymorphic* markers scattered throughout the genome. Each marker consists of a restriction fragment within which are short, identical segments of DNA tandemly arranged head to tail. Differences among individuals result from a great variation in the number of the tandem repeats, called *variable number of tandem repeats*, or VNTRs. Thus when DNA is digested with the

restriction enzyme, fragments of different sizes result because of the VNTRs. VNTR probes have been constructed to analyze the VNTRs so that it is possible, in most cases, to identify an individual unambiguously based on a DNA sample.

DNA fingerprinting proceeds as follows (Figure 13.19): DNA is obtained from the crime scene and the suspect (Figure 13.19[1]). The DNA may be obtained from blood, semen, other bodily tissues or fluids, or hair roots, as appropriate for the crime. Let us consider a rape case. In some instances, PCR may be used to amplify the sample to produce enough DNA for analysis. For example, enough DNA for analysis can be produced by using PCR with DNA from a single hair root. The DNA is cut with the appropriate restriction enzyme(s) (Figure 13.19[1]), and the resulting fragments are separated by electrophoresis (Figure 13.19[2]), Southern blotted to a filter (Figure 13.19[3]), and probed with a radioactive VNTR probe (Figure 13.19[4,5]). The DNA banding pattern on an autoradiogram made from the filter is then analyzed to compare the two samples (Figure 13.19[6]). If the DNA banding pattern for a semen sample taken from a rape victim *matches exactly* the DNA banding pattern for samples taken from a suspect, this is excellent evidence that the suspect is the rapist because the odds of two people having exactly the same DNA banding patterns are extremely low (with the exception of the patterns of identical twins). DNA fingerprinting is being used more and more frequently in U.S. criminal cases, although there remain legal challenges to the method. And in the U.S. military DNA fingerprints are being substituted for the traditional "dog tags." In the United Kingdom DNA fingerprinting has become an accepted method to resolve immigration cases. That is, DNA evidence can more readily prove that a potential immigrant is related to a citizen than can blood typing or other types of evidence.

Gene Therapy

Is it possible to treat genetic diseases? Genetic diseases result from the phenotypes caused by the mutant gene in somatic cells. Theoretically, if a wild-type copy of the gene could be introduced into somatic cells where the gene's activity is required, then the genetic disease could be treated. However, since germ-line cells would not be similarly treated, the mutant gene could still be passed on to the progeny.

Two types of gene therapy are theoretically possible: (1) Somatic cell therapy, in which somatic cells are modified genetically to correct a genetic defect; and (2) germ-line cell therapy, in which the

■ *Figure 13.19*   Procedure for DNA fingerprinting.

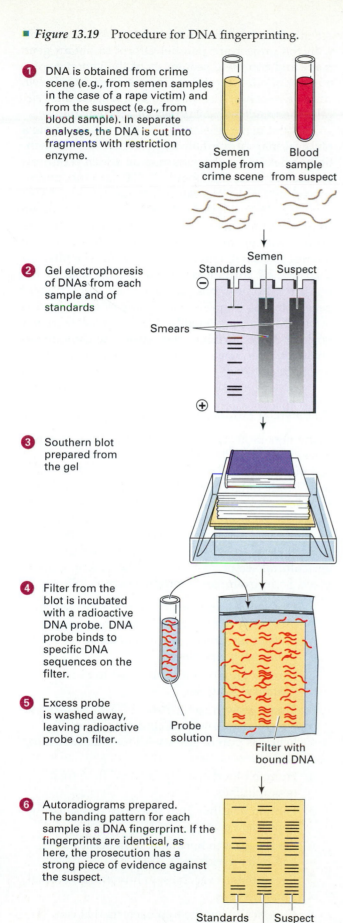

1 DNA is obtained from crime scene (e.g., from semen samples in the case of a rape victim) and from the suspect (e.g., from blood sample). In separate analyses, the DNA is cut into fragments with restriction enzyme.

Semen sample from crime scene   Blood sample from suspect

2 Gel electrophoresis of DNAs from each sample and of standards

Semen
Standards   Suspect

Smears

3 Southern blot prepared from the gel

4 Filter from the blot is incubated with a radioactive DNA probe. DNA probe binds to specific DNA sequences on the filter.

Probe solution

Filter with bound DNA

5 Excess probe is washed away, leaving radioactive probe on filter.

6 Autoradiograms prepared. The banding pattern for each sample is a DNA fingerprint. If the fingerprints are identical, as here, the prosecution has a strong piece of evidence against the suspect.

Standards   Suspect
Semen

germ-line cells are modified to correct a genetic defect. Somatic cell therapy results in a treatment for the genetic disease in the individual, but not a cure, since progeny could still inherit the mutant gene. Germ-line cell therapy, however, would result in a cure since the mutant gene(s) would be replaced by the normal gene. While both somatic cell therapy and germ-line cell therapy have been demonstrated in nonhuman organisms, only somatic cell therapy has been attempted in humans because of the ethical issues raised by germ-line cell therapy.

The most promising candidates for somatic cell therapy are genetic disorders that result from a simple defect of a single gene, and for which the cloned normal gene is available. Gene therapy involving somatic cells proceeds as follows: (1) A sample of the individual's mutant cells is taken; (2) Normal, wild-type copies of the mutant gene are introduced into the cells and the cells reintroduced into the individual where, it is hoped, the cells will function to produce normal gene product and the symptoms of the genetic disease will be treated successfully.

The source of the mutant cells is likely to vary with the genetic disease. For blood disorders, such as thalassemia or sickle-cell disease, modification of blood-line cells isolated from the bone marrow would be required. For genetic diseases affecting circulating proteins, a promising approach is the gene therapy of skin fibroblasts, constituents of the dermis (the lower layer of the skin). The modified fibroblasts can easily be implanted back into the dermis, where the tissue becomes vascularized, allowing gene products to be distributed.

A cell that has had a gene introduced into it by artificial means is called a *transgenic* cell, and the gene involved is called a *transgene*. The introduction of normal genes into a mutant cell poses several problems. First, procedures to introduce DNA into cells (transformation) typically are inefficient; perhaps only 1 in 1,000 or 100,000 cells will receive the gene of interest. Thus a large population of cells must be obtained from the individual in order for gene therapy to be attempted. Present procedures use special viral-related vectors to facilitate the gene uptake. Second, for those cells that take up the cloned gene, the fate of the "foreign" DNA cannot be predicted. In some cases the mutant gene will be replaced by the normal gene, while in others the normal gene will integrate into the chromosome at a site distinct from that of the gene locus of the mutant gene. In the first case, the gene therapy will be successful provided that the gene is expressed. In the second case, successful treatment of the disease will only result if: (1) the introduced gene is expressed; and (2) the resident mutant gene produces no, or very little protein; that is, the mutant phenotype results from no or low gene activity, as is common for recessive mutants. In terms of the mutant gene, if the disease results not from low gene activity, but from the production of an altered protein which is an inhibitor or interferes with other regulatory proteins, then the presence of a normal gene in addition to the mutant gene in the genome may not help treat the disease. The mutants here would likely be known dominant mutants.

Somatic cell therapy has been repeatedly demonstrated in experimental animals such as mice, rats, and rabbits. In recent years a few gene therapy trials have been done with humans. For example, in 1990 a four-year-old girl suffering from a deficiency in adenosine deaminase (ADA), an enzyme required for normal function of the immune system, was subjected to gene therapy involving somatic cells. T cells (cells involved in the immune system) were isolated from the girl, grown in the laboratory, and the normal ADA gene was introduced using a viral vector. The "engineered" cells were then reintroduced into the patient. Since T cells have a finite life in the body, continued infusions of engineered cells have been necessary. It is clear that the introduced ADA gene is expressed, probably throughout the life of the T cell. As a result, the girl's immune system is functioning more normally; she now gets no more than the average number of infections, compared with many more than the average number before the therapy.

It is clear that, with time, many other genetic diseases will become targets for somatic cell therapy. Human genetic diseases that could be early candidates for gene therapy are sickle-cell anemia, thalassemia, phenylketonuria, Lesch-Nyhan Syndrome, and Tay-Sachs syndrome. As the "tricks" are learned to target genes to replace their mutant counterparts and to regulate the expression of the introduced genes, increasing success in treating genetic diseases will undoubtedly result. However, many scientific, ethical, and legal questions will have be addressed before we will see the routine implementation of somatic cell therapy.

Human Genome Project

The **Human Genome Project** is an extensive, collaborative effort to map all of the estimated 50,000 to 100,000 human genes, and to obtain the sequence of the complete 3 billion (3×10^9) nucleotide pairs of the genome. In the United States, the Human Genome Project is being overseen primarily by the National Center of Human Genome Research (a part of the National Institutes of Health) and the Department of Energy. The Human Genome Proj-

ect officially began on October 1, 1990. Associated with the project are parallel efforts to obtain gene maps and complete sequences of the genomes of a number of other model organisms, including *E. coli*, yeast, *Drosophila*, the nematode *Caenorhabditis elegans*, the weed *Arabidopsis thaliani*, and the mouse.

The first major goal of the project is to generate a physical map of the human genome. This is essentially a detailed restriction map on which is indicated the locations of about 30,000 specific probes spaced at DNA intervals of about 100 kb. As of October 1992 detailed physical maps have been completed for human chromosomes Y and 21. The probes will make it possible to localize genes of interest to a small chromosome region. The physical map will be correlated with a linkage map of the human chromosomes achieved by conventional gene mapping analysis. This phase of the project is targeted to be completed by 1995. The second major goal is to determine the entire nucleotide-by-nucleotide sequence of the human genome, by a target date of 2005.

Commercial Products

Many new biotechnology companies have emerged since the mid-1970s, when basic recombinant DNA techniques were developed. The first was Genentech. With older, established pharmaceutical and chemical companies, these companies are focusing on using recombinant DNA technology to make a wide array of commercial products that are useful in businesses in the human health industry and agricultural sector, as well as in the general marketplace. Some examples of products are as follows:

1. Tissue plasminogen activator (TPA): used to prevent or reverse blood clots, therefore preventing strokes, heart attacks, or pulmonary embolisms.
2. Human growth hormone: used to treat pituitary dwarfism.
3. Tissue growth factor-beta (TGF-β): to promote new blood vessel and epidermal growth; hence, is potentially useful for wound healing and burns.
4. Human blood clotting factor VIII: to treat hemophiliacs.
5. Human insulin ("humulin"): to treat insulin-dependent diabetes.
6. Bovine growth hormone: to increase cattle and dairy yields.
7. Recombinant vaccines: to human and animal viral diseases.
8. Genetically engineered bacteria that can accelerate the degradation of oil pollutants or

of certain chemicals in toxic wastes (e.g., dioxin).

9. Genetically engineered bacteria that, when sprayed on fields of crops such as strawberries or potatoes, lower by a degree or two the temperature at which the leaves will freeze, thus providing some protection against frost damage compared with untreated plants.

Genetic Engineering of Plants

For many centuries the traditional genetic engineering of plants involved selective breeding experiments in which plants with desirable traits were used as parents for the next generation in order to reproduce offspring with those traits. As a result, humans have produced hardy varieties of plants (e.g., corn, wheat, oats) and have been successful in breeding varieties with increased yields, all by using standard plant breeding techniques. (Similar experiments have also been done with animals such as dogs and horses to produce desired breeds.)

A number of techniques are presently available for the genetic manipulation of plants using recombinant DNA technology. In the near future we can expect a range of plants to be produced that have been engineered with recombinant DNA techniques. Of particular value will be crop plants that have increased yield (e.g., through more efficient nitrogen fixation), insect pest resistance, and herbicide resistance (to enable fields to be sprayed to kill weeds but not the crop). Already, rice has been genetically engineered to produce strains resistant to the damaging rice stripe virus, and wheat has been engineered to be resistant to a particular herbicide. Such genetically engineered plants must be tested extensively before their future release for commercial use. With more sophisticated approaches, some of which are already available, it will be possible to introduce genes into plants to produce transgenic plants and/or control their expression in different tissues. An example would be controlling the rate at which cut flowers die, or the time at which fruit ripens. Already approved for market is the "Flavr Savr" tomato, genetically engineered by Calgene Inc. in collaboration with the Campbell Soup Co. Tomatoes are picked while in an unripe state so they can be shipped without bruising. Prior to shipping they are exposed to ethylene gas, which initiates the ripening process, so that they arrive in the ripened state at the store. Such prematurely picked and artificially ripened tomatoes do not have the flavor of tomatoes picked when they are ripe. Calgene scientists devised a way to block the tomato from making the normal amount of polygalacturonase (PG), a fruit-softening enzyme. They introduced into the plant a copy of the PG gene that

was backwards in its orientation with respect to the promoter and the terminator. When this gene is transcribed, the mRNA is complementary to the mRNA produced by the normal gene: it is called an *antisense* mRNA. In the cell the antisense mRNA binds to the normal, "sense" mRNA, preventing much of it from being translated. Thus much less PG enzyme is produced and the tomato can remain longer on the vine without getting too soft for handling. Once picked, the Flavr-Savr tomato is also less susceptible to bruising when being packed and traveling to the store or to overripening in the store.

Keynote

Recombinant DNA technology is finding ever-increasing applications in the world. In the basic research laboratory biological processes across all of biology are being studied. With appropriate probes a number of genetic diseases can now be diagnosed using recombinant DNA methods, and DNA fingerprinting is being used in forensic science. In addition, many products in the clinical, veterinary, and agricultural areas can be synthesized in commercial quantities using recombinant DNA procedures. Genetic engineering of plants is also possible using recombinant DNA technology. It is expected that many types of improved crops will result from applications of this new technology.

Summary

In this chapter we have discussed some of the procedures involved in recombinant DNA technology and the manipulation of DNA. Collectively these procedures are also referred to as genetic engineering, at least in the popular press. We have seen how it is possible to cut DNA at specific sites using restriction enzymes, how DNA can be cloned into specially constructed vectors, and how the cloned DNA can be analyzed in various ways. Through the construction of genomic libraries and cDNA libraries, and the application of screening procedures to those libraries, a large number of genes have been cloned and identified from a wide variety of organisms.

Restriction mapping analysis has provided detailed molecular maps of genes and chromosomes analogous to the genetic maps constructed on the basis of recombination analysis. Rapid DNA sequencing methods have given us an enormous amount of information about the DNA coding sequences and regulatory sequences of genes. The amount of DNA sequence information available is growing at an extremely rapid rate and computer databases of such sequences are available for researchers to analyze. For example, when a new gene is sequenced, it is of interest to determine if it has any DNA sequences in common with other sequenced genes that are in the database. In this way, potential families of proteins or parts of proteins have been identified. Such groups possibly have related functions in the organisms from which they derive.

In the mid-1980s a new technique called polymerase chain reaction (PCR) was developed. Given some sequence information about a DNA fragment, synthetic oligonucleotides can be made which can be used to amplify large amounts of the DNA fragment from the genome in a repeated cycle of DNA denaturation, annealing of oligonucleotides to act as primers, and extension of the primers with a special DNA polymerase. Numerous applications have been quickly found for PCR, including cloning rare pieces of DNA, preparing DNA for sequencing without cloning, and genetic disease diagnosis.

Recombinant DNA technology and PCR are being widely applied in both basic research and commercial areas. There is probably not an area of basic biology in which these revolutionary molecular techniques have not been applied to the questions being asked. The Human Genome Project has the mandate to generate a complete map of all of the genes in the human genome, and to obtain the complete sequence of the human genome. The knowledge obtained from this ambitious, long-range project will contribute markedly to our understanding of human genetics. In the commercial area recombinant DNA techniques are being used to develop new pharmaceuticals (including drugs, other therapeutics, and vaccines), to develop new tools for diagnosis of infectious and genetic diseases (as part of genetic counseling efforts), for human gene therapy, in forensic analysis (e.g., analyzing DNA from a crime scene to match with a suspect), and in agricultural areas (e.g., improving disease resistance and yields of livestock and crops). While products generated by genetic engineering have been slower coming to the market than originally expected, there is an undiminished enthusiasm for the genesis and commercialization of a wide array of useful products in the present decade.

Analytical Approaches for Solving Genetics Problems

While this is a rather descriptive area, it is often necessary to interpret data derived from restriction enzyme analysis of DNA fragments in order to generate a restriction map, that is, a map of the locations of restriction enzymes. The logic used for this type of analysis is very similar to that used in generating a genetic map of loci from two-point mapping crosses.

Q13.1   A piece of DNA 900 bp long is cloned and then cut out of the vector for analysis. Digestion of this linear piece of DNA with three different restriction enzymes singly and in all possible combinations of pairs, gave the following restriction fragment size data:

ENZYME(S)	RESTRICTION FRAGMENT SIZES
EcoRI	200 bp, 700 bp
HindIII	300 bp, 600 bp
BamHI	50 bp, 350 bp, 500 bp
EcoRI + HindIII	100 bp, 200 bp, 600 bp
EcoRI + BamHI	50 bp, 150 bp, 200 bp, 500 bp
HindIII + BamHI	50 bp, 100 bp, 250 bp, 500 bp

Construct a restriction map from these data.

A13.1   The approach to this kind of problem is to consider a pair of enzymes and to analyze the data from the single and double digestions. First, let us consider the EcoRI and HindIII data. Cutting with EcoRI produces two fragments, one of 200 bp and the other of 700 bp, while cutting with HindIII also produces two fragments, one of 300 bp and the other of 600 bp. Thus we know that both restriction sites are asymmetrically located along the linear DNA fragment with the EcoRI site 200 bp from an end and the HindIII site 300 bp from an end. When we consider the EcoRI + HindIII data we can determine the positions of these two restriction sites relative to one another. If, for example, the EcoRI site is 200 bp from the fragment end, and the HindIII site is 300 bp from that same end, then we would predict that cutting with both enzymes would produce three fragments of sizes 200 bp (end to EcoRI site), 100 bp (EcoRI site to HindIII site), and 600 bp (HindIII site to other end). On the other end, if the EcoRI site is 200 bp from one fragment end, and the HindIII site is 300 bp from the other fragment end, cutting with both enzymes would produce three fragments of sizes 200 bp (end to EcoRI site), 400 bp (EcoRI site to HindIII site), and 300 bp (HindIII site to end). The actual data support the first model.

Now we pick another pair of enzymes, for example, HindIII and BamHI. Cutting with HindIII produces fragments of 300 bp and 600 bp as we have seen, and cutting with *Bam*HI produces three fragments of sizes 50 bp, 350 bp, and 500 bp, indicating that there are two *Bam*HI sites in the DNA fragment. Again the double digestion products are useful in locating the sites. Double digestion with *Hind*III and *Bam*HI produces four fragments of 50 bp, 100 bp, 250 bp, and 500 bp. The simplest interpretation of the data is that the 300-bp *Hind*III fragment is cut into the 50-bp and 250-bp fragments by *Bam*HI, and that the 600-bp *Hind*III fragment is cut into the 100-bp and 500-bp fragments by *Bam*HI. Thus the restriction map shown in the accompanying figure can be drawn:

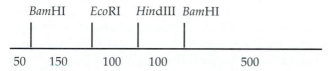

The *Bam*HI + *Eco*RI data are compatible with this model.

Q13.2 The recessive allele *bw*, when homozygous, results in brown eyes in *Drosophila*, in contrast to the wild-type bright-red eye color. A restriction fragment length polymorphism (RFLP) for a particular DNA region results in either two restriction fragments (type I) or one restriction fragment (type II) when *Drosophila* DNA is cut with restriction enzyme C, the fragments separated by DNA electrophoresis, blotted to nitrocellulose, and probed with a particular radioactive DNA probe.

A true-breeding brown-eyed fly with type I DNA was crossed with a true-breeding wild-type fly with type II DNA. The F$_1$ flies had wild-type eye color and exhibited both type I and type II DNA patterns. The F$_1$ flies were crossed with true-breeding brown-eyed flies with type I DNA, and the progeny were scored for eye color and RFLP type. The results were as follows:

CLASS	PHENOTYPES	NUMBER
1	red eyes, type I and II DNA	184
2	red eyes, type I DNA	21
3	brown eyes, type I DNA	168
4	brown eyes, type I and II DNA	27
	total progeny	400

Analyze these data.

A13.2 The eye color mutation is a familiar genetic marker. The restriction fragment length polymorphisms (RFLPs) are also genetic markers and can be analyzed just like any gene marker. If we symbolize the type I DNA as I and the type II DNA as II, the backcross is:

$$\frac{bw^+ \; \text{II}}{bw \;\; \text{I}} \times \frac{bw \; \text{I}}{bw \; \text{I}}$$

(The cross is drawn as if the markers were linked; of course, we have yet to show this.) This is a testcross with the exception that DNA markers do not exhibit dominance or recessiveness. If the eye color and RFLP markers are unlinked, the result would be equal numbers of the four progeny classes. However, the data show a great excess of these two classes: (a) red eyes, type I and II DNA; and (b) brown eyes, type I DNA. Their origin was the pairing of F$_1$ bw^+ II and bw I gametes with bw I gametes to give bw^+ II/bw I and bw I/bw I progeny genotypes, respectively. Similarly, the progeny (a) red eyes, type I DNA and (b) brown eyes, type I and II DNA derive from pairing bw^+ I and bw II gametes with bw I gametes to give bw^+ I/bw I and bw II/bw I progeny, respectively. These two latter classes occur with about equal frequency, that is a frequency much lower than those for the other two classes. The simplest explanation is that the brown eye color gene and the RFLP marker are linked, so the F$_1$ cross was a mapping cross, much like those we analyzed in Chapter 5. Classes 1 and 3 are the parentals, and classes 2 and 4 are the recombinants. The map distribution between *bw* and the RFLP is, therefore:

$$\frac{21 + 27}{\text{Total}} \times 100\%$$

$$= \frac{48}{400} \times 100\%$$

$$= 12 \text{ map units}$$

Questions and Problems

*13.1 A new restriction endonuclease is isolated from a bacterium. This enzyme cuts DNA into fragments that average 4096 base pairs long. Like all other known restriction enzymes, the new one recognizes a sequence in DNA that has twofold rotational symmetry. From the information given, how many base pairs of DNA constitute the recognition sequence for the new enzyme?

13.2 An endonuclease called *Avr*II ("a-v-r-two") cuts DNA whenever it finds the sequence 5'-CCTAGG-3' / 3'-GGATCC-5'. About how many cuts would *Avr*II make in the human genome, which is about 3×10^9 base pairs long and about 40% GC?

13.3 About 40% of the base pairs in human DNA are GC. On the average, how far apart (in terms of base pairs) will the following sequences be?
a. two *Bam*HI sites
b. two *Eco*RI sites
c. two *Not*I sites
d. two *Hae*III sites

13.4 What are the features of plasmid cloning vectors that make them useful for constructing and cloning recombinant DNA molecules?

***13.5** Genomic libraries are important resources for isolating genes of interest and for studying the functional organization of chromosomes. List the steps you would use to make a genomic library of yeast in a lambda vector.

***13.6** Suppose you wanted to produce human insulin (a peptide hormone) by cloning. Assume that this could be done by inserting the human insulin gene into a bacterial host, where, given the appropriate conditions, the human gene would be transcribed and then translated into human insulin. Which do you think it would be best to use as your source of the gene—human genomic insulin DNA or a cDNA copy of this gene? Explain your choice.

13.7 You are given a genomic library of yeast prepared in a bacterial plasmid vector. You are also given a cloned cDNA for human actin, a protein that is conserved in protein sequences among eukaryotes. Outline how you would use these resources to attempt to identify the yeast actin gene.

13.8 Restriction endonucleases are used to construct restriction maps of linear or circular pieces of DNA. The DNA is usually produced in large amounts by recombinant DNA techniques. The generation of restriction maps is similar to the process of putting the pieces of a jigsaw puzzle together. Suppose we have a circular piece of double-stranded DNA that is 5000 base pairs long. If this DNA is digested completely with restriction enzyme I, four DNA fragments are generated: fragment *a* is 2000 base pairs long; fragment *b* is 1400 base pairs long; *c* is 900 base pairs long; and *d* is 700 base pairs long. If, instead, the DNA is incubated with the enzyme for a short time, the result is incomplete digestion of the DNA; not every restriction enzyme site in every DNA molecule will be cut by the enzyme, and all possible combinations of adjacent fragments can be produced. From an incomplete digestion experiment of this type, fragments of DNA were produced from the circular piece of DNA which contained the following combinations of the above fragments: *a-d-b*, *d-a-c*, *c-b-d*, *a-c*, *d-a*, *d-b*, and *b-c*. Lastly, after digesting the original circular DNA to completion with restriction enzyme I, the DNA fragments were treated with restriction enzyme II under conditions conducive to complete digestion. The resulting fragments were: 1400, 1200, 900, 800, 400, and 300. Analyze all the data to locate the restriction enzyme sites as accurately as possible.

***13.9** A piece of DNA 5000 bp long is digested with restriction enzymes A and B, singly and together. The DNA fragments produced were separated by DNA electrophoresis and their sizes were calculated, with the following results:

DIGESTION WITH		
A	B	A + B
2100 bp	2500 bp	1900 bp
1400 bp	1300 bp	1000 bp
1000 bp	1200 bp	800 bp
500 bp		600 bp
		500 bp
		200 bp

Each A fragment was extracted from the gel and digested with enzyme B, and each B fragment was extracted from the gel and digested with enzyme A. The sizes of the resulting DNA fragments were determined by gel electrophoresis, with the following results:

A FRAGMENT	FRAGMENTS PRODUCED BY DIGESTION WITH B	B FRAGMENT	FRAGMENTS PRODUCED BY DIGESTION WITH A
2100 bp →	1900, 200 bp	2500 bp →	1900, 600 bp
1400 bp →	800, 600 bp	1300 bp →	800, 500 bp
1000 bp →	1000 bp	1200 bp →	1000, 200 bp
500 bp →	500 bp		

Construct a restriction map of the 5000 bp DNA fragment.

***13.10** Draw the banding pattern you would expect to see on a DNA sequencing gel if you annealed the primer 5'-C-T-A-G-G-3' to the following single-stranded DNA fragment and carried out a dideoxy sequencing experiment. Assume the dNTP precursors were all labeled with ^{32}P.

3'-G-A-T-C-C-A-A-G-T-C-T-A-C-G-T-A-T-A-G-G-C-C-5'

13.11 DNA was prepared from small samples of white blood cells from a large number of people. Ten different patterns were seen when these DNAs were all digested with *Eco*RI, then subjected to electrophoresis and Southern blotting. Finally, the blot was probed with a radioactively labeled cloned human sequence. The figure below shows the ten DNA patterns taken from ten people.

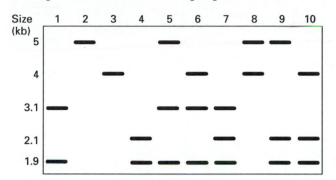

a. Explain the hybridization patterns seen in the ten people in terms of variation in *Eco*RI sites.
b. If the individuals whose DNA samples are in lanes 1 and 6 on the blot were to produce offspring together, what bands would you expect to see in DNA samples from these offspring?

*13.12 Filled symbols in the pedigree (Figure 13.A) indicate people with a rare autosomal dominant genetic disease. DNA samples were prepared from each of the individuals in the pedigree. The samples were digested with a restriction enzyme, electrophoresed, blotted, and probed with a cloned human sequence called DS12–88, with the results shown in Figure 13.B. Do the data in these two figures support the hypothesis that the locus for the disease that is segregating in this family is linked to the region homologous to DS12–88? Make your answer quantitative, and explain your reasoning.

13.13 Imagine that you have been able to clone the structural gene for an enzyme in a catecholamine biosynthetic pathway from the adrenal gland of rats. How could you use this cloned DNA as a probe to determine whether this same gene functions in the brain?

13.14 Imagine that you find an RFLP in the rat genomic region homologous to your cloned catecholamine synthetic gene from Question 13.13, and that in a population of rats displaying this polymorphism there is also a behavioral variation. You find that some of the rats are normally calm and placid, but others are hyperactive, nervous, and easily startled. Your hypothesis is that the behavioral difference is caused by variations in the cloned gene. How could you use your cloned sequence to test this hypothesis?

Figure 13A

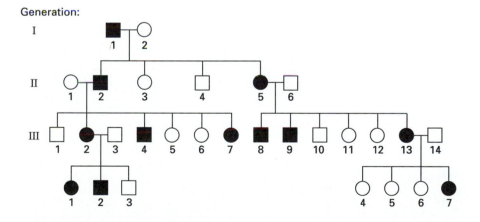

Figure 13B

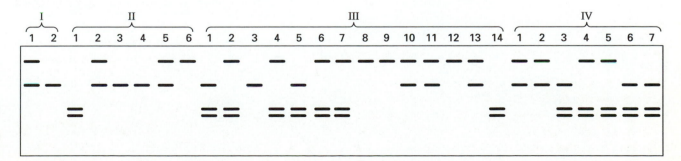

13.15 One application of DNA fingerprinting technology has been to identify stolen children and return them to their parents. Bobby Larson was taken from a supermarket parking lot in New Jersey in 1978 when he was 4 years old. In 1990 a sixteen-year-old boy called Ronald Scott was found in California, living with a couple named Susan and James Scott who claimed to be his parents. Authorities suspected that Susan and James might be the kidnappers, and that Ronald Scott might be Bobby Larson. DNA samples were obtained from Mr. and Mrs. Larson, and from Ronald, Susan, and James Scott. Then DNA fingerprinting was done, using a probe for a particular VNTR family, with the results shown in the figure. From the information in the figure, what can you say about the parentage of Ronald Scott? Explain.

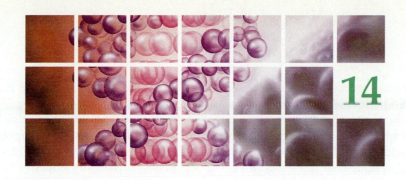

14

Regulation of Gene Expression in Bacteria and Bacteriophages

- A model generated from studies of the synthesis of the lactose-utilizing enzymes of *E. coli* explains the regulation of gene expression in a large number of bacterial and bacteriophage systems. In the lactose system the addition of lactose to the cells brings about a rapid synthesis of three enzymes required for lactose utilization. In the absence of lactose the synthesis of the three enzymes is turned off. The genes for the enzymes are contiguous on the *E. coli* chromosome and are adjacent to a controlling site (an operator) and a single promoter. The genes, the operator, and the promoter constitute an operon, which is transcribed as a single unit. A regulatory gene is associated with an operon. To turn on gene expression in the lactose system, a lactose metabolite binds with a repressor protein (the product of the regulatory gene), inactivating it and preventing it from binding to the operator. As a result RNA polymerase can bind to the promoter and transcribe the three genes as a single polygenic mRNA.

- Expression of a number of bacterial amino acid synthesis operons is accomplished by a repressor-operator system and through attenuation at a second controlling site called an attenuator. The

repressor-operator system functions essentially like that for the *lac* operon, except that the addition of amino acid to the cell activates the repressor, thereby turning the operon off. The attenuator is located downstream from the operator, and acts as a partially effective transcription termination site that allows only a fraction of RNA polymerases to transcribe the rest of the operon. Attenuation involves a coupling between transcription and translation, and the formation of particular RNA secondary structures that signal whether or not transcription can continue.

- Bacteriophages such as lambda are especially adapted for reproducing within a bacterial host. Many of the genes related to the production of progeny phages, or to the establishment or reversal of lysogeny of temperate phages, are organized into operons. These operons, like bacterial operons, are controlled through the interaction of regulatory proteins with operators that are adjacent to clusters of structural genes. Phage lambda has been an excellent model for studying the genetic switch that controls the choice between lytic and lysogenic pathways in a lysogenic phage.

O RGANISMS DO NOT LIVE and reproduce in a constant environment. Through evolutionary processes they have developed ways to compensate for environmental changes and hence to function in a variety of environments. One way that an organism can adjust to a new environment is to alter its gene activity so that gene products appropriate to the new conditions are synthesized. As a result the organism adjusts to grow and reproduce in that environment. This is particularly evident in free-living bacteria.

A change in the set of gene products synthesized in a prokaryotic cell involves regulatory mechanisms that control gene expression. Genes whose activities are controlled in response to the needs of a cell or organism are called **regulated genes**. An organism also possesses a large number of genes whose products are essential to the normal func-

tioning of a growing and dividing cell, no matter what the life-supporting environmental conditions are. These genes are always active in growing cells and are known as **constitutive genes**; examples include genes that code for the components of ribosomes, and enzymes needed for protein synthesis and glucose metabolism. Recognize, though, that all genes are regulated on some level. If the environment suddenly becomes no longer conducive for normal cell function, the expression of all genes, including constitutive genes, will be reduced by regulatory mechanisms. Thus the distinction between regulated and constitutive genes is a somewhat arbitrary one.

In this chapter we examine the mechanisms by which gene expression is regulated in bacteria and bacteriophages. To adapt to alterations in their environments, bacteria have evolved several regulatory mechanisms for turning off the genes whose products are not needed and for turning on the genes whose products are needed. We will learn in detail about some of the basic gene regulation mechanisms in bacteria and how this regulation relates to the organization of genes in the bacterial genome.

Since bacteriophages are parasites, relying on the activities of a host bacterium to reproduce, they must direct their own reproductive cycle in a carefully programmed way; this cycle is directed through the regulation of gene expression. In this chapter we examine the gene regulation events required for controlling the choice between the lytic and lysogenic pathways in bacteriophage lambda.

As we discuss the specific examples of gene regulation in bacteria and their viruses, we will learn that turning genes on and off involves specific interactions between regulatory proteins and DNA sequences. We have already seen many examples of the importance of specific protein-nucleic acid interactions to cell function. The long list includes the interaction of DNA polymerase with DNA origins during replication, and the binding of RNA polymerase with promoters, aminoacyl-tRNA synthetase with tRNAs, ribosomal proteins with rRNAs, restriction enzymes with DNA, and release factors with stop codons on mRNA. The interactions involved in gene regulation are simply further examples of a common theme in cell function. In the next chapter we will see that this theme extends also to gene regulation in eukaryotes.

Gene Regulation of Lactose Utilization in *E. coli*

When gene expression is turned on in a bacterium by the addition of a substance (such as lactose) to the medium, the genes are said to be *inducible.* The regulatory substance that brings about this gene induction is called an *inducer.* The inducer is an example of a class of small molecules, called *effectors* or *effector molecules,* that are involved in the control of expression of many regulated genes.

Figure 14.1 diagrams an inducible gene. Transcription of an inducible gene occurs only in response to a particular regulatory event occurring at a specific DNA sequence near the gene. This DNA sequence is called a *controlling site.* The regulatory event typically involves an inducer—a molecule that causes the cell to turn on the synthesis of specific enzymes—and a regulatory protein. When the regulatory event occurs, RNA polymerase binds to the promoter (usually adjacent to the controlling site) and initiates transcription of the gene. The gene is "turned on," mRNA is made, and the enzyme(s) coded for by the gene(s) is produced.

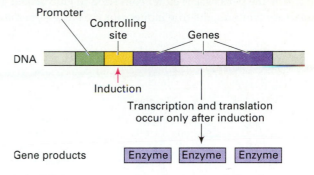

■ *Figure 14.1* General organization of an inducible gene.

Inducible genes are expressed only when induced

The controlling site itself does not code for any product. The phenomenon of producing a gene product only in response to an inducer is called **induction**. We will now discuss the induction of expression of genes required for lactose utilization in *E. coli.*

Lactose as a Carbon Source for *E. coli*

Escherichia coli is able to grow in a simple medium containing salts (including a nitrogen source) and a carbon source such as glucose. The energy for biochemical reactions in the cell comes from the metabolism of the glucose. The enzymes required for the glucose metabolism are coded for by constitutive genes. *E. coli* can use a variety of carbon sources other than glucose for energy. For each carbon source there is a set of genes that encode the necessary enzymes for its metabolism. For the metabolism of lactose (a disaccharide consisting of the two component monosaccharides—glucose and galactose), three enzymes are needed:

1. *β-galactosidase.* This enzyme has two functions (Figure 14.2): (a) It catalyzes the breakdown of lactose into glucose and galactose; (b) It catalyzes the conversion of lactose to a related form called *allolactose,* a compound important in the regulation of expression of the lactose utilization genes.

2. *Lactose permease* (also called *M protein*) is found in the *E. coli* membrane and is needed for the active transport of lactose from the growth medium into the cell.

3. *Transacetylase* is a protein whose function is poorly understood.

In a wild-type *E. coli* growing in a medium containing glucose but no lactose, only a few molecules of each of the three enzymes are produced, indicating a low level of expression of the three genes that

■ *Figure 14.2*   Reactions catalyzed by the enzyme β-galactosidase. Lactose brought into the cell by the permease is either converted to glucose and galactose (top) or to allolactose (bottom), the true inducer for the lactose operon of *E. coli*.

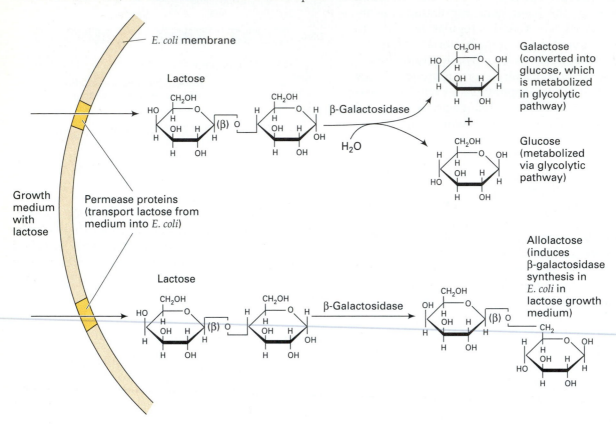

code for the enzymes. If lactose but no glucose is present in the growth medium, however, the number of molecules of each of the three enzymes increases coordinately about a thousandfold. This occurs because the three essentially inactive genes are now being actively transcribed and the resulting mRNA translated. This process is called **coordinate induction**. Further, the mRNAs for the enzymes have a relatively short half-life, so the transcripts must be made continually in order for the enzymes to be produced. So when lactose is no longer present, transcription of the three genes is stopped and any mRNAs present are rapidly broken down, thereby resulting in a drastic reduction in the amounts of the three enzymes in the cell. We will next describe the key features of this gene regulation system.

Experimental Evidence for the Regulation of the lac Genes

Our basic understanding of the organization of the genes, the controlling sites involved in lactose utilization, and the control of expression of the *lac* genes came largely from the genetic experiments of François Jacob and Jacques Monod, for which they received the Nobel Prize.

Mutants of the Protein-Coding Genes. Mutants of the protein-coding (structural) genes were obtained and mapped using standard genetic procedures. From the mutation studies the β-galactosidase gene was named *lacZ*, the permease gene *lacY*, and the transacetylase gene *lacA*. Mapping experiments using mutations of the three genes showed that the structural genes were clustered together in the genome in the order *lacZ-lacY-lacA*. The three structural genes are transcribed onto a single mRNA molecule—called a **polygenic mRNA** or **polycistronic mRNA**—rather than onto three separate mRNAs. That is, RNA polymerase initiates transcription at a single promoter and a polygenic mRNA is synthesized with the gene transcripts in the order 5'-*lacZ*⁺-*lacY*⁺-*lacA*⁺-3'. A ribosome first synthesizes β-galactosidase, then slides along the mRNA until it recognizes the initiation sequence for permease, synthesizes permease, slides along the mRNA until it recognizes the initiation

■ *Figure 14.3*   Translation of the polygenic mRNA encoded by *lac* utilization genes in wild-type *E. coli.*

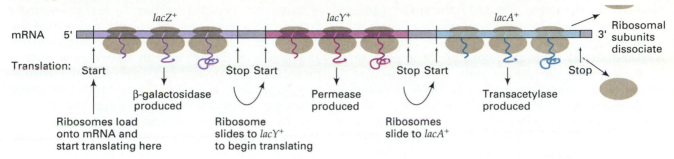

sequence for transacetylase, synthesizes transacetylase, and finally dissociates from the mRNA (Figure 14.3).

Regulatory Mutants.   Of special interest to Jacob and Monod were mutants that affected the regulation of the three lactose utilization genes. In wild-type *E. coli* the three gene products are induced coordinately only when lactose is present. Jacob and Monod isolated a number of mutants in which all gene products of the lactose operon were synthesized *constitutively*; that is, all three enzymes were synthesized whether or not lactose was present. Jacob and Monod hypothesized that the mutants were regulatory mutants that affected the normal mechanisms responsible for controlling the expression of the structural genes. Jacob and Monod identified two classes of constitutive mutants. One class mapped to a small DNA region called the **operator** (*lacO*) that was adjacent to the *lacZ* gene. The other class mapped to a gene-sized DNA region a short distance away, called the *lacI* gene or *lac* **repressor** gene. Figure 14.4 diagrams the organization of the *lac* structural gene cluster and the associated regulatory elements. We will discuss this complex, called the *lac* operon, a little later.

Operator Mutants.   The mutations of the operator were called operator-constitutive, or *lacO*ᶜ, mutations. All *lacO*ᶜ mutants synthesize the lactose utilization enzymes in the presence or absence of lactose. Through the use of partial diploid strains (i.e., F′ strains; see Figure 6.7), Jacob and Monod were able to define better the role of the operator in regulating the expression of the *lac* genes. One such partial diploid was *lacO*⁺ *lacZ*⁻ *lacY*⁺/*lacO*ᶜ *lacZ*⁺ *lacY*⁻ (both gene sets have a normal promoter, and the *lacA* gene is omitted because it is not important for our discussions). This partial diploid was tested for the production of β-galactosidase (from the *lacZ*⁺ gene) and of permease (from the *lacY*⁺ gene), both in the presence and the absence of the inducer lactose.

For this partial diploid, β-galactosidase is synthesized in the absence of the inducer, but permease is not. Only when lactose is added to the culture does permease synthesis occur. The *lacZ*⁺ gene (which is on the same chromosome strand as *lacO*ᶜ) is *constitutively expressed* (the gene is active in the presence or absence of lactose) whereas the *lacY*⁺ gene (which is on the same chromosome strand as *lacO*⁺) is under normal inducible control (the gene is inactive in the absence of lactose and active in the pres-

■ *Figure 14.4*   Organization of the *lac* genes of *E. coli* and the associated regulatory elements, the operator, promoter, and regulatory gene. The region defined as the *lac* operon is shown.

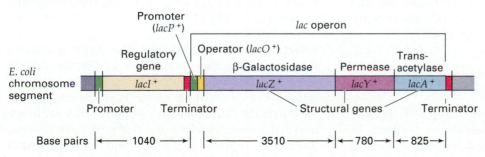

ence of lactose). In other words, a *lacO^c mutation affects only those genes that are adjacent to it on the same chromosome strand*. Similarly, the *lacO^+* region only controls those *lac* structural genes adjacent to it and has no effect on the genes on the other chromosome strand. This phenomenon of a gene or DNA sequence controlling only genes that are on the same, contiguous piece of DNA is called **cis-dominance.** Thus the *lacO^c* mutation is cis-dominant since the defect affects the adjacent genes only and cannot be overcome (complemented) by a normal *lacO^+* region elsewhere in the genome. We know now that the operator is a binding site for the lactose repressor molecule made by the *lacI^+* gene; the operator does not code for a product.

The lacI Gene Regulatory Mutants. The second class of *lac* constitutive mutants defined the *lacI* gene which, we now know, encodes the lactose repressor protein. Again, the use of partial diploid strains illuminated the normal function of the *lacI* gene.

The partial diploid here is *lacI^+ lacO^+ lacZ^- lacY^+/ lacI^- lacO^+ lacZ^+ lacY^-*; both operons have normal operators and promoters. In the absence of lactose, no β-galactosidase or permease was produced; both were synthesized in the presence of lactose. In other words, the expression of *both* genes was inducible. This means that the *lacI^+* gene in the cell can overcome the defect of the *lacI^-* mutation. Since the two *lacI* genes are located on different chromosomes (they are in a *trans* configuration) the *lacI^+* is said to be **trans-dominant** to *lacI^-*.

Since the *lacI^+* gene controlled the genes on the other chromosome strand, the *lacI* gene must produce a product that could move through the cell. Jacob and Monod proposed that the *lacI^+* gene produces a functional **repressor molecule** (hence the *lacI* gene is also called the *lac* **repressor gene**), and that no functional repressor molecules are produced in *lacI^-* mutants. Thus in a haploid bacterial strain that has a *lacI^-* mutation, the *lac* operon is constitutive. In a partial diploid with both a *lacI^+* and a *lacI^-*, however, the functional repressor molecules produced by the *lacI^+* gene control the expression of *both lac* operons present in the cell, making *both* operons inducible. As we will soon see, the repressor encoded by the *lacI^+* gene exerts its regulatory action by binding to the operator.

Jacob and Monod's Operon Model for the Regulation of the *lac* Genes

Based on the results of their studies of genetic mutants affecting the regulation of the synthesis of the lactose utilization enzymes, Jacob and Monod

proposed their now-classical **operon model.** By definition, an **operon** is *a cluster of genes, the expressions of which are regulated together by operator-regulator protein interactions, plus the operator region itself and the promoter.* The order of the controlling elements and genes in the *lac* operon (see Figure 14.4) is promoter-operator-*lacZ-lacY-lacA* and the regulatory gene, *lacI*, is located close to the structural genes, just upstream of the promoter. The *lacI* gene has its own promoter and terminator. The *lac* repressor encoded by *lacI* is made constitutively but its ability to bind to the operator is affected by the presence of lactose.

The following description of the Jacob-Monod model for the regulation of the *lac* operon has been embellished with more up-to-date molecular information. Figure 14.5 diagrams the state of the *lac* operon in wild-type *E. coli* growing in the absence of lactose. In this situation the *lacI^+* gene is expressed to produce a polypeptide, four of which associate together to form the *lac* repressor protein. The repressor binds to the operator (*lacO^+*). When the repressor is bound to the operator, RNA polymerase can bind to the operon's promoter but cannot initiate transcription; hence, transcription of the three protein-coding genes is prevented. The *lac* operon is said to be under *negative control* since the binding of the repressor at the operator site blocks transcription of the structural genes.

When wild-type *E. coli* grows in the presence of lactose as the sole carbon source (Figure 14.6), some of the lactose transported into the cell is converted by existing molecules of β-galactosidase into a metabolite of lactose called *allolactose*, which is the actual *inducer* for the *lac* operon. Allolactose binds to the repressor, causing the shape of the repressor to change. Consequently, the repressor cannot bind to the *lac* operator, and it falls off the operator. Free repressor proteins are also changed in shape so that they cannot bind to the operator. Now, RNA polymerase is no longer blocked and is able to transcribe the operon to produce the polygenic mRNA molecule for the three structural genes. The polygenic mRNA for the *lac* operon is translated by a string of ribosomes to produce the three proteins specified by the operon. This efficient mechanism ensures the coordinate (simultaneous) production of proteins of related function.

Effect of *lacO^c* Mutations. The *lacO^c* mutations lead to constitutive expression of the *lac* operon genes and are *cis*-dominant to *lacO^+*. The explanation of the phenotype of *lacO^c* mutations is that base-pair alterations of the operator DNA sequence make it unrecognizable by the repressor protein. Since the repressor cannot bind, the structural

■ *Figure 14.5* Functional state of the *lac* operon in wild-type *E. coli* growing in the absence of lactose.

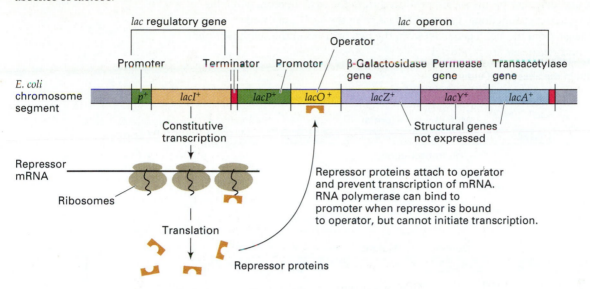

genes become constitutively expressed. The cis-dominance of *lacO*ᶜ is due to the fact that the repressor cannot bind to a *lacO*ᶜ mutant, but can bind to a wild-type operator. The cis-dominance of a *lacO*ᶜ mutant is illustrated for the partial diploid described earlier, *lacI*⁺ *lacO*⁺ *lacZ*⁻ *lacY*⁺ / *lacI*⁺ *lacO*ᶜ *lacZ*⁺ *lacY*⁻, growing in the absence (Figure 14.7[a]) and presence (Figure 14.7[b]) of inducer.

Effects of *lacI*⁻ Mutations. The *lacI*⁻ mutations lead to constitutive expression of the *lac* operon genes and are recessive to *lacI*⁺. The *lacI*⁻ mutations result in a repressor with an altered shape such that it cannot bind to the operator. As a consequence, transcription cannot be prevented, even in the absence of lactose, and there is constitutive expression of the *lac* operon.

■ *Figure 14.6* Functional state of the *lac* operon in wild-type *E. coli* growing in the presence of lactose as the sole carbon source.

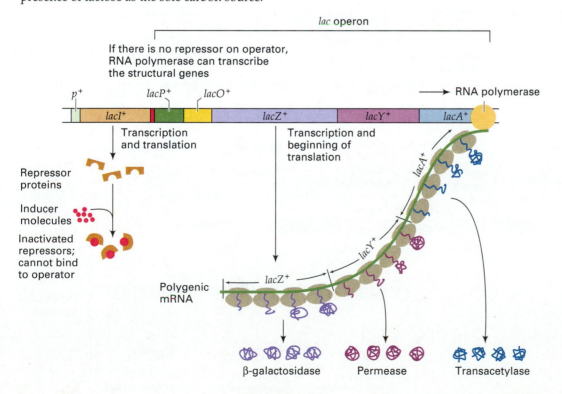

■ **Figure 14.7** *Cis*-dominant effect of *lacO*ᶜ mutation in a partial-diploid strain of *E. coli*: (a) In the absence of the inducer the *lacO⁺* operon is turned off, while the *lacO*ᶜ operon produces functional β-galactosidase from the *lacZ⁺* gene and nonfunctional permease molecules from the *lacY⁻* gene; (b) In the presence of the inducer, the functional β-galactosidase and defective permease are produced from the *lacO*ᶜ operon, while the *lacO⁺* operon produces nonfunctional β-galactosidase from the *lacZ⁻* gene and functional permease from the *lacY⁺* gene. Between the two operons in the cell, functional β-galactosidase and permease are produced.

a) **Partial diploid in the absence of inducer**

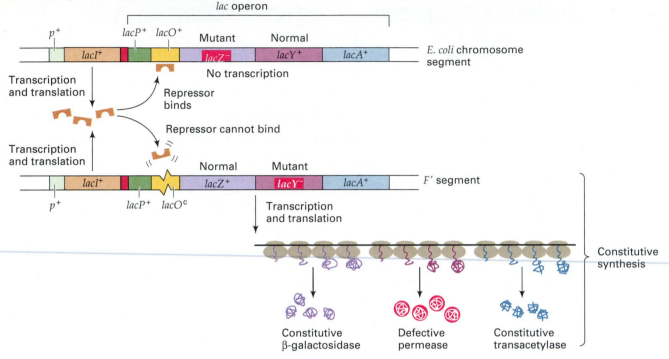

b) **Partial diploid in the presence of inducer**

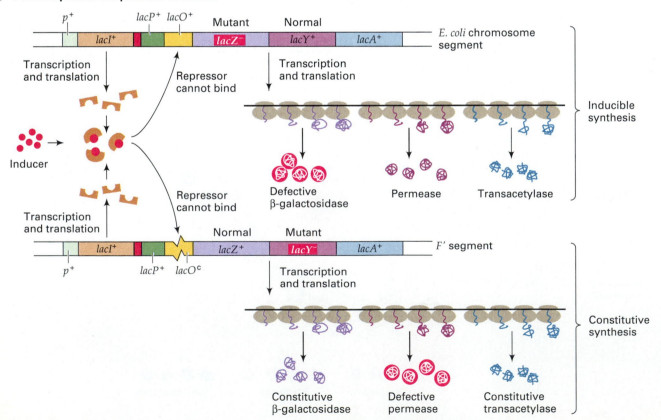

The dominance of the *lacI*⁺ (wild-type) gene over *lacI*⁻ mutants is illustrated for the partial diploid described earlier, *lacI*⁺ *lacO*⁺ *lacZ*⁻ *lacY*⁺ / *lacI*⁻ *lacO*⁺ *lacZ*⁺ *lacY*⁻, in Figure 14.8(a) and (b). When inducer is absent (Figure 14.8[a]) the defective *lacI*⁻ repressor is unable to bind to either normal operator (*lacO*⁺) in the cell. But sufficient normal repressors, produced from the *lacI*⁺ gene, are present that

■ *Figure 14.8* Effect of a *lacI*⁻ mutation in a partial-diploid strain *lacI*⁺ *lacO*⁺ *lacZ*⁻ *lacY*⁺ / *lacI*⁻ *lacO*⁺ *lacZ*⁺ *lacY*⁻ in (a) the absence or (b) the presence of inducer.

a) **Partial diploid in the absence of inducer**

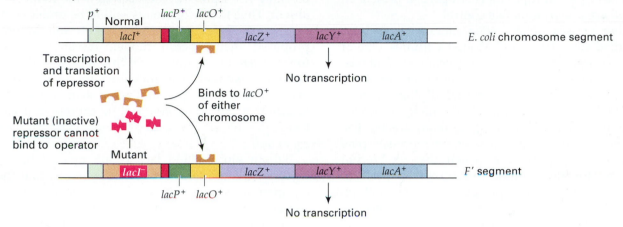

b) **Partial diploid in the presence of inducer**

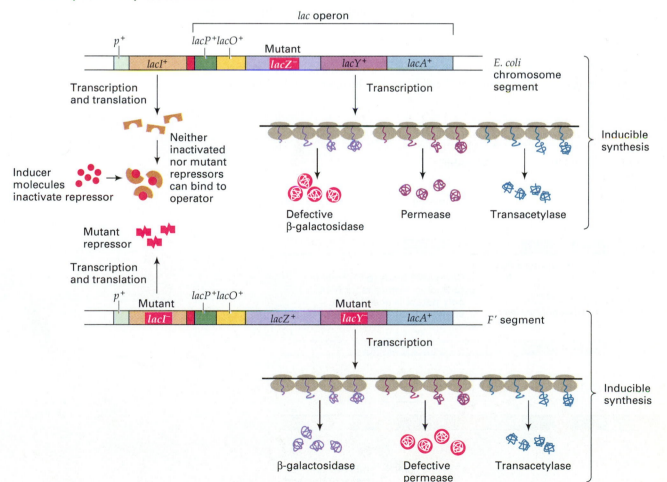

bind to the two operators and block transcription of both operons. When inducer is present (Figure 14.8[b]) the wild-type repressors are inactivated, so both operons are transcribed. One produces a defective β-galactosidase and a normal permease, while the other produces a normal β-galactosidase and a defective permease; between them, active β-galactosidase and permease are produced. Thus in $lacI^+/lacI^-$ partial diploids *both* operons present in the cell are under inducible control.

Other classes of *lacI* gene mutants have been identified since Jacob and Monod's studies. One of these classes, the $lacI^s$ (*superrepressor*) mutants, shows no production of *lac* enzymes in the presence or absence of lactose. In partial diploids with a $lacI^+/lacI^s$ genotype, the $lacI^s$ allele is *trans-dominant*, affecting both chromosomal strands. The explanation here is that the mutant repressor gene produces a superrepressor protein that can bind to the operator but cannot recognize the inducer. Therefore the mutant superrepressors get stuck on the operators, and transcription of the operons can never occur even in the presence of the inducer. The presence of normal repressors in the cell has no effect on this situation, because once a $lacI^s$ repressor is on the operator, the repressor cannot be induced to fall off. Basal levels of transcription will

occur since the superrepressor is not permanently (covalently) bound to the operator. Cells with a $lacI^s$ mutation cannot use lactose as a carbon source.

Positive Control of the *lac* Operon

In the *lac* operon the repressor protein exerts a negative effect on the expression of the *lac* operon by blocking RNA polymerase's action if the inducer is absent. Thus the *lac* operon is said to be under *negative control*. Several years after Jacob and Monod proposed their operon model, researchers also found a *positive control system* that regulates the *lac* operon, a system that functions to turn on the expression of the operon. This system is used to ensure that the *lac* operon will be expressed if lactose is the sole carbon source *but not if glucose is present as well*.

If both glucose and lactose are present in the medium, the glucose is used preferentially and the *lac* operon is not expressed. The *lac* operon is repressed under these conditions because the concentration of a *positive regulator* that binds to the *lac* operon to make transcription possible is reduced in the presence of glucose. Figure 14.9 shows the mechanism involved. First, a protein called *CAP* (catabolite activator protein) binds with *cAMP*

■ *Figure 14.9*   Role of cyclic AMP (cAMP) in the functioning of glucose-sensitive operons such as the lactose operon of *E. coli*. Shown is the operon under inducing conditions.

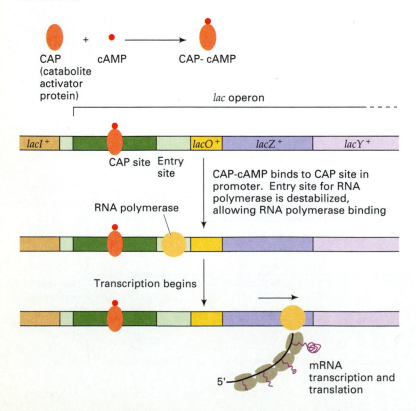

(cyclic AMP, or cyclic adenosine 3',5'-monophosphate; see Figure 14.10), to form a CAP–cAMP complex. This complex is the positive-regulator molecule. Next, the CAP–cAMP complex binds to a specific site in the DNA, called the *CAP site*, and this makes it possible for RNA polymerase to bind to the promoter sequence in the presence of an inducer. The polymerase then moves toward the structural genes and begins transcription of those genes.

In the presence of glucose **catabolite repression (glucose effect)** occurs; that is, the inactivation of an inducible bacterial operon in the presence of glucose even though the operon's inducer is present. Catabolite repression works for the *lac* operon as follows: The presence of glucose in the medium causes the inactivation of adenylate cyclase. As a result, the level of cAMP in the cell drops dramatically. Insufficient CAP–cAMP complex is then available to allow RNA polymerase to bind to the *lac* operon promoter, and transcription is blocked, even though repressors are removed from the operator by the presence of the inducer.

Catabolite repression occurs in a number of other bacterial operons that are involved in the catabolism of sugars other than glucose. In all instances, if glucose is present, catabolite repression occurs and, through a similar mechanism of CAP–cAMP binding to a CAP binding site in DNA, blocks the expression of those operons.

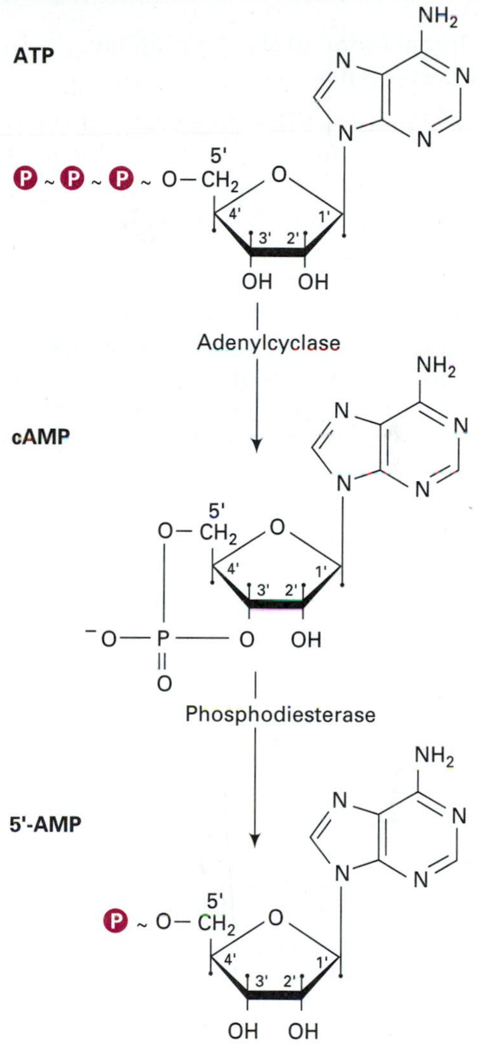

Keynote

Studies of the synthesis of the lactose-utilizing enzymes of *E. coli* generated a model that is the basis for the regulation of gene expression in a large number of bacterial and bacteriophage systems. In the lactose system the addition of lactose to the cells brings about a rapid synthesis of three enzymes required for lactose utilization. The genes for these enzymes are contiguous on the *E. coli* chromosome and are adjacent to a controlling site (an operator) and a single promoter. The genes, the operator, and the promoter constitute an operon, which is transcribed as a single unit. In the absence of lactose the operon is turned off.

Tryptophan Operon of *E. coli*

Just as glucose may not always be available in a bacterial growth medium for use as a carbon source, all necessary amino acids may not be present in a growth medium to enable bacteria to assemble proteins. If an amino acid is missing, a bacterium has

■ *Figure 14.10*   Structure of cyclic AMP (cAMP, or cyclic adenosine 3',5'-monophosphate). cAMP is synthesized from ATP in a reaction catalyzed by adenylcyclase, and is broken down in a reaction catalyzed by phosphodiesterase.

certain operons and other gene systems that enable it to manufacture that amino acid so that the bacterium may grow and reproduce.

When all 20 amino acids are present, the genes encoding the enzymes for all the amino acid biosynthetic pathways are turned off. If an amino acid is not present in the medium, however, the genes must be turned on in order for the biosynthetic enzymes to be made. Unlike the *lac* operon, where gene activity is induced when a chemical (lactose) is added to the medium, in this case there is a repression of gene activity when a chemical (an amino acid) is added. We refer to amino acid biosynthesis operons controlled in this way as *repressible operons*. A repressible operon in *E. coli* that has been extensively studied is the operon for the biosynthesis of the amino acid tryptophan

(Trp). Although the regulation of the trp operon shows some basic similarities to the regulation of the classical *lac* operon, some intriguing differences also appear to be common among similar, repressible, amino acid biosynthesis operons in bacteria.

Gene Organization of the Tryptophan Biosynthesis Genes

Figure 14.11 shows the organization of the controlling sites and of the genes that code for the tryptophan biosynthetic enzymes and how they relate to the biosynthetic steps. Much of the work that we will discuss is that of Charles Yanofsky and his collaborators.

Five structural genes (*trpA* through *trpE*) occur in the tryptophan operon. The promoter and operator regions are upstream from the *trpE* gene. Between the promoter-operator region and *trpE* is a 162-base-pair region called *trpL*, the leader region. Within *trpL*, relatively close to *trpE*, is an *attenuator site* (*att*) that plays an important role in the regulation of the tryptophan operon, as we will see. Tran-

scription of the operon results in the production of a polygenic mRNA containing the transcripts for the five structural genes.

Regulation of the *trp* Operon

Two regulatory mechanisms are involved in controlling the expression of the *trp* operon. One mechanism uses a repressor-operator interaction, and the other determines whether initiated transcripts continue through the structural genes.

Expression of the *trp* Operon in the Presence of Tryptophan. The regulatory gene for the *trp* operon is *trpR*; it is located at some distance from the operon (and therefore does not appear in Figure 14.11). The product of *trpR* is an *aporepressor protein* which, alone, cannot bind to the operator. When tryptophan is abundant in the growth medium, it binds to the aporepressor and converts it to an active repressor. (Tryptophan is an example of an *effector* molecule, just as allolactose is the effector molecule for the *lac* operon.) The active repressor

■ *Figure 14.11* Organization of controlling sites and the structural genes of the *E. coli* tryptophan operon. Also shown are the steps catalyzed by the products of the structural genes *trpA*, *trpB*, *trpC*, *trpD*, and *trpE*.

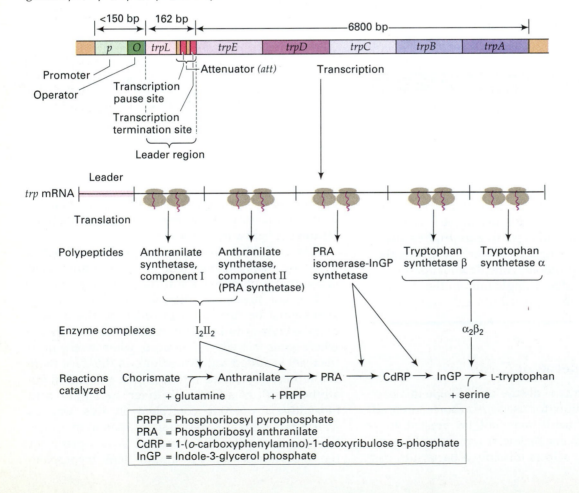

binds to the operator and prevents the initiation of transcription of the *trp* operon protein-coding genes by RNA polymerase. As a result, the tryptophan biosynthesis enzymes are not produced. By repression, transcription of the *trp* operon can be reduced about seventyfold. In short, this part of the *trp* operon regulatory system responds to the amount of tryptophan in the cell.

Expression of the *trp* Operon in the Absence of Tryptophan or in the Presence of Low Amounts of Tryptophan. The second regulatory mechanism is involved in the expression of the *trp* operon under conditions of tryptophan starvation or tryptophan limitation. Under severe tryptophan starvation, the *trp* genes are expressed maximally, while under less severe starvation conditions, the *trp* genes are expressed at less than maximal levels. This is accomplished by a mechanism that controls the proportion of the transcripts that, once started, continue through the five *trp* structural genes. Some RNA transcripts are of the complete operon—the leader region plus the structural genes—while others are short, 140-bp transcripts that have terminated at the attenuator site within the *trpL* region (see Figure 14.11). The short transcripts have terminated by a process called **attenuation**. The proportion of the transcripts that continue through the structural genes is inversely related to the amount of tryptophan in the cell: the more tryptophan there is, the greater the proportion of short transcripts. Attenuation can reduce transcription of the *trp* operon by 8-fold to 10-fold. Thus repression and attenuation together can regulate the transcription of the *trp* operon over a range of about 560-fold to 700-fold.

Molecular Model for Attenuation. The mRNA transcript of the leader region includes a sequence that can be translated to produce a short polypeptide. Near the stop codon are two adjacent codons for tryptophan. As we will see, these Trp codons play an important role in attenuation.

There are four regions of the leader peptide mRNA that can form secondary structures by complementary base pairing (Figure 14.12). Pairing of regions 1 and 2 results in a *pause signal*, pairing of 3 and 4 is a *termination of transcription signal* (similar to the *rho*-independent signal for termination of transcription; see Chapter 11, p. 240), and pairing of 2 and 3 is an *antitermination signal* for transcription. The role of these signals will now be explained.

Crucial to the attenuation model is the fact that transcription and translation are tightly coupled in prokaryotes. In the *trp* regulatory system translation of the leader mRNA transcript is occurring just behind the point at which RNA polymerase is at work extending the transcript. This coupling of transcription and translation is brought about by a pause of the RNA polymerase caused by the pairing of RNA regions 1 and 2 just after they have been synthesized (see Figure 14.12). The pause lasts long enough for the ribosome to load onto the mRNA and to begin translating the leader peptide, and this results in the tight coupling of transcription and translation.

As coupled transcription/translation continues, the position of the ribosome on the leader transcript plays a significant role in the regulation of transcription termination at the attenuator. If the cells are starved for tryptophan, then the amount of Trp-

■ *Figure 14.12* Four regions of the tryptophan operon leader mRNA that can form alternate secondary structures by complementary base pairing.

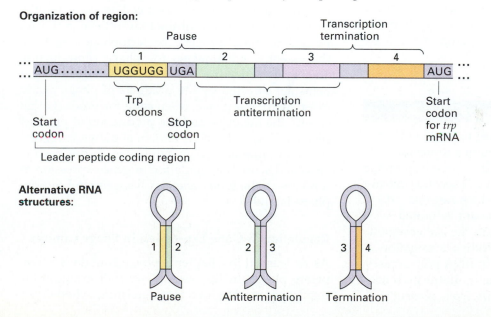

tRNA.Trp molecules (charged tryptophanyl-tRNA) drops dramatically, because very few tryptophan molecules are available for the aminoacylation of the tRNA.Trp molecules. As a result, a ribosome translating the leader transcript stalls at the tandem Trp codons because the next specified amino acid in the peptide cannot be added to the growing chain; the leader peptide cannot be completed (Figure 14.13[a]). The stall allows RNA region 2 to pair with RNA region 3 once region 3 is synthesized. Because region 3 is paired with region 2, region 3 cannot then pair with region 4 when it is synthesized. The 2:3 pairing somehow acts as an antitermination signal, allowing RNA polymerase to continue past the attenuator and transcribe the structural genes.

If, instead, sufficient tryptophan is present so that the ribosome can read the Trp codons (Figure 14.13[b]), the ribosome stalls instead at the stop codon for the leader peptide. Since the ribosome is then covering part of RNA region 2, region 2 is unable to pair with region 3, and region 3 is then able to pair with region 4 when it is synthesized. The bonding of region 3 with region 4 is a termination signal for RNA polymerase to stop transcription. The 3:4 structure is referred to as the attenuator. In other words, the RNA polymerase is "sensing" the secondary structure of the RNA it is making. The key signal for attenuation is the level of Trp-tRNA.Trp in the cell; that level determines whether or not ribosomes will stall on the leader transcript.

Attenuation has been shown to be involved in the genetic regulation of a number of other amino acid biosynthetic operons of *E. coli* and *Salmonella typhimurium*. In every case a leader sequence exists in which there are many codons that specify the amino acid for which the operon specifies the biosynthetic enzymes. For example, the histidine operon of *E. coli* has a string of 7 histidines in the leader peptide, and 7 of the 15 amino acids in the leader for the phenylalanine A operon of *E. coli* are phenylalanine.

Keynote

Expression of the tryptophan (*trp*) operon of *E. coli* is accomplished primarily through a repressor-operator system, which responds to free tryptophan levels, and through attenuation at a second controlling site called an attenuator, which responds to Trp-tRNA.Trp levels. The attenuator is located downstream from the operator in the leader region. The attenuator is a partially effective termination site that allows only a fraction of RNA polymerases to transcribe the rest of the operon. In the presence of tryptophan enough Trp-tRNA.Trp is present so

that the ribosome can move past the attenuator and allow the leader transcript to form a secondary structure that causes transcription to be blocked. In the absence of tryptophan the ribosomes stall at the attenuator, and the leader transcript forms a secondary structure that permits continued transcription activity.

Summary of Operon Function

There are numerous examples of operons in bacteria and their bacteriophages. The following generalizations can be made about the regulation of operons:

1. A regulator protein (e.g., the *lac* repressor) plays a key role in the regulation process since it is able to bind to a controlling site in the operon, the operator.

2. The transcription of a set of clustered structural genes is controlled by an adjacent operator through interaction with a regulator protein which can exert positive or negative control depending on the operon.

3. The trigger for changing the state of an operon from off to on and vice versa is an effector molecule (e.g., the inducer allolactose in the case of the *lac* operon); it controls the conditions under which the regulator protein will or will not bind to the operator.

It should be pointed out, however, that not in all cases are genes for a related function clustered in the prokaryotic genome.

Gene Regulation in Bacteriophages

Because bacteriophages exist by parasitizing bacteria, they themselves do not need to code for the required enzymes for all the biosynthetic pathways needed to make progeny bacteriophages. Instead, the essential components for phage reproduction are provided by the host cell, and those components are directed by the products of phage genes. Most genes of a phage, then, code for products that control the life cycle and the production of progeny phage particles. Much is known about gene regulation in a number of bacteriophages. In this section we will discuss the regulation of gene expression as it relates to the lytic cycle and lysogeny in bacteriophage lambda (λ).

Regulation of Gene Expression in Phage Lambda

As we learned in Chapter 6, phage lambda is a temperate phage. In this section we will examine the regulatory mechanisms that determine whether a λ

■ *Figure 14.13* Models for attenuation in the tryptophan operon of *E. coli;* the taupe structures are ribosomes that are translating the leader transcript: (a) Tryptophan-starved cells; (b) Cells not starved for tryptophan.

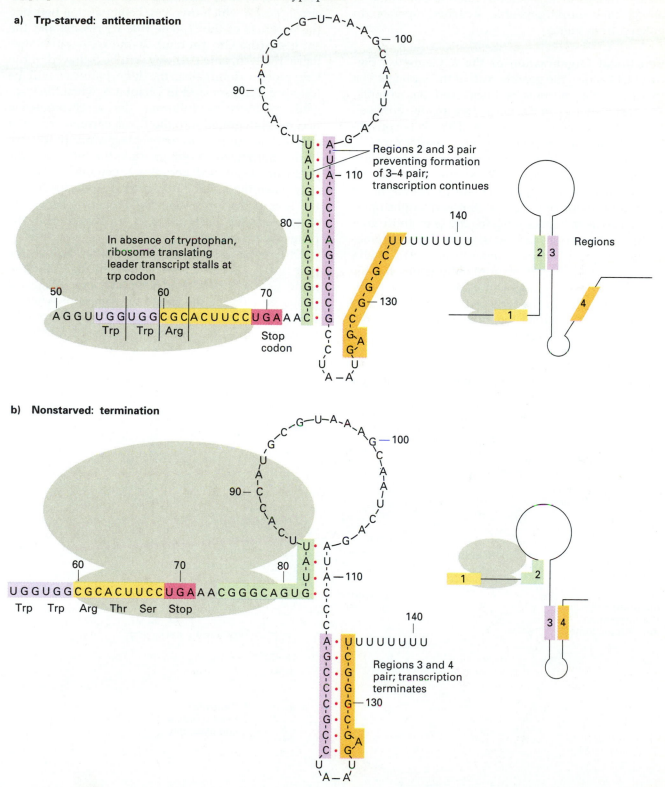

phage enters the lytic or lysogenic cycle. This system is an excellent model for a developmental switch and, as such, has contributed to our thinking about how developmental switches operate in eukaryotic systems.

Functional Organization of the λ Genome. Figure 14.14 shows the genetic map of the λ phage. The mature λ chromosome is linear and has complementary "sticky" ends. Once free in the host cell the λ chromosome circularizes. Thus it is conventional to show the genetic map in a circular form.

In λ, genes with related function are clustered in the genome. The genes for DNA replication that are active during the lytic cycle are clustered in an operon at about one o'clock (the early right operon), whereas the genes for lysogeny (e.g., those controlling integration and excision of the λ chromosome) are clustered in an operon at ten to twelve o'clock (the early left operon). The genes for the various structural components of the phage particles (active only in the lytic cycle)—the heads and the tails—are clustered together in the late operon (from two to seven o'clock).

Early Transcription Events. Soon after λ infects *E. coli* a choice is made between the lytic and lysogenic pathways. This decision depends on a sophisticated *genetic switch* which involves a competition between the products of the *cI* gene (the repressor) and the *cro* gene (the Cro protein). If the repressor dominates, the lysogenic pathway will be followed; if the Cro protein dominates, the lytic pathway will be followed. As discussed in Chapter 6, when the lysogenic pathway is followed the λ chromosome becomes integrated into the *E. coli* chromosome at a specific site and no progeny phages are produced. In this integrated, prophage state the lytic pathway genes are repressed and the λ genome replicates only when the *E. coli* chromosome replicates. When the lytic pathway is followed progeny phages are assembled, the bacterial cells are lysed, and the phages are released.

Different genes are expressed when λ follows the lytic or the lysogenic pathway. When the λ chromosome first infects a cell, however, some stages of phage growth are the same, regardless of whether the lytic or lysogenic pathway is followed. First, the linear chromosome becomes circular through the

■ *Figure 14.14*   A map of phage λ showing the major genes.

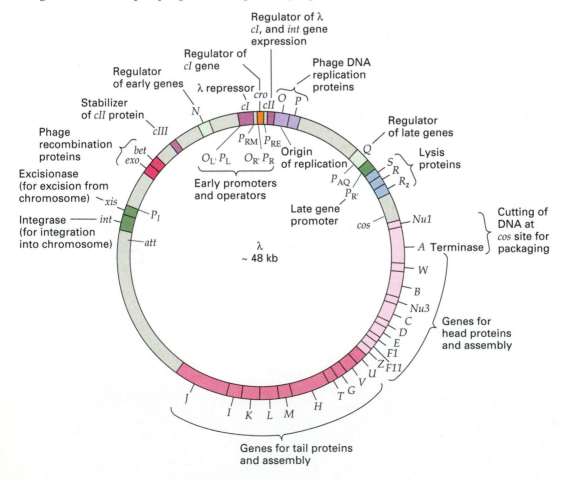

pairing of the complementary ends. Then, transcription begins at promoters P_L and P_R (Figure 14.15[1]). (P_L is the promoter for leftward transcription of the left early operon, and P_R is the promoter for rightward transcription of the right early operon.) Promoter P_L is on a different DNA strand from the P_R promoter, and hence P_L is oriented in an opposite direction from P_R. As a result, transcrip-

tion occurs in opposite directions, counterclockwise for P_L, and clockwise for P_R.

From P_R, the first gene to be transcribed is *cro* (control of *repressor* and *other*), the product of which is the Cro protein. This protein is important for setting the genetic switch to the lytic pathway. From P_L, the first gene to be transcribed is *N*. The resulting N protein is a transcription *antiterminator*

■ *Figure 14.15* Expression of λ genes after infecting *E. coli* and the transcriptional events that occur when either the lysogenic or lytic pathways are followed.

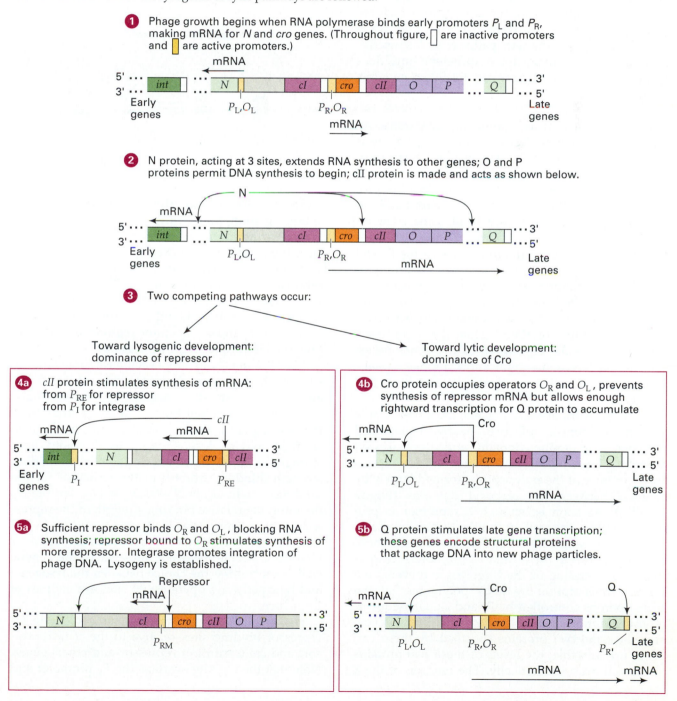

1 Phage growth begins when RNA polymerase binds early promoters P_L and P_R, making mRNA for *N* and *cro* genes. (Throughout figure, ▯ are inactive promoters and ▮ are active promoters.)

2 N protein, acting at 3 sites, extends RNA synthesis to other genes; O and P proteins permit DNA synthesis to begin; cII protein is made and acts as shown below.

3 Two competing pathways occur:

Toward lysogenic development: dominance of repressor

Toward lytic development: dominance of Cro

4a *cII* protein stimulates synthesis of mRNA: from P_{RE} for repressor from P_I for integrase

5a Sufficient repressor binds O_R and O_L, blocking RNA synthesis; repressor bound to O_R stimulates synthesis of more repressor. Integrase promotes integration of phage DNA. Lysogeny is established.

4b Cro protein occupies operators O_R and O_L, prevents synthesis of repressor mRNA but allows enough rightward transcription for Q protein to accumulate

5b Q protein stimulates late gene transcription; these genes encode structural proteins that package DNA into new phage particles.

that allows RNA synthesis to continue past certain transcription terminators. This process is called *antitermination*. N protein acts to allow RNA polymerase to proceed leftward of gene N, and rightward of gene *cro*, thereby including all of the early genes (Figure 14.15[2]). One of the genes transcribed as a result of the action of N protein is *cII*. The cII protein turns on the following genes: *cI* (encodes the λ repressor), O and P (encodes two DNA replication proteins), and Q (encodes a protein needed to turn on late genes for lysis and phage particle proteins). The Q protein functions as another antiterminator protein, permitting transcription to continue into the late genes, which are involved in the lytic pathway. However, only when the switch is set to the lytic pathway and transcription continues from P_R for a sufficient time does enough Q protein accumulate to function effectively.

The Lysogenic Pathway.   After the early transcription events, either the lysogenic or lytic pathway is followed (Figure 14.15[3]). The switch for the lysogenic pathway is set in the following way.

The establishment of lysogeny requires the protein products of the *cII* (right early operon) and *cIII* (left early operon; see Figure 14.14) genes. The cII protein (stabilized by cIII protein) activates transcription of the *cI* gene (located between the P_L and P_R promoters) (Figure 14.15[4a]) from a promoter called P_{RE} (promoter for *r*epressor *e*stablishment), located to the right of the *cI-cro* region. This transcription takes place in a counterclockwise direction. The product of the *cI* gene, the λ repressor protein, blocks the expression of the genes necessary for chromosome replication, progeny phage assembly, and cell lysis, thereby maintaining the lysogenic state.

From the work of Mark Ptashne the λ repressor is known to consist of two nearly equal-sized domains, the amino domain and the carboxyl domain (Figure 14.16[a]). The functional λ repressor is a dimer of the polypeptide, formed largely by contacts between the carboxyl domains (Figure 14.16[b]). As soon as enough λ repressor is produced within the cell, it binds to two operator regions, O_L and O_R (see Figure 14.14), whose sequences overlap the P_L and P_R promoters, respectively. The binding of the λ repressor prevents the further transcription by RNA polymerase of the early operons controlled by P_L and P_R. Consequently, transcription of the N and *cro* genes is blocked, and since the two proteins specified by these two genes are unstable, the levels of these two proteins in the cell drop dramatically. The binding of the λ

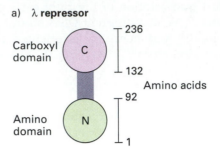

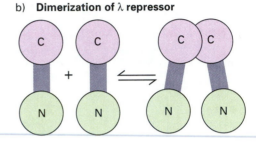

■ *Figure 14.16*   (a) The λ repressor showing the amino (N) and carboxyl (C) domains and the short connecting segment; (b) Repressor monomers from dimers, which can dissociate to monomers.

repressor also blocks the N protein-controlled antitermination transcription of the O, P, and Q genes. Furthermore, a repressor bound to O stimulates the synthesis of more repressor mRNA from a different promoter, P_{RM}, thereby maintaining repressor concentrations in the cell (see Figure 14.15[5a]). Thus if enough λ repressors are present, lysogeny is established by the binding of the repressor to the O_L and O_R operator regions and integration of λ DNA catalyzed by integrase.

The O_L and O_R operators each have three binding sites for the repressor, designated O_{L1}, O_{L2}, and O_{L3}, and O_{R1}, O_{R2}, and O_{R3}, respectively. Interestingly, the repressor does not bind with equal strength to the binding sites. For O_L, the repressor binds most strongly to O_{L1}, less strongly to O_{L2}, and least strongly to O_{L3}. In other words, the relative binding strength (binding affinity) of the repressor for the binding sites in O_L, is $O_{L1} \geqslant O_{L2} \geqslant O_{L3}$. For the O_R operator, the relative binding strength of the repressor for the binding sites is $O_{R1} \geqslant O_{R2} \geqslant O_{R3}$. Let us consider the events that occur at the right operator (O_R) sites, which are major sites where the genetic switch controlling the choice between the lysogenic and lytic pathways operates; O_L sites are not part of the switch.

Figure 14.17 shows the arrangement of the O_R repressor binding sites relative to the *cI* (λ repressor) and *cro* (*c*ontrol of *r*epressor and *o*ther) genes. Note that the O_{R1} site overlaps the P_R promoter and

■ *Figure 14.17* Part of the chromosome, showing the two promoters P_{RM} and P_R and the right operator (O_R), which overlaps them. Transcription from P_{RM} occurs leftward to transcribe the repressor gene (*cI*). Transcription from P_R proceeds rightward to transcribe the *cro* gene.

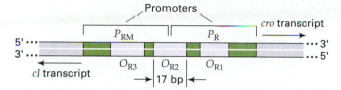

■ *Figure 14.18* A repressor binds to three sites in the right operator (O_R): (a) The affinity of repressor for O_{R1} is about 10 times higher than for O_{R2} or O_{R3}; (b) Once repressor is bound to O_{R1}, a second repressor rapidly binds to O_{R2}; (c) Repressor binds to O_{R3} only at high repressor concentrations.

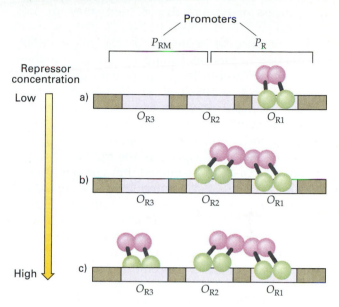

that the O_{R3} site overlaps the P_{RM} (RM = repressor maintenance) promoter. Because of the different binding affinities of the three binding sites for repressor, if we started with a repressor-free operator, a repressor would bind first to O_{R1} (Figure 14.18[a]). Once a repressor is bound to O_{R1}, the affinity of O_{R2} for the second repressor increases because the second repressor dimer not only binds to O_{R2}, but also touches the repressor that is bound at O_{R1} (Figure 14.18[b]). At higher repressor concentrations repressor will also bind to O_{R3} (Figure 14.18[c]). This last repressor to bind does not touch either of the other two repressors already bound.

For setting the switch to the lysogenic state, then, repressor is typically present at concentrations so that repressor is bound to both O_{R1} and O_{R2}. This prevents RNA polymerase from binding to P_R (see Figures 14.14 and 14.15[5a]) and further transcribing the *cro* gene. However, since no repressor is bound to O_{R3}, RNA polymerase can bind to P_{RM}, and the repressor gene *cI* can be transcribed (Figure 14.19). As a result, the amount of repressor in the cell increases and a repressor molecule will bind also to the O_{R3} site. This blocks repressor synthesis by preventing RNA polymerase from binding and transcribing the *cI* gene. This causes the amount of repressor to drop in the cell and the repressor bound to O_{R3} falls off. RNA polymerase can then bind again to P_{RM}, and *cI* transcription can resume. In other words, the λ repressor plays a negative regulatory role by blocking transcription of the *cro* gene, and a positive regulatory role by maintaining optimum repressor levels in the cell by modulating RNA polymerase binding at P_{RM}. As long as enough repressor is present to bind to O_{R1} and O_{R2}, the lysogenic state is maintained.

The Lytic Pathway. The lysogenic pathway is favored when enough λ repressor is made so that early promoters are turned off, thereby repressing all the genes needed for the lytic pathway. One

important lytic pathway gene that is repressed is *Q*; the Q protein is a positive regulatory protein required for the production of lysis proteins and phage coat proteins.

Let us consider the induction of the lytic pathway caused by ultraviolet light irradiation. Inducers such as ultraviolet light typically damage the DNA, and this somehow causes a change in the function of the bacterial protein RecA (product of the *recA* gene). Normally RecA functions in DNA recombi-

■ *Figure 14.19* In the lysogenic state repressors are bound to O_{R1} and O_{R2}, but not to O_{R3}, which allows RNA polymerase to bind to P_{RM} and transcribe the *cI* (repressor) gene. Polymerase cannot bind to P_R since that promoter is covered by repressors; hence the *cro* gene is not transcribed.

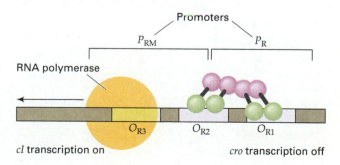

nation events, but when DNA is damaged, RecA cleaves λ repressor monomers, separating the amino and carboxyl domains. This event causes the chain reaction shown in Figure 14.20: the cleaved monomers are unable to form dimers and then, when repressors fall off the operator through the process of normal dissociation, they are not replaced.

The absence of repressor at O_{R1} allows RNA polymerase to bind at P_R and the *cro* gene is then transcribed. The resulting Cro protein then acts to decrease RNA synthesis from P_L and P_R, which reduces synthesis of cII protein, the regulator of λ repressor synthesis, and blocks synthesis of λ repressor mRNA from P_{RM} (Figure 14.15[4b]). At the same time, transcription of the right early operon genes from P_R is decreased, but enough Q proteins are accumulated to stimulate transcription of the genes for starting the lytic pathway (Figure 14.15[5b]).

How does Cro protein work? Cro is a protein with one domain; the functional form of Cro is a dimer, and each dimer can bind to the same operator binding sites to which the repressor can bind. Cro dimers bind independently to the three O_R sites, although the order of affinity for the three sites is *exactly opposite* that of the λ repressor dimer; that is, the strongest affinity of Cro protein is for O_{R3} (Figure 14.21[a]); after that the affinity is approximately equal for O_{R2} and O_{R1} (Figure 14.21[b]). Unlike the λ repressor, which has both a

■ *Figure 14.21* Cro protein binds to the same three sites in the right operator (O_R) as λ repressor, but with different affinities for the three sites: (a) The affinity of Cro protein for O_{R3} is about 10 times higher than for O_{R2} or O_{R1}; (b) Once a Cro protein is bound to O_{R3}, the second dimer binds with equal chance to O_{R2} or O_{R1}; (c) At high Cro dimer concentration, all three sites are occupied.

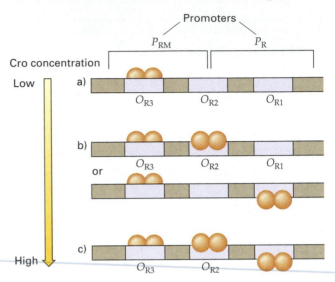

positive and negative regulatory action, Cro acts only as a negative regulator. The first Cro synthesized binds to O_{R3}, blocking the ability of RNA polymerase to bind to P_{RM}, thereby stopping further repressor synthesis. At this point, the switch has been set to the lytic pathway.

As with the repressor protein, the Cro protein controls its own transcription. At high Cro protein concentrations all three O_R binding sites become occupied by Cro protein (Figure 14.21[c]), blocking transcription initiation at the P_R promoter. This results in decreased transcription of the *cro* gene, and of the other early lytic pathway genes.

In summary, lambda uses complex regulatory systems to direct either the lytic or the lysogenic pathway. The choice is made between the two pathways using an elaborate genetic switch. This switch involves two regulatory proteins with opposite affinities for three binding sites within each of the two major operators that control the λ life cycle. Which pathway is followed depends on which regulatory protein "wins" in terms of binding to those two operators.

■ *Figure 14.20* Activated RecA protein cleaves repressor which then cannot dimerize. As a result, when repressors dissociate from operators they are not replaced, *cro* is transcribed, and induction of lytic growth follows.

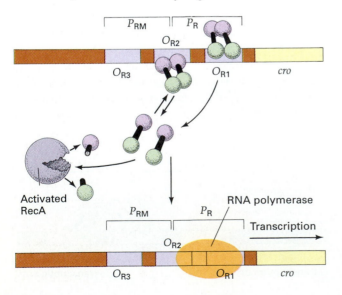

K e y n o t e

Bacteriophages are especially adapted for undergoing reproduction within a bacterial host. Many of the genes related to the production of progeny phages, or to the establishment or reversal of lysogeny in temperate phages, are organized into operons. These operons (like bacterial operons) are controlled through the interaction of regulatory proteins with operators that are adjacent to clusters of structural genes.

Summary

Bacteria typically live in environments that can change rapidly. To function efficiently in the changing environment, bacteria have evolved special systems for regulating gene expression in response to the dictates of conditions. Generally, when a particular function is needed, the genes for that function (e.g., the lactose utilization genes when lactose is present as a carbon source, or the tryptophan biosynthesis genes when tryptophan is absent from the medium) are coordinately transcribed so that the relevant enzymes are coordinately produced. Conversely, when the conditions change so that function is not needed, gene transcription is turned off.

From studies of the regulation of expression of the three lactose-utilizing genes in *E. coli* bacteria, Jacob and Monod developed a model that is the basis for the regulation of gene expression in a large number of bacterial and bacteriophage systems. The genes for the enzymes are contiguous in the chromosome and are adjacent to a controlling site (an operator) and a single promoter. This complex constitutes a transcriptional regulatory unit called an operon. A regulatory gene, which may or may not be nearby, is associated with the operon. The addition of lactose to the cell results in the coordinate induction of the *lac* genes. Induction occurs as follows: lactose binds with a repressor protein that is encoded by the regulatory gene, inactivating it and preventing it from binding to the operator. As a result, RNA polymerase can bind to the promoter and transcribe the three genes onto a single polygenic mRNA. As long as lactose is present, mRNA continues to be produced and the enzymes are made. When lactose is no longer present, the repressor protein is no longer inactivated, and it binds to the operator, thereby preventing RNA polymerase from transcribing the *lac* genes. If glucose is present as well as lactose, the lactose operon is not induced. This is the case because less energy is required to metabolize glucose than lactose. This phenomenon is called catabolite repression.

The genes for a number of bacterial amino acid biosynthesis pathways are arranged in operons. Expression of these operons is accomplished by a repressor-operator system and, in some cases, also through attenuation at a second controlling site called an attenuator. The tryptophan operon is an example of an operon with both types of transcription regulation systems. The repressor-operator system functions essentially like that for the *lac* operon, except that the addition of amino acid to the cell activates the repressor, thereby turning the operon off. This makes sense since the role of the gene products is to make the amino acid. The attenuator is located downstream from the operator in a leader region that is translated. The attenuator is a partially effective transcription termination site that allows only a fraction of RNA polymerases to transcribe the rest of the operon. Attenuation involves a coupling between transcription and translation, and the formation of particular RNA secondary structures that signal whether or not transcription can continue.

Key to the attenuation phenomenon is the presence of multiple copies of codons for the amino acid, the synthesis of which the operon controls. When enough of the amino acid is present in the cell, enough charged tRNAs are produced so that the ribosome can read the key codons in the leader region; this causes the RNA that is being made by the RNA polymerase ahead of the ribosome to assume a secondary structure which signals transcription to stop. However, when the cell is starved for that amino acid, there are insufficient charged tRNAs to be used at the key codons so that the ribosome stalls at that point. As a result, the RNA ahead of that point assumes a secondary structure that permits continued transcription into the structural genes. The combination of repressor-operator regulation and attenuation control permits a fine degree of control of transcription of the operon.

Operons are also extensively used by bacteriophages to coordinate the synthesis of proteins with related functions. As we know, bacteriophages such as lambda are especially adapted for undergoing their reproduction within a bacterial host. Many of the genes related to the establishment or reversal of lysogeny of temperate phages are organized into operons. These operons, like bacterial operons, are controlled through the interaction of regulatory proteins with operators that are adjacent to clusters of structural genes. Bacteriophage lambda has been an excellent model for studying the elaborate genetic switch that controls the choice between lytic and lysogenic pathways in a lysogenic phage. The switch involves two regulatory proteins, the lambda repressor and the cro protein, which have opposite affinities for three binding sites within each of the two major operators that control the lambda life cycle.

In sum, operons are commonly encountered in bacteria and their viruses. They provide a simple way to coordinate the transcription of genes with related function at the transcriptional level. Generally, there is an interaction between an effector molecule (such as lactose or tryptophan) and a repressor molecule. That interaction may inactivate (lactose) or activate (tryptophan) the repressor, causing it to be released from or bind to the operator, respectively. If no repressor is bound to the operator, RNA polymerase can bind to the adjacent promoter and the genes can be transcribed. Operons have not been found in eukaryotic systems.

Analytical Approaches for Solving Genetics Problems

Q14.1 In the laboratory you are given ten strains of *E. coli* with the following lactose operon genotypes:

(1) $lacI^+$ $lacP^+$ $lacO^+$ $lacZ^+$;
(2) $lacI^-$ $lacP^+$ $lacO^+$ $lacZ^+$;
(3) $lacI^+$ $lacP^+$ $lacO^c$ $lacZ^+$;
(4) $lacI^-$ $lacP^+$ $lacO^c$ $lacZ^+$;
(5) $lacI^+$ $lacP^+$ $lacO^c$ $lacZ^-$;
(6) F' $lacI^+$ $lacP^+$ $lacO^c$ $lacZ^-$ /
    $lacI^+$ $lacP^+$ $lacO^+$ $lacZ^+$;
(7) F' $lacI^+$ $lacO^+$ $lacZ^-$ / $lacI^+$ /
    $lacP^+$ $lacO^c$ $lacZ^+$ $lacP^+$;
(8) F' $lacI^+$ $lacP^+$ $lacO^+$ $lacZ^+$ /
    $lacI^-$ $lacP^+$ $lacO^+$ $lacZ^-$
(9) F' $lacI^+$ $lacP^+$ $lacO^c$ $lacZ^-$ /
    $lacI^-$ $lacP^+$ $lacO^+$ $lacZ^+$;
(10) F' $lacI^-$ $lacP^+$ $lacO^+$ $lacZ^-$ /
    $lacI^-$ $lacP^+$ $lacO^c$ $lacZ^+$.

(Note: In the partial-diploid strains (6–10) one copy of the *lac* operon is in the host chromosome and the other copy is in the F plasmid, F'.)

For each strain, predict whether the β-galactosidase enzyme will be produced (a) if lactose is absent from the growth medium, and (b) if lactose is present in the growth medium.

A14.1 The answers are as follows, where + = β-galactosidase produced and − = β-galactosidase not produced:

GENOTYPE	NONINDUCED: LACTOSE ABSENT	INDUCED: LACTOSE PRESENT
(1)	−	+
(2)	+	+
(3)	+	+
(4)	+	+
(5)	−	−
(6)	−	+
(7)	+	+
(8)	−	+
(9)	−	+
(10)	+	+

To answer this question completely, we need a good understanding of how the lactose operon is regulated in the wild type and of the consequences of particular mutations on the regulation of the operon.

Strain (1) is the standard wild-type operon. No enzyme is produced in the absence of lactose, since the repressor produced by the *lacI* gene binds to the operator (*lacO*) and blocks the initiation of transcription. When lactose is added it binds to the repressor, changing its conformation so that it no longer can bind to the *lacO* region, thereby facilitating transcription of the structural genes by RNA polymerase.

Strain (2) is a haploid strain with a mutation in the *lacI* gene (*lacI⁻*). The consequence is that the repressor protein cannot bind to the (normal) operator region, so there is no inhibition of transcription, even in the absence of lactose. This strain, then, is constitutive, meaning that the functional enzyme is produced by the *lacZ⁺* gene in the presence or absence of lactose.

Strain (3) is the other possible constitutive mutant. In this case the repressor gene is a wild type and the β-galactosidase gene *lacZ⁺* is a wild type, but there is a mutation in the operator region (*lacO^c*). Therefore repressor protein cannot bind to the operator, and the transcription occurs in the presence or absence of lactose.

Strain (4) carries both of the regulatory mutations of the previous two strains. Functional repressor is not produced, but even if it were, the operator is changed so that it cannot bind. The consequence is the same: constitutive enzyme production.

Strain (5) produces functional repressor, but the operator (*lacO^c*) is mutated. Therefore transcription cannot be blocked, and lactose polygenic mRNA is produced in the presence or absence of lactose. However, because there is also a mutation in the β-galactosidase gene (*lacZ⁻*), no functional enzyme is generated.

In the partial-diploid strain (6) one lactose operon is completely wild type, and the other carries a constitutive operator mutation and a mutant β-galactosidase gene. In the absence of lactose no functional enzyme is produced. For the wild-type operon the functional repressor binds to the operator and blocks transcription. For the operon with the two mutations the operator region is mutated and cannot bind repressor, so the mRNA for the mutated operon is produced, since transcription cannot be inhibited, and the *lacZ* gene is mutated so that functional enzyme cannot be produced. In the presence of lactose functional enzyme is produced, since repression of the wild-type operon is relieved so that the *lacZ*+ gene can be transcribed. This type of strain provided one of the pieces of evidence that the operator region does not produce a diffusible substance.

In diploid (7) functional enzyme will be produced in the presence or absence of lactose, because one of the operons has a *lacO*c mutation that does not respond to a repressor and that is linked to a wild-type *lacZ*+ gene. That operon is transcribed constitutively. The other operon is inducible, but because there is a *lacZ*− mutation, no functional enzyme is produced.

Diploid (8) has a wild-type operon and an operon with a *lacI*− regulatory mutation and a *lacZ*−

mutation. The *lacI*+ gene product is diffusible so that it can bind to the *lacO*+ region of both operons, thereby putting both operons under inducer control. This strain gives a classic demonstration that the *lacI*+ gene is *trans*-dominant to a *lacI*− mutation. In this case the particular location of the one *lacZ*− mutation is irrelevant: The same result would have been obtained had the *lacZ*+ and *lacZ*− been switched between the two operons. In this diploid β-galactosidase is not produced unless lactose is present.

In strain (9) β-galactosidase is produced only when the lactose is present, because the *lacO*c region controls only those genes that are adjacent to it on the same chromosome (*cis*-dominance), and in this case one of the adjacent genes is *lacZ*−, which codes for a nonfunctional enzyme. The diploid is heterozygous *lacI*+/*lacI*−, but *lacI*+ is *trans*-dominant, as discussed for strain (8). Thus the only normal *lacZ*+ gene is under inducer control.

Diploid (10) has defective repressor protein as well as a *lacO*c mutation adjacent to a *lacZ*+ gene. On the latter ground alone this diploid is constitutive. The other operon is also constitutively transcribed, but because there is a *lacZ*− mutation, no functional enzyme is generated from it.

Questions and Problems

14.1 How does lactose bring about the induction of synthesis of β-galactosidase, permease, and transacetylase? Why does this event not occur when glucose is also in the medium?

14.2 Operons produce polygenic mRNA when they are active. What is a polygenic mRNA? What advantages, if any, do they confer on a cell in terms of its function?

*14.3 If an *E. coli* mutant strain synthesizes β-galactosidase whether or not the inducer is present, what genetic defect(s) might be responsible for this phenotype?

14.4 The elucidation of the regulatory mechanisms associated with the enzymes of lactose utilization in *E. coli* was a landmark in our understanding of regulatory processes in microorganisms. In formulating the operon hypothesis as applied to the lactose system, Jacob and Monod found that results from particular partial-diploid strains were invalu-

able. Specifically, in terms of the operon hypothesis, what information did the partial diploids provide that the haploids could not?

*14.5 For the *E. coli lac* operon, write the partial-diploid genotype for a strain that will produce β-galactosidase constitutively and permease by induction.

14.6 Mutants were instrumental in the elaboration of the model for the regulation of the lactose operon.
a. Discuss why *lacO*c mutants are *cis*-dominant but not *trans*-dominant.
b. Explain why *lacI*s mutants are *trans*-dominant to the wild-type *lacI*+ allele but *lacI*− mutants are recessive.

*14.7 This question involves the lactose operon of *E. coli*. Complete Table 14.A, using + to indicate if the enzyme in question will be synthesized and − to indicate if the enzyme will not be synthesized.

TABLE 14.A

	INDUCER ABSENT		INDUCER PRESENT	
GENOTYPE	β-GALACTOSIDASE	PERMEASE	β-GALACTOSIDASE	PERMEASE
a. $lacI^+ lacO^+ lacZ^+ lacY^+$				
b. $lacI^+ lacO^+ lacZ^- lacY^+$				
c. $lacI^+ lacO^+ lacZ^+ lacY^-$				
d. $lacI^- lacO^+ lacZ^+ lacY^+$				
e. $lacI^s lacO^+ lacZ^+ lacY^+$				
f. $lacI^+ lacO^c lacZ^+ lacY^+$				
g. $lacI^s lacO^c lacZ^+ lacY^+$				
h. $lacI^+ lacO^c lacZ^+ lacY^-$				
i. $lacI^- lacO^+ lacZ^+ lacY^+/$ $lacI^+ lacO^+ lacZ^- lacY^-$				
j. $lacI^- lacO^+ lacZ^+ lacY^-/$ $lacI^+ lacO^+ lacZ^- lacY^+$				
k. $lacI^s lacO^+ lacZ^+ lacY^-/$ $lacI^+ lacO^+ lacZ^- lacY^+$				
l. $lacI^+ lacO^c lacZ^- lacY^+/$ $lacI^+ lacO^+ lacZ^+ lacY^-$				
m. $lacI^s lacO^+ lacZ^+ lacY^+/$ $lacI^+ lacO^c lacZ^+ lacY^+$				

14.8 A new sugar, sugarose, induces the synthesis of two enzymes from the *sug* operon of *E. coli*. Some properties of deletion mutations affecting the appearance of these enzymes are as follows (+ = enzyme induced normally, i.e., synthesized only in the presence of the inducer; C = enzyme synthesized constitutively; 0 = enzyme cannot be detected):

	ENZYME	
MUTATION OF	1	2
Gene A	+	0
Gene B	0	+
Gene C	0	0
Gene D	C	C

a. The genes are adjacent in the order *ABCD*. Which gene is most likely to be the structural gene for enzyme 1?

b. Complementation studies using partial-diploid (F′) strains were made. The plasmid (F′) and chromosome each carried one set of *sug* genes. The results were as follows (symbols are the same as in previous table):

GENOTYPE OF F′	CHROMOSOME	1	2
$A^+ B^- C^+ D^+$	$A^- B^+ C^+ D^+$	+	+
$A^+ B^- C^- D^+$	$A^- B^+ C^+ D^+$	+	0
$A^- B^+ C^- D^+$	$A^+ B^- C^+ D^+$	0	+
$A^- B^+ C^+ D^+$	$A^+ B^- C^+ D^-$	+	+

From all the evidence given, determine whether the following statements are true or false:

1. It is possible that gene *D* is a structural gene for one of the two enzymes.

2. It is possible that gene *D* produces a repressor.

3. It is possible that gene *D* produces a cytoplasmic product required to induce genes *A* and *B*.

4. It is possible that gene *D* is an operator locus for the *sug* operon.

5. The evidence is also consistent with the possibility that gene *C* could be a gene that produces a cytoplasmic product required to induce genes *A* and *B*.

6. The evidence is also consistent with the possibility that gene *C* could be the controlling end of the *sug* operon (end from which mRNA synthesis presumably commences).

*****14.9** What consequences would a mutation in the catabolite activator protein (CAP) gene of *E. coli* have for the expression of a wild-type lac operon?

14.10 The lactose operon is an inducible operon, whereas the tryptophan operon is a repressible operon. Discuss the differences between these two types of operons.

14.11 Bacteriophage λ, upon infecting an *E. coli* cell, has a choice between the lytic and lysogenic pathways. Discuss the molecular events that determine which pathway is taken.

14.12 How do the lambda repressor protein and the Cro protein regulate their own synthesis?

*14.13 If a mutation was induced in the *cI* gene of phage lambda such that the resulting *cI* gene product was nonfunctional, what phenotype would you expect the phage to exhibit?

14.14 Bacteriophage λ can form a stable association with the bacterial chromosome because the virus manufactures a repressor. This repressor prevents the virus from replicating its DNA, making lysozyme and all the other tools used to destroy the bacterium. When you induce the virus with UV light, you destroy the repressor, and the virus goes through its normal lytic cycle. This repressor is the product of a gene called the *cI* gene and is a part of the wild-type viral genome. A bacterium that is lysogenic for λ+ is full of repressor substance, which confirms immunity against any λ virus added to these bacteria. These added viruses can inject their DNA, but the repressor from the resident virus prevents replication, presumably by binding to an operator on the incoming virus. Thus this system has many analogous elements to the lactose operon. We could diagram a virus as shown in the figure. Several mutations of the *cI* gene are known. The c_i mutation results in an inactive repressor.

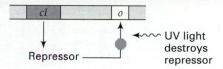

a. If you mix λ containing a c_i mutation, can it lysogenize (form a stable association with the bacterial chromosome)? Why?
b. If you infect a bacterium simultaneously with a wild-type c^+ and a c_i mutant of λ, can you obtain stable lysogeny? Why?
c. Another class of mutants called c^{IN} makes a repressor that is insensitive to UV destruction. Will you be able to induce a bacterium lysogenic for c^{IN} with UV light? Why?

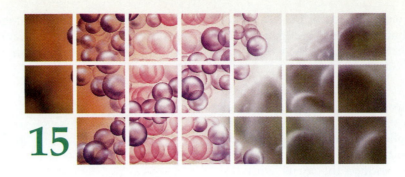

15

Regulation of Gene Expression and Development in Eukaryotes

- There are no operons in eukaryotes. However, genes for related functions are often regulated coordinately.

- In eukaryotes gene expression is regulated at a number of distinct levels. There are regulatory systems for the control of transcription, of precursor-RNA processing, translation of the mRNAs, degradation of the mature RNAs, and degradation of the protein products.

- Eukaryotic protein-coding genes contain both promoter elements and enhancer elements. Some promoter elements are required for transcription to begin. Other promoter elements have a regulatory function; they bind specific regulatory proteins that control expression of the gene. Specific regulatory proteins bind also to the enhancer elements and regulate (activate or repress) transcription through their interaction with proteins bound to the promoter elements.

- Regions of the chromosome that are being transcribed have a more loosened DNA-protein structure than chromosome regions that are transcriptionally inactive.

- Histones repress transcription by assembling nucleosomes on TATA boxes associated with genes. Gene activation occurs when proteins become bound to enhancers and disrupt the nucleosomes on the TATA boxes, thereby allowing the appropriate proteins to bind to the promoter elements to initiate transcription.

- The DNA of most eukaryotes has been shown to be methylated at a certain proportion of the bases. In those eukaryotes transcriptionally active genes generally exhibit lower levels of DNA methylation than transcriptionally inactive genes.

- In higher eukaryotes steroid hormones regulate the expression of particular sets of genes. To function in this short-term regulatory system, a steroid hormone enters a cell. If that cell contains a cytoplasmic receptor molecule specific for the hormone, a hormone-receptor complex is formed which migrates to the nucleus and binds to hormone regulatory elements next to genes, thereby regulating the expression of those genes.

- Regulation at the level of RNA processing operates in the production of mature mRNA molecules from precursor-mRNA molecules. Two regulatory events that exemplify this level of control are choice of poly(A) site and choice of splice site for intron removal. In both cases different types of mRNAs are produced, depending on the choices made.

- Gene expression is regulated also by mRNA translation control and by mRNA degradation control. The latter is believed to be a major control point in the regulation of gene expression.

- Long-term regulation is involved in controlling events that activate and repress genes during development and differentiation. Those two processes result from differential gene activity of a genome that contains a constant amount of DNA, rather than from a programmed loss of genetic information.

- Antibodies are specialized proteins called immunoglobulins, which bind specifically to antigens. Antibody molecules consist of two light (L) chains and two heavy (H) chains. In germ-line DNA the coding regions for L and H chains are scattered in tandem arrays of gene segments. During development somatic recombination occurs to bring particular gene segments together into a functional gene. A large number of different L and H chain genes result from the many possible ways in which the gene segments can recombine.

- *Drosophila* is a model system for studying the genetic control of development. *Drosophila* body structures result from specific gradients in the egg, and the subsequent determination of embryo segments that directly correspond to adult body segments. Both processes are under genetic control as defined by mutations that disrupt the development events. Studies of the mutants indicate that *Drosophila* development is directed by a temporal regulatory cascade.

 S WE LEARNED in Chapter 14, gene expression in prokaryotes is commonly regulated in a unit of protein-coding genes and adjacent controlling sites (promoter and operator) collectively called an operon. Operons function in a simple way to coordinate the synthesis of proteins with related functions. The discovery of operons in prokaryotes stimulated searches for similar regulatory systems in eukaryotes, but no operons have been found. In this chapter we look at some of the well-studied regulatory systems in eukaryotes.

Eukaryotic gene regulation falls into two categories. Short-term regulation involves regulatory events in which gene sets are quickly turned on or off in response to changes in environmental or physiological conditions in the cell's or organism's environment. Long-term gene regulation involves regulatory events other than those required for rapid adjustment to local environmental or physiological changes; that is, those events that are required for an organism to develop and differentiate. Both of these categories of gene regulation are addressed in this chapter.

Levels of Control of Gene Expression in Eukaryotes

There are both unicellular and multicellular eukaryotes. In both eukaryote types the control of gene expression is more complicated than is the case with prokaryotes. Figure 15.1 diagrams some of the levels at which the expression of protein-coding genes can be regulated in eukaryotes: *transcriptional control*, *mRNA processing control*, *mRNA translation control*, *mRNA degradation control*, and *protein degradation control*. We will consider each of these in turn with the exception of the last, since it is less relevant to the study of gene expression than the others.

Transcriptional Control

General Aspects of Transcriptional Control
Transcriptional control regulates whether or not a gene is to be transcribed and the rate at which transcripts are produced.

Protein-coding eukaryotic genes contain both promoter elements and enhancers (see Chapter 11). Recall that particular proteins bind to promoters and enhancers to facilitate the initiation of transcription by RNA polymerase II. The promoter elements are located just upstream of the site at which transcription begins. The enhancers are usually some distance away, either upstream or downstream. We can think of the different promoter elements as modules that function in the regulation of expression of the gene. Certain promoter elements, such as the TATA element, are required to specify where transcription is to begin. Other promoter elements control whether or not transcription of the gene occurs; specific regulatory proteins bind to

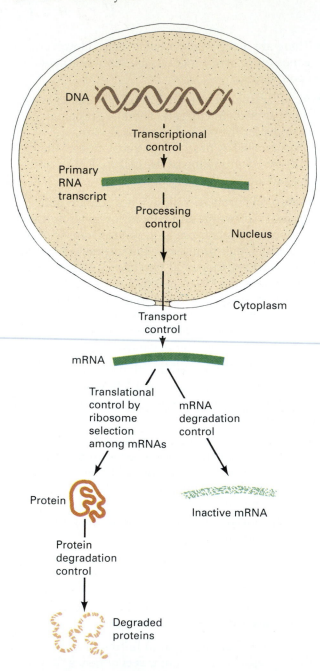

■ *Figure 15.1* Levels at which gene expression can be controlled in eukaryotes.

these elements (Figure 15.2). If transcription is activated (turned on), the promoter element is a *positive regulatory element* and the protein is a *positive regulatory protein*. If transcription is repressed (turned off), the promoter element is a *negative regulatory element* and the protein is a *negative regulatory protein*.

A regulatory promoter element is special to the gene (or genes) it controls because it binds a signaling molecule that is involved in the regulation of that gene's expression. Depending on the particular gene there can be one, a few, or many regulatory

■ *Figure 15.2*   Schematics of (a) Positive regulation of gene transcription; (b) Negative regulation of gene transcription.

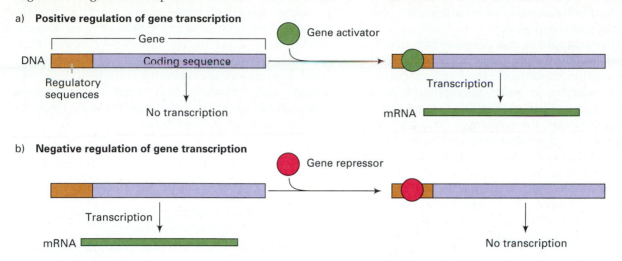

a)   **Positive regulation of gene transcription**

b)   **Negative regulation of gene transcription**

promoter elements, since under various conditions there may be one, a few, or many regulatory proteins that control the gene's expression. The remarkable specificity of regulatory proteins in binding to their specific regulatory element in the DNA and to no others ensures careful control of which genes are turned on and which are turned off.

Both promoters and enhancers are important in regulating transcription of a gene. Each regulatory promoter element and enhancer element binds a special regulatory protein. Some regulatory proteins are found in most or all cell types, while others are found in only a limited number of cell types. Because some of the regulatory proteins activate transcription when they bind to the enhancer or promoter element, while others repress transcription, the net effect of a regulatory element on transcription depends on the combination of different proteins bound. If positive regulatory proteins are bound at both the enhancer and promoter elements, the result is activation of transcription. However, if a negative regulatory protein binds to the enhancer and a positive regulatory protein binds to the promoter element, the result will depend on the interaction between the two regulatory proteins. If the negative regulatory protein has a strong effect, the gene will be repressed.

Enhancer elements and promoter elements appear to bind many of the same proteins. This implies that both types of regulatory elements affect transcription by a similar mechanism, probably involving interactions of the regulatory proteins as described before. Therefore, by combining relatively few regulatory proteins in particular ways, the transcription of different arrays of genes is reg-

ulated, and a large number of cell types is specified. This is called **combinatorial gene regulation.**

Keynote

Eukaryotic protein-coding genes contain both promoter elements and enhancer elements. Some promoter elements are required for transcription to begin. Other promoter elements have a regulatory function; these are specialized for the gene they control, binding specific regulatory proteins that control expression of the gene. Specific regulatory proteins bind also to the enhancer elements and activate or repress transcription through their interaction with proteins bound to the promoter elements. Enhancer elements and promoter elements appear to bind many of the same proteins. This implies that both types of regulatory elements affect transcription by a similar mechanism, probably involving interactions of the regulatory proteins.

Chromosome Changes and Transcriptional Control
As we learned in Chapter 9, the eukaryotic chromosome consists of DNA complexed with histones (to form nucleosomes) and nonhistone chromosomal proteins. In this section we will learn about the relationship between chromosome structure and transcriptional control.

DNase Sensitivity and Gene Expression   The chromosome structure is much looser in regions where genes are being transcribed than in regions where genes are transcriptionally inactive. The evidence for this is that the region of a chromosome with an active gene is much more sensitive to digestion by DNase I (an endonuclease) *in vitro* than is a transcriptionally inactive region.

A specific example of this phenomenon involves globin and ovalbumin synthesis in chickens. In the nuclei of **erythroblasts** (red blood-cell precursors) from chick embryos, in which hemoglobin is made, the genes for globin are susceptible to digestion by DNase I. In contrast, the sequences encoding ovalbumin, which is not produced in erythroblasts, are resistant to DNase digestion. Conversely, in hen oviduct nuclei ovalbumin mRNA is actively transcribed but globin mRNA is not. In hen oviduct nuclei the ovalbumin gene sequences are much more sensitive to DNase I digestion than are the globin gene sequences. From similar experiments with other systems, we have found that almost all transcriptionally active genes have an increased DNase I sensitivity. The extent of the DNase I-sensitive region varies with the particular gene, ranging from a few kilobase pairs around the gene to as much as 20 kb of flanking DNA.

More detailed studies of the regions of DNA around transcriptionally active genes have shown that certain sites, called **hypersensitive sites** or **hypersensitive regions**, are even more highly sensitive to digestion by DNase I. These sites or regions are typically the first to be cut with DNase I. Most, but not all, DNase-hypersensitive sites are in the regions upstream from the start of transcription, probably corresponding to the DNA sequences where RNA polymerase and other gene regulatory proteins bind.

Keynote

Chromosome regions that are transcriptionally active have looser DNA-protein structures than chromosome regions that are transcriptionally inactive. The promoter regions of active genes typically have an even looser DNA-protein structure.

Histones and Gene Regulation In the chromosome DNA is wrapped around histones to form nucleosomes. The histones act as general repressors of transcription because, when they are associated with the histones of a nucleosome core, the promoter elements (particularly the TATA box, which is important for transcription initiation as discussed in Chapter 11) cannot be found by the regulatory proteins and transcription factors that must bind to them to initiate transcription. (These regulatory proteins and transcription factors are in the nonhistone class of chromosomal proteins.)

Some test-tube experiments gave important insights into how histones influence gene expression. These experiments investigated what happened to transcription when histones, proteins that bind to promoter elements (promoter-binding proteins), and proteins that bind to enhancers (enhancer-binding proteins) interact with DNA. In brief the results were as follows:

1. If DNA is mixed simultaneously with histones and promoter-binding proteins, the histones compete more strongly for promoters and form nucleosomes at TATA boxes. As a result the promoter-binding proteins cannot bind and transcription cannot occur;

2. If DNA is mixed first with promoter-binding proteins, they assemble on TATA boxes and other promoter elements and block nucleosome assembly on those sites when histones are subsequently added. As a result transcription occurs;

3. If DNA is mixed *simultaneously* with histones, promoter-binding proteins, and enhancer-binding proteins, the enhancer-binding proteins bind to enhancers and help promoter-binding proteins to bind to TATA boxes by blocking access by the histones. As a result transcription occurs.

These results indicate that histones are effective repressors of transcription, but other proteins can overcome that repression.

How does gene activation occur in the cell? One model is as follows: Histones block transcription by forming nucleosomes on TATA boxes. Promoter-binding proteins are unable to disrupt the nucleosomes on the TATA boxes. However, enhancer-binding proteins bind to the enhancers (displacing any histones that are bound there) and interact with the histones in the TATA-bound nucleosomes. This causes the nucleosome core particle to break up, and promoter-binding proteins can now bind to the freed DNA.

Keynote

Histones repress transcription by assembling nucleosomes on TATA boxes associated with genes. Gene activation occurs when proteins become bound to enhancers and disrupt the nucleosomes on the TATA boxes, thereby allowing the appropriate proteins to bind to the promoter elements to initiate transcription.

■ *Figure 15.3* Production of 5-methylcytosine from cytosine in DNA by the action of the enzyme DNA methylase.

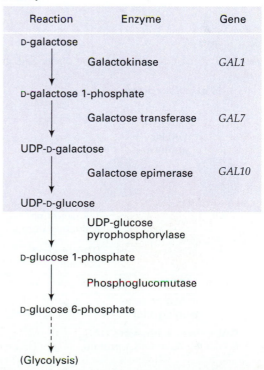

Cytosine (in DNA)   5-Methylcytosine (5mC)

DNA Methylation and Transcriptional Control

Once DNA has been synthesized within the eukaryotic cell, a small proportion of bases become chemically modified through the action of enzymes. One such modified base, 5-methylcyto-sine (5mC), is produced by methylation of cytosine by the enzyme DNA methylase shortly after DNA replication (Figure 15.3).

The percentage of 5mC in eukaryotic DNA varies over a wide range. In mammalian DNA, for example, about 3 percent of cytosines are present as 5mC. The DNAs of less complex eukaryotes, such as *Drosophila* and yeast, appear to contain very little, if any, 5mC. There is no clear relationship, however, between the amount of 5mC and the complexity of an organism.

For at least 30 genes examined a negative correlation exists between DNA methylation and transcription; that is, there is a lower level of methylated DNA in transcriptionally active genes compared with transcriptionally inactive genes. We must exercise caution in taking this broad generalization too far, because not all methylated Cs in a gene region become demethylated when the gene is expressed. And, as was stated earlier, some organisms do not have significant amounts of methylated C in their DNA. Further, the fact that a correlation exists between the level of DNA methylation and the transcriptional state of a gene provides no information about cause and effect. In other words, we do not know whether methylation changes are necessary for the onset of transcriptional activity (in those organisms in which there is significant DNA methylation) or whether the changes are the result of transcriptional activity initiated by other means.

K e y n o t e

The DNA of most eukaryotes has been shown to be methylated at a certain proportion of the bases,

with most methylations occurring on cytosine residues. Transcriptionally active genes exhibit lower levels of DNA methylation than transcriptionally inactive genes.

Transcriptional Control of Gene Expression in Eukaryotes

In this section we shall took at two examples of the *short-term* regulation of gene expression in eukaryotes. This regulation involves rapid responses to changes in environmental or physiological conditions.

Galactose-utilizing Genes of Yeast   In the galactose-utilizing system of yeast, three genes encode enzymes needed for the metabolism of the sugar galactose (Figure 15.4[a]). The enzymes are galactokinase (*GAL1* gene), galactose transferase (*GAL7* gene), and galactose epimerase (*GAL10* gene). (Note that genes are in italics while the proteins they

■ *Figure 15.4* The galactose metabolizing pathway of yeast: (a) The steps catalyzed by the enzymes encoded by the *GAL* genes; (b) The organization of the *GAL* protein-coding genes. The arrows indicate the direction of transcription. (UDP = uridine diphosphate)

a)   Pathway

Reaction	Enzyme	Gene
D-galactose		
	Galactokinase	*GAL1*
D-galactose 1-phosphate		
	Galactose transferase	*GAL7*
UDP-D-galactose		
	Galactose epimerase	*GAL10*
UDP-D-glucose		
	UDP-glucose pyrophosphorylase	
D-glucose 1-phosphate		
	Phosphoglucomutase	
D-glucose 6-phosphate		
(Glycolysis)		

b)   Gene organization

encode are not.) Once D-glucose 6-phosphate is made, it enters the glycolytic pathway, where it is acted upon by other enzymes. In the absence of galactose the *GAL* genes are not transcribed. When galactose is added there is a rapid, coordinate induction of transcription of the *GAL* genes, and therefore a rapid production of the three galactose-utilizing enzymes.

The *GAL1*, *GAL7*, and *GAL10* genes are located near each other, but do not constitute an operon (Figure 15.4[b]). There is an unlinked regulatory gene for the system, *GAL4*, which encodes the GAL4 protein. The GAL4 protein is similar to the lambda repressor in that there are two domains. One is a DNA-binding domain and the other is for turning on transcription. GAL4 binds to a specific sequence in the DNA called *upstream activator sequence-galactose* (UAS$_G$); this sequence is a promoter element. A UAS$_G$ is located between the *GAL1* and *GAL10* genes; it consists of four similar 17-bp sequences, each of which binds GAL4. GAL4 binds to each 17-bp sequence as a dimer (much like lambda repressor does to its target sequences: see Chapter 14). If only one GAL4 dimer is bound, transcription is activated at a low level; maximal expression requires all four sites to be occupied by GAL4 dimers. Transcription occurs in both directions from the UAS$_G$—the *GAL10* gene is transcribed to

the left, and the *GAL1* gene is transcribed to the right (see Figure 15.4).

A model for the regulation of transcription of the GAL genes is presented in Figure 15.5. In the absence of galactose a GAL4 dimer is bound to a UAS$_G$. Another protein, GAL80 (encoded by the *GAL80* gene), is bound to the domain of GAL4 that is needed to turn on transcription. No transcription can occur under these circumstances. When galactose is added, a metabolite of galactose somehow results in GAL4 becoming phosphorylated (i.e., phosphate groups are added to certain amino acids). This changes the shape of the GAL4-GAL80 complex so that GAL4 can then turn on transcription of the *GAL* enzyme-coding genes. In this system, then, the GAL4 protein acts as a positive regulator (activator), and galactose is an effector molecule.

Steroid Hormone Regulation of Gene Expression in Animals   Short-term regulation enables an organism or cell to adapt rapidly to changes in its physiological environment so that it can function as optimally as possible. While lower eukaryotes exhibit some cell specialization, higher eukaryotes are differentiated into a large number of cell types, each of which carries out a specialized function or functions. The cells of higher eukaryotes are not

■ *Figure 15.5*   Model for the activation of the *GAL* genes of yeast.

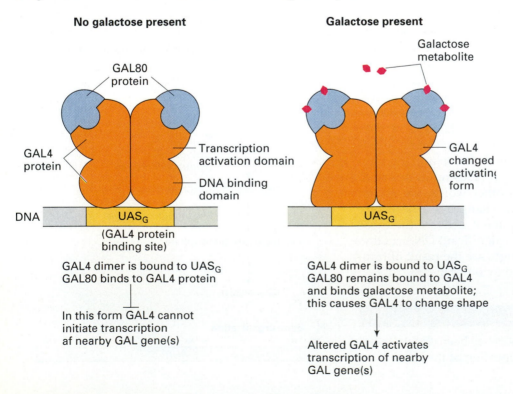

No galactose present

GAL80 protein

GAL4 protein

Transcription activation domain

DNA binding domain

DNA

UAS$_G$

(GAL4 protein binding site)

GAL4 dimer is bound to UAS$_G$
GAL80 binds to GAL4 protein

In this form GAL4 cannot initiate transcription af nearby GAL gene(s)

Galactose present

Galactose metabolite

GAL4 changed activating form

UAS$_G$

GAL4 dimer is bound to UAS$_G$
GAL80 remains bound to GAL4 and binds galactose metabolite; this causes GAL4 to change shape

Altered GAL4 activates transcription of nearby GAL gene(s)

■ *Figure 15.6* Mechanisms of action of polypeptide hormones and steroid hormones.

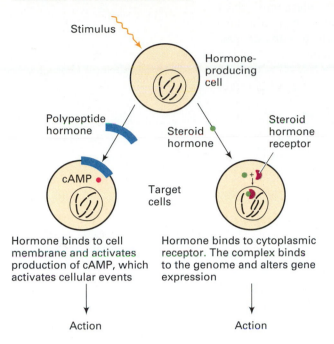

Hormone binds to cell membrane and activates production of cAMP, which activates cellular events

Hormone binds to cytoplasmic receptor. The complex binds to the genome and alters gene expression

exposed to rapid changes in environment as are cells of bacteria and of microbial eukaryotes. The environment to which most cells of higher eukaryotes are exposed, the intercellular fluid, is relatively constant in the nutrients, ions, and other important molecules it supplies. The intercellular fluid, in a sense, is analogous to a bacterium's growth medium. The constancy of the cell's environment is, in part, maintained through the action of chemicals called *hormones*, which are secreted by various cells in response to signals and which circulate in the blood until they stimulate their target cells. Elaborate feedback loops control the amount of hormone secreted, controlling the response so that appropriate levels of chemicals in the blood and tissues are maintained.

A hormone, then, is an effector molecule that is produced in low concentrations by one cell and that causes a physiological response in another cell. As shown in Figure 15.6, steroid hormones act by binding to a receptor, and then the steroid-receptor complex binds to the cell's genome, causing changes in gene expression. Other hormones (e.g., polypeptide hormones) may act at the cell surface to activate a membrane-bound enzyme, adenyl cyclase, which produces cyclic AMP (cAMP) from ATP (see Chapter 14). The cAMP functions as an intracellular signaling compound (called a *second messenger*) to activate the cellular events associated with the hormone.

A key to hormone action is that hormones act on

specific target cells that have receptors capable of recognizing and binding that particular hormone. For most of the polypeptide hormones (e.g., insulin, ACTH, vasopressin), the receptors are on the cell surface, whereas the receptors for steroid hormones are inside the cell.

Since the action of steroid hormones on gene expression has been well studied, we will concentrate on steroid hormones here. Steroid hormones have been shown to be important to the development and physiological regulation of organisms ranging from fungi to humans. Figure 15.7 gives the structures of four of the most common mammalian steroid hormones. All have a common four-ring structure; the variations in the side groups are responsible for their different physiological effects.

Steroid hormones have tissue-specific effects. Estrogen, for example, induces the synthesis of the protein prolactin in the rat pituitary gland; of the protein vitellogenin in frog liver; and of the proteins conalbumin, lysozyme, ovalbumin, and ovomucoid in the hen oviduct. Glucocorticoids induce the synthesis of growth hormone in the rat pituitary gland and of the enzyme phosphoenolpyruvate carboxykinase in the rat kidney. The specificity of the response to steroid hormones is controlled by the hormone receptors. That is, only target tissues that contain receptors for a particular steroid hormone can respond to that hormone, and this directly controls the sites of the hormone's action. With the exception of receptors for the steroid hormone glucocorticoid, which are widely distributed among tissue types, steroid receptors are found in a limited number of target tissues.

Mammalian cells contain between 10,000 and 100,000 steroid receptor molecules. The receptors are cytoplasmic proteins with a very high affinity for their respective steroid hormones. All steroid hormones work in the same general way. For example, in the absence of glucocorticoid hormone, the glucocorticoid hormone receptor is found in the cytoplasm in a complex with a protein called Hsp90. When glucocorticoid enters a cell it binds to its receptor, displacing Hsp90, and forming a glucocorticoid-receptor complex (Figure 15.8). The steroid-receptor complex diffuses to the nucleus where it binds to specific DNA regulatory sequences, activating the transcription of the specific genes controlled by the hormone. The new mRNAs produced appear within minutes after a steroid hormone encounters its target cell, enabling new proteins to be produced rapidly.

All the genes activated by a particular steroid hormone have in common a DNA sequence to which the steroid-receptor complex binds. The

■ *Figure 15.7*　Structures of four mammalian steroid hormones: (a) Hydrocortisone, which helps regulate carbohydrate and protein metabolism; (b) Aldosterone, which regulates salt and water balance; (c) Testosterone, which is used for the production and maintenance of male sexual characteristics; (d) Progesterone, which, with estrogen, prepares and maintains the uterine lining for embryo implantation.

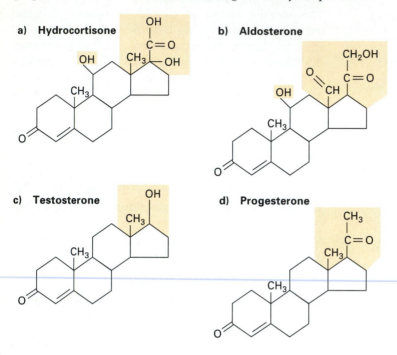

■ *Figure 15.8*　Model for the action of the steroid hormone, glucocorticoid, in mammalian cells.

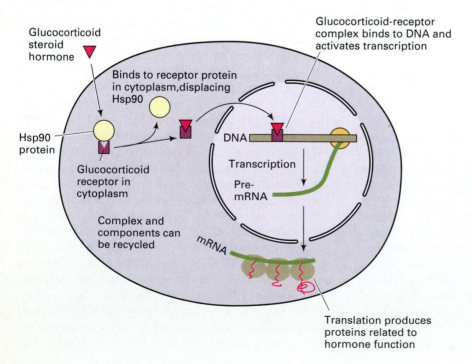

binding regions are called **steroid response elements** or **REs**. An additional letter is added to the acronym to indicate the specific steroid involved. Thus GRE is the glucocorticoid response element and ERE is the estrogen response element. The REs are located, often in multiple copies, in the enhancer region of genes. How the hormone-receptor complexes, once bound to the correct REs, regulate transcription is not completely known. Presumably each steroid hormone regulates its specific transcriptional activation by the same general mechanism. The unique action of each type of steroid results, then, from the different receptor proteins and REs involved.

In different types of cells the same steroid hormone may activate different sets of genes, even though the various cells have the same steroid hormone receptor. This is because many regulatory proteins bind to both promoter elements and enhancers to activate genes (see the discussions earlier in this chapter and Chapter 11). Thus a steroid-receptor complex can activate a gene only if the correct set of other regulatory proteins is present. Since the other regulatory proteins are specific for the cell type, different patterns of gene activation result.

In sum, steroid hormones act as positive effector molecules, and receptors act as regulatory molecules. When the two combine the resulting complex binds to DNA and activates gene transcription. The specific responses characteristic of each steroid hormone result from the fact that receptors are found in only certain cell types, and each of those cell types contains different sets of other cell-type-specific regulatory proteins that interact with the steroid-receptor complex to activate specific genes.

Keynote

In higher eukaryotes one of the well-studied systems of short-term gene regulation is the control of enzyme synthesis by hormones. Some hormones (the steroid hormones) exert their action by binding to a cytoplasmic receptor molecule and then to the cell's genome in a hormone-receptor complex, while others act at the cell surface, activating a system to produce cAMP, which functions as a second messenger to bring about the events controlled by the hormone. The specificity of hormone action is caused by the presence of hormone receptors in only certain cell types and by interactions of steroid-receptor complexes with cell-type-specific regulatory proteins.

RNA Processing Control

RNA processing control regulates the production of mature RNA molecules from precursor-RNA molecules. As we discussed in Chapter 11, mRNA, tRNA, and rRNA can be synthesized as precursor molecules. Perhaps the most significant range of control is exhibited by the mRNA class and involves control of processing of mRNA precursor molecules. We will discuss two types of RNA processing control: choice of alternative poly(A) sites (Figure 15.9[a]) and choice of alternative splice sites (Figure 15.9[b]). Small nuclear ribonucleoprotein particles (snRNPs) may well play a role in the differential processing process.

Choice of a Poly(A) Site   Recall that the poly(A) site is a sequence in a pre-mRNA molecule that specifies the position at which a poly(A) tail is added. One example of regulation by choice of poly(A) site is the production of immunoglobulin M (IgM), one of the immunoglobulin molecules involved in the immune response system (discussed later in this chapter.) Five copies of an IgM monomer aggregate to form the functional IgM molecule. Each monomer consists of two copies of a small protein (the light chain) and two copies of a large protein (the heavy chain) (Figure 15.10, p. 355). The heavy chain is encoded by the μ gene. Transcription of this gene produces a pre-mRNA that can be processed at two different poly(A) addition sites to produce heavy chains of different lengths. If the cell is in an early developmental stage, the longer heavy chain is produced and the resulting IgM associates with the cell membrane. If the cell is in a later developmental stage, the shorter heavy chain is produced and the resulting IgM is secreted from the cell.

Choice of Splice Site   A well-studied example of gene regulation by differential splicing of pre-mRNA involves sex determination in *Drosophila* (see Chapter 3). Recall that sex is determined by the X chromosome : autosome (X:A) ratio. A number of mutations disrupt normal sex determination. Study of these mutations has led to a *regulation cascade model* for sex determination in *Drosophila* (Figure 15.11, p. 356). First, the X:A ratio is read during development in an as-yet-unknown way. For wild-type *Drosophila* the ratio for females (XX) is 2X:2 autosome sets, and the ratio for males (XY) is 1:2. This information is transmitted to the sex-determination genes, which make the choice between the alternative female and male developmental pathways, starting with the master regulatory gene *Sxl* (*sex-lethal*).

■ *Figure 15.9* Models for control of the processing of mRNA precursors in eukaryotes by choice of: (a) poly(A) sites; or (b) Splice sites. (p = transcription initiation site [promoter]; t = transcription termination site.)

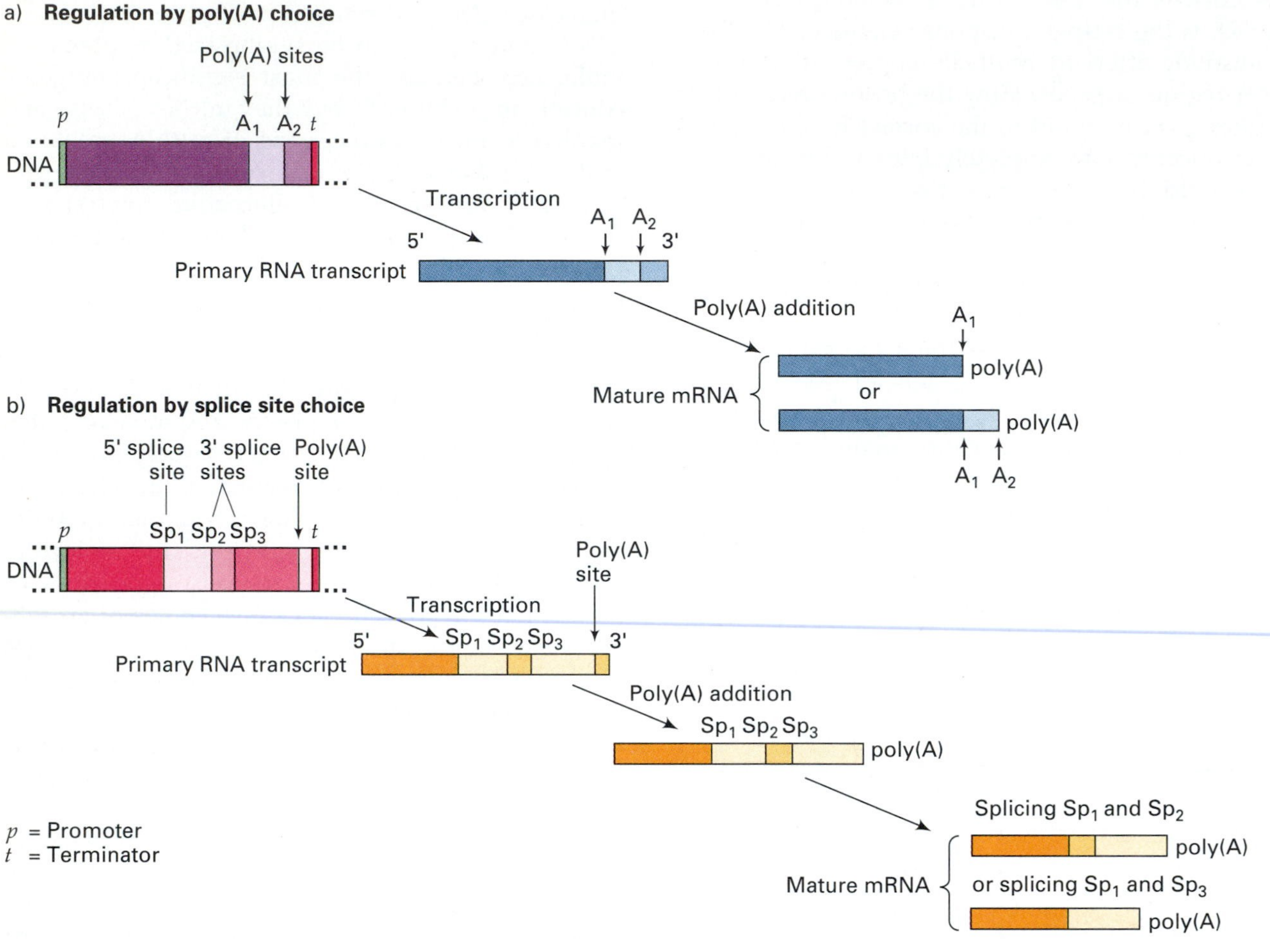

In both males and females *Sxl* is transcribed, but as a result of differential splicing, the male mRNA has an extra exon (the third of ten) compared with the female mRNA. The male-specific exon contains a number of stop codons so only a truncated, nonfunctional protein is produced in males, while a functional protein is produced in females. The mechanism for this *alternative splicing* is not yet understood. The Sxl protein causes the *tra* (*transformer*) gene pre-mRNA to undergo female-specific splicing. Translation of the resulting mRNA produces a protein which, along with the *tra-2* protein, causes the *dsx* (*doublesex*) pre-mRNA to be spliced to produce a female-specific mRNA. The resulting *dsx* protein interacts with the *ix* (*intersex*) protein to turn off the genes for male differentiation and set the switch so that female somatic cells are produced. If no Sxl protein is produced, alternative splicing of *tra* pre-mRNA produces an mRNA from which no functional protein is produced. Since *tra-2*

cannot act alone, alternative splicing of *dsx* premRNA occurs and a male-specific protein is made from the resulting mRNA. This protein is a repressor of female differentiation genes, so the switch is set to produce male somatic cells.

Keynote

Gene expression in eukaryotes can be regulated at the level of RNA processing. This type of regulation operates to determine the production of mature RNA molecules from precursor-RNA molecules. Two regulatory events that exemplify this level of control are choice of poly(A) site and choice of splice site. In both cases different types of mRNAs are produced, depending on the choices made. There are known examples of both types of regulation in developmental systems.

■ *Figure 15.10* Structure of an IgM monomer. Two light (L) chains and two heavy (H) chains are linked by disulfide bridges. Each type of chain has a variable domain (V_L and V_H) characterized by variations in amino acid sequences among IgM molecules. The light chains have one constant domain (C_L), while the heavy chains have four constant domains (C_H). In all IgM molecules constant domains have the same amino acid sequence. The variable domains collectively constitute the antigen-binding sites. Functional IgM is a pentamer of the structure shown here.

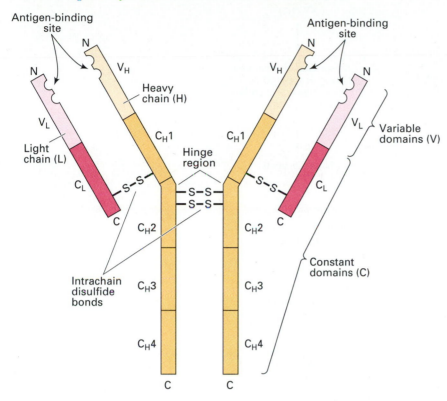

mRNA Translation Control

Messenger RNA molecules are subject to **translational control** by ribosome selection among mRNAs. Thus differential translation can greatly affect gene expression. For example, mRNAs are stored in many vertebrate and invertebrate unfertilized eggs. In the unfertilized state the rate of protein synthesis is very slow; however, protein synthesis increases significantly after fertilization. Since this increase occurs without new mRNA synthesis, it can be concluded that translational control is responsible. The mechanisms involved are not completely understood, but typically stored mRNAs are associated with proteins that serve both to protect the mRNAs and to inhibit their translation. Further, it is well known that the poly(A) tail promotes the initiation of translation. It has been found that stored, inactive messages generally have shorter poly(A) tails than active mRNAs. So to inactivate an mRNA for storage, the poly(A) tail is shortened. To reactivate a stored mRNA, the poly(A) tail is lengthened by cytoplasmic polyadenylation enzymes.

mRNA Degradation Control

Once they are in the cytoplasm, all RNA species are subjected to **degradation control,** in which the rate of RNA breakdown (also called RNA turnover) is regulated. Usually both rRNA (in ribosomes) and tRNA are very stable species. By contrast, mRNA molecules exhibit a diverse range of stability, with some mRNA types known to be stable for many months while others degrade within minutes. The stability of particular mRNA molecules may change in response to regulatory signals. For example, the addition of a regulatory molecule to a cell type can lead to an increase in synthesis of a particular protein or proteins. This is accomplished by an increase in the rate of transcription of the gene(s) involved and/or an increase in the stability of the mRNAs produced.

■ *Figure 15.11* Regulation cascade model for sex determination in *Drosophila*. For details, see text.

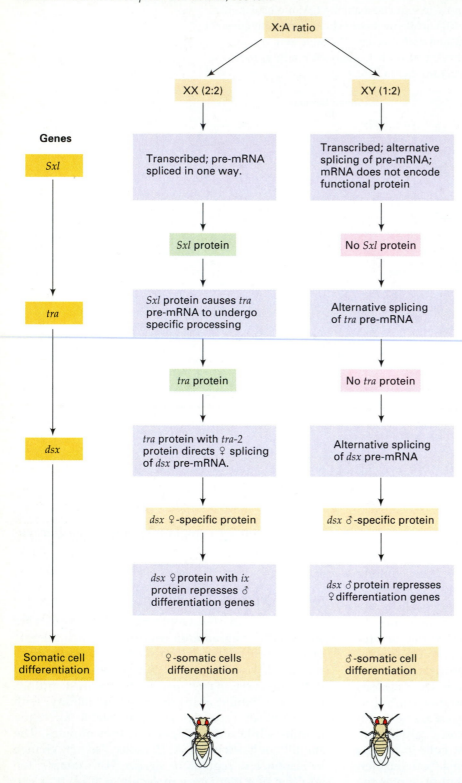

mRNA degradation is believed to be a major control point in the regulation of gene expression. The great diversity of mRNA degradation rates is due to the structural features in individual mRNAs. Specifically, mRNA degradation depends on inter- actions between nucleases and internal mRNA structures. For instance, a group of AU-rich sequences in the 3'-*un*translated *r*egion (UTR) of many short-lived mRNAs is responsible for their metabolic instability. It is not known how the AU-rich sequences act to cause mRNA instability, but possibly they cause destabilization of the poly(A).

In sum, the regulation of gene expression occurs at both transcriptional and post-transcriptional levels. The intertwining of the regulatory events at these different levels leads to the fine tuning of the amount of the controlled protein in the cell.

Keynote

Gene expression is regulated also by mRNA translation control and by mRNA degradation control. The latter is believed to be a major control point in the regulation of gene expression. Structural features of individual mRNAs have been shown to be responsible for the range of mRNA degradation rates, although the precise roles of cellular factors and enzymes have yet to be determined. In prokaryotes control of gene expression occurs mainly at the transcriptional level, in association with rapid turnover of mRNA molecules. In eukaryotes gene expression is regulated at a number of distinct levels. Regulatory systems appear to exist for the control of transcription, precursor-RNA processing, transport out of the nucleus, degradation of mature RNA species, translation of the mRNA, and degradation of the protein product.

Gene Regulation in Development and Differentiation

Most higher eukaryotes have many different cells, tissues, and organs with specialized functions, and yet all cells of the same organism have the same genotype. During an organism's growth throughout its life cycle, cells which were genetically identical become physiologically and phenotypically differentiated in terms of their structure and function.

Two terms are used in the description of long-term gene regulation. **Development** refers to the process of regulated growth that results from the genome's interaction with cytoplasm and the external environment and that involves a programmed sequence of phenotypic events that are typically irreversible. The total of the phenotypic changes constitutes the life cycle of an organism. **Differentiation,** the most spectacular aspect of development, involves the formation of different types of cells, tissues, and organs from a zygote through the processes of specific regulation of gene expression; differentiated cells have characteristic structural and functional properties.

At a very general level the processes in differentiation and development are the result of a highly programmed pattern of gene activation and gene repression. However, while we know a great deal about the molecular events that turn the bacterial *lac* operon on and off, we are a long way from understanding the molecular bases for the differentiation and development events in higher eukaryotes. We can describe these events well at the morphological level and to a certain extent at the biochemical level. We are only in the relatively early stages of knowing the details of how the complex activation-repression patterns are programmed, because of the greater complexity of the processes in comparison with the processes of bacterial operons.

In the following we will discuss only selected aspects of gene regulation in development and differentiation, since much of this area belongs more appropriately in a developmental biology course.

Gene Expression in Higher Eukaryotes

As discussed in Chapter 9 the genomes of higher eukaryotes are much larger than those of prokaryotes, and they have a large amount of highly repetitive and moderately repetitive DNA. About 20 to 40 percent of the genomic DNA of higher eukaryotes is highly repetitive DNA. This DNA does not appear to encode proteins. The remaining 60 to 80 percent of the DNA is distributed between the moderately repetitive and unique-sequence DNA.

What is the function of the moderately repetitive and unique-sequence DNA? In his studies of the sea urchin, Eric Davidson quantified the amount of DNA that encodes proteins by analyzing the mRNAs transcribed at various stages of development. He found that maximal transcription occurs in the oocyte, and that a maximum of about 6 percent of the unique-sequence DNA is transcriptionally active, producing proteins in mature tissue. Thus the majority of the genome in sea urchins (and in other eukaryotes) is not transcribed. By studying the locations of transcribed and nontranscribed regions of the genome, we have discovered that the structural genes (which are transcribed) are located within long stretches of DNA that are not transcribed. The function of the large amount of nontranscribed DNA in higher eukaryotes (less is found in lower eukaryotes) has long been controversial. It could be "junk" DNA left over from evolutionary changes. Or it might have important regulatory roles as yet undetermined, or perhaps a structural role.

Constancy of DNA in the Genome During Development

In early studies an important area of research focused on whether differentiation and development involve a *loss* of genes, or whether these processes involve a programmed sequence of gene activation and repression involving a constant genome; that is, a genome that is the same in the adult as it is in the zygote. The most direct way to decide is to determine whether the genome of a differentiated cell contains the same genetic information found in the zygote. A number of significant studies were done to answer this question. One of these studies will now be described.

Nuclear Transplantation Experiments with *Xenopus*   These experiments were done in 1964 by John Gurdon using the South African clawed toad, *Xenopus laevis*.

Gurdon's experiments (Figure 15.12) tested whether or not a nucleus taken from a tadpole (one differentiated stage of the *Xenopus* life cycle) could direct the development of an egg into a new tadpole or even into an adult toad. Donor nuclei could

■ *Figure 15.12*   Representation of Gurdon's experiments, which showed the totipotency of the nucleus of a differentiated cell of *Xenopus laevis*.

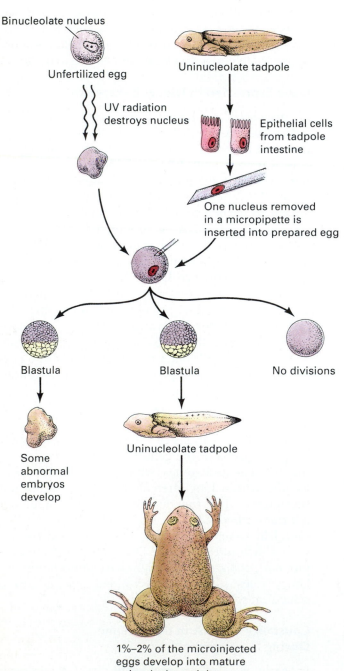

be distinguished from nuclei of the host because they had only one nucleolus instead of the normal two.

Intestinal epithelial cells (cells of the gut lining) were isolated from tadpoles with one nucleolus (uninucleolate tadpoles), and their nuclei were removed. An unfertilized egg from a strain with two nucleoli (binucleolate strain) was then isolated and irradiated with ultraviolet light to destroy the genetic information it contained. Next, the intestinal donor nucleus was introduced into the recipient egg, which was then incubated to see whether development would occur.

The micromanipulations involved in the experiment are very tricky and many failures occurred because of damage to the nucleus or to the cell. Thus in some cases no divisions occurred, in others abnormal embryos developed, and in others normal swimming tadpoles developed that later died. Significantly, in 1 to 2 percent of the cases a fertile, adult uninucleolate toad developed. Because the nuclei of the tadpoles and adult toads produced were always uninucleolate, Gurdon concluded that the donor tadpole nucleus must have contained all the genetic information needed to specify an adult toad. The nucleus from the *Xenopus* intestine is said to be totipotent, where **totipotency** is the capacity of a nucleus to direct a cell through all the stages of development and therefore produce a normal adult.

In sum, differentiation and development in most cases do not involve a loss of genetic information from the genome. Therefore, because there is constancy of DNA in the genome over an organism's life cycle, differentiation and development must result from regulatory processes affecting gene expression. (As always, there are some rare exceptions. Certain organisms have somatic cells that do not contain all of the genetic information that is found in germ-line cells, while other organisms display mature differentiated cells that have genetic material arranged differently from the way it is found in its embryonic precursor cells.)

Keynote

Long-term regulation is involved in controlling the events that activate and repress genes during development and differentiation. Development and differentiation result from differential gene activity of a genome that contains a constant amount of DNA, from the zygote stage to the mature organism stage. Thus these events do not result from a loss of genetic information.

Differential Gene Activity in Tissues and During Development

Specialized cell types have different cell morphologies; for example, nerve cells are clearly different from intestinal epithelial cells. Different organs and tissues can also be distinguished easily. Since researchers have shown that the amount of DNA remains constant during development, the simplest hypothesis is that the phenotypic variations between different cell types reflect differential gene activity. The following examples discuss some of the evidence supporting this hypothesis.

Hemoglobin Types and Human Development

Human adult hemoglobin Hb-A has been examined in this book in many contexts. Hb-A is a tetramer protein made up of two α and two β polypeptides, where each type of polypeptide is coded by a separate globin gene, α and β. The two genes appear to have arisen during evolution by duplication of a single ancestral gene, followed by alteration of the base sequences in each gene.

Hemoglobin Hb-A is only one type of hemoglobin found in humans. Several distinct genes code for α- and β-like globin polypeptides that are assembled in specific combinations to form different types of hemoglobin. The various types are synthesized and function at different times during human development. In the human embryo the hemoglobin initially made in the yolk sac consists of two ζ (zeta) polypeptides and two ε (epsilon) polypeptides. (ζ is an α-like polypeptide, and ε is a β-like polypeptide). After about three months of development, synthesis of embryonic hemoglobin ceases (the ζ and ε genes are no longer transcribed), and the site of hemoglobin synthesis shifts to the liver and the spleen. Here, *fetal hemoglobin* (Hb-F) is made. Hb-F contains two true α polypeptides, and either two β-like γA polypeptides or two β-like γG polypeptides; thus there are two forms of fetal hemoglobin produced. Each type of γ (gamma) chain is coded for by a distinct gene.

Fetal hemoglobin is made until just before birth, when synthesis of the two types of γ chains stops, and the site of hemoglobin switches to the bone marrow. In that tissue α and true β polypeptides are made, along with some β-like δ (delta) polypeptides. In the newborn through adult human, most of the hemoglobin is the familiar $\alpha_2\beta_2$ molecule (Hb-A), with about 1 in 40 molecules being $\alpha_2\delta_2$. Thus globin gene expression switches during human development, and this switch must involve a sophisticated gene regulatory system that turns the appropriate globin genes on and off over a long time period.

■ *Figure 15.13*   Linkage maps of human globin gene clusters. (Map is not to scale.)

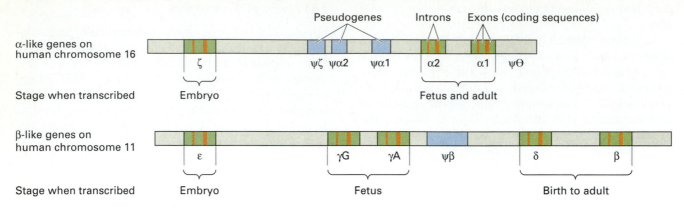

In the genome the α-like genes (two α genes and one ζ gene) are all on chromosome 16, and the β-like genes (ε, γA, γG, δ, and β) are all on chromosome 11 (Figure 15.13). Note that between the ζ and α genes on chromosome 16 is a sequence that closely resembles the base-pair sequence of the α-globin gene. However, this sequence does not have all the features required for producing a functional polypeptide product. Sequences such as this one, which are highly related to known functional genes but which themselves are defective (they cannot produce a functional product), are called *pseudogenes*. Pseudogenes are thought to be nonfunctional relics of gene duplication and gene modification that occurred in the past in evolutionary time. The β-like gene cluster has a pseudogene located between the γ genes and the δ gene.

Significantly, the α-like genes and the β-like genes are each arranged in the chromosome in an order that exactly parallels the timing in which the genes are transcribed during human development. Recall that embryonic hemoglobin consists of ζ and ε polypeptides, which are the first functional genes at the left of the clusters. Next, the α and γ genes are transcribed to produce fetal hemoglobin (Hb-F), and if one moves from left to right, these genes are the next functional genes that can be transcribed from the clusters. Lastly the δ and β polypeptides are produced, and these genes are last in line in the β-like globin gene cluster. Although such an arrangement surely must occur by more than coincidence, there is no insight as yet about how the gene order relates to the regulation of expression of these genes during development.

Immunogenetics

In this section we will discuss how antibodies, the proteins that are of central importance in the immune response, are encoded in the DNA. In this system DNA rearrangements are involved in the regulation of gene products.

Synopsis of the Immune System   The following is a brief list of the important properties of the immune system.

1. All vertebrates have an *immune system*. The immune system provides protection against infectious agents such as viruses, bacteria, fungi, and protozoans.

2. The immune system distinguishes between molecules that are *foreign* from molecules that are *self*.

3. Any substance that elicits an immune response is called an **antigen** (*anti*body *gen*erator).

4. The cells that are responsible for immune specificity are *lymphocytes*, specifically *T cells* and *B cells*. We will focus our discussion on B cells. B cells develop in the adult bone marrow. They secrete **antibodies**—specialized proteins called **immunoglobulins**—which circulate in the blood and lymph and which are responsible for the humoral ("humor" means fluid) immune responses; that is, the antibodies bind to foreign antigens they encounter.

5. A key feature of the immune response is that once an organism has been exposed to a particular antigen, it becomes *immune* to the effects of a new infection. That is, when an individual encounters a foreign antigen, the immune system mounts a response by committing cells to making antibodies against that foreign antigen. The next time the same antigen is encountered, the individual "remembers," and sufficient antibodies can be synthesized rapidly to respond to the new invasion.

6. The establishment of immunity against a particular antigen results from **clonal selection**. This is a process whereby cells that already have antibodies specific to the antigen on their surfaces are stimulated to proliferate and secrete that same antibody. During development each lymphocyte becomes committed to react with a particular antigen, even though the cell has never encountered the antigen. For the humoral response system, there is a population of B cells, *each of which can recognize a single antigen*. A B cell recognizes an antigen because the B cell has made antibody molecules that are attached to the outer membrane of the cell and act as receptor molecules. When an antigen encounters a B cell that has the appropriate antibody receptor capable of binding to it, that B cell is stimulated selectively to proliferate. This produces a clonal population of plasma cells, each of which makes and secretes the identical antibody. It is important to note that *any given cell makes only one specific kind of antibody towards one specific antigen*. The actual immune response, though, may involve the binding of many different antibodies to an array of antigens on such invaders as an infecting virus or bacterium. The end result is inactivation of the infecting agent.

Antibody Molecules   All antibody molecules made by a given B cell are identical: they have the same protein chains and bind the same antigen. Considering the organism as a whole, millions of different antibody types can be produced, each with a different amino acid sequence and a different antigen-binding site.

As a group, antibodies are proteins called *immunoglobulins* (Ig's). A stylized antibody (immunoglobulin) molecule is shown in Figure 15.14[a] (also see Figure 15.10), and a model of an antibody molecule based on X-ray crystallography is shown in Figure 15.14[b]. Both figures show the molecule's two short polypeptide chains, called *light (L) chains*, and two long polypeptide chains, called *heavy (H) chains*. (All antibody molecules also have carbohydrates attached to the regions of H chains not involved in binding with L chains.) The two H chains are held together by disulfide (-S-S-) bonds, and an L chain is bonded to each H chain by disulfide bonds. Other disulfide bonds within each L and H chain cause the chains to fold up into their characteristic shapes.

The overall structure resembles a Y, with the two arms containing the two antigen-binding sites. The two L chains in each Ig molecule are identical, as are the two H chains, so the two antigen-binding sites are identical. The hinge region (see Figure 15.14[a]) allows the two arms to move in space, making it easier for the antibody to bind the antigen. Also, one arm can then bind an antigen on, say, one virus, while the other arm binds the same antigen on a different virus. Such *crosslinking* of antibody molecules in solution helps inactivate infecting agents.

In mammals there are five classes of antibodies: IgA, IgD, IgE, IgG, and IgM. (IgM was diagrammed in Figure 15.10.) They have different H chains— α (alpha), δ (delta), ϵ (epsilon), γ (gamma), and μ (mu), respectively. Two types of L chains are found: κ (kappa) and λ (lambda). Both L chain types are found in all Ig classes, but a given antibody molecule will have either two identical κ chains or two identical λ chains. A complete discussion of the functions of the five Ig classes is beyond the scope of this text. We only need to be aware that the major class of immunoglobulin in the blood is IgG, and IgM plays an important role in the early stages of an antibody response to a previously unseen antigen. We will focus on these two antibody classes from here on.

Each polypeptide chain in an antibody is organized into domains of about 110 amino acids (Figure 15.14[a]). Each L chain (κ or λ) has two domains, and the H chain of IgG (the γ chain) has four domains, whereas the IgM's H chain (the μ chain) has five domains. The N-terminal domains of the H and L chains have highly variable amino acid sequences that constitute the antigen-binding sites. These domains, representing in IgG the N-terminal half of the L chain and the N-terminal quarter of the H chain, are termed the *variable*, or V, regions. The V regions are symbolized generically as V_L for the light chain and V_H for the heavy chain. The amino acid sequence of the rest of the L chain is constant for all antibodies with the same L chain type (i.e., μ or λ) and is termed C_L. Similarly, the amino acid sequence of the rest of the H chain is constant and is termed C_H. For IgG there are three approximately equal domains of C_H called C_H1, C_H2, and C_H3 (Figure 15.14[a]). For IgM there are four approximately equal domains of C_H called C_H1, C_H2, C_H3, and C_H4 (see Figure 15.10). Thus the production of antibody molecules involves synthesizing polypeptide chains, one part of which varies from molecule to molecule, and the other part of which is constant. We will discuss how this occurs in the following section.

Assembly of Antibody Genes from Gene Segments During B Cell Development   A mammal may produce 10^6 to 10^8 different antibodies. Since each antibody molecule consists of one kind of L chain and

■ *Figure 15.14* Antibody molecule: (a) Diagram showing the two heavy and two light chains and the antigen-binding sites. The heavy and light chains are held together by disulfide bonds; (b) Model of antibody molecule based on X-ray crystallography.

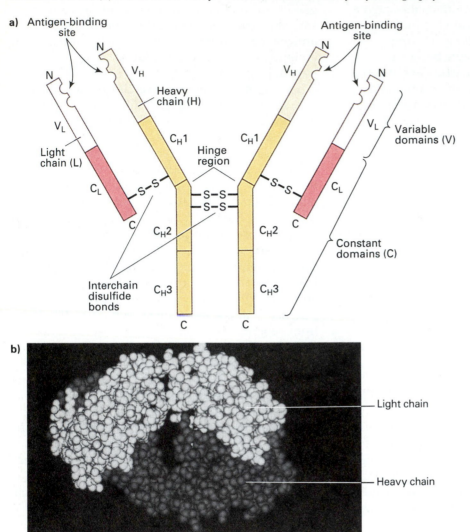

one kind of H chain, then 10^6 to 10^8 antibodies theoretically would require 10^3 to 10^4 different L chains and 10^3 to 10^4 different H chains, if L and H chains paired randomly. The dilemma is that the entire human genome is thought to contain perhaps only 10^5 genes! So, how can the observed diversity be generated? The answer is that variability in L and H chains results from particular DNA rearrangements that occur during B cell development. These rearrangements involve the joining of different gene segments to form a gene that is transcribed to produce an Ig chain. This process is called *somatic recombination* and the many permutations of recombinations that are possible are largely responsible for the diversity of antibody structure. The process will now be illustrated for mouse immunoglobulin chains.

Light Chain Gene Recombination   In mouse germline DNA there are three types of gene segments that encode the κ light chain (Figure 15.15):

1. A number of L-V_κ segments, which consist of a leader sequence (L) and a sequence, V_κ, which varies among L-V_κ segments. The V_κ segment encodes most of the amino acids of the L chain variable domain. The L sequence also encodes a signal sequence (see Chapter 12) required for secretion of the Ig molecule from the cell. The signal sequence is subsequently removed and is not part of the functional antibody molecule;

2. A single C_κ segment, which specifies the constant domain of the kappa L chain; and

3. A number of J_κ segments, which are joining

■ *Figure 15.15* Production of the light chain gene in mouse by recombination of V, J, and C gene segments during development. The rearrangement shown is only one of many possible recombinations.

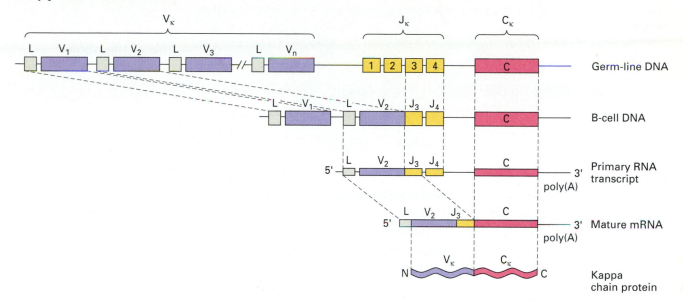

segments used to join V_κ and C_κ segments in the production of a functional κ light chain gene.

In the pre-B cell the L-V_κ, J_κ, and C_κ, segments, in that order, are widely dispersed on the chromosome. Each L-V_κ segment is about 400 nucleotide pairs long; L-V_κ segments are tandemly arranged on a chromosome with about 7 kb of DNA between them. The J_κ segments are about 30 nucleotide pairs long, and they are tandemly arranged about 20 kb apart. An intron of about 2 to 4 kb separates the J_κ segments and the single C_κ segment. As the B cell develops, a particular L-V_κ segment will become associated with one of the J_κ segments and with the C_κ segment (see Figure 15.15). In the example L-$V_{\kappa2}$ has recombined next to $J_{\kappa4}$. After transcription and intron removal, the mature mRNA has the organization L-$V_{\kappa2}J_{\kappa4}C_\kappa$. Translation of this mRNA and removal of the leader from the protein product gives the κ light chain that the B cell is committed to make.

In the mouse there are about 350 L-V_κ gene segments, 4 functional J_κ segments, and 1 C_κ gene segment. Thus the number of possible κ chain variable regions that can be produced by this mechanism is $350 \times 4 = 1400$. Further diversity of κ chains results from imprecise joining of the V_κ and J_κ gene segments. That is, during the joining process a few nucleotide pairs from V_κ and a few nucleotide pairs from J_κ are lost from the DNA at the $V_\kappa J_\kappa$ joint, generating significant diversity in sequence at that

point. Thus diversity of κ light chains results from: (1) variability in the sequences of the multiple V_κ gene segment; (2) variability in the sequences of the four J_κ gene segments; and (3) variability in the number of nucleotide pairs deleted at V_κ-J_κ joints.

A similar mechanism exists for mouse λ light chain gene assembly. In this case there are only two L-V_λ gene segments and four C_λ gene segments, each with its own J_λ gene segment. Fewer λ variable regions can be produced than is the case for κ chains.

Heavy Chain Gene Recombination The immunoglobulin heavy chain gene is also encoded by V_H, J_H, and C_H segments. In this case additional diversity is provided by another gene segment, D (diversity), which is located between the V_H segments and the J_H segments. Figure 15.16 shows the assembly of an IgG heavy chain. In the germ line of mouse there is a tandem array of L-V_H segments, then a gap, then 12 D segments, then a gap, and then 4 J_H segments. After another gap the constant region gene segments are arranged in a cluster which, in mouse, has the order μ, δ, γ (four different ones for four different, but similar, IgG H chain constant domains), ε, and α for the H chain constant domains of IgM, IgD, IgG, IgE, and IgA, respectively. As with L chain gene rearrangements, further antibody diversity results from imprecise joining of the gene segments for the variable region.

■ *Figure 15.16*   Production of heavy chain genes in mouse by recombination of V, D, J, and C gene segments during development. Depending on the C_H segment used, the resulting antibody molecule will be IgM, IgD, IgG, IgE, or IgA. Shown here is the assembly of an IgG heavy chain. This rearrangement is only one of the many thousands possible.

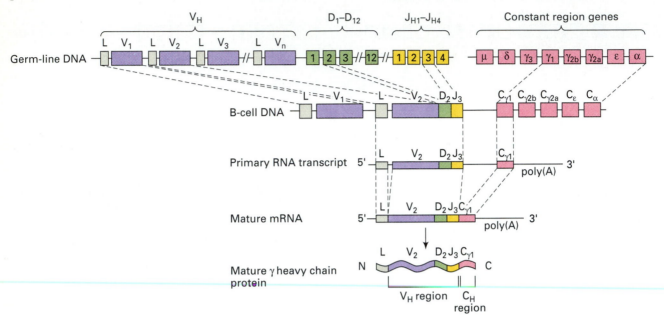

K e y n o t e

Antibodies are specialized proteins called immunoglobulins, which bind specifically to antigens. Immunity against a particular antigen results from clonal selection, in which cells already making the required antibody are stimulated to proliferate. Antibody molecules consist of two light (L) chains and two heavy (H) chains. The amino acid sequence of one domain of each type of chain is variable; this variation is responsible for the different antigen-binding site on different antibody molecules. The other domain(s) of each chain is constant in amino acid sequence. In germ-line DNA the coding regions for immunoglobulin chains are scattered in tandem arrays of gene segments. Thus for light chains there are many variable region (V) gene segments, a few joining (J) gene segments, and one constant region (C) gene segment. During development somatic recombination occurs to bring particular gene segments together into a functional L chain gene. A large number of different L chain genes result from the many possible ways in which the gene segments can recombine. Similar rearrangements occur for H chain genes, but with the addition of several D (diversity) segments that are between V and J, increasing the possible diversity of H chain genes.

Genetic Regulation of Development in *Drosophila*

There are a few systems in which significant progress is being made in relating gene activity to developmental events. One such system is *Drosophila* development, and this section presents a brief overview of what is known.

Drosophila Developmental Stages

The production of an adult *Drosophila* from a fertilized egg involves a well-ordered sequence of developmentally programmed events under strict genetic control (Figure 15.17). About 24 hours after fertilization, a *Drosophila* egg hatches into a larva, which undergoes two molts, after which it is called a pupa. The pupa metamorphoses into an adult fly. The whole process from egg to adult fly takes about ten days at 25°C.

Embryonic Development

Development commences with a single fertilized egg, giving rise to cells that have different developmental fates. What follows is a brief discussion of the information that has been obtained about the relationship between the developmental events in the egg and the determination of adult body parts.

■ *Figure 15.17* Development of an adult *Drosophila* from a fertilized egg.

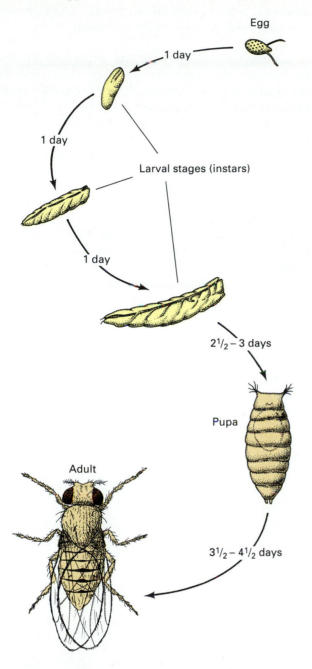

Egg

1 day

1 day

Larval stages (instars)

1 day

2½ – 3 days

Pupa

Adult

3½ – 4½ days

■ *Figure 15.18* Embryonic development in *Drosophila:* (a) The fertilized egg, with the two parental nuclei. Polar cytoplasm indicates the posterior end; (b) The two parental nuclei fuse to produce a diploid zygote nucleus; (c) The nucleus undergoes nine divisions in a common cytoplasm to produce a multinucleate syncytium; (d) Nuclei migrate and divide, producing a layer at the periphery of the egg. This stage is the syncytial blastoderm; (e) Nuclei divide four times and a membrane then forms around each to produce the somatic cells of the cellular blastoderm.

a) **Fertilized egg with two parental nuclei**

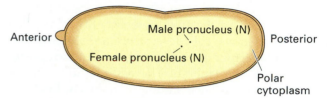

Anterior

Male pronucleus (N)

Female pronucleus (N)

Posterior

Polar cytoplasm

b) **Parental nuclei fuse and produce a diploid zygote nucleus**

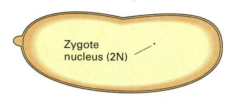

Zygote nucleus (2N)

c) **The nucleus divides for nine divisions in a common cytoplasm to give a multinucleate syncytium**

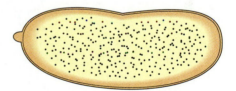

d) **Syncytial blastoderm nuclei migrate and divide, producing a layer at the periphery of the egg**

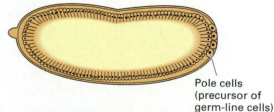

Pole cells (precursor of germ-line cells)

e) **Cellular blastoderm. Nuclei divide four times; membranes form around them and produce somatic cells**

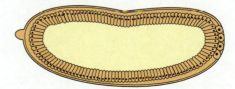

Before a mature egg is fertilized, particular molecular gradients are established within it. The posterior end is indicated by the presence of a region called the *polar cytoplasm* (Figure 15.18[a]). At fertilization the two parental nuclei are roughly centrally located in the egg. The two nuclei fuse to produce a 2N zygote nucleus (Figure 15.18[b]). For the first nine divisions, only the nuclei divide in a common cytoplasm—cytokinesis does not occur—to produce what is called a multinucleate *syncytium* (Figure 15.18[c]). (After seven of those divisions, some nuclei migrate into the polar cytoplasm, where they become precursors to germ-line cells.) Next, the remaining nuclei migrate and divide to form a layer at the surface of the egg, producing the *syncytial blastoderm* (Figure 15.18[d]). After four more divisions, membranes form around the nuclei to produce somatic cells, about 4000 of which make up the *cellular blastoderm* (Figure 15.18[e]).

Subsequent development of body structures depends on two processes (Figure 15.19): (1) Two gradients of molecules form, one along the anterior-posterior axis and the other along the dorsal-ventral axis of the egg. Somehow, a nucleus is aware of its position with reference to the molecular concentration in the two gradients; (2) In the late embryo, regions called *segments* are determined, forming a striped pattern along the anterior-posterior axis of the embryo. The embryonic segments give rise to the corresponding segments of the larva and then the corresponding segments of the adult fly.

Genes involved in regulating *Drosophila* development are defined by mutations that have a lethal phenotype early in development or that result in the development of abnormal structures (e.g., embryos with abnormal striping). Three major classes of such genes have been defined:

1. *Maternal genes*: These are expressed by the mother during oogenesis. Their products specify the gradients in the egg and determine spatial organization in early development. Examples of maternal genes are *bicoid* (*bcd*: required for normal anterior development), *tudor* (*tud*: posterior segment development), and *torso* (*tor*: terminal posterior segment development).

The *bicoid* gene, for example, is transcribed during oogenesis. The bicoid mRNAs become localized at the anterior tip of the embryo. Soon after the egg is laid, the mRNAs are translated. The bicoid protein establishes a gradient along the long axis of the embryo with the highest concentration at the anterior end and the lowest concentration at the posterior end. The concentration of the bicoid protein is correlated with the development of anterior structures. That is, the developmental fate of cells located in

■ *Figure 15.19* *Drosophila* development results from gradients in the egg that define segments in the late embryo, larva (not shown) and adult. The adult segment organization directly reflects the segment pattern of the late embryo.

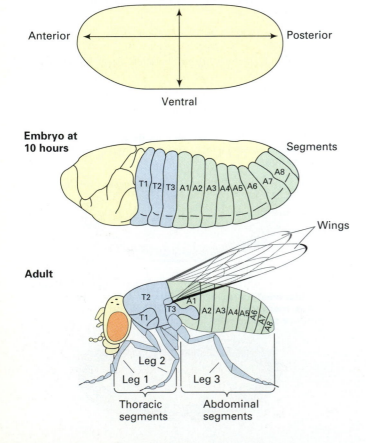

the anterior part of the embryo is determined by the concentration of bicoid protein in which they find themselves.

2. *Segmentation genes*: These are expressed after fertilization. Mutations in these genes affect the number or polarity of body segments. At least 20 gene loci have been shown to affect segmentation. The loci are subclassified on the basis of the size of the unit affected (Figure 15.20). Mutations in *gap genes* (e.g., *Krüppel*, *hunchback*) result in several adjacent segments being deleted; mutations in *pair-rule genes* (e.g., *even-skipped*, *fushi tarazu*) result in the same part of the pattern deleted in every other segment; and mutations in *segment polarity genes* (e.g., *gooseberry*, *engrailed*) have segments replaced by mirror images.

3. *Homeotic genes*: These genes control the *identity* of a segment but do not affect the number, polarity, or size of segments. Homeotic mutants, which act after segmentation genes, cause one body part to develop into a different body part. Examples of homeotic mutants will be described later.

In sum, genes control *Drosophila* development in a temporal regulatory cascade. First, the maternal genes define the anterior-posterior and dorsal-ventral gradients in the egg. Next, the gap genes are transcribed, functioning to define four major regions of the egg. Then pair rule genes are transcribed, and their products are confined to pairs of segments. Following that, segment polarity genes are expressed; their products act at the individual segment level. Last, homeotic genes act to determine the identity of the segments.

Keynote

Development of *Drosophila* body structures results from gradients along the anterior-posterior and dorsal-ventral axes of the egg and from the subsequent determination of regions in the embryo that directly correspond to adult body segments. As defined by mutations, genes control *Drosophila* development in a temporal regulatory cascade. First, maternal genes specify the gradients in the egg, then segmentation genes (gap genes, pair rule genes, and segment genes) determine the segments of the embryo and adult, and finally homeotic genes specify the identity of the segments.

Imaginal Discs

Two types of cells are specified by cellular blastoderm cells: (1) those that will produce larval tissues; and (2) those that will develop into the adult tissues and organs. For the latter, certain groups of undifferentiated cells form larval structures called **imaginal discs** ("imago" means adult), each of which differentiates into a specific organ in the adult fly. A number of *Drosophila* adult structures develop from imaginal discs, which are determined early in larval development. These adult structures include mouth parts, antennae, eyes, wings, halteres, legs, and the external genitalia. Imaginal disc cells remain in an embryonic state throughout larval development, even though larval cells differentiate around them. Other structures, such as the ner-

■ *Figure 15.20* Functions of segmentation genes as defined by mutations.

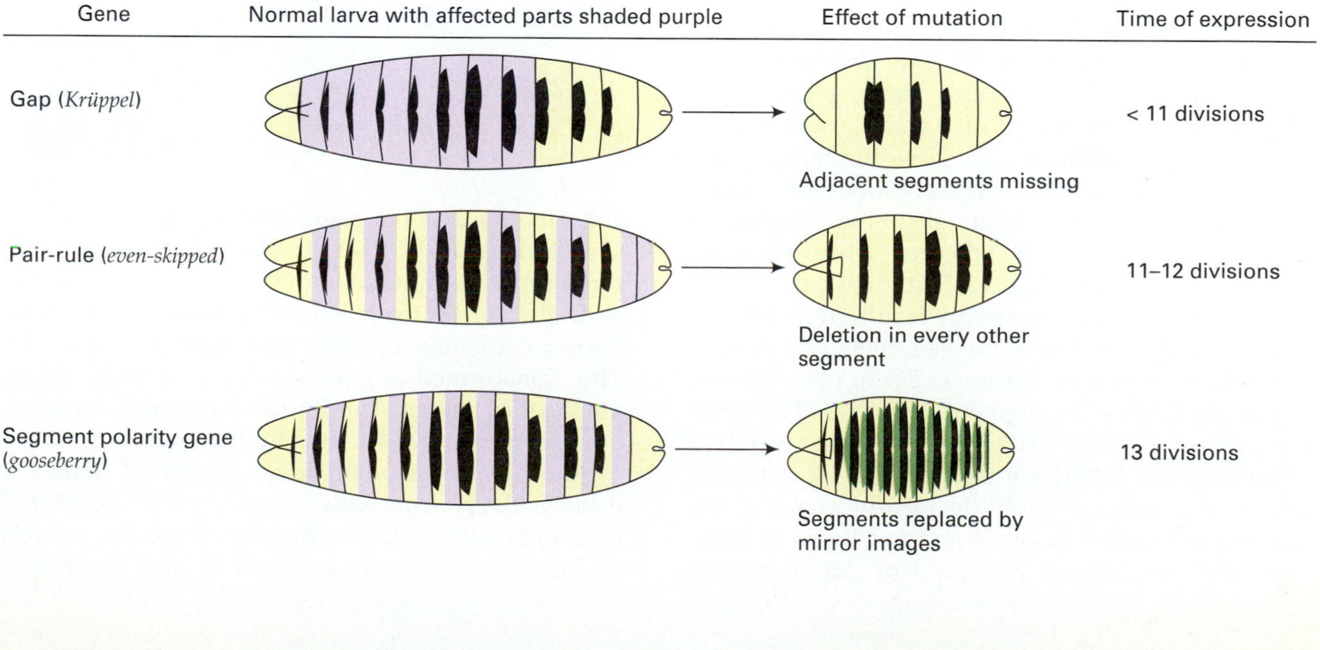

vous system, gut, and cuticle do not develop from imaginal discs.

Each imaginal disc develops during the first larval stage, and when it consists of about 20–50 cells, it is already programmed to specify its given adult structure; in other words, its fate is determined. From then on the number of cells in each disc increases by mitotic division, until by the end of the larval stages, there are many thousand cells per disc.

The nature of the disc determination process is unknown, although at a simple level it probably involves a programming of those genes that are accessible to hormone (ecdysone)-stimulated activation during the pupal stage.

Keynote

A number of *Drosophila* adult structures develop from imaginal discs, which are determined early in larval development. The determination process is a remarkably stable event since all the cells in a particular disc usually remain programmed in the same way.

Homeotic Genes

Once the basic segmentation pattern has been laid down, the homeotic genes give a specific developmental identity to each of the segments. The homeotic genes have been defined by mutations that affect the development of the fly. That is, **homeotic mutations** alter the identity of particular segments, transforming them into copies of other segments. The principal pioneer of genetic studies of homeotic mutants is Edward Lewis.

Lewis's pioneering studies involved a cluster of homeotic genes called the *bithorax* complex (*BX-C*). *BX-C* determines the posterior identity of the fly, namely thoracic segment T3 and abdominal segments A1–A8. *BX-C* contains three genes called *Ultrabithorax* (*Ubx*), *abdominal-A* (*abd-A*), and *Abdominal-B* (*Abd-B*). Mutations in these homeotic genes are often lethal, so the fly typically does not survive past embryogenesis. Some nonlethal mutations have been characterized, however, which allow an adult fly to develop. Figure 15.21 shows the abnormal adult structures that can result from *bithorax* mutations. A diagram of the segments of a normal adult fly is shown in Figure 15.21(a): note that the wings are located on segment Thorax 2 (T2), while the pair of halteres (rudimentary wings used as "balancers" in flight) are on segment T3. A photograph of a normal adult fly clearly showing the wings and halteres is presented in Figure 15.21(b). Figure 15.21(c) shows one type of developmental

■ *Figure 15.21*   (a) Drawing of a normal fly. T = thoracic segment. A = abdominal segment. The haltere (rudimentary wing) is on T3 (see Figure 15.23). (b) Photograph of a normal fly with a single set of wings. (c) Photograph of a fly homozygous for three mutant alleles (*bx³*, *abx*, and *pbx*) that results in the transformation of segment T3 into a structure like T2: namely, a segment with a pair of wings. These flies therefore have two sets of wings, but no halteres.

a)

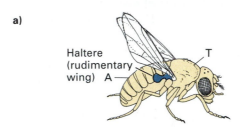

b)

c)

abnormality that can result from nonlethal homeotic mutations in *BX-C*—a fly that is homozygous for three separate mutations in the *Ubx* gene, *abx*, *bx³*, and *pbx*. Collectively these mutations transform segment T3 into an adult structure similar to T2. The transformed segment exhibits a fully developed set of wings. The fly lacks halteres, however, because no normal T3 segment is present.

Another well-studied large cluster of homeotic genes is the *Antennapedia* complex (*ANT-C*). *ANT-C* determines the anterior identity of the fly, namely the head and thoracic segments T1 and T2. *ANT-C* contains at least four genes, called *Deformed* (*Dfd*), *fushi tarazu* (*ftz*), *Sex combs reduced* (*Scr*), and *Antennapedia* (*Antp*). Most *ANT-C* mutations are lethal. Among the nonlethal mutations is a group of mutations in *Antp* that results in leg parts instead of an antenna growing out of the cells near the eye during the development of the eye disc (Figure 15.22[a] and [b]). Note that the leg has a normal structure but is obviously positioned in an abnormal location. A different mutation in *Antp*, called *Aristapedia*, has a different effect: only the distal part of the antenna, the arista, is transformed into a leg (Figure 15.22[c]).

The homeotic genes *ANT-C* and *BX-C*, therefore, encode products that are involved in controlling the normal development of the relevant adult fly structures. The *Antennapedia* complex (*ANT-C*) and the *bithorax* complex (*BX-C*) have both been cloned. Both complexes are very large. In *ANT-C* the *Antp* gene is 103 kb long, with many introns; this gene encodes a mature mRNA of only a few kilobase pairs. *BX-C* covers more than 300 kb of DNA and contains only three protein-coding regions amounting to about 50 kb of that DNA: *Ubx*, *abdA*, and *AbdB* (Figure 15.23, p. 371). At least *Ubx* and *abdA* have introns. Other RNA products are known to be transcribed from *BX-C*, but they do not code for proteins, and they have no known functions. The other 250 kb of DNA in the complex is not transcribed and is believed to consist of regulatory regions of significant size and complexity. The functions of these regulatory regions are to control the expression of the protein-coding genes.

Because the *ANT-C* and *BX-C* protein-coding genes have similar functions but are located in different places in the genome, Lewis predicted that the genes would have related structures. Analysis of the DNA sequences for the genes revealed that each has a similar 180-bp sequence named a **homeobox**. The homeobox is part of the protein-coding sequence of each gene, and the corresponding 60-amino-acid part of each protein is called the **homeodomain**. Homeoboxes have been found in over 20 *Drosophila* genes, most of which regulate development. All homeodomain-containing proteins appear to be located in the nucleus, and there is a similarity in structure between the homeodomain and other known DNA-binding proteins and transcription factors. Thus homeodomain-containing proteins may play a role in transcriptional regulation by binding to specific DNA sequences. This role has been definitely established in a few cases.

Homeoboxes have also been found in many higher eukaryotes, including humans, mice, chickens, and *Xenopus*. They have also been found in arthropods, annelids, ascidians, echinoderms, brachiopods, tapeworms, molluscs, and other chordates. Homeoboxes have not yet been found in coelenterates, sponges, flatworms, slime molds, filamentous fungi, or bacteria. It is hoped that study of the homeoboxes will provide a more general understanding of transcriptional regulation of genes involved in development.

Keynote

Homeotic genes in *Drosophila* determine the developmental identity of a segment, once the basic segmentation pattern has been laid down. Homeotic genes have in common similar DNA sequences called homeoboxes. The homeobox is part of the protein-coding sequence of each gene and the corresponding approximately 60-amino-acid part of the protein is called the homeodomain. Homeoboxes have been found in developmental genes in other organisms, and homeodomains are thought to play a role in regulating transcription by binding specific DNA sequences.

Summary

In this chapter we have considered a number of examples of gene regulation in eukaryotes. The general picture is one of much greater complexity than prokaryotes. Genes are not organized into operons, since genes of related function are often scattered around the genome; nonetheless, genes *are* regulated coordinately. The chromosome organization in eukaryotes also makes for greater complexity in regulating gene expression.

Three main topics were discussed in the chapter: the levels of control of gene expression, gene regulation during development and differentiation, and the specific example of genetic regulation of development in *Drosophila*. Given the amount of information available, each of these topics could be the subject of its own book.

Levels of Control of Gene Expression

Gene expression in eukaryotes is regulated at a number of distinct levels. Regulatory systems have been found for the control of: (1) transcription; (2) precursor-RNA processing; (3) translation of the mRNAs; (4) degradation of the mature RNAs; and (5) degradation of the protein products. All but the last were discussed in this chapter.

The control of transcription itself involves a number of elements. At the DNA level, transcription of protein-coding genes involves the interaction of transcription factors with promoter elements, and of regulatory proteins with both promoter elements and enhancer elements. Depending on

■ *Figure 15.22*   (a) Scanning electron micrograph (left) and drawing (right) of the antennal area of a wild-type fly; (b) Scanning electron micrograph (left) and drawing (right) of the antennal area of the homeotic mutant of *Drosophila Antennapedia*, in which the antenna is transformed into a leg; (c) Scanning electron micrograph (left) and drawing (right) of the homeotic mutant of *Drosophila Aristapedia*, in which the arista is transformed into a leg.

a) Normal

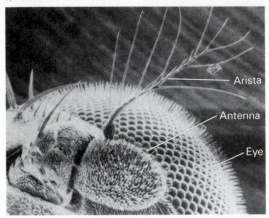
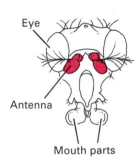

b) Antennapedia

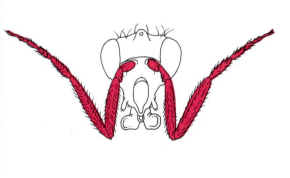

c) Aristapedia

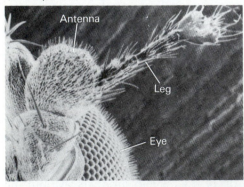
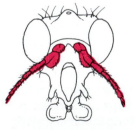

the element and the protein that binds to it, the effect on transcription may be positive or negative. Activation of transcription through protein binding at the distant enhancer elements is thought to involve a looping of the DNA, brought about by interaction of the regulatory proteins bound at the enhancer and promoter elements. While unique regulatory proteins certainly bind to promoter and enhancer elements, some regulatory proteins are shared by the two, indicating that both types of regulatory elements affect transcription by a similar mechanism.

■ *Figure 15.23*   Organization of the *bithorax* complex (BX-C). The DNA spanned by this complex is 300 kb long. The transcription units for *Ubx*, *abdA*, and *AbdB* are shown below the DNA: the exons are shown by colored blocks and the introns by bent, dotted lines. All three genes are transcribed from right to left. Shown above the DNA are regulatory mutants that affect the development of different fly segments.

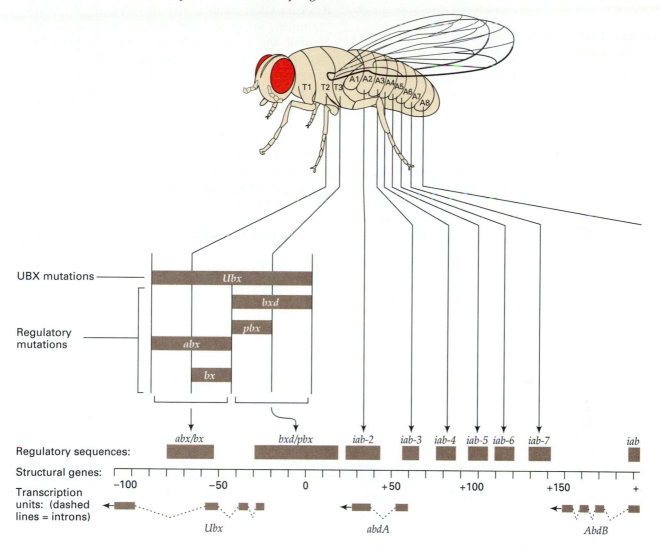

At the chromosome level, the regulation of transcription must deal with the interaction of histones and nonhistones with the DNA. Generally, we can view the eukaryote chromosome as repressed for transcription by this interaction. So activation of transcription can be seen as a derepression phenomenon that results from a loosening of the DNA-protein structure in the region of a gene being activated. The promoter regions of genes have an even looser DNA-protein structure.

Histones play an important role in gene repression by assembling nucleosomes on TATA boxes in promoter regions. Gene activation involves regulatory proteins binding to enhancers and then disrupting the nucleosomes on the TATA boxes. This allows regulatory proteins and transcription factors to bind to promoters to initiate transcription.

Gene activation in many eukaryotes is accompanied in many instances by a decrease in the level of DNA methylation, although the relationship, if any, to chromosome organization changes is not clear.

Gene expression can be regulated at the level of RNA processing. This type of regulation operates to determine the production of mature RNA molecules from precursor-RNA molecules. Two regulatory events that exemplify this level of control are choice of poly(A) site and choice of splice site. In both cases different types of mRNAs are produced, depending on the choice made.

Gene expression is also regulated by mRNA translation control and by mRNA degradation control. The latter is believed to be a major control point in the regulation of gene expression, as evidenced by the wide range of mRNA stabilities found within organisms. It is clear that nucleases are ultimately responsible for the degradation of the RNAs, and the signals for the differential mRNA stabilities seem to be a property of the structural features of individual mRNAs.

Gene Regulation During Development and Differentiation

This topic is a vast one, bringing together material in the domains of the geneticist, the cell biologist, the embryologist, the developmental biologist, and the molecular biologist. Classical experiments showed that development and differentiation result from differential gene activity of a genome that contains a constant amount of DNA, rather than from a programmed loss of genetic information. Development and differentiation, then, must involve regulation of gene expression, using the levels of control we have just discussed. Adding to the complexity is the fact that we must consider the regulation of a large number of genes for each developmental process and communication between differentiating tissues, as well as systems for timing the activa-

tion and repression of genes during those important events. For example, the structure and function of a cell are often determined early, even though the manifestations of this determination process are not seen until later in development. Such early determination events may involve some preprogramming of genes that will be turned on later. Neither the nature of these determination events nor the mechanism of timing in developmental processes are well understood, although progress is being made in understanding the genetic regulation of development in certain organisms such as *Drosophila*.

In sum, a great deal of information has been learned in the past decade or so about gene regulation in eukaryotes. We have merely scratched the surface in this chapter. Thousands of researchers are currently working to elaborate the molecular details of gene regulation in model systems. Much of our advancing knowledge has been made possible by the application of recombinant DNA and related technologies, and we can look forward to significant increases in our understanding of eukaryotic gene regulation by the turn of the century.

Analytical Approaches for Solving Genetics Problems

Q15.1 We learned in this chapter that, in humans, there are several distinct genes that code for α- and β-like globin polypeptides that are assembled in specific combinations to form different types of hemoglobin. The various types synthesized and functioning at different times during human development were discussed on p. 359–360, and Figure Q15.1 summarizes this information in graphical form. That is, human α, β, γ, δ, ε, and ζ globin genes are transcriptionally active at various stages

of development. Fill in the following table, indicating whether the globin gene in question is sensitive (S) or resistant (R) to DNase I digestion at each of the developmental stages listed.

| | | TISSUE | |
GLOBIN GENE	EMBRYONIC YOLK SAC	SPLEEN	ADULT BONE MARROW
α			
β			
γ			
δ			
ζ			
ε			

A15.1 The correct filled-in table is shown at the end of this answer.

The explanation for the answers is as follows: DNase I typically digests regions of DNA that are transcriptionally active, while not digesting regions of DNA that are transcriptionally inactive. This is because transcriptionally inactive DNA is more highly coiled than transcriptionally active DNA. "R" means, then, that the gene was transcriptionally

■ *Figure Q15.1*

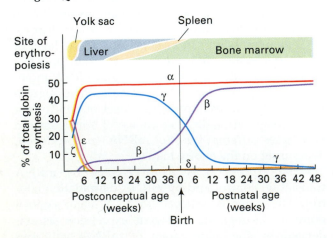

inactive, while "S" means that the gene was transcriptionally active.

To consider each globin gene type in turn, the α gene is transcriptionally inactive in the embryonic yolk sac, but active in spleen and adult bone marrow. That is, in the spleen, fetal hemoglobin (Hb-F) is made; Hb-F contains two α polypeptides, and two γ polypeptides. In the bone marrow, Hb-A is made, which contains two α and two β polypeptides.

The β-globin gene is inactive in the yolk sac and the spleen, and is active in bone marrow, making one of the two polypeptides found in Hb-A the main adult form of hemoglobin. The β-like γ polypeptide is found in Hb-F, which is made only in the liver and the spleen; thus, the γ gene is active in the spleen and inactive in the yolk sac and bone marrow. The β-like δ polypeptide is found in $\alpha_2\delta_2$ hemoglobin, which is a minor class of hemoglobin found in adults; thus, the δ gene is active only in adult bone marrow.

The ζ gene makes an α-like polypeptide found only in the hemoglobin of the embryo. That is, the ζ gene is active in the yolk sac, but inactive in spleen and bone marrow. Finally, the ε gene encodes the β-like polypeptide of the embryo's hemoglobin, so this gene is also active in the yolk sac, but inactive in the spleen and bone marrow.

GLOBIN GENE	TISSUE		
	EMBRYONIC YOLK SAC	SPLEEN	ADULT BONE MARROW
α	R	S	S
β	R	R	S
γ	R	S	R
δ	R	R	S
ζ	S	R	R
ε	S	R	R

Questions and Problems

*15.1 Eukaryotic organisms have a large number of copies (usually more than a hundred) of the genes that code for ribosomal RNA, yet they have only one copy of each gene that codes for each ribosomal protein. Explain why.

15.2 How do hormones participate in the regulation of gene expression in eukaryotes?

*15.3 Figure 15A shows the effect of the hormone estrogen on ovalbumin synthesis in the oviduct of 4-day-old chicks. Chicks were given daily injections of estrogen ("Primary stimulation") and then after 10 days the injections were stopped. Two weeks after withdrawal (25 days), the injections were resumed ("Secondary stimulation").

Provide possible explanations of these data.

15.4 Distinguish between the terms *development* and *differentiation*.

15.5 Discuss some of the evidence for differential gene activity during development.

*15.6 The enzyme lactate dehydrogenase (LDH) consists of four polypeptides (a tetramer). Two genes are known to specify two polypeptides, A and B, which combine in all possible ways (A_4, A_3B, A_2B_2, AB_3, and B_4) to produce five LDH isozymes. If, instead, LDH consisted of three polypeptides (i.e., it was a trimer), how many possible isozymes would be produced by various combinations of polypeptides A and B?

■ *Figure 15A*

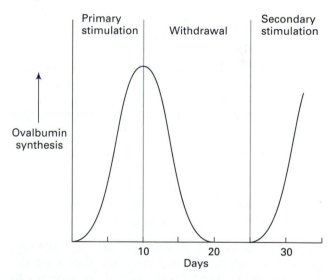

15.7 Discuss the expression of human hemoglobin genes during development.

15.8 In humans β-thalassemia is a disease caused by failure to produce sufficient β-globin chains. In many cases the mutation causing the disease is a deletion of all or part of the β-globin structural gene. Individuals homozygous for certain of the β-thalassemia mutations are able to survive because their bone marrow cells produce γ-globin chains. The γ-globin chains combine with α-globin chains to produce fetal hemoglobin. In these people fetal hemoglobin is produced by the bone marrow cells throughout life, whereas normally it is produced in the fetal liver. Use your knowledge about gene regulation during development to suggest a mechanism by which this expression of γ-globin might occur in β-thalassemia.

*15.9 Figure 15B shows the percentage of ribosomes found in polysomes in unfertilized sea urchin oocytes (0 h) and at various times after fertilization:

■ *Figure 15B*

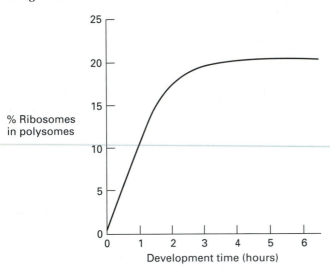

In the unfertilized egg less than 1% of ribosomes are present in polysomes, while at 2 h post-fertilization, about 20% of ribosomes are present in polysomes. It is known that no new mRNA is made during the time period shown. How may the data be interpreted?

15.10 The mammalian genome contains about 10^5 genes. Mammals can produce about 10^6 to 10^8 different antibodies. Explain how it is possible for both of the above sentences to be true.

*15.11 Imagine that you observed the following mutants (*a* through *e*) in *Drosophila*. Based on the characteristics given, assign each of the mutants to one of the following categories: maternal gene, segmentation gene, or homeotic gene.

a. Mutant *a*: In homozygotes phenotype is normal, except wings are oriented backwards.
b. Mutant *b*: Homozygous females are normal but produce larvae that have a head at each end and no distal ends. Homozygous males produce normal offspring (assuming the mate is not a homozygous female).
c. Mutant *c*: Homozygotes have very short abdomens, which are missing segments AB2 through AB4.
d. Mutant *d*: Affected flies have wings growing out of their heads in place of eyes.
e. Mutant *e*: Homozygotes have shortened thoracic regions and lack the second and third pair of legs.

*15.12 If actinomycin D, an antibiotic that inhibits RNA synthesis, is added to newly fertilized frog eggs, there is no significant effect on protein synthesis in the eggs. Similar experiments have shown that actinomycin D has little effect on protein synthesis in embryos up until the gastrula stage. After the gastrula stage, however, protein synthesis is significantly inhibited by actinomycin D, and the embryo does not develop any further. Interpret these results.

*15.13 It is possible to excise small pieces of early embryos of the frog, transplant them to older embryos, and follow the course of development of the transplanted material as the older embryo develops. A piece of tissue is excised from a region of the late blastula or early gastrula that would later develop into an eye and is transplanted to three different regions of an older embryo host; see part a) in Figure 15C. If the tissue is transplanted to the head region of the host, it will form eye, brain, and other material characteristic of the head region. If the tissue is transplanted to other regions of the host, it will form organs and tissues characteristic of those regions in normal development (e.g., ear, kidney, etc.). In contrast, if tissue destined to be an eye is excised from a neurula and transplanted into an older embryo host to exactly the same places as used for the blastula/gastrula transplants, in every case the transplanted tissue differentiates into an eye; see part b) in the figure. Explain these results.

■ *Figure 15C*

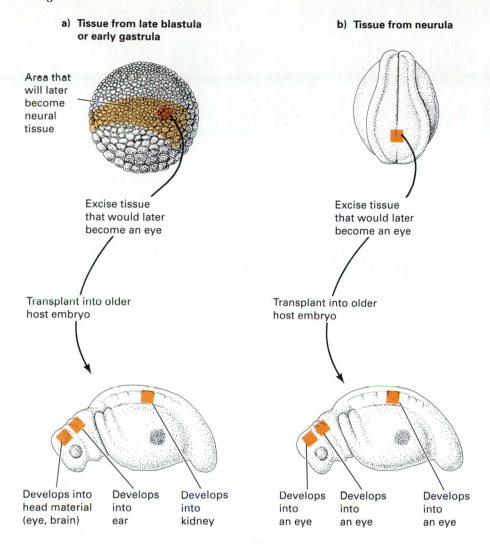

a) **Tissue from late blastula or early gastrula**

Area that will later become neural tissue

Excise tissue that would later become an eye

Transplant into older host embryo

Develops into head material (eye, brain)

Develops into ear

Develops into kidney

b) **Tissue from neurula**

Excise tissue that would later become an eye

Transplant into older host embryo

Develops into an eye

Develops into an eye

Develops into an eye

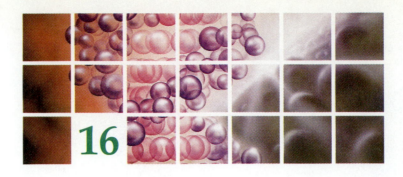

16

Gene Mutation

- Mutation is the process by which a change in DNA base pairs or a change in the chromosomes is produced. A mutation, then, is the DNA base-pair change, or chromosome change resulting from the mutation process. Mutations may occur spontaneously or may be induced experimentally by the application of mutagens.

- Mutations at the level of the chromosome are called chromosomal mutations. Mutations at the level of the base pair, when they affect the expression of genes, are called gene mutations. Gene mutations may occur, for example, by a substitution of one base pair for another or by the addition or deletion of one or more base pairs.

- The consequences of a gene mutation to an organism depend on a number of factors, especially the extent to which the amino acid coding information is changed. For example, missense mutations cause the substitution of one amino acid for another, and nonsense mutations cause premature termination of synthesis of the polypeptide.

- The effects of a gene mutation can be reversed either by reversion of the gene sequence to its original state, or by a mutation at a site distinct from that of the original mutation. The latter is called a suppressor mutation.

- Gene mutations may be caused by exposure to radiation. Radiation may cause genetic damage by breaking chromosomes, by producing chemicals that interact with DNA, or by causing unusual bonds between DNA bases. Mutations result if the genetic damage is not repaired.

- Gene mutations may also be caused by exposure to certain chemicals, called chemical mutagens. Again, mutations result if the genetic damage caused by the mutagen is not repaired. A variety of chemical mutagens is known, and they act in various ways. Base analogs physically replace the proper bases during DNA replication and then shift chemical form so that their base-pairing properties change. Base modifiers cause chemical changes to existing bases, thereby altering their base-pairing properties. Base-pair substitution mutations are the result of treatment with base analogs and base modifiers. Intercalating agents are inserted between existing adjacent bases during replication, resulting in single base-pair additions or deletions, called frameshift mutations.

- Mutations can result from damage to the DNA. Both prokaryotes and eukaryotes have a number of repair systems that deal with different kinds of DNA damage. All of the systems use enzymes to make the correction. Without such repair systems, mutations would accumulate and be lethal to the cell or organism. Not all DNA damage is repaired; hence, mutations do appear, but at relatively low frequencies. At high doses of mutagens, repair systems are unable to correct all of the damage, and cell death results.

- Geneticists have made great progress in understanding how cellular processes take place by studying mutants that have defects in those processes. A number of screening procedures have been developed to enrich selectively for mutants of interest from a population of mutagenized cells or organisms.

E ARLIER IN THIS TEXT we learned that genetic material is stably passed on from mother cell to daughter cell, and from generation to generation through the faithfulness of the DNA replication process. The genetic material is changed, however, through a number of means including spontaneous changes, errors in the replication process, or the action of particular chemicals or radiation. On the broad scale there are two types of changes to the genetic material: changes involving whole chromosomes (the topic of Chapter 17), and changes affecting one or a few base pairs. All organisms have mechanisms for repairing base-pair changes, but not all base-pair changes may be corrected by the repair systems. The changes that are not repaired are base-pair *mutations*.

Mutations of base pairs can occur anywhere in the genome of an organism. Since not all of the genome of an organism consists of genes, a base-pair mutation will have no phenotypic consequences to the organism unless it occurs within a gene or in the sequences regulating it. Thus the mutations that have been of particular interest to geneticists are *gene mutations*—mutations that affect the function of genes. Further, since cell function is the result of protein function, most of the attention has been focused on mutations affecting the expression of protein-coding genes. Figure 16.1 illustrates in general how a gene mutation can alter a phenotype by changing the function of a protein. We have encountered numerous examples throughout this text of mutations affecting protein-coding genes, including white-flowered pea plants, miniature-winged fruit flies, hemophilia in humans, albino mice, *lacZ* mutants of *E. coli*, and so on. Indeed, it is through the comparison of mutant strains with wild-type (normal) strains that geneticists have been able to obtain an understanding of normal cell function and have advanced our knowledge of the genetics of many organisms.

In this chapter we focus on some of the mechanisms that bring about changes in the DNA at the base-pair level, on some of the repair systems that can repair genetic damage, and on some of the methods used to screen for genetic mutants. We will concentrate on gene mutations affecting protein-coding genes and the phenotypic consequences of those changes. As we learn about the specifics of gene mutations, we must be aware that mutations are a major source of genetic variation in a species. Mutations, then, are important elements of the evolutionary process.

Mutations Defined

Mutation is the process by which a DNA base-pair change or a chromosome change is produced. A mutation, then, is the DNA base-pair change or chromosome change resulting from the mutation process. Mutations of the DNA are discussed in this chapter, while mutations of a chromosome set are examined in the next chapter.

A mutation can be transmitted to daughter cells and even to succeeding generations, thereby giving rise to mutant cells or mutant individuals. If a mutant cell gives rise only to somatic cells (in multicellular organisms), a mutant spot or area is produced, but the mutant characteristic is not passed on to the succeeding generation. This type of mutation is called a **somatic mutation**. However, mutations in the germ line of sexually reproducing organisms may be transmitted by the gametes to the next generation, producing an individual with the mutation in both its somatic and germ-line cells. Such mutations are called **germ-line mutations**.

Types of Mutations

A change in the organization of a chromosome or chromosomes is called a **chromosomal mutation** or **chromosomal aberration** (see Chapter 17). A mutation in a gene sequence is called a **gene mutation**, and can involve any one of a number of alterations of the DNA sequence of the gene, including base-pair substitutions and additions or deletions of one or more base pairs. Those gene mutations that affect a single base pair of DNA are called **point mutations**.

Mutations can occur spontaneously, but they can also be induced experimentally by the application of a **mutagen**. A **mutagen** is any physical or chemical agent that significantly increases the frequency of mutational events above the spontaneous mutation rate. Mutations that result from treatment with mutagens are called **induced mutations**; naturally occurring mutations are **spontaneous mutations**. There are no qualitative differences between spontaneous and induced mutations.

With the above background we can now move on to define several other terms. While some of the terms apply generally to mutations throughout the genome, some relate specifically to mutations of genes. These terms are illustrated in Figure 16.2. Keep in mind that since the tertiary structure of a polypeptide is a function of its primary amino acid sequence (which is coded for by a gene), a polypep-

■ *Figure 16.1*   Concept of a gene mutation.

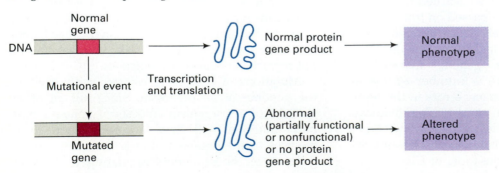

■ *Figure 16.2* Types of base-pair substitution mutations.

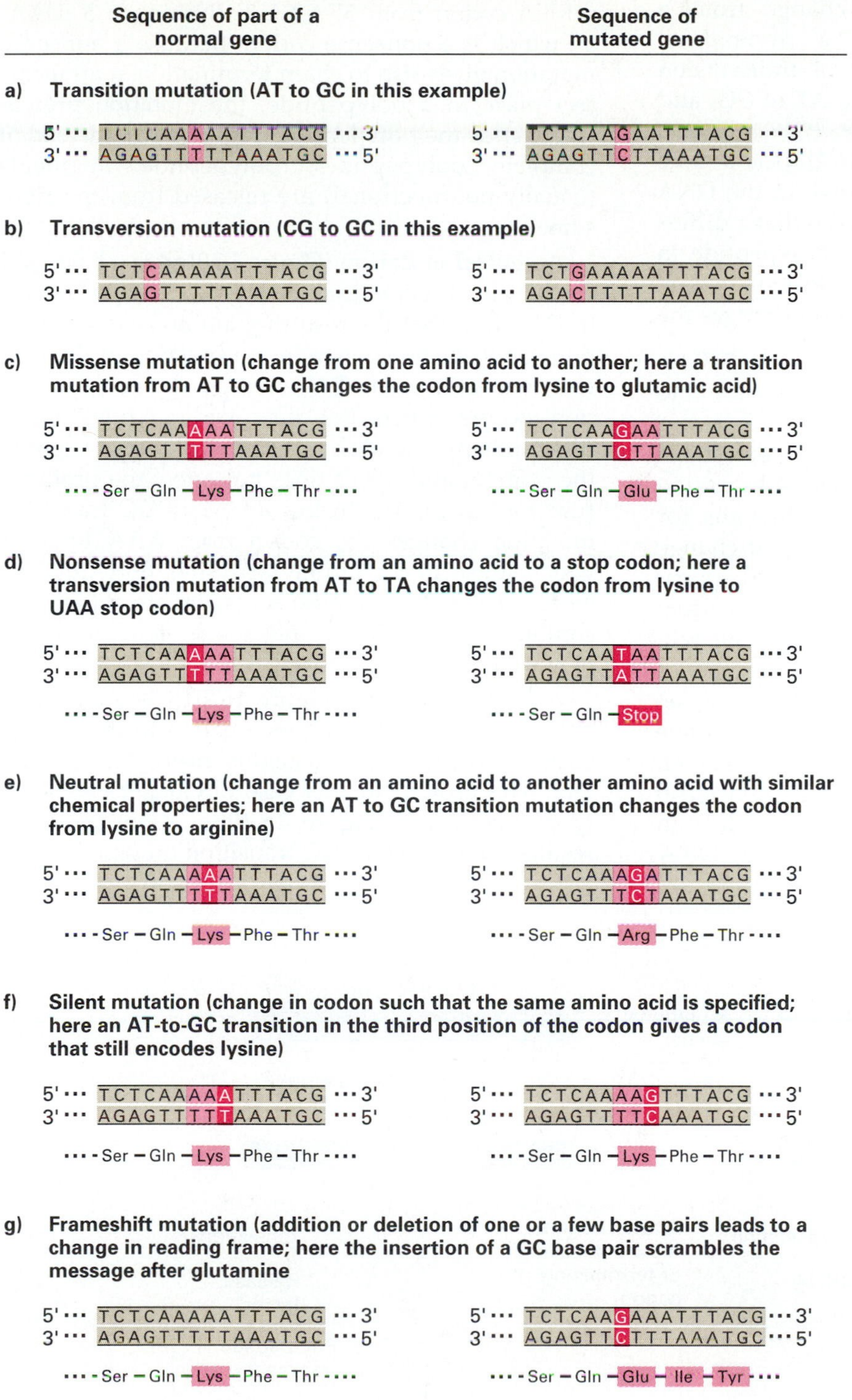

Sequence of part of a normal gene	Sequence of mutated gene

a)   Transition mutation (AT to GC in this example)

```
5'··· TCTCAAAAATTTACG ···3'        5'··· TCTCAAGAATTTACG ···3'
3'··· AGAGTTTTTAAATGC ···5'        3'··· AGAGTTCTTAAATGC ···5'
```

b)   Transversion mutation (CG to GC in this example)

```
5'··· TCTCAAAAATTTACG ···3'        5'··· TCTGAAAAATTTACG ···3'
3'··· AGAGTTTTTAAATGC ···5'        3'··· AGACTTTTTAAATGC ···5'
```

c)   Missense mutation (change from one amino acid to another; here a transition mutation from AT to GC changes the codon from lysine to glutamic acid)

```
5'··· TCTCAAAAATTTACG ···3'        5'··· TCTCAAGAATTTACG ···3'
3'··· AGAGTTTTTAAATGC ···5'        3'··· AGAGTTCTTAAATGC ···5'

···· Ser — Gln — Lys — Phe — Thr ····      ··· Ser — Gln — Glu — Phe — Thr ····
```

d)   Nonsense mutation (change from an amino acid to a stop codon; here a transversion mutation from AT to TA changes the codon from lysine to UAA stop codon)

```
5'··· TCTCAAAAATTTACG ···3'        5'··· TCTCAATAATTTACG ···3'
3'··· AGAGTTTTTAAATGC ···5'        3'··· AGAGTTATTAAATGC ···5'

···· Ser — Gln — Lys — Phe — Thr ····      ···· Ser — Gln — Stop
```

e)   Neutral mutation (change from an amino acid to another amino acid with similar chemical properties; here an AT to GC transition mutation changes the codon from lysine to arginine)

```
5'··· TCTCAAAAATTTACG ···3'        5'··· TCTCAAAGATTTACG ···3'
3'··· AGAGTTTTTAAATGC ···5'        3'··· AGAGTTTCTAAATGC ···5'

···· Ser — Gln — Lys — Phe — Thr ····      ···· Ser — Gln — Arg — Phe — Thr ····
```

f)   Silent mutation (change in codon such that the same amino acid is specified; here an AT-to-GC transition in the third position of the codon gives a codon that still encodes lysine)

```
5'··· TCTCAAAAATTTACG ···3'        5'··· TCTCAAAAGTTTACG ···3'
3'··· AGAGTTTTTAAATGC ···5'        3'··· AGAGTTTTCAAATGC ···5'

···· Ser — Gln — Lys — Phe — Thr ····      ···· Ser — Gln — Lys — Phe — Thr ····
```

g)   Frameshift mutation (addition or deletion of one or a few base pairs leads to a change in reading frame; here the insertion of a GC base pair scrambles the message after glutamine

```
5'··· TCTCAAAAATTTACG ···3'        5'··· TCTCAAGAAATTTACG ···3'
3'··· AGAGTTTTTAAATGC ···5'        3'··· AGAGTTCTTTAAATGC ···5'

···· Ser — Gln — Lys — Phe — Thr ····      ··· Ser — Gln — Glu — Ile — Tyr ···
```

tide synthesized by a mutant strain may be structurally different from the wild-type polypeptide. If so, the mutant polypeptide will be partially functional, nonfunctional, or not produced.

A **base-pair substitution mutation** (point mutation) is a change of one base pair to another base pair; for example, AT to GC.

A **transition mutation** (Figure 16.2[a]) is a specific type of base-pair substitution mutation involving a change from one purine-pyrimidine base pair to the other purine-pyrimidine base pair. The four types of transition mutations are AT to GC, GC to AT, TA to CG, and CG to TA.

A **transversion mutation** (Figure 16.2[b]) is

another specific type of base-pair substitution mutation, this one entailing a change from a purine-pyrimidine base pair to a pyrimidine-purine base pair. The four types of transversion mutations are AT to TA, GC to CG, AT to CG, and GC to TA.

A **missense mutation** (Figure 16.2[c]) is a gene mutation in which a base-pair change in the DNA causes a change in an mRNA codon so that a different amino acid is inserted into the polypeptide in place of the one specified by the wild-type codon. In Figure 16.2(c) an AT-to-GC mutation changes the DNA from $\frac{5'\text{-AAA-}3'}{3'\text{-TTT-}5'}$ to $\frac{5'\text{-GAA-}3'}{3'\text{-CTT-}5'}$ and this changes the mRNA codon from 5'-AAA-3' (lysine) to 5'-GAA-3' (glutamic acid).

Whether a mutant phenotype is easily detectable depends on the particular amino acid substitution produced by the missense mutation. In humans, for example, a particular single nucleotide pair change in the β-globin gene leads to an amino acid substitution in the β-hemoglobin chain. If the individual is homozygous for that mutation, he or she will have sickle-cell anemia, a serious genetic disease.

A **nonsense mutation** (Figure 16.2[d]) is a base-pair change in the DNA that results in the change of an mRNA codon from one that specifies an amino acid to a chain-terminating (nonsense) codon (UAG, UAA, or UGA). In Figure 16.2(d) an AT-to-TA transversion mutation changes the DNA

from $\frac{5'\text{-AAA-}3'}{3'\text{-TTT-}5'}$ to $\frac{5'\text{-TAA-}3'}{3'\text{-ATT-}5'}$ and this changes the mRNA codon from 5'-AAA-3' (lysine) to 5'-UAA-3', which is a nonsense codon. Because a nonsense mutation gives rise to chain termination at an incorrect place in a polypeptide, the mutation prematurely ends the polypeptide (Figure 16.3). Instead of complete polypeptides, polypeptide fragments (usually nonfunctional) are released from the ribosomes.

A **neutral mutation** (Figure 16.2[e]) is a base-pair change in a gene that changes a codon in the mRNA such that the resulting amino acid substitution produces no change in the function of the protein translated from that message. A neutral mutation occurs when the new codon codes for a different amino acid that is chemically equivalent to the original and hence does not affect the protein's function. In Figure 16.2(e) an AT-to-GC transition mutation changes the codon from AAA to AGA which substitutes arginine for lysine. Both arginine and lysine are basic amino acids and are sufficiently similar in properties so that the protein's function may well not be altered significantly.

A **silent mutation** (Figure 16.2[f]) is a base-pair change in a gene that alters a codon in the mRNA such that the *same* amino acid is inserted in the protein. The protein in this case obviously has wild-type function. In Figure 16.2(f) a silent mutation results from an AT-to-GC transition mutation which

■ *Figure 16.3*   Nonsense mutation.

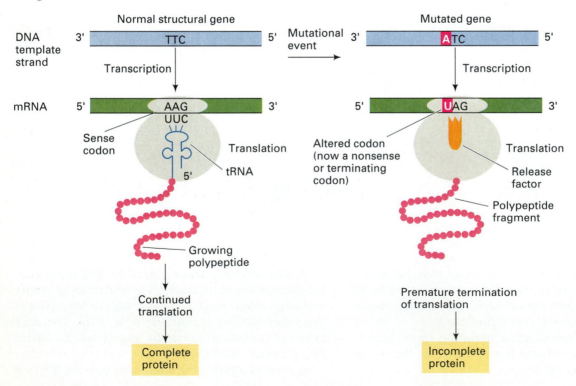

changes the codon from AAA to AAG, both of which specify lysine.

A **frameshift mutation** (Figure 16.2[g]) results from the addition or deletion of one or more base pairs in a gene. (When one base pair is involved, the frameshift mutation is a point mutation.) An addition or deletion of one base pair, for example, shifts the reading frame of the mRNA by one base so that incorrect amino acids are added to the polypeptide chain after the mutation site. A nonfunctional polypeptide results. In Figure 16.2(g) an insertion of a GC base pair scrambles the message after the codon specifying glutamine.

Keynote

Mutation is the process by which DNA base-pair change or a chromosome change occurs. A mutation, then, is the DNA base-pair change or chromosome change resulting from the mutation process. Mutations may occur spontaneously or may be induced experimentally by the application of mutagens. Mutations that affect a single base pair of DNA are called base-pair substitution mutations or point mutations. Mutations that affect gene expression are called gene mutations.

Reverse Mutations and Suppressor Mutations

Point mutations generally fall into two classes in terms of their effects on the phenotype in comparison to the wild type. **Forward mutations** are mutations that occur in the direction wild type to mutant, and **reverse mutations** (or **reversions** or **back mutations**) occur in the direction mutant to wild type.

A reversion can restore a mutant phenotype to wild-type or partially wild-type function. Reversion of a nonsense mutation, for instance, occurs when a base-pair change results in a change of the mRNA nonsense codon to a codon for an amino acid. If the reversion restores the wild-type amino acid, the mutation is a **true reversion**. If the reversion causes some other amino acid to be substituted, the mutation is a *partial reversion*, since complete function is unlikely to be restored. Reversion of missense mutations can occur in the same way.

The effects of a mutation may be diminished or abolished by a **suppressor mutation**, that is, a mutation at a different site from the original mutation (also called a secondary or **second-site mutation**). *A suppressor mutation does not result in a reversal of the original mutation*; instead, it masks or compensates for the effects of the initial mutation.

One major class of suppressor mutations is the **intergenic** ("inter" means between) **suppressor** in which the mutation that restores complete or partial protein function occurs in a different gene from that of the original mutation. Both the original mutation and the suppressor mutation must be present in the cell for suppression to work. Genes that cause suppression of mutations in other genes are called **suppressor genes**.

Intergenic suppressors often work by changing the way the mRNA encoded by the mutant gene is read. Each suppressor gene can suppress the effects of only one type of nonsense, missense, or frameshift mutation; hence, suppressor genes can suppress only a small proportion of the point mutations that theoretically can occur within a gene. On the other hand, a given suppressor gene potentially will suppress all mutations for which it is specific, whatever gene the mutation is in.

The suppressors of nonsense mutations have been well studied. The suppressor genes in this case typically are tRNA genes. That is, particular tRNA genes can mutate so that (in contrast to what occurs with wild-type tRNAs) their anticodons recognize a chain-terminating codon and put an amino acid into the chain. Thus instead of polypeptide chain synthesis being stopped prematurely as a result of a nonsense mutation, the altered (suppressor) tRNA inserts an amino acid at that position, and full or partial function of the polypeptide may be restored. There are three classes of nonsense suppressors: one for each of the nonsense codons UAG, UAA, and UGA. If, for example, a gene for tRNA.Tyr (which has the anticodon 3'-AUG-5') is mutated so that the tRNA has the anticodon 3'-AUC-5', the mutated, suppressor tRNA (which still has tyrosine attached) will compete with the release factor to read the nonsense codon 5'-UAG-3' sometimes (Figure 16.4). Instead of chain termination occurring, tyrosine is inserted at that point in the polypeptide. How functional the complete protein will be then depends on the effects of the inserted tyrosine in the protein. If it is an important part of the protein, then the incorrect amino acid may not restore function to a significant degree. If it is in a less crucial area, the protein may have partial, or complete function.

But we now have a dilemma. If we have changed this particular class of tRNA.Tyr so that its anticodon can now read a nonsense codon, it cannot read the original codon that specifies the amino acid it carries. This is explained by the fact that nonsense suppressor tRNAs are typically produced by mutation of tRNA genes that are redundant in the genome. In other words, there are several different genes all specifying exactly the same tRNA base

■ *Figure 16.4* Mechanism for action of an intergenic nonsense suppressor mutation that results from mutation of a tRNA gene. In this example a tRNA.Tyr gene has mutated so that the tRNA's anticodon is changed from 3'-AUG-5' to 3'-AUC-5', which can read a 5'-UAG-3' nonsense codon (right side of figure), inserting tyrosine in the polypeptide chain at that codon.

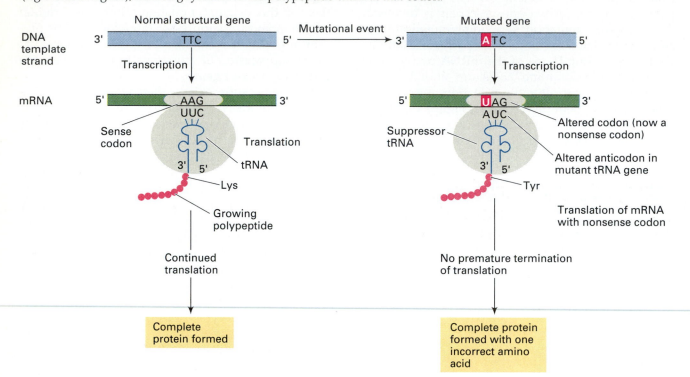

sequence. Therefore if there is a mutation in one of the redundant genes so that the tRNA will read, say, UAG, the other genes specifying the same tRNA produce a tRNA molecule that will read the normal codon.

K e y n o t e

A suppressor mutation is a mutation at a second site that completely or partially restores a function lost because of a mutation at another site. Intergenic suppressors are suppressor mutations that occur in a different gene (called a suppressor gene) than that in which the original mutation occurred. Intergenic suppressors are known for nonsense, missense, and frameshift mutations. Typically, intergenic suppression involves a tRNA with an altered anticodon. Thus nonsense mutations are suppressed by mutated tRNAs which now can read the chain-terminating codon and insert an amino acid at the mutation site.

Causes of Mutation

Mutations can occur spontaneously or they can be induced. **Spontaneous mutations** are mutations that occur without a known cause. They may occur, for example, by errors in cellular processes (such as

DNA replication), or by the action of mutagens in the environment (such as UV irradiation in sunlight). **Induced mutations** are mutations that occur as a result of treatment with known chemical or physical mutagens.

Spontaneous Mutations

Two different terms are often used to quantify the occurrence of mutations. **Mutation rate** is the probability of a particular kind of mutation as a function of time, for instance, number per nucleotide pair per generation, or number per gene per generation. **Mutation frequency**, on the other hand, is the number of occurrences of a particular kind of mutation in a population of cells or individuals, for example, number per 100,000 organisms or number per 1 million gametes.

In humans the spontaneous mutation rate for individual genes varies between 10^{-4} and 4×10^{-6} per gene per generation. For eukaryotes in general the spontaneous mutation rate is 10^{-4} to 10^{-6} per gene per generation, and for bacteria and phages the rate is 10^{-5} to 10^{-7} per gene per generation. The spontaneous mutation rate is known to be affected by the genetic constitution of the organism. For example, male and female *Drosophila* of the same strain have identical mutation rates, while different strains may exhibit different mutation rates.

■ *Figure 16.5*   Examples of mismatched bases in DNA: (a) Mismatched bases resulting from rare forms of pyrimidines; (b) Mismatched bases resulting from rare forms of purines.

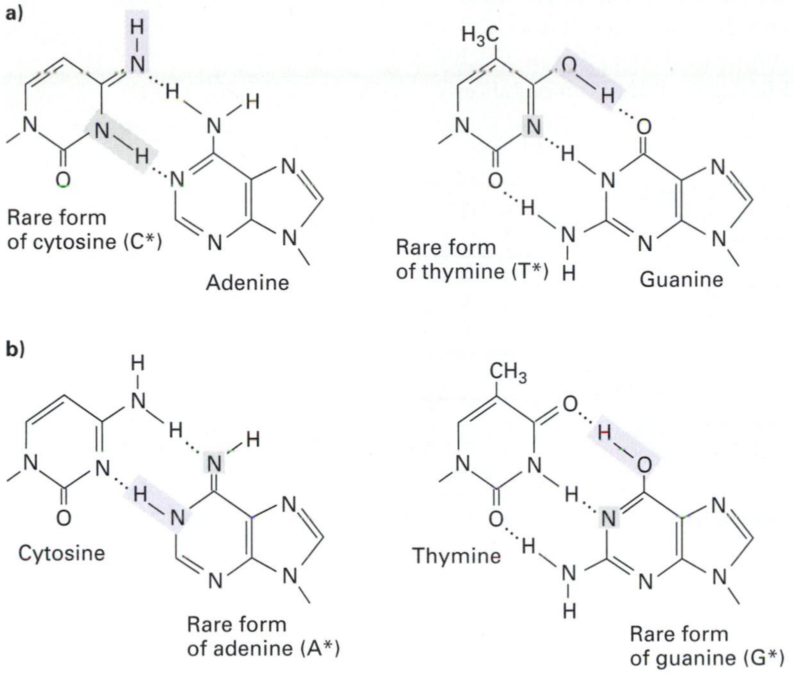

a)

Rare form of cytosine (C*)   Adenine

Rare form of thymine (T*)   Guanine

b)

Cytosine   Rare form of adenine (A*)

Thymine   Rare form of guanine (G*)

Spontaneous mutations can result from any one of a number of events, including errors in DNA replication and spontaneous chemical changes in DNA. All types of point mutations described in the previous section can occur spontaneously.

DNA Replication Errors. Spontaneous mutations can occur if mismatched base pairs occur during DNA replication. How can mismatched base pairs occur? The bases themselves are able to exist in alternate chemical forms, called **tautomers**: one is the normal form and the other is a rare form. The change in the chemical form of a base is called a **tautomeric shift**. In its rare form a base can form different hydrogen bonds, which can lead to mismatched base pairs. As an example, a rare form of cytosine can pair with adenine (Figure 16.5[a]), and a rare form of guanine can pair with thymine (Figure 16.5[b]).

Figure 16.6 illustrates how a GC-to-AT transition mutation can be produced as a result of a tautomeric shift. When the parental DNA (Figure 16.6[a]) is replicating, a guanine could shift to its rare state on a parental strand (top strand, Figure 16.6[b]). As a result, a T will be inserted on the new DNA strand giving a mismatched GT base pair after replication is completed (Figure 16.6[c]). The rare form of guanine in the GT pair is likely to change back to the

normal form so that, in the next round of DNA replication, it acts as a normal G. This will give a progeny DNA molecule with a GC base pair like the original parental DNA (Figure 16.6[d]). The T in the mismatched GT pair specifies an AT in the other progeny DNA molecule (Figure 16.6[d]); this is the GC-to-AT transition mutation. Note that many more mutations would result from tautomeric shifts than are actually observed if it were not for the proofreading activity of DNA polymerases. This activity recognizes many of the mismatches, such as GT, excises them, and replaces them with the correct base pair (see Chapter 10). However, replication of an uncorrected mismatch "fixes" the mutation in the cell. For example, once GT replicates to produce a GC and an AT in the two progeny DNAs, the AT mutation is a normal base pair and cannot be corrected by repair enzymes.

Small additions and deletions can occur spontaneously during replication (Figure 16.7). Such errors occur by erroneous looping out of bases from one or the other DNA strand. If DNA loops out from the template strand, a deletion mutation results; if DNA loops out from the strand being synthesized, an addition mutation results. If they occur in the coding region of a structural gene, such additions or deletions in the DNA will cause frameshift mutations (unless they happen in multiples of three).

■ *Figure 16.6* Production of a mutation as a result of a tautomeric shift in a DNA base:
(a) A guanine on the top strand (b) shifts to its rare form (*G) during DNA replication. In
its rare form thymine on the new DNA strand pairs with it. (c) and (d) The guanine shifts
back to its more stable normal form during the next DNA replication. The mismatched
thymine specifies an adenine on the new strand and the overall result is a GC-to-AT
transition mutation for the DNA strand lineage complementary to that in which the
tautomeric shift occurred. The other DNA strand lineages produce no mutations.

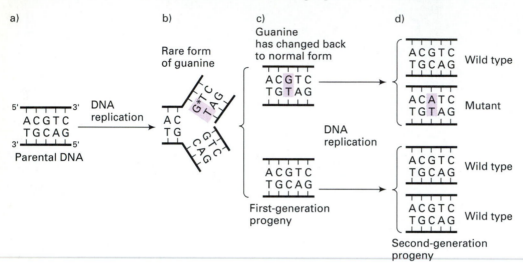

Spontaneous Chemical Changes.

Two of the most common chemical events that occur to produce spontaneous mutations are *depurination* and *deamination* of particular bases. In depurination a purine, either adenine or guanine, is removed from the DNA when the bond breaks between the base and the sugar. Thousands of purines are lost by depurination in a typical generation time of a mammalian cell in tissue culture. If such lesions are not repaired, there is no base to specify a complementary base during DNA replication. Instead, a randomly chosen base is inserted, which may produce

■ *Figure 16.7* Spontaneous generation of addition and deletion mutants by DNA
looping-out errors during replication.

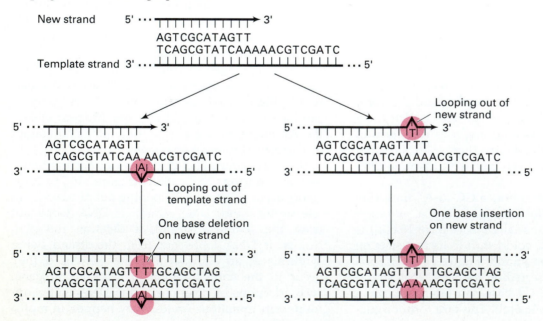

a mismatched base pair. Uncorrected mismatches may lead to a mutation, as described in the previous section.

Deamination is the removal of an amino group from a base. For example, deamination of cytosine produces uracil (Figure 16.8). Uracil is not a normal base in DNA, although it is found in RNA. A repair system described later in this chapter removes most of the uracils produced by deamination of cytosine, reducing the mutational consequences of this event. However, if the uracil is not repaired, an adenine will be incorporated in the new DNA strand opposite it during replication. Ultimately, this will result in the conversion of a CG base pair to a TA base pair; that is, a transition mutation.

DNA of both prokaryotes and eukaryotes contains relatively small amounts of the modified base 5-methylcytosine (5mC) (Figure 16.9; refer also to Figure 15.3) in place of the normal base cytosine. Deamination of 5mC produces thymine (Figure 16.9). Thymine is a normal nucleotide in DNA, so there are no repair mechanisms that can detect and correct such mutations. As a consequence, deamination of 5mC results in 5mCG-to-TA transitions. Because significant proportions of other kinds of mutations are corrected by repair mechanisms but 5mC deamination mutations are not, locations of 5mC in the genome often appear as *mutational hot spots*; that is, nucleotides where a higher-than-average frequency of mutation occurs.

Induced Mutations

Because the rate of spontaneous mutation is so low, geneticists generally use mutagens to increase mutation frequency so that a significant number of organisms have mutations in the gene being studied. Two classes of mutagens are used—radiation and chemical—both of which involve specific mechanisms of action.

Radiation. Both X rays and ultraviolet (UV) light are used to induce mutations. X rays (which are machine produced) are an example of ionizing radiation. Ionizing radiation can penetrate tissues,

■ *Figure 16.8* Deamination of cytosine to uracil.

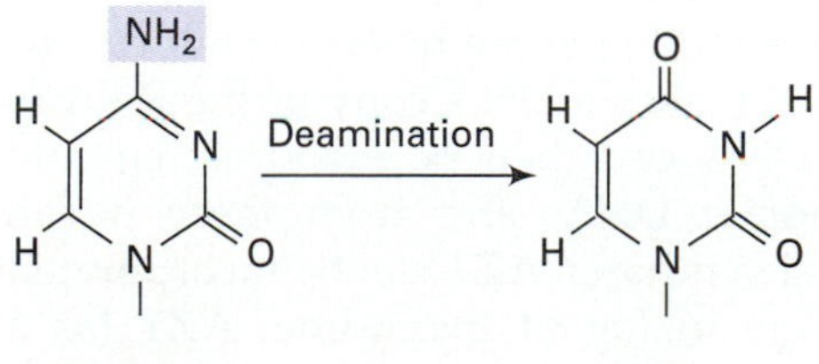

■ *Figure 16.9* Deamination of 5-methylcytosine (5mC) to thymine.

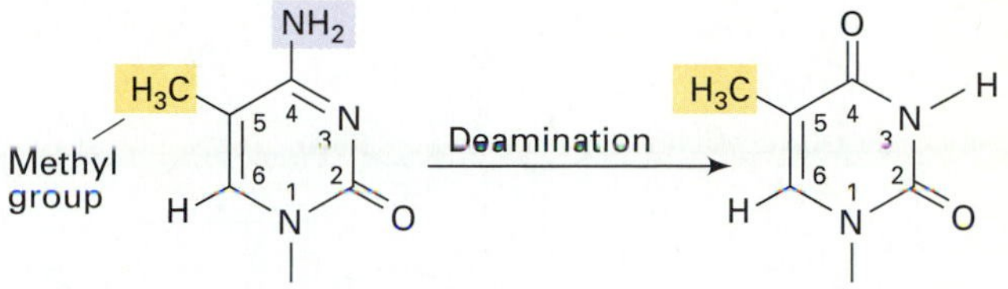

hence the use of X rays as a diagnostic tool. Ionizing radiation produces ions by colliding with atoms and releasing electrons, which in turn collide with other atoms, releasing more electrons, and so on. Therefore, along the track of an X ray a string of ions is formed. These ions can initiate chemical reactions that can induce chromosome breakages, chromosome rearrangements, and point mutations. Since there is a linear relationship between mutation rate and X-ray dosage, people should be concerned about their exposure to X rays or other forms of ionizing radiation. Dentists use X rays as part of routine dental care, and hospitals use X rays for diagnosing bone breaks and certain soft-tissue changes such as tumors (e.g., mammography is an X-ray procedure to detect breast cancer).

Ultraviolet (UV) light rays are nonionizing—they have insufficient energy to induce ionizations. Nonetheless, ultraviolet light is a useful mutagen, and at high enough doses it can kill cells. For geneticists, other scientists, and medical personnel, this property is a useful one: UV light is used as a sterilizing agent in some applications. Ultraviolet light causes mutations because the purine and pyrimidine bases in DNA absorb light very strongly if the light has a wavelength of 254–260 nm, that is, in the ultraviolet range. At this wavelength UV light induces gene mutations primarily by causing photochemical (light-induced chemical) changes in the DNA.

The sun is a very powerful source of UV radiation, but much of the UV light is screened out by ozone in the atmosphere. Nonetheless, significant amounts of UV radiation are present in sunlight, as evidenced by the tanning (or burning) of humans who sunbathe.

One of the effects of UV radiation on DNA is the formation of abnormal chemical bonds between adjacent pyrimidine molecules in the same strand or between pyrimidines on the opposite strands of the double helix. This bonding is induced mostly between adjacent thymines on the same or opposite strands of the DNA, forming what are called *thymine dimers* (Figure 16.10), usually designated T̂T. ĈC and T̂C pairs are also produced by UV radiation.

■ *Figure 16.10*   Production of thymine dimers by ultraviolet light irradiation. The two components of the dimer are covalently linked in such a way that the DNA double helix is distorted at that position.

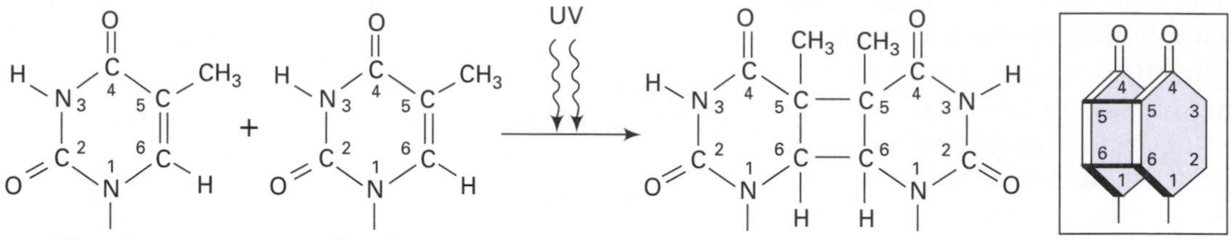

This unusual pairing produces a bulge in the DNA strand and disrupts the normal pairing of the Ts with the corresponding As on the opposite strand. Many of the thymine dimers are repaired (see pp. 392–393); thus it is those that remain or are imperfectly repaired that result in mutations. Indeed, thymine dimers appear to cause mutations by indirect means, for example by interfering with accurate DNA replication. If sufficient dimers remain unrepaired in the cell, cell death may result.

Keynote

Mutations may be induced by the use of radiation or chemical mutagens. Radiation may cause genetic damage by breaking chromosomes, by producing chemicals that affect the DNA (as in the case of X rays), or by causing the formation of unusual bonds between DNA bases such as thymine dimers (as in the case of ultraviolet light). If radiation-induced genetic damage is not repaired, mutations or cell death may result.

Chemical Mutagens.   Many chemicals can induce mutations. They can be grouped into different classes based on their mechanism of action. We will discuss base analogs, base-modifying agents, and intercalating agents and how they can induce mutations.

Base Analog Mutagens.   **Base analogs** have molecular structures that are extremely similar to the bases normally found in DNA. Base analog mutagens cause mutations because they can exist in alternate states (tautomers): a normal state and a rare state. In the two states the base analog pairs with a different base in DNA so that base-pair substitution mutations can be produced. As an illustration we shall consider one of the commonly used base analog mutagens, 5-bromouracil.

5-bromouracil (5BU) has a bromine residue instead of the methyl group of thymine. In the normal state 5BU resembles thymine and will pair only with adenine in DNA (Figure 16.11[a]). In its alternate, rarer state it pairs only with guanine (Figure 16.11[b]). 5BU induces mutations by switching between the two forms once the base analog has been incorporated into the DNA (Figure 16.11[c]). This is essentially similar to the way spontaneous mutations occur by DNA replication errors. The difference here is that base analogs exist much more frequently in the rare state than do normal bases. Consequently, the frequency of mutations induced by base analogs is much higher than the frequency of spontaneous mutations.

If, for example, 5BU is incorporated into DNA in its normal state, it pairs with A (Figure 16.11[c]). If 5BU changes into its rare state during replication, then G will pair with it instead of A. In the next round of replication the G of the G-5BU base pair will specify a GC base pair in the DNA instead of the original AT base pair. In this way a transition mutation is produced, here from AT to GC. 5BU will also induce a mutation from GC to AT if it is first incorporated into DNA in its rare state and then switches to the normal state during replication. Thus 5BU-induced mutations can be reverted by a second treatment of 5BU.

Not all base analogs are mutagens. One of the approved drugs given to AIDS patients, AZT (azidothymidine), is a base analog but it is not a mutagen since it does not result in base-pair changes. Rather, by resembling one of the normal DNA bases, it can act as an inhibitor of DNA replication. AZT works as follows. The AIDS virus HIV-I (human immunodeficiency virus-I) is a retrovirus. That is, its genome is RNA, but when the virus enters a cell, the viral enzyme reverse transcriptase (see Chapter 13) makes a DNA copy of the RNA (a cDNA). That DNA can then be incorporated into the cell's genomic DNA, and from there it can direct new viral synthesis. AZT can be incorporated into DNA as an analog of thymidine. AZT (as a triphosphate derivative) is a substrate for reverse

■ **Figure 16.11** Mutagenic effects of the base analog 5-bromouracil (5BU). (a) In its normal state 5BU pairs with adenine; (b) In its rare state, 5BU (indicated by highlighted 5BU) pairs with guanine; (c) The two possible mutation mechanisms. 5BU induces transition mutations when it incorporates into DNA in one state, then shifts to its alternate state during the next round of DNA replication.

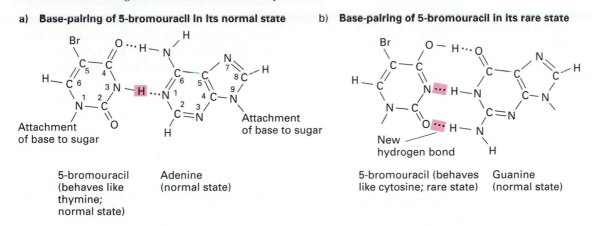

a) **Base-pairing of 5-bromouracil in its normal state**

5-bromouracil (behaves like thymine; normal state)   Adenine (normal state)

b) **Base-pairing of 5-bromouracil in its rare state**

5-bromouracil (behaves like cytosine; rare state)   Guanine (normal state)

c) **Mutagenic action of 5BU**

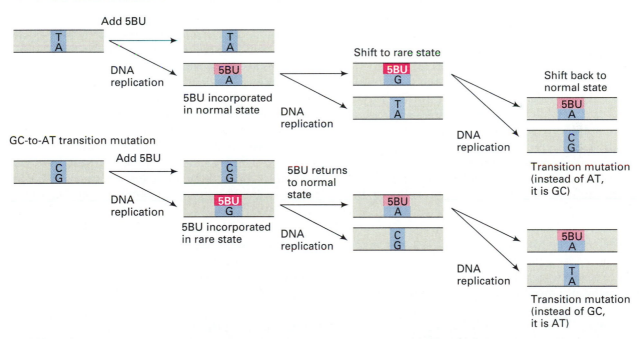

transcriptase in the viral RNA → DNA step. However, AZT is *not* a good substrate for cellular DNA polymerases so host DNA synthesis is not affected. Thus AZT acts as a selective poison by inhibiting the production of the viral cDNA. This blocks new viral synthesis because a cDNA must be incorporated into the host genome in order to code for viral components.

Base-modifying Mutagens. A number of chemicals act as mutagens by directly modifying the chemical structure of the bases so they have different base-pairing properties. Commonly used base-modifying mutagens act as deaminating agents, hydroxylating agents, or alkylating agents. The following is a discussion of how mutations are induced by alkylating agents.

Many alkylating agents can introduce alkyl groups (e.g., -CH₃, -CH₂CH₃) onto the bases at a number of places. An alkylating agent alters one base so that it pairs with the wrong base. For example, the mutagenic action of methylmethane sulfonate (MMS) is shown in Figure 16.12. After treatment with MMS, some guanines and some thymines in the DNA become methylated. A methylated guanine will pair with thymine rather than cytosine, giving GC-to-AT transitions (Figure 16.12[1]), and a methylated thymine will pair with guanine rather than adenine, giving TA-to-CG transitions (Figure 16.12[2]).

Intercalating Agents as Mutagens. The intercalating mutagens (including proflavin, acridine, ethidium bromide, and ICR-170) act by inserting themselves (*intercalating*) between adjacent bases in one or both strands of the DNA double helix. The chemical structures of proflavin and acridine orange are shown in Figure 16.13(a).

If the intercalating agent inserts into a DNA strand that is the template for new DNA synthesis (Figure 16.13[b]), an extra base chosen at random (G in the figure) must be inserted in the new DNA strand. After one more round of replication, during which the intercalating agent is lost, the overall result is a frameshift mutation due to the insertion of one base pair (CG in Figure 16.13[b]). If the intercalating agent inserts into the new DNA strand in place of a base (Figure 16.13[c]), when that DNA double helix replicates after the intercalating agent is lost, a frameshift mutation due to the deletion of one base pair will result (TA in Figure 16.13[c]).

Thus, an intercalating agent produces either a base-pair addition or base-pair deletion. If this occurs in a protein-coding gene, the result is a frameshift mutation. Since intercalating agents can cause either additions or deletions, frameshift mutations induced by intercalating agents can be reverted by a second treatment with these same agents.

Keynote

Mutations may be produced by exposure to chemical mutagens. If the genetic damage caused by the mutagen is not repaired, mutations or cell death result. Chemical mutagens act in a variety of ways. Base analogs physically replace the proper bases during DNA replication, then shift form so that their base-pairing properties change. Base-modifying agents cause chemical changes in existing bases, which alters their base-pairing properties. Base analogs and base-modifying agents result in base-pair substitution mutations. Intercalating agents are inserted between adjacent bases during replication, causing single base-pair additions or deletions, depending on whether the insertion is in the parental or new strand. The result can be a frameshift mutation.

Site-specific *In Vitro* Mutagenesis of DNA. Spontaneous and induced mutations are scattered essentially randomly throughout the genome. However, most geneticists wish to study the effects of mutations in particular genes. By recombinant DNA techniques it is possible to clone genes and produce large amounts of that DNA for analysis and manipulation (see Chapter 13). Further, it is now possible to mutate DNA in the test tube, reinsert the DNA back into the cell by infection or transformation, and examine the effects of the mutation. Such tech-

■ *Figure 16.12* Example of the mutagenic action of an alkylating agent. The alkylating agent methylmethane sulfonate (MMS) methylates some guanine and thymine bases in the DNA. The methylated guanine pairs with thymine giving GC-to-AT transitions (1), and the methylated thymine pairs with guanine giving TA-to-CG transitions (2).

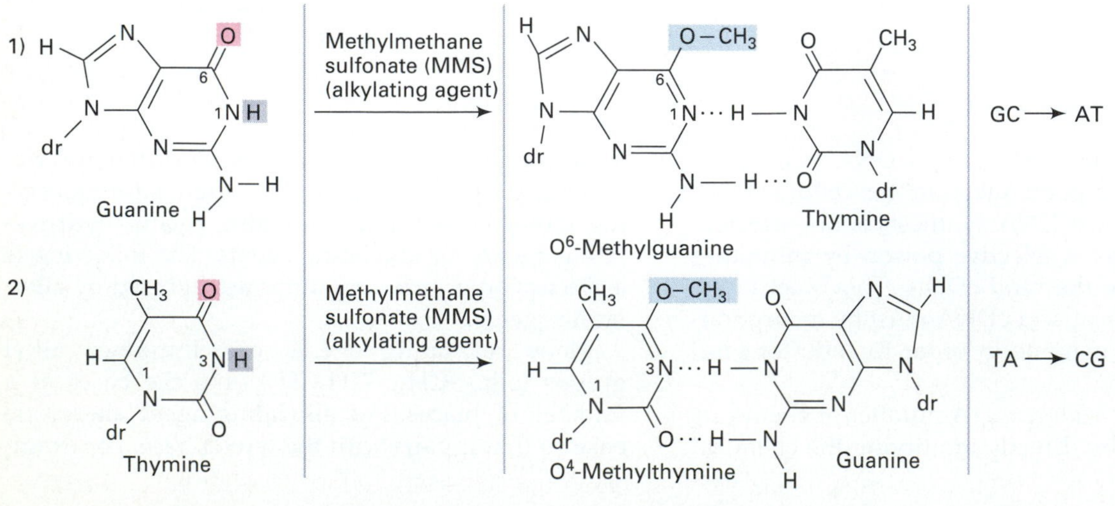

■ *Figure 16.13*   Intercalating mutations: (a) Structures of representative intercalating agents, proflavin and acridine orange; (b) Frameshift mutation by addition, when agent inserts into template strand; (c) Frameshift mutation by deletion, when agent inserts into newly synthesizing strand.

a)   Representative agents

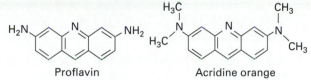

Proflavin   Acridine orange

b)   Mutation by addition

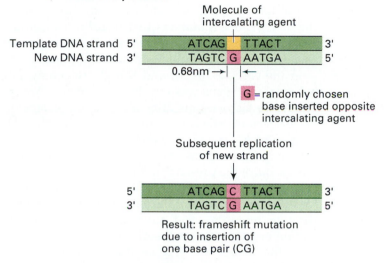

Result: frameshift mutation
due to insertion of
one base pair (CG)

c)   Mutation by deletion

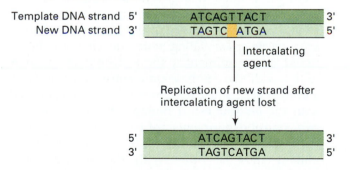

niques enable geneticists to make mutations at specific nucleotide positions within the gene. This procedure is called *site-specific in vitro mutagenesis of DNA.*

Site-specific mutagenesis can be used to create point mutations or small deletions or additions. Figure 16.14 diagrams how a point mutation can be created. A clone of a gene is denatured to single strands. Then an oligonucleotide (a short, synthetic segment of DNA) is annealed to one of the strands. The oligonucleotide is synthesized to be complementary to part of the cloned DNA, except where the point mutation is desired. In the example, a change of a GC base pair to an AT base pair

is desired. Therefore, the oligonucleotide has the sequence TAGACTAT. The italicized A is the mutant site, an incorrect partner for the C on the single-stranded cloned DNA molecule. The annealed oligonucleotide acts as a primer for DNA polymerase to make a complete complementary copy of the circular template strand. DNA ligase seals the gap and makes the new strand circular. We now have a circular, double-stranded molecule with a mismatched AC base pair at one point. The DNA is introduced into *E. coli* cells by transformation. When the DNA replicates, one-half of the molecules are parental with GC at the target site, and the other one-half are mutant with AT at the target

■ *Figure 16.14* Site-specific *in vitro* mutagenesis of DNA to introduce a point mutation at a specific site.

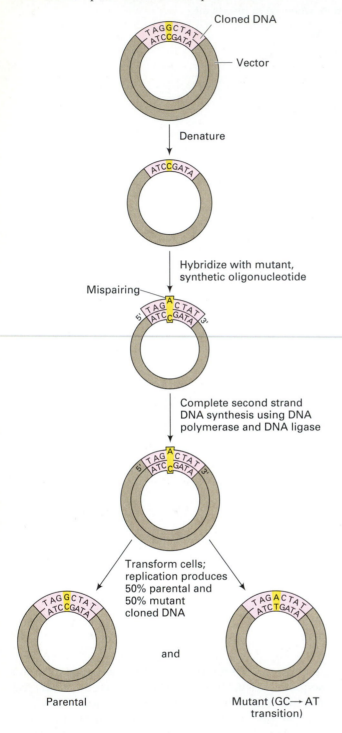

investigate which base pairs are important and which are not. Our understanding of promoter and enhancer elements has come, in part, from site-specific mutagenesis of sequences upstream or downstream from a gene.

The Ames Test: A Screen for Potential Mutagens

All the mutagens (radiation and chemicals) we have examined so far are those commonly used to induce mutations during genetic research. Scientists are aware that many chemicals in our environment might have mutagenic effects. And, since some chemicals cause mutations that result in cancerous growth, there is a strong interest in testing new chemicals for their ability to induce mutations.

In the early 1970s a rapid test for screening for potential mutagens was developed by Bruce Ames. The **Ames test** uses histidine auxotrophic strains of the bacterium *Salmonella typhimurium*. (Recall that an *auxotrophic mutation* requires a particular nutrient in order to grow. Histidine (*his*) auxotrophs, then, require the presence of histidine in the growth medium in order to grow.) We will consider one *his* strain that is mutant because of a base-pair substitution mutation, and another *his* strain that is mutant because of a frameshift mutation.

In the Ames test (Figure 16.15) rat livers are obtained, homogenized, and centrifuged to sediment the cellular debris. The enzymes of the rat liver are collected from the supernatant and are added to a liquid culture of the auxotrophic *Salmonella typhimurium*, along with the chemical being tested. In experiment 1 the potential mutagen is tested with the *his* base-pair substitution mutant, and in experiment 2 the potential mutagen is tested with the *his* frameshift mutant. At the same time two control experiments are run *without* the potential mutagen: one with the *his* base-pair substitution mutant and the other with the *his* frameshift mutant.

Enzymes from the rat liver are used in these experiments because in the living organism, as in humans, the liver enzymes detoxify and, in certain cases, actually toxify various chemicals, including many potential mutagens. The presence of the enzymes allows the investigators to determine whether a chemical that by itself is not mutagenic can become mutagenic when processed in the liver.

The mixture is then plated onto a medium containing no histidine. The control mixture is plated in the same way. Revertants of the two types of *his* mutants are sought. The *his*⁺ revertants are detected because they will produce colonies on the histidine-lacking plates. If there are significantly more *his*⁺

site. To make single base-pair deletions or additions, we carry out this procedure using synthetic oligonucleotides with one fewer or one more base than the complementary cloned sequence.

Site-specific mutagenesis has been used to generate specific changes in key regulatory sequences to

■ *Figure 16.15* Ames test for potential mutagens.

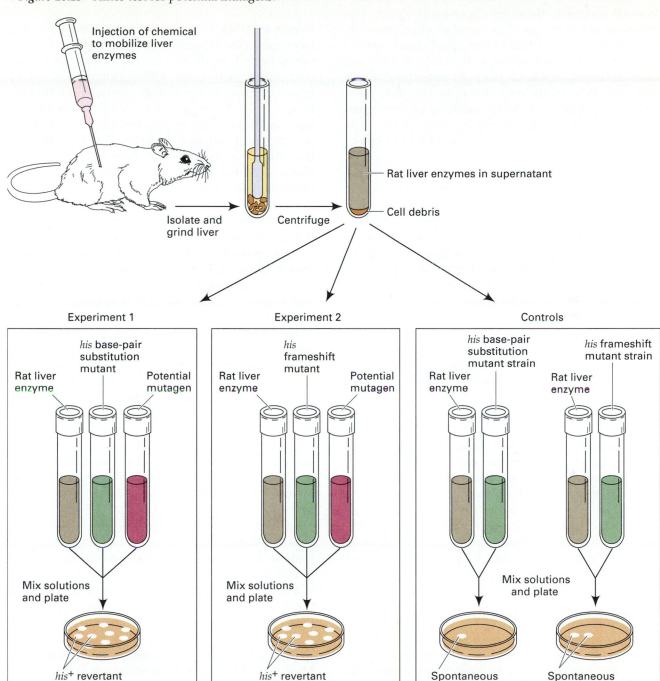

Conclusion: chemical solution is a mutagen

revertants on the plates spread with the chemical-treated mixtures than on the control plate, the chemical is a mutagen. The number of colonies on the control plate indicates the spontaneous reversion rate of the test bacteria. If more colonies are seen on the experimental plates, it indicates that the chemical induced the mutations. To date, the Ames test has identified a large number of mutagens among environmental chemicals such as hair dye additives, vinyl chloride, and particular food colorings, and among natural compounds.

The Ames test can show whether or not a chemical is a mutagen, but it cannot indicate whether or not the chemical is a carcinogen (induces a cancer). That is, carcinogenesis (the induction of cancers) does not necessarily occur because of mutation. In fact, recent data indicate that mitogenesis (induced cell division) plays a dominant role in carcinogenesis. Many carcinogens, then, are *not* mutagens, but cause cancers by inducing mitogenesis.

Several methods are available to test new or old environmental chemicals for carcinogenic effects. One method is to test a chemical directly for carcinogenicity by setting up experiments with inbred lines of mice or rats. The inbred lines are genetically identical so that a statistical analysis of treated versus untreated animals for the incidence of cancers provides information about the probability that a chemical is carcinogenic. Although these studies are expensive and time-consuming, they are performed routinely by research laboratories, often under contract to federal agencies or to private companies. Using animal testing, about one-half of the natural chemicals (those that are normally present in such things as food) and synthetic chemicals (e.g., pesticides, drugs, and food additives) tested have been shown to be carcinogens.

DNA Repair Mechanisms

Spontaneous and induced mutations constitute damage to the DNA of a cell or an organism. Especially with high doses of mutagens, the mutational damage can be considerable. Both prokaryotic and eukaryotic cells have a number of repair systems to deal with damage to DNA. All of the systems use enzymes to make the correction. Some of the systems directly correct the lesion while others first excise the lesion, creating a single-stranded gap, and then synthesize new DNA for the resulting gap. If the repair systems are unable to correct all of the lesions, the result is a mutant cell (or organism) or, if too many mutations remain, death of the cell. Clearly, DNA repair systems are very important for the survival of the cell. The fact that the repair systems are not 100 percent efficient makes it possible to isolate mutants for study.

Direct Correction of Mutational Lesions

Repair by DNA Polymerase Proofreading. In bacterial genes the frequency of base-pair substitutions varies from 1 in 10^7 to 1 in 10^{11} errors per replication event. However, DNA polymerase makes errors in inserting nucleotides perhaps at a frequency of 1 in 10. The discrepancy between the two values is accounted for by proofreading activity of the polymerase itself (see Chapter 10). That is, most bacterial DNA polymerases have a 3'-to-5' exonuclease activity. When an incorrect nucleotide is inserted, the error is most often (but not always) detected by the polymerase, perhaps by the fact that the mismatched base pair results in a bulge in the double helix. Thus DNA synthesis stalls and cannot proceed until the wrong nucleotide is removed and the proper one put in its place. The incorrect nucleotide is removed by the 3'-to-5' exonuclease activity as the polymerase reverses along the template strand. The polymerase then moves forward again, resuming its 5'-to-3' DNA synthesis activity.

The importance of the 3'-to-5' exonuclease activity of DNA polymerase for maintaining a low mutation rate is shown nicely by the existence of *mutator* mutations in *E. coli*. Strains carrying mutator mutations show a much higher-than-normal mutation frequency for all genes. These mutations have been shown to affect proteins whose normal functions are required for accurate DNA replication. For example, the *mutD* mutator of *E. coli* results in an altered ε (epsilon) subunit of DNA polymerase III, the primary replication enzyme in the cell. *MutD* causes a defect in 3'-to-5' proofreading activity, so that many incorrectly inserted nucleotides are unrepaired.

Proofreading does occur in eukaryotes, although the proofreading is done by proteins associated with DNA polymerase, not the polymerase itself.

Photoreactivation of UV-induced Pyrimidine Dimers. Direct correction of lesions can occur also in the repair of ultraviolet-light-induced thymine (or other pyrimidine) dimers. By this process of **photoreactivation** or **light repair** (Figure 16.16), the dimers are reverted directly to the original form by exposure to visible light in the wavelength range of 320 to 370 nm (blue light). Photoreactivation is catalyzed by an enzyme called *photolyase*. When activated by a photon of light, photolyase splits the dimers apart. Presumably, photolyase functions by scavenging along the double helix, seeking the bulges that result when pyrimidine dimers are present. Photolyases are very effective since few pyrimidine dimers (and, hence, potentially lethal damage or mutations) are left after photoreactivation.

■ *Figure 16.16*   Repair of a thymine dimer by photoreactivation.

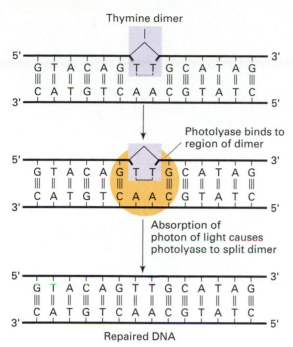

Repaired DNA

Repair of Alkylation Damage.   Alkylating agents transfer alkyl groups (usually methyl or ethyl groups) to certain bases, for example, methylation of the 6-oxygen of guanine by MMS (refer to Figure 16.12). During replication the alkylated guanine pairs with thymine and produces a GC-to-AT transition mutation. Alkylation damage can be corrected by specific DNA repair enzymes. In this particular case an enzyme encoded by the *ada* gene, called *methylguanine methyltransferase*, is used. It recognizes the methylguanine in the DNA and removes the methyl group, changing the guanine back to its original form. Note that in this repair system the modified base is not removed from the DNA.

Repair Involving Excision of Base Pairs

Excision Repair.   Pyrimidine dimers induced by UV light can also be corrected by **excision repair.** Since this repair process does not depend on the presence of light, it is also called **dark repair.** Excision repair can also repair other serious damage-induced distortions of the DNA helix. The discovery of excision repair came through the study of UV-sensitive mutants of *E. coli*. After UV irradiation, UV-sensitive mutants have a higher-than-normal rate of induced mutation in the dark. These mutants can repair dimers only with the input of light; that is, they lack the dark repair system.

The excision repair system in *E. coli* is diagrammed in Figure 16.17. The DNA distortions are recognized by the UvrABC endonuclease, a multi-subunit enzyme encoded by the three genes *uvrA*, *uvrB*, and *uvrC* (*uvr* stands for "UV repair"). This enzyme makes one cut in the damaged DNA strand eight nucleotides to the 5' side of the damage (e.g., a pyrimidine dimer) and four nucleotides to the 3' side of the damage. The cuts release a 12-nucleotide stretch of single-stranded DNA containing the damaged base(s). The 12-nucleotide gap is filled by the 5' → 3' DNA polymerizing activity of DNA polymerase I, and is sealed by DNA ligase. The excision repair system is found in most organisms that have been studied. Its mechanism of action is thought to be essentially like the one found in *E. coli.*

Repair by Glycosylases.   Damaged bases can be excised by other means. For example, cells contain a glycosylase enzyme that can detect an individual unnatural base and remove it from the deoxyribose sugar to which it is attached (Figure 16.18[1], p. 395). This leaves a hole where the base was removed called an *AP site* (for *ap*urinic, where there is no A or G, or *ap*yrimidinic, where there is no C or T). The enzyme *AP endonuclease* recognizes that there is a hole, and cuts the DNA backbone beside the missing base (Figure 16.18[2]). DNA polymerase I (in *E. coli*) then begins repair synthesis, removing a few nucleotides ahead of the missing base with its 5' → 3' exonuclease activity, and filling in the gap with its 5' → 3' DNA polymerizing activity (Figure 16.18[3]). DNA ligase seals the remaining nick in the backbone (Figure 16.18[4]). DNA polymerase I's activity in this repair system, in which the enzyme simultaneously makes new DNA and removes nucleotides ahead of the growing DNA chain, is called *nick translation* (see Box Figure 13.1, p. 297).

Repair by Mismatch Correction.   While proofreading by DNA polymerase is an efficient way of correcting many errors soon after they are made, a significant number of errors still remain uncorrected after replication has been completed. Such errors usually involve mismatched base pairs which, in the next round of replication, will become fixed as spontaneous mutations (see pp. 382–385).

Many mismatched base pairs left after DNA replication may be corrected by another system of repair called *mismatch correction*. This system involves a *mismatch correction enzyme* encoded by three genes in *E. coli*—*mutH*, *mutL*, and *mutS* (Figure 16.19, p. 396). The enzyme searches newly replicated DNA for mismatched base pairs. Once one is found, it removes a single-stranded segment of DNA that includes the mismatched base. DNA polymerase fills in the gap and then DNA ligase seals it.

■ *Figure 16.17* Excision repair of pyrimidine dimer and other damage-induced distortions of DNA initiated by the UvrABC endonuclease.

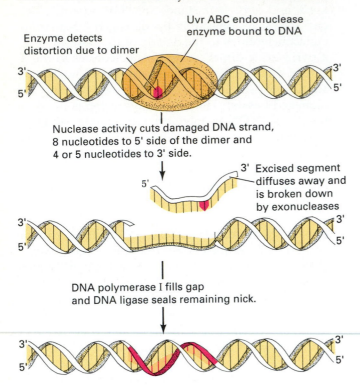

The system is not as simple as it sounds. In particular, how does the system know which base in the mismatched pair is the correct one and which is the erroneous one? The mismatch correction enzyme determines which of the two DNA strands is the parental strand (where the correct base is located) and which is the newly synthesized strand. The signal that it uses is methylation of the A in a 5'-GATC-3' DNA sequence near the mismatch. The GATC sequence is palindromic; that is, the same sequence is present 5' to 3' on both DNA strands to give $\begin{smallmatrix}5'\text{-}GATC\text{-}3'\\3'\text{-}CTAG\text{-}5'\end{smallmatrix}$. In the parental double-stranded DNA both A nucleotides in the DNA segment are methylated. However, after replication the parental DNA strand has a methylated A in the GATC sequence, while the A nucleotide in the GATC sequence of the *new DNA strand* is not methylated until a short time after its synthesis. The mismatch correction enzyme detects the difference and removes the mismatched base only from the newly synthesized strand.

SOS Repair. As we have learned, DNA damage can induce different types of DNA repair enzymes. Perhaps the most important DNA repair enzymes in *E. coli* are those encoded by the *SOS genes*. Expression of the SOS genes is induced when there is sufficient DNA damage to bring DNA replication to a halt. For example, pyrimidine dimers induced

by UV radiation cannot form base pairs. When a replication fork comes to a dimer, the replication process stops and reinitiates some way away along the DNA. This leaves a gap in the new DNA strand. An enzyme involved in genetic recombination, RecA (encoded by the *recA* gene), binds to the gap. Normally, the RecA protein functions in recombinational repair. However, when it binds to single-stranded DNA (in the gap), the RecA protein destroys a repressor that prevents expression of the SOS genes, a group of about 15 genes responsible for the SOS system. As a result, the SOS genes are activated, and the gap is repaired through a variety of mechanisms. This process is called *SOS repair*.

Keynote

Mutations constitute damage to the DNA. Both prokaryotes and eukaryotes have a number of repair systems that deal with different kinds of DNA damage. All of the systems use enzymes to make the correction. Without such repair systems lesions would accumulate and be lethal to the cell or organism. Not all lesions are repaired; thus, mutations do appear, but at relatively low frequencies. At high doses of mutagens repair systems are unable to correct all of the damage, and cell death will result.

■ *Figure 16.18* Excision repair of damaged bases involving the action of glycosylase. Glycosylase detects the damaged base and catalyzes its removal. A few adjacent bases are removed by AP endonuclease, and the resulting single-stranded gap is repaired by DNA polymerase I and DNA ligase.

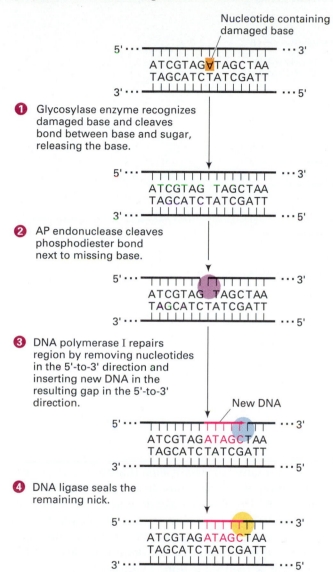

Nucleotide containing damaged base

1 Glycosylase enzyme recognizes damaged base and cleaves bond between base and sugar, releasing the base.

2 AP endonuclease cleaves phosphodiester bond next to missing base.

3 DNA polymerase I repairs region by removing nucleotides in the 5'-to-3' direction and inserting new DNA in the resulting gap in the 5'-to-3' direction.

New DNA

4 DNA ligase seals the remaining nick.

Human Genetic Diseases Resulting from DNA Replication and Repair Errors

Human cells may carry some naturally occurring genetic diseases in which the defect has been attributed to defects in DNA replication. Some of these mutants are listed in Table 16.1. Perhaps the most well-known mutant is *xeroderma pigmentosum* (Figure 16.20), which is caused by homozygosity for a recessive mutation. People with this lethal affliction are photosensitive, and portions of their skin that have been exposed to light show intense pigmenta-tion, freckling, and warty growths that may become malignant. The function affected in these people is excision repair of damage caused by ultraviolet light, X rays, or gamma radiation, or by chemical treatment. Thus individuals with xeroderma pigmentosum are unable to repair radiation damage to DNA and often eventually die as a result of malignancies that arise from the damage. Since the disease is inherited, the defect must result from a mutation in a gene coding for a protein involved in the repair of DNA damage.

Screening Procedures for the Isolation of Mutants

Once mutations have occurred, they must be detected if they are to be studied. Mutations of haploid organisms are readily detectable because there is only one copy of the genome. Mutations of diploid organisms, however, may be more difficult to detect. Let us consider mutations of *Drosophila*, an experimental organism in which genetic crosses can be made as desired. In such an organism dominant mutations will be readily detectable. Recessive mutations are less easily detected. Sex-linked recessive mutations can be detected because they are expressed in one-half of the sons of a mutated, heterozygous female. Autosomal recessive mutations can be detected only if the mutation is homozygous.

The detection of mutations in humans is much more difficult than in *Drosophila* because geneticists cannot make controlled crosses. Dominant mutations can be readily detected, of course, but other types of mutations may be revealed only by pedigree analysis or by direct biochemical or molecular probing. Pedigree analysis can detect sex-linked recessive mutations by mother-son transmission patterns. Rare autosomal recessive mutations, however, will rarely be homozygous, so few pedigrees will have individuals exhibiting such traits.

Thus the detection of mutations is not necessarily simple. Fortunately, for some organisms of genetic interest, particularly microorganisms, screening procedures have been developed to help geneticists obtain particular mutants of interest from among a heterogeneous mixture in a mutagenized population. What follows are descriptions of a few screening procedures for the isolation of particular types of mutations.

Visible Mutations

Visible mutations affect the morphology or physical appearance of an organism (Figure 16.21, p. 398). Examples of visible mutants are eye-color or wing-

■ *Figure 16.19*   Mechanism of mismatch correction repair. The mismatch correction enzyme recognizes which strand the base mismatch is on by reading the methylation state of a nearby GATC sequence. If the sequence is unmethylated, a segment of that DNA strand containing the mismatch is excised and new DNA is inserted.

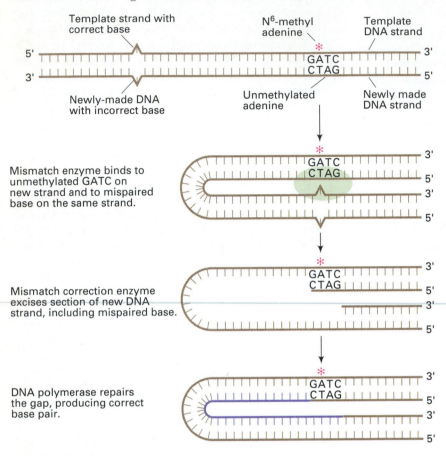

■ *Figure 16.20*   An individual with *xeroderma pigmentosum*.

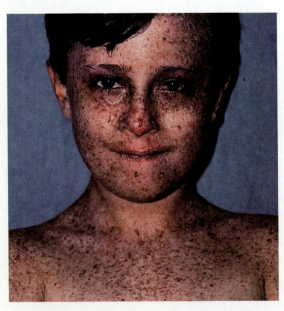

shape mutants of *Drosophila* (Figure 16.21a), coat color mutants of animals (e.g., albino organisms; Figure 16.21b), flower color mutants of plants (Figure 16.21c), colony-size mutants of yeast, and plaque morphology mutants of bacteriophages. Since visible mutations, by definition, are readily apparent, screening is done by inspection.

Auxotrophic Mutations

An **auxotrophic mutation** affects an organism's ability to make a particular molecule essential for growth. Auxotrophic mutations are most readily detected in microorganisms such as *E. coli*, yeast, *Neurospora*, or some unicellular algae that grow on simple growth media from which they synthesize the enzymes used to make all the molecules essential to their growth.

One of the screening procedures for isolating auxotrophic mutants in colony-forming organisms is the **replica-plating** technique (Figure 16.22, p. 399). In replica plating samples from a culture

■ *Figure 16.21* Examples of visible mutants and their wild-type counterparts. a) White-eyed (mutant) (left) and red-eyed (wild-type) (right) *Drosophila*; b) albino (mutant) (left) and agouti (wild type) (right) mice; c) field of white-flowered, pink-flowered and red-flowered snapdragons (flower color is determined by a codominant pair of alleles).

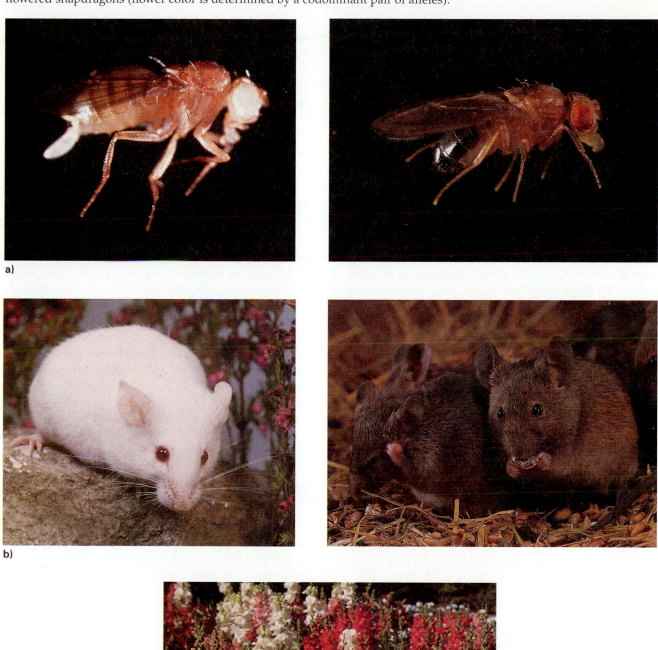

Table 16.1

Examples of Some Naturally Occurring Human Cell Mutants That Are Defective in DNA Replication

DISEASE	SYMPTOMS	FUNCTIONS AFFECTED
Xeroderma pigmentosum (XP)	Skin freckling, cancerous growths on skin, eventually lethal	Repair of DNA damaged by UV irradiation or chemicals
Ataxia telangiectase (AT)	Muscle coordination defect; propensity for respiratory infection; radiation-sensitive; cancer-prone; broken chromosomes	Repair replication of DNA
Fanconi's anemia (FA)	Aplastic anemia[a]; pigmentary changes in skin; malformation of heart, kidney, and extremities; leukemia	Repair replication of DNA, UV dimers and chemical adducts not removed from DNA
Bloom's syndrome (BS)	Dwarfism; sun-sensitive skin disorder; chromosome breaks	Elongation of DNA chains intermediate in replication

[a]Individuals with aplastic anemia make no, or very few, red blood cells.
SOURCE: After R. Sheinin et al., 1978. *Annu . Rev. Biochem.* 47:227–316.

that has or has not been mutagenized are plated onto a complete medium that contains all possible nutrients. The resulting colonies will be a mixture of prototrophic (wild-type) colonies and auxotrophic mutant colonies. The pattern of the colonies is transferred onto sterile velveteen cloth and replicated onto new plates by gently pressing the velveteen onto them. If the new plate contains minimal medium, only prototrophic colonies can grow. So by comparing the patterns on the original master plate with those on the minimal-medium replica plate, researchers can readily identify the potential auxotrophic colonies. If we were looking for a histidine auxotroph, we would replica plate onto another plate containing minimal medium *plus* histidine. The colonies that would grow on this medium but not on the minimal medium are the histidine auxotrophs; that is, the mutants that require histidine in order to grow. They can then be selected from the original master plate and cultured for further study.

Conditional Mutations

The products of many genes—DNA polymerase and RNA polymerase, for example—are important for the growth and division of cells, and most mutations in these genes result in a lethal phenotype. The structure and function of this class of genes may be studied by inducing *conditional mutations* in the genes (see Chapter 10, p. 211 and Figure 10.5). A common type of conditional mutation to study is a heat-sensitive mutation that is characterized by normal function at the normal growth temperature and no or severely impaired function at a higher temperature. In yeast, for instance, normal growth temperature is 23°C; heat-sensitive mutations typically are isolated at 36°C. Heat sensitivity typically

results from the mutation causing a change in the amino acid sequence of a protein such that, at the higher temperature, the protein assumes a shape that is nonfunctional.

Essentially the same procedures are used to screen for heat-sensitive mutations of microorganisms as for auxotrophic mutations. For example, replica plating can select for temperature-sensitive mutants when the replica plate is incubated at a higher temperature than the master plate. Many yeast mutants which disrupt cell division at different steps of the cell cycle (*cdc* mutants) have been identified in this way.

Keynote

Geneticists have made great progress in understanding how cellular processes take place by studying mutants that have defects in those processes. With microorganisms, a number of screening procedures enrich selectively for mutants of interest from a heterogeneous mixture of cells in a mutagenized population of cells.

Summary

In this chapter we have seen that genetic damage can occur to the DNA spontaneously, through replication errors, or through treatment with radiation or chemical mutagens. If the genetic damage is not repaired mutations will result; if there has been too much damage, cell death may result.

Mutations occur spontaneously at a low rate. The mutation rate can be increased through the use of mutagens like irradiation and certain chemicals.

■ *Figure 16.22*   Replica-plating technique to screen for mutant strains of a colony-forming microorganism.

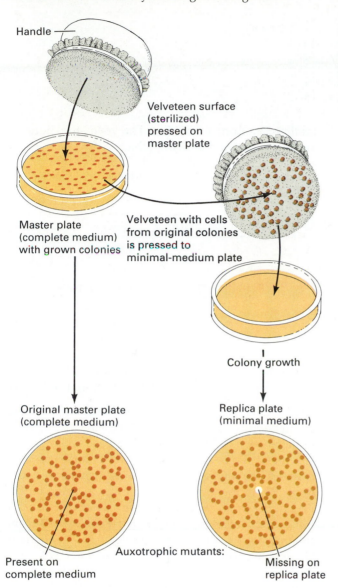

Mutagens are used by researchers so that a mutant of interest is more likely to be found in a population of cells. Chemical mutagens work in a number of different ways, including by acting as base analogs, by modifying bases, or by intercalating into the DNA. The latter results in frameshift mutations, while the other two result in base-pair substitution mutations. Bruce Ames devised a test called the Ames test which has shown that a number of chemicals (e.g., an environmental or commercial chemical) have the potential to cause mutations in humans.

Cells possess a number of repair mechanisms that function to correct at least some damage to DNA. These repair mechanisms include: (1) repair by DNA polymerase proofreading, in which a base-pair mismatch in DNA being synthesized is immediately repaired by 3'-to-5' excision; (2) photoreactivation of pyrimidine dimers induced by UV light; (3) excision repair, in which pyrimidine dimers and other DNA damage that distorts the DNA helix are excised and replaced with new DNA; (4) repair of damaged bases by glycosylases and AP endonuclease; and (5) repair by mismatch correction. Any DNA damage that is not repaired may result in a mutation and may potentially be lethal to the cell. The collective array of repair enzymes, then, serves to reduce mutation rates for spontaneous errors of several orders of magnitude. Such repair mechanisms cannot cope, however, with the extensive amount of DNA damage that arises from the use of chemical mutagens or UV irradiation, and many mutations typically result.

We considered some examples of how a mutagenized population of cells can be screened for particular mutants of interest, for instance, auxotrophs and conditional mutations. Over the decades spontaneous and induced mutants have proved invaluable for the genetic analysis of biological function.

Analytical Approaches for Solving Genetics Problems

Q16.1   Five strains of *E. coli* containing mutations that affect the tryptophan synthetase A polypeptide have been isolated. Figure 16A shows the changes produced in the protein itself in the indicated mutant strains.

In addition A23 can be further mutated to insert Ile, Thr, Ser, or the wild-type Gly into position 210.

a. Using the genetic code (see Figure 12.7), explain how the two mutations *A23* and *A46* can result in two different amino acids being inserted at position 210. Give the nucleotide sequence of the wild-type gene at that position and the two mutants.

b. Can mutants *A23* and *A46* recombine to produce a wild type? Why or why not?

c. From what you can infer of the nucleotide sequence in the wild-type gene, indicate, for the codons specifying amino acids 48, 210, 233, and 234, whether or not a nonsense mutant could be generated by a single nucleotide substitution in the gene.

■ *Figure 16A*

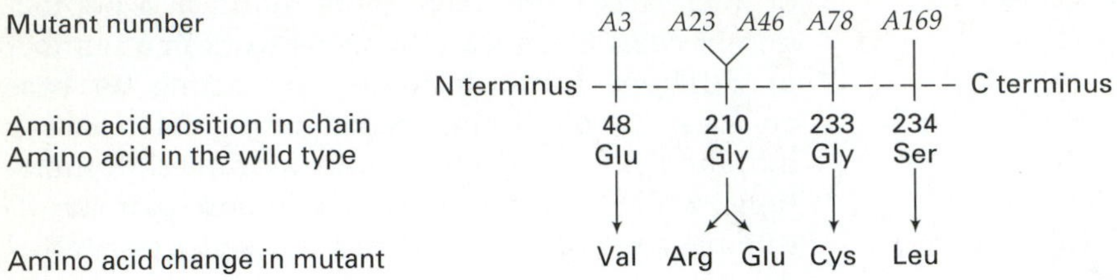

Mutant number

| | A3 | A23 | A46 | A78 | A169 |

N terminus –|– – – –|– – – – – –|– – –|– – C terminus

| | | Y | | | |

Amino acid position in chain   48   210   233   234
Amino acid in the wild type   Glu   Gly   Gly   Ser

Amino acid change in mutant   Val   Arg Glu   Cys   Leu

A16.1 a. There are no simple ways to answer this question. The best approach is to scrutinize the genetic code dictionary and use a pencil and paper to try to define the codon changes that are compatible with all the data. The number of amino acid changes in position 210 of the polypeptide is helpful in this case. The answer is that the original codon for glycine in the wild type was GGA; also *A23* is AGA, and *A46* is GAA. In other words, the two different amino acid substitutions in the two mutants result from different, single base-pair changes in the part of the gene that corresponds to this codon.

b. The answer to this question follows from the answer deduced in part a. Mutants *A23* and *A46* can recombine since the mutations in the two mutant strains are in different base pairs. The results of a single recombination event at position 210 are a wild-type GGA codon (Gly) and a double mutant AAA codon (Lys).

c. Amino acid 48 had a Glu-to-Val change. This change must have involved GAA to GUA or GAG to GUG. In either case the Gly codon can mutate with a single base-pair change to a nonsense codon, that is, UAA or UAG, respectively.

Amino acid 210 in the wild type had a GGA codon, as we have already discussed. This gene could mutate to the UGA nonsense codon with a single base-pair change.

Amino acid 233 had a Gly-to-Cys change. This change must have involved either GGU to UGU or GGC to UGC. In either case the Gly codon cannot mutate to a nonsense codon with one base-pair change.

Amino acid 234 had a Ser-to-Leu change. This change was either UCA to UUA or UCG to UUG. If the Ser codon was UCA, it could change to UGA in one step. But if the Ser codon was UCG, it cannot change to a nonsense codon in one step.

Q16.2 The chemically induced mutations *a*, *b*, and *c* show specific reversion patterns when subjected to treatment by the following mutagens: 5-bromouracil (BU), proflavin (PRO), and methyl-

methane sulfonate (MMS). The reversion patterns are shown in the following table:

	MUTAGENS TESTED IN REVERSION STUDIES		
MUTATION	BU	PRO	MMS
a	–	+	–
b	+	–	+
c	–	–	+

(Note: + indicates many reversions to wild type were found; – indicates no or very few reversions to wild type were found.)

For each original mutation (*a*+ to *a*, *b*+ to *b*, and *c*+ to *c*), indicate the type of mutation that occurred.

A16.2 This question tests knowledge of the base-pair changes that can be induced by the various mutagens used. Mutagen BU is a base analog mutagen that induces mainly GC-to-AT changes and can cause AT-to-GC changes; that is, transition mutations. BU-induced mutations can be reverted by BU. Proflavin causes single base-pair deletions or additions; such changes can be reverted by a second treatment with proflavin. Mutagen MMS alkylates G, which results in its loss from the DNA; it is replaced at random by one of the four bases. If the replacing base is G, there is no mutation, and the effect of MMS is not detected. Substitution with any one of the other three bases, though, results in a mutation. So a third of the time there will be a GC-to-AT change (a transition); a third of the time there will be a GC-to-TA change (a transition); and a third of the time there will be a GC-to-CG change (a transversion). With these mutagen specificities in mind, we can answer the questions for each mutation in turn.

Mutation *a*+ to *a*: The *a* mutation was reverted only by proflavin, indicating that it was a deletion or an addition (a frameshift mutation). Therefore the original mutation was induced by an intercalating agent.

Mutation *b*+ to *b*: The *b* mutation was reverted by BU or MMS. BU can only cause transition mutations while MMS can cause either transition mutations or transversion mutations. The conclusion here is that the reversion occurred by a transition mutation, so the original mutational event must have been a transition mutation.

Mutation c^+ to c: The c mutation was reverted only by MMS but not by 5BU or by PRO. 5BU can only cause transition mutations while PRO can only cause base-pair deletion or base-pair addition mutations. Uniquely among the three mutagens

tested, MMS causes transversions from GC to CG or from CG to GC. Thus the reversion event was most likely a transversion, and this means that the original c^+ to c mutation also was a transversion. For example, if the original mutation was a GC-to-CG transversion, the reversion would be a CG-to-GC transversion. Hence the original mutagen must have been one that could cause a transversion, for instance an alkylating agent such as MMS.

Questions and Problems

*16.1 The following is not a class of mutation (choose the correct answer):
a. Frameshift
b. Missense
c. Transition
d. Transversion
e. None of the above (i.e., all are classes of mutation)

16.2 For the middle region of a particular polypeptide chain, the normal amino acid sequence and the amino acid sequence of several mutants were determined as shown below (.... indicates additional, unspecified amino acids). For each mutant say what DNA level change has occurred, and whether the change is a base-pair substitution mutation (transversion or transition, missense or nonsense) or a frameshift mutation. (Refer to the codon dictionary in Figure 12.7, p. 272.)
a. Normal: ....Phe Leu Pro Thr Val Thr Thr Arg Trp
b. Mutant 1:....Phe Leu His His Gly Asp Asp Thr Val
c. Mutant 2:....Phe Leu Pro Thr Met Thr Thr Arg Trp
d. Mutant 3:....Phe Leu Pro Thr Val Thr Thr Arg
e. Mutant 4:....Phe Pro Pro Arg
f. Mutant 5:....Phe Leu Pro Ser Val Thr Thr Arg Trp

*16.3 In mutant strain X of *E. coli* a Leu tRNA which recognizes the codon 5'-CUG-3' in normal cells has been altered so that it now recognizes the codon 5'-GUG-3'. A missense mutation, which affects amino acid 10 of a particular protein, is suppressed in mutant X cells.
a. What are the anticodons of the two Leu tRNAs, and what mutational event has occurred in mutant X cells?
b. What amino acid would normally be present at position 10 of the protein (without the missense mutation)?
c. What amino acid would be put in at position 10 if the missense mutation is not suppressed (i.e., in normal cells) ?
d. What amino acid is inserted at position 10 if the

missense mutation is suppressed (i.e., in mutant X cells)?

16.4 In any kind of chemotherapy the object is to find a means to kill the invading pathogen or cancer cell without killing the cells of the host. To do this successfully, one must find and exploit biological differences between target organisms and host cells. Explain the nature of the biological difference between host cells and HIV-1 virus that permits the use of azidothymidine (AZT) for chemotherapy.

*16.5 The mutant *lac z-1* was induced by treating *E. coli* cells with acridine, while *lac z-2* was induced with 5BU. What kinds of mutants are these likely to be? Explain. How could you confirm your predictions by studying the structure of the β-galactosidase in these cells?

*16.6 a. The sequence of nucleotides in an mRNA is:

5'-AUGACCCAUUGGUCUCGUUAG-3'

How many amino acids long would you expect the polypeptide chain made with this messenger to be?
b. Hydroxylamine is a base-modifying mutagen that results in the replacement of a GC base pair by an AT base pair in the DNA; that is, it induces a transition mutation. When applied to the organism that made the mRNA molecule shown in part a, a strain was isolated in which a mutation occurred at the 11th position of the DNA that coded for the mRNA. How many amino acids long would you expect the polypeptide made by this mutant to be? Why?

*16.7 Three of the codons in the genetic code are chain-terminating codons for which no naturally occurring tRNAs exist. Just like any other codons in the DNA, though, these codons can change as a result of base-pair changes in the DNA. Confining yourself to single base-pair changes at a time, determine which amino acids could be inserted in a polypeptide by mutation of these chain-terminating codons: (a) UAG; (b) UAA; (c) UGA. (The genetic code is listed in Figure 12.7.)

16.8 The amino acid substitutions in the figure occur in the α and β chains of human hemoglobin. Those amino acids connected by lines are related by single nucleotide changes. Propose the most likely codon or codons for each of the numbered amino acids. (Refer to the genetic code listed in Figure 12.7.)

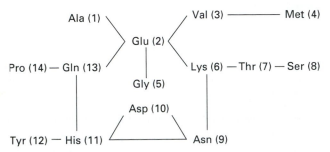

*16.9 Yanofsky studied the tryptophan synthetase of *E. coli* in an attempt to identify the base sequence specifying this protein. The wild type gave a protein with a glycine in position 38. Yanofsky isolated two *trp* mutants, *A23* and *A46*. Mutant *A23* had Arg instead of Gly at position 38, and mutant *A46* had Glu at position 38. Mutant *A23* was plated on minimal medium, and four spontaneous revertants to prototrophy were obtained. The tryptophan synthetase from each of four revertants was isolated, and the amino acids at position 38 were identified. Revertant 1 had Ile, revertant 2 had Thr, revertant 3 had Ser, and revertant 4 had Gly. In a similar fashion three revertants from *A46* were recovered, and the tryptophan synthetase from each was isolated and studied. At position 38 revertant 1 had Gly, revertant 2 had Ala, and revertant 3 had Val. A summary of these data is given in Figure 16B. Using the genetic code in Figure 12.7, deduce the codons for the wild type, for the mutants *A23* and *A46*, and for the revertants, and place each designation in the space provided in Figure 16B.

■ *Figure 16B*

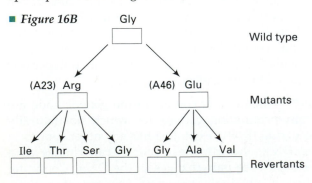

16.10 Consider an enzyme chewase from a theoretical microorganism. In the wild-type cell the chewase has the following sequence of amino acids at positions 39 to 47 (reading from the amino end) in the polypeptide chain:

$$-\text{Met}-\text{Phe}-\text{Ala}-\text{Asn}-\text{His}-\text{Lys}-\text{Ser}-\text{Val}-\text{Gly}-$$
$$394041424344454647$$

A mutant of the organism was obtained that lacks chewase activity. The mutant was induced by a mutagen known to cause single base-pair insertions or deletions. Instead of making the complete chewase chain, the mutant makes a short polypeptide chain only 45 amino acids long. The first 38 amino acids are in the same sequence as the first 38 of the normal chewase, but the last 7 amino acids are as follows:

$$-\text{Met}-\text{Leu}-\text{Leu}-\text{Thr}-\text{Ile}-\text{Arg}-\text{Val}$$
$$39404142434445$$

A partial revertant of the mutant was induced by treating it with the same mutagen. The revertant makes a partly active chewase, which differs from the wild-type enzyme only in the following region:

$$-\text{Met}-\text{Leu}-\text{Leu}-\text{Thr}-\text{Ile}-\text{Arg}-\text{Gly}-\text{Val}-\text{Gly}-$$
$$394041424344454647$$

Using the genetic code given in Figure 12.7, deduce the nucleotide sequences for the mRNA molecules that specify this region of the protein in each of the three strains.

*16.11 After a culture of *E. coli* cells was treated with the chemical 5-bromouracil, it was noted that the frequency of mutants was much higher than normal. Mutant colonies were then isolated, grown, and treated with nitrous acid; some of the mutant strains reverted to wild type. In terms of the Watson-Crick model, diagram a series of steps by which 5BU may have produced the mutants.

16.12 Two mechanisms in *E. coli* were described for the repair of DNA damage (thymine dimer formation) after exposure to ultraviolet light: photoreactivation and excision (dark) repair. Compare and contrast these mechanisms, indicating how each achieves repair.

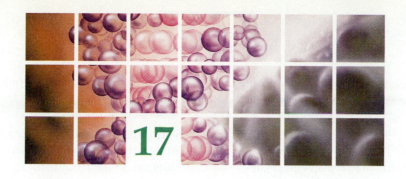

17

Chromosome Mutation

■ Chromosome mutations are variations from the wild-type condition in either chromosome number or chromosome structure. Chromosome mutations can occur spontaneously, or they can be induced by chemical or radiation treatment.

■ Chromosome mutations may involve parts of chromosomes rather than whole chromosomes. The four major types of such mutations are deletions (loss of a DNA segment); duplications (duplication of a DNA segment); inversions (change in orientation of a DNA segment within a chromosome without loss of DNA material); and translocations (movement of a DNA segment to another location in the genome without loss of DNA material).

■ Variations in the chromosome number of a cell or an organism give rise to aneuploidy, monoploidy, and polyploidy. Aneuploidy is the state in which the cell has one, two, or a few whole chromosomes more or less than the basic number of the species involved. Monoploidy is the state in which a normally diploid cell has only one set of

chromosomes, and polyploidy is the state in which there are more than the normal number of sets of chromosomes present.

■ Chromosome number or chromosome structure mutations may have serious, even lethal, consequences to the organism. In eukaryotes the abnormal phenotypes result typically from problems with chromosome segregation during meiosis, with gene disruptions at chromosome breakage sites, or with gene expression levels when gene dosage is altered.

■ Some changes in chromosome organization occur naturally as mechanisms of altering gene expression, often as part of a developmental program. Examples of such changes are amplification of genes, deletion of a chromosome or part of a chromosome, inversion of a chromosome segment that results in a switch in transcription from one gene or set of genes to an alternative set, and transposition of a gene from a silent location in the genome to an active location where transcription then occurs.

C HAPTER 16 FOCUSED on gene mutation. This chapter discusses chromosome mutations, which are changes in normal chromosome structure or chromosome number. In practice, discriminating between gene mutations and chromosomal structural changes, especially minute ones, is somewhat arbitrary. Chromosome mutations affect both prokaryotes and eukaryotes, as well as their viruses. We will focus in this chapter on chromosome mutations in eukaryotes and discuss some of the human disease syndromes that result from chromosome mutations. The association of genetic defects with changes in chromosome structure or chromosome number indicates that not all genetic defects result from simple mutations of single genes. In this chapter we also briefly discuss some examples of changes in chromosome organization that affect gene expression.

Types of Chromosome Mutations

With the exception of gametes, cells of the same eukaryotic organism characteristically have the same number of chromosomes. Further, the organi-

zation and the number of genes on the chromosomes of an organism are the same from cell to cell. These characteristics of chromosome number and gene organization are the same for all members of the same species. Deviations do occur and are known as **chromosome mutations** (also called **chromosome aberrations**). Chromosome mutations are *variations from the wild-type condition in either chromosome structure or chromosome number*. In both prokaryotes and eukaryotes chromosome mutations arise spontaneously or can be induced experimentally by chemical or radiation mutagens. In eukaryotes chromosome mutations can often be detected cytologically during mitosis and meiosis. The ability to detect different kinds of chromosome mutations depends on the size, structure, and number of chromosomes involved and how easily they can be handled.

We tend to believe that reproduction in humans occurs without significant problems affecting chromosome structure or number. After all, the vast majority of babies appear normal, as does the majority of the adult population. However, chromosome mutations are more common than we once thought. In fact they contribute very significantly to spontaneously aborted pregnancies and stillbirths. For example, gross chromosome mutations are present in about half of spontaneous abortions, and a visible chromosome mutation is present in about 6 out of 1000 live births. We know that about one in seven fertilizations ends in a spontaneous abortion, so about 7 percent of all conceptions contain chromosome mutations. Other studies have shown that about 11 percent of men with serious fertility problems and about 6 percent of people institutionalized with mental deficiencies have chromosome mutations. Chromosome mutations are significant causes of developmental disorders.

Variations in Chromosome Structure

This type of chromosome mutation involves changes in parts of individual chromosomes rather than whole chromosomes or sets of chromosomes in a genome. There are four types of such mutations: *deletions* and *duplications* (both of which involve a change in the amount of DNA on a chromosome), *inversions* (which involve a change in the arrangement of a chromosomal segment), and *translocations* (which involve a change in the location of a chromosomal segment). Of these mutations, only deletion mutations cannot revert.

We have learned a lot about changes in chromosome structure from the study of **polytene chromosomes**, a special type of chromosome found in certain insects (e.g., *Drosophila*) that consists of a bundle of chromatids (see Figure 17.8[a]). These chromatid bundles result from repeated cycles of chromosome duplication without nuclear or cell division, and they are easily detectable through microscopic examination. Polytene chromosomes may be a thousand times the size of corresponding chromosomes at meiosis or in the nuclei of ordinary somatic cells. In polytene chromosomes homologous chromosomes are tightly paired, and therefore the number of polytene chromosomes per cell is half the diploid number of chromosomes. The number of chromatids per polytene chromosome is species-specific.

As a result of the intimate pairing of the multiple copies of chromatids, characteristic banding patterns are easily observed, enabling cytogeneticists to identify unambiguously any segment of a chromosome and therefore to see clearly where chromosome changes have occurred. In *Drosophila melanogaster*, for example, over 5000 bands and interbands can be counted in the four polytene chromosomes. Each band contains an average of 30,000 base pairs (30 kb) of DNA, which is enough to encode several average-sized proteins. Many

bands contain a number of genes (up to seven) that are transcribed independently.

Deletion

A **deletion** is a chromosome mutation involving the loss of a segment of a chromosome (Figure 17.1). A deletion starts by breaks occurring in chromosomes. These breaks may be induced by agents such as heat, radiation (especially ionizing radiation; see Chapter 16), viruses, and chemicals, or by errors in recombination. Deletion mutations do not revert.

The consequences of the deletion depend on the genes or parts of genes that have been removed. In diploid organisms the effects may be lessened by the presence in the homologous chromosome of a copy of the set of genes lost from the other chromosome. We are most knowledgeable, then, about the deletions that cause a detectable phenotypic change. If the deletion involves the loss of the centromere, for example, the result is an acentric chromosome, which is usually lost during meiosis. This leads to the deletion of an entire chromosome from the genome. Depending on the organism, this chromosome loss may have very serious or lethal consequences.

In organisms in which karyotype analysis (see Chapter 9, p. 191–196) is practical, deletions can be detected by that procedure, but only if the deficiencies are large enough. In that case, a mismatched pair of homologous chromosomes where one is shorter than the other is seen. In individuals heterozygous for a deletion, unpaired loops are seen when the two homologous chromosomes pair at meiosis. Unpaired loops may also be seen in polytene chromosomes (Figure 17.2).

A number of human syndromes are caused by deletions. In many cases the disorders are found in

■ *Figure 17.1* Deletion, with a chromosome segment (here, *D*) missing.

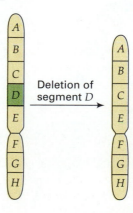

■ *Figure 17.2* Cytological effects at meiosis of heterozygosity for a deletion. Paired *Drosophila* salivary gland polytene chromosomes in a strain with a heterozygous deletion, showing the unpaired region; numbers refer to bands, which are presumed to indicate genes.

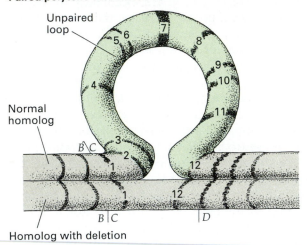

Paired polytene chromosomes

is French for "cry of the cat"). Cri-du-chat syndrome occurs in 1 in 100,000 births.

Another example is *Prader-Willi syndrome* which results from heterozygosity for a deletion of part of the long arm of chromosome 15. Infants with this syndrome are weak because their sucking reflex is poor, making feeding difficult. As a result, growth is poor. By age 5 to 6, Prader-Willi children become compulsive eaters, which results in obesity and related health problems. If untreated, afflicted individuals can eat themselves to death. Other phenotypes associated with the syndrome include poor sexual development in males, behavioral problems, and mental retardation. Many individuals with the syndrome go undiagnosed, so its frequency of occurrence is not accurately known.

Duplication

A **duplication** is a chromosome mutation that results in the doubling of a segment of a chromosome (Figure 17.4). Duplications may or may not be lethal to an organism. Cytologically, heterozygous duplications result in unpaired loops similar to those described for chromosome deletions.

Duplications of particular genetic regions can have unique phenotypic effects, as in the *Bar* mutant on the X chromosome of *Drosophila melanogaster*. In strains carrying the *Bar* mutation (not to be confused with the Barr body discussed in Chapter 3) the number of facets of the compound eye is fewer than for the normal eye (shown in Figure 17.5[a]), giving the eye a bar-shaped (slitlike) rather than an oval appearance. *Bar* acts like an incompletely dominant mutation since females heterozygous for *Bar* have more facets and hence a somewhat larger bar-shaped eye (referred to as

heterozygous individuals; that is, homozygotes for deletions are often lethal if the deletion is large enough. Thus in humans at least, the number of copies of genes is important for normal development and function. One human disorder caused by a deletion is *cri-du-chat syndrome* (Figure 17.3), which results from heterozygosity for a deletion of part of the short arm of chromosome 5, one of the larger human chromosomes. Children with cri-du-chat syndrome are severely mentally retarded, have a number of physical abnormalities, and cry with a sound like the mew of a cat (hence the name, which

■ *Figure 17.3* Photograph of an individual with cri-du-chat syndrome resulting from heterozygosity for a deletion of the short arm of chromosome 5.

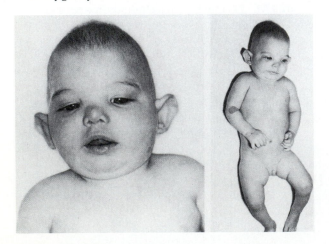

■ *Figure 17.4* Duplication, with a chromosome segment (here, *BC*) repeated.

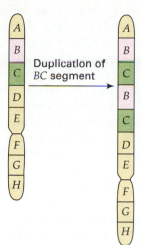

Duplication of *BC* segment

wide-Bar or "kidney" eye) than do females homozygous for *Bar*. Males hemizygous for *bar* have very small eyes like those of homozygous *Bar* females. Cytological examination of polytene chromosomes has shown that the *Bar* trait is the result of a duplication of a small segment of the X chromosome (Figure 17.5[b]). Some flies have been found that have three copies of that segment instead of the normal one copy, a condition called *double-Bar* (Fig-

ure 17.5[c]). These flies have even smaller, slitlike eyes.

A duplication such as *Bar* probably arose as a result of a process called *unequal crossing-over* (Figure 17.6). Unequal crossing-over may result when homologous chromosomes pair inaccurately, perhaps because similar DNA sequences occur in neighboring regions of chromosomes. Crossing-over in the mispaired region will result in gametes with a duplication or a deletion. Figure 17.6 shows how the duplicated 16A segment of the X chromosome of *Drosophila* that is associated with *Bar* (Figure 17.5[b]) could have arisen.

Duplications have played an important role in the evolution of multigene families. For example, the human α-globin and β-globin gene families are each arranged in a cluster (see Chapter 15). The sequences of the genes in the α-globin family are all very similar, as are the sequences of the β-globin family of genes. It is thought that each family evolved from a different ancestral gene by duplication and divergence of the sequences.

Inversion

An **inversion** is a chromosome mutation that results when a segment of a chromosome is excised and then reintegrated in an orientation 180° from the original orientation (Figure 17.7). When the invert-

■ *Figure 17.5* Chromosome segments of *Drosophila* strains, showing the relationship between duplications of region 16A of the X chromosome and the production of reduced-eye size phenotypes: (a) Wild type; (b) *Bar* mutant; (c) *Double-Bar* mutant.

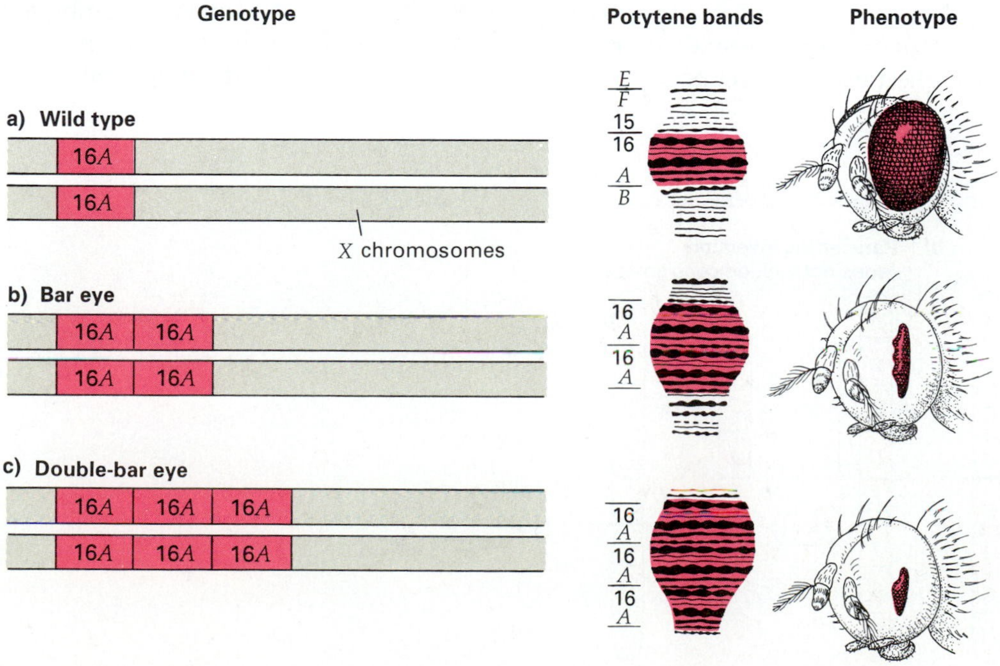

■ *Figure 17.6* Duplications and deletions of chromosome segments that arise because of unequal crossing-over.

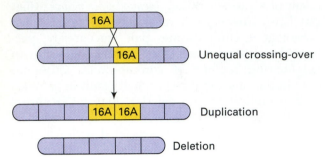

Unequal crossing-over

Duplication

Deletion

ed segment includes the centromere, the inversion is called a **pericentric inversion** (Figure 17.7[a]). When the inverted segment occurs on one chromosome arm and does not include the centromere, the inversion is called a **paracentric inversion** (Figure 17.7[b]).

Genetic material is not generally gained or lost when an inversion takes place, although there can be phenotypic consequences when the break points (inversion points) occur within genes or within regions that control gene expression. Also since gene order (the physical relationships between genes) may affect gene regulation, inversions can disrupt gene regulation. Homozygous inversions can be detected because of the unusual linkage relationships that result for the genes within the inverted segment and the genes that flank the inverted segment. For example, if the normal chromosome is *ABCDEFGH* and the *BCD* segment is inverted, the gene order will now be *ADCBEFGH* (see Figure 17.7[b]).

The meiotic consequences of a chromosome inversion depend on whether the inversion occurs in a homozygote or a heterozygote. If the inversion is homozygous (e.g., *ADCBEFGH/ADCBEFGH*, where the *BCD* segment is the inverted segment), then meiosis will take place normally and there will be no problems related to gene duplications or deletions. However, problems occur if the individual is heterozygous for the inversion (e.g., *ABCDEFGH/ADCBEFGH*, where one chromosome has a normal gene order and the other has an inverted *BCD* segment). In such inversion heterozygotes the homologous chromosomes attempt to pair so that the best possible match of the base-pair sequences is made. Because of the inverted segment on one homolog, pairing of homologous chromosomes requires the formation of loops containing the inverted segments, called inversion loops (Figures 17.8 and 17.9, p. 410). Inversion heterozygotes, then, may be identified by looking for loops in cells undergoing meiosis or, where appropriate, in polytene chromosomes (Figure 17.8[a]).

In inversion heterozygotes the frequency of crossing-over is not diminished in comparison with normal cells, but gametes derived from recombined chromatids are inviable. This is illustrated in Figure 17.8(b), which shows the effects of a single crossover in the inversion loop of an individual heterozygous for a paracentric inversion. During the first meiotic anaphase the two centromeres migrate to opposite poles of the cell. Because of the crossover between genes *B* and *C* in the inversion loop, one recombinant chromatid becomes stretched across the cell as the two centromeres begin to migrate, forming a **dicentric bridge.** As the two centromeres continue to migrate to opposite poles in the cell, the dicentric bridge breaks because of the tension. The other recombinant product of the crossover event is a chromosome without a centromere (an acentric fragment). This

■ *Figure 17.7* (a) Pericentric and (b) paracentric inversions.

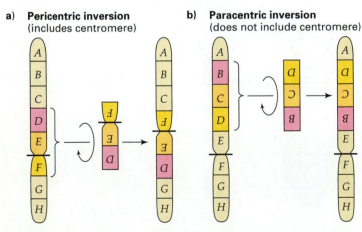

a) **Pericentric inversion**
(includes centromere)

b) **Paracentric inversion**
(does not include centromere)

■ *Figure 17.8*   Consequences of a paracentric inversion: (a) Photomicrograph of an inversion loop in polytene chromosomes of a strain of *Drosophila melanogaster* that is heterozygous for a paracentric inversion; (b) Meiotic products resulting from a single crossover within a heterozygous, paracentric inversion loop. Crossing-over occurs at the four-strand stage involving two nonsister homologous chromatids.

a)

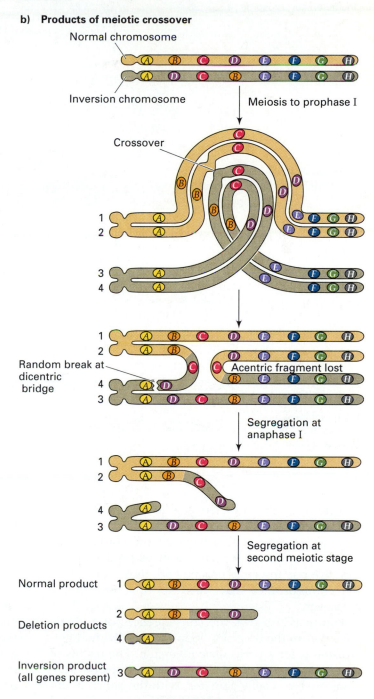

acentric fragment is unable to continue through meiosis and is usually lost (i.e., it is not found in the gametes).

In the second meiotic division the centromeres divide, and the chromosomes are segregated to the four gametes. Two of the gametes have complete sets of genes and are viable: the gamete with the normal order of genes (*ABCDEFGH*) and the

■ *Figure 17.9*   Meiotic products resulting from a single crossover within a heterozygous, pericentric inversion loop. Crossing-over occurs at the four-strand stage, involving two nonsister homologous chromatids.

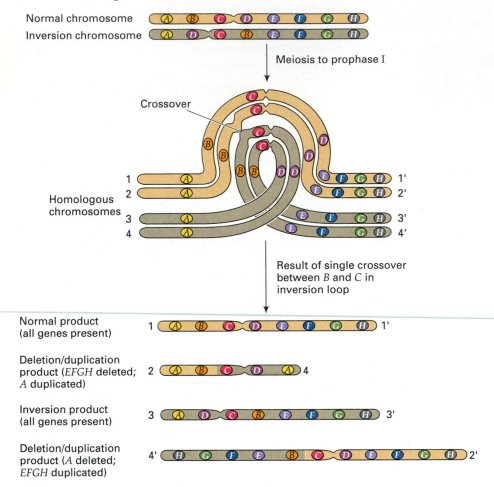

gamete with the inverted segment (*ADCBEFGH*). The other two gametes are inviable because they both have many of the genes of the chromosome deleted. In general the only gametes produced that can give rise to viable progeny are those containing the chromosomes that were not involved in the crossover.

Figure 17.9 shows the consequences of a single crossover in the inversion loop of an individual heterozygous for a pericentric inversion. The results of the crossover event and the ensuing meiotic divisions are two viable gametes, with the nonrecombinant chromosomes *ABCDEFGH* (normal) and *ADCBEFGH* (inversion), and two recombinant gametes that are inviable, each as a result of the deletion of some genes and the duplication of other genes.

Some crossover events within an inversion loop do not affect gamete viability. For example, a double crossover (or an even number of crossovers)

close together and involving the same two chromatids gives four viable gametes.

Translocation

A **translocation** (also called a **transposition**) is a chromosome mutation in which there is a change in position of chromosome segments and the gene sequences they contain (Figure 17.10). There is no gain or loss of genetic material involved with a translocation. Two simple kinds of translocations occur. One involves a change in position of a chromosome segment within the same chromosome: this is called an *intrachromosomal* (within a chromosome) *translocation* (Figure 17.10[a]). The other involves the transfer of a chromosome segment from one chromosome into a nonhomologous chromosome: this is called an *interchromosomal* (between chromosomes) *translocation* (Figure 17.10[b,c]). If this latter translocation is one way, it is a *nonrecipro-*

■ *Figure 17.10*   Translocations: (a) Nonreciprocal intrachromosomal; (b) Nonreciprocal interchromosomal; (c) Reciprocal interchromosomal.

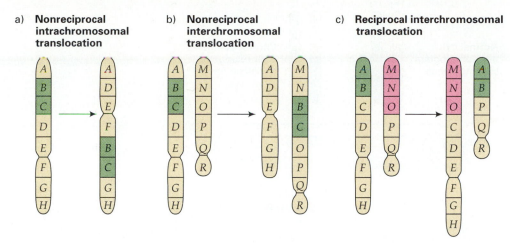

a) **Nonreciprocal intrachromosomal translocation**

b) **Nonreciprocal interchromosomal translocation**

c) **Reciprocal interchromosomal translocation**

cal translocation (Figure 17.10[b]), and if it involves the exchange of segments between the two chromosomes it is a *reciprocal translocation* (Figure 17.10[c]).

In homozygotes for the translocations, the genetic consequence of translocations is an alteration in the linkage relationships of genes, in comparison with strains with normal chromosomes. For example, in the translocation shown in Figure 17.10(a), the *BC* segment has moved to the other chromosome arm and has become inserted between the *F* and *G* segments so that *B* and *C* are now linked to *F* and *G*. As a result of the translocation, genes in the *F* and *G* segments are now farther apart than they are in the normal strain, and genes in the *A* and *D* segments are now more closely linked. Similarly, in reciprocal translocations new linkage relationships are produced.

Translocations typically can affect the production of meiotic products. In many cases some of the gametes produced have duplications and/or deletions; hence in many cases they are inviable. We concentrate here on reciprocal translocations, since they are the most frequent and the most crucial to genetics.

In strains homozygous for a reciprocal translocation, meiosis takes place normally since all the chromosome pairs can pair properly and crossing-over does not produce any abnormal chromatids. In strains heterozygous for a reciprocal translocation, however, all homologous chromosome parts pair as best they can. Because there is one set of normal chromosomes (N) and one set of translocated chromosomes (T) involved, the result is a crosslike configuration in meiotic prophase I (Figure 17.11). The crosslike figures consist of four associated chromosomes, with each chromosome being partially

homologous to two other chromosomes in the group.

Segregation at anaphase I may occur in three different ways, producing six types of gametes. In one way, alternate centromeres segregate to the same pole (Figure 17.11, left: N_1 and N_2 to one pole, T_1 and T_2 to the other pole). This produces two kinds of gametes, each of which is viable because it contains a complete set of genes—no more, no less. One of these gametes has two normal chromosomes and the other has two translocated chromosomes. In the second way, adjacent nonhomologous centromeres migrate to the same pole (Figure 17.11, middle: N_1 and T_2 to one pole, N_2 and T_1 to the other pole). Both gametes produced contain gene duplications and deletions and are often inviable. In the third way, adjacent homologous centromeres migrate to the same pole (Figure 17.11, right: N_1 and T_1 to one pole, N_2 and T_2 to the other pole). This also produces two gametes, each of which has gene duplications and deletions; these gametes are always inviable. In sum, of the six gametes, two are functional and the other four types of gametes are often inviable, since each type contains duplications and deletions.

Therefore, if we assume that the chromosomes in a reciprocal-translocation heterozygote migrate randomly in pairs to opposite poles in meiosis, we expect that two-thirds of the gametes generated will be inviable because of duplications and deletions. Our assumption is not quite valid, though, because not all possible segregation patterns occur with the same frequency: The nonfunctional right pair of gametes diagrammed in Figure 17.11 actually occur infrequently. In reality, then, approximately half the gametes produced from a heterozygous

■ *Figure 17.11*   Meiosis in a translocation heterozygote in which no crossover occurs.

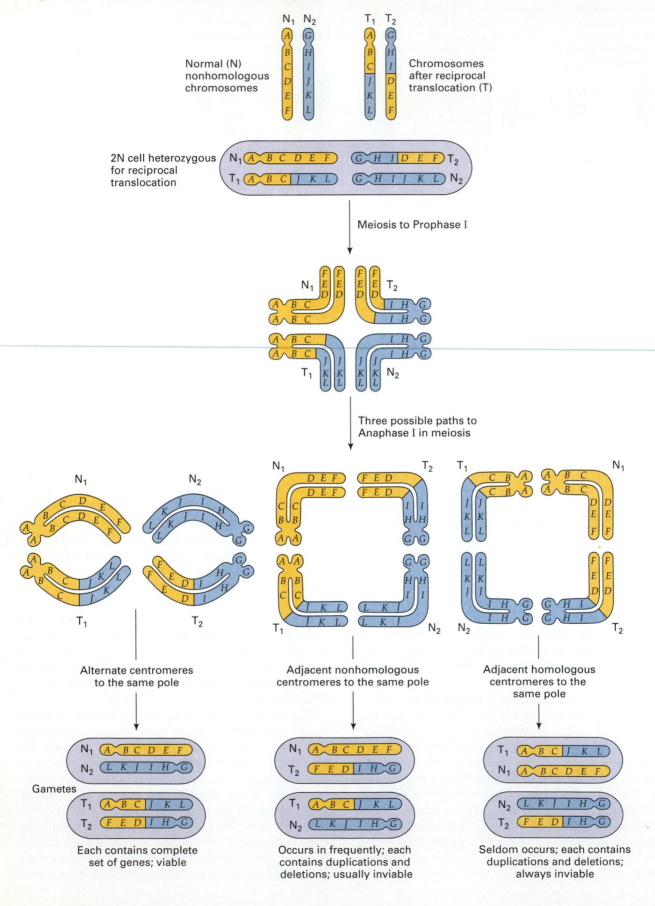

reciprocal translocation are nonfunctional: the term *semisterility* is used for this event. (This term is also used for inversion heterozygotes.)

In practice, animal gametes that have large duplicated or deleted chromosome segments typically are nonfunctional. Gametes may function normally and viable offspring may result if the duplicated and deleted chromosome segments are small. In humans, however, such offspring are phenotypically abnormal. In plants pollen grains with duplicated or deleted chromosome segments typically do not develop completely and are nonfunctional.

Translocations can have serious effects distinct from those associated with meiosis of a translocation heterozygote. For example, some human tumors have been shown to be associated with translocations, such as chronic myelogenous leukemia and Burkitt's lymphoma. Chronic myelogenous leukemia is an invariably fatal cancer involving uncontrolled replication of myeloid stem cells. Ninety percent of chronic myelogenous leukemia patients have a chromosome mutation in the leukemic cells called the *Philadelphia chromosome* (Ph^1). The Philadelphia chromosome results from a reciprocal translocation of part of the long arm of chromosome 22 (the second smallest human chromosome) and a small part of the tip of chromosome 9 (Figure 17.12). This reciprocal translocation event apparently activates genes (called *oncogenes*; see Chapter 18), which initiate the change from a stable differentiated cell to a tumor cell with an uncontrolled pattern of growth. Burkitt's lymphoma is a viral-induced tumor that affects cells of the immune system called B cells. Ninety percent of tumors in Burkitt's lymphoma patients are associated with a reciprocal translocation involving chromosomes 8 and 14. As in the case for chronic myelogenous leukemia and the Philadelphia chromosome, a cancer-causing oncogene (in this case *c-myc* on chromosome 8) is activated by the reciprocal translocation event, which causes tumor development.

■ *Figure 17.12* Origin of the Philadelphia chromosome in chronic myelogenous leukemia by a reciprocal translocation involving chromosomes 9 and 22.

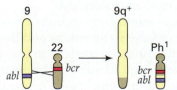

Chromosome mutations may involve parts of individual chromosomes rather than whole chromosomes or sets of chromosomes. The four major types of structural alterations are deletions and duplications (both of which involve a change in the amount of DNA on a chromosome), inversions (which involve no change in the amount of DNA on a chromosome but rather a change in the arrangement of a chromosomal segment), and translocations (which also involve no change in DNA amount but involve a change in location of one or more DNA segments).

Position Effect

Unless inversions or translocations involve breaks within a gene, these chromosomal mutations do not produce mutant phenotypes. Rather, as we have seen, these mutations have significant consequences in meiosis when they are heterozygous with normal sequences. In some cases, however, phenotypic effects of inversions or translocations occur because of a different phenomenon called **position effect**; that is, a change in the phenotypic effect of one or more genes as a result of a change in their position in the genome. For example, position effect may be exhibited if a gene, normally located in euchromatin, is brought near heterochromatin by a chromosomal rearrangement. (Transcription typically occurs in euchromatin but not in heterochromatin.)

An example of position effect involves the X-linked white-eye (*w*) locus in *Drosophila*. One inversion moves the w^+ gene from a euchromatic region near the end of the X chromosome to a position next to the heterochromatin at the centromere of the X. In a w^+ male, or in a w^+/w female where the w^+ is involved in the inversion, the eye exhibits a mottled pattern of red and white rather than being completely red as expected. The explanation is as follows. In flies with the inversion, some eye-cell clones undergo inactivation of the w^+ allele because of the position effect of w^+ near heterochromatin. Those clones give white spots in the eyes. Cell clones in which the w^+ allele is not inactivated give red spots in the eye. Since the inactivation event is variable, the eye ends up with a mottled pattern of red and white spots.

Variations in Chromosome Number

Variations may occur from the wild-type state in the total amount of genetic material in a cell. Such changes may occur spontaneously or may be experimentally induced. Variations in chromosome number, as opposed to variations in parts of chromosomes, give rise to *aneuploidy, monoploidy,* and *polyploidy.*

Changes in One or a Few Chromosomes

We will again focus our discussions on eukaryotes and, in particular, on diploid eukaryotes. Changes in chromosome number can also occur in haploid organisms.

Aneuploidy. One to a few whole chromosomes are lost from or added to the normal set of chromosomes in **aneuploidy** (Figure 17.13). In most cases aneuploidy is lethal in animals, so in mammals it is detected mainly in aborted fetuses. Aneuploidy is encountered more frequently in plants because it is less likely to be lethal. The condition can occur, for example, from the loss of individual chromosomes in mitosis or meiosis. In this case nuclei are formed with less than the normal chromosome numbers. Aneuploidy may also result from nondisjunction, the irregular distribution to the cell poles of sister chromatids in mitosis or of homologous chromosomes in meiosis (see Chapter 3). In nondisjunction one progeny nucleus with more than and one with less than the normal number of chromosomes are produced.

In diploid organisms aneuploid variations take four main forms (see Figure 17.13):

1. **Nullisomy** (a nullisomic cell) involves a loss of one homologous chromosome pair. The nullisomic cell is 2N − 2.
2. **Monosomy** (a monosomic cell) involves a loss of a single chromosome. The monosomic cell is 2N − 1.
3. **Trisomy** (a trisomic cell) involves a single extra chromosome so the cell has three copies of one chromosome type and two copies of every other chromosome type. A trisomic cell is 2N + 1.
4. **Tetrasomy** (a tetrasomic cell) involves an extra chromosome pair, resulting in the presence of four copies of one chromosome type and two copies of every other chromosome type. A tetrasomic cell is 2N + 2.

Aneuploidy may involve the loss or the addition of more than one specific chromosome or a chromosome pair. For example, a *double monosomic* has two chromosomes present in only one copy each; it is 2N − 1 − 1, and a *double tetrasomic* has two chromosomes present in four copies each; it is 2N + 2 + 2, and so forth.

All forms of aneuploidy have serious consequences in meiosis. Monosomics, for instance, will produce two kinds of haploid gametes, N and N − 1. Alternatively, the odd, unpaired chromosome in the 2N − 1 cell may be lost during anaphase and not be included in either daughter nucleus, thereby producing two N − 1 gametes.

Table 17.1 presents a summary of various human aneuploid abnormalities for autosomes and for sex chromosomes. Examples of aneuploidy of the X and Y chromosomes were discussed in Chapter 3. We learned in that chapter that in mammals, aneuploidy of the sex chromosomes is more readily found than aneuploidy of the autosomes because of a dosage compensation mechanism by which excess X chromosomes are inactivated.

Autosomal monosomy in humans is only found rarely in spontaneous abortions and live births. Presumably monosomic embryos do not develop significantly and are lost early in pregnancy. By contrast, autosomal trisomy is seen in about one-half of

■ *Figure 17.13* Normal (theoretical) set of metaphase chromosomes in a diploid (2N) organism (top) and examples of aneuploidy (bottom).

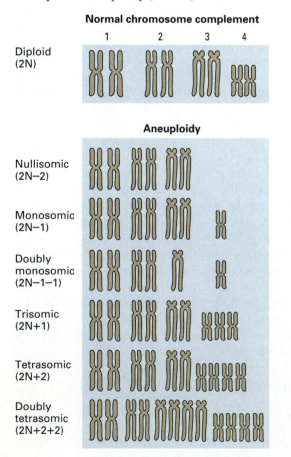

Table 17.1

Aneuploid Abnormalities in the Human Population

CHROMOSOMES	SYNDROME	FREQUENCY AT BIRTH
AUTOSOMES		
Trisomic 21	Down	1/700
Trisomic 13	Patau	1/15,000
Trisomic 18	Edwards	1/7,500
SEX CHROMOSOMES, FEMALES		
XO	Turner	1/5,000
XXX	Triplo-X (Viable, though most are sterile)	1/700
XXXX		
XXXXX		
SEX CHROMOSOMES, MALES		
XYY	XYY (Normal phenotype)	1/1,000
XXY, XXYY, XXXY	Klinefelter	1/500

chromosome abnormalities in fetal deaths. In fact only a few autosomal trisomies are seen in live births. Most of these (trisomy 8, 13, and 18) do not survive long. Only in trisomy 21 (Down syndrome) does survival to adulthood occur.

Trisomy-21. **Trisomy-21** occurs when there are three copies of the chromosome 21. Trisomy-21 occurs with a frequency of about 3510 per 1 million conceptions and about 1500 per 1 million live births. Individuals with trisomy-21 have Down syndrome and exhibit such abnormalities as low IQ, epicanthal folds over the eyes, short and broad hands, sterility, shorter life span, and below-average height (Figure 17.14).

There is a relationship between the age of the mother and the probability of her having a trisomy-21 child (Table 17.2). (The correlation with age of the father is much more tenuous.) During the development of a female child before birth, the eggs in the ovary go through meiosis and arrest in prophase I. In a fertile female, each month at ovulation the nucleus of a secondary oocyte (see Chapter 1) begins the second meiotic division but progresses only to metaphase, when division again stops. If a sperm penetrates the secondary oocyte, the second meiotic division is completed. As an egg matures the probability of nondisjunction increases with the storage time of the egg in the ovary. It is important, then, that older mothers-to-be consider testing with amniocentesis or chorionic villus sampling (see Chapter 13) to determine whether the fetus has a normal complement of chromosomes.

Trisomy-13. Trisomy-13 produces Patau syndrome (Figure 17.15). About 1 in 15,000 live births have trisomy-13. Individuals with trisomy-13 have cleft lip and palate, small eyes, polydactyly (extra fingers and toes), mental and developmental retardation, and cardiac anomalies, among many other abnormalities. Most die before the age of three months.

Trisomy-18. Trisomy-18 produces Edwards syndrome. About 1 in 7500 live births have trisomy-18.

■ *Figure 17.14* An individual with Trisomy-21 (Down syndrome).

Table 17.2

Relationship Between Age of Mother and Risk of Trisomy-21

AGE OF MOTHER	RISK OF TRISOMY-21 IN CHILD
<29	1/3000
30–34	1/600
35–39	1/280
40–44	1/70
45–49	1/40
All mothers combined	1/665

For reasons that are not known, about 80 percent of Edwards syndrome infants are female. Individuals with trisomy-18 are small at birth and have multiple congenital malformations affecting almost every organ system in the body. Clenched fists, elongated skull, low-set malformed ears, mental and developmental retardation, and many other abnormalities are associated with the syndrome. Ninety percent of infants with trisomy-18 die within six months, often from cardiac problems.

Changes in Complete Sets of Chromosomes

Monoploidy and *polyploidy* involve variations from the normal state in the number of complete sets of chromosomes. Monoploidy and polyploidy are lethal for most animal species, but are tolerated more readily by plants. Both have played significant roles in plant speciation.

■ *Figure 17.15*   An individual with Trisomy-13 (Patau syndrome).

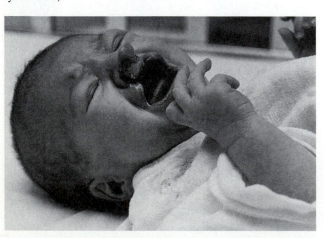

Monoploidy.   If an organism is usually diploid, a monoploid individual has only one set of chromosomes (Figure 17.16[a]). Monoploidy occurs only rarely and, in many organisms, monoploids do not survive. However, certain species have monoploid organisms as a normal part of their life cycle. Some male wasps, ants, and bees, for example, are normally monoploid because they develop from unfertilized eggs.

Monoploids are used extensively in plant-breeding experiments. Monoploid cells can be isolated from the normal haploid products of meiosis in plant anthers and induced to grow for studies of genetic traits of interest.

Cells of a monoploid individual are very useful for mutagenesis because mutants are not masked by a dominant allele on the homologous chromosome. Thus the mutants can be isolated directly.

Polyploidy.   Polyploidy is the situation where a cell or organism has more than its normal number of sets of chromosomes (Figure 17.16[b]). Again the best examples of polyploids come from plants. Polyploids as deviants from normal diploidy are not found in most living animals, although there are some polyploid animal species such as North American suckers, a freshwater fish.

Polyploids may arise spontaneously or be induced experimentally. They often occur as a result of a breakdown in the spindle apparatus in one or more meiotic divisions or in mitotic divisions. For example, the drug colchicine inhibits the formation of the mitotic spindle, so administering this drug can produce somatic tissue with twice the normal sets of chromosomes. Relative to a normal diploid cell, a cell with three sets of chromosomes is called a *triploid* (3N), and one with four sets of chromosomes is called a *tetraploid* (4N).

There are two general classes of polyploids: those that have an *even* number of chromosome sets and those that have an *odd* number of sets. Polyploids with an even number of chromosome sets have a better chance of being at least partially fertile, since there is the potential for chromosomes to come together in pairs during meiosis. However, polyploids with an odd number of chromosome sets always have an unpaired chromosome for each chromosome type, so the probability of a balanced gamete is extremely low. Such organisms are usually sterile.

In humans the most common type of polyploidy is triploidy. Triploidy is always lethal. Triploidy is seen in 15 to 20 percent of spontaneous abortions. About 1 in 10,000 live births is triploid, but most die within one month. Triploid infants have many

■ *Figure 17.16* Variations in number of complete chromosome sets: (a) Monoploidy
(only one set of chromosomes instead of two); (b) Polyploidy (more than the normal
number of chromosomes).

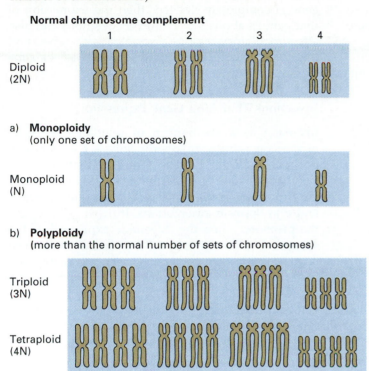

Normal chromosome complement

1   2   3   4

Diploid
(2N)

a) **Monoploidy**
(only one set of chromosomes)

Monoploid
(N)

b) **Polyploidy**
(more than the normal number of sets of chromosomes)

Triploid
(3N)

Tetraploid
(4N)

abnormalities, including a characteristic enlarged head. Tetraploidy in humans is always lethal. It is found very rarely in live births and is seen in about 5 percent of spontaneous abortions.

Plants are more "tolerant" to polyploidy for two main reasons. First, sex determination is less sensitive to polyploidy in plants than in animals. Second, many plants undergo self-fertilization so if a plant is produced with an even polyploid number of chromosome sets (e.g., 4N) it can still produce fertile gametes and reproduce.

Two types of polyploidy are encountered in plants. In **autopolyploidy** all the sets of chromosomes are of the same species. The condition probably results from a defect in meiosis, which leads to diploid or triploid gametes. If a diploid gamete fuses with a normal haploid gamete, the zygote and the organism that develops from it will have three sets of chromosomes: it will be triploid. The cultivated banana is an example of a triploid autopolyploid plant. Because it has an odd number of chromosome sets, the gametes have a variable number of chromosomes and few fertile seeds are set, thereby making the banana a highly palatable, and seedless, fruit to eat. Because of the triploid state, cultivated bananas are propagated vegetative-

ly (by cuttings). In general, the development of seedless fruits relies on odd-number polyploidy.

In **allopolyploidy** the sets of chromosomes involved differ; that is, they come from different, though usually related, species. This situation can arise if two different species interbreed to produce an organism with one haploid set of each parent's chromosomes and then both chromosome sets double. For example, fusion of haploid gametes of two diploid plants that can cross may produce an $N_1 + N_2$ hybrid plant. The hybrid will have a haploid set of chromosomes from plant 1 and a haploid set from plant 2. However, because of the differences between the two chromosome sets, pairing of chromosomes does not occur at meiosis and no viable gametes are produced. As a result, the hybrid plants are sterile. In rare instances, through a division error the two sets of chromosomes may double, producing tissues of $2N_1 + 2N_2$ genotype. That is, the cells in the tissue have a diploid set of chromosomes from plant 1 and a diploid set from plant 2. Each diploid set can function normally in meiosis, so that gametes produced from the $(2N_1 + 2N_2)$ plant are $(N_1 + N_2)$. Fusion of two gametes like this can produce fully fertile, allotetraploid $2N_1 + 2N_2$ plants.

An example of an important allopolyploid is the cultivated bread wheat, *Triticum aestivum*, which is an allohexaploid with 42 chromosomes, an even number of chromosome sets. This plant species is descended from three distinct diploid species, each of which contributed a diploid set of 14 chromosomes. Meiosis is normal because only homologous chromosomes pair, so the plant is fertile, producing gametes with 21 chromosomes.

Keynote

Variations in the chromosome number of a cell or an organism give rise to aneuploidy, monoploidy, and polyploidy. In aneuploidy a cell or organism has one, two, or a few whole chromosomes more or less than the basic number of the species under study. In monoploidy an organism that is usually diploid has only one set of chromosomes. And in polyploidy an organism has more than its normal number of sets of chromosomes. Any or all of these abnormal conditions may have serious consequences to the organism.

Chromosome Rearrangements That Alter Gene Expression

In this section we will describe some chromosomal rearrangements that occur in nature as mechanisms of altering gene expression. Many of these chromosomal rearrangements are part of the normal developmental program of an organism, and thus belong more in the area of developmental biology than genetics. Here we will only give a brief list of the types of rearrangements that alter gene expression.

Amplification or Deletion of Genes

The programmed amplification or deletion of genes is found widely in nature. Amplification may involve a large part or all of the genome, or only a small set of genes. The former is exemplified by polytenization of chromosomes in insect salivary glands (see p. 405), and the latter by the amplification of the 18S, 5.8S, and 28S rRNA transcription units in frog eggs. Programmed deletions also vary in magnitude. They may involve a systematic loss of whole chromosomes (e.g., the loss of an X chromosome in some organisms as a way to achieve the same dosage compensation that other organisms acquire through X chromosome inactivation), or the loss of one or a few genes.

Gene amplification generally leads to the synthesis of more gene product, not to a change in gene expression of the genes involved. The effects of deletions on gene expression vary, however. If a gene is completely deleted, then the expression of that gene is abolished. However, deletions can also affect regulatory regions associated with genes, and this can produce an increase or decrease of gene expression.

Inversions That Alter Gene Expression

Inversion as a mechanism of regulation of gene expression is rare, involving only a few genes in bacteria and their viruses. One way that an inversion can function as a regulator is as follows: A single promoter flanks a piece of DNA with genes *A* and *B* in opposite orientations. If the *A* gene is near the promoter, then the *A* gene is expressed. If the segment of DNA inverts, the *B* gene is brought next to the promoter, and the *B* gene is expressed instead of the *A* gene. Thus the inversion event acts as a genetic switch. This type of inversion control of gene expression is seen in phage Mu, where two types of tail fibers can be made, which enables the phage to infect more than one type of bacterial host.

Transpositions That Alter Gene Expression

As discussed earlier, transpositions are movements of DNA segments from one location to another in the genome. Programmed transpositions that alter gene expression are known for a number of systems, two of which will be mentioned here.

1. *Antigenic variation in Trypanosomes.* Trypanosomes are parasitic protozoa that can multiply for long periods in the bloodstream of mammalian hosts. The parasite evades the host immune defense system by repeatedly altering its surface coat so that it avoids attack by antibodies. The surface coat consists of a single glycoprotein species, and the changes result from transposition of a silent copy of a new surface protein gene to a site in the genome where surface protein genes can be transcribed. This site is called the *expression site*. With at least 10^3 silent surface protein genes in the genome, an extremely large repertoire of possible surface coats is possible.

2. *Mating-type switching in yeast.* Yeast has two mating types, *a* and α (see Chapter 5). Mating type is controlled by the DNA sequence present at the mating types locus, MAT. Under genetic control, haploid cells may frequently switch mating type during growth. Figure 17.17(a) diagrams the organization of mating-type genes on chromosome III of

■ *Figure 17.17* (a) Organization of mating-type genes on chromosome III of a mating-type α yeast cell. The *HMLE* sequence represses transcription of *HMLα*, and the *HMRE* sequence represses transcription of *HMRα*; (b) Mating-type switching from α to *a*; (c) Mating-type switching from *a* to α.

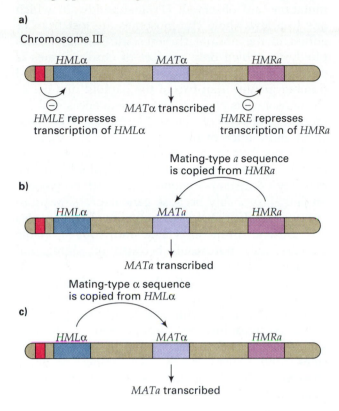

a)

Chromosome III

HMLα   *MATα*   *HMRa*

⊖   *MATα* transcribed   ⊖
HMLE represses   *HMRE* represses
transcription of *HMLα*   transcription of *HMRa*

Mating-type *a* sequence
is copied from *HMRa*

b)

HMLα   *MATa*   *HMRa*

MATa transcribed

Mating-type α sequence
is copied from *HMLα*

c)

HMLα   *MATα*   *HMRa*

MATa transcribed

an α cell. The α phenotype is specified by a 2.5-kb segment of the *MATα* locus. An identical copy of the 2.5-kb, α-specific sequence is present on chromosome III at a locus called *HMLa*. This DNA is not expressed because associated with it is a regulatory (silencer) sequence *HMLE* that represses transcription from the *HMLα* promoter; that is, *HMLα* is a "silent," unexpressed copy of the *MATα* sequence. On the opposite side of the active *MATα* locus is a locus called *HMRa*, which contains all of the DNA necessary for the mating-type *a* phenotypes. *HMRa* also has two promoters, and transcription from them is repressed by an associated silencer, *HMRE*.

Mating-type switching in yeast, then, involves replacing the mating-type sequence at the active *MAT* locus with a *copy* of a sequence with opposite mating-type information. Thus in an α cell, a copy of the *HMRa* sequence will replace the active sequence at *MAT* to produce a mating-type *a* cell (Figure 17.17[b]). The next switch of the cell would replace the active *a* sequence at *MAT* with a copy of the *HMLα* sequence (Figure 17.17[c]).

K e y n o t e

Some chromosome organizational changes occur in nature as mechanisms for altering gene expression. Often these changes function as genetic switches during development. Examples of such changes are amplification of the entire genome, amplification of one or a few genes, deletion of a chromosome or part of a chromosome, inversion of a chromosome segment that switches transcription from one gene to another, and transposition of a gene from a silent location to an active location, where transcription can occur.

Summary

In this chapter we considered chromosome mutations. Chromosome mutations are defined as variations from the wild-type conditions in either chromosome structure or chromosome number. They may occur spontaneously or their frequency can be increased by exposure to radiation or to chemical mutagens. The four major types of chromosome structure mutations are: (1) deletion, in which a chromosome segment is lost; (2) duplication, in which a chromosome segment is present in more copies than normal; (3) inversion, in which the orientation of a chromosome segment is opposite that of the wild type; and (4) translocation, in which a chromosome segment has moved to a new location in the genome. Of these, only deletions cannot revert. The consequences of these kinds of structural mutations depend on the specific mutation involved. Firstly, each kind of mutation involves one or more breaks in a chromosome. If a break occurs within a gene, then a gene mutation has been produced. In the case of deletions, of course, genes may be lost and multiple mutant phenotypes may result. In some cases deletions and duplications result in severe defects or are lethal because the normal gene dosage is altered.

Secondly, chromosome structure mutations in the heterozygous condition can result in production of some gametes that are inviable because of duplications and/or deficiencies. Commonly this is seen following meiotic crossovers that produce inversions and translocations in heterozygotes. Thirdly, inversions and translocations may cause position effects in which a gene normally active in euchromatin becomes inactive in some cells because of

now being located near heterochromatin.

Chromosome number mutations involve departure from the normal diploid (or haploid) state of the organism. For diploids, three classes of such mutations are observed: (1) aneuploidy, in which one to a few whole chromosomes are lost from or added to the normal chromosome set; (2) monoploidy, in which only one set of chromosomes is present; and (3) polyploidy, in which an integral number greater than two of the haploid number of chromosomes is present. The consequences of these mutations depend on the organism. Plants are generally more tolerant than are animals of variations in the number of chromosome sets. While some animal species are naturally polyploid, in the majority of instances monoploidy and polyploidy are lethal, probably because gene expression problems occur when abnormal numbers of gene copies are present. Even in viable individuals, viable gametes may not result because of segregation problems during meiosis.

Finally, we mentioned a few examples of natural systems in which changes in chromosome organization are associated with altered gene expression. These changes include amplification of chromosomes or chromosome segments, deletions, inversions, and transpositions. A number of these changes function as genetic switches during development.

Analytical Approaches for Solving Genetics Problems

Q17.1 Diagram the meiotic pairing behavior of the four chromatids in an inversion heterozygote $a\ b\ c\ d\ e\ f\ g/a'\ b'\ f'\ e'\ d'\ c'\ g'$. Assume that the centromere is to the left of gene a. Next, diagram the early anaphase configuration if a crossover occurred between genes d and e.

A17.1 This question requires a knowledge of meiosis and the ability to draw and manipulate an appropriate inversion loop. Part a of Figure 17A below gives the diagram for the meiotic pairing.

Note that the lower pair of chromatids (a', b', etc.) must loop over in order for all the genes to align; this looping is characteristic of the pairing behavior expected for an inversion heterozygote.

Once the first diagram has been constructed, answering the second part of the question is straightforward. We diagram the crossover, then trace each chromatid from the centromere end until

■ *Figure 17.A*

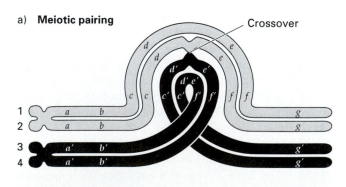

a) **Meiotic pairing**

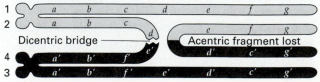

b) **Early anaphase**

the other end is reached. It is convenient to distinguish maternal and paternal genes, perhaps by *a'* versus *a*, and so on, as we did in part a in the figure. The result of the crossover between *d* and *e* is shown in part b.

In anaphase I of meiosis the two centromeres, each with two chromatids attached, migrate toward the opposite poles of the cell. At anaphase the non-crossover chromatids (top and bottom chromatids in the figure) segregate to the poles normally. As a result of the single crossover between the other two chromatids, however, unusual chromatid configurations are produced, and these configurations are found by tracing the chromatids from left to right. If we begin by tracing the second chromatid from the top, we get ○ *a b c d e' f' b' a'* ○ (where ○ is a centromere); in other words, we have a single chromatid attached to two centromeres. This chromatid also has duplications and deletions for some of the genes. Thus during anaphase this so-called dicentric chromosome becomes stretched between the two poles of the cells as the centromeres separate, and the chromosome will eventually break at a random location. The other product of the single crossover event is an acentric fragment that can be traced starting from the *right* with the second chromatid from the top. This chromatid is *g f e d' c' g'*, which contains neither a complete set of genes nor a centromere—it is an acentric fragment that will be lost as meiosis continues.

Thus the consequence of a crossover event within the inversion in an inversion heterozygote is the production of gametes with duplicated or deleted genes. Hence these gametes often will be inviable.

Viable gametes are produced, however, from the noncrossover chromatids: One of these chromatids (1 in part b of the figure) has the normal gene sequence, and the other (3 in part b of the figure) has the inverted gene sequence.

Q17.2 Eyeless is a recessive gene (*ey*) on chromosome 4 of *Drosophila melanogaster*. Flies homozygous for *ey* have tiny eyes or no eyes at all. A male fly trisomic for chromosome 4 with the genotype +/+/*ey* is crossed with a normal, eyeless female of genotype *ey/ey*. What are the expected genotypic and phenotypic ratios that would result from random assortment of the chromosomes to the gametes?

A17.2 To answer this question, we must apply our understanding of meiosis to the unusual situation of a trisomic cell. Regarding the *ey/ey* female, only one gamete class can be produced, namely, eggs of genotype *ey*. Gamete production with respect to the trisomy for chromosome 4 occurs by a random segregation pattern in which two chromosomes migrate to one pole and the other chromosome migrates to the other pole in meiosis I. (This pattern is similar to the meiotic segregation pattern shown in secondary nondisjunction of XXY cells; see Chapter 3.) Three types of segregation are possible in the formation of gametes in the trisomy. This is seen more clearly if we distinguish the two + chromosomes as +$_1$ and +$_2$, as shown in part a of Figure 17B. The union of these sperm at random with eggs of genotype *ey* occurs as shown in part b.

The resulting genotypic and phenotypic ratios are illustrated in part c.

■ *Figure 17.B*

a) **Segregation**

+$_1$/+$_2$ *ey*   +$_1$ +$_2$/*ey*   +$_2$ +$_1$/*ey*

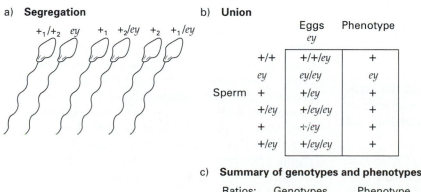

b) **Union**

	Eggs *ey*	Phenotype
Sperm +/+	+/+/*ey*	+
ey	*ey/ey*	*ey*
+	+/*ey*	+
+/*ey*	+/*ey/ey*	+
+	+/*ey*	+
+/*ey*	+/*ey/ey*	+

c) **Summary of genotypes and phenotypes**

Ratios:
Genotypes	Phenotype
1/6 +/+/*ey*	5/6 wild type
1/3 +/*ey/ey*	1/6 eyeless
1/3 +/*ey*	
1/6 *ey/ey*	

Questions and Problems

*17.1 A normal chromosome has the following gene sequence:

$A\ B\ C\ D\ {}_{\circ}\ E\ F\ G\ H$

Determine the chromosome mutation in each of the following chromosomes:

a. $A\ B\ C\ F\ E\ {}_{\circ}\ D\ G\ H$

b. $A\ D\ {}_{\circ}\ E\ F\ B\ C\ G\ H$

c. $A\ B\ C\ D\ {}_{\circ}\ E\ F\ E\ F\ G\ H$

d. $A\ B\ C\ D\ {}_{\circ}\ E\ F\ F\ E\ G\ H$

e. $A\ B\ D\ {}_{\circ}\ E\ F\ G\ H$

*17.2 Define pericentric and paracentric inversions.

17.3 What would be the result, in terms of protein structure, if a small inversion were to occur within the coding region of a structural gene?

*17.4 Inversions are said to affect crossing-over. The following homologs with the indicated gene order are given:

$\bullet\ \underline{A\ B\ C\ D\ E}$

${}_{\circ}\ \underline{A\ D\ C\ B\ E}$

a. Diagram the way these chromosomes would align in meiosis.

b. Diagram what a single crossover between homologous genes *B* and *C* in the inversion would result in.

c. Considering the position of the centromere, what is this sort of inversion called?

17.5 An inversion heterozygote possesses one chromosome with genes in the normal order:

${}_{\circ}\ \underline{a\ b\ c\ d\ e\ f\ g\ h}$

It also contains one chromosome with genes in the inverted order:

${}_{\circ}\ \underline{a\ b\ f\ e\ d\ c\ g\ h}$

A four-strand double crossover occurs in the areas *e-f* and *c-d*. Diagram and label the four strands at synapsis (showing the crossovers) and at the first meiotic anaphase.

*17.6 Mr. and Mrs. Lambert have not yet been able to produce a viable child. They have had two miscarriages and one severely defective child who died soon after birth. Studies of banded chromosomes of father, mother, and child showed all chromosomes were normal except for pair number 6. The number 6 chromosomes of mother, father, and child are shown in Figure 17.C.

■ *Figure 17.C*

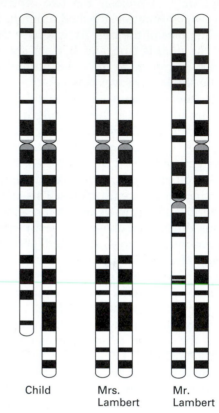

Child   Mrs. Lambert   Mr. Lambert

a. Does either parent have an abnormal chromosome? If so, what is the abnormality?

b. How did the chromosomes of the child arise? Be specific as to what events in the parents gave rise to these chromosomes.

c. Why is the child not phenotypically normal?

d. What can be predicted about future conceptions by this couple?

*17.7 Mr. and Mrs. Simpson have been trying for years to have a child but have been unable to conceive. They consulted a physician, and tests revealed that Mr. Simpson had a very reduced sperm count. His chromosomes were studied, and a testicular biopsy was done as well. His chromosomes proved to be normal, except for pair 12. Figure 17D shows a normal pair of number 12 chromosomes (Mrs. Simpson's) and Mr. Simpson's number 12 chromosomes.

a. What is the nature of the abnormality in Mr. Simpson's chromosome 12s?

b. What abnormal feature would you expect to see in the testicular biopsy, in which cells in various stages of meiosis can be seen?

■ *Figure 17.D*

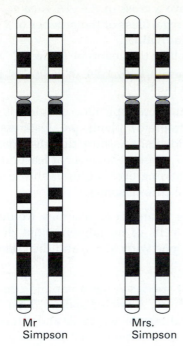

Mr
Simpson

Mrs.
Simpson

c. Why is Mr. Simpson's sperm count low?
d. What can be done about it?

*17.8 The following gene arrangements in a particular chromosome are found in *Drosophila* populations in different geographical regions. Assuming the arrangement in part a is the original arrangement, in what sequence did the various inversion types most likely arise?
a. *ABCDEFGHI*
b. *HEFBAGCDI*
c. *ABFEDCGHI*
d. *ABFCGHEDI*
e. *ABFEHGCDI*

*17.9 Chromosome I in maize has the gene sequence *ABCDEF*, whereas chromosome II has the sequence *MNOPQR*. A reciprocal translocation resulted in *ABCPQR* and *MNODEF*. Diagram the expected meiosis I pachytene configuration of the F_1 of a cross of homozygotes of these two arrangements.

*17.10 Mr. and Mrs. Denton have been trying for several years to have a child. They have experienced a series of miscarriages, and last year they had a child with multiple congenital defects. The child died within days of birth. The birth of this child prompted the Denton's physician to order a chromosome study of parents and child. The results of the study are shown in the figure below. Chromosome banding was done, and all chromosomes were normal in these individuals except some copies of number 6 and number 12. The number 6 and number 12 chromosomes of mother, father, and child are shown in Figure 17E (the number 6 chromosomes are the larger pair).

■ *Figure 17E*

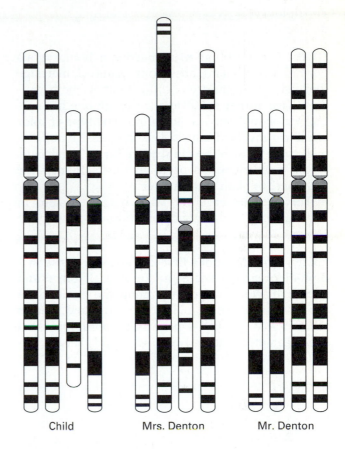

Child   Mrs. Denton   Mr. Denton

a. Does either parent have an abnormal karyotype? If so, which parent has it, and what is the nature of the abnormality?
b. How did the child's karyotype arise (what pairing and segregation events took place in the parents)?
c. Why is the child phenotypically defective?
d. What can this couple expect to occur in subsequent conceptions?
e. What medical help, if any, can be offered to them?

17.11 Define the terms *aneuploidy, monoploidy,* and *polyploidy.*

17.12 If a normal diploid cell is 2N, what is the chromosome content of the following: (a) a nullisomic; (b) a monosomic; (c) a double monosomic; (d) a tetrasomic; (e) a double trisomic; (f) a tetraploid; (g) a hexaploid?

*17.13 In humans how many chromosomes would be typical of nuclei of cells that are (a) monosomic; (b) trisomic; (c) monoploid; (d) triploid; (e) tetrasomic?

*17.14 An individual with 47 chromosomes, including an additional chromosome 15, is said to be:
a. A triplet
b. Trisomic
c. Triploid
d. Tricycle

*17.15 A color-blind man marries a homozygous normal woman and after four years of marriage they have two children. Unfortunately, both children have Turner syndrome, although one has normal vision and one is color-blind. The type of color blindness involved is a sex-linked recessive trait.
a. For the color-blind child with Turner syndrome, did nondisjunction occur in the mother or the father? Explain your answer.
b. For the Turner child with normal vision, in which parent did nondisjunction occur? Explain your answer.

*17.16 From Mendel's first law genes A and a in a diploid organism segregate from each other and appear in equal numbers among the gametes. In the real world, some plants are tetraploid. Let us assume that Mendel's peas were tetraploid, that every gamete contains two alleles, and that the distribution of alleles to the gamete is random. Suppose we have a cross of $AA\ AA \times aa\ aa$, where A is dominant, regardless of the number of a alleles present in an individual.
a. What will be the genotype of the F_1?
b. If the F_1 is selfed, what will be the phenotypic ratios in the F_2?

17.17 What phenotypic ratio of A to a is expected if $AA\ aa$ (tetraploid) plants are testcrossed against $aa\ aa$ individuals? (Assume that the dominant phenotype is expressed whenever at least one A is present, no crossing-over occurs, and each gamete receives two chromosomes.)

*17.18 How many chromosomes would be found in somatic cells of an allotetraploid derived from two plants, one with N = 7 and the other with N = 10?

17.19 Plant species A has a haploid complement of 4 chromosomes. A related species B has N = 5. In a geographical region where A and B are both present, C plants are found that have some characters of both species and somatic cells with 18 chromosomes. What is the chromosome constitution of the C plants likely to be? With what plants would they have to be crossed in order to produce fertile seed?

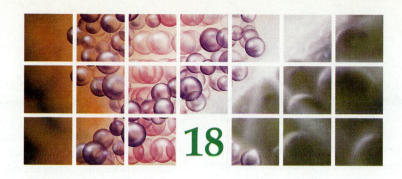

18

Transposable Elements, Tumor Viruses, and Oncogenes

- Transposable elements are DNA segments that can insert themselves at one or more sites in a genome. The presence of transposable elements in a cell is usually detected by the changes they bring about in the expression and activities of the genes at or near the chromosomal sites into which they integrate.

- In bacteria three major types of transposable elements are insertion sequence (IS) elements, transposons (Tn), and some bacteriophages, such as Mu. Insertion sequence elements and transposons have repeated sequences at their ends and encode proteins that are responsible for their transposition. When ISs and Tns integrate into the genome, a short sequence at the target site is duplicated, giving rise to directly repeated sequences flanking the integrated element.

- Transposable elements in eukaryotes resemble bacterial transposons in general structure and transposition properties. While most transposons move using a DNA-to-DNA mechanism, some eukaryotic transposons move via an RNA intermediate (using a transposon-encoded reverse transcriptase). Such transposons resemble retroviruses in genome organization and other properties.

- Some forms of cancer are caused by tumor viruses. Both DNA and RNA tumor viruses are known. RNA tumor viruses, which contain tumor-inducing genes termed oncogenes, are part of the retroviruses. Both normal cells and nonviral-induced cancer cells contain sequences that are related to the viral oncogenes. The model is that tumor-causing retroviruses have picked up certain normal cellular genes—called proto-oncogenes—while simultaneously losing some of their genetic information. In normal cells proto-oncogenes function in various ways to regulate cell division. In the retrovirus these genes have been modified or their expression controlled differently so that the protein product, now synthesized under viral control, is both qualitatively and quantitatively changed, as well as being expressed in cells in which the products are not normally found. These gene products, which include growth factors, are directly responsible for the transformation of cells to the cancerous state.

THE CLASSIC PICTURE OF GENES is one in which the genes are at fixed loci on a chromosome. In the last chapter we learned that segments of chromosomes can change their orientation or position in the genome as a result of chromosome mutations. These changes inform us about ways in which genomes might have evolved. For example, a chromosome segment containing a gene might become duplicated. Not only does this increase genome size, but the two copies may, in turn, mutate independently to produce two distinct genes with related function. Moreover, through translocation, genes can be moved to different places in the genome. Genomic rearrangements can occur in other ways. That is, certain genetic elements in the chromosomes of both prokaryotes and eukaryotes have the capacity to move from one location to another in the genome. These mobile genetic elements have been given a number of names in the literature, including controlling elements, jumping genes, insertion sequences, and transposons. We shall use a generic term that has become fairly widely accepted—**transposable elements**—since this term reflects the *transposition* ("change in position") events associated with these elements. Undoubtedly transposable elements have contributed to the evolution of the genomes of both prokaryotes and eukaryotes through the chromosome rearrangements they cause.

In structure and function transposable elements are similar in both prokaryotes and in eukaryotes and their viruses. In prokaryotes transposable elements can move to new positions on the same chromosome (since there is only one), while in eukaryotes transposable elements have the ability to move to new positions within the same chromosome or to a different chromosome. In both prokaryotes and eukaryotes transposable elements are identified by the changes they cause: they can produce mutations by inserting into genes, they can affect gene expression by inserting into gene regulatory sequences, and they can produce various kinds of chromosome mutations. They have also probably been important in genome evolution. The effects of transposable

Table 18.1

Properties of Some IS Elements of *E. coli*

INSERTION SEQUENCE	LENGTH (BP)	INVERTED TERMINAL REPEATING SEQUENCE (BP)	TARGET SITE DUPLICATIONS (BP)
IS1	768	23	9
IS2	1327	32/41	5
IS10	1329	22	9

elements have been established through genetic, cytological, molecular, and recombinant DNA procedures.

In this chapter we also discuss tumor viruses, oncogenes, and cancer. Some cancers are caused by tumor viruses, and both DNA and RNA tumor viruses are known. Here we focus on retroviruses, which are RNA tumor viruses. Retroviruses have an RNA genome in the virion (virus particle) but replicate via a DNA intermediate that becomes integrated into the host cell's chromosome. Continuing the theme of genes that move, tumor-inducing retroviruses cause cancer because of the activities of normal genes that have been picked up from the host genome, inserted into the retroviral genome, and altered in some way.

Transposable Elements in Prokaryotes

There are three types of transposable elements in prokaryotes: insertion sequence (IS) elements, transposons (Tn), and certain bacteriophages.

Insertion Sequences

An **insertion sequence (IS),** or **IS element,** is the simplest transposable element found in prokaryotes. An IS element contains only genes required for inserting the element into a chromosome and for mobilizing the element to different locations. IS elements are normal constituents of bacterial chromosomes and plasmids.

Properties of Insertion Sequences. IS elements have characteristic lengths and unique nucleotide sequences. Table 18.1 lists the properties of some of the IS elements found in *E. coli*. IS1 is 768 bp long, and is present in 4 to 19 copies on the *E. coli* chromosome (Figure 18.1). IS2 is present in 0 to 12 copies on the *E. coli* chromosome and in one copy on the *F* plasmid, and IS10 is found on a class of plasmids that can replicate in *E. coli* called *R* plasmids.

All IS elements end with perfect or nearly perfect

inverted terminal repeats (IRs) of between 9 and 41 bp. This means that essentially the same sequence is found at each end of an IS but in opposite orientations. The inverted repeats of IS1 consist of 23 bp of not quite identical sequence (Figure 18.2).

Movement of IS elements to new genome locations often causes mutations, either by disrupting the coding sequence of a gene or by disrupting a gene's regulatory region. Promoters within the IS elements themselves may also have effects by altering the expression of nearby genes. Thus depending on the promoter, gene expression might become turned on or turned off. Moreover, the presence of an IS element in the chromosome can cause chromosome mutations such as deletions and inversions in the adjacent DNA.

When transposition of an IS element takes place, a copy of the IS element inserts into a new chromosome location while the original IS element remains in place. Transposition requires the precise replication of the original IS element. This replication uses some enzymes of the host cell. Transposition requires an enzyme encoded by the IS element called **transposase**. The IR sequences are essential for the transposition process; that is, those sequences are recognized by transposase to initiate transposition. The frequency of transposition is characteristic of each IS element, ranging between 10^{-5} and 10^{-7} per generation.

IS elements insert into the chromosome at sites with which they have no sequence homology. These sites are called *target sites*. The process of IS insertion into a chromosome is shown in Figure 18.3. First, a staggered cut is made in the target site and the IS element is then inserted, becoming joined to

■ *Figure 18.1* The insertion sequence transposable element, IS1. The IS element has inverted repeat (IR) sequences at the ends.

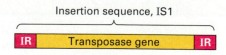

Insertion sequence, IS1

■ *Figure 18.2*   23-bp inverted terminal repeats (IR) of IS1.

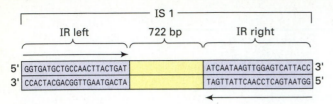

the single-stranded ends. The gaps are filled in by DNA polymerase and DNA ligase, producing an integrated IS element with two direct repeats of the target site sequence flanking the IS element. "Direct" in this case means that the two sequences are repeated in the same orientation (see Figure 18.3). The direct repeats are called *target site duplications*. The sizes of target site duplications vary with the IS element, but tend to be small (4 to 13 bp as shown in Table 18.1).

IS Elements in the F Factor   Recall that the transfer of genetic material between conjugating *E. coli* is the

result of the function of the fertility factor *F* (see Chapter 6). The *F* factor, a circular double-stranded DNA molecule, is one example of a natural, bacterial **plasmid**, an extrachromosomal genetic element capable of self-replication. Plasmids such as *F* that are also capable of integrating into the bacterial chromosome are called **episomes.**

Figure 18.4 shows the genetic organization of an *E. coli F* factor. It consists of 94,500 bp of DNA that code for a variety of proteins. The important genetic elements are:

1. *tra* (for "transfer") genes, required for the conjugal transfer of the DNA from a donor bacterium to a recipient bacterium;
2. Genes that encode proteins required for the plasmid's replication;
3. Four IS elements: two copies of IS3, one of IS2, and one of an element called γδ (gamma-delta).

The *F* factor can integrate into the *E. coli* chromosome at different sites and in different orientations

■ *Figure 18.3*   Schematic of the integration of an IS element into chromosomal DNA. As a result of the integration event, the target site becomes duplicated to produce direct target repeats. Thus the integrated IS element is characterized by its inverted repeat (IR) sequences flanked by direct target repeat sequences. Integration involves making staggered cuts in the host target site. After joining one strand of the IS to the overhangs, the gaps that result are filled in with DNA polymerase and DNA ligase. [Note: The base sequences given for the IR are for illustration only and are not the actual sequences found, either in length or sequences.]

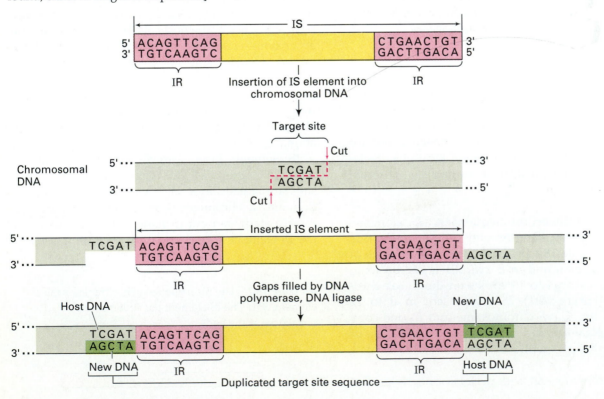

■ *Figure 18.4* Organizational map of the *E. coli F* factor. The map shows the locations of genes required for transfer of the *F* factor during conjugation and for the replication of *F*; also shown are the insertion elements responsible for *F*'s integration into the bacterial chromosome.

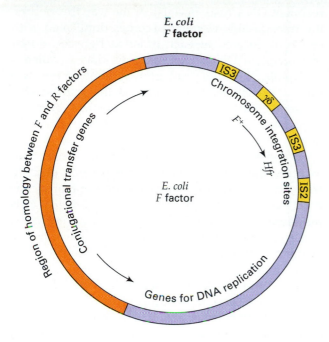

E. coli
F factor

E. coli
F factor

Region of homology between *F* and *R* factors

Conjugational transfer genes

IS3

γδ

Chromosome integration sites

F+

Hfr

IS3

IS2

Genes for DNA replication

because the *E. coli* chromosome has copies of these insertion sequences at various positions. That is, *F* factor integration occurs by conventional genetic recombination between the homologous sequences of the insertion elements. Once integrated into the *E. coli* chromosome, the *tra* genes in the *F* factor direct the conjugal transfer functions in *Hfr* strains.

Transposons

A **transposon (Tn)** is a mobile DNA segment that, like an IS element, contains genes for the insertion of the DNA segment into the chromosome and for the mobilization of the element to other locations on the chromosome. Unlike IS elements, Tns also contain other genes. The properties of some *E. coli* transposons are summarized in Table 18.2.

There are two types of prokaryotic transposons: composite transposons and noncomposite transposons. *Composite transposons* are complex transposons with a central region containing genes (e.g., drug-resistance genes) flanked on both sides by IS elements. Composite transposons may be thousands of base pairs long. The IS elements are both of the same type and are called IS-L (for "left") and IS-R (for "right"). Depending on the transposon, IS-L and IS-R may be in the same or inverted orientation relative to each other. Because the ISs themselves have terminal inverted repeats, the composite transposons also have terminal inverted repeats. Figure 18.5 diagrams the structure of the composite transposon Tn10 to illustrate the general features of such transposons. This transposon is 9300 bp long and consists of 6500 bp of DNA containing the tetracycline resistance gene flanked at each end with a 1400-bp IS element. These IS elements are designated IS10L and IS10R and are arranged in an inverted orientation. Cells containing Tn10 are resistant to tetracycline because of the tetracycline resistance gene.

Transposition of composite transposons occurs because of the function of the IS elements they contain. One or both IS elements supplies the transposase. The IRs of the IS elements at the two ends of the transposon are recognized by transposase to initiate transposition (see IS elements transposition). Like IS elements, composite transposons produce target site duplications after transposition. For example, Tn10 produces a 9-bp target site duplication (see Table 18.2).

Noncomposite transposons, like composite transposons, contain genes such as those for drug resistance. Unlike composite transposons, they do not terminate with IS elements. They do have repeated

---

			Table 18.2		

Properties of Some Transposons Found in *E. coli*.

TRANSPOSON	GENE MARKER(S)	LENGTH (BP)	TERMINAL REPEAT SEQUENCES	IS ORIENTATION	TARGET SITE DUPLICATION
Tn3 (noncomposite)	Resistance to ampicillin	4957	38 bp; not an IS element	—	5 bp
Tn9 (composite)	Resistance to chloramphenicol	2638	IS1L, IS1R	Direct	9 bp
Tn10 (composite)	Resistance to tetracycline	9300	IS10L, IS10R	Inverted	9 bp

■ *Figure 18.5* Structure of the composite transposon Tn10. The general features of composite transposons are evident: a central region carrying a gene or genes such as for drug resistance flanked by either direct or inverted IS elements. The IS elements themselves each have terminal inverted repeats.

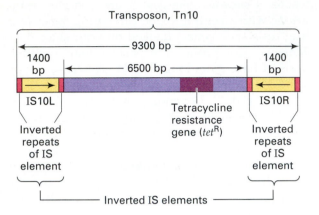

Transposon, Tn10

←—————— 9300 bp ——————→

1400 bp ←—————— 6500 bp ——————→ 1400 bp

IS10L / Tetracycline resistance gene (*tet*ᴿ) / IS10R

Inverted repeats of IS element / Inverted repeats of IS element

└——————— Inverted IS elements ———————┘

sequences at their ends, however, that are required for transposition. Tn3 is a noncomposite transposon (Table 18.2 and Figure 18.6). Tn3 has 38-bp inverted terminal repeats and contains three genes in its central region. One of those genes, *amp*ᴿ, encodes β-lactamase which breaks down ampicillin and therefore makes cells containing Tn3 resistant to ampicillin. The other two genes, *tnpA* and *tnpB* encode the enzymes transposase and resolvase that are needed for transposition of Tn3. The genes for transposition are in the central region for noncomposite transposons, while they are in the terminal IS elements for composite transposons. Like composite transposons, noncomposite transposons cause target site duplications when they move. Tn3, for example, produces a 5-bp target site duplication.

A number of detailed models have been generated for transposition of transposons. Figure 18.7

shows a *cointegration* model involving the transposition of a transposon from one genome to another; e.g., from a plasmid to a bacterial chromosome or vice versa. Similar events can occur between two locations on the same chromosome. First, the donor DNA containing the transposable element and the recipient DNA fuse. Because of the way this occurs, the transposable element becomes duplicated with one copy located at each junction between donor and recipient DNA. This fused product is called a *cointegrate*. Next, the cointegrate is resolved into two products, each with one copy of the transposable element. Since the transposable element becomes duplicated, the process is called *replicative transposition*.

A second type of transposition mechanism involves the movement of a transposable element from one location to another on the same or different DNA *without* replication of the element. This mechanism is called *conservative (nonreplicative) transposition* or *simple insertion*. In other words, the element is lost from the original position when it transposes.

Which transposition mechanism is used depends on the particular transposon. Tn10, for example, transposes by conservative transposition, while Tn3 and related noncomposite transposons, moves by replicative transposition.

Bacteriophage Mu

Mu is a temperate bacteriophage that infects *E. coli*. Recall that *temperate bacteriophages* (e.g., λ) can go through the lytic cycle or enter the lysogenic phase (see Chapter 6). Mu is also a transposon and can cause mutations when it transposes—in fact Mu stands for mutator.

In the phage particle the Mu genome is a 37-kb linear piece of DNA consisting mostly of phage

■ *Figure 18.6* Structure of the noncomposite transposon Tn3. Tn3 has genes in the central region for three enzymes: β-lactamase (destroys antibiotics like penicillin and ampicillin), transposase, and resolvase. Transposase and resolvase are involved in the transposition process. Tn3 has short inverted terminal repeats that are unrelated to IS elements.

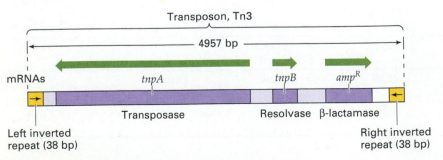

Transposon, Tn3

←—————————— 4957 bp ——————————→

mRNAs

tnpA / *tnpB* / *amp*ᴿ

Transposase / Resolvase / β-lactamase

Left inverted repeat (38 bp) / Right inverted repeat (38 bp)

■ *Figure 18.7*   Transposition of a transposable element by replicative transposition. A donor DNA with a transposable element fuses with a recipient DNA. During the fusion, the transposable element is duplicated so that the product is a cointegrate molecule with one transposable element at each junction between donor and recipient DNA. The cointegrate is resolved into two molecules, each with one copy of the transposable element.

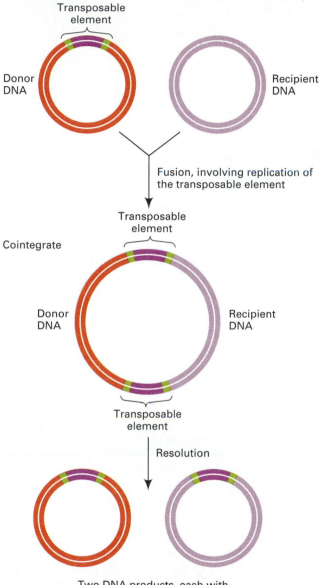

DNA, with unequal lengths of host DNA at the two ends (Figure 18.8[a]). When Mu infects *E. coli* and enters the lysogenic state, the Mu genome integrates into the host chromosome by *conservative* transposition to produce the integrated prophage DNA, flanked by a 5-bp direct repeat of the host target site sequence (Figure 18.8[b]). During integration the flanking host DNA that was present in the particle is lost. In the lysogenic state a phage-encoded repressor prevents most Mu gene expression (see λ repressor in Chapter 6), and the Mu prophage replicates when the *E. coli* chromosome replicates.

Mu's lytic cycle is different from that of λ. The Mu genome integrates into the *E. coli* chromosome in a different way than for lysogeny and remains integrated throughout the lytic phase. Replication of the Mu genome occurs by replicative transposi-

■ *Figure 18.8* Temperate bacteriophage Mu genome shown (a) in phage particles and (b) integrated into the *E. coli* chromosome as a prophage.

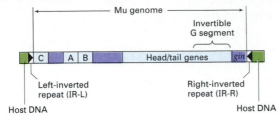

a) Phage DNA present in virus particles

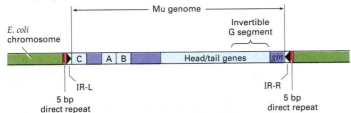

b) Prophage DNA

tion. The details of this transposition remain to be determined.

Transposition by Mu causes mutations of various kinds. Apart from simple insertions, Mu can also cause deletions, inversions, and translocations. For example, by homologous recombination between two identical copies, Mu (and other multicopy transposons) can cause deletions (Figure 18.9[a]) and inversions (Figure 18.9[b]). Deletion occurs if the two Mu prophages or transposons are in the same orientation in the chromosome, while inversion occurs if the two are in opposite orientations.

Keynote

Transposable elements are unique DNA segments that can insert themselves at one or more sites in a genome. The presence of transposable elements in a cell is usually detected by the changes they bring about in the expression and activities of the genes at or near the chromosomal sites into which they integrate. In prokaryotes the three major types of transposable elements are insertion sequence (IS) elements, transposons (Tn), and bacteriophages such as Mu.

Transposable Elements in Eukaryotes

The earliest observations of the phenotypic effects of a transposable element in eukaryotes came from Barbara McClintock's work with corn (*Zea mays*) in the 1940s and 1950s. On the basis of her genetic work, she hypothesized the existence of what she called "controlling elements," which modify or suppress gene activity in corn and which are mobile in the genome. Decades later the controlling elements were shown to be transposable elements. For her original work in deducing the presence of transposable elements, Barbara McClintock was awarded the Nobel Prize in 1983.

Transposable elements have since been identified in many eukaryotes, and they have been studied mostly in yeast, *Drosophila*, corn, and humans. In general their structure and function are very similar to those of prokaryotic transposable elements. Eukaryotic transposable elements can integrate into chromosomes at a number of sites. Thus such elements may be able to affect the function of virtually any gene, turning it on or off, depending on the element involved and how it integrates into or near the gene. Integration of eukaryotic transposable elements, like that of most prokaryotic transposable elements, occurs into sites with which they have no homology. Many of the eukaryotic transposable elements carry genes. However, while some of the genes must code for enzymes required for transposition, in most cases the functions of the genes are unknown.

Transposons in Plants

Some of the transposable elements found in plants are presented in Table 18.3. Like the transposons we have already discussed, plant transposons have inverted repeated (IR) sequences at their ends and generate short direct repeats of the target site DNA when they integrate.

When a plant transposon inserts into a chromosome, the consequence depends on the properties of the transposon. Typically, the effects range from

■ *Figure 18.11* The structure of the *Ac* autonomous transposable element of corn, and of several *Ds* nonautonomous elements derived from *Ac*: (a) The *Ac* element is 4563 bp long and has imperfect 11-bp terminal inverted repeats. *Ac* has a single transcription unit, comprising most of the element, and going from left to right. The five exons of the resulting RNA are shown in green within the *Ac* element itself; (b) Several *Ds* elements are shown in the same way as for the *Ac* element. They are derived from *Ac* by internal deletions of various lengths.

a) **Activator element (*Ac*)**

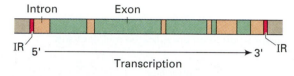

b) **Dissociation elements (*Ds*)**

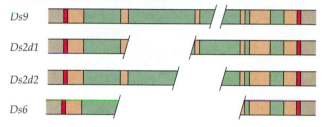

Consider a chromosome with one copy of *Ac* at a site called the *donor site*. When the chromosome region containing *Ac* replicates, two copies of *Ac* result, one on each progeny chromatid. There are two possible outcomes of *Ac* transposition depending on whether transposition occurs to a replicated or unreplicated chromosome site. If one of the two *Ac* elements transposes to a replicated chromosome site (Figure 18.12[a]), an empty donor site is left on one chromatid, while an *Ac* element remains in the homologous donor site on the other chromatid. The transposing *Ac* element inserts into a new, already replicated recipient site, which is often on the same chromosome. In Figure 18.12(a) the site is shown on the same chromatid as the parental *Ac* element. Thus in the case of transposition to an already-replicated site, there is no net increase in the number of *Ac* elements.

Figure 18.12(b) shows transposition of one *Ac* element to a site on the chromosome that has not yet been replicated. As with the first case, only one of the two *Ac* elements transposes, leaving an empty donor site on one chromatid and an *Ac* element in the homologous donor site on the other chromatid. But now the transposing element inserts into a nearby recipient site that has yet to be replicated. When that region of the chromosome replicates, a copy of the transposed *Ac* element will be

on both chromatids, in addition to the one original copy of the *Ac* element at the donor site on one chromatid. Thus in the case of transposition to an unreplicated recipient site, there is a net increase in the number of *Ac* elements.

Transposition of most *Ds* elements occurs in the same way as for *Ac* transposition, with functions for transposition supplied by an *Ac* element in the genome.

K e y n o t e

The mechanism of transposition of plant transposons is quite similar to transposition of bacterial IS elements or transposons. Transposons integrate at a target site by a precise mechanism so that the integrated elements are flanked at the insertion site by a short duplication of target site DNA of characteristic length. Many plant transposons occur in families, the autonomous elements of which are able to direct their own transposition, and the nonautonomous elements of which are able to transpose only when activated by an autonomous element in the same genome. Most nonautonomous elements are derived from autonomous elements by internal deletions or complex sequence rearrangements.

Ty Elements in Yeast

***Ty* Structure and Properties** All *Ty* transposable elements of yeast have a number of structural properties in common with bacterial transposons: they have terminal repeated sequences, they integrate at sites with which they have no homology, and they generate a target site duplication (of 5 bp) upon insertion.

A *Ty* element is diagrammed in Figure 18.13. The element is about 5.9 kb long and includes two 334-bp-long, directly repeated sequences called long terminal repeats (LTR) or deltas (δ), one at each end of the element. Each delta has inverted repeats at its ends that are only 2 bp long and contains a promoter and sequences recognized by transposing enzymes. *Ty* elements encode a single, 5700-nucleotide mRNA that begins at the promoter in the delta at the 5' end of the element. The mRNA transcript contains two open reading frames (ORFs); that is, regions with a start codon in frame with a chain-terminating codon, indicating that two proteins could be produced from the mRNA. The two regions have been designated *TyA* and *TyB*. There are about 35 copies of *Ty* in the haploid

■ *Figure 18.12*   The *Ac* transposition mechanism: (a) Transposition to an already-replicated recipient site results in no net increase in the number of *Ac* elements in the genome; (b) Transposition to an unreplicated recipient site results in a net increase in the number of *Ac* elements when the region of the chromosome containing the transposed element is subsequently replicated.

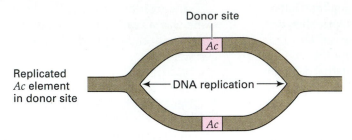

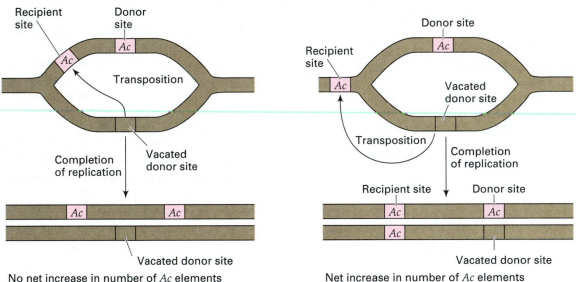

genome of laboratory strains of yeast, representing about 1 percent of the genome.

Ty Elements and Retroviruses   Retroviruses are single-stranded RNA viruses that replicate via dou-

■ *Figure 18.13*   The *Ty* transposable element of yeast. (Courtesy of Dr. Gerald Fink.)

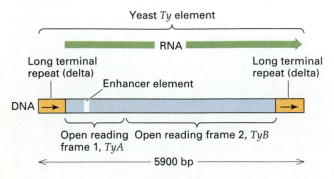

ble-stranded DNA intermediates. When a retrovirus infects a cell, its RNA genome is copied by viral reverse transcriptase to produce a double-stranded DNA. The DNA integrates into the host's chromosome where it can be transcribed to produce progeny RNA viral genomes and viral protein mRNAs.

Yeast *Ty* elements are similar in organization to retroviruses. In fact *Ty* elements transpose using the same mechanisms as those involved in the production and integration of retrovirus DNA into the chromosome. That is, rather than transposing DNA to DNA as is the case with bacterial transposons and most eukaryotic transposons, *Ty* elements transpose by *making an RNA copy* of the integrated DNA sequence and then creating a new *Ty* element by reverse transcription. The new DNA element integrates at a new chromosome location. Because of their similarity with retroviruses, *Ty* elements have been referred to as *retrotransposons*.

Evidence showing that *Ty* transposition occurs

via an RNA intermediate came from experiments in which an intron was placed into the *Ty* element by recombinant DNA techniques (there is none in normal *Ty* elements). The element was then followed through a transposition event. At the new location, the *Ty* element no longer had the intron sequence, indicating that transposition must have occurred via an RNA intermediate. The intron had been removed from the RNA by the usual splicing processes that function to remove introns from pre-mRNAs. The *tyB* reading frame of *Ty* elements encodes the reverse transcriptase that is used to make *Ty* DNA from the RNA transcript.

Drosophila Transposons

A number of classes of transposons have been identified in *Drosophila*. In this organism it is estimated that about 15 percent of the genome is mobile. One example will be considered here.

P Elements and Hybrid Dysgenesis   Hybrid dysgenesis is the appearance of a series of defects, including gene mutations, chromosomal mutations, and sterility, when certain strains of *Drosophila melanogaster* are crossed. For example, hybrid dysgenesis occurs when female laboratory strains are mated to males from natural populations (Figure 18.14[a]). The laboratory strain is said to be of the

M cytotype (maternal contributing cell type) and the naturally occurring strain is said to be of the *P cytotype* (paternal contributing cell type); thus the cross is M♀ × P♂. In the reciprocal P♀ × M♂ cross hybrid dysgenesis does not occur and all progeny are fertile (Figure 18.14[b]).

Hybrid dysgenesis primarily affects germ line-cells. F_1 hybrids of a M♀ × P♂ cross have normal somatic tissues, but gonads do not develop, and the flies are sterile. Hybrid dysgenesis results when chromosomes of the P male parent become exposed to the cytoplasm derived from the M female parent. Cytoplasm from P females does not induce hybrid dysgenesis, so progeny of P♀ × P♂ are fully fertile.

The model for the induction of hybrid dysgenesis in the M♀ × P♂ crosses is as follows. The haploid genome of the P male has about 40 copies of a family of transposons called *P elements*. The M strain does not have any *P* elements. *P* elements vary in length from 500 to 2900 bp; each has 31-bp inverted terminal repeats. The shorter *P* elements (which are nonautonomous elements, like *Ds* in corn) are derived from the longest *P* element by internal deletions. The longest *P* elements are autonomous elements like *Ac* in corn; that is, they encode a transposase which can catalyze its own transposition and the transposition of the shorter *P* elements. In the P cytotype *P* elements are stable in the chromo-

■ *Figure 18.14*   Hybrid dysgenesis, exemplified by the production of sterile flies, results from (a) a cross of M♀ × P♂, but not from (b) a cross of P♀ × M♂.

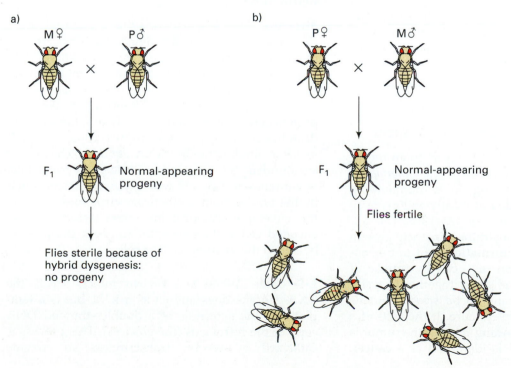

somes because the elements encode a repressor protein that prevents the transcription of the transposase gene, therefore preventing transposition. In the hybrid flies produced from the M♀ × P♂ crosses, the cytoplasm derives from the M cytotype, which lacks *P* repressors. So transposase is produced, causing the activation of *P* element transposition. The transposition events cause the defects associated with hybrid dysgenesis. In hybrid flies produced from the reciprocal P♀ × M♂ crosses, the cytoplasm derives from the P cytotype, so repressors are present; consequently, *P* element transposition is not activated and no hybrid dysgenesis is seen in those crosses.

The mechanisms by which *P* elements are inserted or excised are not known. In the research laboratory *P* elements are used as vectors to introduce genes into the *Drosophila* genome.

Keynote

Transposable elements in eukaryotes are typically transposons. Transposons can transpose to new sites while leaving a copy behind in the original site (replicative transposition), or they may excise themselves from the chromosome (conservative transposition). When the excision is imperfect, deletions can occur, and by various recombination events other chromosomal rearrangements such as inversions and duplications may occur. Most transposons move by using a DNA-to-DNA mechanism. Some eukaryotic transposons, such as yeast *Ty* elements, transpose via an RNA intermediate (using a transposon encoded reverse transcriptase). Such transposons resemble retroviruses in genome organization and other properties.

Tumor Viruses, Oncogenes, and Cancer

Occasionally differentiated cells revert to an undifferentiated state (a process called *dedifferentiation*) in which, instead of remaining in a nondividing mode, they begin to divide mitotically and give rise to tissue masses called *tumors* or *neoplasms*. The failure of cells to remain constrained in their growth properties is called **transformation** (not to be confused with transformation of a cell by uptake of DNA). If the neoplastic cells stay together in a single mass, the tumor is said to be *benign*, and its removal results in a complete cure. If the cells of a tumor can invade surrounding tissues, the tumor is said to be *malignant* and is identified as a **cancer**.

Cells from malignant tumors can also break off and move through the blood system or lymphatic system, forming new tumors at other locations in the body. The spreading of malignant tumor cells throughout the body is called *metastasis*.

The initiation of cancers in an organism is called **oncogenesis** (*onkos* means "mass" or "bulk"; *genesis* means "birth"). There are many different causes of cancer such as spontaneous genetic changes (e.g., spontaneous gene mutations or chromosome mutations); exposure to mutagens and radiation; or cancer-inducing viruses (tumor viruses). There is also hereditary predisposition to cancer.

In this section we will focus on tumor viruses and, in particular, on the class of tumor viruses called retroviruses. We will see that tumor-causing retroviruses carry genes that have moved from the host's genome to the virus's genome.

Tumor Viruses

Tumors can result from infection of cells with **tumor viruses**. Both DNA viruses and RNA viruses can transform cells into cancer cells. An important class of RNA tumor viruses is the retroviruses and studies of these viruses have provided important information about the relationship between viruses, oncogenes, and the host cell. We will focus our attention on retroviruses in the remaining discussions.

Retroviruses

Structure   The RNA tumor viruses exemplified by *Rous sarcoma* virus, feline leukemia virus, mouse mammary tumor virus, and human immunodeficiency virus (HIV-1, the causative agent of AIDS) are all *retroviruses*. A drawing of a stylized retrovirus particle is shown in Figure 18.15. Within a protein core, which often is icosahedral (an icosahedron has 20 faces) in shape, are two copies of the 7-to-10-kb single-stranded RNA genome. The core is surrounded by an envelope derived from host membranes and coated with glycoproteins encoded by the viral genome. When the virus infects a cell, the envelope glycoproteins interact with a host-encoded cell surface receptor to begin the process by which the virus enters the cell.

Life Cycle   When a retrovirus infects a cell, the RNA genome does not act as mRNA but is a template for the synthesis of a double-stranded DNA copy of the retrovirus (the *provirus*). This process is catalyzed by reverse transcriptase, an enzyme

■ *Figure 18.15* Stylized drawing of a retrovirus.

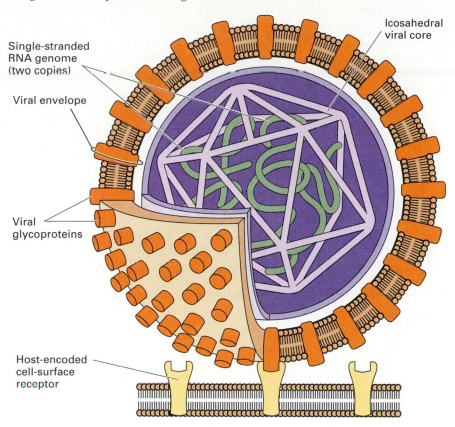

Single-stranded
RNA genome
(two copies)

Viral envelope

Viral
glycoproteins

Host-encoded
cell-surface
receptor

Icosahedral
viral core

brought into the cell as part of the virus particle. The double-stranded DNA then integrates into the host chromosome. Once integrated, the viral DNA is transcribed to produce the various viral mRNAs. Progeny RNA genomes are produced by transcription of the entire integrated viral DNA.

The RNA genome organization of the retrovirus *Rous sarcoma* virus (RSV) is shown in Figure 18.16(a). RSV causes *sarcomas*, cancers of the connective tissue or muscle cells, in chickens. Typical retroviruses have three protein-coding genes required for the virus life cycle: *gag*, *pol*, and *env*. The *gag* gene encodes a precursor protein that, when cleaved, produces virus particle proteins. The *pol* gene encodes a precursor protein that is cleaved to produce reverse transcriptase and an enzyme needed for the integration of the proviral DNA into the host cell chromosome. The *env* gene encodes the precursor to the envelope glycoprotein. The RSV retrovirus also carries an *oncogene* called *src*, which is not involved in the viral life cycle. By definition, an **oncogene** is a gene that transforms a normal cell to the cancerous state. Different retroviruses carry different oncogenes. The ends of the RNA genome consist of the sequences R and U5 (shown at left in Figure 18.16[a]) and U3 and R (shown at right).

When proviral DNA is produced by reverse transcriptase, the end sequences are duplicated to produce long terminal repeats (LTRs) of the sequences U3-R-U5 (Figure 18.16[b]). The LTRs contain many of the transcription regulatory signals for the viral sequence.

The first step in the integration of the proviral DNA into the host's chromosome is ligation of the ends of the linear molecule to produce a circular, double-stranded molecule (Figure 18.16[c]). This brings two LTRs next to each other. Staggered nicks are made in both the viral and cellular DNAs (Figure 18.16[d]). By recombination, the viral ends become joined to the ends of the cellular DNA; at this point integration has occurred (Figure 18.16[e]). Last, the single-stranded gaps are filled in. Like the transposition of transposons, the integration of retrovirus proviral DNA results in a duplication of DNA at the target site, producing short, direct repeats in the host cell DNA flanking the provirus.

Types of Retroviruses   There are two types of retroviruses: (1) *Replication-competent retroviruses* which direct their own life cycle but do not change the growth properties of the cells they infect; that

■ *Figure 18.16*   The *Rous sarcoma* virus (RSV) RNA genome and the integration of the proviral DNA into the host chromosome: (a) RSV genome RNA; (b) RSV proviral DNA produced by reverse transcriptase; (c) Circularization of the proviral DNA; (d) Staggered nicks are made in viral and cellular DNAs; (e) By recombination, the viral ends become joined to the ends of the cell's DNA; (f) The single-stranded gaps are filled in and a complete, double-stranded integrated RSV provirus results.

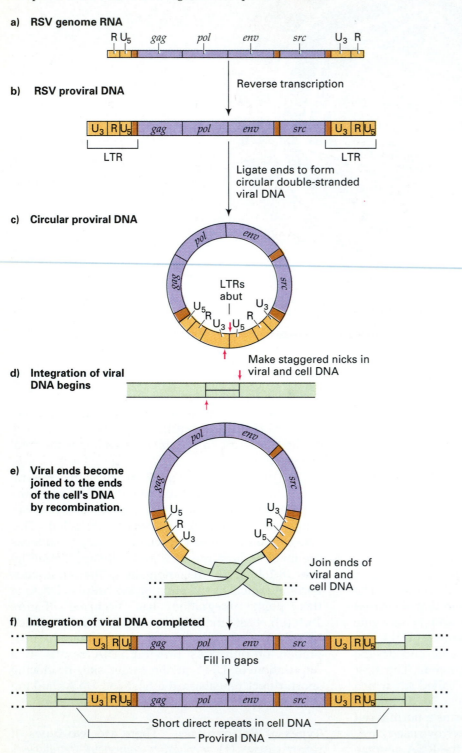

is, they are nontransforming viruses; and (2) *Transformation-competent retroviruses* which have the ability to transform an infected cell to the tumorous state. In short, all RNA tumor viruses are retroviruses, but not all retroviruses induce tumors.

Viral Oncogenes   Cancer induction by RSV results from the activity of the **viral oncogene** *src* in the retroviral genome. Viral oncogenes (which will generically be called v-*onc*s) are responsible for many different cancers. Only retroviruses that contain a v-*onc* gene are tumor viruses. Such retroviruses are called **transducing retroviruses,** because they have picked up an oncogene from the cell's genome. (We will learn how this happens later in this section.) The v-*onc* genes are named for the tumor that the virus causes, with the prefix "v" to indicate that the gene is of viral origin. Thus the v-*onc* gene of RSV is v-*src*. Table 18.4 lists some transducing retroviruses and their viral oncogenes.

Cells infected by RSV rapidly transform into the cancerous state because of the activity of the v-*src* gene. Since RSV contains all the genes necessary for viral replication (*gag*, *env*, and *pol*), a RSV-transformed cell produces progeny RSV particles. RSV is an exception in this ability, however, because all other transducing retroviruses are defective in some of their viral replication genes (Figure 18.17): they can transform cells but are unable to produce progeny viruses because of the absence of one or more genes needed for virus reproduction. These defective retroviruses can produce progeny viral particles, however, if cells containing them are also infected with a normal virus (a *helper virus*) that can supply the missing gene products.

Cellular Proto-Oncogenes   In the mid-1970s it was discovered that normal animal cells contain genes with DNA sequences very closely related to the viral oncogenes. Then in the early 1980s it was

---

Table 18.4

Some Transducing Retroviruses and Their Oncogenes

Onco-Gene	Retrovirus Isolates	v-*onc* Origin	v-*onc* Protein	Virus Disease
src	Rous sarcoma virus (RSV)	Chicken	$pp60^{src}$	Sarcoma
fps[a]	Fujinami sarcoma virus (FuSV)	Chicken	$P130^{gag\text{-}fps}$	Sarcoma
	PRCII-avian sarcoma virus (ASV)	Chicken	$P105^{gag\text{-}fps}$	Sarcoma
fgr	Gardner-Rasheed FeSV	Cat	$P70^{gag\text{-}actin\text{-}fgr}$	Sarcoma
abl	Abelson murine leukemia virus (MLV)	Mouse	$P90\text{-}P160^{gag\text{-}abl}$	Pre-B cell leukemia
	Hardy-Zuckerman (HZ2)-FeSV	Cat	$P98^{gag\text{-}abl}$	Sarcoma
erbA	Avian erythroblastosis virus (AEV)	Chicken	$P75^{gag\text{-}erb\text{-}A}$	Erythroblastosis and sarcoma
sis	Simian sarcoma virus (SSV)	Monkey	$P28^{env\text{-}sis}$	Sarcoma
	Parodi-Irgens FeSV	Cat	$P76^{gag\text{-}sis}$	Sarcoma
myc	MC29	Chicken	$P100^{gag\text{-}myc}$	Sarcoma, carcinoma, and myelocytoma
myb	Avian myeloblastosis virus (AMV)	Chicken	$p45^{myb}$	Myeloblastosis
	AMV-E26	Chicken	$P135^{gag\text{-}myb\text{-}ets}$	Myeloblastosis and erythroblastosis
kit	HZ4-FeSV	Cat	$P80^{gag\text{-}kit}$	Sarcoma
H-*ras*	Harvey MSV (Ha-MSV)	Rat	$pp21^{ras}$	Sarcoma and erythroleukemia
	RaSV	Rat	$P29^{gag\text{-}ras}$	Sarcoma?
K-*ras*	Kirsten MSV (Ki-MSV)	Rat	$pp21^{ras}$	Sarcoma and erythroleukemia

[a] *fps*, the oncogene of several chicken sarcoma viruses, is derived from a chicken gene, c-*fps*, that is the chicken equivalent of c-*fes*, a gene of cats first discovered as the oncogene *fes* in several feline sarcoma viruses. It was not realized that *fes* and *fps* were homologous when they were assigned different names. Note that AMV-E26 has two oncogenes.
Source: R. Weiss, N. Teich, H. Varmus, and J. Coffin (eds.), 1985. *RNA Tumor Viruses*, 2d ed. Cold Spring Harbor Laboratory, Cold Spring Harbor, N.Y.

■ *Figure 18.17* Structure of three defective transducing viruses (not drawn to scale): (a) Avian myeloblastosis virus (AMV) contains the v-*myb* oncogene, which replaces some of the 3' end of *pol* and most of *env*; (b) Avian defective leukemia virus (DLV) contains the v-*myc* oncogene, which replaces the 3' end of *gag*, all of *pol*, and the 5' end of *env*; (c) Abelson murine leukemia virus (AbMLV) contains the v-*abl* oncogene, which replaces the 3' end of *gag*, and all of *pol* and *env*.

a) Avian myeloblastosis virus (AMV) genomic RNA

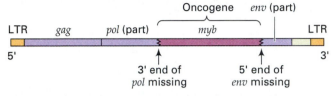

b) Avian defective leukemia virus (DLV) genomic RNA

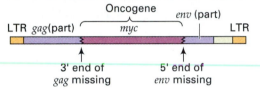

c) Abelson murine leukemia virus (AbMLV) genomic RNA

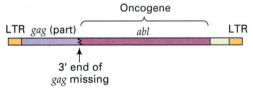

shown that a variety of human tumor cells contain oncogenes. These genes, when introduced into other cells growing in culture, transformed those cells into cancer cells. The human oncogenes were found to be very similar to viral oncogenes that had been characterized earlier, even though viruses did not induce the human cancers involved. Moreover, the human oncogenes were shown also to be closely related to genes found in normally growing cells.

For human and other animal oncogenes, the related genes in normal cells were not identical in sequence but there were long sequences showing high degrees of similarity. That is, all cells contain normal, nontransforming copies of oncogenes called **proto-oncogenes**. When a proto-oncogene is mutated or altered into a transforming oncogene, it is called a **cellular oncogene**, or a *c-onc* (the "c" stands for "cellular," and the *onc* is replaced by the three-letter sequence of the related viral oncogene; e.g., c-*src*). A transducing retrovirus carries a signif-

icantly altered form of a cellular proto-oncogene (now a v-*onc*). When the transducing retrovirus infects a normal cell, the hitchhiking oncogene transforms the cell into a cancer cell. Normal cells can also become transformed into cancer cells even if a tumor virus does not infect them if (1) a proto-oncogene becomes mutated to a cellular oncogene, or (2) chromosomal mutations occur (see Chapter 17) to move the proto-oncogene next to the regulatory region of another gene.

A cellular proto-oncogene differs in a number of ways from its viral oncogene v-*onc* counterpart:

1. Most proto-oncogenes contain introns that are not present in the corresponding v-*onc*. For example, the chicken *src* proto-oncogene gene is over 7 kb long with 12 exons separated by introns (Figure 18.18[a]). In the RSV RNA genome the v-*src* oncogene is 1.7 kb long with no introns. The absence of introns is not surprising since viral propagation involves the production of RNA from the integrated proviral DNA. That is, during RNA production any introns present in the proto-oncogene are removed. Genomic RNA produced by the full-length transcription of the proviral DNA is packaged into virus particles.

Figure 18.18(b) shows the three mRNAs transcribed from the RSV proviral DNA. For each mRNA, transcription starts at a promoter in the left U3 sequence, and the addition of the poly(A) tail is signaled by a sequence in the right R sequence.

2. When a normal cellular proto-oncogene becomes a retroviral oncogene, the retroviral genome is altered significantly (Figure 18.19). After a retrovirus integrates near a cellular proto-oncogene, a deletion may occur which fuses retroviral sequences with proto-oncogene sequences. Usually the deletion event causes the loss of some or all of the *gag*, *pol*, and *env* genes. Then, by transcription (and splicing if introns are present in the proto-oncogene) an mRNA is produced of the retrovirus-oncogene fusion. If this mRNA is packaged into a virus particle along with a normal retrovirus genome, reverse transcriptase can produce a new defective transducing virus by switching RNA templates during RNA replication (Figure 18.19, p. 444).

3. The v-*onc* is transcribed in a different range of cells, and results in larger amounts of mRNA than that produced by the corresponding proto-oncogene. This is because the gene is now under viral control, using the retroviral promoter. Thus there is a significant increase in the amount of protein encoded by the oncogene. The proto-oncogene has usually become mutated when picked up by the retroviruses (e.g., by point mutations, additions,

■ *Figure 18.18* (a) Top: Molecular organization of the chicken *src* proto-oncogene. The gene contains twelve exons (shown as purple boxes). Below is the molecular organization of the Rous sarcoma virus RNA genome to indicate the relationship of nucleotide sequences in the cellular *src* proto-oncogene and v-*src*; v-*src* was produced mostly by intron removal from the cellular *src* proto-oncogene; (b) mRNAs produced by transcription of RSV proviral DNA genome. (The angled lines indicate that the sequences on either side are actually connnected.)

a) Chicken c-*src* proto-oncogene

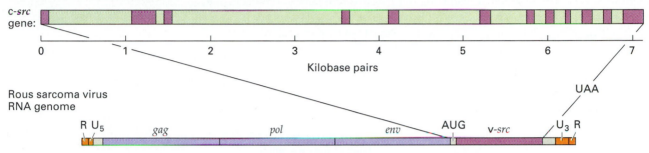

b) RSV proviral DNA transcripts

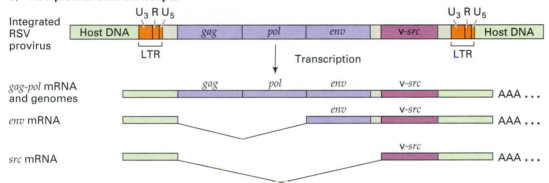

deletions, rearrangements), so there is typically a change in function of the proto-oncogene product.

Protein Products of Proto-Oncogenes There are at least 30 known proto-oncogenes. They code for products that are essential for normal development and cell function. Based on DNA sequence similarities and similarities in amino acid sequences of the protein products, proto-oncogenes fall into several distinct classes, each with a characteristic type of protein product (Table 18.5, p. 445). The major classes of protein products are growth factors (e.g., *sis* product), receptor and nonreceptor protein tyrosine kinases (e.g., *src* product), receptors lacking kinase activity (e.g., *mas* product), membrane-associated G proteins (e.g., H-*ras* product), cytoplasmic protein serine kinases (e.g., *pim-1* product), cytoplasmic regulators (e.g., *crk* product), and nuclear transcription factors (e.g., *myc* product). All of these proteins are involved with cell growth control. Let us consider three examples in brief.

Firstly, let us consider the growth factors. Based on the effect of oncogenes on cell growth and division, it was hypothesized early that proto-oncogenes might be regulatory genes involved with the control of cell multiplication during differentiation. Evidence supporting this hypothesis came when the product of the viral oncogene v-*sis* was shown to be identical to part of platelet-derived growth factor (PDGF, a factor found in blood platelets in mammals), which is released after tissue damage. PDGF affects only one type of cell, fibroblasts, causing them to grow and divide. The fibroblasts are part of the wound-healing system. PDGF itself consists of two polypeptides, one of which is encoded by the proto-oncogene *sis*. The link between PDGF and tumor induction was demonstrated in an experiment in which the cloned PDGF gene was introduced into a cell that normally does not make PDGF; that cell was transformed into a cancer cell.

We can generalize and say that cancer cells can result from the synthesis of growth factors. Such

■ *Figure 18.19*   Model for the formation of a transducing retrovirus. First a wild-type provirus integrates near a cellular proto-oncogene (*c-onc*). Next, a hypothetical deletion-fusion event fuses a *c-onc* exon into the *gag* region of the retrovirus. The LTR in this fusion directs the synthesis of a transcript that undergoes splicing to produce an mRNA for the *gag-onc* fusion protein. If the cell is also infected with a wild-type (helper) retrovirus, as part of the virus life cycle a wild-type RNA and a *gag-onc* RNA can be packaged into one virus particle. In a virus particle containing the two RNAs, reverse transcriptase can start copying the *gag-onc* RNA, then switch to copying the wild-type RNA to produce a new defective transducing retrovirus. (From James D. Watson, et al., *Molecular biology of the gene*, 4th ed. Copyright 1965, 1970, 1976, 1987 by The Benjamin/Cummings Publishing Company, Inc. Reprinted by permission.)

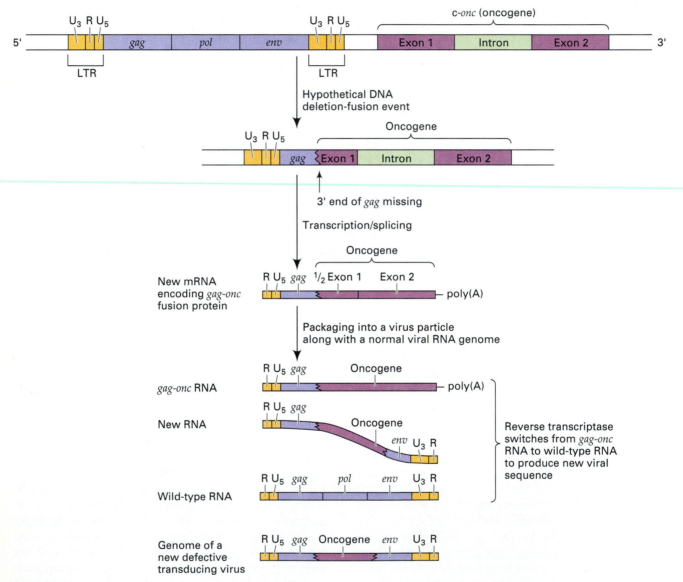

growth factors are not normally produced in those cells, but the introduction of a certain class of transducing retrovirus can cause the growth factor genes to come under viral, not cellular, control.

Secondly, the protein kinases phosphorylate (add phosphate groups) to particular target proteins. Some of the target proteins are presumed to be involved in pathways that regulate the cell divi-

sion cycle. The oncogene form of protein kinases stimulate cell division in conditions in which the normal form would not have.

Thirdly, G proteins participate in a cascade of reactions that transmit a signal from a surface receptor to the cytoplasm. The oncogene forms of G proteins have a limited ability to self-deactivate (as do normal G proteins) so signal transmission con-

Table 18.5
Examples of the Functions of Oncogene Products

CLASS 1—GROWTH FACTORS

sis	PDGF B-chain growth factor
int-2	FGF-related growth factor

CLASS 2—RECEPTOR AND NONRECEPTOR PROTEIN-TYROSINE KINASES

src	Membrane-associated nonreceptor protein-tyrosine kinase
fgr	Membrane-associated nonreceptor protein-tyrosine kinase
fps/fes	Nonreceptor protein-tyrosine kinase
abl/bcr-abl	Nonreceptor protein-tyrosine kinase
kit	(W locus) Truncated stem cell receptor protein-tyrosine kinase

CLASS 3—RECEPTORS LACKING PROTEIN KINASE ACTIVITY

mas	Angiotensin receptor

CLASS 4—MEMBRANE-ASSOCIATED G PROTEINS

Hras	Membrane-associated GTP-binding/GTPase
K-ras	Membrane-associated GTP-binding/GTPase
gsp	Mutant activated form of $G_s \alpha$

CLASS 5—CYTOPLASMIC PROTEIN-SERINE KINASES

pim-1	Cytoplasmic protein-serine kinase
mos	Cytoplasmic protein-serine kinase (cytostatic factor)

CLASS 6—CYTOPLASMIC REGULATORS

crk	SH-2/3 protein that binds to (and regulates?) phosphotyrosine-containing proteins

CLASS 7—NUCLEAR TRANSCRIPTION FACTORS

myc	Sequence-specific DNA-binding protein
fos	Combines with c-jun product to form AP-1 transcription factor
jun	Sequence-specific DNA-binding protein; part of AP-1
erbA	Dominant negative mutant thyroxine (T_3) receptor
ski	Transcription factor?

From T. Hunter, 1991. Cooperation between oncogenes. *Cell* 64:249–270.

tinues abnormally, and uncontrolled cell division results.

Keynote

Some forms of cancer are caused by tumor viruses. The tumor-causing RNA retroviruses contain cancer-inducing genes called oncogenes. Both normal cells and nonviral-induced cancer cells contain sequences that are homologous to the viral oncogenes. Retroviruses that induce tumors have picked up certain normal cellular genes, called proto-oncogenes, while (in most cases) they simultaneously lose part of their genetic information. The cellular proto-oncogenes function in normal cells in various ways to regulate cell division. In the retrovirus these genes have become modified so that the cell's protein product, now produced under viral control, is both quantitatively and qualitatively changed. In addition the cell's protein product is expressed in cells that normally do not contain that particular protein product. The oncogene products, which include growth factors, are directly responsible for the transformation of cells to the cancerous state.

HIV—The AIDS Virus

Acquired Immunodeficiency Syndrome (AIDS) may be induced by the retrovirus *HIV* (*human immunodeficiency virus*) (Figure 18.20[a], p. 446). The capsid (protein coat) of HIV is bullet-shaped. HIV contains complete *gag*, *pol*, and *env* genes, so HIV can produce progeny viruses on its own. HIV also contains five or six other genes that are not oncogenes (Figure 18.20[b]). The exact functions of these genes are not completely understood. HIV infects T lymphocytes, which are part of the immune system, but instead of transforming the cells to the tumorous state, HIV kills them. As a consequence, the immune system is damaged severely or totally destroyed, leading to immunodeficiency. In the

■ *Figure 18.20*   (a) Schematic drawing of a cross-section through an HIV particle. Note that the capsid of this particular retrovirus is bullet-shaped; (b) Organization of the HIV genome.

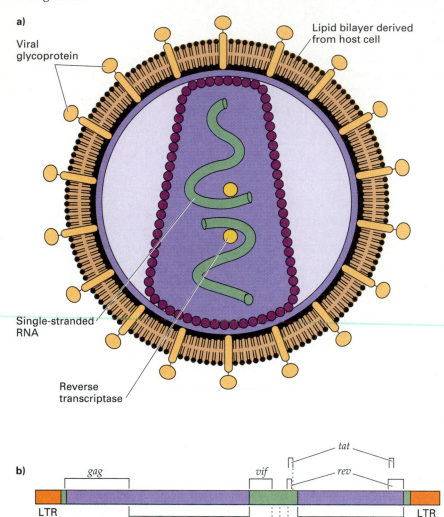

a)

Viral glycoprotein

Lipid bilayer derived from host cell

Single-stranded RNA

Reverse transcriptase

b)

gag

vif

tat

rev

LTR

pol

vpu

vpr

env

LTR

absence of a well-functioning immune system, the patient becomes vulnerable to infections (e.g., bacterial, viral, fungal) and becomes susceptible to numerous cancers. Most patients die from the infections. To date, at least 10 million people in the world are infected with HIV, and a cure is many years away.

Summary

Bacteria and eukaryotic cells contain a variety of transposable elements that have the property of moving from one site to another in the genome. In bacteria there are three major types of transposable elements: insertion sequence (IS) elements, transposons (Tn), and certain bacteriophages. The sim-

plest type of transposable element is an IS element. An IS element typically consists of inverted terminal repeat sequences flanking a coding region, the products of which provide transposition activity. Tn elements are more complex; unlike IS elements, Tns also contain other genes. There are two types of prokaryotic transposons. Composite transposons consist of a central region flanked on both sides by IS elements. The central region contains genes, such as those for drug resistance. The IS elements contain the genes encoding the proteins required for transposition. Noncomposite transposons consist of a central region containing genes (e.g., for drug resistance), but they do not end with IS elements. Instead short repeated sequences are found at their ends that are required for transposition. In these transposons the transposition functions are encoded by genes in the central region.

Transposable elements in eukaryotes resemble bacterial transposons in general structure and transposition properties. Corn transposons often occur as families, each family containing an autonomous element (an element capable of transposing by itself) and one or more nonautonomous elements (elements that can only transpose if the autonomous element of the family is also present in the genome). Interestingly, and perhaps uniquely, the timing and frequency of transposition of corn transposons is developmentally regulated. Some eukaryotic transposons, such as yeast *Ty* elements, transpose via an RNA intermediate (using a transposon-encoded reverse transcriptase). These types of transposons resemble retroviruses in genome organization and other properties, and hence have been called retrotransposons.

The presence of bacterial transposons (IS and Tn elements) and eukaryotic transposons in a cell is usually detected by the changes they bring about in the expression and activities of the genes at or near the chromosomal sites into which they integrate. Gene expression may be increased or decreased if the element inserts into a promoter or other regulatory sequence, mutant alleles of a gene can be produced if an element inserts within the coding sequence of the gene, and various chromosome rearrangements or chromosome breakage events can occur as a result of the transposition. In addition genes can be turned on next to a transposon because of the action of promoters in the transposons themselves. Generally, mutant alleles produced by insertion of a transposable element are unstable, since they revert when the transposable element undergoes a new transposition event.

Some forms of cancer are caused by tumor viruses. The RNA tumor retroviruses are single-stranded RNA viruses that may or may not contain cancer-inducing genes termed oncogenes. When a cell is infected by a retrovirus, a double-stranded DNA copy of the RNA genome is produced by reverse transcriptase, an enzyme brought into the cell by the virus. That DNA integrates into the nuclear genome of the cell, where it is known as a provirus. The provirus directs the synthesis of progeny retroviruses that are released from the cell without causing the death of the cell.

Both normal cells and nonviral-induced cancer cells contain sequences that are related to the viral oncogenes. The model is that tumor-causing retroviruses have picked up certain normal cellular genes—called proto-oncogenes—while simultaneously losing some of their genetic information. In normal cells proto-oncogenes function in various ways to regulate cell differentiation. In the retrovirus these genes have been modified or their expression controlled differently so that the protein product, now synthesized under viral control, is both qualitatively and quantitatively changed, as well as being expressed in cells in which the products are not normally found. These gene products, which include growth factors, are directly responsible for the transformation of cells to the cancerous state.

Analytical Approaches for Solving Genetics Problems

Q18.1 Imagine that you are a corn geneticist and are interested in a gene you call *zma*, which is involved in formation of the tiny hairlike structures on the upper surfaces of leaves. You have a cDNA clone of this gene. In a particular strain of corn that contains many copies of *Ac* and *Ds*, but no other transposable elements, you observe a mutation of the *zma* gene. You want to figure out whether this mutation does or does not involve the insertion of a transposable element into the *zma* gene. How would you proceed? Suggest at least two approaches, and say how your expectations for an inserted transposable element would differ from your expectations for an ordinary gene mutation.

A18.1 One approach would be to make a detailed examination of leaf surfaces in mutant plants. Since there are many copies of *Ac* in the strain, if a transposable element has inserted into *zma* it should be able to leave again, so that the mutation of *zma* would be unstable. The leaf surfaces should then show a patchy distribution of regions with and without the hairlike structures. A simple point mutation would be expected to be stable.

A second approach would be to digest the DNA from mutant plants and DNA from normal plants with a particular restriction endonuclease, run the digested DNAs on a gel and prepare a Southern blot, and probe the blot using the cDNA. If a transposable element has inserted into the *zma* gene in the mutant plants, then the probe should bind to different molecular weight fragments in mutant as compared to normal DNAs. This would not be the case if a simple point mutation had occurred.

Questions and Problems

18.1 Compare and contrast the types of transposable elements in bacteria.

18.2 What are the properties in common between bacterial and eukaryotic transposable elements?

18.3 An IS element became inserted into the *lacZ* gene of *E. coli*. Later, a small deletion occurred in this gene which removed 40 base pairs, starting to the left of the IS element. Ten *lacZ* base pairs were removed, including the left copy of the target site, and the 30 left-most base pairs of the IS element were removed. What will be the consequence of this deletion?

18.4 A geneticist was studying glucose metabolism in yeast, and had deduced both the normal structure of the enzyme glucose-6-phosphatase (G6Pase) and the DNA sequence of its coding region. She had been using a wild-type strain called A to study another enzyme for many generations, when she noticed a morphologically peculiar mutant had arisen from one of the strain A cultures. She grew the mutant up into a large stock and found that the defect in this mutant involved a markedly reduced G6Pase activity. She isolated the G6Pase protein from these mutant cells and found it was present in normal amounts, but had an abnormal structure. The N-terminal 70% of the protein was normal. The C-terminal 30% was present but altered in sequence by a frame shift reflecting the insertion of 1 base pair. The N-terminal 70% and the C-terminal 30% were separated by 111 new amino acids unrelated to normal G6Pase. These amino acids represented predominantly the AT-rich codons (Phe, Leu, Asn, Lys, Ile, Tyr). There were also two extra amino acids added at the C-terminal end. Explain these results.

*18.5 Consider two theoretical yeast transposons, A and B. Each contains an intron. Each transposes to a new location in the yeast genome and then is examined for the presence of the intron. In the new locations, you find that A has no intron, while B does. What can you conclude about the mechanisms of transposon movement for A and B from these facts?

*18.6 An investigator has found a retrovirus capable of infecting human nerve cells. This is a complete virus, capable of reproducing itself, and it contains no oncogenes. People who are infected suffer a debilitating encephalitis. The investigator has shown that when he infects nerve cells in culture with the complete virus, the nerve cells are killed as the virus reproduces, but if he infects cultured nerve cells with a virus in which he has created deletions in the *env* or *gag* genes, no cell death occurs. The investigator is interested in finding ways to bring about nerve cell growth or regeneration in people who have suffered nerve damage. For example, in a patient with a severed spinal cord, nerve regeneration might relieve paralysis. The investigator has cloned the human nerve growth factor gene, and wants to insert it into the genome of his retrovirus from which he has deleted parts of the *env* and *gag* genes. He would then use the engineered retrovirus to infect cultured nerve cells. Adult nerve cells do not normally produce large amounts of nerve growth factor. If he is successful in inducing growth in them without causing any cell death, he would like to move on to clinical trials on injured patients. When the investigator applied for grant support to do this work, his application was denied on grounds that there were inadequate safeguards in the plan. Why might this work be dangerous? What comparisons can you draw between the virus the investigator wants to create and, for example, Avian myeloblastosis virus?

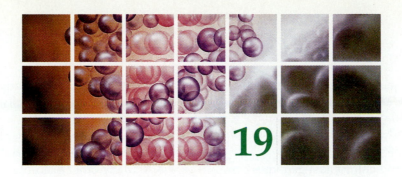

Extranuclear Genetics

449

- Both mitochondria and chloroplasts contain their own DNA genomes. The DNA in most species' mitochondria and all chloroplasts is circular, double-stranded, and supercoiled. The mitochondrial DNA of some species is linear.

- The mitochondrial and chloroplast genomes contain genes for the rRNA components of the ribosomes of these organelles, for many (if not all) of the tRNAs used in organellar protein synthesis, and for a few proteins that remain in the organelles and perform functions specific to the organelles. All other proteins are nuclear-encoded, synthesized on cytoplasmic ribosomes, and imported into the organelles. At least in the mitochondria of some organisms, the genetic code is different from that found in nuclear protein-coding genes.

- The inheritance of extranuclear genes follows rules different from those for nuclear genes: no meiotic segregation is involved, uniparental (and often maternal) inheritance is generally seen, extranuclear genes are nonmappable to the known nuclear linkage groups, and a phenotype resulting from an extranuclear mutation persists after nuclear substitution.

- Not all cases of extranuclear inheritance result from genes on mitochondrial DNA or chloroplast DNA. Many other examples in eukaryotes result from infectious heredity, in which symbiotic, cytoplasmically located bacteria or viruses are transmitted when cytoplasms mix.

- Maternal effect is defined as the predetermination of gene-controlled traits by the maternal *nuclear* genotype prior to the fertilization of the eggs. Maternal effect is different from extranuclear inheritance. The maternal inheritance pattern of extranuclear genes occurs because the zygote obtains most of its organelles (containing the extranuclear genes) from the female parent. In contrast, in maternal effect the inherited trait is controlled by the maternal *nuclear* genotype before the fertilization of the egg and does not involve extranuclear genes.

I N OUR DISCUSSION of eukaryotic genetics up to now, we have analyzed the structure and expression of the genes located on chromosomes in the nucleus and have defined rules for the segregation of nuclear genes. DNA is also found outside the nucleus in two principal organelles, the mitochondrion (found in both animals and plants) and the chloroplast (found only in green plants). The genes in these mitochondrial and chloroplast genomes have become known as extrachromosomal genes, cytoplasmic genes, non-Mendelian genes, organellar genes, or *extranuclear genes*.

Although *extranuclear* is used in the discussions that follow, the term *non-Mendelian* is also useful because extranuclear genes do not follow the rules of Mendelian inheritance like nuclear genes. The application of modern molecular biology techniques has led to rapid advances in knowledge about the organization of extranuclear genomes. In this chapter we first examine some of these advances and then discuss the inheritance patterns of extranuclear genes.

Organization of Extranuclear Genomes

Mitochondrial Genome

Mitochondria, organelles found in the cytoplasm of all aerobic animal and plant cells, are the principal sources of energy in the cell (see Figure 1.6, p. 6). They contain the enzymes of the Krebs cycle, carry out oxidative phosphorylation, and are involved in fatty acid biosynthesis. Some of the properties of mitochondrial DNA, or mtDNA, will now be described.

Structure   Many mitochondrial (mt) genomes are circular, double-stranded, supercoiled DNA molecules. Linear mitochondrial genomes are found in some protozoa and fungi. In many cases the GC content of mtDNA differs greatly from that of the nuclear DNA, which allows mtDNA to be isolated by CsCl equilibrium density gradient centrifugation. No structural proteins are associated with the mtDNA.

The gene content is very similar among mitochondrial genomes in both number and function.

Despite that, the size of the genome varies tremendously from organism to organism. In animals the circular mitochondrial genome is less than 20 kb; in humans it is 16,569 bp. By contrast, the circular yeast mitochondrial genome is about 80 kb. The mitochondrial genomes of plants are even larger, ranging from 250,000 to 2 million base pairs. What is the basis for the difference in size between mitochondrial genomes? In animals the mitochondrial genome is very efficiently organized with very little DNA between genes, whereas in yeast and in plants most of the extra DNA consists of noncoding sequences.

While the informational content of the mitochondrial genome is very small in comparison to the nuclear genome, the relative amount of mitochondrial DNA is actually quite large. Within mitochondria are *nucleoid regions* (similar to those of bacterial cells), each of which contains several copies of the mitochondrial chromosome. Yeast has between 4 and 5 mtDNA molecules per nucleoid, and each mitochondrion has 10 to 30 nucleoids. Since each yeast cell has between 1 and 45 mitochondria per cell, there are between 40 and 6,750 mtDNA molecules per cell. Each mtDNA molecule is 80 kb so there is between 3,200 and 540,000 kb of mitochondrial DNA in the cell, compared with 17,500 kb of genomic DNA in the haploid nucleus.

Replication Replication of mtDNA uses DNA polymerases specific to the mitochondrion. As is the case for replication of nuclear DNA and other DNA, RNA primers are synthesized for initiation.

The mitochondrial DNA replication process occurs throughout the cell cycle, without any preference for the S phase of the cell cycle, which is when nuclear DNA replicates. Some differences appear in the details of mtDNA replication among eukaryotes, and the displacement loop (D loop) model (deduced from observing animal mitochondria *in vivo*) is useful as a general scheme (Figure 19.1). This model is as follows:

In most animals the two strands of mtDNA have different densities, so they are called the H (heavy) and L (light) strands. In the D loop model the synthesis of the new H strand is started at a replication origin for the H strand and forms a D loop structure. When the new H strand has extended about halfway around the molecule, synthesis of the new L strand starts at a second origin. Both strands are completed by continuous replication. The circular DNAs are each converted to a supercoiled form, with approximately 100 superhelical twists.

As far as duplication of the entire organelle is concerned, all the evidence indicates that mitochondria (and chloroplasts) grow and divide rather than assemble from simple components.

Gene Organization of mtDNA Our present-day understanding of the gene organization of mtDNA derives mainly from DNA sequencing experiments. Today the sequence of human (and several other organisms') mtDNA is completely known. Figure 19.2 presents the map of the genes of human mtDNA.

Mitochondrial DNA contains information for a

■ *Figure 19.1* Model for mitochondrial DNA replication that involves the formation of a D loop structure.

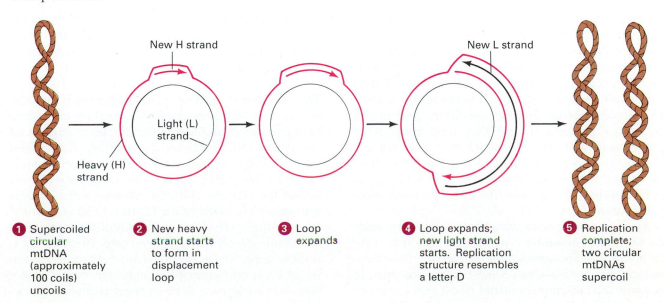

1 Supercoiled circular mtDNA (approximately 100 coils) uncoils

2 New heavy strand starts to form in displacement loop

3 Loop expands

4 Loop expands; new light strand starts. Replication structure resembles a letter D

5 Replication complete; two circular mtDNAs supercoil

■ *Figure 19.2* Map of the genes of human mitochondrial DNA. The outer circle shows the genes transcribed from the H (heavy) strand, and the inner circle shows the genes transcribed from the L (light) strand. The origins of replication (*ori*) for the H and L strands are indicated. 16S and 12S (blue) are the rRNA genes. tRNA genes are shown in purple and protein-coding genes in yellow. Key to abbreviations: ATPase 6 and 8: components of the mitochondrial ATPase complex. COI, COII, and COIII: cytochrome c oxidase subunits. cyt b: cytochrome b. ND1–6: NADH dehydrogenase components (previously URF1–6, where URF = unidentified reading frame).

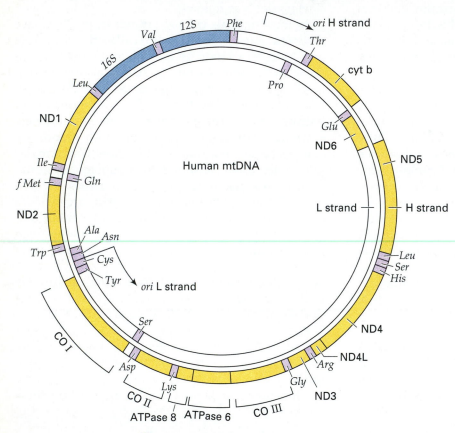

number of mitochondrial components, such as tRNAs, rRNAs, and some of the polypeptide sub-units of the proteins cytochrome oxidase, NADH-dehydrogenase and ATPase. Other components found in the mitochondria are encoded by nuclear genes and must be imported into the mitochondria. These components include the DNA polymerase and other proteins for mtDNA replication, RNA polymerase and other proteins for transcription, ribosomal proteins, protein factors for translation, the aminoacyl tRNA synthetases, and the other polypeptide subunits for the three proteins men-tioned above.

The polypeptides in the mitochondria are made on ribosomes assembled within the organelle. Like cytoplasmic ribosomes and bacterial ribosomes, mitochondrial ribosomes consist of two subunits. In humans the 60S mitochondrial ribosomes consist of 45S and 35S subunits. There are only two rRNAs in

the mitochondrial ribosome: 16S rRNA in the human large subunit and 12S rRNA in the small subunit. There is one gene in the mitochondrial genome for each kind of rRNA.

Transcription of mtDNA   The transcription of mam-malian mtDNA is unusual. Rather than each gene being transcribed independently, both strands of the entire mitochondrial genome are transcribed into two single strands. The origins for both strands are near the DNA replication origin for the H strand (see Figure 19.2). In addition some shorter transcripts are made of the H strand that stop at the far end of the 16S rRNA gene. All of the transcripts are made by an RNA polymerase encoded by a nuclear gene. The exact mechanism of transcription initiation is not understood, although the promoter elements are known to be in short sequences flank-ing the transcription start sites.

How are the large RNA transcripts processed to produce the mature mRNAs, rRNAs, and tRNAs? Notice in the organization map shown in Figure 19.2 that most of the genes encoding the rRNAs and the mRNAs are punctuated by tRNA genes. A current model for processing has the tRNAs playing a crucial role. The tRNA sequences fold up into characteristic cloverleaf shapes that are recognized by specific enzymes that cut the tRNAs out of the transcript. Since there are essentially no gaps between genes, the removal of the tRNAs liberates essentially complete mRNAs and rRNAs. The processed transcripts are then modified to produce the mature RNAs: a poly(A) tail is added to the 3' ends of the mRNAs, and CCA is added to the 3' ends of the tRNAs.

Interestingly, the DNA sequences for some of the mitochondrial mRNAs do not encode complete chain termination codons. Instead, the processed transcripts end with either U or UA. The subsequent addition of a poly(A) tail completes the missing part(s) of a UAA stop codon.

The transcription and RNA processing mechanisms just described are characteristic only of animal mitochondria. The much larger mitochondrial genomes of yeast and plants do not have tRNA genes separating other genes. Rather, transcription starts from many promoters around the mitochondrial genome and the long transcripts are then processed by enzymes to give the individual mature RNAs.

Translation in the Mitochondria The mRNAs in animal mitochondria have no 5' cap and the start codon is very near the 5' end of the molecule so that there is virtually no 5' leader sequence. As a result, the initiation of translation must be quite different from that for cytoplasmic mRNAs where both of those features are present. In other words, mitochondrial ribosomes in animals must bind to the mRNAs and orient themselves to start translation in a unique way. Again, yeast and plant mitochondrial mRNAs are more "traditional" in sequence; although there is no 5' cap, there is a 5' leader so translation initiation may be like that in the cytoplasm. In all mitochondria a mitochondrial tRNA.fMet is used in the initiation of protein synthesis and special mitochondrial initiation factors (IFs), elongation factors (EFs), and release factors (RFs) are used for the translation process.

For protein synthesis, only plant mitochondria use the universal nuclear genetic code. Mitochondria of other organisms have differences from that universal code, although there is no one pattern for the differences. For example, the differences in the human mitochondrial code from the nuclear code are as follows (see Figure 19.3): (1) AUA and AUG both encode methionine instead of only AUG as in the nuclear code. Both of these codons can function as initiation codons; (2) UGA in mitochondria specifies tryptophan while in the nucleus UGA is a stop codon; and (3) AGA and AGG are stop codons in the mitochondria and arginine codons in the nucleus.

Interestingly, the mitochondrial genetic codes are not the same in all organisms. Further, differences in the nuclear genetic code have also been found in ciliated protozoans. The genetic code is not as universal as was originally thought.

There are also differences in mitochondria with respect to the tRNAs and how they read the mRNAs. We learned in Chapter 12 about base-pairing wobble, in which one tRNA may read more than one codon (see Table 12.1, p. 273). The maximum number of codons any particular cytoplasmic tRNA could recognize with wobble is three. Wobble is "more liberal" in the mitochondria, so some tRNAs are able to recognize four different codons. Figure 19.3 shows as shaded boxes the tRNAs that read the mitochondrial code. As a result of the extended wobble, only 22 tRNA genes are needed in mammalian mitochondria to read all sense codons, while 32 tRNAs are needed in the cytoplasm.

Chloroplast Genome

Chloroplasts are cellular organelles found only in green plants and are the site of photosynthesis in the cells containing them. Chloroplasts are characterized by a double membrane surrounding an internal, chlorophyll-containing lamellar structure embedded in a protein-rich stroma (refer to Figure 1.8, p. 8). Like mitochondria, chloroplasts contain their own genomes, although we do not know as much about the chloroplast (cp) genome as we do about the mitochondrial genome.

Structure and Replication The structure of the chloroplast (cp) genome is generally like the structure of mitochondrial genomes. In all cases the DNA is double-stranded, circular, devoid of structural proteins, and supercoiled. An electron micrograph of cpDNA is shown in Figure 19.4, p. 455. In many cases the GC content of cpDNA differs greatly from that of the nuclear and mitochondrial DNA, which allows cpDNA to be isolated by CsCl equilibrium density gradient centrifugation.

Chloroplast DNA is much larger than animal mtDNA, with a size between 120 kb and 180 kb. The DNA sequences of the chloroplast genomes of liverwort, tobacco, and rice have been completely

■ Figure 19.3 Genetic code of the human mitochondrial genome compared with the "universal" code. Each shaded box indicates that one tRNA is used to read those codons within. Highlighted are the codons that have different coding properties in the human mitochondrion than in the nuclear genome of all eukaryotic organisms or in the genomes of prokaryotic organisms.

First letter	Second letter: U	Univ.	Human mito.	C	Univ.	Human mito.	A	Univ.	Human mito.	G	Univ.	Human mito.	Third letter
U	UUU	Phe (1)	Phe (1)	UCU	Ser (3)	Ser (1)	UAU	Tyr (1)	Tyr (1)	UGU	Cys (1)	Cys (1)	U
	UUC			UCC			UAC			UGC			C
	UUA	Leu (2)	Leu (1)	UCA			UAA	Stop	Stop	UGA	Stop	**Trp (1)**	A
	UUG			UCG			UAG	Stop	Stop	UGG	Trp (1)	Trp	G
C	CUU	Leu (3)	Leu (1)	CCU	Pro (3)	Pro (1)	CAU	His (1)	His (1)	CGU	Arg (2)	Arg (1)	U
	CUC			CCC			CAC			CGC			C
	CUA			CCA			CAA	Gln (2)	Gln (1)	CGA			A
	CUG			CCG			CAG			CGG			G
A	AUU	Ile (2)	Ile (1)	ACU	Thr (3)	Thr (1)	AAU	Asn (1)	Asn (1)	AGU	Ser (1)	Ser (1)	U
	AUC			ACC			AAC			AGC			C
	AUA		**Met (1)**	ACA			AAA	Lys (1)	Lys (1)	AGA	Arg (1)	**Stop**	A
	AUG	Met (1)	Met	ACG			AAG			AGG		**Stop**	G
G	GUU	Val (2)	Val (1)	GCU	Ala (2)	Ala (1)	GAU	Asp (1)	Asp (1)	GGU	Gly (3)	Gly (1)	U
	GUC			GCC			GAC			GGC			C
	GUA			GCA			GAA	Glu (1)	Glu (1)	GGA			A
	GUG			GCG			GAG			GGG			G

determined. The tobacco genome, for instance, is 155,844 bp long. The chloroplast genome contains a lot of noncoding DNA sequences.

The number of copies of cpDNA per chloroplast varies from species to species. In all cases there are multiple copies per chloroplast and, like in mitochondria, these copies are found in nucleoid regions that are also present in multiple copies. For example, leaf cells of the garden beet have between 4 and 8 cpDNA molecules per nucleoid, from 4 to 18 nucleoids per chloroplast, and about 40 chloroplasts per cell, giving almost 6000 cpDNA molecules per cell.

It is not yet known exactly how replication of cpDNA occurs. Chloroplasts themselves grow and divide in the same way as do mitochondria.

Gene Organization of cpDNA   The chloroplast genome contains genes for the chloroplast rRNAs and may contain genes for some ribosomal proteins, for a number of tRNAs, and for some chloroplast enzymes and membrane proteins that are needed for electron transport in the process of photosynthesis. All of the mRNAs transcribed from chloroplast genes are translated by chloroplast ribosomes. Other proteins found in the chloroplast are encoded by nuclear genes; for instance, ribosomal proteins not encoded by chloroplast genes, translation factors, and so on.

Chloroplast Ribosomes and Protein Synthesis
Chloroplast protein synthesis uses ribosomes that are completely distinct from mitochondrial ribo-

■ *Figure 19.4*   Electron micrograph of a chloroplast (cp) DNA molecule.

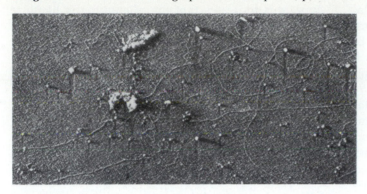

somes and cytoplasmic ribosomes. They have a sedimentation coefficient of 70S and consist of two unequal-sized subunits, 50S and 30S.

The large subunit of the chloroplast ribosome has two rRNAs, 23S and 5S. The small subunit has a 16S rRNA. Many of the ribosomal proteins are made on cytoplasmic ribosomes and are therefore encoded in the nuclear genome. A few ribosomal proteins are made in the chloroplast.

Protein synthesis in chloroplasts is similar to the process in prokaryotes. Formylmethionyl tRNA is used to initiate all proteins, and the chloroplast uses its own initiation factors (IFs), elongation factors (EFs), and release factors (RFs).

K e y n o t e

Both mitochondria and chloroplasts contain their own DNA genomes. The DNA in most species' mitochondria and all chloroplasts is circular, double-stranded, and supercoiled. The mitochondrial DNA of some species is linear. The mitochondrial and chloroplast genomes contain genes for the rRNA components of the ribosomes of these organelles, for many (if not all) of the tRNAs used in organellar protein synthesis, and for a few proteins that remain in the organelles and perform functions specific to the organelles. All other proteins are nuclear-encoded, synthesized on cytoplasmic ribosomes, and imported into the organelles. At least in the mitochondria of some organisms, the genetic code is different from that found in nuclear protein-coding genes.

The Origin of Mitochondria and Chloroplasts

How might mitochondria and chloroplasts have arisen? The widely accepted **endosymbiont hypoth-**esis is that mitochondria and chloroplasts originated as free-living prokaryotes that invaded primitive eukaryotic cells and established a mutually beneficial (symbiotic) relationship. According to this hypothesis, eukaryotic cells started out as anaerobic organisms that lacked mitochondria and chloroplasts. At some point in their evolution, over a billion years ago, the eukaryotic cells established a symbiotic relationship with a bacterium. Then over time, the oxidative phosphorylation activities of the bacterium became used for the benefits of the eukaryotic cell. Eventually the eukaryotic cell became dependent on the intracellular bacterium for survival. The chloroplasts of plants and algae are hypothesized to have occurred later by the ingestion by a eukaryotic cell of an oxygen-producing photosynthetic bacterium.

As we have just discussed, many proteins found in mitochondria and chloroplasts are encoded by nuclear genes. Thus further evolution of mitochondria and chloroplasts must have involved the extensive transfer of genes from the organelles to the nuclear DNA.

Rules of Extranuclear Inheritance

Since the pattern of inheritance shown by genes located in organelles differs strikingly from the pattern shown by nuclear genes, the term **non-Mendelian inheritance** is used when we are discussing extranuclear genes. In fact if results are obtained from genetic crosses that do not conform to predictions based on the inheritance of nuclear genes, extranuclear inheritance is a good possibility.

Here are the four main characteristics of *extranuclear inheritance*:

1. In higher eukaryotes the results of reciprocal crosses involving extranuclear genes are not the

same as reciprocal crosses involving nuclear genes. (Recall from Chapters 2 and 3 that in a reciprocal cross, one selects a pair of contrasting genotypes, then conducts two crosses, reversing the sexes of the parent in each cross. For example, if A and B represent contrasting genotypes, A♀ × B♂ and B♀ × A♂, would be a pair of reciprocal crosses.)

Extranuclear genes usually show **uniparental inheritance** from generation to generation. In uniparental inheritance all progeny (both males and females) have the phenotype of only one parent. For animals the mother's phenotype is expressed exclusively, a phenomenon called **maternal inheritance.** Maternal inheritance occurs because the amount of cytoplasm in the female gamete, the egg, usually greatly exceeds that in the male gamete, the sperm. Therefore, the zygote receives most of its cytoplasm (containing the extranuclear genes in the organelles; i.e., the mitochondria and, where applicable, the chloroplasts) from the female parent and a negligible amount from the male parent.

In contrast, the results of reciprocal crosses between a wild-type and a mutant strain are identical if the genes are located on nuclear chromosomes. One exception for nuclear genes occurs when sex-linked genes are involved (see Chapter 3), but even then the results are distinct from those for extranuclear inheritance. In the case of a recessive sex-linked, nuclear gene, a cross between a wild-type female and a mutant male will produce F_1s that are all wild type. In the F_1s of the reciprocal cross between a mutant female and a wild-type male, however, the females will be wild type and the males will be mutant.

2. Extranuclear genes cannot be mapped to the chromosomes in the nucleus. If the nuclear chromosomes of an organism are well mapped, any new mutations of nuclear genes can be mapped by standard genetic mapping crosses. If a new mutation does not show linkage to any of the nuclear genes, it is probably an allele of an extranuclear gene.

3. Ratios typical of Mendelian segregation are not found. For nuclear gene mutations the rules of Mendelian segregation predict that all F_1 progeny of a cross between a homozygous mutant (recessive) and a wild type will be wild type in phenotype and that a 3:1 ratio of wild type: mutant phenotypes will be seen in the F_2. Such a segregation ratio is not characteristic of extranuclear genes—the uniparental inheritance pattern characteristic of extranuclear genes is clearly a deviation from Mendelian segregation.

4. Extranuclear inheritance is not affected by substituting a nucleus with a different genotype. When a particular phenotype persists after the nucleus is replaced with one with a different genotype, this indicates that the phenotype is likely to be controlled by an extranuclear genome.

Keynote

The inheritance of extranuclear genes follows rules different from those for nuclear genes. No meiotic segregation is involved, uniparental (and often maternal) inheritance is generally seen, extranuclear genes are not mappable to the known nuclear-linkage groups, and the phenotype persists even after nuclear substitution.

Examples of Extranuclear Inheritance

In this section we shall discuss the properties of a selected number of mutations in extranuclear chromosomes in order to illustrate the principles of extranuclear inheritance.

Leaf Variegation in the Higher Plant *Mirabilis jalapa*

A leaf variegation trait in a strain of the plant *Mirabilis jalapa* (also called the four o'clock, or the marvel of Peru) involves extranuclear inheritance. The strain, called *albomaculata*, has shoots that have green leaves with yellowish-white patches (variegated leaves) as well as occasional shoots that are wholly green or wholly yellow-white (see Figure 19.5).

All types of shoots (variegated, green, white) give rise to flowers, so crosses can be made by taking pollen from one type of flower and fertilizing a flower on the same or different type of shoot. The flowers on green shoots give only green progeny, regardless of whether the pollen is from green, white, or variegated shoots. Flowers on white shoots give only white progeny, regardless of the source of the pollen. Flowers on variegated shoots all produce three types of progeny—completely green, completely white, and variegated—regardless of the source of the pollen.

These results indicate there is maternal inheritance of the leaf variegation trait. That is, the *progeny phenotype in each case was the same as that of the maternal parent* (the color of the progeny shoots resembled the color of the parental flower shoot), a property indicative of maternal inheritance.

The basis for the green color of higher plants is the presence of the green pigment chlorophyll in large numbers of chloroplasts. Green shoots in *Mirabilis* have a normal complement of chloroplasts.

■ *Figure 19.5*   Leaf variegation in the four o'clock, *Mirabilis jalapa*. Shoots that are all green, all white, and variegated are found on the same plant, and flowers may form on any of these shoots.

White shoots have abnormal, colorless chloroplasts called leukoplasts. Leukoplasts lack chlorophyll and hence are incapable of carrying out photosynthesis.

The simplest explanation for the inheritance of leaf color in the *albomaculata* strain of *Mirabilis jalapa* is that the abnormal chloroplasts are defective as a result of a mutant gene in the cpDNA. During plant growth the two types of organelles—chloroplasts and leukoplasts—segregate so that a particular cell and its progeny cells may receive only chloroplasts (leading to green tissues), only leukoplasts (leading to white tissues), or a mixture of chloroplasts and leukoplasts (leading to variegation). This model is shown in Figure 19.6.

In a variegated plant, then, the white shoots derive from cells in which, through segregation, only leukoplasts are present. Green shoots are similarly derived from cells containing only chloroplasts. Variegated shoots are generated from cells that contain both chloroplasts and leukoplasts so, when they segregate later, patches of white tissue are produced on the shoots and leaves.

This simple model has three assumptions. One is that the pollen contributes essentially no cytoplasmic information (no chloroplasts or leukoplasts) to the egg. This is reasonable since the egg is much larger than the pollen. Thus in the zygote the extranuclear genetic determinants in the egg over-

whelm those in the pollen. The second assumption is that the chloroplast genome replicates autonomously and, by growth and division of plastids (the general term for photosynthetic organelles), the wild-type and mutant cpDNA molecules segregate randomly to the new plastids so that pure plastid lines can be generated from a mixed line. The third is that segregation of plastids to daughter cells is random, so that some daughters receive chloroplasts, some receive leukoplasts, and some receive mixtures.

Keynote

The leaf color phenotypes in a variegated strain of the four o'clock, *Mirabilis jalapa*, show maternal inheritance. The abnormal chloroplasts in white tissue are the result of a mutant gene in the cpDNA, and the observed inheritance patterns follow the segregation of cpDNA.

Yeast *Petite* Mutants

Yeast grows as single cells. Thus on solid medium, yeast forms discrete colonies consisting of many thousands of individual cells clustered together. Yeast can grow with or without oxygen. In the absence of oxygen yeast obtains energy for cell growth and cellular metabolism through fermentation metabolism, in which the mitochondria are not involved.

Characteristics of *Petite* Mutants   In the late 1940s Boris Ephrussi and his colleagues studied the growth of yeast cells on a solid medium that allowed growth to occur either by aerobic respiration or by fermentation. Occasionally, they noticed a colony that was much smaller than wild-type colonies. Since Ephrussi was French, the small colonies were called *petites* (French for "small"), and the wild-type colonies were called *grandes* (French for "big"). Ephrussi found that *petite* colonies were small not because the individual cells were small but because the growth rate of the mutant *petite* strain is significantly slower than that of the wild-type. Thus there are fewer cells in the *petite* colonies. The *petites* are essentially incapable of carrying out aerobic respiration; they must obtain their energy primarily from fermentation, which is a relatively inefficient process.

Nuclear, Neutral, and Suppressive *Petite* Mutants   Some *petites* have as their genetic basis a mutation in the nuclear genome—a characteristic that is not

■ *Figure 19.6*   Model for the inheritance of leaf color in *Mirabilis jalapa*.

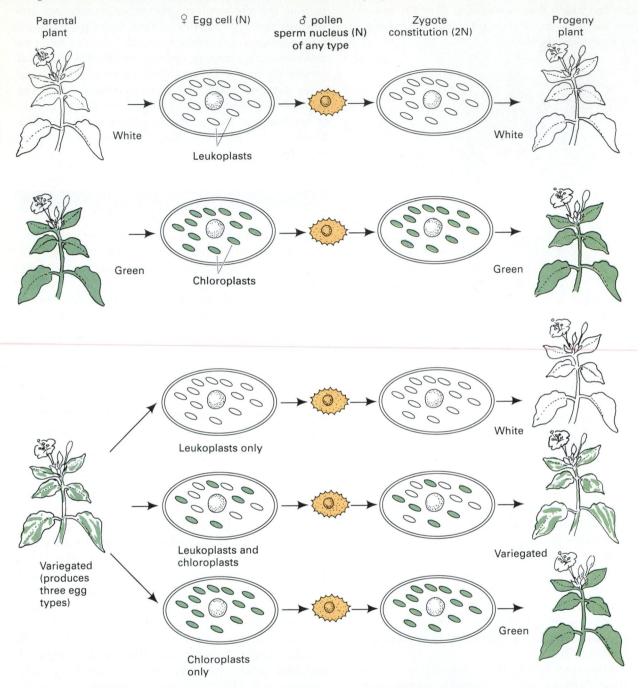

Parental plant	♀ Egg cell (N)	♂ pollen sperm nucleus (N) of any type	Zygote constitution (2N)	Progeny plant

White — Leukoplasts — White

Green — Chloroplasts — Green

Variegated (produces three egg types)

Leukoplasts only — White

Leukoplasts and chloroplasts — Variegated

Chloroplasts only — Green

surprising since some subunits of some mitochondrial proteins are encoded by nuclear genes. The mutations in these *nuclear petites* (also called segregational *petites*) are called *pet⁻*. When a *pet⁻* mutant is crossed with the wild type (*pet⁺*), the diploid is *pet⁺/pet⁻* which produces *grande* colonies. When a *pet⁺/pet⁻* cell goes through meiosis, half the ascospores produced give rise to *grande* colonies and the other half give rise to *petite* colonies. This 1:1 ratio of colony phenotypes is characteristic of Mendelian inheritance.

One class of *petites* that exhibits the traits of extranuclear inheritance is the *neutral petite* class (symbolized [*rho⁻N*], where the brackets indicate a cytoplasmic genotype). Figure 19.7(a) shows the inheritance pattern of neutral *petites*. When a neutral *petite* is crossed with wild type ([*rho⁺N*]), the resulting [*rho⁺N*]/[*rho⁻N*] diploids all produce *grande* colonies. When these diploids go through meiosis, the ascospores produced all give rise to *grande* colonies; that is, no *petite* colonies are seen in the progeny. (The name *neutral*, then, refers to that

■ *Figure 19.7* Extranuclear inheritance pattern of (a) *neutral petites* and (b) suppressive petites in yeast.

a) Inheritance of neutral *petites*

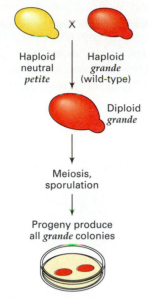

b) Inheritance of suppressive *petites*

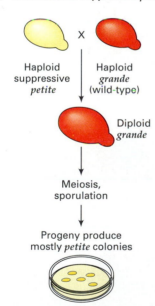

found in the cytoplasm. In genetic crosses of neutral *petites* with wild-type yeasts, normal (wild-type) mitochondria form a population from which new, normal (wild-type) mitochondria are produced in all progeny, and hence the *petite* trait is lost after one generation.

The second class of *petites* that shows extranuclear inheritance is the *suppressive petites* (symbolized [rho⁻S]). The suppressive *petites* are different from the neutrals because they *do* have an effect on the wild type. Most *petite* mutants are of the suppressive type. Like the neutral and the nuclear *petites*, suppressive *petites* are deficient in mitochondrial protein synthesis.

The inheritance pattern of suppressive *petites* is different from those of nuclear and neutral *petites* (Figure 19.7[b]). While there is variation from one suppressive *petite* strain to another, generally most of the ascospores produced by meiosis of the diploid give rise to *petite* colonies. In other words, somehow the suppressive *petite* mitochondria are favored over *grande* mitochondria. Clearly, suppressive *petites* show non-Mendelian inheritance.

The suppressive *petite* mutations start out as deletions of part of the mtDNA. Then, by some correction mechanism, sequences that are not deleted become duplicated until the normal amount of mtDNA is restored. During these events rearrangements of the mtDNA sometimes occur. Since the protein-coding genes in the mitochondrial genome are widely scattered, these deletions and rearrangements lead to deficiencies in the enzymes involved in aerobic respiration, and a *petite* colony results. Suppressive *petites* could have a suppressive effect over normal mitochondria in one of two ways: (1) Suppressive mitochondria could replicate faster than normal mitochondria and simply overrun them in the cell. The variations seen in suppressiveness between strains would then reflect the relative competitiveness of suppressive mitochondria in replication; (2) Suppressive and normal mitochondria could fuse and recombination between rearranged suppressive mtDNA and normal mtDNA could severely alter the latter's gene organization and produce a *petite* phenotype.

fact that this class of *petites* does not affect the wild type.) This result is a classical example of *uniparental inheritance*, in which all progeny have the phenotype of only one parent. *This is not maternal inheritance, however, since the two haploid cells that fuse to produce the diploid are the same size and contribute equally to the cytoplasm.*

Essentially 99–100 percent of the mtDNA is missing in *neutral petites*. Not surprisingly, the neutral *petites* cannot make mitochondrion-encoded proteins and, therefore, cannot perform mitochondrial functions. They survive, however, because yeast fermentation processes are carried out by enzymes

K e y n o t e

Yeast *petite* mutants grow slowly and have various deficiencies in mitochondrial functions as a result of alterations in mitochondrial DNA. The *petite* mutants show extranuclear inheritance, with particular patterns of inheritance varying with the type of *petite* involved.

Extranuclear Genetics of *Chlamydomonas*

The unicellular alga *Chlamydomonas reinhardi* is a motile, haploid eukaryotic organism. Motility is the result of the activity of two flagella. Of interest to us in this section is the single chloroplast *Chlamydomonas* contains.

Chlamydomonas has two mating types, + and −. Mating occurs by the fusion of two equal-sized cells (which therefore contribute an equal amount of cytoplasm), one of each mating type. A thick-walled cyst develops around the resulting diploid zygote. After meiosis four haploid progeny cells are produced, and since mating type is determined by a nuclear gene, a 2:2 segregation of +:− mating types results.

The erythromycin resistance ([ery^{-r}]) trait is inherited in an extranuclear manner in *Chlamydomonas*. Wild-type *Chlamydomonas* cells are erythromycin-sensitive ([ery^{-s}]). If we cross + [ery^{-r}] × − [ery^{-s}],

■ *Figure 19.8* Uniparental inheritance in *Chlamydomonas*: (a) From a cross of + [ery^{-r}] × − [ery^{-s}], 95 percent of the zygotes give tetrads that segregate 2:2 for the nuclear mating-type genes, and 4:0 for the extranuclear gene carried by the + parent (here, [ery^{-r}]); (b) From the reciprocal cross of − [ery^{-r}] × + [ery^{-s}], 95 percent of the zygotes give tetrads segregating 2+:2− and 0 [ery^{-r}]:4 [ery^{-s}], again showing uniparental inheritance for the extranuclear trait of the + parent.

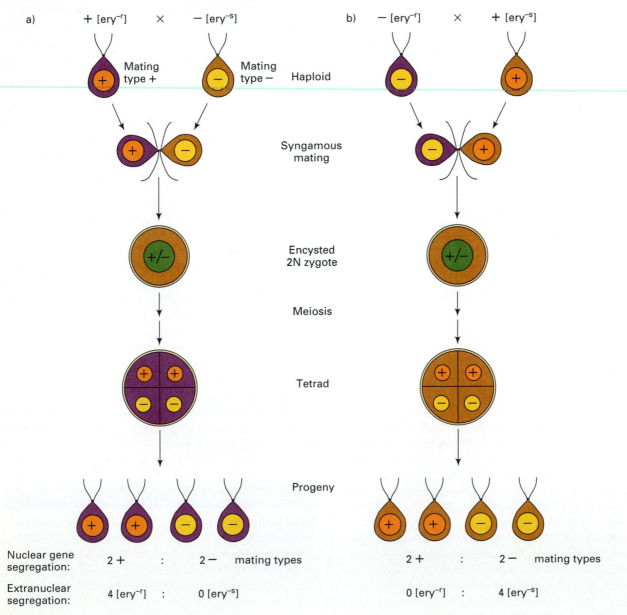

all of the progeny are erythromycin-resistant (Figure 19.8[a]). This is uniparental inheritance, an attribute we have come to expect of extranuclear traits. From the reciprocal cross − [ery⁻ʳ] × + [ery⁻ˢ] (Figure 19.8[b]), all of the progeny are erythromycin sensitive; again, this is uniparental inheritance. Thus in this system, even though both parents contribute equal amounts of cytoplasm to the zygote, the progeny *always* resemble the + parent.

A number of *Chlamydomonas* mutants in addition to ery⁻ʳ show uniparental inheritance, with progeny of crosses *always* resembling the phenotype of the + parent. These mutations are in genes on the chloroplast genome. Some of the evidence is as follows. The density of the chloroplast DNA is different from that of nuclear DNA and from that of mitochondrial DNA, so it is possible to study cpDNA selectively. By using density labels, it is possible to make the cpDNA of the + and − parents different in density (c.f., Meselson and Stahl experiment, Chapter 10, pp. 208–210). When +/− zygotes are examined for their cpDNA, the cpDNA from the − parent has always disappeared. This loss of cpDNA from the − parent clearly parallels the loss of uniparental genes (such as the erythromycin genes) of the − parent in the progeny of the genetic crosses.

Keynote

A number of genes have been identified and mapped in the *Chlamydomonas* chloroplast genome. These genes are inherited in an extranuclear manner.

Human Genetic Diseases and Mitochondrial DNA Defects

A number of human genetic diseases have been shown to be the result of mtDNA gene mutations. These diseases show maternal inheritance as would be expected for defects involving mitochondria. The following are some brief, selected examples of these diseases.

Leber's hereditary optic neuropathy (LHON): This disease usually affects young adults, resulting in complete or partial blindness from optic nerve degeneration. Men are affected more frequently than women. Missense mutations in the mitochondrial genes for the proteins ND1, ND2, ND4, ND 5, cytb, and COI (see Figure 19.2) can lead to LHON. Those proteins are included in mitochondrial electron transport chain enzyme complexes. The electron transport chain drives cellular ATP production by oxidative phosphorylation. It appears that death of the optic nerve in LHON is a common result of oxidative phosphorylation defects, brought about here by the inhibition of the electron transport chain.

Kearns-Sayre syndrome: People with this syndrome have encephalomyopathy, a brain disease. The cause of the syndrome is large deletions at various positions in the mtDNA. One model is that each deletion removes one or more tRNA genes whose products are essential for mitochondrial protein synthesis. In some unknown way, this leads to development of the syndrome.

Myoclonic epilepsy: People with this disease exhibit dementia, deafness, and seizures. The mitochondria of these individuals are abnormal in appearance. The disease is caused by a single nucleotide substitution in the gene for tRNA.Lys. The mutated tRNA has a detrimental effect on mitochondrial protein synthesis, and somehow this gives rise to the various phenotypes of the disease.

In most diseases resulting from mtDNA defects, the cells of affected individuals have a mixture of mutant and normal mitochondria. This condition is known as **heteroplasmy**. Characteristically the proportions of the two mitochondrial types vary from tissue to tissue and from individual to individual within a pedigree. The severity of the disease symptoms correlates approximately with the relative amount of mutant mitochondria.

Infectious Heredity—Killer Yeast

The examples of extranuclear inheritance we have discussed so far have all resulted from mutations in the organelles—in mtDNA or cpDNA. There are other examples of eukaryotic extranuclear inheritance that are due to the presence of cytoplasmic bacteria or viruses coexisting with the eukaryotes in a symbiotic relationship. One example is the killer phenomenon in yeast. Some yeast strains secrete a killer toxin that will kill sensitive strains of yeast. Killer strains are immune to their own toxin. The killer phenomenon results from the presence in the cell's cytoplasm of two types of viruses, L and M (Figure 19.9[a]). Neither cause deleterious effects on the host cell.

The L virus consists of a protein capsid within which is the double-stranded (ds) RNA genome called L-dsRNA. L-dsRNA encodes the capsid proteins for both L and M viruses and the viral polymerase required for viral RNA replication. So since all viral particles are encoded by an L-dsRNA, M viruses are only found in cells if L viruses are also

■ *Figure 19.9* The killer phenomenon in yeast: (a) Killer yeast contains two virus types, L and M, each of which contains a double-stranded RNA genome. L-dsRNA encodes both virus particles and the replication enzyme required for L and M virus replication. M-dsRNA encodes the killer toxin; (b) Sensitive yeast, which can be killed by killer toxin, either have L viruses but no M viruses, or have neither virus type.

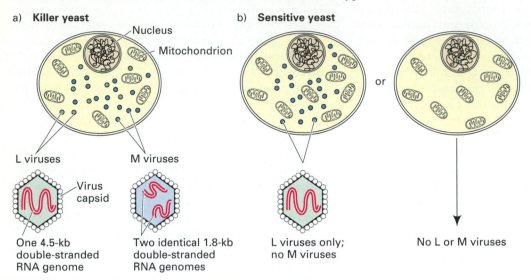

present. The M virus consists of a virus particle encoded by L-dsRNA, and two copies of a double-stranded RNA genome called M-dsRNA. M-dsRNA encodes the killer toxin protein, which is secreted from the cell. The same protein confers immunity on the killer cell.

Sensitive yeast cells are cells that can be killed by the killer toxin. There are two types as shown in Figure 19.9(b). One type has only L viruses; the other has neither L nor M viruses. In both of these types no immunity function is produced because killer toxin is not made.

Unlike most viruses, the yeast L and M viruses are not found outside of the cell, so sensitive yeast cells cannot be infected by viruses that invade from outside. Rather, virus transmission from yeast to yeast occurs whenever there is cytoplasmic mixing, most commonly when two yeast cells mate. All progeny of the mating will inherit copies of the viruses in the parental cells, illustrating an infectious mechanism of cytoplasmic inheritance. For example, if a killer yeast mates with a sensitive yeast that lacks viruses (obviously before killing it), the resulting diploid will be a killer because of the presence of both L and M viruses. When ascospores are produced, both virus types will be distributed to them. Thus the ascospores will all give rise to cells that are killers. The killer phenomenon in yeast, then, is a case of extranuclear, uniparental inheritance resulting from infectious heredity of virus particles.

Keynote

Not all cases of extranuclear inheritance result from genes on mtDNA or cpDNA. Many other examples in eukaryotes result from infectious heredity in which symbiotic, cytoplasmically located bacteria or viruses are transmitted when cytoplasms mix. The killer phenomenon in yeast is such an example: it results from the infectious heredity of cytoplasmically located viruses.

Maternal Effect

The maternal inheritance pattern of extranuclear genes is distinct from the phenomenon of **maternal effect,** which is defined as the predetermination of gene-controlled traits by the maternal *nuclear* genotype prior to the fertilization of the egg. That is, in maternal inheritance the progeny always have the maternal phenotype, whereas in maternal effect the progeny always have the phenotype specified by the maternal nuclear *genotype. Maternal effect does not involve any extranuclear genes and is discussed here to make the distinction from extranuclear inheritance clear.*

Maternal effect is seen, for instance, in the inheritance of the direction of coiling of the shell of the snail *Limnaea peregra* (Figure 19.10). The shell-coiling trait is determined by a single pair of *nuclear* alleles: *D* for coiling to the right (dextral coiling)

■ *Figure 19.10*   The snail, *Limnaea peregra*.

and *d* for coiling to the left (sinistral coiling). The *D* allele is completely dominant to the *d* allele and the shell-coiling phenotype is *always determined by the genotype of the mother*. The latter is shown by the results of reciprocal crosses between a true-breeding, dextral-coiling, and a sinistral-coiling snail (Figure 19.11). All the F₁s have the same genotype since a nuclear gene is involved, yet the *phenotype* is different for the reciprocal crosses.

In the cross of a dextral (*D/D*) female with a sinistral (*d/d*) male (Figure 19.11[a]), the F₁s are all *D/d* in genotype and dextral in phenotype. Selfing the F₁ produces F₂s with a 1:2:1 ratio of *D/D*, *D/d*, and *d/d* genotypes. *All* of the F₂s are dextral, even the *d/d* snails whose genotype would seem to indicate sinistral phenotype. Here is our first encounter with maternal effect; the *d/d* snails have a coiling phenotype not specified by their genotype, but one specified by their mother's genotype (*D/d*). Selfing the F₂ snails gives F₃ progeny, 3/4 of which are dextral and 1/4 of which are sinistral. The latter are the *d/d* progeny of the F₂ *d/d* snails; these F₃ snails are

■ *Figure 19.11*   Inheritance of the direction of shell coiling in the snail *Limnaea peregra* is an example of maternal effect: (a) Cross between a true-breeding dextral-coiling female (*D/D*) and a sinistral-coiling male (*d/d*); (b) Cross between a sinistral-coiling female (*d/d*) and a dextral-coiling male (*D/D*).

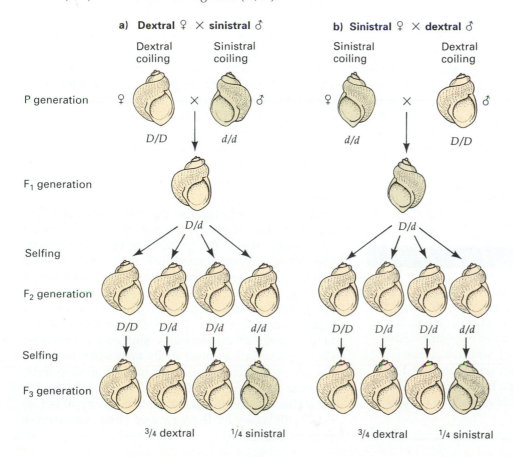

sinistral because their phenotype reflects their mother's (the F_2) genotype.

Similar results are seen in the reciprocal cross of a sinistral (d/d) female with a dextral (D/D) male (Figure 19.11[b]). The F_1s are all D/d in genotype yet they are sinistral in phenotype because the mother was d/d in genotype. Selfing the F_1 produces F_2s, all of which are dextral for the same reason as the reciprocal cross already described. The genotypes and phenotypes of the F_2 and F_3 generations are the same as for the reciprocal cross (see Figure 19.11[b]) and for the same reasons.

The above results do *not* fit our criteria for extranuclear inheritance. That is, if the coil direction phenotype were controlled by an extranuclear gene, the progeny would always exhibit the *phenotype* of the mother, owing to maternal inheritance. Here, the coiling phenotype is governed directly by the nuclear *genotype* of the mother and is an example of maternal effect.

We encountered cases of maternal effect earlier in our discussion of the genetic control of *Drosophila* development (Chapter 15, pp. 364–369). The class of genes called *maternal genes* are nuclear genes that are expressed by the mother during oogenesis. The products of these genes are deposited in the egg and function to specify the gradients in the egg that control spatial organization in early development. These genes were identified by studying the properties of mutants which did not develop normally. For example, mothers homozygous for a mutated maternal effect gene, *bicoid* (*bcd*), produce mutant embryos that have no heads or thoraxes, only abdomens. On the basis of that result, the *bcd* gene was deduced to play an important role in determining normal anterior development. Relevant to our discussion here is that the *phenotype* (mutant) of the progeny embryos reflects the *genotype* (*bcd/bcd*) and not the phenotype (normal) of the mothers; that is, maternal effect is involved. The role of the *bcd* protein in development was discussed in Chapter 15 (p. 366).

Keynote

Maternal effect is different from extranuclear inheritance. The maternal inheritance pattern of extranuclear genes occurs because the zygote contains most of its organelles (containing the extranuclear genes) from the female parent, whereas in the maternal effect the trait inherited is controlled by the maternal *nuclear* genotype before the fertilization of the egg and does not involve extranuclear genes.

Summary

Both mitochondria and chloroplasts contain DNA, the length of which varies from organism to organism. The genomes of both organelles are naked, circular, double-stranded DNAs. The two genomes contain genes that are not duplicated in the nuclear genome; hence the organelle genes are contributing different information for the function of the cell. The organellar genes encode the rRNA components of the ribosomes that are assembled and function in the organelles, for many (if not all) of the tRNAs used in organellar protein synthesis, and for a few proteins that remain in the organelles and perform functions specific to them. Many of the organellar proteins function in multi-protein complexes. Some of these proteins are encoded by the organelle; the remainder are nuclear-encoded. Nuclear-encoded proteins found in organelles are synthesized on cytoplasmic ribosomes, then imported into the appropriate organelles. These proteins include the ribosomal proteins for the organellar ribosomes and, in the case of mitochondria, component proteins of cytochromes.

Given that organelles contain genetic material, it is not surprising that there are many examples of mitochondrial and chloroplast mutants. The inheritance of these extranuclear genes follows rules different from those for nuclear genes, and this is how extranuclear genes were originally identified. For extranuclear genes no meiotic segregation is involved, uniparental (and often maternal) inheritance is generally exhibited, a trait resulting from an extranuclear mutation persists after nuclear substitution, and extranuclear genes cannot be mapped to the known nuclear linkage groups. Not all cases of extranuclear inheritance result from genes on mtDNA or cpDNA. A number of other examples result from infectious heredity, in which cytoplasmically located bacteria or viruses are transmitted when cytoplasms mix.

A specific inheritance pattern sometimes observed is the phenomenon in which the mother specifically affects the phenotype of the offspring. This phenomenon, called the maternal effect, is defined as the predetermination of gene-controlled traits by the maternal *nuclear* genotype prior to the fertilization of the eggs. Maternal effect is different from extranuclear inheritance. The maternal inheritance pattern of extranuclear genes occurs because the zygote receives most of its organelles (containing the extranuclear genes) from the female parent, whereas in the maternal effect the trait inherited is controlled by the maternal *nuclear* genotype before the fertilization of the egg and does *not* involve extranuclear genes.

Analytical Approaches for Solving Genetics Problems

Q19.1 In *Neurospora* the sexual phase of the life cycle is initiated following a fusion of nuclei from mating type *A* and *a* parents. A sexual cross can be made in one of two ways: either by putting both parents on the crossing medium simultaneously or by inoculating the medium with one strain, and after three or four days at 25°C, adding the other parent. In the latter case the first parent on the medium produces all the *protoperithecia*, the bodies that will give rise to the true fruiting bodies in which are the asci with the sexual spores. The protoperithecia have a tremendous amount of cytoplasm and hence can be considered the female parent in much the same way as an egg of a plant or an animal is the female parent. Now by adding conidia of a strain of the opposite mating type to the crossing medium, we have what is called a *controlled cross* in which one strain acts as the female and the other as the male parent. Using a strain to produce the protoperithecia as the female parent and conidia of another strain as a male parent, geneticists can make reciprocal crosses to determine whether any trait shows extranuclear inheritance.

Four slow-growing mutant strains of *Neurospora crassa*, coded *a*, *b*, *c*, and *d*, were isolated. All have an abnormal system of respiratory mitochondrial enzymes. The inheritance patterns of these mutants were tested in controlled crosses with the wild type (+), with the following results:

PROTO-PERITHECIAL (FEMALE) PARENT		CONIDIAL (MALE) PARENT	PROGENY (ASCOSPORES)	
			WILD TYPE	SLOW GROWING
+	×	*a*	847	0
a	×	+	0	659
+	×	*b*	1113	0
b	×	+	0	2071
+	×	*c*	596	590
+	×	*d*	1050	1035

Give a genetic interpretation of these results.

A19.1 This question asks us to consider the expected transmission patterns for nuclear genes and for extranuclear genes. The nuclear genes will have a 1:1 segregation in the offspring, since this organism is a haploid organism and hence should exhibit no differences in the segregation patterns, whichever strain is the maternal parent. On the other hand, a distinguishing characteristic of extranuclear genes is a difference in the results of reciprocal crosses. In *Neurospora* this characteristic is usually manifested by all progeny having the phenotype of the maternal parent. With these ideas in mind we can analyze each mutant in turn.

Mutant *a* shows a clear difference in its segregation in reciprocal crosses and is, in fact, a classic case of maternal inheritance. In each case, all of the progeny resemble the female parent. The interpretation here is that the gene is extranuclear, hence, the gene must be in the mitochondrion.

By the same reasoning used to analyze mutant *a*, the mutation in strain *b* must also be extranuclear.

Mutants *c* and *d* segregate 1:1, indicating that the mutations involved are in the nuclear genome. In these cases we need not consider the reciprocal cross since there is no evidence for maternal inheritance. In fact, the actual mutations that are the basis for this question are female sterile, so the reciprocal cross cannot be done. We can confirm that the mutations are in the nuclear genome by doing mapping experiments, using known nuclear markers. Evidence of linkage to such markers would confirm that the mutations are not extranuclear.

Questions and Problems

19.1 What genes are present in the human mitochondrial genome?

19.2 What conclusions can you draw from the fact that most nuclear-encoded mRNAs and all mitochondrial mRNAs have a poly(A) tail at the 3′ end?

*19.3 Discuss the differences between the universal genetic code of the nuclear genes of most eukaryotes and the code found in human mitochondria. Is there any advantage to the mitochondrial code?

19.4 When the DNA sequences for most of the

mRNAs in human mitochondria are examined, no nonsense codons are found at their termini. Instead, either U or UA is found. Explain this result.

*19.5 What features of extranuclear inheritance distinguish it from the inheritance of nuclear genes?

19.6 Distinguish between maternal effect and extranuclear inheritance.

19.7 Distinguish between nuclear (segregational), neutral, and suppressive *petite* mutants of yeast.

*19.8 In yeast a haploid nuclear (segregational) *petite* is crossed with a neutral *petite*. Assuming that both strains have no other abnormal phenotypes, what proportion of the progeny ascospores are expected to be *petite* in phenotype if the diploid zygote undergoes meiosis?

19.9 When grown on a medium containing acriflavin, a yeast culture produces a large number of very small (*tiny*) cells that grow very slowly. How would you determine whether the slow-growth phenotype was the result of a cytoplasmic factor or a nuclear gene?

19.10 *Drosophila melanogaster* has a sex-linked, recessive, mutant gene called *maroon-like* (*ma-l*). Homozygous *ma-l* females or hemizygous *ma-l* males have light-colored eyes, due to the absence of the active enzyme xanthine dehydrogenase, which is involved in the synthesis of eye pigments. When heterozygous *ma-l$^+$/ma-l* females are crossed with *ma-l* males, all the offspring are phenotypically wild type. However, half the female offspring from this cross, when crossed back to *ma-l* males, give all *ma-l* progeny. The other half of the females, when crossed to *ma-l* males, give all phenotypically wild-type progeny. What is the explanation for these results?

*19.11 When females of a particular mutant strain of *Drosophila melanogaster* are crossed to wild-type males, all the viable progeny flies are females. Hypothetically, this result could be the consequence of either a sex-linked, lethal mutation or a maternally inherited factor that is lethal to males. What crosses would you perform in order to distinguish between these alternatives?

19.12 Reciprocal crosses between two *Drosophila* species, *D. melanogaster* and *D. simulans*, produce the following results:

melanogaster ♀ × *simulans* ♂ → females only
simulans ♀ × *melanogaster* ♂ → males, with few or no females

Propose a possible explanation for these results.

19.13 Some *Drosophila* flies are very sensitive to carbon dioxide—they become anesthetized when it is administered to them. The sensitive flies have a cytoplasmic particle called *sigma* that has many properties of a virus. Resistant files lack *sigma*. The sensitivity to carbon dioxide shows strictly maternal inheritance. What would be the outcome of the following two crosses: (a) sensitive female × resistant male and (b) sensitive male × resistant female?

19.14 Reciprocal crosses between two types of the evening primrose, *Oenothera hookeri* and *Oenothera muricata*, produce the following effects on the plastids:

O. *hookeri* female × O. *muricata* male → yellow plastids
O. *muricata* female × O. *hookeri* male → green plastids

Explain the difference between these results, noting that the chromosome constitution is the same in both types.

*19.15 A form of male sterility in corn is maternally inherited. Plants of a male-sterile line crossed with normal pollen give male-sterile plants. Some lines of corn carry a dominant, so-called restorer (*Rf*) gene, which restores pollen fertility in male-sterile lines.

a. If a male-sterile plant is crossed with pollen from a plant homozygous for gene *Rf*, what will be the genotype and phenotype of the F$_1$?

b. If the F$_1$ plants of part a are used as females in a test cross with pollen from a normal plant (*rf/rf*), what would be the result? Give genotypes and phenotypes, and designate the type of cytoplasm.

*19.16 A few years ago the political situation in Chile was such that very many young adults were kidnapped, tortured, and killed by government agents. When abducted young women had young children or were pregnant, those children were often taken and given to government supporters to raise as their own. Now that the political situation has changed, grandparents of stolen children are trying to locate and reclaim their grandchildren. Imagine that you are a judge in a trial centering on the custody of a child. Mr. and Mrs. Escobar believe Carlos Mendoza is the son of their abducted, murdered daughter. If this is true, then Mr. and Mrs. Sanchez are the paternal grandparents of the child, as their son (also abducted and murdered) was the husband of the Escobars' daughter. Mr. and Mrs. Mendoza claim Carlos is their natural child. The attorney for the Escobar and Sanchez couples informs you that scientists have discovered a series of RFLPs in human mitochondrial DNA. He tells you his clients are eager to be tested, and asks that you order that Mr. and Mrs. Mendoza and Carlos be tested also.

a. Can mitochondrial RFLP data be helpful in this case? In what way?

b. Do all seven parties need to be tested? If not, who actually needs to be tested in this case? Explain your choices.

c. Assume the critical people have been tested, and you have received the results. How would the results determine your decision?

19.17 The pedigree in the figure below shows a family in which an inherited disease called Leber's optic atrophy is segregating. This condition causes blindness in adulthood. Studies have recently shown that the mutant gene causing Leber's optic atrophy is located in the mitochondrial genome.

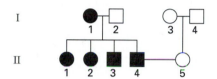

a. Assuming II-4 marries a normal person, what proportion of his offspring should inherit Leber's optic atrophy?

b. What proportion of the sons of II-2 should be affected?

c. What proportion of the daughters of II-2 should be affected?

*19.18 Imagine you have discovered a new genus of yeast. In the course of your studies on this organism, you isolate DNA and subject it to CsCl density gradient centrifugation. You observe a major peak at a density of 1.75 g/cm^3 and a minor peak at a density of 1.70 g/cm^3. How could you determine whether the minor peak represents organellar (presumably mitochondrial) DNA, as opposed to a relatively AT-rich repeated sequence in the nuclear genome?

*19.19 The inheritance of the direction of shell coiling in the snail *Limnaea peregra* has been studied extensively. A snail produced by a cross between two individuals has a shell with a right-hand twist (dextral-coiling). This snail produces only left-hand (sinistral) progeny on selfing. What are the genotypes of the F_1 snail and its parents?

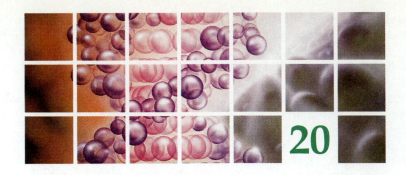

20

Quantitative Genetics

THE NATURE OF CONTINUOUS TRAITS
   Why Some Traits Have Continuous
     Phenotypes

STATISTICS
   Samples and Populations
   Distributions
   Binomial Theorem
   The Mean
   The Variance and the Standard Deviation

POLYGENIC INHERITANCE
   Inheritance of Ear Length in Corn
   Polygene Hypothesis for Quantitative
     Inheritance

HERITABILITY
   Components of the Phenotypic Variance
   Broad-Sense and Narrow-Sense Heritability
   Understanding Heritability
   How Heritability Is Calculated
   Identification of Genes Influencing a
     Quantitative Trait

RESPONSE TO SELECTION
   Estimating the Response to Selection
   Types of Selection

- Discontinuous traits exhibit only a few distinct phenotypes. Continuous traits display a range of phenotypes.

- Continuous traits have many phenotypes because they are encoded by genotypes at many loci (are polygenic) and/or because environmental factors cause each genotype to produce a range of phenotypes.

- The polygene or multiple-gene hypothesis of inheritance proposes that quantitative traits are determined by effects of numerous genes.

- The broad-sense heritability of a trait is the proportion of the phenotypic variance that results from genetic differences among individuals; it does not indicate the extent to which a trait is genetic. Narrow-sense heritability is the proportion of the phenotypic variance that results from additive genetic variance.

- The amount that a trait changes in one generation as a result of selection for the trait is called the response to selection. The magnitude of the response to selection depends on the selection differential and the narrow-sense heritability.

- Directional selection is selection in which individuals with one extreme phenotype survive and reproduce best. Stabilizing selection is selection in which individuals of intermediate phenotype survive and reproduce best. Disruptive selection is selection in which individuals at both ends of the phenotype survive and reproduce best.

IN PREVIOUS CHAPTERS much of our attention focused on the molecular aspects of gene structure, function, and expression. By isolating mutants that affect a particular process and comparing the mutants with the wild type, geneticists can often describe the molecular basis for the mutant phenotype, piecing together an understanding of how the events for normal cells and organisms proceed. The mutations used in these molecular studies, and in fact most of the traits we have studied up to this point, have been characterized by the presence of a few distinct phenotypes. The seed coats of pea plants, for example, were either gray or white, the seed pods were green or yellow, and the plants were tall or short. In each trait the different phenotypes were distinct, and each phenotype was easily separated from all other phenotypes. Such a trait, with only a few distinct phenotypes, is called a **discontinuous trait.** The phenotypes of a discontinuous trait can be described in qualitative terms, and all individuals can be placed into a few phenotypic categories, as shown in Figure 20.1. Some additional examples of discontinuous traits are the ABO blood types, coat colors in mice, the prototrophic versus auxotrophic mutants of bacteria, and the presence or absence of extra digits on the human hand.

A simple relationship usually exists between the genotype and the phenotype for discontinuous traits. In most cases each genotype produces a single phenotype; frequently each phenotype results from a single genotype. When dominance or epistasis occurs the same phenotype may be produced by

■ *Figure 20.1* Discontinuous distribution of shell color in the snail *Cepaea nemoralus* from a population in England.

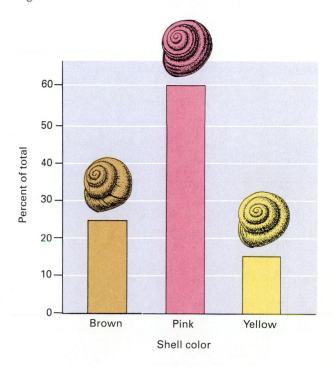

several different genotypes, but the relationship between the genes and the trait remains simple. Because of this uncomplicated relationship, genotypes can be inferred by studying the phenotypes of parents and offspring. Such study enabled Mendel to work out the basic principles of heredity.

Not all traits, however, exhibit phenotypes that fall into a few distinct categories. Many traits, such as birth weight and adult height in humans, protein content in corn, and number of eggs laid by *Drosophila*, exhibit a wide range of possible phenotypes. Traits with a continuous distribution of phenotypes are called **continuous traits.** The distribution of a continuous trait is illustrated in Figure 20.2. Since the phenotypes of continuous traits must be described in quantitative terms (e.g., centimeters for height), such traits are also known as **quantitative traits.** The study of the inheritance of quantitative traits comprises the field of **quantitative genetics.**

The Nature of Continuous Traits

Biologists began developing statistical techniques for the study of continuous traits during the latter part of the nineteenth century, even before they were aware of Mendel's principles of heredity. Francis Galton and his associate Karl Pearson studied a number of continuous traits in humans, such as height, weight, and mental traits. By demonstrating that the traits of parents and their offspring are statistically associated, they were able to show that these traits are inherited, but they were not successful in determining how genetic transmission occurs. After the rediscovery of Mendel's work, considerable controversy arose over whether continuous traits also follow Mendel's principles or whether they are inherited in some different fash-

ion. Around 1903 Wilhelm Johannsen conducted a series of experiments on the inheritance of seed weight of beans, and he demonstrated that variation in continuous traits is partly genetic and partly environmental. Several years later Herman Nilsson-Ehle, working with wheat, proposed that continuous traits are determined by multiple genes, each of which segregates according to Mendelian principles.

Why Some Traits Have Continuous Phenotypes

Continuous traits have many phenotypes. To understand the inheritance of continuous traits, we must first determine why some traits have many phenotypes.

Multiple phenotypes of a trait arise in several ways. Frequently a range of phenotypes occurs because numerous genotypes exist among the individuals of a group—numerous genotypes occur when the trait is influenced by a large number of loci. For example, when a single locus with two alleles determines a trait, three genotypes are present: AA, Aa, and aa. With two loci, each with two alleles, the number of genotypes is $3^2 = 9$ ($AA\ BB$, $Aa\ BB$, $AA\ Bb$, $Aa\ Bb$, $AA\ bb$, $aa\ BB$, $aa\ Bb$, $Aa\ bb$, and $aa\ bb$). In general, the number of genotypes is 3^n, where n equals the number of loci with two alleles (see Chapter 2, p. 40). If more than two alleles are present at a locus, the number of genotypes is even greater. As the number of loci influencing a trait increases, the number of genotypes quickly becomes large. Traits that are encoded by many loci are referred to as **polygenic traits.** If each genotype in a polygenic trait encodes a separate phenotype, many phenotypes will be present. And because many phenotypes are present, and the differences between phenotypes are slight, the trait appears to be continuous.

A second reason that a trait may have a range of phenotypes is that environmental factors also affect the trait. When environmental factors exert an important influence on the phenotype, each genotype is capable of producing a range of phenotypes. Which phenotype is expressed depends both on the genotype and on the specific environment in which the genotype is found. For most continuous traits, both multiple genotypes and environmental factors influence the phenotype; such a trait is **multifactorial** (influenced by multiple genes and environmental factors).

When multiple genes and environmental factors influence a trait, one does not find the simple relationship between genotype and phenotype that exists in discontinuous traits. Therefore, the simple rules of inheritance that we learned in classical

■ *Figure 20.2* Continuous distribution of height for male students at the University of Minnesota in 1924–25.

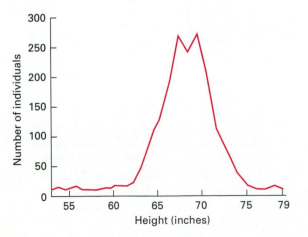

genetics provide us with little information about the genes involved in continuous traits. For example, in transmission genetics we frequently determined the probability of inheriting a discontinuous trait, such as 3/4 or 1/4. With most continuous traits, however, numerous genes are involved and no simple relationship exists between the genotype and the phenotype; thus we do not know the genotypes of the parents and we cannot make precise predictions about the probability of inheriting a continuous trait, as we did for simple discontinuous traits. In order to understand the genetic basis of these traits and their inheritance, we must employ special concepts and analytical procedures.

Keynote

Discontinuous traits exhibit only a few distinct phenotypes and can be described in qualitative terms. Continuous traits, on the other hand, display a spectrum of phenotypes and must be described in quantitative terms. Many phenotypes are present in a continuous trait because the trait is encoded by many loci, producing many genotypes, and because environmental factors cause each genotype to produce a range of phenotypes.

Statistics

With discontinuous traits we described the phenotypes of a group of individuals by listing the numbers or frequencies of each phenotype. For instance, we might state that three-fourths of the progeny were yellow and one-fourth were green. With continuous traits such a simple description of the phenotypes is often impossible because a spectrum of phenotypes is present, and the phenotype can usually only be described in terms of a measurement. Furthermore, the relationship between genotype and phenotype is complex, so the simple ratios predicted by Mendelian principles cannot be applied. Consequently, quantitative genetics requires statistical procedures for describing continuous traits and for understanding their inheritance.

Samples and Populations

Suppose we want to describe some aspect of a trait in a large group of individuals. For example, we might be interested in the average birth weight of infants born in New York City during the year 1987. Since thousands of babies are born in New York City every year, collecting information on each baby's weight might not be practical. An alternative

method would be to collect information on a subset of the group, say birth weights on 100 infants born in New York City during 1987, and then use the average obtained on this subset as an estimate of the average for the entire city. Biologists and other scientists commonly employ this sampling procedure in data collection, and statistics are necessary for analyzing such data. The group of ultimate interest (in our example, all infants born in New York City during 1987) is called the **population,** and the subset used to give us information about the population (our set of 100 babies) is called a **sample.** For a sample to give us reliable information about the population, it must be large enough so that chance differences between the sample and the population are not misleading. If our sample consisted of only a single baby, and that infant was unusually large, then our estimate of the average birth weight of all babies would not be very accurate. The sample must also be a random subset of the population. If all the babies in our sample came from Hope Hospital for Premature Infants, then we would grossly underestimate the true birth weight of the population.

Keynote

To describe and study a large group of individuals, scientists frequently examine a subset of the group. This subset, called a sample, provides information about the larger group, which is termed the population. The sample must be of reasonable size and must be a random subset of the larger group to provide accurate information about the population.

Distributions

When we studied discontinuous traits, we were able to describe the phenotypes found among a group of individuals by stating the proportion of individuals falling into each phenotypic class, such as 3/4 yellow and 1/4 green. As we discussed earlier, continuous traits exhibit a range of phenotypes, so describing the phenotypes found within a group of individuals is more complicated. One means of summarizing the phenotypes of a continuous trait is with a **frequency distribution,** which is a description of the population in terms of the proportion of individuals that exhibit each phenotype.

To make a frequency distribution classes are constructed to consist of individuals falling within a specified range of the phenotype; then the number of individuals in each class is counted. Table 20.1 presents a frequency distribution constructed from

Table 20.1

Weight of 5494 F$_2$ Beans (Seeds of *Phaseolus vulgaris*) Observed by Johannsen in 1903

Weight (cg)	5–15	15–25	25–35	35–45	45–55	55–65	65–75	75–85	85–95
(Midpoint of Range)	(10)	(20)	(30)	(40)	(50)	(60)	(70)	(80)	(90)
Number of Beans	5	38	370	1676	2255	928	187	33	2

the data in Johannsen's study of the inheritance of weight in the dwarf bean, *Phaseolus vulgaris*. As shown in the table, Johannsen weighed 5494 beans from the F$_2$ progeny of a cross and classified them into nine groups or classes, each of which covered a 10-centigram (cg) range of weight. A frequency distribution such as this can be displayed graphically by plotting the phenotypes in a frequency histogram, as shown in Figure 20.3. In the histogram the phenotypic classes are indicated along the horizontal axis and the number present in each class is plotted on the vertical axis. If a curve is drawn tracing the outline of the histogram, the curve assumes a shape that is characteristic of the frequency distribution.

Many continuous phenotypes exhibit a symmetrical, bell-shaped distribution similar to the curve shown in Figure 20.3. This type of distribution is called a **normal distribution**. The normal distribution is produced when a large number of independent factors influence the measurement. Since many continuous traits are multifactorial, observing a normal distribution for these traits is not surprising.

Binomial Theorem

Another distribution that is particularly useful for understanding the inheritance of polygenic traits is the binomial distribution. The **binomial distribu-**tion is the theoretical frequency distribution of events that have two possible outcomes. For example, among humans two sexes are present, male and female. Suppose that we want to know what the probability is that in a family of three children, two will be boys and one will be a girl. The probability is not $1/2 \times 1/2 \times 1/2 = 1/8$, but rather $3/8$. This is because there are three different ways, or three permutations, in which a family of three children can have two boys and one girl: The first two children might be boys and the third child a girl (BBG). Alternatively, the first child might be a boy, followed by a girl and then another boy (BGB), or the first child might be a girl followed by two boys (GBB). In each case, the probability is $1/2 \times 1/2 \times 1/2 = 1/8$ (using the product rule; see Chapter 2), so the total probability is the sum of these individual probabilities (using the sum rule; see Chapter 2), or $1/8 + 1/8 + 1/8 = 3/8$.

In a family with three children, four different combinations of the sexes are possible: three boys, two boys and a girl, one boy and two girls, and three girls. Three boys can occur in only one way: BBB. Therefore the probability that a family of three children has all boys is $1/2 \times 1/2 \times 1/2 = 1/8$. As we have seen, the probability of two boys and one girl is $3/8$. Following the same reasoning, the probability of two girls and one boy is $3/8$, because three permutations of this combination are possible (GGB, GBG, and BGG). Since three girls can be present in only one way (GGG), the probability of this combination is $1/2 \times 1/2 \times 1/2 = 1/8$. In summary, we have probabilities of $1/8:3/8:3/8:1/8$ for three boys:two boys and one girl:two girls and one boy:three girls.

Such computations get rather laborious when the number of combinations is large. A convenient, efficient way around the tedium is to use the *binomial theorem*. According to this theorem, the frequencies (or probabilities of occurrence) of the various combinations correspond to the terms of the *binomial expansion*.

The general expression for the binomial expansion is $(a + b)^n$, where a is the probability of one event (having a boy) and b is the probability of the alternative event (having a girl); $a + b$ must equal 1. The term n is the number of trials (number of chil-

■ *Figure 20.3* Frequency histogram for bean weight in *Phaseolus vulgaris* plotted from the data in Table 20.1. A normal curve has been fitted to the data and is superimposed on the frequency histogram.

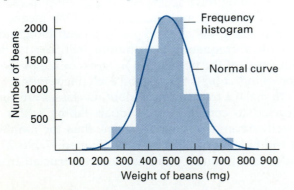

dren). The first three binomial expansions are as follows:

$$(a + b)^2 = a^2 + 2ab + b^2$$

$$(a + b)^3 = a^3 + 3a^2b + 3ab^2 + b^3$$

$$(a + b)^4 = a^4 + 4a^3b + 6a^2b^2 + 4ab^3 + b^4$$

Let us consider the first expansion. The exponent (2) indicates that we are dealing with a two-child family. The term a is the probability of having a boy, namely 1/2, and b is the probability of having a girl, also 1/2. To the right of the equals sign are the terms for the various two-child families that can result, and from these terms the probability of each combination can be determined. The a^2 term, for example, is the probability of having two boys, which is $(1/2)^2 = 1/4$. The probability of having a boy and a girl is $2ab = 2(1/2)(1/2) = 2/4$; and so on.

To answer questions of probability, we must be able to expand the binomial to any power. Consider the example $(a + b)^6$.

$$(a + b)^6 = a^6 + 6a^5b^1 + 15a^4b^2 + 20a^3b^3 + 15a^2b^4$$
$$+ 6a^1b^5 + b^6$$

Let's see how this binomial expansion can be obtained. First, note that the number of terms in the expansion is always one more than the power to which the binomial is raised ($n + 1$). In our example the binomial is raised to the 6th power and thus there will be $6 + 1 = 7$ terms in the expansion. The first term in the expansion is always a raised to the power of the binomial, which in this case is 6. In each successive term the exponent of a decreases by one. Thus the exponent of a in the second term is 5, the exponent of a in the third term is 4, and so on. At the same time that the exponents of a decrease,

the exponents of b increase by one in each successive term, going from 0 to 6.

Next, we must figure the coefficients of the terms. The first coefficient in the binomial expansion is always 1: the first term in our example is $1a^6$. (For simplicity, we usually do not write the coefficient 1, writing instead a^6.) For the second term, the coefficient is always the same power to which the binomial is raised (n); in our example the second coefficient would be 6. To determine the coefficient of the third term, we look at the preceding term ($6a^5b^1$), take the coefficient of this term (6) and multiply it by the exponent of a (5), and then divide by the number of that term (the second term or 2). Thus the coefficient of the third term is $(6 \times 5)/2 = 15$. We can use this same procedure for each successive term. The coefficients for the terms of the binomial expansion can also be obtained from Pascal's Triangle (Figure 20.4).

A more direct way to determine the coefficient is to use the general formula

$$\frac{n!}{r!s!} a^r b^s$$

for a specific item in the expansion, where $n!/r!s!$ is the coefficient, r is the number of "a"s, s is the number of "b"s, $n = r + s$, and ! stands for "factorial," meaning that the number is multiplied by every successive decreasing integer down to 1. 0! is defined as 1. For example, the coefficient for a^4b^2 in the expansion

$$(a + b)^6 = \frac{6!}{4!2!} = \frac{6 \times 5 \times 4 \times 3 \times 2 \times 1}{(4 \times 3 \times 2 \times 1)(2 \times 1)} = \frac{6 \times 5}{2} = 15$$

Later in this chapter we will use the binomial expansion for determining the phenotypic ratios in crosses involving a polygenic trait (see p. 477).

■ *Figure 20.4* Pascal's Triangle. The numbers in each row are the coefficients for terms in the binomial expansion $(a + b)^n$ for values of n from 1 to 8.

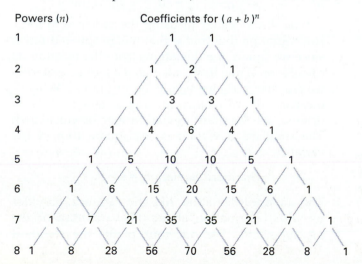

Powers (n)   Coefficients for $(a + b)^n$

Powers (n)																
1								1		1						
2							1		2		1					
3						1		3		3		1				
4					1		4		6		4		1			
5				1		5		10		10		5		1		
6			1		6		15		20		15		6		1	
7		1		7		21		35		35		21		7		1
8	1		8		28		56		70		56		28		8	1

The Mean

A distribution of phenotypes can be summarized in the form of two convenient statistics, the mean and the variance. The **mean,** which is also known as the average, gives us information concerning the center of the distribution of the phenotypes in a sample. The mean of a sample $(\bar{x})$ is calculated by adding up all the individual measurements $(x_1, x_2, x_3,$ etc.) and dividing by the number of measurements we added (n). We can represent this calculation with the following formula:

$$\text{Mean} = \bar{x} = \frac{x_1 + x_2 + x_3 + ...x_n}{n}$$

This equation can be abbreviated by using the symbol Σ, which means the summation, and x_i, which means the ith (or individual) value of x:

$$\text{Mean} = (\bar{x}) = \frac{\Sigma x_i}{n}$$

Suppose we measured the length of four flowers and obtained the following measurements: 36 mm, 40 mm, 44 mm, and 48 mm. To calculate the mean length $(\bar{x})$ of these flowers, we sum the individual lengths $(\Sigma x_i = 36 + 40 + 44 + 48 = 168)$ and divide the total by the number of measurements $(\Sigma x_i/n = 168/4 = 42$ mm).

The mean is frequently used in quantitative genetics to summarize the phenotypes of a group of individuals. For instance, in an early study of continuous variation, Edward M. East examined the inheritance of flower length in several strains of the tobacco plant. He crossed a short-flowered strain of tobacco with a long-flowered strain. Within each strain, however, flower length varied somewhat, so East reported that the mean phenotype of the short strain was 40.4 mm and the mean phenotype of the long strain was 93.1 mm. The F_1 progeny, which consisted of 173 plants, had a mean flower length of 63.5 mm. The mean provides a convenient way to quickly summarize the phenotypes of parents and offspring.

The Variance and the Standard Deviation

A second statistic that provides key information about a distribution is the **variance.** The variance is a measure of how much the individual measurements vary from the mean. Two populations may have the same mean but very different variances, as shown in Figure 20.5. The variance of a sample (s^2) is defined as the average squared deviation from the mean:

■ *Figure 20.5* Graphs showing three distributions with the same mean but different variances.

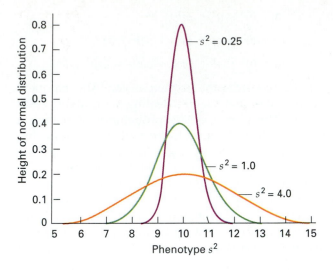

$$\text{Variance} = s^2 = \frac{\Sigma(x_i - \bar{x})^2}{n - 1}$$

The variance can be calculated by first subtracting each individual measurement from the mean $(x_i - \bar{x})$. Each value obtained from this subtraction is squared $(x_i - \bar{x})^2$, and all the squared values are added up: $\Sigma(x_i - \bar{x})^2$. The sum of this calculation is then divided by the number of original measurements minus one $(n - 1)$.

Suppose we wanted to calculate the variance for the four flowers we used in our calculation of the mean. Recall that these four flowers had lengths of 36 mm, 40 mm, 44 mm, and 48 mm, and that the mean flower length was 42 mm. The sum of the squared deviations from the mean is: $\Sigma(x_i - \bar{x})^2 = (36 - 42)^2 + (40 - 42)^2 + (44 - 42)^2 + (48 - 42)^2 = 36 + 4 + 4 + 36 = 80$ mm^2. To calculate the variance, we divide the sum of the squared deviations by the number of original measurements minus one $(s^2 = \frac{\Sigma(x_i - \bar{x})^2}{n - 1})$, obtaining $80/3 = 26.7$ mm^2.

One disadvantage of using the variance to measure variation in a trait is that it is in squared units, since we square the deviations from the mean in its calculation. Thus if we calculated a variance of seed weight, the variance would be in grams^2. When a measure of variability in the original units is desired, the **standard deviation** may be calculated. The standard deviation is the square root of the variance:

$$\text{Standard deviation} = s = \sqrt{s^2}$$

The standard deviation, like the variance, provides us with information about the variability of the measurements. For our four plants the standard deviation is the square root of the variance or $\sqrt{26.7} = 5.16$ mm.

The variance and the standard deviation can provide us with valuable information about the distribution of phenotypes among a group of individuals. In our discussion of the mean, we saw how East used the mean to describe flower length of parents and offspring in crosses of the tobacco plant. When East crossed a strain of tobacco with short flowers to a strain with long flowers, the F_1 offspring had a mean flower length of 63.5 mm, which was intermediate to the phenotypes of the parents. When he intercrossed the F_1, the mean flower length of the F_2 offspring was 68.8 mm, approximately the same as the mean phenotype of the F_1. However, the F_2 progeny differed from the F_1 in a significant attribute that is not apparent if we only examine the means of the phenotype—the F_2 were more variable in phenotype than the F_1. The variance in the flower length of the F_2 was 42.4, whereas the variance in the F_1 was only 8.6. This finding indicated that more genotypes were present among the F_2 progeny than in the F_1. Thus the mean and the variance are both necessary for fully describing the distribution of phenotypes among a group of individuals.

Polygenic Inheritance

In the examples of polygenic inheritance described in the following sections, geneticists did not know at first how these traits were inherited, although it was apparent that their pattern of inheritance differed from that of discontinuous traits.

Inheritance of Ear Length in Corn

An organism that has been the subject of genetic and cytological studies for many years is corn, *Zea mays*. Ear length is one of the traits of the corn plant that came under scrutiny in a classic study that demonstrated quantitative inheritance for that trait. In this study, reported in 1913, Rollins Emerson and Edward East started their experiments with two pure-breeding strains of corn, each of which displayed little variation in ear length. The two varieties were Tom Thumb popcorn (which had short ears of mean length 6.63 cm) and Black Mexican sweet corn (which had long ears of mean length 16.80 cm).

Emerson and East crossed the two strains and then interbred the F_1 plants. Figure 20.6 presents the results in both photographs and a histogram. Note that the F_1 ears have a mean length of 12.12 cm, which is intermediate between the mean ear lengths of the two parental lines. The parental plants are pure breeding, so we can assume that

each is homozygous for whatever genes control the lengths of their ears. However, since the two parental plants differ in ear length, each must be genetically different. When two pure-breeding strains are crossed, the F_1 plants will be heterozygous for all the genes that controlled the differences in the ear lengths of the pure-breeding strains, and all F_1 plants should have the same genotype. Therefore the range of ear length phenotypes seen in the F_1 plants must be due to factors other than genetic differences; these other factors are probably environmental, since it is impossible to grow plants in exactly identical conditions.

In the F_2 the mean ear length of 12.89 cm is about the same as the mean for the F_1 population, but the F_2 population has a much larger variation around the mean than the F_1 population has. This variation is easy to see in Figure 20.6(b); it can also be shown by calculating the standard deviation s. The standard deviation of the long-eared parent is 0.816, and that of the short-eared parent is 1.887. In the F_1, $s = 1.519$, and in the F_2, $s = 2.252$. The larger standard deviation in the F_2 confirms that the F_2 has greater variability, which could also be gathered just by looking at the data.

Is this variation the result of the effects of environmental factors? Certainly, if the environment were responsible for variation in the parental and the F_1 generations, we would have every reason to believe that it would have a similar effect on the F_2. However, we have no reason to suppose that the environment would have a greater influence on the F_2 than on the other two generations, and so there must be another explanation for the greater variation in ear length in the F_2 generation. A more reasonable hypothesis is that the increased variability of the F_2 results from the presence of greater genetic variation.

Setting aside the environmental influence for the moment, the data reveal four observations that apply generally to quantitative-inheritance studies:

1. The mean value of the quantitative trait in the F_1 is approximately intermediate between the means of the two true-breeding parental lines.

2. The mean value for the trait in the F_2 is approximately equal to the mean for the F_1 population.

3. The F_2 shows more variability around the mean than the F_1 does.

4. The extreme values for the quantitative trait in the F_2 extend further into the distribution of the two parental values than do the extreme values of the F_1.

The data presented cannot be explained in terms of principles that govern the inheritance of simply

■ *Figure 20.6* Inheritance of ear length in corn: (a) Representative corn ears from the parental, F₁, and F₂ generations from an experiment in which two pure-breeding corn strains that differ in ear length were crossed and then the F₁s interbred; (b) Histograms of the distributions of ear length (in centimeters) of ears of corn from the experiment represented in part (a). The vertical axes represent the percentages of the different populations found at each ear length.

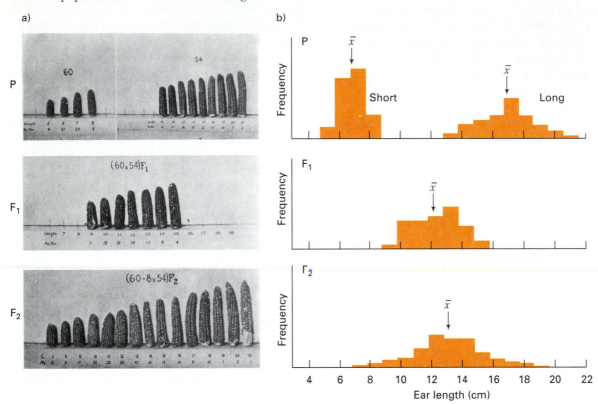

inherited discontinuous traits. That is, if a single gene were responsible for the two phenotypes of the original parents (AA = homozygous, long; aa = homozygous, short), then the F₁ data could be explained if we assume incomplete dominance. However, crossing the F₁ heterozygote (Aa) should produce a 1:2:1 ratio of AA, Aa, and aa, or long, intermediate, and short phenotypes. Clearly, the data do not fall into such discrete classes.

Keynote

For a quantitative trait the F₁ progeny of a cross between two phenotypically distinct, pure-breeding parents has a phenotype intermediate between the parental phenotypes. The F₂ shows more variability than the F₁, with a mean phenotype close to that of the F₁. The extreme phenotypes of the F₂ extend well beyond the range of the F₁ and into the ranges of the two parental values.

Polygene Hypothesis for Quantitative Inheritance

The simplest explanation for the data obtained from Emerson and East's experiments on corn ear length and from other experiments with quantitative traits is that quantitative traits are controlled by not one but by many genes. This explanation is called the **polygene,** or **multiple-gene, hypothesis for quantitative inheritance**.

The polygene hypothesis can be traced back to 1909 and the classic work of Hermann Nilsson-Ehle, who studied the color of the wheat kernel. Like Mendel and Emerson and East, Nilsson-Ehle began by crossing true-breeding lines of plants— wheat with red kernels and wheat with white kernels. The F₁ had kernels that were all the same shade of an intermediate color between red and white. At this point he could not rule out incomplete dominance as the basis for the F₁ results. However, when he intercrossed the F₁s, a number of the F₂ progeny showed a ratio of approximately 15 red (all shades):1 white kernels, clearly a deviation from a 3:1 ratio expected for a monohybrid cross.

Nilsson-Ehle recognized four discrete shades of red, in addition to white, among the progeny. He counted the relative number of each class and found a 1:4:6:4:1 phenotypic ratio of wheat with dark red, medium red, intermediate red, light red, and white kernels. Note that 1/16 of the F_2 has a kernel phenotype as extreme as the original red parent, and 1/16 has a kernel phenotype as extreme as the original white parent.

What genetic explanation can be given for these data? Recall from Chapter 4 (pp. 84; 93–94) that a 15:1 ratio of two alternative characteristics resulted from the interaction of the products of two genes, each of which affects the same trait; the two genes are known as duplicate genes. The explanation for the 15:1 ratio was that two allelic pairs are involved in determining the phenotypes segregating in the cross. Since several of the F_2 populations from the wheat crosses exhibited a 15 red:1 white ratio, we will apply that explanation to the kernel trait.

Let us hypothesize that two independently segregating genes control the production of red pigment: alleles R (red) and C (crimson) of the two genes result in red pigment and alleles r and c result in the lack of pigment. Nilsson-Ehle's parental cross and the F_1 genotypes can then be shown as follows:

$$\text{P} \quad RR\ CC \quad \times \quad rr\ cc$$
$$\text{(dark red)} \qquad \text{(white)}$$

$$F_1 \qquad Rr\ Cc$$
$$\text{(intermediate red)}$$

When the F_1 is interbred, the distribution of genotypes in the F_2 is typical of dihybrid inheritance; 1/16 RR CC + 2/16 Rr CC + 1/16 rr CC + 2/16 RR Cc + 4/16 Rr Cc + 2/16 rr Cc + 1/16 RR cc + 2/16 Rr cc + 1/16 rr cc. If R and C are dominant to r and c, the 9:3:3:1 phenotypic ratio characteristic of dihybrid inheritance should result.

For the wheat kernel color phenotype, then, dominance is not the simple answer, since the observed phenotypic ratio approximates 1:4:6:4:1.

Note that these numbers in the phenotypic ratio are the same as the coefficients in the binomial expansion of $(a + b)^4$, or $1a^4 + 4a^3b^1 + 6a^2b^2 + 4a^1b^3 + 1b^4$. A simple explanation is that each dose of an allele controlling pigment production allows the synthesis of a certain amount of pigment. Therefore the intensity of red coloration is a function of the number of R or C alleles in the genotype; RR CC would be dark red and rr cc would be white. Table 20.2 summarizes this situation with regard to the five phenotypic classes observed by Nilsson-Ehle. In other words, the alleles of the genes represented by capital letters code for products that add to the phenotypic characteristic; for example, each allele, R or C, causes more red pigment to be added to the wheat kernel color phenotype. Alleles that contribute to the phenotype of a quantitative trait are called *contributing alleles*. The alleles that do not add to the phenotype of the quantitative trait, such as the r and c alleles in the wheat kernel color trait, are called noncontributing alleles. Thus the inheritance of red kernel color in wheat is an example of a multiple-gene (polygene) series of as many as four contributing alleles.

We must be cautious in interpreting the genetic basis of this particular quantitative trait, though. Some F_2 populations show only three phenotypic classes with a 3:1 ratio of red to white, while other F_2 populations show a 63:1 ratio of red to white, with discrete classes of color between the dark red and the white. These results indicate that the genetic basis for the quantitative trait can vary with the strain of wheat involved. The 3:1 case could be explained by a single-gene system, while the 63:1 case could indicate a polygene series with six contributing alleles. The number of discrete classes in

Table 20.2

Genetic Explanation for the Number and Proportions of F_2 Phenotypes for the Quantitative Trait Red Kernel Color in Wheat

GENOTYPE	NUMBER OF CONTRIBUTING ALLELES FOR RED	PHENOTYPE	FRACTION OF F_2
RR CC	4	Dark red	1/16
RR Cc or Rr CC	3	Medium red	4/16
RR cc or rr CC or Rr Cc	2	Intermediate red	6/16
rr Cc or Rr cc	1	Light red	4/16
rr cc	0	White	1/16

the latter case would be seven, with the proportion of each class following the coefficients in the binomial expansion of $(a + b)^6$, or, 1:6:15:20:15:6:1.

In its basic form the multiple-gene hypothesis proposes that a number of the attributes of quantitative inheritance can be explained on the basis of the action and segregation of a number of alleles that have equal additive effects on the phenotype and that involve no complete dominance. The assumptions of the hypothesis are as follows:

1. In no allelic pair does one allele exhibit dominance over another allele. Instead a series of contributing and noncontributing alleles is involved.
2. Each contributing allele in a series has an equal effect.
3. The effect of each contributing allele is additive.
4. No genetic interaction occurs among the genes at different loci (no epistasis) in a polygenic series.
5. No genetic linkage is exhibited between the genes in a polygenic series, and therefore they assort independently.
6. There are no environmental effects.

For many quantitative traits, all of these assumptions may not be valid. For example, some of the genes affecting a quantitative trait may be linked; others may exhibit various degrees of dominance and epistasis. Environmental effects are also common with quantitative traits. Nevertheless, the multiple-gene hypothesis is satisfactory as a working hypothesis for interpreting many quantitative traits, and quantitative inheritance provides an explanation for the inheritance of continuous traits that is compatible with Mendel's laws. In reality, however, the analysis of quantitative traits is quite complicated. There are still many gaps in our understanding of quantitative inheritance.

Keynote

Quantitative traits are based on genes in a multiple-gene series, or polygenes. The multiple-gene hypothesis assumes that contributing and noncontributing alleles in the series operate so that as the number of contributing alleles increases, there is an additive effect on the phenotype, that no genetic interaction occurs between the gene loci, that the loci are unlinked, and that there are no environmental effects.

Heritability

Heritability is the proportion of a population's phenotypic variation that is attributable to genetic factors. The concept of heritability is used to examine the relative contributions of genes and environment to *variation* in a specific trait. As we have seen, continuous traits are frequently influenced by multiple genes and environmental factors. One crucial question addressed in quantitative genetics is: How much of the variation in phenotype results from genetic differences among individuals and how much is a product of environmental variation? For example, multifactorial traits such as weight of cattle, number of eggs laid by chickens, and amount of wool produced by sheep are significant for breeding programs and agricultural management. Many ecologically important traits, such as variation in body size, fecundity, and developmental rate, are also multifactorial, and determining the genetic contribution to this variation is important for understanding how natural populations evolve. The extent to which genetic and environmental factors contribute to human variation in traits like blood pressure and birth weight is essential for health care. Hence the study of heritability is crucial in a variety of scientific disciplines.

The following section deals with two types of heritability: *broad-sense heritability* and *narrow-sense heritability*. To assess heritability we must first measure the variation in the trait and then we must partition that variance into components attributable to different causes.

Components of the Phenotypic Variance

The **phenotypic variance,** represented by V_P, is a measure of the variability of a trait. It is calculated by computing the variance of the trait for a group of individuals, as discussed in the earlier section on statistics. Differences among individuals arise from several factors; therefore, we can divide the phenotypic variance into a number of components attributable to different sources. First, some of the phenotypic variation arises because of genetic differences among individuals (different genotypes within the group). This contribution to the phenotypic variation is called **genetic variance** and is represented by the symbol V_G. As noted, additional variation often results from environmental differences among the individuals; in other words, different environments experienced by individuals may contribute to the differences in their phenotypes. The **environmental variance** is symbolized by V_E, and by definition, includes any nongenetic source of variation. Temperature, nutrition, and parental care are examples of obvious environmental factors that may cause differences among individuals. Environmental variance also includes random factors that occur during development, factors that are sometimes referred to as *developmental noise.* A third source of phenotypic variance is genetic-environmental interaction, represented by V_{GE}. Genetic-environmental interaction occurs when the relative effects of the genotypes differ among environments. For example, in a cold environment, genotype AA of a plant may be 40 cm in height and genotype Aa may be 35 cm in height. However, when the genotypes are moved to a warm climate, genotype Aa is now 60 cm, as compared with genotype AA, which is only 50 cm in height. In this example both genotypes grow taller in the warm environment, but the relative performance of the genotypes switches in the two environments. Therefore, both environmental differences (temperature) and genetic differences (genotypes) contribute to the phenotypic variance. However, the effects of genotype and environment cannot simply be added together. An additional component, V_{GE}, that accounts for how genotype and environment interact must be considered.

The phenotypic variance, composed of differences arising from genetic variation, environmental variation, and genetic-environmental variation, can be represented by the following equation:

$$V_P = V_G + V_E + V_{GE}$$

The relative contributions of these three factors to the phenotypic variance depend on the genotypes present in the population, the specific environment, and the manner in which the genes interact with the environment.

The genetic variance V_G can be further subdivided into components arising from different types of interactions between genes. Some of the genetic variance occurs as a result of the additive effects of the different genes on the phenotype, and is called **additive genetic variance.** You may recall that the genes studied by Nilsson-Ehle, which determine kernel color in wheat, are strictly additive in this way. Some genes contribute to the pigment of the kernel while others do not; the added effects of all the individual contributing genes determine the phenotype of the kernel. Thus the genotypes $AA\ bb$, $aa\ BB$, and $Aa\ Bb$ all produce the same phenotype, since each genotype has two contributing alleles. The phenotypic variance arising from the additive effects of genes is the additive genetic variance and is symbolized by V_A.

Some genes may exhibit dominance, which comprises another source of genetic variance, the **dominance variance** (V_D). Finally, epistatic interactions may occur among genes. Recall that in epistasis genes at different loci interact to determine the phenotype. The presence of epistasis adds another source of genetic variation, called epistatic or **interaction variance** (V_I). So we can partition the genetic variance as follows:

$$V_G = V_A + V_D + V_I$$

The total phenotypic variance can then be summarized as

$$V_P = V_A + V_D + V_I + V_E + V_{GE}$$

Partitioning the phenotypic variance into these components is useful for thinking about the contribution of different factors to the variation in phenotype.

Broad-Sense and Narrow-Sense Heritability

Geneticists frequently estimate the different variance components of a trait to determine the extent to which variation among individuals results from genetic differences. Thus they are interested in what proportion of the phenotypic variance V_P can be attributed to genetic variance V_G. This quantity, the proportion of the phenotypic variance that consists of genetic variance, is called **the broad-sense heritability.** It is expressed as follows:

$$\text{Broad-sense heritability} = H^2 = \frac{V_G}{V_P}$$

The heritability of a trait can range from 0 to 1. A broad-sense heritability of 0 indicates that none of the variation in phenotype among individuals results from genetic differences. A heritability of 0.5 means that 50 percent of the phenotypic variation arises from genetic differences among individuals, and a heritability of 1 would suggest that all the phenotypic variance is genetically based.

Broad-sense heritability includes genetic variation from all types of genes. We are often more interested in the proportion of the phenotypic variation that results only from *additive* genetic effects. If the genes are additive, the offspring should be exactly intermediate between the parents. However, if dominant genes are important, the offspring may not be intermediate to the parental phenotypes. In a similar fashion epistatic genes will not always contribute to the resemblance between parents and offspring.

Because the additive genetic variance allows one to make accurate predictions about the resemblance between offspring and parents, quantitative geneticists frequently determine the proportion of the phenotypic variance that results from additive genetic variance, a quantity referred to as the **narrow-sense heritability.** The additive genetic variance is also that variation that responds to selection in a predictable way; thus the narrow-sense heritability provides information about how a trait will evolve. The narrow-sense heritability is represented mathematically as

$$\text{Narrow-sense heritability} = h^2 = \frac{V_A}{V_P}$$

Understanding Heritability

Despite their widespread use, heritability estimates have a number of significant limitations. These limitations are often ignored and consequently heritability is one of the most misunderstood and misused concepts in genetics. Before we discuss how heritability is determined, we must list some of the important qualifications and limitations of heritability.

1. Heritability does not indicate the extent to which a trait is genetic. What heritability does measure is the *proportion of the phenotypic variance* among individuals in a population that results from genetic differences. Genes often influence the development of a trait; thus the trait may be said to be genetic. However, the differences in phenotype among individuals may not be genetic at all—they may be entirely the result of environmental differences. If all individuals in a population have identical genes at the loci that control the trait, then the genetic variance is zero ($V_G = 0$). Because heritability equals the proportion of the phenotypic variance that results from genetic variance ($H^2 = V_G/V_P$), and V_G is zero, the heritability must be zero. Although the heritability in this case is zero, it would be incorrect to assume that genes play no role in the development of the trait; it just means that all individuals are genetically uniform. Similarly, a high heritability does not negate the significance of environmental factors influencing a trait; a high heritability might simply mean that the environmental factors that influence the trait are relatively uniform among the individuals studied.

2. Heritability does not indicate what proportion of an individual's phenotype is genetic. Since it is based on the variance, which can only be calculated on a group of individuals, heritability is characteristic of a population. An individual does not have heritability—a population does.

3. Heritability is not fixed for a trait. Thus there is no universal heritability for a trait like human stature. Rather, the heritability value for a trait depends on the genetic makeup and the specific

environment of the population. We could calculate the heritability of stature among Hopi Indians living in Arizona, for example, but heritability calculated for other populations and other environments might be very different. If the genetic composition of the group or the environment is different, the original heritability value is no longer valid. Changing groups or environments does not alter the way in which genes affect the trait, but it may change the amount of genetic and environmental variance for the trait, which would then alter the heritability.

4. Even if heritability is high in each of two populations, and the populations differ markedly in a particular trait, one cannot assume that the populations are genetically different. Heritability cannot be used to draw conclusions about the nature of differences between populations.

5. Traits shared by members of the same family do not necessarily have high heritability. When members of the same family share a trait, the trait is said to be **familial.** Familial traits may arise because family members share genes or because they are exposed to the same environmental factors. Familiality is not the same as heritability.

Keynote

Broad-sense heritability of a trait represents the proportion of the phenotypic variance that results from genetic differences among individuals. The narrow-sense heritability is more restricted, measuring the proportion of the phenotypic variance that results from additive genetic variance. Narrow-sense heritability allows quantitative geneticists to make predictions about the resemblance between parents and offspring, and it represents that part of the phenotypic variance that responds to selection in a predictable manner.

How Heritability Is Calculated

A number of different methods are available for calculating heritability. Many of the methods involve comparing related and unrelated individuals, or comparing individuals with different degrees of relatedness. If genes are important in determining the phenotypic variance, then closely related individuals should be more similar in phenotype than unrelated individuals, since they have more genes in common. Alternatively, if environmental factors are responsible for determining differences in the trait, then related individuals should be no more similar in phenotype than unrelated individuals. A crucial point to remember is that *the related individu-*

als studied must not share a more common environment than unrelated individuals. We assume that if related individuals are more similar in phenotype, it is because they share similar genes. If related individuals also share a more common environment than unrelated individuals, separating the effects of genes and environment is much more difficult and frequently impossible. The elimination of common environmental factors among related individuals can often be achieved in domestic plants and animals, and common environments may not exist among family members in the wild; however, control of the environment is very difficult to obtain in humans, where family structure and extended parental care create common environments for many related individuals.

Some of the methods used to calculate heritability include comparison of parents and offspring, comparison of full and half siblings, comparison of identical and nonidentical twins, and response-to-selection experiments. Heritability values for a number of traits in different species are given in Table 20.3.

Identification of Genes Influencing a Quantitative Trait

One of the limitations of the traditional approach of quantitative genetics is that it tells us little about the genes involved in quantitative traits. While heritability may indicate how much of the phenotypic variance in a trait is due to genetic differences, we learn nothing about how many genes are involved, where they are located, or what proteins the genes produce. Identification of individual genes that affect quantitative traits like yield in corn would allow more efficient design of breeding programs to enhance these traits; finding genes in humans that affect quantitative traits such as high blood pressure would allow identification of those at risk for the disease and could permit early application of preventive treatments.

Table 20.3
Heritability Values for Some Traits in Humans, Domesticated Animals, and Natural Populations

ORGANISM	TRAIT	HERITABILITY	REF*
Humans	Stature	0.65	1
	Serum immunoglobulin (IgG) level	0.45	1
Cattle	Milk yield	0.35	1
	Butterfat content	0.40	1
	Body weight	0.65	1
Pigs	Back-fat thickness	0.70	1
	Litter size	0.05	1
Poultry	Egg weight	0.50	1
	Egg production (to 72 weeks)	0.10	1
	Body weight (at 32 weeks)	0.55	1
Mice	Body weight	0.35	1
Drosophila	Abdominal bristle number	0.50	1
Jewelweed	Germination time	0.29	2
Milkweed bugs	Wing length (females)	0.87	3
	Fecundity (females)	0.50	3
Spring peeper (frog)	Size at metamorphosis	0.69	4
Wood frog	Development rate (mountain population)	0.31	5
	Size at metamorphosis (mountain population)	0.62	5

NOTE: The estimates given in this table apply to particular populations in particular environments; heritability values for other individuals may differ.

*REFERENCES: (1) Falconer, D. S. 1981. *Introduction to Quantitative Genetics,* 2d ed. New York; Longman; (2) Mitchell-Olds, T. 1986. *Evolution* 40:107–116; (3) Palmer, J. O., and H. Dingle. 1986. *Evolution* 40:767–777; (4) Travis, J., et al. 1987. *Evolution* 41:145–156; (5) Berven, K. A. 1987. *Evolution* 41:1088–1097.

Using genetic crosses and rather complex statistical procedures, it is possible to identify and map genes that affect quantitative traits, just as genes for Mendelian traits are mapped by linkage analysis (see Chapter 5). The approach for quantitative traits makes use of **genetic markers**. A genetic marker is a gene with an observable phenotypic effect and a known chromosome location. In specially designed genetic crosses, or in pedigrees with humans, one looks to see if a genetic marker is inherited along with the quantitative trait of interest. If a genetic marker and the trait are inherited together, then one or more genes affecting the quantitative trait (called *quantitative trait loci* or *QTLs*) are located close to the genetic marker on the same chromosome; in other words, the marker and the QTL are linked. Using this approach, J. M. Thoday and his colleagues mapped genes that affected bristle number in *Drosophila*, a trait that had been shown to exhibit quantitative inheritance. They identified five genes that accounted for almost 90 percent of the phenotypic variation in bristle number.

Similar studies have been carried out on a few other organisms, but past attempts to identify QTLs were limited by the paucity of known genetic markers. To identify QTLs a number of genetic markers of known chromosome location are required. The set of markers should ideally span the lengths of all the chromosomes, so that QTLs located anywhere in the genome can be identified. The markers must also be variable (each marker locus must possess two or more common alleles) and code for easily observable phenotypes.

The recent development of methods to identify restriction fragment length polymorphisms (RFLPs; see Chapter 13) provides the technology to identify large numbers of genetic markers that can be used for mapping genes that affect quantitative traits. For example, hundreds of new genetic markers in corn and tomatoes have been uncovered with RFLP methods. These markers now provide the means to map QTLs for a number of agriculturally important traits in these plants. Work on mapping human genes has also provided hundreds of RFLPs with known chromosomal locations; these can be used for mapping QTLs that affect significant quantitative traits in humans. Undoubtedly, these new developments will greatly enhance our understanding of how numerous loci interact to determine quantitative traits.

Response to Selection

Two fields of study in which quantitative genetics has played an important role are plant and animal breeding and evolutionary biology. Both fields are concerned with genetic change within groups of organisms: in the case of plant and animal breeding genetic change can lead to improvement in yield, hardiness, size, and other agriculturally important qualities; in the case of evolutionary biology genetic change occurs in natural populations as a result of the forces of nature. **Evolution** can be defined as genetic change that takes place over time within a group of organisms. Therefore, both evolutionary biologists and plant and animal breeders are interested in the process of evolution, and both use the methods of quantitative genetics to predict the rate and magnitude of genetic change that will occur.

Evolution results from a variety of forces, many of which will be explored more thoroughly in the next chapter. Here we concentrate on genetic change that results from selection. **Natural selection** is undoubtedly one of the most powerful forces of evolution in natural populations. First fully described by Charles Darwin, the concept of natural selection can be summarized in four statements:

1. More individuals are produced every generation than can survive and reproduce.

2. Much phenotypic variation exists among these individuals. Some of this variation is heritable.

3. Individuals with certain traits survive and reproduce better than others; these individuals leave more offspring in the next generation.

4. Since individuals with certain traits leave more offspring, and since those traits are heritable, traits that allowed increased survival and reproduction will be more common in the next generation.

The essential element of natural selection is that individuals with certain genotypes leave more offspring than others. In this way groups of individuals change or evolve over time and become better adapted to their particular environment.

Humans bring about evolution in domestic plants and animals through the similar process of **artificial selection.** In artificial selection humans, not nature, select the individuals that are to survive and reproduce. If the selected traits have a genetic basis, they too will change over time and evolve, just as traits in natural populations evolve as a result of natural selection. Artificial selection can be a powerful force in bringing about rapid evolutionary change, as evidenced by the extensive variation observed among domesticated plants and animals. For example, all breeds of domestic dogs are derived from one species that was domesticated some 10,000 years ago. The large number of breeds that exist today, encompassing a tremendous variety of sizes, shapes, colors, and even behaviors, has

■ *Figure 20.7*   All breeds of dogs have been produced by artificial selection, a demonstration of the tremendous power of this method for bringing about change.

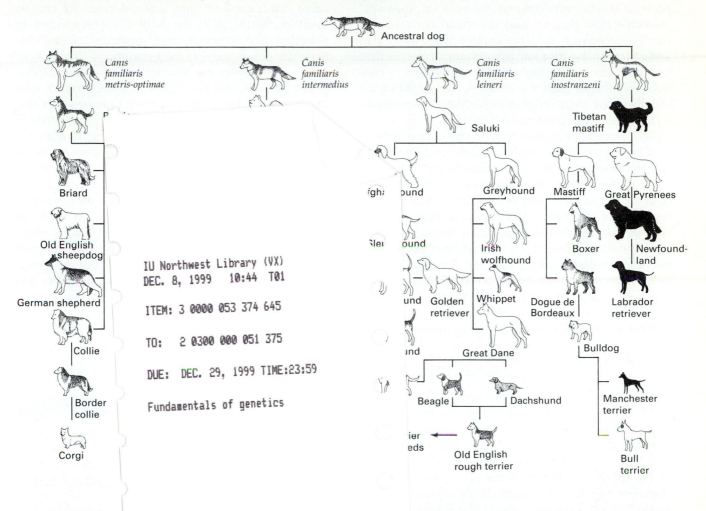

been produced b[...]
breeding (Figure 2[...]

Both the pr[...] described by Cha[...] tion practiced by p[...] on the presence of [...] variation is presen[...] uals can that pop[...] evolve. Furthermore, the amount and the type of genetic variation present is crucial in determining how fast evolution will occur. Both evolutionary biologists and breeders are interested in the question of how much genetic variation for a particular trait exits within a population, which is essentially the question of heritability. Quantitative genetics is often employed to answer this question.

Estimating the Response to Selection

When natural or artificial selection is imposed upon a phenotype, the phenotype will change from one generation to the next, provided that genetic variation underlying the trait is present in the pop-

[...] The amount that a continuous trait changes [...] eneration is termed the **selection response.** [...] rate the concept of selection response, sup- [...] geneticist wishes to produce a strain of [...] *la melanogaster* with large body size. In [...] increase body size in fruit flies, the geneti- [...] ld first examine flies from a genetically diverse culture and would measure the body size of these *unselected flies*. Suppose that our geneticist found the mean body weight of the unselected flies to be 1.3 mg. After determining the mean body weight of this population, the geneticist would select flies that were endowed with large bodies (assume that the mean body weight of these selected flies was 3.0 mg). He would then place the large, selected flies in a separate culture vial and allow them to interbreed. After the F_1 offspring of these selected parents emerged, the geneticist would measure the body weights of the F_1 flies and compare them with the body weights of the original, unselected population.

What our geneticist has done in this procedure is to apply selection for large body size to the population of fruit flies. If additive genetic variation underlies the variation in body size of the original population, the offspring of the selected flies will resemble their parents and the mean body size of the F$_1$ generation will be greater than the mean body size of the original population. If the F$_1$ flies have a mean body weight of 2.0 mg, which is considerably larger than the mean body weight of 1.3 mg observed in the original, unselected population, a response to selection has occurred.

The amount of change that occurs in one generation, or the selection response, is dependent on two things: the narrow-sense heritability and the **selection differential.** The selection differential is defined as the difference between the mean phenotype of the selected parents and the mean phenotype of the unselected population. In our example of body size in fruit flies, the original population had a mean weight of 1.3 mg, and the mean weight of the selected parents was 3.0 mg, so the selection differential is 3.0 mg − 1.3 mg = 1.7 mg. The selection response is related to the selection differential and the heritability by the following formula:

Selection response = narrow-sense heritability
× selection differential

When the geneticist applied artificial selection to body size in fruit flies, the difference in the mean body weight of the F$_1$ flies and the original population was 2.0 mg − 1.3 mg = 0.7 mg, which is the response to selection. We now have values for two of the three parameters in the above equation: the selection response (0.7 mg) and the selection differential (1.7 mg). By rearranging the formula for the selection response, we can solve for the narrow-sense heritability.

Narrow-sense heritability
= h^2
= selection response/selection differential

$$h^2 = 0.7 \text{ mg}/1.7 \text{ mg} = 0.41$$

Measuring the response to selection provides another means for determining the narrow-sense heritability. Heritability determined from the selection response is called the **realized heritability,** and heritabilities for many traits are determined in this way.

A trait will continue to respond to selection, generation after generation, as long as genetic variation for the trait remains within the population. The results from an actual, long-term selection experiment on phototaxis in *Drosophila pseudoobscura* are presented in Figure 20.8. Phototaxis is a behavioral response to light. In this study flies were scored for the number of times they moved toward light in a total of 15 light-dark choices. Two different

response-to-selection experiments were carried out. In one, attraction to light was selected, and in the other, avoidance of light was selected. As can be seen in Figure 20.8, the fruit flies responded to selection for positive and negative phototactic behavior for a number of generations. Eventually, however, the response to selection tapered off (after generation 15), and no consistent directional change in phototactic behavior occurred. One possible reason for this lack of response in later generations is that no more genetic variation for phototactic behavior exists within the population. In other words, all flies at this point are homozygous for all the alleles affecting the behavior. If this were the case, phototactic behavior could not undergo further evolution in this population unless input of additional genetic variation occurred. More often, however, some variation still exists for the trait even after the selection response levels off, but the population fails to respond to selection because the genes for the selected trait have detrimental effects on other traits. For example, flies that exhibit stronger phototaxis might not respond properly to other important stimuli such as attraction to food and thus would not survive well.

Keynote

The amount that a trait changes in one generation as a result of selection for the trait is called the selection response or the response to selection. The magnitude of selection response depends on both

■ *Figure 20.8* Selection for phototaxis in *Drosophila pseudoobscura.* The upper graph is the line selected for avoidance of light. The lower graph is the line selected for attraction to light. The phototactic score is the number of times the fly moved toward the light out of a total of 15 light-dark choices. (After Dobzhansky and Spassky, 1969. *Proc. Natl. Acad. Sci.* 62: 75–80.)

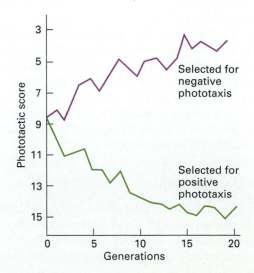

■ *Figure 20.9*   (a) Directional selection; (b) Stabilizing selection, and (c) Disruptive selection. At the top of the graph are the original distributions of phenotypes. At the bottom are the distributions of phenotypes after several generations of selection. The shaded areas represent those phenotypes that survive and reproduce best. (After Audesirk, G. and T. Audesirk, 1993. *Biology Life on Earth.* New York: Macmillan.

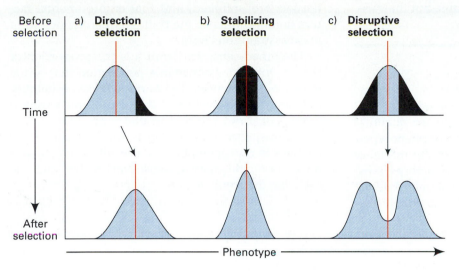

Types of Selection

Figure 20.9 shows three different types of selection acting on a continuous trait. In **directional selection** the individuals that survive and reproduce best are those that have one of the extreme phenotypes. These individuals are at one end of the phenotypic distribution (Figure 20.9[a]). If genes determine the differences in phenotype—if the trait has high heritability—then a response to selection will occur and the population mean will shift in the direction of the extreme phenotype. For instance, perhaps individuals with large body size produced more offspring than those with small body size. If directional selection for large body size occurred, the mean body size of the population would increase over time.

A second type of selection is **stabilizing selection** (Figure 20.9[b]). In stabilizing selection individuals with the extreme phenotypes are at a disadvantage and individuals with intermediate phenotype survive and reproduce best. Stabilizing selection will maintain the same mean phenotype over time, but the variance in the trait will decrease, be-

cause genes that produce the extreme phenotypes are eliminated from the population. Stabilizing selection would occur if individuals with intermediate body size produced more offspring. If stabilizing selection for body size occurred, the average body size of the population would stay the same, but there would be fewer very small and fewer very large individuals in the population. **Disruptive selection** (Figure 20.9[c]) occurs when individuals at both ends of the phenotypic distribution survive and reproduce best and individuals with intermediate phenotype are at a disadvantage. Disruptive selection will maintain the same mean phenotype over time, but because genes for extreme phenotypes are favored, the variance in the trait increases. If disruptive selection for body size occurred, small and large individuals would leave more offspring than those of intermediate size. Over time, the mean body size of the population would remain the same, but the variance in body size would increase.

K e y n o t e

In directional selection individuals with one extreme phenotype survive and reproduce best; over time, the mean phenotype of the population will shift in the direction of the selected phenotype. Individuals of intermediate phenotype survive and reproduce best with stabilizing selection; over time, the mean phenotype will remain the same but the variance in the phenotype will decrease. In disruptive selection individuals at both ends of the phenotype survive and reproduce best; over time, the mean phenotype will remain the same but the variance in the phenotype will increase.

Summary

Quantitative genetics is the field of genetics that studies the inheritance of continuous, or quantitative, traits—those traits with a range of phenotypes. Continuous traits usually result from the influence of multiple genes and environmental factors. Statistics such as the mean, variance, and standard deviation can be used to describe continuous traits.

Polygenic traits are caused by genes at many loci, each of which follows the principles of Mendelian inheritance. The multiple-gene hypothesis assumes that the effects of the genes are additive, that no genetic interaction occurs between gene loci, that the loci are unlinked, and that environmental factors do not influence the trait. Frequently, not all of these assumptions hold.

The broad-sense heritability is the proportion of the phenotypic variance in a population that is due to genetic differences. The narrow-sense heritability indicates the proportion of the phenotypic variance that results from additive genetic variance.

The response to selection is the amount a trait changes in one generation as a result of selection. It depends on the narrow-sense heritability and the selection differential.

Analytical Approaches for Solving Genetics Problems

Q20.1 Assume that the four genes *A*, *B*, *C*, and *D* are members of a multiple-gene series that control a quantitative trait. Each gene assorts independently. Each capital-letter allele of these genes has a duplicate, cumulative effect in that it contributes 3 cm of height to the organism when it is present. In addition allele *L* of a fifth gene is always present in the homozygous state, and the *LL* genotype contributes a constant 40 cm of height. The alleles *a*, *b*, *c*, and *d* do not contribute anything to the height of the organism. If we ignore height variation caused by environmental factors, an organism with genotype *AA BB CC DD LL* would be 64 cm high, and one with genotype *aa bb cc dd LL* would be 40 cm. A cross is made of *AA bb CC DD LL* × *aa BB cc DD LL* and is carried into the F_2 by selfing of the F_1.

a. How does the size of the F_1 individuals compare with the size of each of the parents?

b. Compare the mean of the F_1 with the mean of the F_2, and comment on your findings.

c. What proportion of the F_2 population would show the same height as the *AA bb CC DD LL* parent?

d. What proportion of the F_2 population would show the same height as the *aa BB cc DD LL* parent?

e. What proportion of the F_2 population would breed true for the height shown by the *aa BB cc DD LL* parent?

f. What proportion of the F_2 population would breed true for the height characteristic of F_1 individuals?

A20.1 This question explores our understanding of the basic genetics involved in a multiple-gene series that in this case controls a quantitative trait. The approach we will take is essentially the same as the approach used with a series of independently assorting genes that control distinctly different traits. That is, we make predictions on the basis of genotypes and relate the results to phenotypes, or we make predictions on the basis of phenotypes and relate the results to genotypes.

a. Each allele represented by a capital letter contributes 3 cm of height to the base height of 40 cm, which is controlled by the ever-present *LL* homozygosity. Therefore the *AA bb CC DD LL* parent, which has six contributing alleles from the *A*-through-*D*, multiple-gene series, is 40 + (6 × 3) = 58 cm high. Similarly, the *aa BB cc DD LL* parent has four contributing alleles and therefore is 40 + 12 = 52 cm high. The F_1 from a cross between these two individuals would be heterozygous for the *A*, *B*, and *C* loci and homozygous for *D* and *L*, that is, *Aa Bb Cc DD LL*. This progeny has five contributing alleles apart from *LL* and hence is 40 + 15 = 55 cm high.

b. The F_2 is derived from a self of the *Aa Bb Cc DD LL* F_1. All the F_2 individuals will be *DD LL*, making them at least 40 + 6 = 46 cm high. Now we must deal with the heterozygosity at the other three loci. What we need to calculate is the relative proportions of individuals with all the various possible numbers of contributing alleles. This calculation is equivalent to determining the relative distribution of three independently assorting traits, each showing incomplete dominance. In other words, we must calculate directly the relative frequencies of all possible genotypes for the three loci and collect those with no, one, two, three, four, five, and six contributing alleles. Thus the probability of getting an individual with two contributing alleles for each locus is 1/4, the probability of getting an individual with one contributing allele for each locus is 1/2, and the probability of getting an individual with no contributing alleles for each locus is 1/4. So the probability of getting an F_2 individual with six contributing alleles for the A, B, and C loci is $(1/4)^3 = 1/64$, and the same probability is obtained for an individual with no contributing alleles. This analysis gives us a clue about how we should consider all the possible combinations of genotypes that have the other numbers of contributing alleles. That is, the simplest approach is to compute the coefficients in the binomial expansion of $(a + b)^6$. The expansion gives a 1:6:15:20:15:6:1 distribution of zero, one, two, three, four, five, and six contributing alleles, respectively. Now since each contributing allele in the A, B, and C set contributes 3 cm of height over the 46-cm height given by the DD LL genotype common to all, then the F_2 individuals would fall into the following distribution:

NUMBER OF CONTRIBUTING ALLELES	HEIGHT ADDED TO BASIC HEIGHT OF 46 CM FOR COMMON DD LL GENOTYPE (CM)	HEIGHT OF INDIVIDUALS (CM)	FREQUENCY
6	18	64	1/64
5	15	61	6/64
4	12	58	15/64
3	9	55	20/64
2	6	52	15/64
1	3	49	6/64
0	0	46	1/64

The distribution is clearly symmetrical, giving an average of 55 cm, the same height shown in F_1 individuals.

c. The AA bb CC DD LL parent was 58 cm, so we can read the proportion of F_2 individuals that show this same height directly from the table in part b. The answer is 15/64.

d. The aa BB cc DD LL parent was 52 cm, and from the table in part b the proportion of F_2 individuals that show this same height is 15/64.

e. We are asked to determine the proportion of the F_2 population that would breed true for the height shown by the aa BB cc DD LL parent, which was 52 cm. To breed true, the organism must be homozygous. We have also established that DD LL is a constant genotype for the F_2 individuals, giving a basic height of 46 cm. Therefore, for a height of 52 cm, two additional, active, contributing alleles must be present, apart from those at the D and L loci. With the requirement for homozygosity, there are only three genotypes that give a 52-cm height; they are AA bb cc DD LL, aa BB cc DD LL, and aa bb CC DD LL. The probability of each combination occurring in the F_2 is 1/64, so the answer to the problem is $1/64 + 1/64 + 1/64 = 3/64$. (Note that the individual probability for each genotype may be calculated. That is, probability of $AA = 1/4$, probability of bb 1/4, probability of $cc = 1/4$, probability of DD $LL = 1$, giving an overall probability for AA bb cc DD LL of 1/64.)

f. We are asked to determine the proportion of the F_2 population that would breed true for the height characteristic of F_1 individuals. Again, the basic height given by DD LL is 46 cm. The F_1 height is 55 cm, so three contributing alleles must be present in addition to DD LL to give that height, since (3×3) cm = 9 cm, and 9 cm + 46 cm = 55 cm. However, since an individual must be homozygous to be true-breeding, the answer to this question is none, since 3 is an odd number, meaning that at least one locus must be heterozygous in order to get the 55-cm height.

Questions and Problems

20.1 Using methods described in this chapter, answer the following questions.
a. In a family of 6 children, what is the probability that 3 will be girls and 3 will be boys?
b. In a family of 5 children, what is the probability that 1 will be a boy and 4 will be girls?
c. What is the probability that in a family of 6 children, all will be boys?

*20.2 In flipping a coin, there is a 50 percent chance of obtaining heads and a 50 percent chance of obtaining tails on each flip. If you flip a coin 10 times, what is the probability of obtaining exactly 5 heads and 5 tails?

*20.3 The F_1 generation from a cross of two pure-breeding parents that differ in a size character is usually no more variable than the parents. Explain.

20.4 If two pure-breeding strains, differing in a size trait, are crossed, is it possible for F_2 individuals to have phenotypes that are more extreme than either grandparent (i.e., be larger than the largest or smaller than the smallest in the parental generation)? Explain.

20.5 Two pairs of genes with two alleles each, A/a and B/b, determine plant height additively in a population. The homozygote $AA\ BB$ is 50 cm tall, the homozygote $aa\ bb$ is 30 cm tall.
a. What is the F_1 height in a cross between the two homozygous stocks?
b. What genotypes in the F_2 will show a height of 40 cm after an $F_1 \times F_1$ cross?
c. What will be the F_2 frequency of the 40-cm plants?

*20.6 Three independently segregating loci (A, B, C), each with two alleles, determine height in a plant. Each capital-letter (contributing) allele adds 2 cm to a base height of 2 cm.
a. What are the heights expected in the F_1 progeny of a cross between homozygous strains $AA\ BB\ CC$ (14 cm) $\times$ $aa\ bb\ cc$ (2 cm)?
b. What is the distribution of heights (frequency and phenotype) expected in an $F_1 \times F_1$ cross?
c. What proportion of F_2 plants will have heights equal to the heights of the original two parental strains?
d. What proportion of the F_2 will breed true for the height shown by the F_1?

20.7 Repeat Problem 20.6, but assume that each capital-letter (contributing) allele acts to double the existing height; for example, $Aa\ bb\ cc = 4$ cm, $AA\ bb\ cc = 8$ cm, $AA\ Bb\ cc = 16$ cm, and so on.

20.8 Assume two equally and additively contributing pairs of alleles control flower length in nasturtiums. A completely homozygous plant with 10 mm flowers is crossed to a completely homozygous plant with 30 mm flowers. F_1 plants all have flowers about 20 mm long. F_2 plants show a range of lengths from 10 to 30 mm, with about 1/16 of the F_2 having 10 mm flowers and 1/16 having 30 mm flowers. What distribution of flower length would you expect to see in the offspring of a cross between an F_1 plant and the 30-mm parent?

*20.9 In a particular experiment the mean internode length in spikes of the barley variety *asplund*

was found to be 2.12 mm. In the variety *abed binder* the mean internode length was found to be 3.17 mm. The mean of the F_1 of a cross between the two varieties was approximately 2.7 mm. The F_2 gave a continuous range of variation from one parental extreme to the other. Analysis of the F_3 generation showed that in the F_2 8 out of the total 125 individuals were of the *asplund* type, giving a mean of 2.19 mm. Eight other individuals were similar to the parent *abed binder*, giving a mean internode length of 3.24 mm. Is the internode length in spikes of barley a discontinuous or a quantitative trait? Why?

20.10 From the information given in Problem 20.9, determine how many gene pairs involved in the determination of internode length are segregating in the F_2.

20.11 Assume that the difference between a type of oats yielding about 4 g per plant and a type yielding 10 g is the result of three equal and cumulative multiple-gene pairs, $AA\ BB\ CC$. If you cross the type yielding 4 g with the type yielding 10 g, what will be the phenotypes of the F_1 and the F_2? What will be their distribution?

*20.12 Assume that in squashes the difference in fruit weight between a 3-lb type and a 6-lb type is due to three allelic pairs, A/a, B/b, and C/c. Each capital-letter (contributing) allele contributes a half pound to the weight of the squash. From a cross of a 3-lb plant ($aa\ bb\ cc$) with a 6-lb plant ($AA\ BB\ CC$), what will be the phenotypes of the F_1 and the F_2? What will be their distribution?

20.13 Refer to the assumptions stated in Problem 20.12. Determine the range in fruit weight of the offspring in the following squash crosses:
(a) $Aa\ Bb\ CC \times aa\ Bb\ Cc$; (b) $AA\ bb\ Cc \times Aa\ BB\ cc$; (c) $aa\ BB\ cc \times AA\ BB\ cc$.

*20.14 Assume that the difference between a corn plant 10 dm (decimeters) high and one 26 dm high is due to four pairs of equal and cumulative multiple alleles, with the 26-dm plants being $AA\ BB\ CC\ DD$ and the 10-dm plants being $aa\ bb\ cc\ dd$.
a. What will be the size and genotype of an F_1 from a cross between these two true-breeding types?
b. Determine the limits of height variation in the offspring from the following crosses:
(1) $Aa\ BB\ cc\ dd \times Aa\ bb\ Cc\ dd$;
(2) $aa\ BB\ cc\ dd \times Aa\ Bb\ Cc\ dd$;
(3) $AA\ BB\ Cc\ DD \times aa\ BB\ cc\ Dd$;
(4) $Aa\ Bb\ Cc\ Dd \times Aa\ bb\ Cc\ Dd$.

20.15 Refer to the assumptions given in Problem 20.14. But for this problem two 14-dm corn plants, when crossed, only give 14-dm offspring (case A).

Two other 14-dm plants give one 18-dm, four 16-dm, six 14-dm, four 12-dm, and one 10-dm offspring (case B). Two other 14-dm plants, when crossed, give one 16-dm, two 14-dm, and one 12-dm offspring (case C). What genotypes for each of these 14-dm parents (cases A, B, and C) would explain these results? Would it be possible to get a plant taller than 18 dm by selection in any of these families?

20.16 A quantitative geneticist determines the following variance components for leaf width in a population of wildflowers growing along a roadside in Kentucky:

Additive genetic variance (V_A)	= 4.2
Dominance genetic variance (V_D)	= 1.6
Interaction genetic variance (V_I)	= 0.3
Environmental variance(V_E)	= 2.7
Genetic-environmental variance (V_{GE})	= 0.0

a. Calculate the broad-sense heritability and the narrow-sense heritability for leaf width in this population of wildflowers.
b. What do the heritabilities obtained in part a indicate about the genetic nature of leaf width variation in this plant?

*__20.17__ On his farm in Kansas a farmer is growing a variety of wheat called TK138. He calculates the narrow-sense heritability for yield (the amount of wheat produced per acre) and finds that the heritability of yield for TK138 is 0.95. The next year he visits a farm in Poland and observes that a Russian variety of wheat, UG334, growing there has only about 40 percent as much yield as TK138 grown on his farm in Kansas. Since he found the heritability of yield in his wheat to be very high, he concludes that the American variety of wheat (TK138) is genetically superior to the Russian variety (UG334), and he tells the Polish farmers that they can increase their yield by using TK138. What is wrong with his conclusion?

*__20.18__ A scientist wishes to determine the narrow-sense heritability of tail length in mice. He measures tail length among the mice of a population and finds a mean tail length of 9.7 cm. He then selects the ten mice in the population with the longest tails; mean tail length in these selected mice is 14.3 cm. He interbreeds the mice with the long tails and examines tail length in their progeny. The mean tail length in the F_1 progeny of the selected mice is 13 cm.

Calculate the selection differential, the response to selection, and the narrow-sense heritability for tail length in these mice.

*__20.19__ The narrow-sense heritability of egg weight in a particular flock of chickens is 0.60. A farmer selects for increased egg weight in this flock. The difference in the mean egg weight of the unselected chickens and the selected chickens is 10 g. How much should egg weight increase in the offspring of the selected chickens?

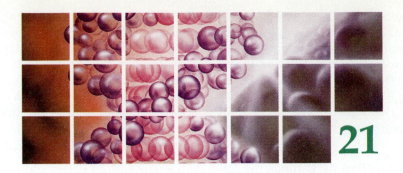

21

Population Genetics

- The gene pool of a population is the total of all genes within a group of sexually interbreeding individuals. The gene pool can be described in terms of allelic and genotypic frequencies.

- The Hardy-Weinberg law states that in a large, randomly mating population, free from evolutionary forces, the allelic frequencies do not change, and the genotypic frequencies stabilize after one generation in the proportions p^2, $2pq$, and q^2, where p and q equal the allelic frequencies of the population.

- The classical, balance, and neutral models are hypotheses to explain how much genetic variation should exist within natural populations and what forces are responsible for the variation observed.

- Mutation, genetic drift, migration, and natural selection are forces that can alter allelic frequencies of a population.

- Recurrent mutation changes allelic frequencies, and at equilibrium the relative rates of forward and reverse mutations determine allelic frequencies of a population in the absence of other forces.

- Genetic drift is random change in allelic frequencies due to chance. Genetic drift produces genetic change within populations, genetic differentiation among populations, and loss of genetic variation within populations.

- Migration, also termed gene flow, involves movement of genes between populations. Migration can alter the allelic frequencies of a population, and it tends to reduce genetic divergence among populations.

- Natural selection is differential reproduction of genotypes. It is measured by fitness, which is the relative reproductive ability of genotypes.

- Nonrandom mating affects the genotypic frequencies of a population, inbreeding leads to an increase in homozygosity, and outbreeding leads to an increase in heterozygosity.

- Principles of population genetics can be applied to the management of rare and endangered species. Genetic diversity is best maintained by establishing a population with adequate founders, expanding the population rapidly, avoiding inbreeding, and maintaining equal sex ratio and equal family size.

- Rates of evolution can be measured by comparing DNA or RNA sequences. Different genes and even different parts of the same gene usually evolve at different rates.

- In eukaryotic organisms genes frequently occur in multiple copies with identical or similar sequences. A group of such genes is termed a multiple-gene family.

- The mitochondrial DNA of some organisms evolves at a faster rate than the nuclear DNA.

- Concerted evolution is a process that maintains sequence uniformity among multiple copies of the same sequence within a species.

- Evolutionary relationships among organisms can be revealed by study of DNA and RNA sequences.

POPULATION GENETICS is the field of genetics that studies heredity in groups of individuals. Population geneticists investigate the patterns of genetic variation found within groups and the genetic basis of evolutionary change. In this discipline the perspective shifts away from the individual and the cell, and focuses instead on a **Mendelian population.** A Mendelian population is a group of *interbreeding* individuals, who through their interbreeding share a common set of genes over evolutionary time. The genes shared by the individuals of a Mendelian population are called the **gene pool.** To understand the genetics of the evolutionary process, we study the gene pool of a Mendelian population, rather than the genotypes of its individual members. Except for somatic mutations, individuals are born and die with the same set of genes; what changes genetically over time (evolves) is the hereditary makeup of the Mendelian population.

Population geneticists frequently develop mathematical models and equations to describe what happens to the gene pool of a population under various conditions. An example is the set of equations that describes the influence of random mating on the allelic and genotypic frequencies of an infinitely large population. This model is called the **Hardy-Weinberg law** and will be discussed later in the chapter. Students often have difficulty grasping the significance of such mathematical models, because the models are frequently simplistic and require numerous assumptions that are unlikely to be met by organisms in the real world. Beginning with simple models is useful, however, because we can use them to examine the effects of individual evolutionary factors on the genetic structure of a population. Once we understand the results of the noncomplex models, we can incorporate more realistic conditions into the equations. In the case of the Hardy-Weinberg law, the assumptions of infinitely large size and random mating may seem unrealistic, but these assumptions are necessary for simplifying the mathematical analysis. We cannot begin to understand the impact of nonrandom mating and limited population size on allelic frequencies until we first know what happens under the simpler conditions of random mating and large population size.

Genotypic and Allelic Frequencies

Genotypic Frequencies

To study the genetic composition of a Mendelian population, population geneticists must first quantitatively describe the gene pool of the group. This is done by calculating genotypic frequencies and allelic frequencies within the population. A frequency is a proportion and always ranges between 0 and 1. If 43 percent of the people in a group have red hair, the frequency of red hair in the group is 0.43. To calculate the **genotypic frequencies** at a specific locus, we count the number of individuals with one particular genotype and divide this number by the total number of individuals in the population. We do this for each of the genotypes at the locus, and all the genotypic frequencies should add up to 1. Consider a locus that determines the pattern of spots in the scarlet tiger moth, *Panaxia dominula* (Figure 21.1). Three genotypes are present in most populations, and each genotype produces a different phenotype. E. B. Ford collected moths at one locality in England and found the following numbers of genotypes: 452 *BB*, 43 *Bb*, and 2 *bb*, out of a total of 497 moths. The genotypic frequencies are:

$$f(BB) = 452/497 = 0.909$$
$$f(Bb) = 43/497 = 0.087$$
$$f(bb) = 2/497 = \underline{0.004}$$
$$\text{Total} \quad 1.000$$

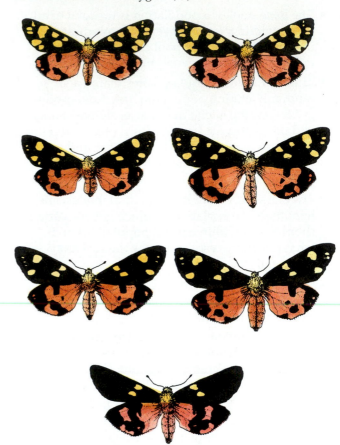

■ *Figure 21.1 Panaxia dominula,* the scarlet tiger moth. The top two moths are normal homozygotes (*BB*), those in rows 2 and 3 are heterozygotes (*Bb*), and the bottom moth is the rare homozygote (*bb*).

Allelic Frequencies

While genotypic frequencies are useful for examining the effects of certain evolutionary forces on a population, in most cases population geneticists use frequencies of alleles to describe the genetic structure of the gene pool. The frequencies of alleles at a locus are called the **allelic frequencies** or the **gene frequencies.** The use of allelic frequencies offers several advantages over the genotypic frequencies. First, there are always fewer alleles than genotypes, so the gene pool can be described with fewer parameters when allelic frequencies are used. For example, if a locus has three alleles (A^1, A^2, and A^3), six genotypic frequencies (frequencies for A^1A^1, A^1A^2, A^2A^2, A^1A^3, A^2A^3, A^3A^3) must be calculated to describe the gene pool, whereas use of the allelic frequencies requires the calculation of only three allelic frequencies (frequencies for A^1, A^2, and A^3). In general the number of genotypes is $n(n+1)/2$, thus as the number of alleles increases, the number of genotypes becomes very large. Furthermore, in sexually reproducing organisms the alleles comprising the genotype separate when gametes are formed, and alleles, not genotypes, are passed from one generation to the next. Consequently, only alleles have continuity over time, and the gene pool evolves through changes in the frequencies of alleles. Thus allelic frequencies better represent the gene pool of a population than do the genotypic frequencies.

Allelic frequencies may be calculated in either of two ways: from the observed numbers of different genotypes, or from the genotypic frequencies. We can calculate the allelic frequencies directly from the *numbers* of individuals with different genotypes. In this method we count the number of alleles of one type and divide it by the total number of alleles in the population:

$$\frac{\text{Allelic}}{\text{frequency}} = \frac{\substack{\text{number of copies of a given allele} \\ \text{in the population}}}{\text{sum of all alleles in the population}}$$

Imagine a population of 1000 diploid individuals with 353 *AA*, 494 *Aa*, and 153 *aa* individuals. Each *AA* individual has two *A* alleles, while each *Aa* heterozygote possesses only a single *A* allele. Therefore, the number of *A* alleles in the population is (2 × the number of *AA* homozygotes) + (the number of *Aa* heterozygotes), or (2 × 353) + 494 = 1200. Because every diploid individual has two alleles, the total number of alleles in the population will be twice the number of individuals, or 2 × 1000. Using the equation given above, the frequency of allele *A* is 1200/2000 = 0.60.

The frequencies of two alleles, *f(A)* and *f(a)*, are commonly symbolized as *p* and *q*. When two alleles are present at a locus, we can use the following formulas for calculating allelic frequencies:

$$p = f(A) = \frac{(2 \times \text{number of } AA \text{ homozygotes}) + (\text{number of } Aa \text{ heterozygotes})}{(2 \times \text{total number of individuals})}$$

$$q = f(a) = \frac{(2 \times \text{number of } aa \text{ homozygotes}) + (\text{number of } Aa \text{ heterozygotes})}{(2 \times \text{total number of individuals})}$$

The allelic frequencies for a locus should always add up to 1. Therefore, once *p* is calculated, *q* can be easily obtained by subtraction: $1 - p = q$. It is important to note that the calculation of allelic frequencies requires the ability to identify all the genotypes. If dominance is present, and two genotypes have the same phenotype (e.g., *AA* and *Aa* are both yellow), it is impossible to calculate allelic frequencies directly. Later in the chapter we will discuss how allelic frequencies can be estimated indirectly when dominance is present.

Allelic Frequencies with Multiple Alleles   Suppose we have three alleles—A^1, A^2, and A^3—at a locus, and we want to determine the allelic frequencies. Here we employ the same rule that we used with two alleles: add up the number of alleles of each type and divide by the total number of alleles in the population:

$$p = f(A^1) = \frac{(2 \times A^1A^1) + (A^1A^2) + (A^1A^3)}{(2 \times \text{total number of individuals})}$$

$$q = f(A^2) = \frac{(2 \times A^2A^2) + (A^1A^2) + (A^2A^3)}{(2 \times \text{total number of individuals})}$$

$$r = f(A^3) = \frac{(2 \times A^3A^3) + (A^1A^3) + (A^2A^3)}{(2 \times \text{total number of individuals})}$$

Allelic Frequencies from Genotypic Frequencies
A second method for calculating allelic frequencies is from the *genotypic frequencies*. This calculation may be quicker if we have already determined the frequencies of the genotypes. For two alleles the allelic frequencies are

$$p = f(A) = (\text{frequency of the } AA \text{ homozygote}) + (1/2 \times \text{frequency of the } Aa \text{ heterozygote})$$

$$q = f(a) = (\text{frequency of the } aa \text{ homozygote}) + (1/2 \times \text{frequency of the } Aa \text{ heterozygote})$$

The frequency of the homozygote is added to half of the heterozygote frequency because half of the heterozygote's alleles are *A* and half are *a*. If three alleles (A^1, A^2, and A^3) are present in the population, the allelic frequencies are

$$p = f(A^1) = f(A^1A^1) + 1/2f(A^1A^2) + 1/2f(A^1A^3)$$

$$q = f(A^2) = f(A^2A^2) + 1/2f(A^1A^2) + 1/2f(A^2A^3)$$

$$r = f(A^3) = f(A^3A^3) + 1/2f(A^1A^3) + 1/2f(A^2A^3)$$

Allelic Frequencies at an X-Linked Locus   Calculation of allelic frequencies at an X-linked locus is slightly more complicated, because males have only a single X chromosome and thus only a single X-linked allele. Consider two X-linked alleles, X^A and X^a. Females may be homozygous (X^AX^A or X^aX^a) or heterozygous (X^AX^a); males are hemizygous and carry only one copy of the X-linked allele (X^AY or X^aY). To calculate the frequency of an X-linked allele, we can use the same rules we used for autosomal loci. Each female homozygous for the allele carries two copies of it; heterozygous females and hemizygous males have only one copy of the allele. We first total the number of copies of the allele by multiplying the number of homozygous females by 2, then adding the number of heterozygous females and the number of hemizygous males. We next divide by the total number of alleles in the population. When determining the total number of alleles, we add twice the number of females (because each female has two X-linked alleles) to the number of males (who have a single allele at X-linked loci). Using this reasoning, the frequencies of two alleles at an X-linked locus (X^A and X^a) are determined with the following equations:

$$p = f(X^A) = \frac{(2 \times X^AX^A\ \text{females}) + (X^AX^a\ \text{females}) + (X^AY\ \text{males})}{(2 \times \text{number of females}) + (\text{number of males})}$$

$$q = f(X^a) = \frac{(2 \times X^aX^a\ \text{females}) + (X^AX^a\ \text{females}) + (X^aY\ \text{males})}{(2 \times \text{number of females}) + (\text{number of males})}$$

These equations calculate the allelic frequencies from the *numbers of individuals* with different genotypes. We can also determine allelic frequencies at an X-linked locus from the *frequencies of the genotypes* using the following equations:

$$p = f(X^A) = f(X^AX^A) + 1/2f(X^AX^a) + f(X^AY)$$

$$q = f(X^a) = f(X^aX^a) + 1/2f(X^AY) + f(X^aY)$$

Students should strive to understand the logic behind these calculations, not just memorize the formulas. If you fully understand the basis of the calculations, you will not need to remember the exact equations and will be able to determine allelic frequencies for any situation.

The Hardy-Weinberg Law

The Hardy-Weinberg law is the most important principle in population genetics. It explains how random mating influences allelic and genotypic frequencies of a population. The Hardy-Weinberg law is named after the two individuals who independently discovered the relationship between allele and genotypic frequencies in the early 1900s (Box 21.1).

The Hardy-Weinberg law is divided into three parts—a set of assumptions and two major results—as follows:

Part 1: In an infinitely large, randomly mating population, free from evolutionary forces (mutation, migration, natural selection);

Part 2: The frequencies of the alleles do not change over time; and

Part 3: After one generation of random mating,

BOX 21.1

Hardy, Weinberg, and the History of Their Contribution to Population Genetics

Godfrey H. Hardy (1877–1947), a mathematician at Cambridge University, often met R. C. Punnett, the Mendelian geneticist, at the faculty club. One day in 1908 Punnett told Hardy of a problem in genetics that he attributed to a strong critic of Mendelism, G. U. Yule (Yule later denied having raised the problem). Supposedly, Yule said that if the gene for short fingers (brachydactyly) was dominant (which it is) and its allele for normal-length fingers was recessive, then short fingers ought to become more common with each generation. Eventually virtually everyone in Britain would have short fingers. Punnett believed the argument was incorrect, but he could not prove it.

Hardy was able to write a few equations showing that, given any particular frequency of alleles for short fingers and alleles for normal fingers in a population, the relative number of people with short fingers and people with normal fingers will stay the same generation after generation providing no natural selection is involved that favors one phenotype or the other in producing offspring. Hardy published a short paper describing the relationship between genotypes and phenotypes in populations, and within a few weeks a paper was published by Wilhelm Weinberg (1862–1937), a German physician from Stuttgart, that clearly stated the same relationship. The Hardy-Weinberg law signaled the beginning of modern population genetics.

We should note that in 1903 the American geneticist W. E. Castle of Harvard University was actually the first to recognize the relationship between allele and genotypic frequencies, but it was Hardy and Weinberg who clearly described the relationship in mathematical terms. The law is sometimes referred to as the Castle-Hardy-Weinberg law.

the genotypic frequencies will remain in the proportions p^2 (frequency of AA), $2pq$ (frequency of Aa), and q^2 (frequency of aa), where p is the allelic frequency of A and q is the allelic frequency of a. The sum of the genotypic frequencies should be equal to 1 (that is, $p^2 + 2pq + q^2 = 1$).

The Hardy-Weinberg law explains what happens to a population's allelic and genotypic frequencies as the genes are passed from generation to generation in the absence of evolutionary forces.

Assumptions of the Hardy-Weinberg Law

Part 1 of the Hardy-Weinberg law presents certain conditions, or assumptions, which must be present for the law to apply. First, the law indicates that the population must be infinitely large. If a population is limited in size, chance (random) deviations from expected ratios can cause changes in allelic frequency. Changes in allelic frequency due to chance deviations are called **genetic drift.** It is true that the assumption of infinite size in part 1 is unrealistic— no population has an infinite number of individuals. However, chance has a significant effect on allelic frequencies only in "fairly small populations." (We will discuss this phenomenon later when we examine genetic drift in more detail. At this point, we merely want to stress that populations need not be infinitely large for the Hardy-Weinberg law to hold true.)

A second condition of the Hardy-Weinberg law is that mating must be random. **Random mating** refers to matings between genotypes occurring in proportion to the frequencies of the genotypes in the population. The requirement of random matings for the Hardy-Weinberg law is often misinterpreted. Many students assume, incorrectly, that the population must be interbreeding randomly for *all* traits for the Hardy-Weinberg law to hold. If this were true, human populations would never obey the Hardy-Weinberg law, because humans do not mate randomly. Humans mate preferentially for height, IQ, skin color, socioeconomic status, and other traits. However, while mating may be nonrandom for some traits, most humans still mate randomly for other traits such as the *M-N* blood types; few of us even know what our *M-N* blood types are. The principles of the Hardy-Weinberg law apply to any locus for which random mating occurs, even if mating is nonrandom for other loci.

A third condition of the Hardy-Weinberg law is that the population must be free from evolutionary forces. In the model law we are interested in whether heredity alone changes allelic frequencies and how reproduction influences genotypic frequencies. Therefore, the influence of evolutionary factors must be excluded. Later we will discuss these other evolutionary forces and their effects on the gene pool of a population. This assumption (that no evolutionary forces act on the population) applies only to the locus in question: a population may be subject to evolutionary forces acting on some genes, while still meeting the Hardy-Weinberg conditions at other loci.

Predictions of the Hardy-Weinberg Law

If the conditions in part 1 of the Hardy-Weinberg law are met, two results are expected. First, the frequencies of the alleles do not change from one generation to the next. Second, the genotypic frequencies will be in the proportions p^2, $2pq$, and q^2 after one generation of random mating. To understand the basis of these frequencies, consider a hypothetical population in which the frequency of allele A is p and the frequency of allele a is q. In producing gametes each genotype passes on both alleles that it possesses with equal frequency; therefore, the frequencies of A and a are p and q in the gametes as well. Table 21.1 shows the combinations of gametes when mating is random. It illustrates the relationship between the allelic frequencies and the genotypic frequencies, forming the basis of the Hardy-Weinberg law. To obtain the probabilities of the different genotypes, we use the product rule of probability (Chapter 2), multiplying the probabilities of the two gametes that join to produce the genotype. For example, the probability of genotype AA is the probability of the first gamete containing A (p) × the probability of the second gamete containing A (p), which equals $p \times p = p^2$. As shown in Table 21.1, the genotypes occur in the proportions p^2 (AA), $2pq$ (Aa), and q^2 (aa). These genotypic proportions result from the expansion of the square of the allelic frequencies $(p + q)^2 = p^2 + 2pq + q^2$, and are reached after one generation of random mating. The genotypic frequencies will remain in these pro-

Table 21.1

Possible Combinations of A and a Gametes from Gametic Pools for a Population

	GAMETES	♂	
		$p\,A$	$q\,A$
♀	$p\,A$	$p^2\,AA$	$pq\,Aa$
	$q\,a$	$pq\,Aa$	$q^2\,aa$

In sum, $p^2\,AA + 2pq\,Aa + q^2\,aa = 1.00$

portions as long as all the conditions required by the Hardy-Weinberg law are met. When the genotypes are in these proportions, the population is said to be in **Hardy-Weinberg equilibrium**. The Hardy-Weinberg law provides an important mechanism for determining the genotypic frequencies from the allelic frequencies when the population is in equilibrium.

To summarize, the Hardy-Weinberg law makes several predictions about the allelic frequencies and the genotypic frequencies of a population when certain conditions are satisfied. The necessary assumptions are that the population is large, randomly mating, and free from evolutionary forces. When these conditions are met, the Hardy-Weinberg law indicates that allelic frequencies will not change. Furthermore, the genotypic frequencies will be determined by the allelic frequencies, occurring in the proportions p^2, $2pq$, and q^2 after one generation of random mating.

According to the Hardy-Weinberg law, at equilibrium, the genotypic frequencies depend on the frequencies of the alleles. This relationship between allelic frequencies and genotypic frequencies for a locus with two alleles is represented graphically in Figure 21.2. Several aspects of this relationship should be noted: (1) The maximum frequency of the heterozygote is 0.5, and occurs only when the frequencies of A and a are both 0.5; (2) If allelic frequencies are between 0.33 and 0.66, the heterozygote is the most numerous genotype; (3) When the frequency of one allele is low, the homozygote for that allele is the rarest of the genotypes. For instance, when $f(a) = q = 0.1$ and $f(A) = p = 0.9$, the frequencies of the genotypes are $f(AA) = 0.81$ (p^2), $f(Aa) = 0.18$ $(2pq)$, and $f(aa) = 0.01$ (q^2).

This last aspect is also illustrated by the distribution of genetic diseases in humans, which are often rare and recessive. For a rare recessive trait, the frequency of the allele causing the trait will be much higher than the frequency of the trait itself, because most of the rare alleles are in nonaffected heterozygotes. Albinism, for example, is a rare recessive condition in humans. One form of albinism is *tyrosinase-negative albinism*. In this type of albinism affected individuals have no tyrosinase activity, which is required for normal production of pigment. Among North American whites, the frequency of tyrosinase-negative albinism is roughly 1 in 40,000, or 0.000025. Since albinism is a recessive condition, the genotype of affected individuals is *aa*. According to the Hardy-Weinberg law, the frequency of the *aa* genotypes equals q^2. If $q^2 = 0.000025$, then $q = 0.005$ and $p = 1 - q = 0.995$. The heterozygote frequency is therefore $2pq = 2 \times 0.995 \times 0.005 = 0.00995$ (almost 1 percent). Thus while the frequency of albinos is low (1 in 40,000), individuals heterozygous for albinism are much more common (almost 1 in 100). Heterozygotes for recessive traits can be common, even when the trait is rare.

■ *Figure 21.2*   Relation of the frequencies of the genotypes AA, Aa, and aa, and the frequencies of alleles A and a (in values of p and q, respectively) in a randomly mating population, as predicted by the Hardy-Weinberg formula. It is assumed that all three genotypes have equal reproductive success.

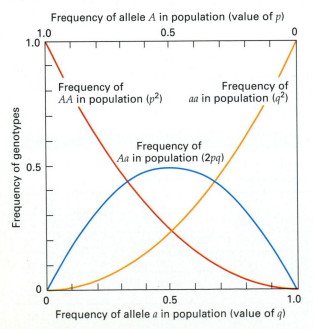

Keynote

The Hardy-Weinberg law describes what happens to allelic and genotypic frequencies of a large population as a result of random mating, assuming that no evolutionary forces act on the population. If these conditions are met, allelic frequencies do not change from generation to generation, and the genotypic frequencies stabilize after one generation in the proportions p^2, $2pq$, and q^2, where p and q equal the frequencies of the alleles in the population.

Testing for Hardy-Weinberg Proportions

To determine whether the genotypes of a population are in Hardy-Weinberg proportions, we first compute p and q from the observed frequencies of the genotypes. Once we have obtained these allelic frequencies, we can calculate the expected genotypic frequencies (p^2, $2pq$, and q^2) and compare these frequencies with the actual observed frequencies of the genotypes using a chi-square test (see Chapter 5, pp. 101–103). The chi-square test gives the probability that the difference between what is observed and what is expected under the Hardy-Weinberg law is due to chance.

To illustrate this procedure, consider a locus that codes for transferrin (a blood protein) in the red-backed vole, *Clethrionomys gapperi*. Three genotypes are found at the transferrin locus: *MM*, *MJ*, and *JJ*. In a population of voles trapped in the Northwest Territories of Canada in 1976, 12 *MM* individuals, 53 *MJ* individuals, and 12 *JJ* individuals were found. To determine if the genotypes are in Hardy-Weinberg proportions, we first calculate the allelic frequencies for the population using our familiar formula:

$$p = \frac{(2 \times \text{number of homozygotes}) + (\text{number of heterozygotes})}{(2 \times \text{total number of individuals})}$$

Therefore:

$$p = f(M) = \frac{(2 \times 12) + (53)}{(2 \times 77)} = 0.50$$

$$q = 1 - p = 0.50$$

Using p and q calculated from the observed genotypes, we can now compute the expected Hardy-Weinberg proportions for the genotypes: $f(MM) = p^2 = (0.50)^2 = 0.25$; $f(MJ) = 2pq = 2(0.50)(0.50) = 0.50$; and $f(JJ) = q^2 = (0.50)^2 = 0.25$. However, for the chi-square test, actual numbers of individuals are needed, not the proportions. To obtain the expected numbers, we simply multiply each expected proportion times the total number of individuals counted (N), as shown below.

	EXPECTED	OBSERVED
$f(MM) = p^2 \times N$		
$= 0.25 \times 77 =$	19.3	12
$f(MJ) = 2pq \times N$		
$= 0.50 \times 77 =$	38.5	53
$f(JJ) = q^2 \times N$		
$= 0.25 \times 77 =$	19.3	12

With observed and expected numbers, we can compute a chi-square value to determine the probability that the differences between observed and expected numbers could be the result of chance. The chi-square is computed using the same formula that we employed for analyzing genetic crosses—d, the deviation, is calculated for each class as (observed—expected); d^2, the deviation squared, is divided by the expected number e for each class; and chi-square (χ^2) is computed as the sum of all d^2/e values. In this example $\chi^2 = 10.98$. We now need to find this value in the chi-square table (see Table 5.2, p. 102) under the appropriate degrees of

freedom. This step is not as straightforward as in our previous χ^2 analyses for genetic crosses. In those examples the number of degrees of freedom was the number of classes in the sample minus 1. Here, however, while there are three classes, there is only one degree of freedom; this is because the frequencies of alleles in a population have no theoretically expected values like the expected values in a genetic cross between two individuals. Thus p must be estimated from the observed data. So one degree of freedom is lost for every parameter (p in this case) that must be calculated from the data. Another degree of freedom is lost because, for the fixed number of individuals, once all but one of the classes have been determined, the last class has no degree of freedom and is set automatically. Therefore with three classes (*MM*, *MJ*, and *JJ*), two degrees of freedom are lost, leaving one degree of freedom.

Under the column for one degree of freedom in the chi-square table, the chi-square value of 10.98 indicates a P value less than 0.05. Thus the probability that the differences between the observed and expected values is due to chance is very low (less than 5 percent). That is, the observed numbers of genotypes are not the expected numbers under the Hardy-Weinberg law, and it is unlikely that the population is in Hardy-Weinberg equilibrium.

Using the Hardy-Weinberg Law to Estimate Allelic Frequencies

An important application of the Hardy-Weinberg law is the calculation of allelic frequencies when one or more alleles is recessive. For example, we have seen that albinism in humans results from an autosomal recessive gene. Normally this trait is rare, but among the Hopi Indians of Arizona, albinism is remarkably common. Charles M. Woolf and Frank C. Dukepoo visited Hopi villages in 1969 and observed 26 cases of albinism in a total population of about 6000 Hopis (Figure 21.3). This gave a frequency for the trait of 26/6000, or 0.0043, which is much higher than the frequency of albinism in most populations. Although we have calculated the frequency of the trait, we cannot directly determine the frequency of the albino gene because we cannot distinguish between heterozygous individuals and those individuals homozygous for the normal allele. Recall that our computation of the allelic frequency involves counting the number of alleles:

$$p = \frac{(2 \times \text{number of homozygotes}) + (\text{number of heterozygotes})}{(2 \times \text{total number of individuals})}$$

But since heterozygotes for a recessive trait such as albinism cannot be identified, this is impossible.

■ *Figure 21.3*   Three Hopi girls, photographed about 1900. The middle child is an albino. Albinism, an autosomal recessive disorder, occurs with high frequency among the Hopi Indians of Arizona.

Nevertheless, we can estimate the allelic frequency from the Hardy-Weinberg law if the population is in equilibrium. At equilibrium, the frequency of the homozygous recessive genotype is q^2. For albinism among the Hopis, $q^2 = 0.0043$; q can be obtained by taking the square root of the frequency of the trait. Therefore, $q = \sqrt{0.0043} = 0.065$, and $p = 1 - q = 0.935$. Following the Hardy-Weinberg law, the frequency of heterozygotes in the population will be $2pq = 2 \times 0.935 \times 0.065 = 0.122$. Thus one out of eight Hopis, on the average, carries an allele for albinism.

We should not forget that this method of calculating allelic frequency rests on the assumptions of the Hardy-Weinberg law. If the conditions of the Hardy-Weinberg law do not apply, then our estimate of allelic frequency will be inaccurate. Also, once we calculate allelic frequencies using these assumptions, we cannot then test the population to determine if the genotypic frequencies are in the Hardy-Weinberg expected proportions. To do so would involve circular reasoning, for we assumed Hardy-Weinberg proportions in the first place to calculate the allelic frequencies.

Genetic Variation in Natural Populations

One of the most significant questions addressed in population genetics is how much genetic variation exists within natural populations. The amount of heritable variation within populations is important for several reasons. First, it determines the potential for evolutionary change and adaptation. The amount of variation also provides important clues about the relative significance of various evolutionary forces, since some forces increase variation and others decrease it. Furthermore, the manner in which new species arise may depend on the amount of genetic variation harbored within populations. For all these reasons population geneticists are interested in measuring genetic variation and in attempting to understand the evolutionary forces that control it.

Models of Genetic Variation

During the 1940s and 1950s, population geneticists developed two opposing hypotheses about the amount of genetic variation within natural populations. These hypotheses, or models of genetic variation, have become known as the balance model and the classical model. The **classical model** emerged primarily from the work of laboratory geneticists, who proposed that most natural populations possess little genetic variation. According to this model, within each population one allele functions best, and this allele is strongly favored by natural selection. As a consequence, almost all individuals in the population are homozygous for this best or wild-type allele. New alleles arise from time to time through mutation, but almost all are deleterious and are kept in low frequency by selection. Once in a long while, a new mutation arises that is better than the wild-type allele. This new allele increases the survival and reproduction of the individuals that carry it, so the frequency of the allele increases over time because of its selective advantage. Eventually the new allele reaches high frequency and becomes the new wild type. In this way a population evolves, but little genetic variation is found within the population at any one time.

In contrast, the **balance model** was developed by geneticists whose backgrounds were in natural history and who studied wild populations. This model predicts that no single allele is best and widespread. The gene pool of a population consists of many alleles at each locus; therefore individuals in the populations are heterozygous at numerous loci. To account for the presence of this variation, the proponents of the balance model suggest that natural selection actively maintains genetic variation within a population through balancing selection. **Balancing selection** is selection that maintains a balance between alleles, preventing any single allele from reaching high frequency. One form of balancing selection is overdominance, in which the heterozygote has higher fitness than either homozygote.

Measuring Genetic Variation with Protein Electrophoresis

For many years population geneticists were unable to establish how much variation existed within natural populations. It was difficult to determine which model of genetic variation was correct because no general procedure was available for assessing the amount of genetic variation in nature. To distinguish between the classical and balance models, population geneticists required data on genotypes at many loci of many individuals from multiple species.

In 1966 population geneticists began to apply the principle of protein electrophoresis to the study of natural populations (Box 21.2). Electrophoresis provided population geneticists with a technique for quickly determining the genotypes of many individuals at many loci. This procedure has now been used to examine genetic variation in hundreds of plant and animal species. The results of these studies are unambiguous—most species possess large amounts of genetic variation in their proteins, and the classical model is clearly wrong. Actually the technique of standard electrophoresis misses a large proportion of the genetic variation that is present, because only genetic variants that cause a change in the movement of the protein on a gel will be observed. For example, some differences in the amino acid sequences do not significantly alter the charge of the protein, which largely determines the rate of movement in the gel. Therefore, the true amount of genetic variation is even greater than that revealed by this technique.

Studies of electrophoretic variation in natural populations showed that the classical model was incorrect, because populations possess large amounts of genetic variation. However, this did not prove that the balance model was correct. The balance model predicts that much genetic variation is found within natural populations, and this turns

---

BOX 21.2

Analysis of Genetic Variation with Protein Electrophoresis

Gel electrophoresis has been widely used to study genetic variation in natural populations. This process is a biochemical technique that separates large molecules on the basis of size, shape, and charge. To examine genetic variation in proteins by electrophoresis, separate tissue samples, taken from a number of individuals, are ground up, releasing the proteins into an aqueous solution. The individual solutions are then inserted into a gel made of starch, agar, polyacrylamide, or some other porous substance. Electrodes are placed at the ends of the gel, as shown in Box Figure 21.1, and a direct electrical current is supplied, setting up an electrical field across the gel.

Because proteins are charged molecules, they will migrate within the electrical field. Molecules with different shape, size, and/or charge will migrate at different rates and will separate from one another within the gel over time. If two individuals have different genotypes at a locus coding for a protein, they will produce slightly different molecular forms of the protein, which can be separated and identified with electrophoresis.

After a current has been supplied to the gel for several hours and the proteins allowed to separate, the current is turned off and the proteins visualized. The gel now contains a large number of different enzymes and proteins; to identify the product of a single locus, one must stain for a specific enzyme or protein. This is frequently accomplished by having the enzyme in the gel carry out a specific biochemical reaction. The substrate for the reaction is added to the gel, along with a dye that changes color when the reaction takes place. A colored band will appear on the gel wherever the enzyme is located.

If two individuals have different molecular forms of the enzyme, indicating differences in genotypes, bands for those individuals will appear at different locations on the gel. A diagram of the banding pattern is shown in Box Figure 21.1. Histochemical stains are available for dozens of enzymes, so genotypes at a number of loci in many individuals may be quickly determined.

Continued ⟶

Box 21.2 **Continued**

■ *Box Figure 21.1* The technique of protein electrophoresis is used to measure genetic variation in natural populations: (a) Solutions of proteins from different individuals are inserted into the slots in the gel, and a direct current is supplied to separate the proteins; (b) After electrophoresis, enzyme-specific staining reactions are used to reveal the proteins. The pattern of bands on the gel indicates the genotype of each individual.

a)

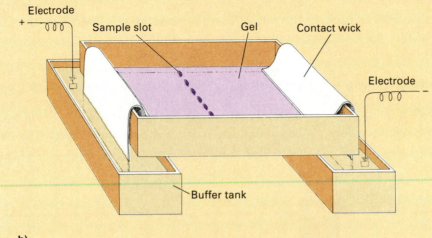

b)

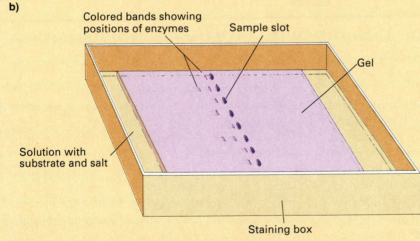

out to be the case. But the balance model also proposes that large amounts of genetic variation are maintained by balancing selection, and this was not demonstrated by electrophoretic studies.

At this point, the nature of the controversy changed. Previously, the central question was how much genetic variation exists. Then, emphasis shifted to the question of what maintains the extensive variation observed. A new hypothesis was proposed to replace the classical model. The **neutral-mutation hypothesis** acknowledges the presence of extensive genetic variation in proteins but proposes that this variation is neutral with regard to natural selection. This does not mean that the proteins detected by electrophoresis have no function, rather that the different genotypes are physiologically equivalent. Therefore, natural selection does not differentiate between neutral alleles, and random processes such as mutation and genetic drift shape the patterns of genetic variation that we see in natural populations. The neutral-mutation hypothesis proposes that different alleles at some loci do affect fitness, but natural selection eliminates variation at these loci.

Distinguishing between the balance model and the neutral-mutation hypothesis has been difficult. Population geneticists still argue over the relative merits of the two, and no clear consensus has emerged as to which model of genetic variation is correct. In reality both models may be partly correct; at some loci genetic variation may be essentially neutral, while at others it may be acted upon by natural selection.

K e y n o t e

Two hypotheses were developed during the 1940s and 1950s to explain the amount of variation in natural populations. The classical model predicted that little variation would be present in natural populations, and the balance model proposed that much genetic variation would occur. Through the use of electrophoresis, population geneticists eventually demonstrated that natural populations harbor much genetic variation, thus disproving the classical model. However, the forces responsible for keeping this variation within populations are still controversial. The neutral-mutation hypothesis proposes that the genetic variation detected by electrophoresis is neutral with regard to natural selection, whereas the balance model proposes that this variation is favored by balancing selection.

Measuring Genetic Variation with RFLPs and DNA Sequencing

Techniques in molecular genetics now provide the means to directly examine genetic variation in the DNA and to unambiguously determine the amount of genetic variation present in natural populations. One such technique employs restriction enzymes, which were discussed in Chapter 13. You will recall that restriction enzymes make double-stranded cuts in DNA at specific base sequences. Most restriction enzymes recognize a sequence of four bases or six bases (see Table 13.1, p. 288.) For example, the restriction enzyme *BamHI* recognizes the sequence $\frac{GGATCC}{CCTAGG}$. Whenever this sequence appears in the DNA, *BamHI* will cut the DNA. The resulting fragments can be separated by agarose gel electrophoresis and observed by staining the DNA or by using probes for specific genes (see Chapter 13).

Suppose that two individuals differ in one or more nucleotides at a particular DNA sequence and that the differences occur at a site recognized by a restriction enzyme (Figure 21.4). One individual has a DNA molecule with the restriction site, but the other individual does not, because the sequences of DNA nucleotides differ. If the DNA from these two individuals is mixed with the restriction enzyme and the resulting fragments are separated on a gel, the two individuals produce different patterns of fragments, as shown in Figure 21.4. The different patterns on the gel are termed restriction fragment length polymorphisms, or RFLPs (see Chapter 13). They indicate that the DNA sequences of the two individuals differ. RFLPs are inherited in the same way that alleles coding for traits are inherited, because alleles for a trait and RFLPs represent heritable differences in the DNA sequences. However, RFLPs usually do not produce any outward phenotypes; the phenotype of the RFLP is the fragment patterns produced on a gel when the DNA is cut by the restriction enzyme. Because they are inherited, RFLPs can be used as genetic markers for mapping genes, as discussed in Chapter 13. They can also provide information about how DNA sequences differ among individuals. Such differences involve only a small part of the DNA, specifically those few nucleotides recognized by the restriction enzyme. However, if we assume that restriction sites occur randomly in the DNA, which is not an unreasonable assumption since the sites are not expressed as traits, the presence or absence of restriction sites can be used to estimate the amount of variation in DNA sequence.

To illustrate the use of RFLPs for estimating genetic variation, suppose we isolate DNA from five wild mice, cut the DNA with the restriction enzyme *BamHI*, and separate the fragments with agarose gel electrophoresis. We then transfer the DNA to nitrocellulose, using Southern blotting (see Chapter 13), and add a probe that will detect the gene for β-hemoglobin. A typical set of restriction patterns that might be obtained is shown in Figure 21.5, p. 503. Remember that each mouse carries two copies of the β-hemoglobin gene, one on each homologous chromosome. Thus a mouse could be +/+ (the restriction site is present on both chromosomes), +/− (the restriction site is present on one chromosome and is absent on the other), or −/− (the

■ *Figure 21.4* DNA from individual 1 and individual 2 differ in one nucleotide, found within the sequence recognized by the restriction enzyme *Bam*HI. Individual 1's DNA contains the *Bam*HI restriction sequence and is cleaved by the enzyme. Individual 2's DNA lacks the *Bam*HI restriction sequence and is not cleaved by the enzyme. Flanking *Bam*HI sites are the same in the two individuals. When placed on an agarose gel and separated by electrophoresis, the DNAs from 1 and 2 produce different patterns on the gel. This variation is called a restriction fragment length polymorphism (RFLP).

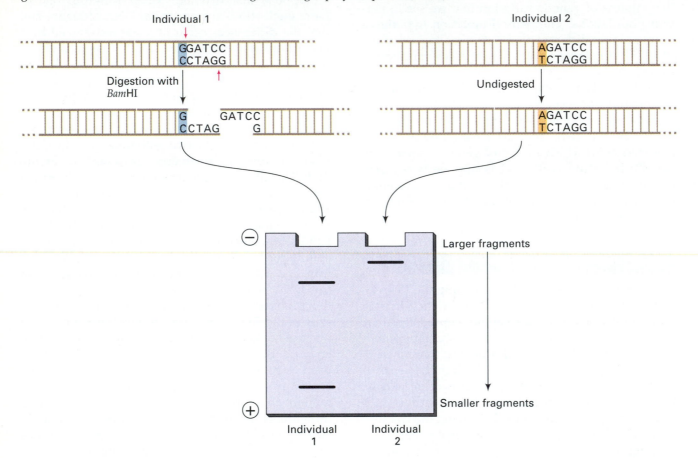

restriction site is absent on both chromosomes). For the ten chromosomes present among these particular five mice, four have the restriction site and six do not.

The presence and absence of restriction sites can be used to estimate the heterozygosity of nucleotide sequence (H_{nuc}), which is the probability that different nucleotides will occur at the same position on the two homologous chromosomes. To calculate the H_{nuc}, we use the formula:

$$H_{nuc} = \frac{n(\Sigma c_i) - \Sigma c_i^2}{j(\Sigma c_i)(n-1)}$$

In this equation n equals the number of homologous DNA molecules examined. In our example we looked at five mice, each with two homologous chromosomes, so $n = 10$. The quantity j equals the number of nucleotides in the restriction site; in our example this is 6, because *Bam*HI recognizes a six-base sequence. For each restriction site i, c_i represents the number of molecules in the sample that were cleaved at that restriction site. We examined only a single restriction site, so we have a single value of c_i which is 4. If additional restriction enzymes were used, we would have a value of c_i for each site recognized by an enzyme. The symbol Σc_i is the total number of cuts at all cleavage sites in all the chromosomes. Since we examined only a single restriction site, and that site was cut in four of the chromosomes, $\Sigma c_i = 4$. Thus we obtain

$$H_{nuc} = \frac{10(4) - (4)^2}{6(4)(10-1)} = \frac{40 - 16}{24(9)} = \frac{24}{216} = 0.11$$

The use of this equation assumes that each RFLP results from a single nucleotide difference.

Nucleotide heterozygosity has been studied for a number of different organisms through the use of

■ *Figure 21.5* Restriction patterns from five mice. The patterns differ in the presence (+) or absence (−) of a restriction site. Each mouse has two homologous chromosomes, each of which potentially carries the restriction site. Thus a mouse may be +/+ (has the restriction site on both chromosomes), +/− (has the restriction site on one chromosome), or −/− (has the restriction site on neither chromosome). When the restriction site is present, the DNA is broken into two fragments after digestion with the restriction enzyme and separation with electrophoresis.

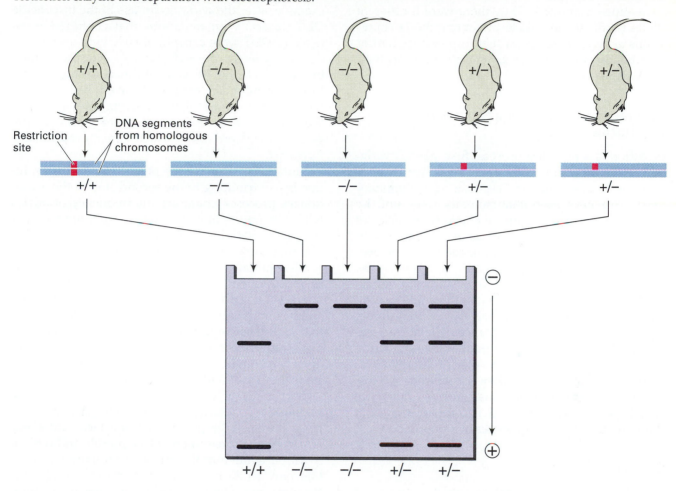

restriction enzymes. A few examples are shown in Table 21.2. Nucleotide heterozygosity typically varies from 0.002 to 0.02 in eukaryotic organisms. This means that an individual is heterozygous at about 1 in every 50 to 500 nucleotides. Another way to interpret nucleotide heterozygosity is that two randomly chosen chromosomes from a population will differ at about 1 in every 50 to 500 nucleotides.

One disadvantage of using RFLPs for examining genetic variation is that this method reveals variation at only a small subset of the nucleotides that make up a gene. With RFLP analysis we are taking a small sample of the nucleotides (those recognized by the restriction enzyme) and using this sample to estimate the overall level of variation. DNA sequencing methods, which were described in Chapter 13, provide a method for detecting all

nucleotide differences that exist among a set of DNA molecules. Martin Krietman used this

Table 21.2

Estimates of Nucleotide Heterozygosity for DNA Sequences

DNA SEQUENCES	ORGANISM	H_{nuc}
β-globin genes	Humans	0.002
Growth hormone gene	Humans	0.002
Alcohol dehydrogenase gene	Fruit fly	0.006
Mitochondrial DNA	Humans	0.004
H4 gene region	Sea urchin	0.019

method to examine variation in the alcohol dehydrogenase gene in *Drosophila melanogaster*. He sequenced 11 copies of a 2659 base-pair segment of this gene from different fruit flies. Among the 11 copies he found different nucleotides at 43 positions within the 2659 base-pair segment. Furthermore, only 3 of the 11 copies were identical at all nucleotides examined. Thus there were 8 different alleles (at the nucleotide level) among the 11 copies of this gene. This suggests that populations harbor a tremendous amount of genetic variation in their DNA sequences.

Changes in Allelic Frequencies of Populations

Evolution is a two-step process: first, genetic variation arises; second, different alleles increase or decrease in frequency in response to evolutionary forces. We have seen that heredity does not, by itself, generate changes in the allelic frequencies of a population. When the population is large, randomly mating, and free from outside forces, no evolution occurs. For many populations, however, the conditions required by the Hardy-Weinberg law do not hold. Populations are frequently small, mating may be nonrandom, and other evolutionary forces may occur. In these circumstances allelic frequencies do change, and the gene pool of the population evolves in response to the interplay of different evolutionary factors. In the following sections we discuss the role of four evolutionary forces—mutation, genetic drift, migration, and natural selection—in changing the allelic frequencies of a population.

Mutation

Mutation plays an essential role in the process of evolution. As we discussed in Chapter 16, gene mutations consist of heritable changes in the DNA that occur within a locus. A mutation converts one allelic form of the gene to another. The rate at which mutations arise is generally low, but varies between loci and among species. Certain genes affect overall mutation rates, and many environmental factors, such as chemicals, radiation, and infectious agents, may increase the number of mutations.

Mutation is the source of new genetic variation; new combinations of genes may arise through recombination, but new alleles occur primarily as a result of mutation. Thus mutation provides the raw genetic material upon which evolution acts. Many mutations will be detrimental and will be eliminated from the population. Other mutations (called neutral mutations) may have no effect on survival

or reproduction. A few mutations, however, will convey some advantage to the individuals that possess them and will spread through the population. Whether a mutation is detrimental, neutral, or advantageous depends on the specific environment: if the environment changes, previously neutral or harmful mutations may become beneficial. For example, after the widespread use of the insecticide DDT, insects with mutations that conferred resistance to DDT were capable of surviving and reproducing, whereas the wild-type insects were killed by the insecticide. Because of this advantage, many insect populations quickly evolved resistance to DDT. Mutations are of fundamental importance to the process of evolution, since they provide the genetic variation upon which other evolutionary forces act.

Mutations also have the potential to affect evolution by contributing to the second step in the evolutionary process—changing the frequency of alleles within a population. Consider a population that consists of 50 individuals, all with the genotype AA. The frequency of $A(p)$ is 1.00 $[(2 \times 50)/100]$. If one A allele mutates to A', the population now consists of 49 AA individuals and 1 AA' individual. The frequency of p is now $[(2 \times 49) + 1]/100 = 0.99$. When another mutation occurs, the frequency of A drops to 0.98. If these mutations continue to occur at a low but steady rate over long periods of time, the frequency of A will eventually decline to zero, and the frequency of A' will reach a value of 1.00.

This effect of the mutational process on allelic frequencies is analogous to the following situation: If we have a large jar of black marbles, and every year or so we take out one black marble and replace it with a white marble, the jar eventually will contain only white marbles. Similarly, in populations the slow, persistent pressure of recurrent mutation may lead to significant changes in allelic frequency, assuming other evolutionary forces are not present.

The mutation of A to A' is referred to as a *forward mutation*. For most genes, mutations also occur in the reverse direction; A' may mutate to A. These mutations are called *reverse mutations*. To include these reverse mutations in our marble jar analogy, suppose we start with a jar of black marbles, but instead of taking out a black marble at each dip and replacing it with a white one, we take out any marble at random and replace it with one of the opposite color. With this sampling system there is no selection against either marble type. At the start we are more likely to pull out black marbles and replace them with white ones, but as white marbles become more common in the jar, we are more and more likely to pull out a white one and replace it with black. Eventually, an **equilibrium** is reached, at which point no further changes in black and white marble frequencies will occur, as long as the same method of marble replacement is applied. At equilibrium the black and white marbles will occur in equal frequency.

In our example with marbles the forward and reverse mutation rates were equal, since every marble drawn from the jar, whether black or white, was replaced with one of the opposite color. In reality, forward and reverse mutation rates usually differ. The forward mutation rate (the rate at which A mutates to A', or $A \rightarrow A'$) is symbolized with μ; the reverse mutation rate (the rate at which A' mutates to A, or $A' \rightarrow A$) is symbolized with v. Let q equal the frequency of A' and p equal the frequency of A. As a result of mutation, the amount that A decreases in one generation is equal to the increase in A alleles due to reverse mutations minus the decrease in A alleles due to forward mutations. That is, the change in the frequency of A, Δp, is

$$\Delta p = vq - \mu p$$

As we have seen, when p is high, the number of A alleles mutating to A' is relatively large, but as more and more A alleles mutate to A', the change in p decreases. At the same time, when q is small, few alleles will be available to undergo reverse mutation to A, but as q increases, the number of alleles undergoing reverse mutation increases. Eventually, the population achieves equilibrium, in which the number of alleles undergoing forward mutation is exactly equal to the number of alleles undergoing reverse mutation. At this point, no further change in allelic frequency occurs, in spite of the fact that forward and reverse mutations continue to take place.

When equilibrium is reached, the value of q (symbolized by q') is

$$q' = \frac{\mu}{\mu + v}$$

The equilibrium value for p, p', is

$$p' = \frac{v}{\mu + v}$$

Consider a population in which the initial allelic frequencies are $p = 0.9$ and $q = 0.1$ and the forward and reverse mutation rates are $\mu = 5 \times 10^{-5}$ and $v = 2 \times 10^{-5}$. (These values are similar to forward and reverse mutation rates observed for many eukaryotic genes.) In the first generation the change in allelic frequency is

$$\Delta p = vq - \mu p$$
$$= (2 \times 10^{-5} \times 0.1) - (5 \times 10^{-5} \times 0.9)$$
$$\Delta p = -0.000043$$

The frequency of A decreases by only four-thousandths of 1 percent. At equilibrium the frequency of the A' allele, q', equals

$$q' = \frac{\mu}{\mu + v}$$

$$q' = \frac{5 \times 10^{-5}}{(5 \times 10^{-5}) + (2 \times 10^{-5})} = 0.714$$

If no other forces act on a population, after many generations the alleles will reach equilibrium. Therefore, mutation rates determine the allelic frequencies of the population in the absence of other evolutionary forces. However, because mutation rates are so low, the change in allelic frequency due to mutation pressure is exceedingly slow.

Keynote

Recurrent mutation can alter the allelic frequencies of a population over time, provided that other evolutionary forces are not active. Eventually, a mutational equilibrium is reached in which the allelic frequencies of the population remain constant in spite of continuing mutation; the allelic frequencies at this equilibrium are a function of the forward and reverse mutation rates.

Genetic Drift

A major assumption of the Hardy-Weinberg law is that the population is infinitely large. Real populations are not infinite in size, but frequently they are large enough that expected ratios are realized and chance factors have insignificant effects on allelic frequencies. Some populations are small, however, and in these groups chance factors may produce random changes in allelic frequencies. Random change in allelic frequency due to chance is called **genetic drift,** or *drift* for short.

Chance Changes in Allelic Frequency   Random changes in allelic frequency resulting from chance can be a significant evolutionary force in small populations. Imagine a small group of humans inhabiting a South Pacific island. Suppose that this population consists of only ten individuals, five of whom have green eyes and five of whom have brown eyes. For this example we assume that eye color is determined by a single locus (actually, eye color is polygenic) and that the allele for green eyes is recessive to brown (*BB* and *Bb* codes for brown eyes and *bb* codes for green). The frequency of the allele for green eyes is 0.7 in the island population. A typhoon strikes the island, killing 50 percent of the population. Just by chance, the five individuals who die all have brown eyes. Eye color in no way affects the probability of surviving; the fact that only those with green eyes survive is strictly the result of chance. After the typhoon, the allelic frequency for green eyes is 1.0. Evolution has occurred in this population—the frequency of the green-eye allele has changed from 0.7 to 1.0—simply as a result of chance.

Now imagine the same scenario, but this time with a population of 1000 individuals. As before, 50 percent of the population has green eyes and 50 percent has brown eyes. A typhoon strikes the island and kills half the population. How likely is it that, just by chance, all 500 people who perish will have brown eyes? In a population of 1000 individuals the probability of this occurring by chance is extremely remote. This example illustrates an important characteristic of genetic drift, that chance factors are likely to produce significant changes in allelic frequencies only in small populations.

Random factors producing mortality in natural populations, such as the typhoon in our example, is only one of several ways in which genetic drift arises. Chance deviations from expected ratios of gametes and zygotes also produce genetic drift. Suppose that a population produces an infinitely large pool of gametes, with alleles in the proportions p and q. If random mating occurs and all the gametes unite to form zygotes, the proportions of the genotypes will be equal to p^2, $2pq$, and q^2, and the frequencies of the alleles in these zygotes will remain p and q, as indicated by the Hardy-Weinberg law. If the number of progeny is limited, however, the gametes that unite to form the progeny constitute a sample from the infinite pool of potential gametes. By chance, this sample may deviate from the larger pool; the smaller the sample, the larger the potential deviation.

Measuring Genetic Drift   Genetic drift is random, and thus we cannot predict what the allelic frequencies will be after drift has occurred. However, since the amount of genetic drift is related to the size of the population, we can make predictions about the magnitude of genetic drift. Ecologists often measure population size by counting the number of individuals, but not all individuals contribute gametes to the next generation. To determine the magnitude of genetic drift, we must know the **effective population size**, which equals the equivalent number of adults contributing gametes to the next generation. If the sexes are equal in number and all individuals have an equal probability of producing offspring, the effective population

size equals the number of breeding adults in the population. However, when males and females are not present in equal numbers, all individuals are not genetically equivalent, and the effective population size is

$$N_e = \frac{4 \times N_f \times N_m}{N_f + N_m}$$

where N_f equals the number of breeding females and N_m equals the number of breeding males. (The derivation of this equation is beyond the scope of this book.)

Students often have difficulty understanding why this equation must be used—why the effective population size is not simply the number of breeding adults. The reason is that males, as a group, contribute half of all genes to the next generation and females, as a group, contribute the other half. Therefore, in a population of 70 females and 2 males, the two males are not genetically equivalent to two females. Because there are only two males, each contributes half of the male genes, which are half of all genes in the progeny, so each male contributes $1/2 \times 1/2 = 0.25$ of the genes to the next generation. However, because there are 70 females, each female contributes $1/70$ of the female genes, which are half of all the genes in the progeny, so each female contributes $1/70 \times 1/2 = 0.007$ of all genes. The small number of males disproportionately influences what alleles are present in the next generation. Using the above equation, the effective population size is $N_e = (4 \times 70 \times 2)/(70 + 2) = 7.8$, or approximately 8 breeding adults. What this means is that in a population of 70 females and 2 males, genetic drift will occur as if the population had only four breeding males and four breeding females. Therefore, genetic drift will have a much greater effect in this population than in one with 72 breeding adults equally divided between males and females.

The amount of variation among populations resulting from genetic drift is measured by the **variance of allelic frequency,** which equals

$$s_p^2 = \frac{pq}{2N_e}$$

where N_e is the effective population size and p and q are the allelic frequencies. Notice that the product of the allelic frequencies (pq) is divided by the effective population size (N_e) in this equation, indicating that genetic differences among populations increase as effective population size decreases.

Causes of Genetic Drift   There are several ways in which sampling error occurs in natural populations. Genetic drift arises when population size remains continuously small over many generations. This situation is frequent, particularly where populations occupy marginal habitats, or when competition limits population growth. In such populations genetic drift has a crucial impact on the evolution of allelic frequencies. Another way in which genetic drift arises is through **founder effect.** Founder effect occurs when a population is initially established by a small number of individuals. Although the population may subsequently grow in size and later consists of a large number of individuals, the gene pool of the population is derived from the genes present in the original founders. Chance may play a significant role in determining which genes were present among the founders, which has a profound effect on the gene pool of subsequent generations.

Many excellent cases of founder effect come from the study of human populations. Consider the inhabitants of Tristan da Cunha, a small, isolated island in the South Atlantic. This island was first permanently settled by William Glass, a Scotsman, and his family in 1817. (Several earlier attempts at settlement failed.) They were joined by a few additional settlers, some shipwrecked sailors, and a few women from the distant island of St. Helena, but for the most part the island remained a genetic isolate. In 1961 a volcano on Tristan da Cunha erupted, and the population of almost 300 inhabitants was evacuated to England. During their two-year stay in England, geneticists studied the islanders and reconstructed the genetic history of the population. These studies revealed that the current gene pool of Tristan da Cunha has been strongly influenced by genetic drift.

Three forms of genetic drift occurred in the evolution of the island's population. First, founder effect took place at the initial settlement. By 1855 the population of Tristan da Cunha consisted of about 100 individuals, but 26 percent of the genes of that population originated from William Glass and his wife. Even in 1961, 14 percent of all the genes in the 300 individuals of the population originated from these initial two settlers. The particular genes that Glass and other original founders carried heavily influenced the subsequent gene pool of the population. Second, population size remained small throughout the history of the settlement, and sampling error continually occurred.

A third form of sampling error, called **bottleneck effect,** also had an important effect on the population of Tristan da Cunha. Bottleneck effect occurs when a population is drastically reduced in size. During such a population reduction, some genes may be lost from the gene pool as a result of chance. Recall our earlier example of the popula-

tion consisting of ten individuals inhabiting a South Pacific island. When a typhoon struck the island, the population size was reduced to five, and by chance, all individuals with brown eyes perished in the storm, changing the frequency of green eyes from 0.6 to 1.0. This is an example of bottleneck effect. Bottleneck effect can be viewed as a type of founder effect, since the population is refounded by those few individuals that survive the reduction.

Two severe bottlenecks occurred in the history of Tristan da Cunha. One took place around 1856 and was precipitated by two events: the death of William Glass and the arrival of a missionary who encouraged the inhabitants to leave the island. At this time many islanders emigrated to America and to South Africa, and the population dropped from 103 individuals at the end of 1855 to 33 in 1857. The second bottleneck occurred in 1885. The island of Tristan da Cunha has no natural harbor, and the islanders intercepted passing ships for trade by rowing out in small boats. On November 28, 1885, 15 of the adult males on the island put out in a small boat to make contact with a passing ship. In full view of the entire island community, the boat capsized and all 15 men drowned. Following this disaster, only four adult males were left on the island, one of whom was insane and two of whom were old. Many of the widows and their families left the island during the next few years, and the population size dropped from 106 to 59. Both bottlenecks had a major effect on the gene pool of the population. All the genes contributed by several settlers were lost, and the relative contributions of others were altered by these events. Thus the gene pool of Tristan da Cunha has been influenced by genetic drift in the form of founder effect, small population size, and bottleneck effect.

As we shall see later when we discuss migration, gene flow among populations increases the effective population size and reduces the effects of genetic drift. Small breeding units that lack gene flow are genetically isolated from other groups and often experience considerable genetic drift, even though surrounded by much larger populations. A good example is a religious sect, known as the Dunkers, found in eastern Pennsylvania. Between 1719 and 1729, 50 Dunker families emigrated from Germany and settled in the United States. Since that time, the Dunkers have remained an isolated group, rarely marrying outside of the sect, and the number of individuals in their communities has always been relatively small.

During the 1950s geneticists studied one of the original Dunker communities in Franklin County, Pennsylvania. At the time of the study, this population had about 300 members, and the population

Table 21.3

Frequencies of Alleles Controlling the ABO Blood Group System in Three Human Populations

| Population | ALLELE FREQUENCIES | | | PHENOTYPE (BLOOD GROUP) FREQUENCIES | | | |
	I^A	I^B	I^O	A	B	AB	O
Dunker	0.38	0.03	0.59	0.593	0.036	0.023	0.348
United States	0.26	0.04	0.70	0.431	0.058	0.021	0.490
West Germany	0.29	0.07	0.64	0.455	0.095	0.041	0.410

size had remained relatively constant for many generations. The investigators found that some of the allelic frequencies in the Dunkers were very different from the frequencies found among the general population of the United States. The Pennsylvania frequencies were also different from the frequencies of the West German population from which the Dunkers descended. Table 21.3 presents some of the allelic frequencies at the ABO blood group locus. The ABO allele frequencies among the Dunkers are not the same as those in either the U.S. population or the West German population. Nor are the Dunker frequencies intermediate between the U.S. and West German frequencies. (Intermediate frequencies might be expected if intermixing of Dunkers and Americans had occurred.) The most likely explanation for the unique Dunker allelic frequencies observed is that genetic drift has produced random change in the gene pool. Founder effect probably occurred when the original 50 families emigrated from Germany, and genetic drift has most likely continued to influence allelic frequencies each generation since 1729 because the population size has remained small.

Effects of Genetic Drift   Genetic drift produces changes in allelic frequencies. These changes have several effects on the genetic structure of populations. Genetic drift causes the allelic frequencies of a population to change over time. This is illustrated in Figure 21.6. The different lines represent allelic frequencies in several populations over a number of generations. Although all populations begin with an allelic frequency equal to 0.50, the frequencies in each population change over time as a result of sampling error. In each generation the allelic frequency may increase or decrease, and over time the frequencies wander randomly or drift (hence the name "genetic drift"). Sometimes, just by chance, the allelic frequency reaches a value of 0.0 or 1.0. At this point, one allele is lost from the population and the population is said to be *fixed* for the remaining

allele. Once an allele has reached fixation, no further change in allelic frequency can occur, unless the other allele is reintroduced through mutation or migration. The probability of fixation in a population increases with time. If the initial allelic frequencies are equal, which allele becomes fixed is strictly random. If, on the other hand, initial allelic frequencies are not equal, the rare allele is more likely to be lost. During this process of genetic drift and fixation, the number of heterozygotes in the population also decreases; after fixation the population heterozygosity is zero. As heterozygosity decreases and alleles become fixed, populations lose genetic variation; thus the second effect of genetic drift is a reduction in genetic variation within populations.

Since genetic drift causes random change in

■ *Figure 21.6*   The effect of genetic drift on the frequency (*q*) of allele *a* in four populations. Each population begins with *q* equal to 0.5, and the effective population size for each is 20. The mean frequency of allele *a* for the four replicates is indicated by the magenta line. These results were obtained by a computer simulation.

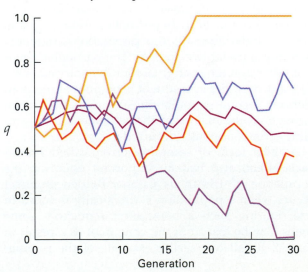

allelic frequency, the allelic frequencies in separate, individual populations need not change in the same direction. Therefore, populations can diverge in their allelic frequencies through genetic drift. This is shown in Figure 21.6 where all the populations begin with p and q equal to 0.5. After a few generations, the allelic frequencies of the populations diverge, and this divergence increases over generations. The maximum divergence in allelic frequencies is reached when all populations are fixed for one or the other allele. If allelic frequencies are initially equal to 0.5, approximately half of the populations will become fixed for one allele, and half will be fixed for the other.

Keynote

Genetic drift, or chance changes in allelic frequency, can be an important evolutionary force in small populations. Genetic drift leads to loss of genetic variation within populations, genetic divergence among populations, and random fluctuation in the allelic frequencies of a population over time.

Migration

One of the assumptions of the Hardy-Weinberg law is that the population is closed and not influenced by outside evolutionary forces. Many populations are not completely isolated, however, and exchange genes with other populations of the same species. Individuals migrating into a population may introduce new alleles to the gene pool, altering the frequencies of existing alleles. Thus **migration** has the potential to disrupt Hardy-Weinberg equilibrium and may influence the evolution of allelic frequencies within populations.

The term *migration* usually implies movement of organisms. In population genetics, however, we are interested in the movement of genes from one gene pool to another, which may or may not occur when organisms move. Movement of genes takes place only when organisms reproduce with members of another population, adding their genes to the gene pool of the other population. This process is also referred to as **gene flow.**

Gene flow has two major effects on a population. First, it introduces new alleles to the population. Because mutation is generally a rare event, a specific mutant allele may arise in one population and not in another. Gene flow spreads unique alleles to other populations and, like mutation, is a source of genetic variation for the population. Second, when the allelic frequencies of migrants and the recipient population differ, gene flow changes the allelic frequencies within the recipient population. Through exchange of genes, populations remain similar; thus migration is a homogenizing force that tends to prevent populations from evolving genetic differences.

To illustrate the effect of migration on allelic frequencies, we will consider a simple model in which gene flow occurs in only one direction, from population I to population II. Suppose that the frequency of allele A in population I (p_I) is 0.8 and the frequency of A in population II (p_{II}) is 0.5. In each generation some individuals migrate from population I to population II, and these migrants are a random sample of the genotypes in population I. After migration population II actually consists of two groups of individuals: the migrants with $p_I = 0.8$, and the residents with $p_{II} = 0.5$. The migrants now make up a proportion of population II, which we will designate m, and the residents make up the rest of the population ($1 - m$). The frequency of A in population II after migration (p'_{II}) is

$$p'_{II} = mp_I + (1 - m)p_{II}$$

We see that the frequency of A after migration is determined by the proportion of A alleles in the two groups that now comprise population II. The first component, mp_I, represents the A alleles in the migrants: we multiply the proportion of the population that consists of migrants (m) by the allelic frequency of the migrants (p). The second component represents the A alleles in the residents, and equals the proportion of the population consisting of residents ($1 - m$) multiplied by the allelic frequency in the residents (p_{II}). Adding these two components together gives us the allelic frequency of A in population II after migration.

With continued migration from population I to population II, p_I and p_{II} become increasingly similar, and, as a result, the change in allelic frequency due to migration decreases. Eventually, allelic frequencies in the two populations will be equal, and no further change will occur. This is only true, however, when other factors besides migration do not influence allelic frequencies.

An additional point to be noted is that migration among populations tends to reduce the effects of genetic drift. As we have seen, genetic drift causes populations to diverge. Migration, on the other hand, reduces divergence among populations. For example, extensive gene flow has been shown to reduce divergence among populations of the monarch butterfly (Figure 21.7). Even a small amount of gene flow can reduce the effect of genetic drift. Calculations have shown that a single migrant moving between two populations each generation will prevent the two populations from becoming fixed for different alleles.

■ *Figure 21.7* (a) Extensive gene flow occurs among populations of the monarch butterfly; (b) The butterflies overwinter in Mexico and then migrate north during spring and summer to breeding grounds as far away as northern Canada. Extensive gene flow occurs during the migration period; consequently monarch populations display relatively little genetic divergence.

a)

b)

K e y n o t e

Migration of individuals into a population may alter the makeup of the population gene pool if the genes carried by the migrants differ from those of the residents of the population. Migration, also termed gene flow, tends to reduce genetic divergence among populations and increases the effective size of the population.

Natural Selection

We have now examined three major evolutionary forces capable of changing allelic frequencies and producing evolution—mutation, genetic drift, and migration. These forces alter the gene pool of a population and influence the evolution of a species. However, mutation, migration, and genetic drift do not result in adaptation. *Adaptation* is the process by which traits evolve that make organisms more suited to their immediate environment; these traits increase the organism's chances of surviving and reproducing. Adaptation is responsible for the many extraordinary traits seen in nature—wings that enable a hummingbird to fly backward, leaves of the pitcher plant that capture and devour insects, brains that allow humans to speak, read, and love. These biological features and countless other remarkable traits are the product of adaptation. Genetic drift, mutation, and migration all influence the pattern and process of adaptation, but adaptation arises chiefly from natural selection. Natural selection is the dominant force in the evolution of many traits and has shaped much of the phenotypic variation observed in nature.

Natural selection can be defined as differential reproduction of genotypes. This means that individuals with certain genes produce more offspring than others; therefore those genes increase in frequency in the next generation. Through natural selection, traits that contribute to survival and reproduction increase over time. In this way organisms adapt to their environment.

Selection in Natural Populations A classic example of natural selection is the evolution of melanic (dark) forms of moths in association with industrial pollution, a phenomenon known as "industrial melanism." Melanic phenotypes have appeared in a number of different species of moths found in the industrial regions of Europe, North America, and England. One of the best studied cases involves the peppered moth, *Biston betularia*. The common phenotype of this species, called the *typical* form, is a grayish white color with black mottling over the body and wings.

Prior to 1848 all peppered moths collected in England possessed this *typical* phenotype, but in 1848 a single black moth was collected near Manchester, England. This new phenotype, called *carbonaria*, presumably arose by mutation, and rapidly increased in frequency around Manchester and in other industrial regions. By 1900 the *carbonaria* phenotype had reached a frequency of more than 90 percent in several populations. High frequencies of *carbonaria* appeared to be associated with industrial regions, whereas the *typical* phenotype remained common in more rural districts. Laboratory studies by a number of investigators, including E. B. Ford and R. Goldschmidt, demonstrated that the *carbonaria* phenotype was dominant to the *typical* phenotype.

H. B. D. Kettlewell investigated color polymorphism in the peppered moth, demonstrating that the increase in the *carbonaria* phenotype occurred as a result of strong selection against the *typical* form in polluted woods. Peppered moths are nocturnal; during the day they rest on the trunks of lichen-covered trees. Birds frequently prey on the moths during the day, but because the lichens that cover the trees are naturally light in color, the *typical* form of the peppered moth is well camouflaged against this background (Figure 21.8[a]). In industrial areas, however, extensive pollution beginning with the Industrial Revolution in the mid-nineteenth century had killed most of the lichens and covered the tree trunks with black soot. Against this black background, the *typical* phenotype was quite conspicuous and was readily consumed by birds. In contrast, the *carbonaria* form was well camouflaged against the blackened trees and had a higher rate of survival than the *typical* phenotype in polluted areas (Figure 21.8[b]). Because *carbonaria* survived better in polluted woods, more *carbonaria* genes were transmitted to the next generation; thus the *carbonaria* phenotype increased in frequency in industrial areas. In rural areas, where pollution was absent, the *carbonaria* phenotype was conspicuous and the *typical* form was camouflaged; in these regions the frequency of the *typical* form remained high.

Kettlewell demonstrated that selection affected the frequencies of the two phenotypes by conducting a series of mark-and-recapture experiments involving dark and light moths in smoky, industrial Birmingham, England, and in nonindustrialized Dorset. As predicted, the *typical* phenotype was favored in Dorset, and *carbonaria* was favored in Birmingham.

Fitness and Coefficient of Selection   Darwin described natural selection primarily in terms of survival, and even today, many nonbiologists think of natural selection in terms of a struggle for survival. However, what is critical in the process of natural selection is the relative number of genes that are contributed to future generations. Certainly the ability to survive is important, but survival alone will not ensure that genes are passed on. Reproduction must also occur. Therefore, we measure natural selection by assessing reproduction. Natural selection is measured by **fitness,** which is defined as the relative reproductive ability of a genotype.

Fitness is usually symbolized as W. Since fitness is a measure of the relative reproductive ability, population geneticists usually assign a fitness of 1 to a genotype that produces the most offspring. The fitnesses of the other genotypes are assigned relative to this. For example, suppose that the genotype G^1G^1 on the average produces 8 offspring, G^1G^2 produces an average of 4 offspring, and G^2G^2 produces an average of 2 offspring. The G^1G^1 genotype has the highest reproductive efficiency, so its fitness is 1 ($W_{11} = 1.0$). Genotype G^1G^2 produces on the average 4 offspring for the 8 produced by the most fit genotype, so the fitness of G^1G^2 (W_{12}) is $4/8 = 0.5$. Similarly, G^2G^2 produces 2 offspring for the 8 produced by G^2G^2, so the fitness of G^2G^2 (W_{22}) is $2/8 = 0.25$.

Fitness tells us how well a genotype is doing in

■ *Figure 21.8* *Biston betularia*, the peppered moth, and its dark form *carbonaria*: (a) On the light-colored trunk of a tree in the unpolluted countryside, and (b) on the trunk of a tree blackened by industrial pollution in Birmingham, England. On the light-colored tree, the dark form is readily seen, whereas the light form is well camouflaged. On the polluted tree, the dark form is well camouflaged.

a)

b)

terms of natural selection. A related measure is the **selection coefficient,** which is a measure of the relative intensity of selection against a genotype. The selection coefficient is symbolized by s and equals $1 - W$. In our example, the selection coefficients for G^1G^1 are $s = 0$; for G^1G^2, $s = 0.5$; for G^2G^2, $s = 0.75$.

Effect of Selection on Allelic Frequencies Natural selection produces a number of different effects. Natural selection can eliminate genetic variation, maintain variation, or increase variation; it can change allelic frequencies or prevent allelic frequencies from changing; it can produce genetic divergence among populations or produce genetic uniformity. Which of these effects occurs depends primarily on the relative fitness of the genotypes and on the frequencies of the alleles in the population.

The change in allelic frequency that results from natural selection can be calculated by constructing a table such as Table 21.4. This "table method" can be used for any type of single-locus trait, whether the trait is dominant, codominant, recessive, or overdominant. To use the table method we begin by listing the genotypes (A^1A^1, A^1A^2, and A^2A^2) and their initial frequencies. If random mating has just taken place, the genotypes are in Hardy-Weinberg proportions and the initial frequencies are $p^2, 2pq$,

and q^2. We then list the fitnesses for each of the genotypes, W_{11}, W_{12}, and W_{22}. Now suppose that selection occurs and only some of the genotypes survive. The contribution of each genotype to the next generation will be equal to the initial frequency of the genotype multiplied by its fitness. For A^1A^1 this will be $p^2 \times W_{11}$. The sum of the contributions equals $p^2W_{11} + 2pqW_{12} + q^2W_{22} = \overline{W}$, which is called the *mean fitness of the population*. Next, we normalize the relative contributions by dividing each by the mean fitness of the population, which gives us the frequencies of the genotypes after selection. We then calculate the new allelic frequencies (p' and q') from the genotypes after selection, using our familiar formula, $p' =$ (frequency of A^1A^1) + (1/2 × frequency of A^1A^2) and $q' = 1 - p'$. Finally, the change in allelic frequency resulting from selection equals $p' - p$. A sample calculation using some actual allelic frequencies and fitness values is presented in Table 21.4.

Keynote

Natural selection involves differential reproduction of genotypes and is measured in terms of fitness, the relative reproductive contribution of a genotype.

Table 21.4

General Method of Determining Change in Allelic Frequency Due to Natural Selection When Initial Allelic Frequencies Are $p = 0.6$ and $q = 0.4$

	GENOTYPES		
	A^1A^1	A^1A^2	A^2A^2
Initial genotypic frequencies	p^2 $(0.6)^2 = 0.36$	$2pq$ $2(0.6)(0.4) = 0.48$ 0.48	q^2 $(0.4)^2 = 0.16$
Fitness	$W_{11} = 0$	$W_{12} = 0.4$	$W_{22} = 1$
Frequency after selection	$p^2W_{11} =$ $(0.36)(0) = 0$	$2pqW_{12} =$ $(0.48)(0.4) =$ 0.19	$q^2W_{22} =$ $(0.16)(1)$ $= 0.16$
Genotypic frequency after selection	$P' = \dfrac{p^2W_{11}}{\overline{W}}$[a] $P' = 0/0.35 = 0$	$H' = \dfrac{2pqW_{12}}{\overline{W}}$ $H' = 0.19/0.35$ $= 0.54$	$Q' = \dfrac{q^2W_{22}}{\overline{W}}$ $Q' = 0.16/0.35$ $= 0.46$

Allelic frequency after selection $p' = P' + 1/2(H')$
$$p' = 0 + 1/2(0.54) = 0.27$$
$$q' = 1 - p' = 1 - 0.27 = 0.73$$
Change in allelic frequency due to selection $= \Delta p = p' - p$
$$\Delta p = 0.27 - 0.6 = -0.33$$

[a]$\overline{W} = p^2W_{11} + 2pqW_{12} + q^2W_{22}$
$\overline{W} = 0 + 0.19 + 0.16$
$\overline{W} = 0.35$

Mutation and Selection

Selection against a deleterious recessive allele depends on the actual frequencies of the alleles in a population. When the frequency of a recessive allele is high, many homozygous recessive individuals are present in the population and will have low fitness, causing a large change in the allelic frequency. When the allelic frequency is low, however, the homozygous recessive genotype is rare, and little change in allelic frequency occurs. Opposing this reduction in the allele's frequency due to selection is mutation pressure, which will continually produce new alleles and tend to increase the frequency. Eventually a balance, or equilibrium, is reached, in which the input of new alleles by recurrent mutation is exactly counterbalanced by the loss of alleles thorugh natural selection. When equilibrium is obtained, the allele's frequency remains stable, in spite of the fact that selection and mutation continue, unless perturbed by some other evolutionary force.

Consider a population in which selection occurs against a deleterious recessive allele, a. If a is rare, then the decrease in frequency caused by selection is given by the formula below. (The derivation of the equation is beyond the scope of this book.)

$$Dq = -spq^2$$

In this equation, s equals the selection coefficient for the homozygous recessive genotype. At the same time that a is decreasing due to selection, its frequency increases as a result of mutation from A to a. Provided the frequency of a is low, the reverse mutation of a to A essentially can be ignored. Equilibrium between selection and mutation occurs when the decrease in gene frequency produced by selection (spq^2) is the same as the increase produced by mutation (μp; see p. 505):

$$spq^2 = \mu p$$

We can predict what the frequency of a at equilibrium (q) will be by rearranging this equation:

$$sq^2 = \mu$$
$$q^2 = \mu/s$$

and

$$q = \sqrt{\mu/s}$$

If the recessive homozygote is lethal ($s = 1$), the equation becomes

$$q = \sqrt{\mu}$$

As an example of the balance between mutation and selection, consider a recessive gene for which the mutation rate is 10^{-6} and s is 0.1. At equilibri-

um, the frequency of the gene will be $q = \sqrt{10^{-6}/0.1} = 3.2 \times 10^{-3}$. Most recessive deleterious traits remain within a population at low frequency because of equilibrium between mutation and selection.

For a dominant allele A, the frequency at equilibrium (q) is

$$q = \mu/s$$

If the mutation rate is 10^{-6} and s is 0.1, as in the above example, the frequency of the dominant gene at equilibrium will be $10^{-6}/0.1 = 0.00001$, which is considerably less than the equilibrium frequency for a recessive allele with the same fitness and mutation rate. This is because selection cannot act on a recessive allele in the heterozygote state, whereas both the homozygote and the heterozygote for a dominant allele have reduced fitness. For this reason, detrimental dominant alleles generally are less common than recessive ones.

Overdominance

An equilibrium of allelic frequencies arises when the heterozygote has higher fitness than either of the homozygotes. This situation is called **overdominance** or **heterozygote advantage**. In overdominance both alleles are maintained in the population, because both are favored in the heterozygote genotype. Allelic frequencies will change as a result of selection until the equilibrium point is reached and then will remain stable. The allelic frequencies at which the population reaches equilibrium depend on the relative fitnesses of the two homozygotes. If the selection coefficient of AA is s and the selection coefficient of aa is t, it can be shown algebraically that at equilibrium

$$p = f(A) = \frac{t}{s + t}$$

and

$$q = f(a) = \frac{s}{s + t}$$

An example of overdominance in humans is sickle-cell anemia. Sickle-cell anemia results from a mutation in the gene coding for beta hemoglobin. In some populations there are three hemoglobin genotypes: $Hb\text{-}A/Hb\text{-}A$, $Hb\text{-}A/Hb\text{-}S$, and $Hb\text{-}S/Hb\text{-}S$. Individuals with the $Hb\text{-}A/Hb\text{-}A$ genotype have completely normal red blood cells; $Hb\text{-}S/Hb\text{-}S$ individuals have sickle-cell anemia; and $Hb\text{-}A/Hb\text{-}S$ individuals have sickle-cell trait, a mild form of sickle-cell anemia. In an environment in which malaria is common, the heterozygotes are at a selective advantage over the two homozygotes. Apparently, the abnormal hemoglobin mixture in the heterozygotes provides an unfavorable environment for the growth or maintenance of the malarial parasite in the red blood cell. The heterozygotes therefore have greater resistance to malaria and thus higher fitness than *Hb-A/Hb-A* individuals. The *Hb-S/Hb-S* individuals are at a serious selective disadvantage because they have sickle-cell anemia, which causes major medical problems. As a result, in malaria-infested areas in which the *Hb-S* gene is also found, an equilibrium state is established in which a significant number of *Hb-S* alleles are found in the heterozygotes because of the selective advantage of this genotype. The distributions of malaria and the *Hb-S* allele are illustrated in Figure 21.9.

Nonrandom Mating

A fundamental assumption of the Hardy-Weinberg law is that members of the population mate randomly. However, many populations do not mate randomly for some traits, and when nonrandom mating occurs, the genotypes will not exist in the proportions predicted by the Hardy-Weinberg law.

One form of nonrandom mating is **positive assortative mating,** which occurs when individuals with similar phenotypes mate preferentially. Positive assortative mating is common in natural populations. For example, humans mate assortatively for height; tall men and tall women marry more frequently and short men and short women marry more frequently than would be expected on a random basis. **Negative assortative mating** occurs when phenotypically dissimilar individuals mate more often than randomly chosen individuals. If humans exhibited negative assortative mating for height, tall men and short women would marry preferentially and short men and tall women would marry preferentially. Neither positive nor negative assortative mating affects the allelic frequencies of a population, but it may influence the genotypic frequencies if the phenotypes for which assortative mating occurs are genetically determined.

■ *Figure 21.9*   The distribution of malaria caused by the parasite *Plasmodium falciparum* coincides with distribution of the *Hb-S* allele for sickle-cell anemia. The frequency of the *Hb-S* is high in areas where malaria is common, because *Hb-A/Hb-S* heterozygotes are resistant to malarial infection. (After A. C. Allison. 1961. Abnormal hemoglobin and erythrocyte enzyme-deficiency traits. In G. A. Harrison, ed., *Genetic Variation in Human Populations*. Pergamon, New York: pp. 16–40.)

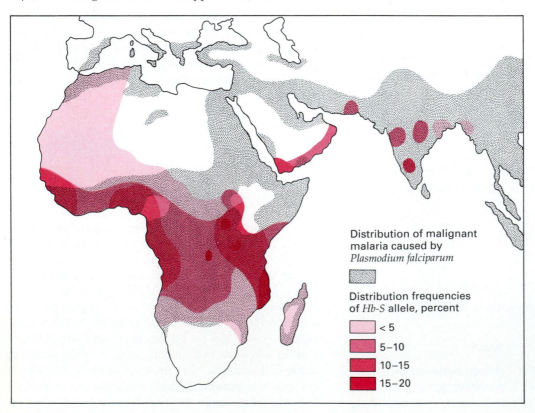

Distribution of malignant malaria caused by *Plasmodium falciparum*

Distribution frequencies of *Hb-S* allele, percent

< 5
5–10
10–15
15–20

■ *Figure 21.10*   Consequences of continued inbreeding of three different lines of corn, A, B, and C.

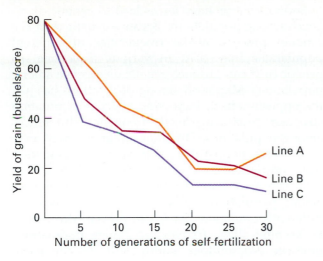

Inbreeding

One departure from nonrandom mating that is of special interest to population geneticists is inbreeding. **Inbreeding**, or mating among related individuals, is really positive assortative mating for relatedness. The overall result of inbreeding is to increase homozygosity and decrease heterozygosity of the population. Many people assume that inbreeding is always harmful. Indeed, close inbreeding in humans generally has unfavorable consequences; so laws in many countries prohibit marriages between close relatives. Close relatives have a high proportion of genes in common, as they are descended from the same ancestors. For example, first cousins share on the average one eighth of the same genes; siblings share on the average one half of the same

genes. Because close relatives share many of the same genes, there is a greater chance of bringing together two deleterious recessive mutations to produce a homozygous offspring when close relatives mate.

A substantial amount of data from plant and animal breeding also indicates deleterious effects of inbreeding in these organisms. As an example, Figure 21.10 shows the consequences of 30 generations of inbreeding on the yield of three different lines of corn. Note that in each line yield gradually declines as the number of generations of self-fertilization (the most extreme form of inbreeding) increases. This type of reduction in fitness due to inbreeding is referred to as *inbreeding depression*.

Although inbreeding has negative effects in humans and many domesticated plants and animals, a number of organisms reproduce predominantly by self-fertilization; many of these species are quite successful. For example, the land snail, *Rumina decollata*, was introduced into the United States from its native habitat in Europe sometime before 1822. By 1915 it had spread to Florida, Texas, Oklahoma, and Mexico, and is common today across the southern United States. Genetic studies indicate that *Rumina decollata* is completely homozygous in North America and reproduces only through self-fertilization. This species illustrates that inbreeding is not always harmful.

To demonstrate how inbreeding leads to a reduction in heterozygosity, let us consider the effects of self-fertilization, the most extreme type of inbreeding. Assume that we begin with a population consisting entirely of heterozygotes and that all individuals in this population reproduce by self-fertilization (Table 21.5). After one generation of self-fertilization, the progeny will consist of 1/4 *AA*,

Table 21.5

Relative Genotype Distributions Resulting from Self-Fertilization over Several Generations Starting with an *Aa* Individual

GENERATION	PROPORTIONS OF POPULATION		
	AA	*Aa*	*Aa*
0		1	
1	1/4	1/2	1/4
2	1/4 + 1/8 = 3/8	1/4	1/4 + 1/8 = 3/8
3	3/8 + 1/16 = 7/16	1/8	3/8 + 1/16 = 7/16
4	7/16 + 1/32 = 15/32	1/16	7/16 + 1/32 = 15/32
5	15/32 + 1/64 = 31/64	1/32	15/32 + 1/64 = 31/64
n	$[1 - (1/2)^n]/2$	$(1/2)^n$	$[1 - (1/2)^n]/2$
∞	1/2	0	1/2

1/2 *Aa*, and 1/4 *aa*. Now only half of the population consists of heterozygotes. When this generation undergoes self-fertilization, the *AA* homozygotes will produce only *AA* progeny, and the *aa* homozygotes will produce only *aa* progeny. When the heterozygotes reproduce, however, only half of their progeny will be heterozygous like the parents, and the other half will be homozygous (1/4 *AA* and 1/4 *aa*). This means that in each generation of self-fertilization, the percentage of heterozygotes decreases by 50 percent. After an infinite number of generations, there will be no heterozygotes and the population will be divided equally between the two homozygous genotypes.

The result of continued self-fertilization is to increase homozygosity at the expense of heterozygosity. The frequencies of alleles *A* and *a* remain constant, while the frequencies of the three genotypes change significantly. When less intensive inbreeding occurs, similar, but less pronounced, effects occur. On the other hand, outbreeding (preferential mating with unrelated individuals) increases heterozygosity.

Keynote

Inbreeding involves preferential mating between close relatives. Continued inbreeding increases homozygosity within a population. Outbreeding, or preferential mating with unrelated individuals, increases heterozygousity.

Summary of the Effects of Evolutionary Forces on the Gene Pool of a Population

Let us now review the major effects of the different evolutionary forces (mutation, migration, genetic drift, and natural selection) on (1) changes in allelic frequency within a population, (2) genetic divergence among populations, and (3) increases and decreases in genetic variation within populations.

Mutation, migration, genetic drift, and selection all have the potential to change the allelic frequencies of a population over time. Mutation, however, usually occurs at such a low rate that the change resulting from mutation pressure alone is frequently negligible. Genetic drift will produce substantial changes in allelic frequency only when the population size is small. Furthermore, mutation, migration, and selection may lead to equilibrium, where evolutionary forces continue to act but the allelic frequencies no longer change. Nonrandom mating does not change *allelic frequencies*, but it does affect the *genotypic frequencies* of a population: inbreeding leads to increases in homozygosity and outbreeding produces an excess of heterozygotes.

Several evolutionary forces lead to genetic divergence among populations. Because genetic drift is a random process, allelic frequencies in different populations may drift in various directions, so genetic drift can produce genetic divergence among populations. Migration among populations has just the opposite effect, increasing effective population size and equalizing allelic frequency differences among populations. Different mutations may arise in different populations, and, therefore, mutation may contribute to population differentiation. Natural selection can increase genetic differences among populations by favoring different alleles in different populations, or it can prevent divergence by keeping allelic frequencies among populations uniform. Nonrandom mating will not, by itself, generate genetic differences among populations, although it may contribute to the effects of other forces by increasing or decreasing effective population size.

Gene flow and mutation tend to increase genetic variation within populations by introducing new alleles to the gene pool. Genetic drift produces the opposite effect, decreasing genetic variation within small populations through loss of alleles. Because inbreeding leads to increases in homozygosity, it also diminishes genetic variation within populations; outbreeding, on the other hand, increases genetic variation by increasing heterozygosity. Natural selection may increase or decrease genetic variation. If one particular allele is favored, other alleles decrease in frequency and may be eliminated from the population by selection. Alternatively, natural selection may increase genetic variation within populations through overdominance and other forms of balancing selection.

In practice, these evolutionary forces never act in isolation but combine and interact in complex ways. In most natural populations the combined effects of these forces and their interaction determine the pattern of genetic variation observed in the gene pool over time.

Conservation Genetics

Although extinction is a natural process and ultimately the fate of every species, human activities in the past 200 years have greatly accelerated the pace of extinction among life on Earth. Tropical forests contain more than 50 percent of all species, and until recently, were relatively undisturbed. However, humans are now exploiting these areas on an unprecedented scale. Every year 200,000 square kilometers of tropical forest are destroyed or impoverished, leading to the extinction of hundreds of species. In temperate regions, where the total number of species is much less than in the tropics, thousands of plants and animals are still endangered or threatened. If these trends continue, humans will probably exterminate 25 percent of all species in the world during the next 50 years. To preserve as much of the remaining biodiversity as possible, biologists have begun applying principles of population genetics to the management of rare and endangered species, leading to the development of the new field of **conservation genetics**.

Many endangered plants and animals exist in small, isolated populations. As we have seen, genetic drift becomes an important evolutionary force when population size is small. For example, small populations are often forced into inbreeding. One consequence of genetic drift is the loss of genetic variation and increases in homozygosity. For many organisms, the loss of genetic variation is undesirable for several reasons. First, the evolutionary potential of a population to respond to environmental change is directly related to the amount of genetic variation within the species. Species without significant variation may be unable to adapt to new environmental conditions and therefore may become extinct. Second, populations frequently harbor a number of deleterious recessive mutations, whose effects are masked by the presence of dominant alleles in the heterozygous state. As genetic drift reduces heterozygosity, these deleterious mutations are more likely to be expressed. Consequently, recessive diseases and disorders are more common in populations with high homozygosity. For example, studies of cheetahs reveal that most of these cats are totally lacking in genetic variation—they are completely homozygous at over 200 loci. As a result, cheetahs experience high juvenile mortality, low fertility, and extreme susceptibility to diseases. Reduced genetic variation in this species threatens its long-term survival in the wild.

The projected amount of heterozygosity of a population in the next generation (H_1) is given by the formula

$$H_1 = \left(1 - \frac{1}{2N_e}\right) H_o$$

where N_e is the effective population size, and H_o is the initial heterozygosity. Suppose that the initial heterozygosity in a population of 50 adult black rhinoceroses is 0.3. In the next generation the heterozygosity is expected to be $H_1 = (1 - 1/100)(0.3) = 0.297$. In this case the heterozygosity has been reduced very little, only 1 percent. Even if the number of breeding adults is only five, the heterozygosity in the next generation will still be $H_1 = (1 - 1/10)(0.3) = 0.27$; only 10 percent of the heterozygosity has been lost. This illustrates an important point: Even when population size is quite small, relatively little genetic variation is lost in a single generation. However, if drift occurs for a number of generations, the loss of genetic variation can be quite substantial. The heterozygosity remaining after t generations (H_t) is given by

$$H_t = \left(1 - \frac{1}{2N_e}\right)^t H_o$$

If our population of black rhinoceroses consists of only 5 breeding individuals for 20 generations, then the amount of heterozygosity would be reduced to $H_t = (1 - 1/10)^{20}(0.3) = 0.036$, a reduction of 88 percent.

From the equations given, it is clear that the reduction in heterozygosity is inversely proportional to the effective population size N_e; the smaller the effective population size the larger the reduction in heterozygosity. Remember that the N_e represents the equivalent number of breeding adults. We have already mentioned one factor that affects N_e—the sex ratio. The effective population size is higher when there are equal numbers of breeding males and females in the population, and it declines as one of the sexes becomes rare. Thus for preserving genetic diversity it is desirable to maintain equal numbers of adult males and females. Several additional factors also influence N_e. One is the family size. If some breeding individuals produce many offspring while others produce few, N_e is smaller than if all breeding individuals produce equal numbers of offspring. Therefore equal-size families will preserve the most genetic diversity.

These principles suggest practices that wildlife managers and zoo personnel can use to preserve genetic diversity in small populations:

1. Acquire an adequate number of founders when establishing a captive or wild population. Founder effect can produce genetic drift and loss of genetic diversity when the population is established, but as we have discussed, the amount of genetic variation lost in one generation is not great. Thus relatively few founders (five to ten pairs) can be used without great loss of genetic variation, provided that the founders have substantial heterozygosity to begin with and that the population size does not remain small for long. It is essential that the founders not be collected from an inbred group so that their initial heterozygosity (H_o in the above equations) is high.

2. Expand the population rapidly. Most genetic diversity is lost through drift when the population remains small over several generations. Thus a small number of founders

may be sufficient if the population is expanded rapidly following founding.

3. Prevent inbreeding by controlled breeding. Inbreeding (mating among close relatives) also leads to loss of genetic variation. Careful recording of pedigrees and managed breeding can be useful to prevent matings between close relatives. This is frequently possible in captive populations, such as those existing in zoos. Breeding adults should be exchanged between populations to reduce inbreeding within populations. Many zoos now routinely exchange adults of rare and endangered species for breeding purposes.

4. Maximize effective population size by maintaining an equal sex ratio. If only a limited number of adults can be maintained, it is best to keep the number of males and females equal, as unequal sex ratios lower N_e and reduce genetic diversity.

5. Maintain equal family size. Large numbers of offspring produced by some adults and few offspring produced by others tend to reduce N_e and lead to loss of genetic diversity.

These practices, which are based on population genetics principles, are now being used by many zoo personnel, wildlife managers, and animal breeders to prevent the loss of genetic diversity in a number of rare and endangered species.

Keynote

Conservation genetics uses principles of population genetics to help manage rare and endangered species. Genetic diversity can be maintained best by establishing populations with adequate numbers of founders, expanding the population rapidly, avoiding inbreeding, and maintaining equal sex ratio and equal family size.

Molecular Evolution

In recent years population geneticists have begun to apply molecular genetic techniques to studies of genetic variation within populations and to questions about the molecular basis of evolution. By using restriction mapping and DNA sequencing methods (see Chapter 13), biologists can now examine evolution at the most basic genetic level, the level of the DNA. These studies have not altered basic principles of population genetics, but they have provided a more complete and detailed pic-

ture of how the forces of nature produce evolutionary change.

DNA Sequence Variation

An important question in the study of evolution is how the patterns and rates of evolution differ among genes and among different parts of the same gene. Rates of evolution of given DNA sequences can be measured by comparing sequences in two different organisms that diverged from a common ancestor at some known time in the past. We assume that the common ancestor had a single DNA sequence. Then, after the two organisms diverged from this ancestor, their DNA sequences underwent independent evolutionary changes, producing the differences we see in their sequences today. For example, it is believed that most major groups of mammals diverged from a common ancestor some 65 million years ago. Suppose we examined a particular DNA sequence, perhaps the gene for growth hormone, in mice and humans. We might find that the sequences for this gene in mouse and human differ in 20 nucleotides. These 20 nucleotides must have changed during the past 65 million years following the split between the mouse and human evolutionary lines. To calculate the rate of evolutionary change in this gene, we first compute the number of nucleotide substitutions (changes) that occurred to produce the 20 nucleotide differences we observe today. There are various mathematical methods for estimating this number, which is usually expressed as nucleotide substitutions per nucleotide site, so that the rate of evolution is independent of the length of the sequences compared. To obtain rates of change, we then divide our value of nucleotide substitutions per nucleotide site by the number of years of evolutionary time that separate the two organisms. The rate of change obtained is then expressed as substitutions per nucleotide site per year. In our example with the growth hormone gene, we might find that the rate of change was 4×10^{-9} substitutions per site per year.

Studies of nucleotide sequences in numerous genes have revealed that different parts of genes evolve at different rates. Recall from our discussion of molecular genetics that a typical eukaryotic gene is made up of some nucleotides that specify the amino acid sequence of a protein (coding sequences) and other nucleotides that do not code for amino acids in a protein (noncoding sequences). Noncoding sequences include introns, leader regions, trailer regions which are transcribed but not translated, and 5′ and 3′ flanking sequences that are not transcribed. Other noncoding sequences include pseudogenes, which are nucleotide sequences that no longer produce functional gene products because of mutations. Even within the coding regions of a functional gene, not all nucleotide substitutions produce a corresponding change in the amino acid sequence of a protein. In particular, many mutations occurring at the third position of the codon have no effect on the amino acid sequence of the protein, because such mutations produce a synonymous codon—one that codes for the same amino acid as the unmutated codon (see Chapter 12).

Table 21.6 shows relative rates of change in different parts of mammalian genes. Within functional genes, notice that the highest rate of change involves synonymous changes in the coding sequences. The rate of synonymous nucleotide change is about five times greater than the observed rate of nonsynonymous changes. Synonymous changes do not alter the amino acid sequence of the protein. Thus the high rate of evolutionary change seen there is not unexpected, because these changes do not affect a protein's functioning. Synonymous and nonsynonymous mutations probably arise with equal frequency. However, many nonsynonymous changes that arise within coding sequences are detrimental to fitness and consequently are eliminated by natural selection, whereas synonymous mutations are rarely detrimental and are tolerated.

You will notice in Table 21.6 that high rates of evolutionary change also occur in the 3' flanking regions of functional genes. Like synonymous changes, sequences in the 3' flanking regions have no known effect on the amino acid sequence and usually have little effect on gene expression; most mutations that occur here will be tolerated by natural selection. Rates of change in introns are also high, but not as high as the synonymous changes and those in the 3' flanking regions. Although the sequences in the introns do not code for proteins, the intron must be properly spliced out for the mRNA to be translated into a functional protein. A few sequences within the intron are critical for proper splicing; these include the consensus sequences at the 5' and 3' ends of the intron and the sequence at the branch point (see Chapter 11). As a result, not all changes in introns will be tolerated, so the overall rate of evolution is a bit lower than that seen in synonymous coding sequences and in the 3' flanking region.

Lower rates of evolutionary change are seen in the 5' flanking region. Although this region is neither transcribed nor translated, it does contain the promoter for the gene; thus sequences in the 5' flanking region are significant for gene expression. Next in evolutionary rate are the leader and trailer

Table 21.6

Relative Rates of Evolutionary Change in DNA Sequences of Mammalian Genes

SEQUENCE	NUCLEOTIDE SUBSTITUTIONS PER SITE PER YEAR ($\times 10^{-9}$)
Functional Genes	
5' Flanking region	2.36
Leader	1.74
Coding sequence— synonymous	4.65
Coding sequence— nonsynonymous	0.88
Intron	3.70
Trailer	1.88
3' Flanking region	4.46
Pseudogenes	4.85

From Li, W., C. Luo, and C. Wu, 1985. Evolution of DNA sequences. In R. J. MacIntyre, ed. *Molecular Evolutionary Genetics*. New York: Plenum Press.

regions (Table 21.6), which have somewhat lower rates than the 5' flanking region. Although leaders and trailers are not translated, they are transcribed, and they provide important signals for processing the mRNA and for attachment of the ribosome to the mRNA. Nucleotide substitutions in these regions are therefore limited. The lowest rate of evolution is seen in the nonsynonymous coding sequences. Alteration of these nucleotides changes the amino acid sequence of the protein, and most mutations that occur here are eliminated by natural selection.

One final thing to note in Table 21.6 is that the highest rate of evolution is seen in nonfunctional pseudogenes. Among human globin pseudogenes, for example, the rate of nucleotide change is approximately 10 times that observed in the coding sequences of functional globin genes. The high rate of evolution observed in these sequences occurs because pseudogenes no longer code for proteins; changes in these genes do not affect an individual's fitness, so they are not eliminated by natural selection. Thus we observe the highest rates of evolutionary change in those sequences that have the least effect on function. Different functional genes also evolve at different rates (Table 21.7). For instance, the rate of nonsynonymous nucleotide substitution in the prolactin gene of mammals is over 300 times higher than that found in the histone H4 gene of mammals.

Table 21.7

Relative Rates of Evolutionary Change in DNA Sequences of Different Mammalian Genes[a]

GENE	NONSYNONYMOUS RATE	SYNONYMOUS RATE
Histone H4	0.004	1.43
Insulin	0.16	5.41
Prolactin	1.29	5.59
α-globin	0.56	3.94
β-globin	0.87	2.96
Albumin	0.92	6.72
α-fetoprotein	1.21	4.90

[a]All rates are in nucleotide substitutions per site per year $\times 10^{-9}$. Data from Li, W. et al., 1985. *Mol. Bio. Evol.* 2:150–174.

DNA Length Polymorphisms

In addition to evolution of nucleotide sequences through nucleotide substitution, variation frequently occurs in the number of nucleotides found within a gene. These variations, called DNA length polymorphisms, arise through deletions and insertions of relatively short stretches of nucleotides. For example, DNA length polymorphisms have been observed in the alcohol dehydrogenase gene of *Drosophila melanogaster*. In a study mentioned earlier, Martin Krietman sequenced 11 copies of this gene in fruit flies. In addition to extensive variation in nucleotide sequence, Krietman found six insertions and deletions in the 11 copies of the gene he examined. All of these were confined to introns and flanking regions of the DNA: none were found within exons. Insertions and deletions within exons usually alter the reading frame, so they will be selected against. As a result, insertions and deletions are most commonly found in noncoding regions of the DNA. However, some insertions and deletions have been found in the coding regions of certain genes. Another class of DNA length polymorphisms involves variation in the number of copies of a particular gene. Among individual fruit flies, for instance, the number of copies of rRNA genes varies extensively. The number of copies of transposons also varies extensively among individuals and is responsible for some DNA length polymorphisms.

Evolution of Multigene Families Through Gene Duplication

In eukaryotic organisms we frequently find multiple copies of genes, all having identical or similar sequences. A group of such genes is termed a **multigene family,** which is defined as a set of related genes that have evolved from some ancestral gene through the process of gene duplication. Members of a multigene family may be clustered together or dispersed on different chromosomes.

An example of a multigene family is the globin gene family, which consists of the genes that code for the polypeptide chains making up the hemoglobin molecule. The organization of this multigene family in humans was discussed in Chapter 15. The globin multigene family in humans is comprised of seven α-like genes found on chromosome 16 and six β-like genes found on chromosome 11. Globin genes are also found in other animals, and globin-like genes are even found in plants, suggesting that this is a very ancient gene family. Almost all functional globin genes in animal species have the same general structure, consisting of three exons separated by two introns. However, the numbers of globin genes and their order varies among species, as is shown for the β-like genes in Figure 21.11. Because all globin genes have similarities in structure and sequence, it appears that an ancestral globin gene (perhaps related to the present-day myoglobin gene) duplicated and diverged to produce an ancestral α-like gene and an ancestral β-like gene. These two genes then underwent repeated duplications, giving rise to the various α-like and β-like genes found in vertebrates today. Repeated gene duplication, such as that giving rise to the globin gene family, appears to be a frequent evolutionary occurrence. Indeed, the number of copies of globin genes varies in some human populations. For example, most humans have two α-globin genes on chromosome 16 (see Figure 15.13, p. 360). However, some individuals have a single α-globin gene on chromosome 16, while other individuals have three or even

■ *Figure 21.11* Organization of the β-globin gene families in several mammalian species. (ψ = Pseudogenes.)

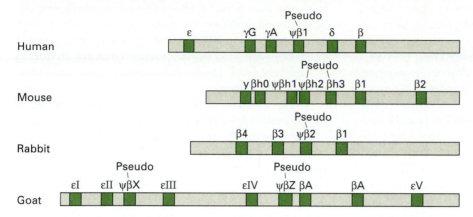

four copies of the α-globin gene on one of their chromosomes. These observations indicate that duplication and deletion of genes in a multigene family are ongoing processes.

Evolution in Mitochondrial DNA Sequences

In Chapter 19 we discussed the mitochondrial genome. We saw that animal mitochondrial DNA (mtDNA) is comprised of approximately 15,000 base pairs and that this DNA encodes 2 rRNAs, 22 to 23 tRNAs, and 10 to 12 proteins. A number of recent studies have examined sequence variation in mtDNA. These studies reveal that mtDNA evolution differs from evolution typically observed in nuclear DNA. For example, nucleotide sequences within vertebrate mtDNA evolve at a faster rate than coding sequences in nuclear genes. In fact the mtDNA of mammals undergoes evolutionary change at a rate that is five to ten times faster than that typically observed in mammalian nuclear DNA. However, the rapid rate of evolution observed in vertebrate mtDNA is not universal for all eukaryotes. For unknown reasons, plant mtDNA appears to evolve at an even slower rate than plant nuclear DNA. It is not entirely clear why animal mtDNA undergoes rapid evolutionary change, but the reason may be related to a higher mutation rate in mtDNA sequences and/or the lack of repair mechanisms.

Whatever the cause of accelerated evolution of mtDNA, the fact that these sequences change rapidly makes them ideal for assessing evolutionary relationships among groups of closely related organisms. Analyses of mtDNA sequences have been used to investigate the relationships among humans, chimpanzees, gorillas, orangutans, and gibbons, a group whose precise evolutionary relationships have been controversial.

Mitochondrial DNA also differs from nuclear DNA in that mtDNA is usually inherited from the mother; mitochondria are located in the cytoplasm and in most cases, only the egg cell contributes cytoplasm to a zygote. Furthermore, mtDNA does not undergo meiosis, and all offspring inherit the maternal genotype for mtDNA sequences (the offspring are clones for mtDNA genes). This pattern of inheritance allows matriarchal lineages (descendants from one female) to be traced, and provides a means for examining family structure in some populations. Recent studies of coyotes and wolves illustrates the utility of mtDNA for addressing evolutionary and ecological questions.

At one time gray wolves ranged over the entire North American continent, but during the past 200 years wolf populations have declined dramatically, primarily due to habitat destruction and direct extermination by humans. During this same time, coyotes have greatly expanded their range in North America, perhaps as a result of the same habitat destruction and wolf decline. Although coyotes and wolves belong to different species, observations of large coyotes in some areas and coyotelike characteristics in the nearly extinct red wolf led some experts to suggest that wolves and coyotes occasionally hybridize (interbreed) in nature. In captivity they occasionally mate, and such matings produce fertile offspring. Until recently, however, direct evidence for hybridization in nature was lacking.

To examine the issue of hybridization between coyotes and wolves, Niles Lehman and his colleagues collected DNA samples from 276 gray wolves and 240 coyotes from localities across North America. They cut the mitochondrial DNA (mtDNA) with 21 restriction enzymes and then separated the resulting fragments on a gel; differences in the DNA sequences were revealed by the presence of different restriction patterns or RFLPs (refer to Figure 21.5). This analysis showed clear differences in the mtDNA of wolves and coyotes. Surprisingly, some wolves possessed coyote mtDNA, indicating that hybridization has occurred between these two species. However, no coyotes possessed wolf mtDNA. Since mtDNA is inherited only from the female parent, this indicates that wolf-coyote hybridization has been unidirectional—female coyotes have contributed their mtDNA to the wolf population but female wolves have not contributed their mtDNA to the coyote population. This suggests that wolf-coyote hybridization occurs through matings between female coyotes and male wolves. This is actually the type of cross one might expect based on the larger size of the wolves: smaller male coyotes may be physically unable to mate with larger female wolves.

Hybridization between wolves and coyotes appears to be restricted to northern Minnesota, southern Ontario, southern Quebec, and Isle Royale (an island in Lake Superior). Interestingly, this is an area where coyotes have recently extended into the wolf distribution. The investigators proposed that repeated matings between female coyotes and male wolves produce hybrid offspring, which then backcross with wolves, allowing the coyote mtDNA to enter wolf populations. This hybridization threatens to contaminate the gene pool of the North American gray wolf with coyote genes, contributing further to the destruction of this already endangered species.

Concerted Evolution

One of the surprising findings from studies of DNA sequence variation is the observation that some molecular force or forces maintains uniformity of sequence in multiple copies of a gene. This phenomenon has been termed **concerted evolution** or **molecular drive.** As we discussed in previous chapters, some genes in eukaryotes exist in multiple copies. Ribosomal RNA genes in complex organisms, for example, typically exist in hundreds or thousands of copies. Undoubtedly, these multiple copies arose through duplication. Following duplication, individual copies of a gene might be expected to acquire mutations and diverge. Selection might limit mutations in coding regions, but if many copies exist, some divergence would be expected to occur, especially in the noncoding sequences. Contrary to this expectation, numerous studies have revealed that nucleotide sequences in the different copies of a gene are frequently

quite homogeneous. Furthermore, the noncoding sequences are also homogeneous. This suggests that selection is not responsible for maintaining the homogeneity. When the same genes are examined in a second, closely related species, that group's sequences are also homogeneous, but are frequently different from the homogeneous sequence found in the first species.

These observations have led to the conclusion that a molecular process continually maintains uniformity among multiple copies of the same sequence within a species. At the same time, the process allows for rapid differentiation among species. The mechanism of concerted evolution is not understood, but concerted evolution has important consequences for how species undergo genetic divergence, and represents an evolutionary force that was unknown before the application of modern molecular techniques to population genetics.

Evolutionary Relationships Revealed by RNA and DNA Sequences

Within the past few years, molecular genetics has provided powerful tools for deciphering the evolutionary history of life. Because evolution is defined as genetic change, genetic relationships are of primary importance in the construction of evolutionary trees. However, until recently, it was impossible to examine the genes directly. In the past evolutionary biologists were forced to rely entirely on comparison of phenotypes to infer genetic similarities and differences. They assumed that if the phenotypes were similar, the genes coding for the phenotypes were similar; if the phenotypes were different, the genes were different. Thus phenotypes were used for evolutionary studies. Originally the phenotypes examined consisted largely of gross anatomical features. Later, behavioral, ultrastructural, and biochemical characteristics were also studied. Comparisons of such traits were successfully used to construct evolutionary trees for many groups of plants and animals, and, indeed, are still the basis of most evolutionary studies today.

However, relying on the study of such traits has limitations. Similar phenotypes can sometimes evolve in organisms that are distantly related. For example, if a naive biologist tried to construct an evolutionary tree on the basis of whether wings were present or absent, he might place birds, bats, and insects in the same evolutionary group, since all have wings. It is fairly obvious that these three organisms are not closely related—they differ in many features besides presence of wings, and the wings themselves are very different. But this extreme example shows that phenotypes can sometimes be misleading about evolutionary relationships, and phenotypic similarities do not necessarily reflect genetic similarities.

Another problem with relying on a comparison of phenotypes is that not all organisms have a number of easily studied phenotypic features. For instance, the study of evolutionary relationships among bacteria has always been problematic, because bacteria have few obvious traits that correlate with the degree of genetic relatedness. A third problem arises when we try to compare distantly related organisms. How do we compare bacteria and mammals, which have very few traits in common?

We have seen how DNA sequencing methods and analysis of restriction fragment length polymorphisms provide information about DNA sequences. DNA sequences provide the most accurate and reliable information upon which to infer evolutionary relationships. They allow direct comparison of the genetic differences among organisms, are easily quantified, and are found in all organisms (at least all organisms have some genes in common such as tRNA genes, rRNA genes, and genes for a few proteins). Because of these considerable advantages, many evolutionary biologists have turned to DNA sequences for assessing evolutionary relationships and for constructing evolutionary trees.

One case where sequence data have provided new information about evolutionary relationships is in our understanding of the primary divisions of life. Many years ago, biologists divided all of life into two major groups, the plants and the animals. As more organisms were discovered and their features examined in more detail, it became clear that this simple dichotomy was inappropriate. It was later recognized that organisms can be divided into prokaryotes and eukaryotes on the basis of cell structure. More recently, several primary divisions of life have been recognized, such as the five kingdoms (prokaryotes, protista, plants, fungi, and animals) proposed by Whittaker.

Within the past 10 years, RNA and DNA sequences have been used to uncover the primary lines of evolutionary history among all organisms. In one study Norman Pace and his colleagues constructed an evolutionary tree of life based on the sequences found in the 16S rRNA, which all organisms possess. As illustrated in Figure 21.12, their evolutionary tree revealed three major evolutionary groups: the eubacteria (the traditional prokaryotes), the eukaryotes, and the archaebacteria (a relatively little-known group of bacteria). The eubacteria and the archaebacteria, although both prokaryotic, were

■ *Figure 21.12*   An evolutionary tree of life revealed by comparison of 16S rRNA sequences. (From Pace et al., 1986. *Cell* 45:325–326.

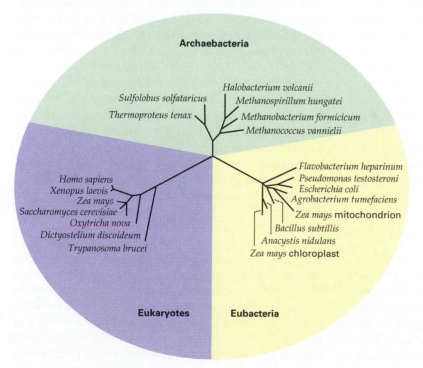

found to be as different genetically as were the eubacteria and eukaryotes. The deep evolutionary differences that separate the eubacteria and the archaebacteria were not obvious on the basis of phenotype; this became clear only after their nucleotide sequences were compared. Sequences of other genes, including 5S rRNAs, large rRNAs, and genes for some basic proteins, support the idea that three major evolutionary groups exist among living organisms.

Another field where DNA sequences are being used to study evolutionary relationships is in human evolution. Despite the extensive variation that is observed in size, body shape, facial features, skin color, and so on, genetic differences among human populations are relatively small. For example, analysis of mtDNA sequences shows that the mean difference in sequence between two human populations is about one-third of 1 percent. Other primates exhibit much larger differences. The two subspecies of orangutan, for instance, differ by as much as 5 percent. DNA sequence analysis indicates that all human groups are closely related. Nevertheless, some genetic differences do occur among different human groups. Surprisingly, the greatest differences are not found between populations located on different continents, but among human populations residing in Africa. All other human populations show fewer differences than those found among the African populations. Many experts interpret these findings to mean that humans experienced their origin and early evolutionary divergence in Africa. After a number of genetically differentiated populations had evolved in Africa, it is hypothesized that a small group of humans may have migrated out of Africa and given rise to all other human populations. This hypothesis has been called the "out-of-Africa" theory. Sequence data from both mitochondrial DNA and nuclear genes are consistent with this hypothesis. Although the out-of-Africa theory is not universally accepted, DNA sequence data are playing an increasingly important role in the study of human evolution and indeed in the study of the evolution of many groups.

Summary

Population genetics is the study of patterns of genetic variation found within groups and of how these patterns change or evolve over time. The gene pool of a population is the total of all genes within a Mendelian population, and is described in terms of the allelic and genotypic frequencies. The Hardy-Weinberg law describes what happens to allelic and genotypic frequencies of a large, randomly mating population free from evolutionary forces. When these conditions are met allelic frequencies do not change, genotypic frequencies stabilize after one generation in the proportions p^2, $2pq$, q^2, where p and q equal the allelic frequencies of the population.

The classical, balance, and neutral-mutation models are hypotheses to explain how much genetic variation should exist within natural populations and what forces are responsible for the variation observed. Protein electrophoresis showed that most populations of plants and animals contain large amounts of genetic variation, proving that the classical model was wrong. However, attempts to determine whether most genetic variation is maintained by natural selection (the balance model) or by the neutral forces of genetic drift and mutation (the neutral-mutation model) have provided no clear answer.

Mutation, genetic drift, migration, and natural selection are forces that can alter allelic frequencies of a population. Recurrent mutation changes the allelic frequencies of a population; the relative rates of forward and reverse mutation will determine allelic frequencies of a population in the absence of other forces. Genetic drift, chance change in allelic frequencies due to small effective population size, leads to a loss of genetic variation within a population, genetic divergence among populations, and change of allelic frequency within a population. Migration tends to reduce genetic divergence among populations and increase effective population size. Natural selection is differential reproduction of genotypes; it is measured by fitness, the relative reproductive contribution of genotypes. The effects of natural selection depend on the fitness of the genotypes, the degree of dominance, and the frequencies of the alleles in the population. Nonrandom mating affects only the genotypic frequencies of a population; the allelic frequencies are unaffected. One type of nonrandom mating, inbreeding, leads to an increase in homozygosity.

Current techniques of molecular genetics, including analysis of restriction fragment length polymorphisms and RNA and DNA sequences, have provided new insight into evolutionary processes. Different parts of a gene are found to evolve at different rates, and those parts of the gene that have the least effect on fitness appear to evolve at the highest rates. Multigene families evolve through repeated duplication of genes, followed by genetic divergence of their sequences. Mitochondrial DNA of animals appears to evolve at a faster rate than the DNA of nuclear genes. RNA and DNA sequences can be used for inferring evolutionary rates and relationships among organisms.

Analytical Approaches for Solving Genetics Problems

Q21.1 In a population of 2000 gaboon vipers a genetic difference with respect to venom exists at a single locus. The alleles are incompletely dominant. The population shows 100 individuals homozygous for the *t* allele (genotype *tt*, nonpoisonous), 800 heterozygous (genotype *Tt*, mildly poisonous), and 1100 homozygous for the *T* allele (genotype *TT*, deadly poisonous).

a. What is the frequency of the *t* allele in the population?

b. Are the genotypes in Hardy-Weinberg equilibrium?

A21.1 This question addresses the basics of calculating allelic frequencies and relating them to the genotype frequencies expected of a population in Hardy-Weinberg equilibrium.

a. The *t* frequency can be calculated from the information given, since the trait is an incompletely dominant one. There are 2000 individuals in the population under study, meaning a total of 4000 alleles at the *T/t* locus. The number of *t* alleles is given by

$$(2 \times tt \text{ homozygotes}) + (1 \times Tt \text{ heterozygotes})$$
$$= (2 \times 100) + (1 \times 800) = 1000$$

This calculation is straightforward, since both alleles in the nonpoisonous snakes are *t*, while only one of the two alleles in the mildly poisonous snakes is *t*. Since the total number of alleles under study is 4000, the frequency of *t* alleles is 1000/4000 = 0.25. This system is a two-allele system, so the frequency of *T* must be 0.75.

b. For the genotypes to be in Hardy-Weinberg equilibrium, the distribution must be $p^2\, TT + 2pq\, Tt + q^2\, tt$ genotypes, where *p* is the frequency of the *T* allele and *q* is the frequency of the *t* allele. In part a we established that the frequency of *T* is 0.75 and the frequency of *t* is 0.25. Therefore *p* = 0.75 and *q* = 0.25. Using these values, we can determine the genotype frequencies expected if this population is in Hardy-Weinberg equilibrium:

$$(0.75)^2\, TT + 2(0.75)(0.25)\, Tt + (0.25)^2\, tt$$

This expression gives 0.5625 *TT* + 0.3750 *Tt* + 0.0625 *tt*. Thus with 2000 individuals in the population we would expect 1125 *TT*, 750 *Tt*, and 125 *tt*. These values are close to the values given in the question, suggesting that the population is indeed in genetic equilibrium.

To check this result, we should perform a chi-square analysis (see Chapter 5), using the given numbers (not frequencies) of the three genotypes as the observed numbers and the calculated numbers as the expected numbers. The chi-square analysis is as follows, where $d = (\text{observed} - \text{expected})$:

GENOTYPE	OBSERVED	EXPECTED	d	d^2	d^2/e
TT	1100	1125	−25	625	0.556
Tt	800	750	+50	2500	3.334
tt	100	125	−25	625	5.000
Totals	2000	2000	0		8.890

Thus the the chi-square value (i.e., the sum of all the d^2/e values) is 8.89. For the reasons discussed in the text for a similar example, there is only one degree of freedom. Looking up the chi-square value in the chi-square table (Table 5.2), we find a *P* value of approximately 0.0025. So about 25 times out of 10,000 we would expect chance deviations of the magnitude observed. In other words, our hypothesis that the population is in Hardy-Weinberg equilibrium is not substantiated. [In this case our guess that it was in equilibrium was inaccurate.] Nonetheless, the population is not greatly removed from an equilibrium state.

Q21.2 About one normal allele in 30,000 mutates to the X-linked recessive allele for hemophilia in each human generation. Assume for the purposes of this problem that one *h* allele in 300,000 mutates to the normal alternative in each generation. (Note that in reality it is difficult to measure the reverse mutation of a human recessive allele that is essentially lethal, like the allele for hemophilia.) The mutation frequencies are indicated in the following equation:

$$h^+ \underset{\upsilon}{\overset{\mu}{\rightleftarrows}} h$$

where $\mu = 10\upsilon$. What allelic frequencies would prevail at equilibrium under mutation pressures alone in these circumstances?

A21.2 This question seeks to test our understanding of the effects of mutation on allelic frequencies. In the chapter we discussed the consequences of mutation pressure. The conclusion was that if *A* mutates to *A'* at *n* times the frequency that *A'* mutates back to *A*, then at equilibrium the value of

q will be $q = \mu/(\mu + v)$ or $q = nv/(n + 1)v$. Applying this general derivation to this particular problem, we simply use the values given. We are told that the forward mutation rate is 10 times the reverse mutation rate, or $\mu = 10v$. At equilibrium the value of q will be $q = \mu/(\mu + v)$. Since $\mu = 10v$, this equation becomes $q = 10v/11v$, so $q = 10/11$, or 0.909. Therefore, at equilibrium brought about by mutation pressures the frequency of h (the hemophilia allele) will be 0.909, and the frequency of h^+ (the normal allele) will be p, that is, $(1 - q) = (1 - 0.909) = 0.091$.

Questions and Problems

21.1 The following numbers of genotypes were collected from a population of the scarlet tiger moth, *Panaxia dominula*: 52 *BB*, 40 *Bb*, and 8 *bb*. Calculate the frequencies of the *B* and *b* alleles in this population.

21.2 Three alleles are found at a locus coding for malate dehydrogenase (MDH) in the spotted chorus frog. Chorus frogs were collected from a breeding pond, and each frog's genotype at the MDH locus was determined with electrophoresis. The following numbers of genotypes were found:

M^1M^1	8
M^1M^2	35
M^2M^2	20
M^1M^3	53
M^2M^3	76
M^3M^3	62
Total	254

a. Calculate the frequencies of the M^1, M^2, and M^3 alleles in this population.
b. Using a chi-square test, determine whether the MDH genotypes in this population are in Hardy-Weinberg proportions.

21.3 In a large interbreeding population 81 percent of the individuals are homozygous for a recessive character. In the absence of mutation or selection, what percentage of the next generation would be homozygous recessives? Homozygous dominants? Heterozygotes?

*21.4 Let *A* and *a* represent dominant and recessive alleles whose respective frequencies are p and q in a given interbreeding population at equilibrium (with $p + q = 1$).
a. If 16 percent of the individuals in the population have recessive phenotypes, what percentage of the total number of recessive genes exists in the heterozygous condition?
b. If 1.0 percent of the individuals were homozygous recessive, what percentage of the recessive genes would occur in heterozygotes?

*21.5 A population has eight times as many heterozygotes as homozygous recessives. What is the frequency of the recessive gene?

21.6 In a large population of range cattle the following ratios are observed: 49 percent red (*RR*), 42 percent roan (*Rr*), and 9 percent white (*rr*).
a. What percentage of the gametes that give rise to the next generation of cattle in this population will contain allele *R*?
b. In another cattle population only 1 percent of the animals are white and 99 percent are either red or roan. If this population is in Hardy-Weinberg equilibrium, what is the percentage of *r* alleles in this case?

21.7 In a pool of gametes, the alleles *A* and *a* have initial frequencies of p and q. Explain why the frequencies of the genotypes that result from these gametes will be p^2 (*AA*), $2pq$ (*Aa*), and q^2 (*aa*).

*21.8 The *S-s* antigen system in humans is controlled by two codominant alleles, *S* and *s*. In a group of 3146 individuals the following genotypic frequencies were found: 188 *SS*, 717 *Ss*, and 2241 *ss*.
a. Calculate the frequency of the *S* and *s* alleles.
b. Determine whether the genotypic frequencies conform to the Hardy-Weinberg equilibrium by using the chi-square test.

21.9 Refer to Problem 21.8. A third allele is sometimes found at the *S* locus. This allele S^u is recessive to both the *S* and the *s* alleles and can only be detected in the homozygous state. If the frequencies of the alleles *S*, *s*, and S^u are p, q, and r, respectively, what would be the expected frequencies of the phenotypes *S-*, *Ss*, *s-*, and S^uS^u?

21.10 In a large randomly mating human population, 60 percent of individuals belong to blood group O (genotype I^O/I^O). Assuming negligible mutation and no selective advantage of one blood type over another, what percentage of the grandchildren of the present population will be type O?

*21.11 A selectively neutral, X-linked recessive character appears in 0.40 of the males and in 0.16 of the females in a randomly interbreeding population. What is the frequency of the X-linked recessive gene that codes for the trait? How many females are heterozygous for it? How many males are heterozygous for it?

21.12 Suppose you found two distinguishable types of individuals in wild populations of some organism in the following frequencies:

	TYPE 1	TYPE 2
Female	99%	1%
Male	90%	10%

The difference is known to be inherited. What is its genetic basis?

21.13 List some of the basic differences in the classical, balance, and neutral-mutation models of genetic variation.

*21.14 Two alleles of a locus, A and A', can be interconverted by mutation:

$$A \xrightarrow[\;\;v\;\;]{\;\;\mu\;\;} A'$$

and μ is a mutation rate of 6.0×10^{-7}, and v is a mutation rate of 6.0×10^{-8}. What will be the frequencies of A and A' at mutational equilibrium, assuming no selective difference, no migration, and no random fluctuation caused by genetic drift?

*21.15 a. Calculate the effective population size (N_e) for a breeding population of 50 adult males and 50 adult females.
b. Calculate the effective population size (N_e) for a breeding population of 60 adult males and 40 adult females.
c. Calculate the effective population size (N_e) for a breeding population of 10 adult males and 90 adult females.
d. Calculate the effective population size (N_e) for a breeding population of 2 adult males and 98 adult females.

21.16 The land snail *Cepaea nemoralis* is native to Europe but has been accidentally introduced into North America at several localities. These introductions occurred when a few snails were inadvertently transported on plants, building supplies, soil, or other cargo. The snails subsequently multiplied and established large, viable populations in North America.

Assume that today the average size of *Cepaea nemoralis* populations found in North America is equal to the average size of *Cepaea nemoralis* populations in Europe. What predictions can you make about the amounts of genetic variation present in

European and North American populations of *Cepaea nemoralis*? Explain your reasoning.

*21.17 A population of 80 adult squirrels resides on campus, and the frequency of the Est^1 allele among these squirrels is 0.70. Another population of squirrels is found in a nearby woods, and there, the frequency of Est^1 allele is 0.5. During a severe winter, 20 of the squirrels from the woods population migrate to campus in search of food and join the campus population. What will be the allelic frequency of Est^1 in the campus population after migration?

*21.18 A completely recessive gene, owing to changed environmental circumstances, becomes lethal in a certain population. It was previously neutral, and its frequency was 0.5.
a. What was the genotype distribution when the recessive genotype was not selected against?
b. What will be the allelic frequency after one generation in the altered environment?
c. What will be the allelic frequency after two generations?

21.19 Human individuals homozygous for a rare recessive autosomal gene die before reaching reproductive age. In spite of this removal of all affected individuals, the trait still appears generation after generation. Why doesn't this trait rapidly decrease in frequency, since the affected individuals do not pass on the gene?

*21.20 A completely recessive gene (Q^1) has a frequency of 0.7 in a large population, and the Q^1Q^1 homozygote has a relative fitness of 0.6.
a. What will be the frequency of Q^1 after one generation of selection?
b. If there is no dominance at this locus (the fitness of the heterozygote is intermediate to the fitnesses of the homozygotes), what will the allelic frequency be after one generation of selection?
c. If Q^1 is dominant, what will the allelic frequency be after one generation of selection?

21.21 As discussed earlier in this chapter, the gene for sickle-cell anemia exhibits overdominance. An individual who is an *Hb-A/Hb-S* heterozygote has increased resistance to malaria and therefore has greater fitness than the *Hb-A/Hb-A* homozygote, who is susceptible to malaria, and the *Hb-S/Hb-S* homozygote, who has sickle-cell anemia. Suppose that the fitness values of the genotypes in Africa are as presented below:

$$Hb\text{-}A/Hb\text{-}A = 0.88$$
$$Hb\text{-}A/Hb\text{-}S = 1.00$$
$$Hb\text{-}S/Hb\text{-}S = 0.14$$

Give the expected equilibrium frequencies of the sickle-cell gene (*Hb-S*).

21.22 The frequencies of the L^M and L^N blood group alleles are the same in each of the populations I, II, and III, but the genotypes' frequencies are not the same, as shown below. Which of the populations is most likely to show each of the following characteristics: random mating, inbreeding, genetic drift. Explain your answers.

	$L^M L^M$	$L^M L^N$	$L^N L^N$
I	0.50	0.40	0.10
II	0.49	0.42	0.09
III	0.45	0.50	0.05

*21.23 DNA was collected from 100 people randomly sampled from a given human population and was digested with the restriction enzyme *Bam*HI, electrophoresed, and Southern blotted. The blots were probed with a particular cloned sequence. Three different patterns of hybridization were seen on the blots. Some DNA samples (56 of them) showed a single band of 6.3 kb, others (6) showed a single band at 4.1 kb, and yet others (38) showed both the 6.3 and the 4.1 kb bands.

a. Interpret these results in terms of *Bam*HI sites.

b. What are the frequencies of the restriction site alleles?

c. Does this population appear to be in Hardy-Weinberg equilibrium for the relevant restriction site(s)?

21.24 What factors cause genetic drift?

21.25 What are the primary effects of the following evolutionary forces on the allelic and genotypic frequencies of a population?

a. Mutation

b. Migration

c. Genetic drift

d. Inbreeding

21.26 Explain how overdominance leads to a relatively high frequency of sickle-cell anemia in areas where malaria is widespread.

*21.27 Suppose we examine the rates of nucleotide substitution in two 300-nucleotide sequences of DNA isolated from humans. In the first sequence (sequence A), we find a nucleotide substitution rate of 4.88×10^{-9} substitutions per site per year. The substitution rate is the same for synonymous and nonsynonymous substitutions. In the second sequence (sequence B), we find a synonymous substitution rate of 4.66×10^{-9} substitutions per site per year and a nonsynonymous substitution rate of 0.70×10^{-9} substitutions per site per year. Referring to Table 21.6, what might you conclude about the possible functions of sequence A and sequence B?

21.28 What are some of the characteristics of mitochondrial DNA evolution in animals?

21.29 What is concerted evolution?

21.30 What are some of the advantages of using DNA sequences to infer evolutionary relationships?

Glossary

[Numbers in parentheses indicate the page(s) on which the term is defined.]

acrocentric chromosome A chromosome with a centromere near the end such that it has one long arm plus a stalk and a satellite. (8)

additive genetic variance Genetic variance that arises from the additive effects of genes on the phenotype. (479)

adenine (A) A purine base found in RNA and DNA; in double-stranded DNA adenine pairs with the pyrimidine thymine. (177)

allele One of two or more alternative forms of a single gene locus. Different alleles of a gene each have a unique nucleotide sequence, and are all concerned with the same biochemical or developmental process, although their individual phenotypes may differ. (31; 34)

allelic frequencies (gene frequencies) The frequencies of alleles at a locus occurring among individuals in a population. (492)

allelomorph (allele) A term coined by William Bateson; literally means "alternative form"; later shortened to *allele*. (40)

allopolyploidy Polyploidy involving two or more genetically distinct sets of chromosomes. (417)

alternation of generations The two distinct reproductive phases of green plants in which stages alternate between haploid cells and diploid cells (gametophyte cells and sporophyte cells). (21)

Ames test A test developed by Bruce Ames in the early 1970s that investigates new or old environmental chemicals for carcinogenic effects. It uses the bacterium *Salmonella typhimurium* as a test organism for mutagenicity of compounds. (390)

amino acids The building blocks of polypeptides. There are 20 different amino acids used to make proteins. (265)

aminoacyl-tRNA A tRNA molecule covalently bound to an amino acid. This complex brings the amino acid to the ribosome so that it can be used in polypeptide synthesis. (273)

aminoacyl-tRNA synthetase An enzyme that catalyzes the addition of a specific amino acid to a tRNA molecule. Since there are 20 amino acids, there are also at least 20 synthetases. (273)

amniocentesis A procedure in which a sample of amniotic sac fluid is withdrawn from the amniotic sac of a developing fetus. Fetal cells can be tested for DNA defects, or cultured and examined for biochemical defects. (307)

anaphase The stage in mitosis or meiosis during which the sister chromatids (mitosis) or homologous chromosomes (meiosis) separate and migrate toward the opposite poles of the cell. (11, 12)

anaphase II The second stage of meiosis during which the centromeres (and therefore the chromatids) are pulled to the opposite poles of the spindle. The separated chromatids are now referred to as chromosomes in their own right. (18)

aneuploidy The abnormal condition in which one or more whole chromosomes of a normal set of chromosomes either are missing or are present in more than the usual number of copies. (58; 414)

antibody A protein molecule that recognizes and binds to a foreign substance introduced into the organism. (360)

anticodon A three-nucleotide sequence on a tRNA that pairs with a codon in mRNA by complementary base pairing. (257)

antigen Any large molecule that stimulates the production of specific antibodies or binds specifically to an antibody. (360)

artificial selection Human determination as to which individuals will survive and reproduce. If the selected traits have a genetic basis, they will change and evolve. (482)

asexual (vegetative) reproduction Reproduction in which a new individual develops from either a single cell or from a group of cells in the absence of any sexual process. (8)

attenuation A regulatory mechanism in certain bacterial biosynthetic operons that controls gene expression by causing RNA polymerase to terminate transcription. (331)

autopolyploidy Polyploidy involving more than two chromosome sets of the same species. (417)

autosome A chromosome other than a sex chromosome. (8; 56)

auxotroph A mutant strain of a given organism that is unable to synthesize a molecule required for growth and therefore must have the molecule supplied in the growth medium in order for it to grow. (131)

auxotrophic mutation (nutritional, biochemical) A mutation that affects an organism's ability to make a particular molecule essential for growth. (396)

back mutation *See* reversion (reverse mutation). (381)

bacteria Spherical, rod-shaped, or spiral-shaped unicellular prokaryotic microorganisms lacking a defined nucleus and organelles. (3)

balance model A hypothesis of genetic variation proposing that balancing selection maintains large amounts of genetic variation within populations. (498)

balancing selection Selection that maintains balance between alleles, preventing any single allele from reaching high frequency. (498)

Barr body   A highly condensed mass of chromatin found in the nuclei of normal females, but not in the nuclei of normal male cells. It represents a cytologically condensed and inactivated X chromosome. (61)

base analog   A chemical whose molecular structure is extremely similar to the bases normally found in DNA. (386)

base-pair substitution mutation   A change in a gene such that one base pair is replaced by another base pair, e.g., an AT is replaced by a GC pair. (379)

bidirectional replication   The DNA synthesis that takes place in both directions away from the origin of the replication point. (215)

binary fission   The process in cell reproduction in which a parental cell divides into two daughter cells of approximately equal size. (3)

binomial distribution   The theoretical frequency distribution of events that have two possible outcomes. (472)

biochemical mutation   *See* auxotrophic mutation. (396)

bivalent   A pair of homologous, synapsed chromosomes during the first meiotic division. (15)

bottleneck effect   A form of genetic drift that occurs when a population is drastically reduced in size. Some genes may be lost from the gene pool as a result of chance. (507)

branch-point sequence   The consensus sequence in mammalian cells, YNCURAY (where Y is a pyrimidine, R is a purine, and N is any base), to which the free 5' end of the intron loops and binds to the A nucleotide in the sequence during intron splicing. (247)

broad-sense heritability   A quantity representing the proportion of the phenotypic variance that consists of genetic variance. (479)

C value   The amount of DNA found in the haploid set of chromosomes. (7; 196)

CAAT element   One of the eukaryotic promoter elements found approximately 80 base pairs upstream of the initiation site. It can also function at a number of other locations and in either orientation with respect to the start point. The consensus sequence is 5'-GGCCA-ATCT-3'. (241)

cancer   Diseases characterized by the uncontrolled and abnormal division of eukaryotic cells to produce a malignant tumor, and by the spread of the disease (metastasis) to disparate sites in the organism. (438)

5'-capping   The addition of a methylated guanine nucleotide (a "cap") to the 5' end of a premessenger RNA molecule; the cap is retained on the mature mRNA molecule. (243)

catabolite repression (glucose effect)   The inactivation of an inducible bacterial operon in the presence of glucose even though the operon's inducer is present. (329)

cDNA   DNA copies made from an RNA template catalyzed by the enzyme reverse transcriptase. (293)

cDNA library   The collection of molecular clones that contains cDNA copies of the entire mRNA population of a cell. *See* cDNA. (293)

cell cycle   The cyclical process of growth and cellular reproduction in unicellular and multicellular eukaryotes. The cycle includes nuclear division, or mitosis, and cell division, or cytokinesis. (9)

cell division   A process whereby one cell divides to produce two cells. (9)

cell-free, protein-synthesizing system   A system, isolated from cells, which contains ribosomes, mRNA, tRNAs with amino acids attached, and all the necessary protein factors for the *in vitro* synthesis of polypeptides. (270)

cellular oncogene (c-*onc*)   The genes, present in a functional state in cancerous cells, which are responsible for the cancerous state. (442)

centi-Morgan (cM)   The map unit, so named in honor of T. H. Morgan. (103)

Central Dogma   Crick's name for the two-step process of transcription (DNA → RNA) followed by translation (RNA → protein) for the synthesis of proteins encoded by DNA. (234)

centromere (kinetochore)   A specialized region of a chromosome seen as a constriction under the microscope. This region is important in the activities of the chromosomes during cellular division. (8; 200)

chain-terminating codon   One of three codons for which no normal tRNA molecule exists with an appropriate anticodon. A nonsense codon in an mRNA specifies the termination of polypeptide synthesis. (272)

character   An observable phenotypic feature of the developing or fully developed organism that is the result of gene action. (26)

charged tRNA   The product of an amino acid added to a tRNA. (275)

charging   The act of adding the amino acid to the tRNA. (275)

chiasma   A cross-shaped structure formed during crossing-over and visible during the diplonema stage of meiosis. (17)

chiasma interference (chromosomal interference)   The physical interference caused by the breaking and rejoining chromatids that reduces the probability of more than one crossing-over event occurring near another one in one part of the meiotic tetrad. (112)

chi-square test   A statistical procedure that determines what constitutes a significant difference between observed results and results expected on the basis of a particular hypothesis; a goodness-of-fit test. (33)

chloroplast   The cellular organelle found only in green plants that is the site of photosynthesis in the cells containing it. (453)

chorionic villus sampling   A procedure in which a sample of chorionic villus tissue of a developing fetus is collected and examined for chromosomal abnormalities. (308)

chromatid   One of the two visibly distinct longitudinal subunits of all replicated chromosomes that becomes visible between early prophase and metaphase of mitosis. (11)

chromatin   The complex of DNA, chromosomal proteins, and RNA that comprises eukaryotic chromosomes. (196)

chromosomal interference   *See* chiasma interference. (112)

chromosome   The genetic material of the cell, complexed with protein and organized into a number of linear structures. Chromosome literally means "colored body," because the threadlike structures are visible under the microscope only after they are stained with dyes. (4)

chromosome mutation (chromosome aberration)   A change from the wild type in the structure or number of chromosomes. (378; 404)

chromosome theory of heredity   The theory that the chromosomes are the carriers of the genes. The first clear formulation of the theory was made by both Sutton and Boveri, who independently recognized that the transmission of chromosomes from one generation to the next closely paralleled the pattern of transmission of genes from one generation to the next. (51)

***cis*-dominance**   The phenomenon of a gene or DNA sequence controlling only genes that are on the same contiguous piece of DNA. (324)

***cis-trans* (complementation) test**   *See* complementation test. (146)

classical model   A hypothesis of genetic variation proposing that natural populations contain little genetic variation as a result of strong selection for one allele. (498)

clonal selection   A process whereby cells that already have antibodies specific to an antigen on their surfaces are stimulated to proliferate and secrete that antibody. (361)

cloning   The generation of many copies of a DNA molecule (e.g., a recombinant DNA molecule) by replication in a suitable host. (286)

cloning vector (cloning vehicle)   A DNA molecule that is able to replicate autonomously in a host cell and with which a DNA fragment (or fragments) can be bonded to form a recombinant DNA molecule for cloning. (289)

clustered repeated sequences   DNA sequences that are repeated one after another in a row. (202)

coding sequence   The part of a gene or an mRNA molecule that specifies the amino acid sequence of a polypeptide during translation. (243)

codominance   The situation in which the heterozygote exhibits the phenotypes of both homozygotes. (78)

codon   A group of three adjacent nucleotides in an mRNA molecule that specifies either one amino acid in a polypeptide chain or the termination of polypeptide synthesis. (268)

coefficient of coincidence   A number that expresses the extent of chiasma interference throughout a genetic map; the ratio of observed double-crossover frequency to expected double-crossover frequency. Interference is equal to 1 minus the coefficient of coincidence. (114)

combinatorial gene regulation   Transcriptional control (i.e., whether a gene is active or inactive) achieved by combining relatively few regulatory proteins (negative and positive) binding to particular DNA sequences. (347)

complementary base pairing   The hydrogen bonding between a particular purine and a particular pyrimidine in double-stranded nucleic acid molecules (DNA-DNA, DNA-RNA, or RNA-RNA). The major specific pairings are guanine with cytosine and adenine with thymine or uracil. (182)

complementary DNA   *See* cDNA. (293)

complementation test   A test used to determine whether two different mutations are within the same cistron (gene). (146)

complete dominance   The case in which one allele is dominant to the other so that at the phenotypic level the heterozygote is essentially indistinguishable from the homozygous dominant. (77)

complete recessiveness   The situation in which an allele is phenotypically expressed only when it is homozygous. (77)

concerted evolution (molecular drive)   A poorly understood evolutionary process that produces uniformity of sequence in multiple copies of a gene. (522)

conditional mutant   A mutant organism that is normal under one set of conditions but becomes seriously impaired or dies under other conditions. Temperature-sensitive mutants are examples of conditional mutants. (211)

conjugation   A process having a unidirectional transfer of genetic information through direct cellular contact between a donor ("male") and a recipient ("female") bacterial cell. (131)

consensus sequence   The sequence showing the nucleotide found most frequently at each position for a number of sequences examined. Examples of usage in the text are consensus sequences for promoters and for splice sites for intron removal from pre-mRNA molecules. (238)

conservation genetics   Application of the principles of population genetics to the management of rare and endangered species. (517)

conservative model   A DNA replication scheme in which the two parental strands of DNA remain together and serve as a template for the synthesis of a new daughter double helix. (207)

constitutive gene   A gene that is always active in growing cells. (320)

constitutive heterochromatin   Condensed chromatin that is constant in all cells of the organism. Constitutive heterochromatin is always genetically inactive, and is found at homologous sites on chromosome pairs. (196)

continuous trait   *See* quantitative (continuous) traits. (470)

contributing allele   An allele of a gene that contributes to the phenotype of a quantitative trait. (477)

controlling site   A specific sequence of nucleotide pairs adjacent to the gene where the transcription of a gene occurs in response to a particular molecular event. (321)

coordinate induction   The simultaneous transcription and translation of two or more genes brought about by the presence of an inducer. (322)

core enzyme   The portion of the *E. coli* RNA polymerase that is the active enzyme; it consists of four polypeptides. (238)

cosmids   Cloning vectors derived from plasmids that are capable of cloning large fragments of DNA. In addition to the features of plasmid cloning vectors (i.e., origin of replication and selectable marker(s) for growth in bacteria), cosmids contain phage lambda *cos* sites, which permit recombinant DNA molecules that are constructed to be packaged into the lambda phage head *in vitro*. (292)

cotransduction   The simultaneous transduction of two or more bacterial genes; a good indication that the bacterial genes are closely linked. (141)

cotranslational transport   The movement of a protein into the endoplasmic reticulum (ER) simultaneously with its synthesis. (280)

coupling   An arrangement in which the two wild-type alleles are on one homologous chromosome and the two recessive mutant alleles are on the other. (103)

crisscross inheritance   A type of gene transmission passed from a male parent to a female child to a male grandchild. (55)

cross   *See* cross-fertilization. (28; 34)

cross-fertilization (cross)   A term used for the fusion of male and female gametes from different individuals; the bringing together of genetic material from different individuals for the purpose of genetic recombination. (28)

crossing-over   A term introduced by Morgan and E. Cattell in 1912 to describe the process of reciprocal chromosomal interchange by which recombinants arise. (15)

cytokinesis   A term that refers to the division of the cytoplasm. The two new nuclei compartmentalize into separate daughter cells, and the mitotic cell division process is completed. (9)

cytoplasm   The semisolid matrix of cell material in the surrounding cell body. The term was coined by E. Strasburger. (4)

cytosine (C)   A pyrimidine base found in RNA and DNA. In double-stranded DNA cytosine pairs with the purine guanine. (177)

dark repair   *See* excision repair. (393)

daughter chromosome   The chromatid after anaphase of mitosis or anaphase II of meiosis. (11)

degeneracy   A multiple coding; more than one codon per amino acid. (272)

degradation control   Regulation of the RNA breakdown rate in the cytoplasm. (355)

deletion (deficiency)   A chromosome mutation resulting in the loss of a segment of the genetic material and the genetic information contained therein from a chromosome. (405)

deletion mutant   A mutation that arises by the loss of a segment of DNA. (145)

deoxyribonuclease (DNase)   An enzyme that catalyzes the degradation of DNA to nucleotides. (174)

deoxyribonucleic acid (DNA)   A polymeric molecule consisting of deoxyribonucleotide building blocks. In a double-stranded, double-helical form DNA is the genetic material of most organisms. (173)

deoxyribonucleotide   The basic building block of DNA, consisting of a sugar (deoxyribose), a base, and a phosphate. (177)

deoxyribose   The pentose (five-carbon) sugar found in DNA. (177)

development   The process of regulated growth that results from the interaction of the genome with cytoplasm and the environment. It involves a programmed sequence of phenotypic changes that are typically irreversible. (357)

dicentric bridge   A chromosome with two centromeres stretched across the cell in anaphase I of meiosis. For example, as a result of the crossover between genes in the loop in a paracentric inversion heterozygote, one recombinant chromatid becomes stretched across the cell as the two centromeres begin to migrate, forming a dicentric bridge. (408)

dideoxy nucleotide   A modified nucleotide which has a 3'-H on the deoxyribose sugar rather than a 3'-OH. If a dideoxy nucleoside triphosphate (ddNTP) is used in a DNA synthesis reaction, the ddNTP can be incorporated into the growing chain. However, no further DNA synthesis can then occur because no phosphodiester bond can be formed with an incoming DNA precursor. (302)

dideoxy (Sanger) sequencing   A method of rapid sequencing of DNA molecules developed by Fred Sanger. This technique incorporates the use of dideoxy nucleotides in a DNA polymerase-catalyzed DNA synthesis reaction. (302)

differentiation   An aspect of development that involves the formation of different types of cells, tissues, and organs from a zygote through the processes of specific regulation of gene expression. (357)

dihybrid cross   A cross between individuals heterozygous at the same two loci. (37)

dioecious   A term referring to plant species that have male and female sex organs on different individuals. (63)

diploid   A eukaryotic cell with two sets of chromosomes. (7)

directional selection   Selection in which individuals at one end of the phenotypic distribution survive and reproduce best. (485)

discontinuous DNA replication   A DNA replication involving the synthesis of short DNA segments, which are subsequently linked to form a long polynucleotide chain. (215)

discontinuous trait   A heritable trait in which the mutant phenotype is sharply distinct from the alternative, wild-type phenotype. (469)

disjunction   The process in anaphase during which sister chromatid pairs undergo separation. (12)

dispersive model   A DNA replication scheme in which the parental double helix is cleaved into double-stranded DNA segments that act as templates for the synthesis of new double-stranded DNA sgements. Somehow, the segments reassemble into complete DNA double helices, with parental and progeny DNA segments interspersed. (207)

disruptive selection   Selection in which individuals at both ends of the phenotypic distribution survive and reproduce best. (485)

DNA   *See* deoxyribonucleic acid. (173)

DNA fingerprinting   The use of DNA restriction fragment length polymorphisms to identify an individual. (310)

DNA helicase   An enzyme that catalyzes the unwinding of the DNA double helix during replication in *E. coli*; a product of the *rep* gene. (214)

DNA ligase (polynucleotide ligase)   An enzyme that catalyzes the formation of a covalent bond between free single-stranded ends of DNA molecules during DNA replication and DNA repair. (215; 217)

DNA polymerase   An enzyme that catalyzes the synthesis of DNA. (210)

DNA polymerase I   An *E. coli* enzyme that catalyzes DNA synthesis, originally called the Kornberg enzyme. (210)

DNA primase   The enzyme in DNA replication that catalyzes the synthesis of a short nucleic acid primer. (214)

docking protein   An integral protein membrane of the endoplasmic reticulum (ER) to which the nascent polypeptide-signal recognition particle (SRP)-ribosome complex binds to facilitate the binding of the polypeptide's signal sequence and associated ribosome to the ER. (280)

dominance variance (represented by V_D)   Genetic variance that arises from the dominance effects of genes. (479)

dominant   An allele or phenotype that is expressed in either the homozygous or the heterozygous state. (32)

dominant lethal allele   An allele that will exhibit a lethal phenotype when present in the heterozygous condition. (85)

dosage compensation   A mechanism in mammals which compensates for extra copies of genes on the X chromosome. *See* Barr body. (61)

double crossover   Two crossovers occurring in a particular region of a chromosome in meiosis. (106)

Down syndrome (trisomy-21)   A human clinical condition characterized by various abnormalities. It is caused by the presence of an extra copy of chromosome 21. (415)

duplication   A chromosome mutation that results in the doubling of a segment of a chromosome. (406)

effective population size   The effective number of adults contributing gametes to the next generation. (506)

effector   A small molecule involved in the control of expression of many regulated genes. (321)

endoplasmic reticulum (ER)   A system of membranes in the cytoplasm of many eukaryotic cells. In places the ER is continuous with the plasma membrane or the outer membrane of the nuclear envelope. ER to which ribosomes are bound is called rough ER, and ER without ribosomes is called smooth ER. The ribosomes on rough ER are engaged in the synthesis of polypeptides that are transported into the lumen of the ER to be secreted from the cell or to be localized in the cell membrane or in particular vacuoles within the cell. (4)

endosymbiont hypothesis   The hypothesis that mitochondria and chloroplasts originated as free-living prokaryotes that invaded primitive eukaryotic cells and established a mutually beneficial (symbiotic) relationship. (455)

enhancer (enhancer element)   In eukaryotes, a type of DNA sequence element having a strong, positive effect on transcription by RNA polymerase II. (241)

environmental variance (represented by V_E)   Any nongenetic source of phenotypic variation among individuals. (478)

episome   An autonomously replicating plasmid (a circular, double-stranded DNA molecule) that is capable of integrating into the host cell's chromosome. (428)

epistasis   A form of gene interaction in which one gene interferes with the phenotypic expression of another nonallelic gene so that the phenotype is governed by the former gene and not by the latter. (80)

erythroblasts   Red blood cell precursors. (348)

essential genes   Genes which, when they are mutated, can result in a lethal phenotype. (85)

euchromatin   Chromatin that stains normally. It is uncoiled during interphase, but becomes condensed during mitosis. Most of the eukaryotic genome consists of euchromatin. (196)

eukaryote   A term that literally means "true nucleus." Eukaryotes are organisms that have cells in which the genetic material is located in a membrane-bound nucleus. Eukaryotes can be unicellular or multicellular. (3)

eukaryotic initiation factor   In eukaryotes, proteins that associate with the small ribosomal subunit at the initiation stage of translation. (276)

evolution   Genetic change that takes place over time within a group of organisms. (482)

excision repair (dark repair)   An enzyme-catalyzed, light-independent process of repair of, e.g., ultraviolet light–induced thymine dimers in DNA that involves removal of the dimers and synthesis of a new piece of DNA complementary to the undamaged strand. (393)

exon   *See* coding sequence. (243)

exonuclease   An enzyme that digests a nucleic acid from a free end. For example, a $5' \rightarrow 3'$ exonuclease digests from the 5' end. (211)

expressivity   The degree to which a particular genotype is expressed in the phenotype. (86)

F_1 generation   The first filial generation produced by crossing two parental strains. (29)

F_2 generation   The second filial generation produced by selfing the F_1. (29)

***F*-pili (sex pili)**   Hairlike cell surface components produced by bacterial cells containing the *F* factor, which allow the physical union of F^+ and F^- cells, or *Hfr* and F^- cells, to take place. (131)

facultative heterochromatin   Chromatin that has the potential to become condensed throughout the cell cycle and may contain genes that are inactivated when the chromatin becomes condensed. (196)

familial trait   A trait shared by members of a family. (480)

fine-structure mapping   A high-resolution mapping of allelic sites within a gene. (144)

first filial generation   *See* F₁ generation; the offspring that result from the first experimental crossing of animals or plants. (29)

first law   *See* principle of segregation. (33)

fitness   The relative reproductive ability of a genotype. (511)

formylmethionine (fMet)   A specially modified amino acid involving the addition of a formyl group to the amino group of methionine. It is the first amino acid incorporated into a polypeptide chain in prokaryotes and in eukaryotic cellular organelles. (275)

forward mutation   A mutational change from a wild-type allele to a mutant allele. (381)

founder effect   A change in allele frequency due to migration of a small number of individuals from a large population. (507)

frameshift mutation   A mutational addition or deletion of a base pair in a gene that disrupts the normal reading frame of an mRNA, which is read in groups of three bases. (381)

frequency distribution   A means of summarizing the phenotypes of a continuous trait whereby the population is described in terms of the proportion of individuals that fall within a defined range. (471)

gametes   Mature reproductive cells that are specialized for sexual fusion. Each gamete is haploid and fusion of two gametes produces a diploid zygote. (7)

gametogenesis   The formation of male and female gametes by meiosis. (15)

gametophyte   The haploid sexual generation in the life cycle of plants that produces the gametes. (21)

G banding   A technique for staining human metaphase chromosomes in order to distinguish clearly each chromosome type. In G banding, metaphase chromosomes are treated with proteolytic enzymes and then stained with Giemsa stain (a permanent DNA dye) to produce the G banding pattern. G bands reflect regions of DNA that are rich in adenine and thymine. (194)

GC element   A eukaryotic promoter element with the consensus sequence 5'-GGGCGGG-3' that can be found in either orientation upstream of the transcription initiation site. The GC elements appear to help the RNA polymerase locate the transcription start point. (241)

gene (Mendelian factor)   The determinant of a characteristic of an organism. Genetic information is coded in the DNA, which is responsible for species and individual variation. A gene's nucleotide sequence specifies a polypeptide or RNA and is subject to mutational alteration. (26; 34)

gene flow   The movement of genes that takes place when organisms migrate and then reproduce, contributing their genes to the gene pool of the recipient population. (509)

gene frequency (allele frequency)   The proportion of one particular type of allele to the total of all alleles at this genetic locus in a Mendelian breeding population. (492)

gene locus   *See* locus. (33; 34)

gene mutation   An alteration of the DNA sequence of a gene. (378)

gene pool   The total genetic information present in a breeding population at a given time. (491)

gene redundancy   A situation in which two or more copies of a gene are present in a genome. (258)

gene regulatory elements   *See* transcription-controlling sequence. (234)

gene segregation   *See* principle of segregation (first law). (33)

generalized transduction   A type of transduction in which any gene may be transferred between bacteria. (139)

genetic code   The base-pair information that specifies the amino acid sequence of a polypeptide. (265)

genetic counseling   The procedures whereby the risks of prospective parents having a child who expresses a genetic disease are evaluated and explained to the couple. The genetic counselor typically makes predictions about the probabilities of particular traits (deleterious or not) occurring among children of a couple. (41; 306)

genetic drift   Any change in gene frequency due to chance in a population. (495; 505)

genetic engineering   The alteration of the genetic constitution of cells or individuals by directed and selective modification, insertion, or deletion of an individual gene or genes. In some cases novel gene combinations are made by joining DNA fragments from different organisms. (286)

genetic map (linkage map)   A representation of the order and relative location of genes. (101)

genetic mapping   The use of genetic crosses to locate genes on chromosomes relative to one another. (101)

genetic marker   Any allele used as an experimental probe to mark a gene, chromosome, or nucleus. (482)

genetic recombination   A process by which new combinations of alleles are produced. For example, from *A B* and *a b* the recombinants *A b* and *a B* are produced. (15)

genetic variance (represented by V_G)   Genetic sources of phenotypic variation among individuals of a population. Includes dominance genetic variance, additive genetic variance, and epistatic genetic variance. (478)

genetics   The science of heredity, involving study of the structure and function of genes and the way genes are passed from one generation to the next. (2)

genome   The total amount of genetic material in a cell; in eukaryotes the haploid set of chromosomes of an organism. (7)

genomic library   The collection of molecular clones that contains at least one copy of every DNA sequence in the genome. (293)

genotype   The genetic constitution of an organism. (26; 34)

genotypic frequencies   The frequencies or percentages of different genotypes found within a population. (492)

genotypic sex determination   The process by which the sex chromosomes play a decisive role in the inheritance and determination of sex. (59)

germ-line mutations   Mutations in the germ line of sexually reproducing organisms may be transmitted by the gametes to the next generation, giving rise to an individual with the mutant state in both its somatic and germ-line cells. (378)

glucose effect   *See* catabolite repression. (329)

Goldberg-Hogness box (TATA box, or TATA element)   *See* TATA element. (241)

Golgi apparatus   A membranous organelle in eukaryotic cells that plays a key role in the processing and transport of proteins that are secreted from the cell or that are packaged into special vesicles called lysosomes. (7)

guanine (G)   Purine base found in RNA and DNA. In double-stranded DNA guanine pairs with the pyrimidine cytosine. (177)

haploid   A cell or an individual with one copy of each nuclear chromosome. (7)

Hardy-Weinberg law (Hardy-Weinberg equilibrium, Hardy-Weinberg law of genetic equilibrium)   An extension of Mendel's laws of inheritance that describes the expected relationship between allelic frequencies in natural populations and the frequencies of individuals of various genotypes in the same populations. (492; 494; 496)

helix-destabilizing proteins   *See* single-strand DNA-binding proteins. (217)

hemizygous   The condition of X-linked genes in males. Males that have an X chromosome with an allele for a particular gene but do not have another allele of that gene in the gene complement are hemizygous. (54)

hereditary trait   A characteristic under the control of the genes that is transmitted from one generation to another. (26)

heritability   The proportion of phenotypic variation in a population attributable to genetic factors. (478)

hermaphroditic   For animals (e.g., nematode), the species in which each individual has both testes and ovaries. In plants, the species that have both stamens and pistils on the same flower. (63)

heterochromatin   Chromatin that remains condensed throughout the cell cycle and is genetically active. (196)

heterogametic sex   The sex that has sex chromosomes of different types (e.g., XY) and therefore produces two kinds of gametes with respect to the sex chromosomes. (52)

heterogeneous nuclear RNA (hnRNA)   The RNA molecules of various sizes that exist in a large population in the nucleus. Some of the RNA molecules are precursors to mature mRNAs. (246)

heteroplasmy   The condition in which cells have a mixture of normal and mutant mitochondria. (461)

heterozygosity   The proportion of individuals heterozygous at a locus.

heterozygote advantage   *See* overdominance. (513)

heterozygous   A term describing a diploid organism having different alleles of one or more genes and therefore producing gametes of different genotypes. (33; 34)

Hfr (high-frequency recombination) cell   A male cell in *E. coli*, with the *F* factor integrated into the bacterial chromosome. When the *F* factor promotes conjugation with a female (*F⁻*) cell, bacterial genes are transferred to the female cell with high frequency. (132)

highly repetitive sequence   A DNA sequence that is repeated between 10^5 and 10^7 times in the genome. (202)

histone   One of a class of basic proteins that are complexed with DNA in chromosomes and that play a major role in determining the structure of eukaryotic nuclear chromosomes. (196)

homeobox   A 180-bp consensus sequence found in the protein-coding sequences of genes that regulate development. (369)

homeodomain   The 60-amino-acid part of proteins that corresponds to the homeobox sequence of genes. All homeodomain-containing proteins appear to be located in the nucleus. (369)

homoeotic mutations   Mutations that alter the identity of particular segments, transforming them into copies of other segments. (368)

homogametic sex   The gender in the species (in mammals the female, in birds the male) which has the sex chromosome (X in mammals, Z in birds). (52)

homolog   Each individual member of a pair of homologous chromosomes. (7)

homologous chromosomes   The members of a chromosome pair that are identical in the arrangement of genes they contain and in their visible structure. (7)

homozygous   A term describing a diploid organism having the same alleles at one or more genes and therefore producing gametes of identical genotypes. (33; 34)

homozygous dominant   A diploid organism that has the same dominant allele for a given gene locus on both members of a homologous pair of chromosomes. (34)

homozygous recessive   A diploid organism that has the same recessive allele for a given gene locus on both members of a homologous pair of chromosomes. (34)

human genome project   A project to obtain the sequence of the complete 3 billion (3×10^9) nucleotide pairs of the human genome, and to map all of the estimated 50,000 to 100,000 human genes. (312)

hybrid dysgenesis   The appearance of a series of defects, including mutations, chromosomal aberrations, and sterility, when certain strains of *Drosophila melanogaster* are crossed. (437)

hypersensitive sites (hypersensitive regions)   Sites in the regions of DNA around transcriptionally active genes that are highly sensitive to digestion by DNase I. (348)

imaginal discs   In the *Drosophila* blastoderm, undifferentiated cells which will develop into adult tissue and organs. (367)

immunoglobulins   Specialized proteins (antibodies) secreted by B cells that circulate in the blood and lymph and are responsible for humoral immune responses. (360)

inbreeding   Mating between close relatives. (515)

incomplete (partial) dominance   The condition resulting when one allele is not completely dominant to another allele so that the heterozygote has a phenotype between that shown in individuals homozygous for either individual allele involved. An example of partial dominance is the frizzle chicken. (77)

induced mutations   Mutations that occur as a result of treatment with known chemical or physical mutagens. (378; 382)

inducer   A chemical or environmental agent for bacterial operons that brings about the transcription of an operon. (321)

induction   The synthesis of a gene product (or products) in response to the action of an inducer, i.e., a chemical or environmental agent. (321)

initiation factor   In prokaryotes, proteins that associate with the small ribosomal subunit at the initiation stage of translation. (275)

insertion sequence (IS) element   The simplest transposable element found in prokaryotes. It is a mobile segment of DNA that contains genes required for the process of insertion of the DNA segment into a chromosome and for the mobilization of the element to different locations. (427)

interaction variance (represented by V_I)   Genetic variance that arises from epistatic interactions among genes. (479)

intergenic suppressor   A mutation whose effect is to suppress the phenotypic consequences of another mutation in a gene distinct from the gene in which the suppressor mutation is located. (381)

internal control region (ICR)   Promoter sequence, recognized by RNA polymerase III, that is located within the gene sequence, e.g., in tRNA genes and 5S rRNA genes of eukaryotes. (256)

intron   A nucleotide sequence in eukaryotes that must be excised from a structural gene transcript in order to convert the transcript into a mature messenger RNA molecule containing only coding sequences that can be translated into the amino acid sequence of a polypeptide. (243)

inversion   A chromosome mutation that results when a segment of a chromosome is excised and then reintegrated in an orientation 180° from the original orientation. (407)

IS element   *See* insertion sequence (IS) element. (427)

karyotype   A complete set of all the metaphase chromatid pairs in a cell (literally, "nucleus type"). (191)

kinetochore   *See* centromere. (8)

Klinefelter syndrome   A human clinical syndrome that results in a 47,XXY male. Many of the affected males are mentally deficient, have underdeveloped testes, and are taller than average. (60)

lagging strand   In DNA replication, the DNA strand that is synthesized discontinuously in the 5'-to-3' direction away from the replication fork. (215)

leader sequence   One of three main parts of the mRNA molecule. The leader sequence is located at the 5' end of the mRNA molecule and contains the coded information that the ribosome and special proteins read to tell it where to begin the synthesis of the polypeptide. (243)

leading strand   In DNA replication, the DNA strand synthesized continuously in the 5'-to-3' direction towards the replication fork. (215)

lethal allele   An allele that results in the death of an organism. (85)

light repair   *See* photo-reactivation. (392)

linkage   A term describing genes located on the same chromosome. (99)

linkage map   *See* genetic map. (101)

linked genes   Genes that are located less than 50 map units apart on the same chromosome. (99)

linker   *See* restriction site linker. (295)

locus (*plural:* **loci**)   The position of a gene on a genetic map; the specific place on a chromosome where a gene is located. (33; 34)

lyonization   A mechanism in mammals that allows them to compensate for more than one X chromosome of the normal complement. The excess X chromosomes are cytologically condensed and inactivated, and do not play a role in much of the development of the individual. The name derives from the discoverer of the phenomenon, Mary Lyon. (61)

lysogenic   A term describing a bacterium that contains a temperate phage in the prophage state. The bacterium is said to be lysogenic for that phage. Upon induction phage reproduction is initiated, progeny phages are produced, and the bacterial cell lyses. (138)

lysogenic pathway   A path, besides the lytic cycle, that a phage can follow. The phage chromosome does not replicate; instead, it inserts itself physically into a specific region of the host cell's chromosome in a way that is essentially the same as *F* factor integration. (138)

lysogeny   The phenomenon of the insertion of a temperate phage chromosome into a bacterial chromosome, where it replicates when the bacterial chromosome replicates. In this state the phage genome is repressed and is said to be in the prophage state. (138)

lytic cycle   A type of phage life cycle in which the phage takes over the bacterium and directs its growth and reproductive activities to express the phage's genes and to produce progeny phages. (138)

macromolecule   A large molecule (such as DNA, RNA, and proteins) that has a molecular weight of at least a few thousand daltons. (177)

map unit (mu)   A unit of measurement used for the distance between two gene pairs on a genetic map. A crossover frequency of 1 percent between two genes equals 1 map unit. *See* centi-Morgan. (103)

maternal effect   The predetermination of gene-controlled traits by the maternal nuclear genotype prior to the fertilization of the egg. (462)

maternal inheritance   A phenomenon in which the mother's phenotype is expressed exclusively. (456)

mating types   A genic system in which two sexes are morphologically indistinguishable but carry different alleles and will mate. (63)

Maxam-Gilbert sequencing   A method of rapid sequencing of DNA molecules developed by Allan Maxam and Walter Gilbert. The technique uses specific chemical reactions to break DNA at specific nucleotides. (302)

mean   The average of a set of numbers, calculated by adding all the values represented and dividing by the number of values. (474)

meiosis   Two successive nuclear divisions of a diploid nucleus that result in the formation of haploid gametes or of meiospores having one-half the genetic material of the original cell. (14)

meiosis I   The first meiotic division that results in the reduction of the number of chromosomes. This division consists of four stages: prophase I, metaphase I, anaphase I, and telophase I. (15)

meiosis II   The second meiotic division, resulting in the separation of the chromatids. (18)

Mendelian factor   *See* gene. (34)

Mendelian population   An interbreeding group of individuals sharing a common gene pool; the basic unit of study in population genetics. (491)

messenger RNA (mRNA)   The RNA molecule that contains the coded information for the amino acid sequence of a protein. (235)

metacentric chromosome   A chromosome that has the centromere approximately in the center of the chromosome. (8)

metaphase   A stage in mitosis or meiosis in which chromosomes become aligned along the equatorial plane of the spindle. (11)

metaphase II   The stage of meiosis II during which the centromeres line up on the equator of the second-division spindles (in each of two daughter cells formed from meiosis I). (18)

metaphase plate   The plane where the chromosomes become aligned during metaphase. (11)

migration   Movement of organisms from one location to another. (509)

missense mutation   A gene mutation in which a base-pair change in the DNA causes a change in an mRNA codon, with the result that a different amino acid is inserted into the polypeptide in place of one specified by the wild-type codon. (380)

mitochondria   Organelles found in the cytoplasm of all aerobic animal and plant cells; the principal sources of energy in the cell. (450)

mitosis   The process of nuclear division in haploid or diploid cells producing daughter nuclei which contain identical chromosome complements and which are genetically identical to one another and to the parent nucleus from which they arose. (9)

moderately repetitive sequence   A DNA sequence that is reiterated from a few to as many as 10^3 to 10^5 times in the genome. (202)

molecular cloning   *See* cloning. (286)

molecular drive   *See* concerted evolution. (522)

molecular genetics   A subdivision of the science of genetics involving how genetic information is encoded within the DNA and how biochemical processes of the cell translate the genetic information into the phenotype.

monoecious   A term referring to plants in which male and female gametes are produced in the same individual. (63)

monohybrid cross   A cross between two individuals that are both heterozygous for the same pair of alleles. (e.g., $Aa \times Aa$). By extension, the term also refers to crosses involving the pure-breeding parents that differ with respect to the alleles of one locus (e.g., $AA \times aa$). (29)

monoploidy   Having one set of chromosomes (haploidy). (416)

monosomy   An aberrant, aneuploid state in a normally diploid cell or organism in which one chromosome is missing, leaving one chromosome with no homolog. (414)

mRNA splicing   A process whereby an intervening sequence between two coding sequences in an RNA molecule is excised and the coding sequences ligated (spliced) together. (247)

multifactorial trait   A trait influenced by multiple genes and environmental factors. (470)

multigene family   A set of related genes that have evolved from some ancestral gene through the process of gene duplication. (520)

multiple alleles   Many alternative forms of a single gene. (74)

multiple cloning site   *See* polylinker. (290)

multiple crossovers   More than one crossover occurring in a particular region of a chromosome in a meiosis. (106)

mutagen   Any physical or chemical agent that significantly increases the frequency of mutational events above a spontaneous mutation rate. (378)

mutant allele   Any alternative to the wild-type allele of a gene. Mutant alleles may be dominant or recessive to wild-type alleles. (53; 74)

mutation   (1) The process by which a DNA base-pair change or a chromosome change is produced. (2) DNA base-pair change or chromosome change resulting from the mutation process. (2; 378)

mutation frequency   The number of occurrences of a particular kind of mutation in a population of cells or individuals. (382)

mutation rate   The probability of a particular kind of mutation as a function of time or generations. (382)

narrow-sense heritability   The proportion of the phenotypic variance that results from additive genetic variance. (480)

natural selection   Differential reproduction of genotypes. (482; 510)

negative assortative mating   The situation in which matings occur between dissimilar individuals more often than would be expected by chance. (514)

neutral mutation   A base-pair change in the gene that changes a codon in the mRNA such that there is no change in the function of the protein translated from that message. (380)

neutral-mutation hypothesis   A hypothesis that replaced the classical model by acknowledging the presence of extensive genetic variation in proteins, but proposing that this variation is neutral with regard to natural selection. (500)

nick translation   A process for labeling a double-stranded DNA molecule radioactively using DNase I and DNA polymerase to produce a labeled probe. (297)

nitrogenous base   A nitrogen-containing base that, along with a pentose sugar and a phosphate, is one of the three parts of a nucleotide, the building block of RNA and DNA. (177)

noncontributing alleles   The alleles that do not have any effect on the phenotype of the quantitative trait.

nondisjunction   A failure of homologous chromosomes or sister chromatids to separate at anaphase. (57)

nonhistone   A type of acidic or neutral protein found in chromatin. (196)

nonhomologous chromosomes   The chromosomes containing dissimilar genes that do not pair during meiosis. (7)

non-Mendelian inheritance (cytoplasmic inheritance)   The inheritance of characters determined by genes not located on the nuclear chromosomes but on mitochondrial or chloroplast chromosomes. Such genes show inheritance patterns distinctly different from those of nuclear genes. (455)

nonparental ditype (NPD)   One of three types of tetrads possible when two genes are segregating in a cross. The NPD tetrad contains four nuclei, all of which have recombinant (nonparental) genotypes, i.e., two of each possible type. (119)

nonsense codon   *See* chain-terminating codon. (272)

nonsense mutation   A gene mutation in which a base-pair change in the DNA causes a change in an mRNA codon from an amino acid-coding codon to a chain-terminating (nonsense) codon. As a result, polypeptide chain synthesis is terminated prematurely and the product is therefore either nonfunctional or, at best, partially functional. (380)

nontranscribed spacer (NTS) sequence   Sequences, which are not transcribed, found between transcription units in rDNA. Important sequences that control transcription of the rDNA are within the NTS. (253)

normal distribution   A probability distribution in statistics, graphically displayed as a bell-shaped curve. (472)

norm of reaction   The extent to which the phenotype produced by a genotype varies with the environment. (91)

northern blot analysis   A similar technique to Southern blotting except that RNA rather than DNA is separated and transferred to a filter for hybridization with a probe. (301)

nuclease   An enzyme that catalyzes the degradation of a nucleic acid by breaking phosphodiester bonds. Nucleases specific for DNA are termed deoxyribonucleases (DNases), and nucleases specific for RNA are termed ribonucleases (RNases). (174)

nucleofilament   A fiber seen in chromatin. It is approximately 10 nm in diameter and consists of DNA wrapped around nucleosome cores. (198)

nucleoside phosphate   *See* nucleotide. (178)

nucleosome   The basic structural unit of eukaryotic nuclear chromosomes. It consists of about 140 base pairs of DNA wound almost two times around an octamer of histones. It is connected to adjacent nucleosomes by about 60 base pairs of DNA that is complexed with another histone. (197)

nucleotide   A monomeric molecule of RNA and DNA that consists of three distinct parts: a pentose (ribose in RNA, deoxyribose in DNA), a nitrogenous base, and a phosphate group. (177)

nucleus   A discrete structure within the cell that is bounded by a nuclear membrane. It contains most of the genetic material of the cell. (3)

nullisomy   The aberrant, aneuploid state in a normally diploid cell or organism in which there is a loss of one pair of homologous chromosomes. (414)

nutritional mutation   *See* auxotrophic mutation. (396)

Okazaki fragments   The relatively short, single-stranded DNA fragments which are synthesized in discontinuous DNA replication and which are subsequently covalently joined to make a continuous strand. (215)

oncogene   A gene that transforms a normal cell to a tumorous (cancerous) state. (439)

oncogenesis   Tumor (cancer) initiation in an organism. (438)

one gene-one enzyme hypothesis   The hypothesis, based on Beadle and Tatum's studies in biochemical genetics, that each gene controls the synthesis of one enzyme. (159)

oogenesis   The development in the gonad of the female germ cell (egg cell) of animals. (20)

operator   The controlling site, which is adjacent to a promoter, that is responsible for controlling the transcription of genes that are contiguous to the promoter. (323)

operon   A cluster of genes whose expressions are regulated together by operator-regulator protein interactions, plus the operator region itself and the promoter. (324)

ordered tetrad   A structure resulting from meiosis in which the four meiotic products are in an order that exactly reflects the orientation of the four chromatids at the metaphase plate in meiosis I. (119)

origin   A specific site on the chromosome at which the double helix denatures into single strands and continues to unwind as the replication fork(s) migrates. (132; 214)

origin of replication   A specific DNA sequence that is required for the initiation of DNA replication in prokaryotes. (213)

overdominance (heterozygote advantage)   Condition in which the heterozygote has higher fitness than either of the homozygotes. (513)

ovum   A mature egg cell. In the second meiotic division the secondary oocyte produces two haploid cells; the large cell rapidly matures into the ovum. (20)

P generation   The parental generation, i.e., the immediate parents of an F_1. (29)

paracentric inversion   An inversion in which the inverted segment occurs on one chromosome arm and does not include the centromere. (408)

parental ditype (PD)   One of three types of tetrads possible when two genes are segregating in a cross. The PD tetrad contains four nuclei, all of which are parental genotypes, with two of one parental type and two of the other parental type. (119)

parental genotypes (parental classes, parentals)   Individuals among progeny of crosses that have combinations of genetic markers like one or other of the parents in the parental generation. (99)

partial dominance   *See* incomplete dominance. (77)

particulate factors   The term Mendel used to describe the factors that carried hereditary information and were transmitted from parents to progeny through the gametes. We now know these factors by the name *genes*. (31)

pedigree analysis   A family tree investigation that involves the careful compilation of phenotypic records of the family over several generations. (40)

penetrance   The frequency with which a dominant or homozygous recessive allele manifests itself in the phenotype of an individual. (86)

pentose sugar   A 5-carbon sugar that, along with a nitrogenous base and a phosphate group, is one of the three parts of a nucleotide, the building block of RNA and DNA. (177)

peptide bond   A covalent bond in a polypeptide chain that joins the α-carboxyl group of one amino acid to the α-amino group of the adjacent amino acid. (266)

peptidyl transferase   The RNA enzyme (the rRNA of the large ribosomal subunit) which catalyzes the formation of the peptide bond in protein synthesis. (278)

pericentric inversion   An inversion in which the inverted segment includes the parts of both chromosome arms and therefore includes the centromere. (408)

phage lysate   The progeny phages released following lysis of phage-infected bacteria. (138)

phage vector   A phage that carries pieces of bacterial DNA between bacterial strains in the process of transduction. (138)

phenocopy   An abnormal individual resulting from special environmental conditions. It mimics a similar phenotype caused by gene mutation. (90)

phenotype   The physical manifestation of a genetic trait that results from a specific genotype and its interaction with the environment. (26; 34)

phenotypic sex determination   The process by which the environment plays a major role in determining the sex of an organism. (63)

phenotypic variance   (represented by V_P) A measure of a trait's variability. (478)

phosphate group   A component, along with a pentose sugar and a nitrogenous base, of a nucleotide, the building block of RNA and DNA. Because phosphate groups are acidic in nature, DNA and RNA are called nucleic acids. (177)

phosphodiester bond   A covalent bond in RNA and DNA between two sugars and a phosphate. Phosphodiester bonds form the repeating sugar-phosphate array of the backbone of DNA and RNA. (178; 211)

photoreactivation (light repair)   One way by which thymine dimers can be repaired. The dimers are reverted directly to the original form by exposure to visible light in the wavelength range 320–370 nm. (392)

plaque   A round, clear area in a lawn of bacteria on solid medium that results from the lysis of cells by repeated cycles of phage lytic growth. (142)

plasma membrane   The lipid bilayer surrounding the cytoplasm of both animal and plant cells. (4)

plasmid   An extrachromosomal genetic element consisting of double-stranded DNA that replicates autonomously from the host chromosome. (131; 428)

point mutants   Organisms whose phenotypes result from an alteration of a single nucleotide pair. (144)

point mutation   A mutation caused by a substitution of one base pair for another or by the addition or deletion of one base pair. (378)

poly(A) addition site   The 3' end of mRNA to which 50 to 250 adenine nucleotides are added as part of mRNA posttranscriptional modification. (245)

poly(A) polymerase   The enzyme which catalyzes the production of the 3' poly(A) tail. (245)

poly(A) tail   A sequence of 50 to 250 adenine nucleotides that is added as a posttranscriptional modification at the 3' ends of most eukaryotic mRNAs. (243)

polygene (multiple-gene) hypothesis for quantitative inheritance   The hypothesis that quantitative traits are controlled by many genes. (476)

polygenic mRNA (polycistronic mRNA)   A single mRNA transcript in prokaryotic operons of two or more adjacent structural genes that specifies the amino acid sequences of the corresponding polypeptides. (322)

polygenic traits   Traits determined by many loci. (470)

polylinker (multiple cloning site)   A region of clustered unique restriction sites in a cloning vector. (290)

polymerase chain reaction (PCR)   A method used to replicate defined DNA sequences selectively and repeatedly between two defined primers. (304)

polynucleotide   A linear sequence of nucleotides in DNA or RNA. (179)

polypeptide   A polymeric, covalently bonded linear arrangement of amino acids joined by peptide bonds. (265)

polyploidy   The condition of a cell or organism that has more than its normal number of sets of chromosomes. (416)

polyribosome (polysome)   The complex between an mRNA molecule and all the ribosomes that are translating it simultaneously. (278)

polytene chromosome   A special type of chromosome representing a bundle of numerous chromatids that have arisen by repeated cycles of replication of single chromatids without nuclear division. This type of chromosome is characteristic of various tissues of Diptera. (405)

population   A group of interbreeding individuals that share a set of genes. (471)

population genetics   A branch of genetics that describes in mathematical terms the consequences of Mendelian inheritance on the population level. (491)

position effect   A change in the phenotypic expression of one or more genes as a result of a change in their position in the genome. (413)

positive assortative mating   The situation in which matings occur more frequently between individuals who are phenotypically similar than would be expected by chance. (514)

precursor mRNA (primary transcripts; pre-mRNA)   The initial transcript of a gene that is modified and/or processed to produce the mature, functional mRNA molecule. In eukaryotes, for example, the transcript is modified at both the 5' and the 3' ends, and in a number of cases RNA sequences that do not code for amino acids are present and must be excised. (240)

precursor RNA molecule (primary transcripts; pre-RNA)   The initial transcript whose processing may involve the addition and/or removal of bases, the chemical modification of some bases, or the cleavage of sequences from the precursors. (234)

precursor rRNA (pre-rRNA)   A primary transcript of adjacent rRNA genes (16S, 23S, and 5S rRNA genes in prokaryotes; 18S, 5.8S, and 28S rRNA genes in eukaryotes) plus flanking and spacer DNA that must be processed to release the mature rRNA molecules. (251)

precursor tRNA (pre-tRNA)   A primary transcript of a tRNA gene whose bases must be extensively modified and that must be processed to remove extra RNA sequences in order to produce the mature tRNA molecule. In some cases the primary transcript may contain the sequences of two or more tRNA molecules. (256)

Pribnow box   A part of the promoter sequence in prokaryotic genomes that is located at about 10 base pairs upstream from the transcription starting point. The consensus sequence for the Pribnow box is TATAAT. The Pribnow box is often referred to as the TATA box. (238)

primary nondisjunction (nondisjunction)   A rare event in which sister chromatids (in mitosis) or chromosomes contained in pairing configurations (in meiosis) fail to be distributed to opposite poles. (57)

primary transcripts   *See* precursor RNA molecules (pre-RNAs). (234)

primer   A preexisting polynucleotide chain in DNA replication to which new nucleotides can be added. (214)

primosome   A complex of *E. coli* primase, helicase, and perhaps other polypeptides that together become functional in catalyzing the initiation of DNA synthesis. (214)

principle of independent assortment (second law)   The law that the factors (genes) for different traits assort independently of one another. In other words, genes on different chromosomes behave independently in the production of gametes. (37)

principle of segregation (first law)   The law that two members of a gene pair (alleles) segregate (separate) from each other during the formation of gametes. As a result, one-half the gametes carry one allele and the other half carry the other allele. (33)

probability   The ratio of the number of times a particular event occurs to the number of trails during which the event could have happened. (35)

probe   In molecular genetics a probe is any molecule (usually labeled radioactively or in a nonradioactive way that permits easy detection) that is used to identify or isolate a gene, gene product, or a protein. For example, DNA or RNA probes may be used to identify clones in a library, or to study RNA transcripts produced by a cell. (295)

product rule   The rule that the probability of two independent events occurring simultaneously is the product of each of their probabilities. (35)

prokaryote   A cellular organism whose genetic material is not located within a membrane-bound nucleus (*cf.* **eukaryote**). (3)

promoter elements   Consensus sequences found in the promoter region of the transcription initiation site. The elements are the TATA box (or Goldberg-Hogness box), CAT element, and the GC element. (241)

promoter site (promoter sequence, promoter)   A specific regulatory nucleotide sequence in the DNA to which RNA polymerase binds for the initiation of transcription. (236)

proofreading   In DNA synthesis, the process of recognizing a base-pair error during the polymerization events and correcting it. Proofreading is a property of the DNA polymerase in prokaryotic cells. (213)

prophage   A temperate bacteriophage integrated into the chromosome of a lysogenic bacterium. It replicates with the replication of the host cell's chromosome. (138)

prophase   The first stage in mitosis or meiosis during which the chromosomes (already replicated) condense and become visible under the microscope. (11)

prophase I   The first stage of meiosis. There are several stages of prophase I, including leptonema, zygonema, pachynema, diplonema, and diakinesis. (15)

prophase II   The stage of meiosis II during which there is chromosome contraction. (18)

propositus (male), proposita (female) (proband)   The affected person, in human genetics, with whom the study of a character in a family begins. (41)

protein   One of a group of high-molecular-weight polymers of amino acids that can assume complex shape and composition. (265)

proto-oncogene   A gene which, in normal cells, functions to control the normal proliferation of cells, and which, when mutated or changed in any other way, becomes an oncogene. (442)

prototroph   A strain that is a wild type for all nutritional requirement genes and thus requires no supplements in its growth medium. (131)

Punnett square   A matrix that describes all the possible gametic fusions that will give rise to the zygotes that will produce the next generation. (33)

pure-breeding   *See* true-breeding. (29)

purine   A type of nitrogenous base. In DNA and RNA the purines are adenine and guanine. (177)

pyrimidine   A type of nitrogenous base. Cytosine is a pyrimidine in DNA and RNA; thymine is a pyrimidine in DNA; and uracil is a pyrimidine in RNA. (177)

Q banding   A staining technique in which metaphase chromosomes are stained with quinacrine mustard to produce temporary fluorescent Q bands on the chromosomes. (194)

quantitative genetics   Study of the inheritance of quantitative traits. (470)

quantitative (continuous) traits   Traits that show a continuous variation in phenotype over a range. (470)

quantitative trait loci   Genes affecting a quantitative trait.

random mating   Matings between genotypes occurring in proportion to the frequencies of the genotypes in the population. (495)

rDNA repeat units   The tandem arrays of rRNA genes, 18S-5.8S-28S, repeated many times along the chromosome. (252)

realized heritability   Heritability of a trait determined from the response to selection. (484)

recessive   An allele or phenotype that is expressed only in the homozygous state. (32)

recessive lethal allele   An allele that causes lethality when it is homozygous. (85)

reciprocal cross   A pair of crosses to determine if the sex of the parent with a trait has an effect on the outcome. In the garden pea example a reciprocal cross for smooth and wrinkled seeds is smooth female × wrinkled male and wrinkled female × smooth male. (30)

recombination   The exchange of genetic material between chromosomes. (2)

recombinant chromosome   A chromosome that emerges from meiosis with a combination of alleles different from a parental combination of alleles. (15)

recombinant DNA molecule   A new type of DNA sequence that has been constructed or engineered in the test tube from two or more distinct DNA sequences. (286)

recombinant DNA technology   A collection of experimental procedures that allows molecular biologists to splice a DNA fragment from one organism into DNA from another organism and to clone the new recombinant DNA molecule. It includes the development and the application of particular molecular techniques such as biotechnology or genetic engineering. This technology is important, for example, in the production of antibiotics, hormones, and other medical agents used in the diagnosis and treatment of certain genetic diseases. (286)

recombinants   The individuals or cells that have nonparental combinations of alleles as a result of the processes of genetic recombination. (99)

regulated gene   A gene whose activity is controlled in response to the needs of a cell or organism. (320)

regulatory factors   Proteins active in the activation or repression of transcription of the gene. (241)

release factors   *See* termination factors. (279)

replica plating   The procedure for transferring the pattern of colonies from a master plate to a new plate. In this procedure a velveteen pad on a cylinder is pressed lightly onto the surface of the master plate, thereby picking up a few cells from each colony to inoculate onto the new plate. (396)

replication bubble   Opposing replication forks found with the local denaturing of DNA during replication. (214)

replication fork   A Y-shaped structure formed when a double-stranded DNA molecule unwinds to expose the two single-stranded template strands for DNA replication. (215)

replication machine (replisome)   The complex formed by the close association of the key proteins used during DNA replication. (217)

replicon (replication unit)   The stretch of DNa in eukaryotes from the origin of replication to the two termini of replication on each side of the origin. (222)

repressor   *See* repressor gene. (323)

repressor gene   A regulatory gene whose product is a protein that controls the transcriptional activity of a particular operon. (323; 324)

repressor molecule   The protein product of a repressor gene. (324)

repulsion   An arrangement in which each homologous chromosome carries the wild-type allele of one gene and the mutant allele of the other gene. (103)

restriction endonucleases (restriction enzymes)   Enzymes important for analyzing DNA and for constructing recombinant DNA molecules because of their ability to cleave double-stranded DNA molecules at specific nucleotide pair sequences. (286)

restriction fragment length polymorphisms (RFLPs)   The different restriction maps which result from different patterns of distribution of restriction sites. They are detected by the presence of restriction fragments of different lengths on gels. (309)

restriction map   A genetic map of DNA showing the relative positions of restriction enzyme cleavage sites. (298)

restriction site linker (linker)   A relatively short, double-stranded oligodeoxyribonucleotide about 8 to 12 nucleotide pairs long which is synthesized by chemical means and which contains the cleavage site for a specific restriction enzyme within its sequence. (295)

retroviruses Single-stranded RNA viruses that replicate via double-stranded DNA intermediates. The DNA integrates into the host's chromosome where it can be transcribed. (436)

reverse transcriptase An enzyme (an RNA-dependent DNA polymerase) that makes a complementary DNA copy of an mRNA strand. (294)

reversion (reverse mutation) A mutational change from a mutant allele back to a wild-type allele. (269; 381)

ribonuclease (RNase) An enzyme that catalyzes the degradation of RNA to nucleotides. (174)

ribonucleic acid (RNA) A usually single-stranded polymeric molecule consisting of ribonucleotide building blocks. RNA is chemically very similar to DNA. The three major types of RNA in cells are ribosomal RNA (rRNA), transfer RNA (tRNA), and messenger RNA (mRNA), each of which performs an essential role in protein synthesis (translation). In some viruses RNA is the genetic material. (173)

ribonucleotide The basic building block of RNA consisting of a sugar (ribose), a base, and a phosphate. (177)

ribose The pentose sugar component of the nucleotide building block of RNA. (177)

ribosomal DNA (rDNA) The regions of the DNA that contain the genes for the rRNAs in prokaryotes and eukaryotes. (251)

ribosomal proteins The proteins that, along with rRNA molecules, comprise the ribosomes of prokaryotes and eukaryotes. (249)

ribosomal RNA (rRNA) The RNA molecules of discrete sizes that, along with ribosomal proteins, comprise ribosomes of prokaryotes and eukaryotes. (235)

ribosome A complex cellular particle composed of ribosomal protein and rRNA molecules that is the site of amino acid polymerization during protein synthesis. (235)

R looping (R loops) A technique developed by M. Thomas, R. White, and R. Davis in which molecules of double-stranded DNA are incubated at temperatures below their denaturing temperature to open up short stretches of the DNA double helix so that single-stranded RNA molecules can begin to form DNA/RNA hybrids where the two are complementary. The DNA/RNA hybrid forms an R loop by displacing a single-stranded section of DNA. (245)

RNA *See* ribonucleic acid. (173)

RNA ligase An enzyme that splices together the RNA pieces once the intervening sequence is removed from the pre-tRNA. (258)

RNA polymerase An enzyme that catalyzes the synthesis of RNA molecules from a DNA template in a process called transcription. (214; 234)

RNA polymerase I An enzyme in eukaryotes located in the nucleolus that catalyzes the transcription of the 18S, 5.8S, and 28S rRNA genes. (236)

RNA polymerase II An enzyme in eukaryotes found only in the nucleoplasm of the nucleus. It catalyzes the transcription of mRNA-coding genes. (236)

RNA polymerase III An enzyme in eukaryotes found only in the nucleoplasm. It catalyzes the transcription of the tRNA and 5S rRNA genes. (236)

RNA primer *See* primer. (214)

RNA processing control The second level of control of gene expression in eukaryotes. This level involves regulating the production of mature RNA molecules from precursor-RNA molecules. (353)

RNA synthesis *See* transcription. (233)

sample The subset used to give information about a population. It must be of reasonable size and it must be a random subset of the larger group in order to provide accurate information about the population. (471)

second law *See* principle of independent assortment. (37)

secondary oocyte A large cell produced by the primary oocyte. In the ovaries of female animals the diploid primary oocyte goes through meiosis I and unequal cytokinesis to produce two cells; the large cell is called the secondary oocyte. (20)

second-site mutation *See* suppressor mutation. (381)

selection The favoring of particular combinations of genes in a given environment. (2)

selection coefficient A measure of the relative intensity of selection against a genotype. (512)

selection differential In natural and artificial selection, the difference between the mean phenotype of the selected parents and the mean phenotype of the unselected population. (484)

selection response The amount that a phenotype changes in one generation when selection is applied to a group of individuals. (483)

self-fertilization (selfing) The union of male and female gametes from the same individual. (28)

self-splicing The excision of introns from some precursor RNA molecules which occurs by a protein-independent reaction in some organisms. (255)

semiconservative replication model A DNA replication scheme in which each daughter molecule retains one of the parental strands. (207)

semidiscontinuous In DNA replication, when one new strand is synthesized continuously and the other discontinuously. *See* discontinuous DNA replication. (215)

sense codon A sense codon, as opposed to a nonsense codon, in an mRNA molecule specifies an amino acid in the corresponding polypeptide. (272)

sex chromosome A chromosome in eukaryotic organisms that is represented differently in the two sexes. In many organisms one sex possesses a pair of visibly different chromosomes. One is an X chromosome, and the other is a Y chromosome. Commonly, the XX sex is female and the XY sex is male. (7; 52)

sex-influenced traits The traits that appear in both sexes but either the frequency of occurrence in the two sexes is different or there is a different relationship between genotype and phenotype. (88)

sex-limited trait   A genetically controlled character that is phenotypically exhibited in only one of the two sexes. (88)

sex-linked   *See* X-linked. (56)

sexual reproduction   Reproduction involving the fusion of haploid gametes produced by meiosos. (9)

Shine-Dalgarno sequence   In prokaryotes, a sequence to the 5' side of an mRNA's initiation codon which base pairs with the 3' end of 16S rRNA in the small ribosomal subunit to allow the ribosome to locate the true initiator region of the mRNA. (275)

shuttle vector   A cloning vector that can replicate in two or more host organisms. Shuttle vectors are used for experiments in which recombinant DNA is to be introduced into organisms other than *E. coli.* (291)

signal peptidase   An enzyme in the cisternal space of the ER that catalyzes removal of the signal sequence from the polypeptide. (280)

signal peptide   *See* signal sequence. (280)

signal recognition particle (SRP)   In eukaryotes, a complex of a small RNA molecule with six proteins, which can temporarily halt protein synthesis by recognizing the signal sequence of a nascent polypeptide destined to be translocated through the ER, binding to it, and thereby blocking further translationof the mRNA. (280)

signal sequence   The hydrophobic, amino terminal extension found on proteins that are secreted from a cell. The amino terminus (extension) is removed and degraded in the cisternal space of the endoplasmic reticulum. (280)

silencer   *See* silencer element. (419)

silencer element   In eukaryotes, a transcriptional regulatory element that decreases RNA transcription rather than stimulating it like other enhancer elements. (241; 419)

silent mutation   A mutational change resulting in a protein with a wild-type function because of an unchanged amino acid sequence. (380)

simple telomeric sequences   Simple, tandemly repeated DNA sequences at, or very close to, the extreme ends of the chromosomal DNA molecules. (201)

single-strand DNA binding (SSB) proteins (helix-destabilizing proteins)   Proteins that help the DNA unwinding process by stabilizing the single-stranded DNA. (217)

sister chromatid   A chromatid derived from replication of one chromosome during interphase of the cell cycle. (11)

small nuclear ribonucleoprotein particles (snRNP's)   The complexes formed by small nuclear RNAs and proteins in which the processing of pre-mRNA molecules occurs. (247)

small nuclear RNA (snRNA)   Found only in eukaryotes, one of four major classes of RNA molecules produced by transcription. snRNAs are used in the processing of pre-mRNA molecules. (235)

somatic cell hybridization   The fusion of two genetically different somatic cells of the same or different species to generate a somatic hybrid for genetic analysis. (114)

somatic mutation   A mutation in a cell which produces a mutant spot or area, but the mutant characteristic is not passed on to the succeeding generation. (378)

Southern blot technique   A technique invented by E. M. Southern and used in analyzing genes and gene transcripts, in which DNA fragments are transferred from a gel to a nitrocellulose filter. (299)

spacer sequences   Transcribed sequences which are found between, and flanking, coding RNA sequences. Spacer sequences are removed during processing of pre-rRNA and pre-tRNA to produce mature molecules. (253)

specialized transduction   A type of transduction in which only specific genes are transferred. (139)

spermatogenesis   Development of the male animal germ cell within the male gonad. (20)

sperm cells (spermatozoa)   The male gametes; the spermatozoa produced by the testes in male animals. (20)

spliceosomes   The splicing complexes formed by the association of several snRNPs bound to the pre-mRNA. (247)

spontaneous mutations   The mutations that occur without a known cause. (378; 382)

sporophyte   The haploid, asexual generation in the life cycle of plants that produces haploid spores by meiosis. (21)

stabilizing selection   Selection in which individuals of intermediate phenotype survive and reproduce best. (485)

standard deviation   The square root of the variance. It measures the extent to which each measurement in the data set differs from the mean value and is used as a measure of the extent of variability in a population. (474)

steroid response elements (REs)   The DNA sequence to which steroid hormones will bind to activate a gene. (353)

stop codon   *See* chain-terminating codon. (272)

structural gene   A gene that codes for an mRNA molecule and hence for a polypeptide chain. (236)

submetacentric chromosome   A chromosome which has the centromere nearer one end than the other. Such chromosomes appear J-shaped at anaphase. (8)

sum rule   The rule that the probability of any mutually exclusive events occurring is the sum of their individual probabilities. (35)

suppressor gene   A gene that causes suppression of mutations in other genes. (381)

suppressor mutation   A mutation at a second site that totally or partially restores a function lost because of a primary mutation at another site. (381)

Svedberg units (S values)   The conversions for sedimentation rates in sucrose density centrifugation. Svedberg units are used as a rough indication of relative sizes of the components being analyzed. (237)

synapsis   The specific pairing of homologous chromosomes during the zygonema stage of meiosis. (15)

synaptinemal complex   A complex structure spanning the region between meiotically paired (synapsed) chromosomes that is concerned with crossing-over rather than with chromosome pairing. (15)

synkaryon   A fusion nucleus produced following the fusion of cells with genetically different nuclei. (116)

syntenic   The genes that are localized to a particular chromosome by using an experimental approach (literally "together thread"). Such genes may or may not be genetically linked. (106)

TATA box (TATA element)   Found approximately at position –30 from the transcription initiation site. The Goldberg-Hogness sequence is considered to be a key element of eukaryotic promoter sequences. The consensus sequence for the Goldberg-Hogness box is TATAAAAA. (241)

TATA factor   A key transcription factor (TFIID) that binds to the TATA element. (242)

tautomeric shift   The change in the chemical form of a DNA (or RNA) base. (383)

tautomers   Alternate chemical forms in which DNA (or RNA) bases are able to exist. (383)

telocentric chromosome   A chromosome that has the centromere more or less at one end. (8)

telomere   The region of DNA at each end of a linear eukaryotic chromosome. (200)

telomere-associated sequences   Repeated, complex DNA sequences extending from the molecular gene of chromosomal DNA. Suspected to mediate many of the telomere-specific interactions. (200)

telophase   A stage during which the migration of the daughter chromosomes to the two poles is completed. (11; 12)

telophase II   The last stage of meiosis II during which a nuclear membrane forms around each set of chromosomes, and cytokinesis takes place. (18)

temperate phages   Bacteriophages that have a choice of following the lytic or lysogenic pathways. (138)

template strand   The unwound single strand of DNA upon which new strands are made (following complementary base-pairing rules). (214)

termination factors (release factors, RF)   The specific proteins in polypeptide synthesis (translation) that read the chain termination codons and then initiate a series of specific events to terminate polypeptide synthesis. (279)

terminator   *See* transcription terminator sequence. (236)

testcross   A cross of an individual of unknown genotype, usually expressing the dominant phenotype, with a homozygous recessive individual in order to determine the genotype of the individual. (36)

tetrad analysis   Genetic analysis of all the products of a single meiotic event. Tetrad analysis is possible in those organisms in which the four products of a single nucleus that has undergone meiosis are grouped together in a single structure. (117)

tetrasomy   The aberrant, aneuploid state in a normally diploid cell or organism in which an extra chromosome pair results in the presence of four copies of one chromosome type and two copies of every other chromosome type. (414)

tetratype (T)   One of the three types of tetrads possible when two genes are segregating in a cross. The T tetrad contains two parental and two recombinant nuclei, one of each parental type and one of each recombinant type. (119)

three-point testcross   A test involving three genes within a relatively short section of the chromosome. It is used to map genes for their order in the chromosome and for the distance between them. (108)

thymine (T)   A pyrimidine base found in DNA but not in RNA. In double-stranded DNA thymine pairs with adenine. (177)

topoisomerases   A class of enzymes that catalyze the supercoiling of DNA. (190)

totipotency   The capacity of a nucleus to direct events through all the stages in development and therefore produce a normal adult. (359)

trailer sequence   The sequence of the mRNA molecule beginning at the end of the amino acid-coding sequence and ending at the 3' end of the mRNA. The trailer sequence is not translated and varies in length from molecule to molecule. (243)

transconjugants   In bacteria, the recipients inheriting donor DNA in the process of conjugation. (131)

transcription   The transfer of information from a double-stranded DNA molecule to a single-stranded RNA molecule. It is also called RNA synthesis. (233)

transcription factors (TFs)   Specific proteins which are required for the initiation of transcription by each of the three eukaryotic RNA polymerases. Each polymerase uses its own set of TFs. (241)

transcription terminator sequence (terminator)   A transcription regulatory sequence located at the distal end of a gene that signals the termination of transcription. (236)

transcriptional control   The first level of control of gene expression in eukaryotes. This level involves regulating whether or not a gene is to be transcribed and the rate at which transcripts are produced. (346)

trans-**dominant**   The phenomenon of a gene or DNA sequence controlling genes that are on a different piece (strand) of DNA. (324)

transducing phage   The phage that is the vehicle by which genetic material is shuttled between bacteria. (139)

transducing retroviruses   Retroviruses that have picked up an oncogene from the cellular genome. (441)

transductants   In bacteria, the recipients inheriting donor DNA in the process of transduction. (137)

transduction   A process by which bacteriophages mediate the transfer of bacterial genetic information from one bacterium (the donor) to another (the recipient); a process whereby pieces of bacterial DNA are carried between bacterial strains by a phage. (137)

transfer RNA (tRNA)   One of the classes of RNA molecules produced by transcription and involved in protein synthesis; molecules that bring amino acids to the ribosome, where they are matched to the transcribed message on the mRNA. (235)

transformant   The genetic recombinant generated by the transformation process. (129)

transformation   (1) A process in which genetic information is transferred by means of extracellular pieces of DNA in bacteria. (129). (2) The failure of cells to remain constrained in their growth properties and give rise to tumors. (438)

transition mutation   A specific type of base-pair substitution mutation that involves a change from one purine-pyrimidine base pair to the other purine-pyrimidine base pair at a particular site in the DNA (e.g., AT to GC). (379)

translation (protein synthesis)   The conversion in the cell of the mRNA base sequence information into an amino acid sequence of a polypeptide. (233)

translational control   The regulation of protein synthesis by limiting the entry or movement of ribosomes along an mRNA. (355)

translocation   (1) A chromosome mutation involving a change in position of a chromosome segment (or segments) and the gene sequences it contains (also called transposition). (410). (2) In polypeptide synthesis, the movement of the ribosome, one codon at a time, along the mRNA toward the 3' end. (278)

transmission genetics (classical genetics)   A subdivision of the science of genetics primarily dealing with how genes are passed from one individual to another.

transport control   Regulating the number of transcripts that exit the nucleus to the cytoplasm.

transposable element   A genetic element of chromosomes of both prokaryotes and eukaryotes that has the capacity to mobilize itself and move from one location to another in the genome. (426)

transposase   An enzyme encoded by the IS element of a tranposon which catalyzes transposition activity of a transposable element. (427)

transposon (Tn)   A mobile DNA segment that contains genes for the insertion of the DNA segment into the chromosome and for mobilization of the element to other locations on the chromosomes. (429)

transversion mutation   A specific type of base-pair substitution mutation that involves a change from a purine-pyrimidine base pair to a pyrimidine-purine base pair at a particular site in the DNA (e.g., AT to TA or GC to TA). (379)

trihybrid cross   A cross between individuals of the same type that are heterozygous for three pairs of alleles at three different loci. (39)

trisomy   An aberrant, aneuploid state in a normally diploid cell or organism in which there are three copies of a particular chromosome instead of two copies. (414)

trisomy-21   *See* Down syndrome. (415)

true-breeding (pure-breeding) strain   A strain allowed to self-fertilize for many generations to ensure that the traits to be studied are inherited and unchanging (i.e., the alleles for the traits are homozygous). (29)

true reversion   A point mutation from mutant back to wild type in which the change codes for the original amino acid of the wild type. (381)

tumor viruses   Viruses that induce cells to dedifferentiate and to divide to produce a tumor. (438)

Turner syndrome   A human clinical syndrome that results from monosomy for the X chromosome in the female, which gives a 45,X female. These females fail to develop secondary sexual characteristics, tend to be short, have weblike necks, have poorly developed breasts, are usually infertile, and exhibit mental deficiencies. (60)

uniparental inheritance   A phenomenon, usually exhibited by extranuclear genes, in which all progeny have the phenotype of only one parent. (456)

unique (single-copy) sequence   A class of DNA sequences that has one to a few copies per genome. (202)

unordered tetrad   A structure resulting from meiosis in which the four meiotic products are randomly arranged. (119)

uracil (U)   A pyrimidine base found in RNA but not in DNA. (177)

variance   A statistical measure of how values vary from the mean. (474)

variance of allelic frequency   The variance in the frequency of an allele among a group of populations. (506)

vegetative reproduction   *See* asexual reproduction. (9)

viral oncogene   A viral gene that transforms a cell it infects to a cancerous state. *See* oncogenesis; cellular oncogene. (441)

virulent phage   A phage like T4, which always follows the lytic cycle when it infects bacteria. (138)

virus   A noncellular organism that can produce only within a host cell. It contains genetic material within a membrane or protein coat. Once a virus is within a cell, its genetic material causes the cellular machinery to produce progeny viruses.

visible mutation   A mutation that affects the morphology or physical appearance of an organism. (395)

wild type   A strain, organism, or gene of the type that is designated as the standard for the organism with respect to genotype and phenotype. (53)

wild-type allele   The allele designated as the standard ("normal") for a strain of organism. (43; 74)

wobble hypothesis   A theory proposed by Francis Crick which proposes that the base at the 5' end of the anticodon (3' end of the codon) is not as constrained as the other two bases. This feature allows for less exact base pairing so that the 5' end of the anticodon can potentially pair with one of three different bases at the 3' end of the codon. (272)

X chromosome   A sex chromosome present in two copies in the homogametic sex and in one copy in the heterogametic sex. (52)

X chromosome-autosome balance system   A genotypic sex determination system. The main factor in sex determination is the ratio between the numbers of X chromosomes and autosomes. Sex is determined at the time of fertilization, and sex differences are assumed to be due to the action during development of two sets of genes, one located on the X chromosomes and the other on the autosomes. (59)

X chromosome nondisjunction   An event occurring when the two X chromosomes fail to separate in meiosis so that eggs are produced either with two X chromosomes or with no X chromosomes, instead of the usual one X chromosome. (57)

X-linked   Referring to genes located on the X chromosome. (56)

X-linked dominant trait   A trait due to a dominant mutant gene carried on the X chromosome. (64)

X-linked recessive trait   A trait due to a recessive mutant gene carried on the X chromosome. (64)

Y chromosome   A sex chromosome that, when present, is found in one copy in the heterogametic sex, along with an X chromosome, and is not present in the homogametic sex. Not all organisms with sex chromosomes have a Y chromosome. (52)

Y chromosome mechanism of sex determination   Sex determination mechanism seen, for example, in mammals, where the Y chromosome determines the sex of the individual. If a Y chromosome is present, the individual is male; if it is absent, the individual is a female. (59)

Y-linked (holandric ["wholly male"]) trait   A trait due to a mutant gene carried on the Y chromosome, but with no counterpart on the X. (65)

zygote   The cell produced by the fusion of the male and female gametes. (7; 34)

Suggested Reading

Chapter 1: Viruses, Cells, and Cellular Reproduction

Alberts, B., Bray, D., Lewis, J., Raff, M., Roberts, K., and Watson, J. D. 1989. *Molecular Biology of the Cell,* 2nd ed. New York: Garland Publishing Co.

Chapter 2: Mendelian Genetics

Bateson, W. 1909. *Mendel's principles of heredity.* Cambridge: Cambridge University Press.

Mendel, G. 1866. Experiments in plant hybridization (translation). In *Classic papers in genetics,* edited by J. A. Peters, 1959. Englewood Cliffs, NJ: Prentice-Hall.

Peters, J. A., ed. 1959. *Classic papers in genetics.* Englewood Cliffs, NJ: Prentice-Hall.

Sandler, I., and Sandler, L. 1985. A conceptual ambiguity that contributed to the neglect of Mendel's paper. *Hist. Phil. Life Sci.* 7:3–70.

Sturtevant, A. H. 1965. *A history of genetics.* New York: Harper & Row.

Tschermak-Seysenegg, E. von. 1951. The rediscovery of Mendel's work. *J. Hered.* 42:163–171.

Chapter 3: Chromosomal Basis of Inheritance, Sex Determination, and Sex Linkage

Baker, B., Nagoshi, R. N., and Burtin, K. C. 1987. Molecular genetic aspects of sex determination in *Drosophila. BioEssays* 6:66–70.

Barr, M. L. 1960. Sexual dimorphism in interphase nuclei. *Am. J. Hum. Genet.* 12:118–127.

Bodmer, W. F., and Cavalli-Sforza, L. L. 1976. *Genetics, evolution, and man.* San Francisco: W. H. Freeman.

Bridges, C. B. 1916. Nondisjunction as a proof of the chromosome theory of heredity. *Genetics* 1:1–52, 107–163.

———. 1925. Sex in relation to chromosomes and genes. *Am. Naturalist* 59:127–137.

Cummings, M. R. 1988. *Human Heredity: Principles and Issues.* St. Paul, MN: West Publishing Co.

Dice, L. R. 1946. Symbols for human pedigree charts. *J. Hered.* 37:11–15.

Eicher, E. M., and Washburn, L. L. 1986. Genetic control of primary sex determination in mice. *Annu. Rev. Genet.* 20:327–360.

Hodgkin, J. 1985. Males, hermaphrodites, and females: sex determination in *Caenorhabditis elegans. Trends Genet.* 1:85–88.

Hodgkin, J. 1989. Drosophila sex determination: a cascade of regulated splicing. *Cell* 56:905–906.

Koopman, P., Gubbay, J., Vivian, N., Goodfellow, P., and Lovell-Badge, R. 1991. Male development of chromosomally female mice transgenic for *Sry. Nature* 351:117–121.

Lyon, M. F. 1962. Sex chromatin and gene action in the mammalian X-chromosome. *Am. J. Hum. Genet.* 14:135–148.

McClung, C. E. 1902. The accessory chromosome—sex determinant? *Biol. Bull.* 3:43–84.

McKusick, V. A. 1965. The royal hemophilia. *Sci. Am.* 213:88–95.

Morgan, T. H. 1910. Sex-limited inheritance in *Drosophila. Science* 32:120–122.

———. 1911. An attempt to analyze the constitution of the chromosomes on the basis of sex-limited inheritance in *Drosophila. J. Exp. Zool.* 11:365–414.

Stern, C., Centerwall, W. P., and Sarkar, Q. S. 1964. New data on the problem of Y-linkage of hairy pinnae. *Am. J. Hum. Genet.* 16:455–471.

Sutton, W. S. 1903. The chromosomes in heredity. *Biol. Bull.* 4:231–251.

Wilson, E. B. 1905. The chromosomes in relation to the determination of sex in insects. *Science* 22:500–502.

Wolfner, M. F. 1988. Sex-specific gene expression in somatic tissues of *Drosophila melanogaster. Trends Genet.* 4:333–337.

Chapter 4: Extensions of Mendelian Genetic Analysis

Ginsburg, V. 1972. Enzymatic basis for blood groups. *Methods Enzymol.* 36:131–149.

Landauer, W. 1948. Hereditary abnormalities and their chemically induced phenocopies. *Growth Symposium* 12:171–200.

Landsteiner, K., and Levine, P. 1927. Further observations on individual differences of human blood. *Proc. Soc. Exp. Biol. Med.* 24:941–942.

Reed, T. E., and Chandler, J. H. 1958. Huntington's chorea in Michigan. I. Demography and genetics. *Am. J. Hum. Genet.* 10:201–225.

Chapter 5: Linkage, Crossing-over, and Gene Mapping in Eukaryotes

Bateson, W., Saunders, E. R., and Punnett, R. G. 1905. Experimental studies in the physiology of heredity. *Rep. Evol. Committee R. Soc.* II:1–55, 80–99.

Blixt, S. 1975. Why didn't Mendel find linkage? *Nature* 256:206.

Creighton, H. S., and McClintock, B. 1931. A correlation of cytological and genetical crossing-over in *Zea mays. Proc. Natl. Acad. Sci. USA.* 17:492–497.

Ephrussi, B., and Weiss, M. C. 1969. Hybrid somatic cells. *Sci. Am.* 220:26–35.

Fincham, J. R. S., Day, P. R., and Radford, A. 1979. *Fungal genetics*, 3rd ed. Oxford: Blackwell Scientific.

McKusick, V. A. 1971. The mapping of human chromosomes. *Sci. Am.* 224:104–113.

Morgan, T. H. 1910. Sex-limited inheritance in *Drosophila*. *Science* 32:120–122.

———. 1910. The method of inheritance of two sex-limited characters in the same animal. *Proc. Soc. Exp. Biol. Med.* 8:17.

———. 1911. An attempt to analyze the constitution of the chromosomes on the basis of sex-limited inheritance in *Drosophila*. *J. Exp. Zool.* 11:365–414.

———. 1911. Random segregation versus coupling in Mendelian inheritance. *Science* 34:384.

———, Sturtevant, A. H., Muller, H. J., and Bridges, C. B. 1915. *The mechanism of Mendelian heredity*. New York: Henry Holt.

Ruddle, F. H., and Kucherlapati, R. S. 1974. Hybrid cells and human genes. *Sci. Am.* 231:36–44.

Sturtevant, A. H. 1913. The linear arrangement of six sex-linked factors in *Drosophila*, as shown by their mode of association. *J. Exp. Zool.* 14:43–59.

Sutton, W. S. 1903. The chromosomes in heredity. *Biol. Bull.* 4:231–251.

Chapter 6: Genetic Analysis in Bacteria and Bacteriophages

Benzer, S. 1959. On the topology of the genetic fine structure. *Proc. Natl. Acad. Sci. USA* 45:1607–1620.

———. 1961. On the topography of the genetic fine structure. *Proc. Natl. Acad. Sci. USA* 47:403–415.

———. 1962. The fine structure of the gene. *Sci. Am.* 206:70–84.

Delbruck, M. 1940. The growth of bacteriophage and lysis of the host. *J. Gen. Physiol.* 23:643–660.

Ellis, E. L., and Delbruck, M. 1939. The growth of bacteriophage. *J. Gen. Physiol.* 22:365–384.

Fincham, J. 1966. *Genetic complementation*. Menlo Park, CA: Benjamin.

Hayes, W. 1968. *The genetics of bacteria and their viruses*, 2nd ed. New York: Wiley.

Hershey, A. D., and Rotman, R. 1949. Genetic recombination between host-range and plaque-type mutants of bacteriophage in single bacterial cells. *Genetics* 34:44–71.

Hotchkiss, R. D., and Gabor, M. 1970. Bacterial transformation with special reference to recombination processes. *Annu. Rev. Genet.* 4:193–224.

Jacob, F., and Wollman, E. L. 1951. *Sexuality and the genetics of bacteria*. New York: Academic Press.

Susman, M. 1970. General bacterial genetics. *Annu. Rev. Genet.* 4:135–176.

Vielmetter, W., Bonhoeffer, F., and Schutte, A. 1968. Genetic evidence for transfer of a single DNA strand during bacterial conjugation. *J. Mol. Biol.* 37:81–86.

Wollman, E. L., Jacob, F., and Hayes, W. 1962. Conjugation and genetic recombination in *E. coli K-12*. *Cold Spring Harbor Symp. Quant. Biol.* 21:141–162.

Zinder, N., and Lederberg, J. L. 1952. Genetic exchange in *Salmonella*. *J. Bacteriol.* 64:679–699.

Chapter 7: The Beginnings of Molecular Genetics: Gene Function

Beadle, G. W., and Tatum, E. L. 1942. Genetic control of biochemical reactions in *Neurospora*. *Proc. Natl. Acad. Sci. USA* 27:499–506.

Galjaard, H. 1986. Biochemical diagnosis of genetic diseases. *Experientia* 42:1075–1085.

Garrod, A. E. 1909. *Inborn errors of metabolism*. New York: Oxford University Press.

Gilbert, F., Kucherlapati, R., Creagan, R. P., Murnane, M. J., Darlington, G. J., and Ruddle, F. H. 1975. Tay-Sachs' and Sandhoff's diseases: The assignment of genes for hexosaminidase A and B to individual human chromosomes. *Proc. Natl. Acad. Sci. USA* 72:263–267.

Gusella, J. F., Wexler, N. S., Conneally, P. M., Naylor, S. L., Anderson, M. A., Tanzi, R. E., Watkins, P. C., Ottina, K., Wallace, M. R., Sakaguchi, A. Y., Young, A. B., Shoulson, I., Bonilla, E., and Martin, J. B. 1983. A polymorphic DNA marker genetically linked to Huntington's disease. *Nature* 306:234–238.

Guttler, F., and Woo, S. L. C. 1986. Molecular genetics of PKU. *J. Inherited Metab. Dis. 9* Suppl. 1:58–68.

Ingram, V. M. 1963. *The hemoglobins in genetics and evolution*. New York: Columbia University Press.

Kan, Y. W., and Dozy, A. M. 1978. Polymorphism of DNA sequence adjacent to human β-globin structural gene: Relationship to sickle mutation. *Proc. Natl. Acad. Sci. USA* 75:5631–5635.

Maniatis, T., Fritsch, E. F., Lauer, J., and Lawn, R. M. 1980. The molecular genetics of human hemoglobins. *Annu. Rev. Genet.* 14:145–178.

Milunsky, A. 1976. Prenatal diagnosis of genetic disorders. *N. Engl. J. Med.* 295:377–380.

Motulsky, A. G. 1964. Hereditary red cell traits and malaria. *Am. J. Trop. Med. Hyg.* 13:147–158.

Neel, J. V. 1949. The inheritance of sickle-cell anemia. *Science* 110:64–66.

Pauling, L., Itano, H. A., Singer, S. J., and Wells, J. C. 1949. Sickle-cell anemia, a molecular disease. *Science* 110:543–548.

Srb, A. M., and Horowitz, N. H. 1944. The ornithine cycle in *Neurospora* and its genetic control. *J. Biol. Chem.* 154:129–139.

Woo, S. L. C., Lidsky, A. S., Guttler, F., Chandra, T., and Robson, K. J. H. 1983. Cloned human phenylalanine hydroxylase gene allows prenatal diagnosis and carrier detection of classical phenylketonuria. *Nature* 300:151–155.

Chapter 8: The Structure of Genetic Material

Avery, O. T., MacLeod, C. M., and McCarty, M. 1944. Studies on the chemical nature of the substance inducing transformation of pneumococcal types. Induction of transformation by a deoxyribonucleic acid fraction isolated from pneumococcus type III. *J. Exp. Med.* 79:137–158.

Azorin, F., and Rich, A. 1985. Isolation of Z-DNA binding proteins from SV40 minichromosomes: Evidence for binding to the viral control region. *Cell* 41:365–374.

Chargaff, E. 1951. Structure and function of nucleic acids as cell constituents. *Fed. Proc.* 10:654–659.

Dickerson, R. E. 1983. The DNA helix and how it is read. *Sci. Am.* 249 (Dec.):94–111.

Fraenkel-Conrat, H., and Singer, B. 1957. Virus reconstitution: Combination of protein and nucleic acid from different strains. *Biochim. Biophys. Acta* 24:540–548.

Geis, I. 1983. Visualizing the anatomy of A, B and Z-DNAs. *J. Biomol. Struct. Dynam.* 1:581–591.

Gierer, A., and Schramm, G. 1956. Infectivity of ribonucleic acid from tobacco mosaic virus. *Nature* 177:702–703.

Griffith, F. 1928. The significance of pneumococcal types. *J. Hyg.* (Lond) 27:113–159.

Hershey, A. D., and Chase, M. 1952. Independent functions of viral protein and nucleic acid in growth and bacteriophage. *J. Gen. Physiol.* 36:39–56.

Jaworski, A., Hsieh, W.-T., Blaho, J. A., Larson, J. E., and Wells, R. D. 1988. Left-handed DNA in vivo. *Science* 238:773–777.

Krishna, P., Kennedy, B. P., van de Sande, J. H., and McGhee, J. D. 1988. Yolk proteins from nematodes, chickens, and frogs bind strongly and preferentially to left-handed Z-DNA. *J. Biol. Chem.* 263:19066–19070.

Pauling, L., and Corey, R. B. 1956. Specific hydrogen-bond formation between pyrimidines and purines in deoxyribonucleic acids. *Arch. Biochem. Biophys.* 65:164–181.

Rich, A., Nordheim, A., and Wang, A. H.-J. 1984. The chemistry and biology of left-handed Z-DNA. *Annu. Rev. Biochem.* 53:791–846.

Wang, A. H.-J., Quigley, G. J., Kolpak, F. J., Crawford, J. L., van Boom, J. H., van der Marel, G., and Rich, A. 1979. Molecular structure of a left-handed double helical DNA fragment at atomic resolution. *Nature* 282:680–686.

Wang, J. C. 1982. DNA topoisomerases. *Sci. Am.* 247:94–109.

Watson, J. D. 1968. *The double helix.* New York: Atheneum.

———, and Crick, F. H. C. 1953. Genetical implications of the structure of deoxyribonucleic acid. *Nature* 171:964–969.

———. 1953. Molecular structure of nucleic acids. A structure for deoxyribose nucleic acid. *Nature* 171:737–738.

Wilkins, M. H. F., Stokes, A. R., and Wilson, H. R. 1953. Molecular structure of deoxypentose nucleic acids. *Nature* 171:738–740.

Wittig, B., Dorbic, T., and Rich, A. 1989. The level of Z-DNA in metabolically active, permeabilized mammalian cell nuclei is regulated by torsional strain. *J. Cell Biol.* 108:755–764.

Chapter 9: The Organization of DNA in Chromosomes

Amati, B. B., and Gasser, S. M. 1988. Chromosomal ARS and CEN elements bind specifically to the yeast nuclear scaffold. *Cell* 54:967–978.

Blackburn, E. H. 1984. Telomeres: Do the ends justify the means? *Cell* 37:7–8.

———, and Szostak, J. W. 1984. The molecular structure of centromeres and telomeres. *Annu. Rev. Biochem.* 53:163–194.

Bloom, K. S., Amaya, E., Carbon, J., Clarke, L., Hill, A., and Yeh, E. 1984. Chromatin conformation of yeast centromeres. *J. Cell Biol.* 99:1559–1568.

Boy de la Tour, E., and Laemmli, U. K. 1988. The metaphase scaffold is helically folded: sister chromatids have predominantly opposite helical handedness. *Cell* 55:937–944.

Britten, R. J., and Kohne, D. E. 1968. Repeated sequences in DNA. *Science* 161:529–540.

Burlingame, R. W., Love, W. E., Wang, B.-C., Hamlin, R., Xuang, N.-H., and Moudranakis, E. N. 1985. Crystallographic structure of the octameric histone core of the nucleosome at a resolution of 33 Å. *Science* 228:546–553.

Cai, M., and Davis, R. W. 1990. Yeast centromere binding protein CBFI of the helix-loop-helix protein family, is required for chromosome stability and methionine prototrophy. *Cell* 61:437–446.

Carbon, J. 1984. Yeast centromeres: structure and function. *Cell* 37:351–353.

Comings, D. 1978. Mechanisms of chromosome banding and implications for chromosome structure. *Annu. Rev. Genet.* 12:25–46.

DuPraw, E. J. 1970. *DNA and chromosomes.* New York: Holt, Rinehart and Winston.

Eisenberg, J. C., Cartwright, I. L., Thomas, G. H., and Elgin, S. C. R. 1985. Selected topics in chromatin structure. *Annu. Rev. Genet.* 19:485–536.

Freifelder, D. 1978. *The DNA molecule. Structure and properties.* San Francisco: W. H. Freeman.

Gellert, M. 1981. DNA topoisomerases. *Annu. Rev. Biochem.* 50:879–910.

Jelinek, W. R., and Schmid, C. W. 1982. Repetitive sequences in eukaryotic DNA and their expression. *Annu. Rev. Biochem.* 51:813–844.

Kornberg, R. D. 1977. Structure of chromatin. *Annu. Rev. Biochem.* 46:931–954.

———, and Klug, A. 1981. The nucleosome. *Sci. Am.* 244 (2):52–64.

Mirkovich, J., Mirault, M.-E., and Laemmli, U. K. 1984. Organization of the higher-order chromatin loop: specific DNA attachment sites on nuclear scaffold. *Cell* 39:223–232.

Morse, R. H., and Simpson, R. T. 1988. DNA in the nucleosome. *Cell* 54:285–287.

Moyzis, R. K., Buckingham, J. M., Cram, L. S., Dani, M., Deaven, L.L., Jones, M. D., Meyne, J., Ratliff, R. L., and Wu, J.-R. 1988. A highly conserved repetitive DNA sequence, (TTAGGG)$_n$, present at the telomeres of human chromosomes. *Proc. Natl. Acad. Sci. USA* 85:6622–6626.

Olins, A. L., Carlson, R. D., and Olins, D. E. 1975. Visualization of chromatin substructure: nu-bodies. *J. Cell Biol.* 64:528–537.

Pettijohn, D. E. 1988. Histone-like proteins and bacterial chromosome structure. *J. Biol. Chem.* 263:12793–12796.

Richard, T. J., Finch, J. T., Rushton, B., Rhodes, D., and Klug, A. 1984. Structure of the nucleosome core particle at 7 Å resolution. *Nature* 311:532–537.

Richards, E. J., and Ausubel, F. M. 1988. Isolation of a higher eukaryotic telomere from *Arabidopsis thaliana*. *Cell* 53:127–136.

Singer, M. F. 1982. Highly repeated sequences in mammalian genomes. *Int. Rev. Cytol.* 76:67–112.

———. 1982. SINEs and LINEs: Highly repeated short and long interspersed sequences in mammalian genomes. *Cell* 28:133–134.

———, and Skowronski, J. 1985. Making sense out of LINES: Long interspersed repeat sequences in mammalian genomes. *Trends Biochem. Sci.* (Mar.): 119–121.

van Holde, K. E. 1988. *Chromatin.* Springer-Verlag: New York.

Walmsley, R. W., Chan, C. S. M., Tye, B.-K., and Petes, T. D. 1984. Unusual DNA sequences associated with the ends of yeast chromosomes. *Nature* 310:157–160.

Williamson, J. R., Raghuraman, M. K., and Cech, T. R. 1989. Monovalent cation-induced structure of telomeric DNA: the G-quartet model. *Cell* 59:871–880.

Woodcock, C. L. F., Frado, L.-L. Y., and Rattner, J. B. 1984. The higher-order structure of chromatin: Evidence for a helical ribbon arrangement. *J. Cell Biol.* 99:42–52.

Worcel, A. 1978. Molecular architecture of the chromatin fiber. *Cold Spring Harbor Symp. Quant. Biol.* 42:313–324.

———, and Benyajati, C. 1977. Higher order coiling of DNA in chromatin. *Cell* 12:83–100.

———, and Burgi, E. 1972. On the structure of the folded chromosome of *Escherichia coli. J. Mol. Biol.* 71:127–147.

Yokoyama, R., and Yao, M.-C. 1986. Sequence characterization of *Tetrahymena* macronuclear ends. *Nucleic Acids Res.* 14:2109–2122.

Zimmerman, S. B. 1982. The three-dimensional structure of DNA. *Annu. Rev. Biochem.* 51:395–427.

Chapter 10: DNA Replication and Recombination

Alberts, B. M. 1985. Protein machines mediate the basic genetic processes. *Trends Genet.* 1:26–30.

Bramhill, D., and Kornberg, A. 1988. Duplex opening by *dnaA* protein at novel sequences in initiation of replication at the origin of the *E. coli* chromosome. *Cell* 52:743–755.

Campbell, J. L. 1986. Eukaryotic DNA replication. *Annu. Rev. Biochem.* 55:733–772.

Cox, M. M., and Lehman, I. R. 1987. Enzymes of general recombination. *Annu. Rev. Biochem.* 56:229–262.

Cozzarelli, N. R. 1980. DNA gyrase and the supercoiling of DNA. *Science* 207:953–960.

DeLucia, P., and Cairns, J. 1969. Isolation of an *E. coli* strain with a mutation affecting DNA polymerase. *Nature* 224:1164–1166.

DePamphilis, M. S. 1988. Transcriptional elements as components of eukaryotic origins of DNA replication. *Cell* 52:635–638.

DePamphilis, M. L., and Wassarman, P. M. 1980. Replication of eukaryotic chromosomes: A close-up of the replication fork. *Annu. Rev. Biochem.* 49:627–666.

Gilbert, W., and Dressler, D. 1968. DNA replication: The rolling circle model. *Cold Spring Harbor Symp. Quant. Biol.* 33:473–484.

Holliday, R. 1964. A mechanism for gene conversion in fungi. *Genet. Res.* 5:282–304.

Huberman, J. A. 1987. Eukaryotic DNA replication: A complex picture partially clarified. *Cell* 48:7–8.

———, and Riggs, A. D. 1968. On the mechanism of DNA replication in mammalian chromosomes. *J. Mol. Biol.* 32:327–341.

Kornberg, A. 1960. Biologic synthesis of deoxyribonucleic acid. *Science* 131:1503–1508.

———, and Baker, T. A. 1992. *DNA replication,* 2nd ed. San Francisco: W. H. Freeman.

Meselson, M., and Radding, C. M. 1975. A general model for genetic recombination. *Proc. Natl. Acad. Sci. USA* 72:358–361.

Meselson, M., and Stahl, F. W. 1958. The replication of DNA in *Escherichia coli. Proc. Natl. Acad. Sci. USA* 44:671–682.

Modrich, P. 1987. DNA mismatch correction. *Annu. Rev. Biochem.* 56:435–466.

Ogawa, T., Baker, T. A., van der Ende, A., and Kornberg, A. 1985. Initiation of enzymatic replication at the origin of the *Escherichia coli* chromosome: Contributions of RNA polymerase and primase. *Proc. Natl. Acad. Sci. USA* 82:3562–3566.

———, and Okazaki, T. 1980. Discontinuous DNA replication. *Annu. Rev. Biochem.* 49:424–457.

Okazaki, R. T., Okazaki, K., Sakobe, K., Sugimoto, K., and Sugino, A. 1968. Mechanism of DNA chain growth. I. Possible discontinuity and unusual secondary structure of newly synthesized chains. *Proc. Natl. Acad. Sci. USA* 59:598–605.

Radding, C. 1982. Homologous pairing and strand exchange in genetic recombination. *Annu. Rev. Genet.* 16:405–437.

Recombination at the DNA level. 1984. Cold Spring Harbor Symposium on Quantitative Biology. Vol. 49. Cold Spring Harbor Laboratory, Cold Spring Harbor, NY.

Reynolds, A. E., McCarroll, E. M., Newlon, C. S., and Fangman, W. L. 1989. Time of replication of ARS elements along yeast chromosome III. *Mol. Cell. Biol.* 9:4488–4494.

Szostak, J., Orr-Weaver, T., Rothstein, R., and Stahl, F. 1983. The double-strand break repair model for recombination. *Cell* 33:25–35.

Taylor, J. H. 1970. The structure and duplication of chromosomes. In *Genetic organization,* edited by E. Caspari and A. Ravin, vol. 1, pp. 163–221. New York: Academic Press.

Van der Ende, A., Baker, T. A., Ogawa, T., and Kornberg, A. 1985. Initiation of enzymatic replication at the origin of the *Escherichia coli* chromosome: Primase as the sole priming enzyme. *Proc. Natl. Acad. Sci. USA* 82:3954–3958.

Wang, J. C., and Liu, L. F. 1979. DNA topoisomerases; enzymes that catalyze the concerted breakage and rejoining of DNA backbone bonds. In *Molecular genetics,* edited by J. H. Taylor, part 3, pp. 65–88. New York: Academic Press.

Wickner, S. H. 1978. DNA replication proteins of *Escherichia coli. Annu. Rev. Biochem.* 47:1163–1191.

Zyskind, J. W., and Smith, D. W. 1986. The bacterial origin of replication, *oriC Cell* 46:489–490.

Chapter 11: Transcription, RNA Molecules, and RNA Processing

Baker, S. M., and Platt, T. 1986. Pol I transcription: Which comes first, the end or the beginning? *Cell* 47:839–840.

Banerjee, A. K. 1980. 5'-terminal cap structure in eucaryotic messenger ribonucleic acids. *Microbiol. Rev.* 44:175–205.

Bell, S. P., Pikaard, C. P., Reeder, R. H., and Tjian, R. 1989. Molecular mechanisms governing species-specific transcription of ribosomal RNA. *Cell* 59:489–497.

Birnsteil, M. L., Busslinger, M., and Struhl, K. 1985. Transcription termination and 3' processing: The end is in site! *Cell* 41:349–359.

Bogenhagen, D. F., Sakonju, S., and Brown, D. D. 1980. A control region in the center of the 5S RNA gene directs specific initiation of transcription II: The 3' border of the region. *Cell* 19:27–35.

Brand, A. H., Breeden, L., Abraham, J., Sternglanz, R., and Nasmyth, K. 1987. Characterization of a "silencer" in yeast: A DNA sequence with properties opposite to those of a transcriptional enhancer. *Cell* 41:41–48.

Breathnach, R., and Chambon, P. 1981. Organization and expression of eucaryotic split genes coding for proteins. *Annu. Rev. Biochem.* 50:349–383.

———, Mandel, J. L., and Chambon, P. 1977. Ovalbumin gene is split in chicken DNA. *Nature* 270:314–318.

Brody, E., and Abelson, J. 1985. The "spliceosome": Yeast premessenger RNA associates with a 40S complex in a splicing dependent reaction. *Science* 228:963–967.

Cech, T. R. 1983. RNA splicing: Three themes with variations. *Cell* 34:713–716.

———. 1985. Self-splicing RNA: Implications for evolution. *Int. Rev. Cytol.* 93:3–22.

———. 1986. Ribosomal RNA gene expression in *Tetrahymena:* Transcription and RNA splicing. *Mol. biol. ciliated protozoa.* pp. 203–225. New York: Academic Press.

———. 1986. The generality of self-splicing RNA: Relationship to nuclear mRNA splicing. *Cell* 44:207–210.

Chabot, B., and Steitz, J. A. 1987. Multiple interactions between the splicing substrate and small nuclear ribonucleoproteins in spliceosomes. *Mol. Cell. Biol.* 7:281–293.

Chambliss, G., Craven, G. R., Davies, J., Davis, K., Kahan, L., and Nomura, M., eds. 1980. *Ribosomes. Structure, function, and genetics.* Baltimore: University Park Press.

Choi, Y. D., Grabowski, P. J., Sharp, P. A., and Dreyfuss, G. 1986. Heterogeneous nuclear ribonucleoproteins: Role in RNA splicing. *Science* 231:1534–1539.

Crick, F. H. C. 1979. Split genes and RNA splicing. *Science* 204:264–271.

Dynan, W. S. 1989. Modularity in promoters and enhancers. *Cell* 58:1–4.

Geiduschek, E. P., and Tocchini-Valentini, G. P. 1988. Transcription by RNA polymerase III. *Annu. Rev. Biochem.* 57:873–914.

Grabowski, P. J., Seiler, S. R., and Sharp, P. A. 1985. A multicomponent complex is involved in the splicing of messenger RNA precursors. *Cell* 42:355–367.

Green, M. R. 1986. Pre-mRNA splicing. *Annu. Rev. Genet.* 20:671–708.

Guarente, L. 1988. UASs and enhancers: common mechanism of transcriptional activation in yeast and mammals. *Cell* 52:303–305.

Guthrie, C. 1986. Finding functions for small nuclear RNAs in yeast. *Trends Biochem. Sci.* (Oct.): 430–434.

Jeffreys, A. J., and Flavell, R. A. 1977. The rabbit beta-globin gene contains a large insert in the coding sequence. *Cell* 12:1097–1108.

Konarska, M. M., and Sharp, P. A. 1987. Interactions between small nuclear ribonucleoprotein particles in the formation of spliceosomes. *Cell:* 49:763–771.

Marmur, J., Greenspan, C. M., Palecek, E., Kahan, F. M., Levine, J., and Mandel, M. 1963. Specificity of the complementary RNA formed by *Bacillus subtilis* infected with bacteriophage SP8. *Cold Spring Harbor Symp. Quant. Biol.* 28:191–199.

Moore, P. B. 1988. The ribosome returns. *Nature* 331:223–227.

Nomura, M. 1973. Assembly of bacterial ribosomes. *Science* 179:864–873.

Padgett, R. A., Grabowski, P. J., Konarska, M. M., and Sharp, P. A. 1985. Splicing messenger RNA precursors: Branch sites and lariat RNAs. *Trends Biochem. Sci.* (April): 154–157.

Reeder, R. H. 1989. Regulatory elements of the generic ribosomal gene. *Curr. Opinion Cell Biol.* 1:466–474.

Reznikoff, W. S., Siegele, D. A., Cowing, D. W., and Gross, C. A. 1985. The regulation of transcription initiation in bacteria. *Annu. Rev. Genet.* 19:355–387.

Schmidt, F. J. 1985. RNA splicing in prokaryotes: Bacteriophage T4 leads the way. *Cell* 41:339–340.

Sharp, P. A. 1985. On the origin of RNA splicing and introns. *Cell* 42:397–400.

Sollner-Webb, B. 1988. Surprises in polymerase III transcription. *Cell* 52:153–154.

Starzyk, R. M. 1986. Prokaryotic RNA processing. *Trends Biochem. Sci.* (Feb.):60.

Struhl, K. 1987. Promoters, activator proteins, and the mechanism of transcriptional initiation in yeast. *Cell* 49:295–297.

Tilghman, S. M., Curis, P. J., Tiemeier, D. C., Leder, P., and Weissman, C. 1978. The intervening sequence of a mouse β-globin gene is tanscribed within the 15S β-globin mRNA precursor. *Proc. Natl. Acad. Sci. USA* 75:1309–1313.

————, Tiemeier, D. C., Seidman, J. G., Peterlin, B. M., Sullivan, M., Maizel, J. V., and Leder, P. 1978. Intervening sequence of DNA identified in the structural portion of a mouse beta-globin gene. *Proc. Natl. Acad. Sci. USA* 78:725–729.

Voss, S. D., Schlokat, U., and Gruss, P. 1986. The role of enhancers in the regulation of cell-type-specific transcriptional control. *Trends Biochem. Sci.* (July): 287–289.

Zaug, A. J., and Cech, T. R. 1986. The intervening sequence RNA of Tetrahymena is an enzyme. *Science* 231:470–475.

Chapter 12: The Genetic Code and the Translation of the Genetic Message

Bachmair, A., Finley, D., and Varshavsky, A. 1986. In vivo half-life of a protein is a function of its amino-terminal residue. *Science* 234:179–186.

Blobel, G., and Dobberstein, B. 1975. Transfer of proteins across membranes. I. Presence of proteolytically processed and unprocessed nascent immunoglobulin light chains on membrane-bound ribosomes of murine myeloma. *J. Cell Biol.* 67:835–851.

Brenner, S., Jacob, F., and Meselson, M. 1961. An unstable intermediate carrying information from genes to ribosomes for protein synthesis. *Nature* 190:576–581.

Colman, A., and Robinson, C. 1986. Protein import into organelles: hierarchical targeting signals. *Cell* 46:321–322.

Crick, F. H. C. 1966. Codon-anticodon pairing: The wobble hypothesis. *J. Mol. Biol.* 19:548–555.

————, Barnett, L., Brenner, S., and Watts-Tobin, R. J. 1961. General nature of the genetic code for proteins. *Nature* 192:1227–1232.

Dahlberg, A. 1989. The functional role of ribosomal RNA in protein synthesis. *Cell* 57:525–529.

Garen, A. 1968. Sense and nonsense in the genetic code. *Science* 160:149–159.

Griffiths, G., and Simons, K. 1986. The trans Golgi network: Sorting at the exit site of the Golgi complex. *Science* 234:438–442.

Jackson, R. J., and Standart, N. 1990. Do the poly(A) tail and 3' untranslated region control mRNA translation? *Cell* 62:15–24.

Kozak, M. 1983. Comparison of initiation of protein synthesis in procaryotes, eucaryotes, and organelles. *Microbiol. Rev.* 47:145.

————. 1989. Context effects and inefficient initiation at non-AUG codons in eucaryotic cell-free translation systems. *Mol. Cell. Biol.* 9:5073–5080.

Lingappa, V. R. 1991. More than just a channel: provocative new features of protein traffic across the ER membrane. *Cell* 65:527–530.

Moldave, K. 1985. Eukaryotic protein synthesis. *Annu. Rev. Biochem.* 54:1109–1150.

Moore, P. B. 1988. The ribosome returns. *Nature* 331:223–227.

Morgan, A. R., Wells, R. D., and Khorana, H. G. 1966. Studies on polynucleotides. LIX. Further codon assignments from amino acid incorporation directed by ribopolynucleotides containing repeating trinucleotide sequences. *Proc. Natl. Acad. Sci. USA* 56:1899–1906.

Nirenberg, M., and Leder, P. 1964. RNA code words and protein synthesis. *Science* 145:1399–1407.

————, and Matthaei, J. H. 1961. The dependence of cell-free protein synthesis in *E. coli* upon naturally occurring or synthetic polyribonucleotides. *Proc. Natl. Acad. Sci. USA* 47:1588–1602.

Pfeffer, S. R., and Rothman, J. E. 1987. Biosynthetic protein transport and sorting by the endoplasmic reticulum and Golgi. *Annu. Rev. Biochem.* 56:829–852.

Rogers, S., Wells, R., and Rechsteiner, M. 1986. Amino acid sequences common to rapidly degraded proteins: The PEST hypothesis. *Science* 234:364–368.

Shine, J., and Delgarno, L. 1974. The 3'-terminal sequence of *Escherichia coli* 16S ribosomal RNA: Complementarity to nonsense triplet and ribosome binding sites. *Proc. Natl. Acad. Sci. USA* 71:1342–1346.

Silver, P. A. 1991. How proteins enter the nucleus. *Cell* 64:489–497.

Verner, K., and Schatz, G. 1988. Protein translocation across membranes. *Science* 241:1307–1313.

Watson, J. D. 1963. The involvement of RNA in the synthesis of proteins. *Science* 140:17–26.

Chapter 13: Gene Cloning and Recombinant DNA Technology

Anderson, W. F. 1992. Human gene therapy. *Science* 256:808–813.

Antonarakis, S. E. 1989. Diagnosis of genetic disorders at the DNA level. *New England J. Med.* 320:153–163.

Bobrow, M. 1988. Prenatal diagnosis. *J. Chem. Tech. Biotechnol.* 43:285–291.

Boyer, H. W. 1971. DNA restriction and modification mechanisms in bacteria. *Annu. Rev. Microbiol.* 25:153–176.

Collins, F. S. 1992. Cystic fibrosis: molecular biology and therapeutic implications. *Science* 256:774–779.

Donis-Keller, H., Barker, D., Knowlton, R., Schumm, J., and Braman, J. 1986. Applications of RFLP probes to genetic mapping and clinical diagnosis in humans. In *Current communications in molecular biology: DNA probes-Applications in genetic and infectious disease and cancer,* edited by L. S. Lerman, pp. 73–81. Cold Spring Harbor Laboratory, Cold Spring Harbor, NY.

Drayna, D., and White, R. 1985. The genetic linkage map of the human X chromosome. *Science* 230:753–758.

Eisenstein, B. I. 1990. The polymerase chain reaction: a new method of using molecular genetics for medical diagnosis. *New England J. Med.* 322:178–183.

Foote, S., Vollrath, D., Hilton, A., and Page, D. C. 1992. The human Y chromosome: overlapping DNA clones spanning the euchromatic region. *Science* 258:60–66.

Freifelder, D. 1978. *Recombinant DNA: Readings from Scientific American.* San Francisco: W. H. Freeman.

Hood, L. 1988. Biotechnology and medicine of the future. *J. Am. Med. Assoc.* 259:1837–1844.

Klee, H., Horsch, R., and Rogers, S. 1987. *Agrobacterium*-mediated plant transformation and its further applications to plant biology. *Annu. Rev. Plant Physiol.* 38:467–486.

Maxam, A. M., and Gilbert, W. 1977. A new method for sequencing DNA. *Proc. Natl. Acad. Sci. USA* 74:560–564.

Moores, J. C. 1987. Current approaches to DNA sequencing. *Anal. Biochem.* 163:1–8.

Mullis, K. B. 1990. The unusual origin of the polymerase chain reaction. *Sci. Am.* (April):56–65.

Mullis, K. B., and Faloona, F. A. 1987. Specific synthesis of DNA *in vitro* via a polymerase-catalyzed chain reaction. *Meth. Enzymol.* 155:335–350.

NIH/CEPH Collaborative Mapping Group. 1992. A comprehensive genetic linkage map of the human genome. *Science* 258:67–86.

Sambrook, J., Fritsch, E. F., and Maniatis, T. 1989. *Molecular cloning: A laboratory manual,* 2nd ed. Cold Spring Harbor Laboratory, Cold Spring Harbor, NY.

Sanger, F., and Coulson, A. R. 1975. A rapid method for determining sequences in DNA by primed synthesis with DNA polymerase. *J. Mol. Biol.* 94:441–448.

Schell, J., and Van Montagu, M. 1983. The Ti plasmids as natural and practical gene vectors for plants. *BioTechnology* 1:175–180.

Southern, E. M. 1975. Detection of specific sequences among DNA fragments separated by gel electrophoresis. *J. Mol. Biol.* 98:503–517.

Vollrath, D., Foote, S., Hilton, A., Brown, L. G., Beer-Romero, P., Bogan, J. S., and Page, D. C. 1992. The human Y chromosome: a 43-interval map based on naturally occurring deletions. *Science* 258:52–59.

Watson, J. D., Gilman, J., Witkowski, J., and Zoller, M. 1992. *Recombinant DNA,* 2nd ed. Scientific American Books.

White, T., and Lalouel, J.-M. 1988. Chromosome mapping with DNA markers. *Sci. Am.* 258:40–48.

White, T. J., Arnheim, N., and Erlich, H. A. 1989. The polymerase chain reaction. *Trends Genet.* 5:185–188.

Gilbert, W., Maizels, N., and Maxam, A. 1974. Sequences of controlling regions of the lactose operon. *Cold Spring Harbor Symp. Quant. Biol.* 38:845–855.

———, and Muller-Hill, B. 1966. Isolation of the *lac* repressor. *Proc. Natl. Acad. Sci. USA* 56:1891–1898.

Jacob, F., and Monod, J. 1961. Genetic regulatory mechanisms in the synthesis of proteins. *J. Mol. Biol.* 3:318–356.

Lee, F., and Yanofsky, C. 1977. Transcription termination at the trp operon attenuators of *Escherichia coli* and *Salmonella typhimurium:* RNA secondary structure and regulation of termination. *Proc. Natl. Acad. Sci. USA* 74:4365–4369.

Maizels, N. 1974. *E. coli* lactose operon ribosome binding site. *Nature (New Biol.)* 249:647–649.

Miller, J. H., and Reznikoff, W. S. 1978. *The operon.* Cold Spring Harbor Laboratory, Cold Spring Harbor, NY.

Platt, T. 1981. Termination of transcription and its regulation in the tryptophan operon of *E. coli. Cell* 24:10–23.

Ptashne, M. 1967. Isolation of the λ phage repressor. *Proc. Natl. Acad. Sci. USA* 57:306–313.

———. 1984. Repressors. *Trends Biochem. Sci.* 9:142–145.

———. 1989. How gene activators work. *Sci. Am.* 260(Jan.):24–31.

———. 1992. *A genetic switch,* 2nd ed. Oxford: Cell Press and Blackwell Scientific Publications.

———, and Gilbert, W. 1970. Genetic repressors. *Sci. Am.* 222(June):36–44.

Ullmann, A. 1985. Catabolite repression 1985. *Biochimie* 67:29–34.

Winkler, M. E., and Yanofsky, C. 1981. Pausing of RNA polymerase during in vitro transcription of the tryptophan operon leader region. *Biochemistry* 20:3738–3744.

Yanofsky, C. 1981. Attenuation in the control of expression of bacterial operons. *Nature* 289:751–758.

Yanofsky, C. 1987. Operon-specific control by transcription attenuation. *Trends Genet.* 3:356–360.

———, and Kolter, R. 1982. Attenuation in amino acid biosynthetic operons. *Annu. Rev. Genet.* 16:113–134.

Chapter 14: Regulation of Gene Expression in Bacteria and Bacteriophages

Aloni, Y., and Hay, N. 1985. Attenuation may regulate gene expression in animal viruses and cells. *CRC Crit. Rev. Biochem.* 18:327–383.

Beckwith, J. R., and Zipser, D. 1970. *The lactose operon.* Cold Spring Harbor Laboratory, Cold Spring Harbor, NY.

Dickson, R. C., Abelson, J., Barnes, W. M., and Reznikoff, W. S. 1975. Genetic regulation: The *lac* control region. *Science* 187:27–35.

Fisher, R. F., Das, A., Kolter, R., Winkler, M. E., and Yanofsky, C. 1985. Analysis of the requirements for transcription pausing in the tryptophan operon. *J. Mol. Biol.* 182:397–409.

Chapter 15: Regulation of Gene Expression and Development in Eukaryotes

Bachvarova, R. F. 1992. A maternal tail of poly(A): the long and the short of it. *Cell* 69:895–897.

Baker, W. 1978. A genetic framework for *Drosophila* development. *Annu. Rev. Genet.* 12:451–470.

Beato, M. 1989. Gene regulation by steroid hormones. *Cell* 56:335–344.

Blackwell, T. K., and F. W. 1988. Immunoglobin genes. In *Molecular immunology,* edited by B. D. Homes and D. M. Glover, pp. 1–60. Washington, D.C.: IRL Press.

Boggs, R. T., Gregor, P., Idriss, S., Belote, J. M., and McKeown, M. 1987. Regulation of sexual differentiation in *D. melanogaster* via alternative splicing of RNA from the *transformer gene. Cell* 50:739–747.

Brawerman, G. 1989. mRNA decay: finding the right targets. *Cell* 57:9–10.

Breitbart, R. E., Andreadis, A., and Nadal-Ginard, B. 1987. Alternative splicing: A ubiquitous mechanism for the generation of multiple protein isoforms from single genes. *Annu. Rev. Biochem.* 56:467–495.

Cedar, H. 1988. DNA methylation and gene activity. *Cell* 53:3–4.

Davidson, E. H. 1976. *Genetic activity in early development.* 2nd ed. New York: Academic Press.

De Robertis, E. M., and Gurdon, J. B. 1977. Gene activation in somatic nuclei after injection into amphibian oocytes. *Proc. Natl. Acad. Sci. USA* 74:2470–2474.

Efstratiadis, A., Posakony, J. W., Maniatis, T., Lawn, R. M., O'Connell, C., Spritz, R. A., DeRiel, J. K., Forget, B. G., Weissman, S. M., Slighton, J. L., Blechtl A. E., Smithies, O., Baralle, F. E., Shoulders, C. C., and Proudfoot, N. J. 1980. The structure and evolution of the human β-globin gene family. *Cell* 21:653–668.

Eissenberg, J. C., Cartwright, I. L., Thomas, G. H., and Elgin, S. C. R. 1985. Selected topics in chromatin structure. *Annu. Rev. Genet.* 19:485–536.

Green, M. R. 1989. Pre-mRNA processing and mRNA nuclear export. *Curr. Opinion Cell Biol.* 1:519–525.

Gross, D. S., and Garrard, W. T. 1987. Poising chromatin for transcription. *Trends Biochem. Sci.* (Aug.):293–297.

———. 1988. Nuclease hypersensitive sites in chromatin. *Annu. Rev. Biochem.* 57:159–197.

Grunstein, M. 1992. Histones as regulators of genes. *Sci. Am.* 267 (Oct.):68–74B.

Gurdon, J. B. 1968. Transplanted nuclei and cell differentiation. *Sci. Am.* 219:24–35.

Hodgkin, J. 1989. *Drosophila* sex determination: a cascade of regulated splicing. *Cell* 56:905–906.

Karlsson, S., and Nienhuis, A. W. 1985. Development regulation of human globin genes. *Annu. Rev. Biochem.* 54:1071–1078.

Keyes, L. N., Cline, T. W., and Schedl, P. 1992. The primary sex determination signal of Drosophila acts at the level of transcription. *Cell* 68:933–943.

McGinnis, W., Levine, M. S., Hafen, E., Kuroiwa, A., and Gehring, W. J. 1984. A conserved DNA sequence homeotic genes of the *Drosophila Antennapedia* and *bithorax* complexes. *Nature* 308:428–433.

O'Malley, B. W., and Schrader, W. T. 1976. The receptors of steroid hormones. *Sci. Am.* 234:32–43.

O'Malley, B., Towle, H., and Schwartz, R. 1977. Regulation of gene expression in eucaryotes. *Annu. Rev. Genet.* 11:239–275.

Parthun, M. R., and Jaehning, J. A. 1992. A transcriptionally active form of GAL4 is phosphorylated and associated with GAL80. *Mol. Cell. Biol.* 12:4981–4987.

Peifer, M. F., Karch, F., and Bender, W. 1987. The bithorax complex: control of segmental identity. *Genes Dev.* 1:891–898.

Reeves, R. 1984. Transcriptionally active chromatin. *Biochim. Biophys. Acta* 782:343–393.

Rogers, J. O., Early, H., Carter, C., Calame, K., Bond., M., Hood, L., and Wall, R. 1980. Two mRNAs with different 3' ends encode membrane-bound and secreted forms of immunoglobin chain. *Cell* 20:303–312.

Scott, M. P., and Carroll, S. B. 1987. The segmentation and homeotic gene network in early *Drosophila* development. *Cell* 51:689–698.

Scott, M. P., Tamkun, J. W., and Hartzell III, G. W. 1989. The structure and function of the homeodomain. *Biochim. Biophys. Acta* 989:25–48.

Scott, M. P., Weiner, A. J., Hazelrigg, T. I., Polisky, B. A., Pirotta, V., Scalenghe, F., and Kaufman, T. C. 1983. The molecular organization of the *Antennapedia* locus of *Drosophila. Cell* 35:763–776.

Spelsberg, T. C., Littlefield, B. A., Seelke, R., Dani, G. M., Toyoda, H., Boyd-Leinen, P., Thrall, C., and Kon, O. I. 1983. Role of specific chromosomal proteins and DNA sequences in the nuclear binding sites for steroid receptors. *Recent Prog. Horm. Res.* 39:425–517.

Wang, T. Y., Kostraba, N. C., and Newman, R. S. 1976. Selective transcription of DNA mediated by nonhistone proteins. *Prog. Nucleic Acid Res. Mol. Biol.* 19:447–462.

Yamamoto, K. R. 1985. Steroid receptor regulated transcription of specific genes and gene networks. *Annu. Rev. Genet.* 19:209–252.

Chapter 16: Gene Mutation

Ames, B. N., Durston, W, E, Yamasaki, E, and Lee, F. 1973. Carcinogens are mutagens: A simple test system combining liver homogenates for activation and bacteria for detection. *Proc. Natl. Acad. Sci. USA* 70:2281–2285.

———, and Gold, L. S. 1990. Too many rodent carcinogens: mitogenesis increases mutagenesis. *Science* 249:970–971.

Cairns, J., Overbaugh, J., and Miller, S. 1988. The origin of mutants. *Science* 335:142–145.

Collins, A., Johnson, R. T., and Boyle, J. M., eds. 1987. Molecular biology of DNA repair. Supplement 6, *J. Cell Sci.*

Devoret, R. 1979. Bacterial tests for potential carcinogens. *Sci. Am.* 241:40–49.

Haseltine, W. A. 1983. Ultraviolet light repair and mutagenesis revisited. *Cell* 33:13–17.

Lederberg, J., and Lederberg, E. M. 1952. Replica plating and indirect selection of bacterial mutants. *J. Bacteriol.* 63:399–406.

Modrich, P. 1987. DNA mismatch correction. *Annu. Rev. Biochem.* 56:435–466.

Radman, M., and Wagner, R. 1986. Mismatch repair in *Escherichia coli. Annu. Rev. Genet.* 20:523–538.

Setlow, R. B., and Carrier, W. L. 1964. The disappearance of thymine dimers from DNA: An error-correcting mechanism. *Proc. Natl. Acad. Sci. USA* 51:226–231.

Walker, G. C., Marsh, L., and Dodson, L. A. 1985. Genetic analyses of DNA repair: inference and extrapolation. *Annu. Rev. Genet.* 19:103–126.

Chapter 17: Chromosome Mutation

Bloom, A. D. 1972. Induced chromosome aberrations in man. *Adv. Hum. Genet.* 3:99–153.

Borst, P., and Greaves, D. R. 1987. Programmed gene rearrangements altering gene expression. *Science* 235:658–667.

Dalla-Favera, R., Martinotti, S., Gallo, R., Erickson, J., and Croce, C. 1983. Translocation and rearrangements of the *c-myc* oncogene locus in human undifferentiated B-cell lymphomas. *Science* 219:963–997.

DeKlein, A., van Kessel, A. G., Grosveld, G., Bartram, C. R., Hagemeijer, A., Bootsma, D., Spurr, N. K., Heisterkamp, N., Groffen, J., and Stephenson, J. R. 1982. A cellular oncogene is translocated to the Philadelphia chromosome in chronic myelocytic leukemia. *Nature* 300:765–767.

Herskowitz, I. 1985. Master regulatory loci in yeast and lambda. *Cold Spring Harbor Symp. Quant. Biol.* 50:565–574.

Nasmyth, K. A. 1982. Molecular genetics of yeast mating type. *Annu. Rev. Genet.* 16:439–500.

Penrose, L. S., and Smith, G. F. 1966. *Down's anomaly.* Boston: Little, Brown.

Rowley, J. D. 1973. A new consistent chromosomal abnormality in chronic myelogenous leukemia identified by quinacrine fluorescence and Giemsa staining. *Nature* 243:290–293.

Chapter 18: *Transposable Elements, Tumor Viruses, and Oncogenes*

Adams, S. E., Mellor, J., Gull, K., Sim, R. B., Tuite, M. F., Kingsman, S. M., and Kingsman, A. J. 1987. The functions and relationships of Ty-VLP proteins in yeast reflect those of mammalian retroviral proteins. *Cell* 49:111–119.

Baltimore, D. 1985. Retroviruses and retrotransposons: The role of reverse transcription in shaping the eukaryotic genome. *Cell* 40:481–482.

Berg, D. E., and Howe, M. M. 1989. *Mobile DNA.* Amer. Soc. Microbiol., Washington, D.C.

Bishop, J. M. 1983. Cancer genes come of age. *Cell* 32:1018–1020.

———. 1987. The molecular genetics of cancer. *Science* 235:305–311.

Boeke, J. D., and Corces, V. G. 1989. Transcription and reverse transcription of retrotransposons. *Annu. Rev. Microbiol.* 43:403–434.

Boeke, J. D., Garfinkel, D. J., Styles, C. A., and Fink, G. R. 1985. Ty elements transpose through an RNA intermediate. *Cell* 40:491–500.

Bucheton, A. 1990. I transposable elements and I-R hybrid dysgenesis in *Drosophila. Trends Genet.* 6:16–21.

Cohen, S. N., and Shapiro, J. A. 1980. Transposable genetic elements. *Sci. Am.* 242:40–49.

Cold Spring Harbor Symposia for Quantitative Biology. 1980. Vol. 45. *Movable genetic elements.* Cold Spring Harbor Laboratory, Cold Spring Harbor, NY.

Doring, H.-P., and Starlinger, P. 1984. Barbara McClintock's controlling elements: Now at the DNA level. *Cell* 39:253–259.

———. 1986. Molecular genetics of transposable elements in plants. *Annu. Rev. Genet.* 20:175–200.

Engles, W. R. 1983. The P family of transposable elements in *Drosophila. Annu. Rev. Genet.* 17:315–344.

Finnegan, D. J. 1985. Transposable elements in eukaryotes. *Int. Rev. Cytol.* 93:281–326.

Foster, T. J., Davis, M. A., Roberts, D. E., Takashita, K., and Kleckner, N. 1981. Genetic organization of transposon Tn10. *Cell* 23:201–213.

Frederoff, N. V. 1989. About maize transposable elements and development. *Cell* 56:181–191.

Garfinkel, D. J., Boeke, J. D., and Fink, G. R. 1985. Ty element transposition: Reverse transcriptase and virus-like particles. *Cell* 42:507–517.

Iida, S., Meyer, J., and Arber, W. 1983. Prokaryotic IS elements. In *Mobile Genetic Elements,* edited by J. A. Shapiro, pp. 159–221. New York: Academic Press.

Kingsman, A. J., and Kingsman, S. M. 1988. Ty: A retroelement moving forward. *Cell* 53:333–335.

Kingston, R. E., Baldwin, A. S., and Sharp, P. A. 1985. Transcription control by oncogenes. *Cell* 41:3–5.

Kleckner, N. 1981. Transposable elements in prokaryotes. *Annu. Rev. Genet.* 15:341–404.

Krontiris, T. G. 1983. The emerging genetics of human cancer. *N. Engl. J. Med.* 309:404–409.

McClintock, B. 1961. Some parallels between gene control systems in maize and in bacteria. *Am. Naturalist* 95:265–277.

———. 1965. The control of gene action in maize. *Brookhaven Symp. Biol.* 18:162 ff.

Mount, S. M., and Rubin, G. M. 1985. Complete nucleotide sequence of the *Drosophila* transposable element copia: Homology between copia and retroviral proteins. *Mol. Cell. Biol.* 5:1630–1638.

Ratner, L., Gallo, R. C., and Wong-Staal, F. 1985. Cloning of human oncogenes. In *Recombinant DNA research and virus,* edited by Y. Becker. Boston: Martinus Nijhoff.

———, Josephs, S. F., and Wong-Staal, F. 1985. Oncogenes: Their role in neoplastic transformation. *Annu. Rev. Microbiol.* 39:419–449.

Slamon, D. J., deKernion, J. B., Verma, I. M., and Cline, M. J. 1984. Expression of cellular oncogenes in human malignancies. *Science* 224:256–262.

Spradling, A. C., and Rubin, G. M. 1981. *Drosophila* genome organization: Conserved and dynamic aspects. *Annu. Rev. Genet.* 15:219–264.

Varmus, H. E. 1982. Form and function of retroviral proviruses. *Science* 216:812–821.

Weiss, R. A., Teich, N., Varmus, H. E., and Coffin, J. M., eds. 1985. *Molecular Biology of Tumor Viruses: RNA Tumor Viruses,* 2nd ed. (2 vols.). Cold Spring Harbor Laboratory, Cold Spring Harbor, NY.

Chapter 19: *Extranuclear Genetics*

Ashwell, M., and Work, T. S. 1970. The biogenesis of mitochondria. *Annu. Rev. Biochem.* 39:251–290.

Birky, C. W. 1978. Transmission genetics of mitochondria and chloroplasts. *Annu. Rev. Genet.* 12:471–512.

Borst, P., and Grivell, L. A. 1978. The mitochondrial genome of yeast. *Cell* 15:705–723.

Brown, M. D., Voljavec, A. S., Lott, M. T., MacDonald, I. and Wallace, D. C. 1992. Leber's hereditary optic neuropathy: a model for mitochondrial neurodegenerative diseases. *FASEB J.* 6:2791–2799.

Clayton, D. A. 1982. Replication of animal mitochondrial DNA. *Cell* 28:693–705.

Gellissen, G., and Michaelis, G. 1987. Gene transfer: Mitochondria to nucleus. *Ann. N.Y. Acad. Sci.* 503:391–401.

Gillham, N. W. 1978. *Organelle heredity.* New York: Raven Press.

Grivell, L. 1983. Mitochondrial DNA. *Sci. Am.* 225(Mar.):78–89.

Horowitz, S., and Gorovsky, M. A. 1985. An unusual genetic code in nuclear genes of *Tetrahymena. Proc. Natl. Acad. Sci. USA* 82:2452–2455.

Kirk, J. T. O. 1971. Chloroplast structure and biogenesis. *Annu. Rev. Biochem.* 40:161–196.

Kuroiwa, T., Suzuki, T., Ogawa, K., and Kawano, S. 1981. The chloroplast nucleus: Distribution, number, size, and shape, and a model for the multiplication of the chloroplast genome during chloroplast development. *Plant Cell Physiol.* 22:381–396.

Lander, E. S., and Lodish, H. 1990. Mitochondrial diseases: gene mapping and gene therapy. *Cell* 61:925–926.

Levings, C. S. 1983. The plant mitochondrial genome and its mutants. *Cell* 32:659–661.

Linnane, A. W., and Nagley, P. 1978. Mitochondrial genetics in perspective: The derivation of a genetic and physical map of the yeast mitochondrial genome. *Plasmid* 1:324–345.

Margulis, L. 1981. *Symbiosis in cell evolution.* San Francisco: W. H. Freeman.

Rochaix, J. D. 1978. Restriction endonuclease map of the chloroplast DNA of *Chlamydomonas reinhardi. J. Mol. Biol.* 126:597–617.

Rush, M. G., and Misra, R. 1985. Extrachromosomal DNA in eukaryotes. *Plasmid* 14:177–191.

Schmidt, R. J., Richardson, C. B., Gillham, N. W., and Boynton, J. E. 1983. Sites of synthesis of chloroplast ribosomal proteins in *Chlamydomonas. J. Cell Biol.* 96:1451–1463.

Umesono, K., and Ozeki, H. 1987. Chloroplast gene organization in plants. *Trends Genet.* 3:281–287.

Van Winkle-Swift, K. P., and Birky, C. W. 1978. The nonreciprocality of organelle gene recombination in *Chlamydomonas reinhardi* and *Saccharomyces cerevisiae. Mol. Gen. Genet.* 166:193–209.

Chapter 20: *Quantitative Genetics*

Blumer, M. G. 1980. *The mathematical theory of quantitative genetics.* Oxford: Clarendon Press.

Darwin, C. 1860. *On the origin of species by means of natural selection, or the preservation of favoured races in the struggle for life.* New York: Appleton.

Dobzhansky, T., and Pavlovsky, O. 1969. Artificial and natural selection for two behavioral traits in *Drosophila pseudoobscura. Proc. Nat. Acad. Sci. USA* 62:75–80.

East, E. M. 1910. A Mendelian interpretation of variation that is apparently continuous. *Am. Natur.* 44:65–82.

———. 1916. Studies on size inheritance in *Nicotiana. Genetics* 1:164–176.

———, and Jones, D. F. 1919. *Inbreeding and outbreeding.* Philadelphia: Lippincott.

Falconer, D. S. 1989. *Introduction to quantitative genetics.* New York: Wiley.

Lander, E. S., and Botstein, D. 1989. Mapping Mendelian factors underlying quantitative traits using RFLP linkage maps. *Genetics* 121:185–199.

Mather, K. 1943. Polygenic inheritance and natural selection. *Biol. Rev.* 18:32–64.

Paterson, A. H., Lander, E. S., Hewitt, J. D., Person, S., Lincoln, S. E., and Tanksley, S. D. 1988. Resolution of quantitative traits into Mendelian factors by using a complete RFLP linkage map. *Nature* 335:721–726.

Sokal, R. R., and Rohlf, F. J. 1981. *Biometry,* 2nd ed. San Francisco: W. H. Freeman.

Thoday, J. M. 1961. Location of polygenes. *Nature* 191:368–370.

Thompson, J. N., Jr., and Thompson, J. M., eds. 1979. *Quantitative genetic variation.* New York: Academic Press.

Weir, B. S., Eisen, E. J., Goodman, M. M., and Namkoong, G., eds. 1988. *Proceedings of the second international conference on quantitative genetics.* Sunderland, MA: Sinauer.

Chapter 21: *Population Genetics*

Avise, J. C. 1986. Mitochondrial DNA and the evolutionary genetics of higher animals. *Phil. Trans. Roy. Soc. London.* Ser. B 321:325–342.

Ayala, F. J., ed. 1976. *Molecular evolution.* Sunderland, MA: Sinauer.

Clarke, C. A., and Sheppard, P. M. 1966. A local survey of the distribution of industrial melanic forms in the moth *Biston betularia* and estimates of the selective values of these in an industrial environment. *Proc. R. Soc. Lond. [Biol.]* 165:424–439.

Crow, J. F. 1986. *Basic concepts in population, quantitative, and evolutionary genetics.* New York: W. H. Freeman.

Dobzhansky, T. 1951. *Genetics and the origin of species.* 3rd ed. New York: Columbia University Press.

Fisher, R. A. 1930. *The genetical theory of natural selection.* Oxford: Clarendon Press.

Ford, E. B. 1971. *Ecological genetics.* 3rd ed. London: Chapman and Hall.

Glass, B., Sacks, M. S., Jahn, E. F., and Hess, C. 1952. Genetic drift in a religious isolate: an analysis of the causes of variation in blood group and other gene frequencies in a small population. *Am. Naturalist* 86:145–159.

Gray, M. W. 1989. Origin and evolution of mitochondrial DNA. *Annu. Rev. Cell Biol.* 5:25–50.

Hardy, G. H. 1908. Mendelian proportions in a mixed population. *Science* 28:49–50.

Hartl, D. L., and Clark, A. G. 1988. *Principles of population genetics.* 2nd ed. Sunderland, MA: Sinauer.

Hedrick, P. H. 1983. *Genetics of populations.* Boston: Science Books International.

Hillis, D. M. and Moritz, C. 1990. *Molecular systematics.* Sunderland, MA: Sinauer.

———, Moritz, C., Porter, C. A., and Baker, R. J. 1991. Evidence for biased gene conversion in concerted evolution of ribosomal DNA. *Science* 251:308–309.

Kettlewell, H. B. D. 1961. The phenomenon of industrial melanism in the Lepidoptera. *Annu. Rev. Entomol.* 6:245–262.

Koehn, R., and Hilbish, T. J. 1987. The adaptive importance of genetic variation. *Amer. Sci.* 75:134–141.

Kreitman, M. 1983. Nucleotide polymorphism at the alcohol dehydrogenase locus of *Drosophila melanogaster. Nature* 304:412–417.

Lehman, N., Eisenhawer, A., Hansen, K., Mech, L. D., Peterson, R. O., Gogan, P. J., and Wayne, R. K. 1991. Introgression of coyote mitochrondrial DNA into sympatric North American gray wolf populations. *Evolution* 45:104–119.

Lewontin, R. C. 1974. *The genetic basis of evolutionary change.* New York: Columbia University Press.

———, Moore, J. A., Provine, W. B., and Wallace, B. 1981. *Dobzhansky's genetics of natural populations I-XLIII.* New York: Columbia University Press.

———. 1985. Population genetics. *Annu. Rev. Genet.* 19:81–102.

Li, W-H. 1991. *Fundamentals of molecular evolution.* Sunderland, MA: Sinauer.

———, Luo, C. C., and Wu, C. I. 1985. Evolution of DNA sequences, pp. 1—94. *Molecular evolutionary genetics,* edited by R. J. MacIntyre. New York: Plenum.

MacIntyre, R. J., ed. 1985. *Molecular evolutionary genetics.* New York: Plenum.

Maniatis, T., Fritsch, E. F., Lauer, L., and Lawn, R. M. 1980. The molecular genetics of human hemoglobin. *Annu. Rev. Genet.* 14:145–178.

Maynard Smith, J. 1989. *Evolutionary genetics.* Oxford: Oxford University Press.

Nei, M. 1987. *Molecular evolutionary genetics.* New York: Columbia University Press.

Nei, M., and Koehn, R. K. 1983. *Evolution of genes and proteins.* Sunderland, MA: Sinauer.

Pace, N. R., Olsen, G. J., and Woese, C. R. 1986. Ribosomal RNA phylogeny and the primary lines of evolutionary descent. *Cell* 45:325–326.

Schonewald-Cox, C. M., Chambers, S. M., MacBryde, B., and Thomas, L. 1983. *Genetics and conservation: a reference for managing wild animal and plant populations.* Menlo Park, CA: Benjamin/Cummings.

Selander, R. K., and Kaufman, D. W. 1975. Self fertilization and genetic population structure in a colonizing land snail. *Proc. Natl. Acad. Sci. USA* 70:1186–1190.

Soule, M. E., ed. 1986. *Conservation biology: the science of scarcity and diversity.* Sunderland, MA: Sinauer.

Speiss, E. 1977. *Genes in populations.* New York: Wiley.

Stringer, C. B. 1990. The emergence of modern humans. *Sci. Am.* (December):98–104.

Stringer, C. B., and Andrews, P. 1988. Genetic and fossil evidence for the origin of modern humans. *Science* 239:1263–1268.

Wallace, B. 1981. *Basic population genetics.* New York: Columbia University Press.

Woese, C. R. 1981. Archaebacteria. *Sci. Am.* 244:98–122.

Solutions to Selected Questions and Problems

Chapter 1

1.1   c

1.3   c

1.5   **a.** Yes, if a sexual mating system exists in that species. In that case two haploid cells can fuse to produce a diploid cell, which can then go through meiosis to produce haploid progeny. The fungi *Neurospora crassa* and *Saccharomyces cerevisiae* exemplify this positioning of meiosis in the life cycle.

   b. No, because a diploid cell cannot be formed and meiosis only occurs starting with a diploid cell.

1.7   **a.** metaphase

   b. anaphase

1.11 a. 1/8. The probability of a gamete receiving A is 1/2, because the choice is A or A′. Similarly, the probability of receiving B is 1/2 and of receiving C is 1/2. Therefore, the probability of receiving A, B, and C is $1/2 \times 1/2 \times 1/2 = 1/8$ (like the probability of tossing three heads in a row, the individual probabilities are multiplied together).

   b. 6/8 (= 3/4). The probability of gametes containing chromosomes from both maternal and paternal origin is 1 minus the probability that the gametes will contain either all maternal or all paternal chromosomes. The probability of a gamete containing all maternal chromosomes is 1/8 (part a) and that of a gamete containing all paternal chromosomes is also 1/8. Therefore the answer is $1 - 1/8 - 1/8 = 6/8 = 3/4$.

1.12 One of the long chromosomes and the short chromosome might be members of a heteromorphic pair, i.e., X and Y chromosomes, respectively.

1.13 False. Owing to the randomness of independent assortment and to crossing-over, both of which characterize meiosis, the probability of any two sperm cells being genetically identical is extremely remote.

Chapter 2

2.1   **a.** red

   b. 3 red, 1 yellow

   c. all red

   d. 1/2 red, 1/2 yellow

2.4   The F_2 genotypic ratio (if C is colored, c is colorless) is 1/4 CC:1/2 Cc:1/4 cc. If we consider just the colored plants, there is a 1:2 ratio of CC homozygotes to Cc heterozygotes. Therefore, if a colored plant is picked at random, the probability that it is CC is 1/3 (i.e., 1/3 of the *colored* plants are homozygous) and the probability that it is Cc is 2/3. Only if a Cc plant is selfed will more than one

phenotypic class be found among its progeny; therefore, the answer is 2/3.

2.5   **a.** Parents are *Rr* (rough) and *rr* (smooth); F_1 are *Rr* (rough) and *rr* (smooth).

   b. $Rr \times Rr \rightarrow$ 3/4 rough, 1/4 smooth

2.7   Progeny ratio approximates 3:1, and so the parent is heterozygous. Of the dominant progeny there is a 1:2 ratio of homozygous to heterozygous, so 1/3 will breed true.

2.8   Black is dominant to brown. If *B* is the allele for black and *b* for brown, then female X is *Bb* and female Y is *BB*. The male is *bb*.

2.10 a. The black rabbits of cross (a) could be either *BB* or *Bb*, but you can't tell which is which by looking at them. The following combinations of matings could occur: $BB \times BB$, $BB \times Bb$, and $Bb \times Bb$. The last is the only one that could produce white offspring. The probability that any given black rabbit is a heterozygote is

$$\frac{2Bb}{1\ BB + 2\ Bb} = \frac{2}{3}$$

Accordingly, the probability of selecting two heterozygotes by chance alone is $2/3 \times 2/3$, and the probability that this mating will produce a white offspring is 1/4. Therefore, the probability that in all the random F_1 matings, a white offspring is produced is $2/3 \times 2/3 \times 1/4 = 4/36 = 1/9$ So 1/9 of the $F_1 \times F_1$ progeny will be white.

   b. The only way available to him to produce a white offspring is to pick a heterozygous (*Bb*) male, the probability of which is 2/3, and mating it to the parental female, which we already know is *Bb*. The probability of getting a white rabbit from this cross is $2/3 \times 1 \times 1/4 = 1/6$.

   c. His best strategy would be to backcross a white male obtained from this cross to the parental females, thereby getting a 50% yield of white progeny.

2.13 a. $WW\ Dd \times ww\ dd$

   b. $Ww\ dd \times Ww\ dd$

   c. $ww\ DD \times WW\ dd$

   d. $Ww\ Dd \times ww\ dd$

   e. $Ww\ Dd \times Ww\ dd$

2.14 The cross is $Aa\ Bb\ Cc \times Aa\ Bb\ Cc$.

   a. Considering the *A* gene alone, the probability of an offspring showing the *A* trait from $Aa \times Aa$ is 3/4. Similarly, the probability of showing the *B* trait from $Bb \times Bb$ is 3/4 and the *C* trait from $Cc \times Cc$ is 3/4. Therefore, the probability of a given progeny being phenotypically ABC (using the product rule) is $3/4 \times 3/4 \times 3/4 = 27/64$.

b. Considering the A gene, the probability of an AA offspring from $Aa \times Aa$ is 1/4. The same probability is the case for a BB offspring and for a CC offspring. Therefore, the probability of an AA BB CC offspring (from the product rule) is $1/4 \times 1/4 \times 1/4 = 1/64$.

2.17 a. The F_1 has the genotype $Aa\ Bb\ Cc^h$ and is all agouti, black. The F_2 is 27/64 agouti, black; 9/64 agouti, black, Himalayan; 9/64 agouti, brown; 9/64 black; 3/64 agouti, brown, Himalayan; 3/64 black, Himalayan; 3/64 brown; 1/64 brown, Himalayan.

b. 27/64 of the F_2 are agouti, black. From the $F_1 \times$ F1 cross $Aa\ Bb\ Cc^h \times Aa\ Bb\ Cc^h$, the proportion of F_2 mice with the genotype $Aa\ BB\ Cc = 1/2 \times 1/4 \times 1/2 = 1/16 = 4/64$. Therefore, the proportion of F_2 agouti, blacks that are $Aa\ BB\ Cc^h = 4/64$ divided by $27/64 = 4/64 \times 64/27 = 4/27$.

c. F_2 Himalayans = 9/64 (agouti, black, Himalayan) + 3/64 (agouti, brown, Himalayan) + 3/64 (black, Himalayan) + 1/64 (brown, Himalayan) = 16/64. There are 3/64 + 1/64 = 4/64 brown, Himalayans; hence, the proportion of F_2 Himalayans that are brown = 4/64 divided by $16/64 = 4/64 \times 64/16 = 4/16 = 1/4$

d. F_2 agoutis = 27/64 (agouti, black) + 9/64 (agouti, black, Himalayan) + 9/64 (agouti, brown) + 3/64 (agouti, brown, Himalayan) = 48/64. There are 27/64 + 9/64 = 36/64 F_2 black, agoutis. Therefore, the proportion of F_2 agoutis that are black = 36/64 divided by $48/64 = 36/64 \times 64/48 = 36/48 = 3/4$

2.19 a. Mother must be heterozygous Aa in order to have children who have the trait.

b. Father is homozygous aa since he expresses the trait.

c. The cross is $Aa \times aa$, so children will be aa if they have the trait (II.2 and II.5) and Aa if they do not have the trait (II.1, II.3, and II.4).

d. From the cross $Aa \times aa$, the prediction is that 1/2 of the progeny will be Aa (normal) and 1/2 will be aa (expressing the trait). There are 5 children, two of whom have the trait and three of whom are normal. Thus the ratio fits as well as it could for 5 children.

Chapter 3

3.2 The woman is heterozygous for the recessive color-blindness allele, let us say c^+c. The man is not color-blind and thus has the genotype c^+Y. (It does not matter what phenotype his mother and father have because males are hemizygous for sex-linked genes; thus the genotype can be assigned directly from the phenotype.) All the female offspring will have normal color vision; half the male offspring will be color-blind, and half will have normal color vision.

3.3 If c is the red-green color-blind allele, and a is the albino allele, the parents' genotypes are c^+c^+ aa♀ and $cY\ a^+a^+$♂. All children will be normal visioned (c^+c^+♀ and c^+Y♂ and normally pigmented (a^+a, both sexes).

3.4 The parentals are $ww\ vg^+vg^+$♀ and $w^+Y\ vgvg$♂.

a. The F_1 males are all $wY\ vg^+vg$, white eyes, long wings. The F_1 females are all $w^+w\ vg^+vg$, red eyes, long wings.

b. The F_2 females are 3/8 red, long; 3/8 white, long; 1/8 red, vestigial; 1/8 white, vestigial; the same ratios of the respective phenotypes apply for the males.

c. The cross of an F_1 male with the parental female is $wY\ vg^+vg \times ww\ vg^+vg^+$. All progeny have white eyes and long wings. The cross of an F_1 female with the parental male is $w^+w\ vg^+vg \times w^+Y$ $vgvg$. Female progeny: All have red eyes, half have long wings and half have vestigial wings. Male progeny: 1/4 red, long; 1/4 red, vestigial; 1/4 white, long; 1/4 white, vestigial.

3.7 The simplest hypothesis is that brown-colored teeth are determined by a sex-linked dominant mutant allele. Man A was BY and his wife was bb. All sons will be bY normals and cannot pass on the trait. All the daughters receive the X chromosome from their father, and so they are Bb with brown enamel. Half their sons will have brown teeth because half their sons receive the X chromosome with the B mutant allele.

3.9 The answer is given in the following Punnett squares.

$$ww \times w^+Y$$

	Sperm	
	w^+	Y
Eggs: ww	www^+	wwY
Eggs: O	w^+ O	YO

Survivors are ww Y (white ♀) and w^+O (sterile, red ♂)

Backcross wwY $\times\ w^+$Y

		Sperm	
		w^+	Y
Normal	w Y	$w^+\ wY$ Red ♀	wYY White ♂
	w	$w^+\ w$ Red ♀	wY White ♂
Secondary nondisjunction	ww	$w^+\ ww$ Triplo-X Usually dies	wwY White ♀
	Y	w^+ Y Red ♂	YY Dies

Eggs

3.12

	PEDIGREE A	PEDIGREE B	PEDIGREE C
Autosomal recessive	Yes	Yes	Yes
Autosomal dominant	Yes	Yes	No
X-linked recessive	Yes	Yes	No
X-linked dominant	No	No	No

3.14 a. X-linked recessive can be excluded, for example, because individual I.2 would be expected to pass on the trait to all sons and that is not the case. Also, the II.1, II.2 pairing would be expected to produce offspring, all of whom express the trait.

Autosomal recessive can also be excluded because the II.1, II.2 pairing would be expected to produce offspring, all of whom express the trait.

b. The remaining two mechanisms of inheritance are X-linked dominant and autosomal dominant. Genotypes can be written to satisfy both mechanisms of inheritance. X-linked dominance is perhaps more likely, for the II.6, II.7 pairing shows exclusively father-daughter inheritance; that is, all the daughters and none of the sons exhibit the trait: this is a characteristic of the segregation of X-linked dominant alleles. If the trait were autosomal dominant, then half the sons and half the daughters would be expected to exhibit the trait.

3.16 a is false: the father passes on his X to his daughters so *none* of the sons will be affected.

b is false: since the disease is rare, the mother would be heterozygous and only 50% of her daughters would receive the X with the dominant mutant allele.

c is true: all daughters receive the father's X chromosome.

d is false: only 50% of her sons would receive the X with the dominant mutant allele.

3.19 Since hemophilia is an X-linked trait, the most likely explanation is that random inactivation of X chromosomes (lyonization, see p. 61) produces individuals with different proportions of cells with the normal allele. Thus if some women had only 40% of their cells with an active h^+ allele, and 60% with the h (hemophilia) allele, but other women had 60% of their cells with an active h^+ allele, and 40% with the h allele, there would be a significant difference in the amount of clotting factor these two individuals would make.

3.20 a. Females are XY, and males are XX. (The data also fit the model that males are homozygous for a recessive sex-determining allele while females are heterozygous for that and for a dominant, female-determining allele.) In the mating of estrogen-raised females to normal males, half of the matings were the usual XY♀ × XX♂, and gave the expected 50% XX and 50% XY offspring. The other half of matings were XX sex-reversed (♀) × normal XX (♂), and could produce only XX (♂) offspring.

In the matings of androgen-raised males to normal females, half of the matings were the usual XX × XY, but the other half were XY sex-reversed (♂) × XY (♀). These yielded 25% XX (♂), 50% XY (♀), and 25% YY (♀).

b. There would be two kinds. 75% of the matings would be XX × XY and yield 1 female:1 male. 25% of the matings would be XX × YY and would yield all XY (♀) offspring.

Chapter 4

4.2 6 possible genotypes: w/w, $w/w1$, $w/w2$, $w1/w2$, $w2/w2$.

4.4 The woman's genotype is I^A/I^B and the man's genotype is I^A/I^O.

   a. $1/2 \times 1/2 = 1/4$

   b. Zero. A blood group O baby is not possible.

   c. 1/2 (probability of male) × 1/4 (probability of AB) × 1/2 (probability of male) × 1/4 (probability of B) = 1/64 probability that all four conditions will be fulfilled.

4.6   a. The cross is $F/F\ G^N/G^N \times f/f G^O/G^O$. The F_1 is $F/f\ G^O/G^N$, which is fuzzy with round leaf glands.

   b. Interbreeding the F_1 gives 3/16 fuzzy, oval-glanded; 6/16 fuzzy, round-glanded; 3/16 fuzzy, no-glanded; 1/16 smooth, oval-glanded; 2/16 smooth, round-glanded; 1/16 smooth, no-glanded.

   c. Cross is $F/f\ G^O/G^N \times f/f\ G^O/G^O$. The progeny are 1/4 fuzzy, oval-glanded; 1/4 fuzzy, round-glanded; 1/4 smooth, oval-glanded; 1/4 smooth, round-glanded.

4.11 To show no segregation among the progeny, the chosen plant must be homozygous. The genotypes comprising the 9/16 colored plants are: 1 A/A B/B:2 A/a B/B:2 A/A B/b:4 A/a B/b. Only one of these genotypes is homozygous; the answer is 1/9.

4.13 a. If Y governs yellow and y governs agouti, then Y/Y are lethal, Y/y are yellow, and y/y are agouti. Let C determine colored coat and c determine albino. The parental genotypes, then, are Y/y C/c (yellow) and Y/y c/c (white).

   b. The proportion is 2 yellow:1 agouti:1 albino. None of the yellows breed true because they are all heterozygous, with homozygous YY individuals being lethal.

4.15 a. $Y/Y\ R/R$ (crimson) × $y/y\ r/r$ (white) gives $Y/y\ R/r$ F_1 plants, which have magenta-rose flowers. Selfing the F_1 gives an F_2 as follows: 1/16 crimson ($Y/Y\ R/R$), 2/16 orange-red ($Y/Y\ R/r$), 1/16 yellow ($Y/Y\ r/r$), 2/16 magenta ($Y/y\ R/R$), 4/16 magenta-rose ($Y/y\ R/r$), 2/16 pale yellow ($Y/y\ r/r$), and 4/16 white ($y/y\ R/R$, $y/y\ R/r$, and $y/y\ r/r$). Progeny of the F_1 backcrossed to the crimson parent are 1/4 crimson, 1/4 orange-red, 1/4 magenta, and 1/4 magenta-rose.

b. Y/Y R/r × Y/y r/r gives 1/4 orange-red (Y/Y R/r), 1/4 yellow (Y/Y r/r), 1/4 magenta-rose (Y/r R/r), 1/4 pale yellow (Y/y r/r).

c. Y/Y r/r (yellow) × y/y R/r (white) gives 1/2 magenta-rose (Y/y R/r) and 1/2 pale yellow (Y/y r/r).

4.17 a. The simplest approach is to calculate the proportion of progeny that will be black and then subtract that answer from 1. The black progeny have the genotype $A/-$ $B/-$ $C/-$, and the proportion of these progeny is $(3/4)^3$. Therefore the proportion of colorless progeny is $1 - (3/4)^3 = 1 - 27/64 = 37/64$.

b. Black is produced only when c/c $A/-$ $B/-$ results, and this offspring occurs with the frequency $(1/4)(3/4)(3/4) = 9/64$ black, which gives $55/64$ colorless.

4.19 In males H/H and H/h are horned, and h/h is hornless; in females H/H is horned, and H/h and h/h are hornless. The cross is a H/H W/W male × h/h w/w female. The F_1 is H/h W/w, which gives horned white males and hornless white females. Interbreeding the F_1 gives the following F_2:

	MALE	FEMALE
3/16 H/H $W/-$	horned, white	horned, white
6/16 H/h $W/-$	horned, white	hornless, white
3/16 h/h $W/-$	hornless, white	hornless, white
1/16 H/H w/w	horned, black	horned, black
2/16 H/h w/w	horned, black	hornless, black
1/16 h/h w/w	hornless, black	hornless, black

In sum, the ratios are 9/16 horned, white:3/16 hornless, white:3/16 horned, black:1/16 hornless, black males and 3/16 horned, white:9/16 hornless, white:1/16 horned, black:3/16 hornless, black females.

4.20 Ewe A is H/h w/w; B is H/h W/w or h/h W/w; C is H/H w/w; D is H/h W/w; the ram is H/h W/w.

Chapter 5

5.4 You were lucky in that a double mutant hatched. Based on the presence of one double mutant, you can eliminate linkage between vg and m. This is because there is no crossing-over in the *Drosophila* male, and there is no way to produce a recombinant vg m gamete in the male F_1. The crosses performed here were

$$P \qquad \frac{vg^+}{vg^+} \frac{m}{m} \, ♀ \times \frac{vg}{vg} \, ♂$$

$$\text{All } F_1 \qquad \frac{vg^+}{vg} \frac{m}{m^+} \text{ (wild type)}$$

The results of this cross told you that m is *not* X-linked.

Therefore the F_1 cross was a dihybrid cross and we expect a 9:3:3:1 F_2 (vg^+ m^+ [wild type]:vg m^+:vg^+ m:vg m) ratio. In the small sample, by chance, a double mutant appeared. This tells us that vg and m are segregating independently. If vg and m had been on the same chromosome, the crosses performed would have been

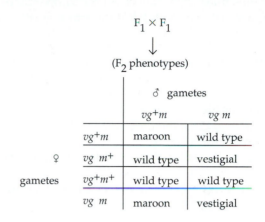

$$F_1 \times F_1$$
$$\downarrow$$
$$(F_2 \text{ phenotypes})$$

		♂ gametes	
		vg^+m	$vg\,m$
	vg^+m	maroon	wild type
♀	$vg\,m^+$	wild type	vestigial
gametes	vg^+m^+	wild type	wild type
	$vg\,m$	maroon	vestigial

The phenotypic ratio would be 4:2:2:0 (wild:vestigial:maroon:vestigial-maroon).

There is no way to get a double mutant from this cross *regardless* of the distance between the two linked genes (0-100 percent recombination). So if the two genes were linked one would never find a double mutant progeny in the F_2. Since a double mutant did hatch, this was enough to allow the preliminary (but correct) conclusion that m was not on chromosome 2. While m could be on chromosome 3 or 4, this experiment does not distinguish between these two possibilities.

5.5

$$
\begin{array}{ccccc}
c & e & a & d & b
\end{array}
$$

←7→←8→←——— 38 ———→←13→

5.6 45% a b^+, 45% a^+b, 5% a^+b^+, 5% a b

5.7 a. $0.035 + 0.465 = 0.50$

b. All the daughters have a^+b^+ phenotype.

5.8 a. The genotype of the F_1 is A B/a b. The gametes produced by the F_1 are 40% A B; 40% a b; 10% A b; 10% a B. From a testcross of the F_1 with a b/a b we get 40% A B/a b; 40% a b/a b, 10% A b/a b; 10% a B/a b.

b. The F_1 genotype is A b/a B. The gametes produced by the F_1 are 40% A b; 40% a B; 10% A B; 10% a b. From a testcross of the F_1 with a b/a b we get 40% A b/a b; 40% a B/ab; 10% A B/a b; 10% a b/a b. This question illustrates that map distance is computed between sites in the chromosome, and it does not matter whether the genes are in coupling or in repulsion, the percentage of recombinants will be the same, even though the phenotypic classes constituting the recombinants differ.

5.10 Each chromosome pair segregates independently. We can compute the relative proportions of gametes produced for each homologous pair of chromosomes separately from the known map distances (P = parental; R = recombinant):

P	R	P	R	P	R
A b 0.4	*A b* 0.1	*C D* 0.45	*C d* 0.05	*E F* 0.35	*E f* 0.15
a b 0.4	*a B* 0.1	*c d* 0.45	*c D* 0.05	*e f* 0.35	*e F* 0.15

To answer the questions, simply multiply the probabilities of getting the particular gamete from the F_1 multiple heterozygote.

    a. *A B C D E F* = 0.4 × 0.45 × 0.35 = 0.063 (6.3%)

    b. *A B C d e f* = 0.4 × 0.05 0.35 = 0.007 (0.7%)

    c. *A b c D E f* = 0.1 × 0.05 × 0.15 = 0.00075 (0.075%)

    d. *a B C d e f* = 0.1 × 0.05 × 0.35 = 0.00175 (0.175%)

    e. *a b c D e F* = 0.4 × 0.05 × 0.15 = 0.003 (0.3%)

5.11 a. 47.5% each of *DPh* and *dph*; 2.5% each of *Dph* and *dPh*.

    b. 23.75% each of *DpH*, *Dph*, *dPH*, and *dPh*; 1.25% each of *DPH*, *DPh*, *dpH*, and *dph*.

5.14 a. By doing a three-point mapping analysis as described in the chapter, we find that the order of genes in the chromosome is *dp-p-hk*, with 35.5 mu between *dp* and *b*, and 5.4 mu between *b* and *hk*.

    b. (1) The frequency of observed double crossovers is 1.4%, and the frequency of expected double crossovers is 1.9%. The coefficient of coincidence, therefore, is 1.4/1.9 = 0.73. (2) The interference value is given by 1 − coefficient of coincidence, 0.27.

5.16 a. $a^+ c b/a c^+ b^+$

    b. $a^+ c^+ b^+/Y$

    c.

5.17 a. *a*, *b*, *c*, and *d* are linked on the same chromosome, since the percentage of recombinations is less than 50. *e* is on a separate chromosome because it segregates independently from the four other genes. The map is constructed by calculating the recombination frequency for all possible pairs of *a*,

b, *c*, and *d*. The map distances allows an order to be determined. The map is shown below.

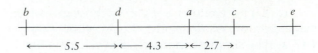

    b. To get an $a^+ b^+ c^+ d^+ e^+$ fly, there must be a crossover between *b* and *d*, *d* and *a*, and *a* and *c*. 1/2 of the progeny from those crossovers are $a^+ b^+ c^+ d^+$, and the other 1/2 are *a b c d*. The 1/2 are e^+. The answer is $1.6 × 10^{-5}$, that is, 0.055 × 0.043 × 0.027 × 0.5 (for the half of the progeny produced by the triple crossover that have all the wild-type alleles) × 0.5 (to give the proportion that are e^+).

5.22 From applying the tetrad analysis formula,

$$\text{map distance} = \frac{1/2T + \text{NPD}}{\text{Total}} × 100\%$$

the *a-b* distance is 19.6 map units, the *b-c* distance is 11 map units, and the *a-c* distance is 14 map units. Thus the gene order is *a-c-b* and the map is as shown:

Chapter 6

6.1 Strain A is *thr leu*$^+$ and B is *thr*$^+$ *leu*. Transformed B should be *thr*$^+$ *leu*$^+$, and its presence should be detected on a medium containing neither threonine nor leucine.

6.2

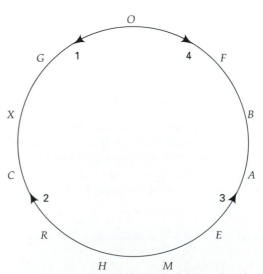

6.5 The *trpB* marker cotransduced with the *cysB*$^+$ selected marker more frequently than did the *trpE* marker; thus *trpB* is closer to *cysB*.

6.7 Only the volume of phage suspension plated is relevant to the calculation. 0.1 ml of the diluted phage suspension produced 20 plaques, meaning there were 200 phages/ml of the diluted suspension. The dilution factor was 10, and so the concentration in the original suspension was 200×10^8/mL or 2×10^{10}/mL.

6.8 0.07 mu. The plaques produced on $K12 (\lambda)$ are wild-type r^+ phages, and in the undiluted lysate there are 470×5 per milliliter (since 0.2 mL was plated), or 2350/mL. The r^+ phages were generated by recombination between the two rII mutations. The other product of the recombination event is the double mutant, and it does not grow on $K12(\lambda)$. Therefore the true number of recombinants in the population is actually equivalent to twice the number of r^+ phages, since for every wild-type phage produced, there ought to be a doubly mutant recombinant produced. Thus there are 4700 recombinants/mL. The total number of phages in the lysate is $627 \times$ (dilution factor) $\times$ (1 mL divided by the sample size plated) per milliliter, or $672 \times 1000 \times 10 = 6,720,000$/mL. The map distance between the mutations is $(4700/6,720,000) \times 1000 = 0.07\%$.

6.10 Two answers are compatible with the data:

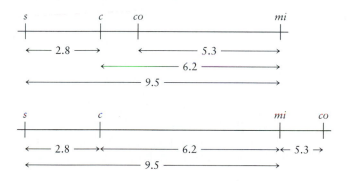

6.12 0.5% recombination (the numbers of plaques counted are so small that this value is a rather rough approximation).

6.15

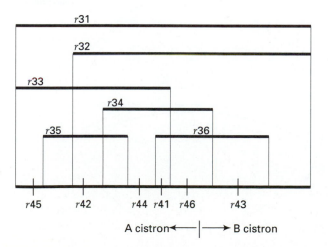

6.17 a. 3 genes

b. A, D, and F are in one; B and G are in the second; and C and E are in the third.

Chapter 7

7.2 The double homozygote should have PKU but not AKU. The PKU block should prevent most homogentisic acid from being formed, so that it could not accumulate to high levels.

7.4 a. The simplest approach is to calculate the proportion of the F_2 that are colored and to subtract that answer from one. In this case the noncolorless are the brown or black progeny. The proportion of progeny that make at least the brown pigment is given by the probability of having the following genotype: $a^+/- b^+/- c^+/- (d^+/d^+, d^+/d, or d/d)$. The answer is $3/4 \times 3/4 \times 3/4 = 27/64$. Therefore the proportion of colorless is $1 - 27/64 = 37/64$.

b. The brown progeny have the following genotype: $a^+/- b^+/- c^+/- d/d$. The probability of getting individuals with this genotype is $3/4 \times 3/4 \times 3/4 \times 1/4 = 27/256$.

7.6 a. The colorless F_2s are $a/a \ b/b \ c/c$ in genotype, the probability for which is $1/4 \times 1/4 \times 1/4 = 1/64$.

b. 67/256.

7.8 a. Half have white eyes (the sons) and half have fire-red eyes (the daughters).

b. All have fire-red eyes.

c. All have brown eyes.

d. All are w^+; a fourth are $bw^+/- st^+/-$, fire-red; a fourth are $bw^+/- st/st$, scarlet; a fourth are $bw/bw \ st^+/-$, brown; a fourth are $bw/bw \ st/st$, the color of 3-hydroxykynurenine plus the color of the precursor to biopterin, or colorless.

7.9 Wild-type T4 will produce progeny phages at all three temperatures. Let us suppose that model (1) is correct. If cells infected with the double mutant are first incubated at 17°C and then shifted to 42°C, progeny phages will then be produced and the cells will lyse. The explanation is as follows: The first step, A to B, is controlled by a gene whose product is heat-sensitive. At 17°C the enzyme works and A is converted to B, but B cannot then be converted to mature phages because that step is cold-sensitive. When the temperature is raised to 42°C, the A-to-B step is now blocked, but the accumulated B can be converted to mature phages for the enzyme involved with that step is cold-sensitive, and so the enzyme is functional at the high temperature. If model 2 is the correct pathway, then progeny phages should be produced in a 42°-to-17°C temperature shift, but not vice versa. In general, two gene product functions can be ordered by this method whenever one temperature shift allows phage production and the reciprocal shift does not, according to the following rules: (1) If a low-to-high temperature results in phages but a high-to-low temperature does not, then the *hs* step precedes the *cs* step (model 1); (2) If a high-to-low temperature results in phages but a low-to-high temperature does not, then the *cs* step precedes the *hs* step (model 2).

7.11 A mutant cell can take up a substance and grow on it only if that substance occurs in the metabolic pathway at a point after the mutant's own block. Since *trpE* can be fed by *C, D, F* and *A, E* is blocked earlier in the pathway than the others. *E* is also earlier than *B*, since *B* can feed *F*. Thus anthranilate synthetase is the first enzyme in the pathway. *F* can feed *C* and *C* can feed *D*. Since *A* and *B* both feed *F*, the order of the steps is *E, D, C, F, [AB]*. After anthranilate synthetase, the enzymes are, in order, PPA transferase, PRA isomerase, IGP synthetase, and tryptophan synthetase.

7.12 a. *c d$^+$* and *c$^+$ d*

b. The genes are not linked because parental ditype (PD) and nonparental ditype (NPD) tetrads occur in equal frequencies.

c. The pathway is Y to X to Z, with *d* blocking the synthesis of Y and *c* blocking the synthesis of X and Y.

Chapter 8

8.2 d

8.3 3′ and 5′ carbons are connected by a phosphodiester bond.

8.7 a. 3′ TCAATGGACTAGCAT 5′
    b. 3′ AAGAGTTCTTAAGGT 5′

8.8 b, c, and d

8.11 C = 17% therefore G = 17% and the % GC = 34. Therefore, the % AT = 66 and the % A = 66/2 = 33.

8.17 a. 200,000
    b. 10,000
    c. 3.4×10^4 nm

8.18 The probability any group of four bases would be GUUA is $(0.3)(0.25)(0.25)(0.2) = 0.00375$. A molecule 10^6 nucleotides long contains very nearly 10^6 groups of four bases. (The first group of four is bases 1, 2, 3 and 4, the second group of four is bases 2, 3, 4 and 5, etc.) Thus the number of occurrences of GUUA should be about $(0.00375)(10^6) = 3750$.

Chapter 9

9.2 a. 9/24 = 0.375. Since S is 37.5% of the cycle, 37.5% of cells should be in S at any instant, and these are the ones that would incorporate label.

b. A little over 4 h. Such cells would have to

take up some ^{3}H (a few minutes), pass through G$_2$ (4 h), and pass through mitotic prophase (several minutes).

c. Both. Each chromatid is a double-stranded DNA molecule containing one old and one newly synthesized DNA strand.

d. A little more than 13 h. Such cells would have to pass through nearly all of S, all of G$_2$, and through prophase.

9.6

DNA Type	Chromatin Type
Barr body (inactivated DNA)	Facultative heterochromatin
Centromere	Constitutive heterochromatin
Telomere	Constitutive heterochromatin
Most expressed genes	Euchromatin

9.11 Most protein-coding genes are found in unique-sequence DNA.

Chapter 10

10.2 Key: ^{15}N-^{15}N DNA = HH; ^{15}N-^{14}N DNA = HL; ^{14}N-^{14}N DNA = LL.

a. Generation 1: all HL; 2: 1/2 HL, 1/2 LL; 3: 1/4 HL, 3/4 LL; 4: 1/8 HL, 7/8 LL; 6: 1/32 HL, 31/32 LL; 8: 1/128 HL, 127/128 LL

b. Generation 1: 1/2 HH, 1/2 LL; 2: 1/4 HH, 3/4 LL; 3: 1/2 HH, 7/8 LL; 4: 1/16 HH, 15/16 LL; 6: 1/64 HH, 63/64 LL; 8: 1/256 HH, 255/256 LL.

10.7 a. One base pair is 0.34 nm, and the chromosome is 1100 μm. Thus the number of base pairs is $(1100/0.34) \times 1000 = 3.24 \times 10^6$.

b. There are 10 base pairs per turn in a normal DNA double helix; therefore it has a total of 3.24×10^5 turns.

c. 3.24×10^5 turns and 60 min for unidirectional synthesis; therefore $(3.24 \times 10^5/60)$ turns per minute = 5400 revolutions per minute.

d.

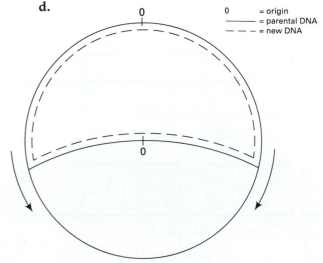

0 = origin
—— = parental DNA
– – – = new DNA

10.9 See text, pp. 213–218.

10.13 Assuming these cells have the typical 4 h G_2 period, you would add ^{3}H thymidine to the medium, wait 4.5 h, and prepare a slide of metaphase chromosomes. Autoradiography would then be done on this chromosome preparation. Regions of chromosomes displaying silver grains are then the late-replicating regions. Cells which were at earlier stages in the S period when they began to take up ^{3}H will not have had sufficient time to reach metaphase.

10.14 See text, pp. 226–227. Hybrid DNA resulting from branch migration may contain mismatched sequences. Mismatched sequences are recognized by special enzymes that remove a short segment of one DNA strand and then fill in the gap. The mismatch repair does not recognize which is the correct sequence.

10.15 In each case there is evidence that the segregation of one of the alleles in the tetrad has resulted from gene conversion caused by mismatch repair of heteroduplex DNA.

For the $a1$ $a2^+$ × $a1^+$ $a2$ cross the tetrad shows 2:2 segregation of $a1^+$:$a1$ but 3:1 segregation of $a2^+$:$a2$, indicating gene conversion of an $a2$ allele to its wild-type counterpart. Similarly, the $a1$ $a3^+$ × $a1^+$ $a3$ cross shows 2:2 segregation of $a3^+$:$a3$ and 3:1 segregation of $a1^+$:$a1$ resulting from gene conversion of an $a1$ allele to $a1^+$. In the $a2$ $a3^+$ × $a2^+$ $a3$ cross, the $a2$ allele segregates in a Mendelian fashion while the $a3$ allele segregates 3:1 $a3^+$:$a3$ as a result of gene conversion of one $a3$ allele to $a3^+$.

Chapter 11

11.1 The DNA contains deoxyribose and thymine, whereas RNA contains ribose and uracil, respectively. Also, DNA is usually double-stranded, but RNA is usually single-stranded.

11.5 See text, pp. 236; 240–242.

11.6 RNA polymerase I transcribes the major rRNA genes that code for 18S, 5.8S, and 28S rRNAs; RNA polymerase II transcribes the protein-coding genes to produce mRNA molecules and some of the genes for small nuclear RNA (snRNA) molecules; and RNA polymerase III transcribes the 5S rRNA genes, the tRNA genes, and the other genes for snRNA molecules.

In the cell the 18S, 5.8S, 28S, and 5S rRNAs are structural components of ribosomes. The mRNAs are translated to produce proteins, the tRNAs bring amino acids to the ribosome to donate to the growing polypeptide chain during protein synthesis, and at least some of the snRNAs are involved in RNA processing events.

11.8 An enhancer element is defined as a DNA sequence that somehow, without regard to its position relative to the gene or its orientation in the DNA, increases the amount of DNA synthesized from the gene it controls (discussed in the chapter on p. 241).

11.11 The three major classes of RNA are mRNA (pp. 243–248), tRNA, (pp. 257–259) and rRNA (pp. 248–257).

11.13 For mRNA: 5′ capping and 3′ polyadenylation in eukaryotes. For tRNA: modification of a number of the bases, removal of 5′ leader and 3′ trailer sequences (if present), and addition of a three-nucleotide sequence (5′-CCA-3′) at the 3′ end of the tRNA. For rRNA in prokaryotes a pre-rRNA molecule is processed to the mature 16S, 23S, and 5S rRNAs. In eukaryotes a precursor molecule is processed to the 18S, 5.8S, and 28S rRNA molecules; 5S rRNA is made elsewhere. The large rRNAs in both types of organisms are methylated. Additionally, the same rRNAs in eukaryotes are pseudouridylated.

11.16 a. This would prevent splicing out of any sequences corresponding to introns, and would result in abnormal sequences of most proteins. This should be lethal.

b. If splicing is not affected, this should have no phenotypic consequence.

c. This would prevent the splicing out of intron 2 material. Since the 5′ cut could still be made, it is likely this mutation would lead to the absence of functional mRNA and the absence of β-globin. (This should produce a phenotype similar to that seen when the β-globin gene is deleted. In that case there is anemia compensated by increased production of fetal hemoglobin. The condition is called β-thalassemia.)

11.18 a. This would be possible because the 5S genes occur in clusters apart from the other rRNA genes.

b. This would not be possible, because the 18S gene copies are interspersed among 5.8S and 28S genes.

c. This would be possible. These three occur in tandem arrays.

d. This would not be possible, because the 5S genes are located separately within the chromosomes.

Chapter 12

12.1 b mRNA

12.9 a. 4 G:1 C

GGG = (4/5)(4/5)(4/5) = 0.512, or 51.2% Gly
GGC = (4/5)(4/5)(1/5) = 0.128, or 12.8% Gly
CCG = (4/5)(1/5)(4/5) = 0.128, or 12.8% Ala
CGG = (1/5)(4/5)(4/5) = 0.128, or 12.8% Arg
CCC = (1/5)(1/5)(1/5) = 0.008, or 0.8% Pro
CCG = (1/5)(1/5)(4/5) = 0.032, or 3.2% Pro
CGC = (1/5)(4/5)(1/5) = 0.032, or 3.2% Arg
GCC = (4/5)(1/5)(1/5) = 0.032, or 3.2% Ala

In sum, 64.0% Gly, 16.0% Ala, 16.0% Arg, and 4.0% Pro.

   b. 1 A:3 U:1 C; the same logic is followed here, using 1/5 as the fraction for A, 3/5 for U, and 1/5 for C.

AAA = 0.008, or 0.8% Lys
AAU = 0.024, or 2.4% Asn
AUA = 0.024, or 2.4% Ile
UAA = 0.024, or 2.4% Chain terminating
AUU = 0.072, or 7.2% Ile
UAU = 0.072, or 7.2% Tyr
UUA = 0.072, or 7.2% Leu
UUU = 0.216, or 21.6% Phe
AAC = 0.008, or 0.8% Asn
ACA = 0.008, or 0.8% Thr
CAA = 0.008, or 0.8% Gln
ACC = 0.008, or 0.8% Thr
CAC = 0.008, or 0.8% His
CCA = 0.008, or 0.8% Pro
CCC = 0.008, or 0.8% Pro
UUC = 0.072, or 7.2% Phe
UCU = 0.072, or 7.2% Ser
CUU = 0.072, or 7.2% Leu
UCC = 0.024, or 2.4% Ser
CUC = 0.024, or 2.4% Leu
CCU = 0.024, or 2.4% Pro
UCA = 0.024, or 2.4% Ser
UAC = 0.024, or 2.4% Tyr
CUA = 0.024, or 2.4% Leu
CAU = 0.024, or 2.4% His
AUC = 0.024, or 2.4% Ile
ACU = 0.024, or 2.4% Thr

In sum, 0.8% Lys, 3.2% Asn, 12.0% Ile, 2.4% Chain terminating, 9.6% Tyr, 19.2% Leu, 28.8% Phe, 4.0% Thr, 0.8% Gln, 3.2% His, 4.0% Pro, and 12.0% Ser. The likelihood is that the chain would not be long because of the chance of the chain-terminating codon.

12.10

WORD SIZE	NUMBER OF COMBINATIONS
a. 5	2^5 5 32
b. 3	3^3 5 27
c. 2	5^2 5 25

(The minimum word sizes must uniquely designate 20 amino acids.)

12.12 a. AAA ATA AAA ATA etc.
   b. TTT TAT TTT TAT etc.
   c. AAA for Phe and AUA for Tyr

12.14 No chain-terminating codons can be produced from only As and Gs. On the other hand, the stop codon UAA can be made from As and Us. Therefore the population A proteins will be longer than those from population B. Most of the population B proteins will be free in solution rather than attached to ribosomes.

12.16 Met-Val-Ser-Ser-Pro-Ile-Gly-Ala-Ala-Ile-Ser . . . (In fact either of the Ile residues might be replaced by Met in a particular molecule.) The normal tRNA recognizes the AUC codon, and therefore carries Ile. The mutant tRNA will recognize the codon AUG and insert Ile there (although the normal tRNA-Met will compete for these sites). The N terminal Met will not be replaced by Ile since it requires the special tRNA-fMet for initiation to occur.

12.18

Normal: Met-Phe-Ser-Asn-Tyr- · · · · -Met-Gly-Trp-Val.
Mutant a: Met-Phe-Ser-Asn.
Mutant b: Starts at later AUG to give: Met-Gly-Trp-Val.
Mutant c: Met-Phe-Ser-Asn-Tyr- · · · · -Met-Gly-Trp-Val.
Mutant d: Met-Phe-Ser-Lys-Tyr- · · · · -Met-Gly-Trp-Val.
Mutant e: Met-Phe-Ser-Asn-Ser- · · · · -Trp-Gly-Gly-Cys . . .
                    (no stop codon; protein continues)
Mutant f: Met-Phe-Ser-Asn-Tyr- · · · · -Met-Gly-Trp-Val-Trp . . .
                    (no stop codon; protein continues)

Chapter 13

13.1 At any one position in the DNA, there are four possibilities for the base pair: A-T, T-A, G-C, and C-G. Therefore the length of the base-pair sequence that the enzyme recognizes is given by the power to which four must be raised to equal (or approximately equal) the average size of the DNA fragment produced by enzyme digestion. The answer in this case is 6; that is, 4 to the power of 6 = 4,096. If the enzyme instead recognized a four base-pair sequence, the average size of the DNA fragment would be 4 to the power of 4 = 256.

13.5 Genomic libraries were discussed on p. 293, and cloning in a lambda vector by replacing a central section of the λ chromosome with foreign DNA (see pp. 291–292). The steps to make a yeast genomic library are:

1. Isolate high-molecular-weight DNA from yeast nuclei;
2. Perform a partial digest of the DNA with a restriction enzyme that cuts frequently (e.g., *Sau*3A), and isolate DNA fragments of the correct size range for cloning in the λ vector by sucrose gradient centrifugation;
3. Remove the central section of an appropriate λ vector by digestion with *Bam*HI;
4. Ligate the left and right λ arms to the yeast DNA fragments—the *Sau*3A and *Bam*HI sticky ends are complementary (see p. 000);
5. Package the recombinant DNA molecules *in vitro* into λ phage particles;
6. Infect *E. coli* cells with the λ phage population, and collect progeny phages produced by cell lysis. These phages represent the yeast genomic library.

13.6 Use cDNA. Human genomic DNA contains introns. mRNA transcribed off genomic DNA needs to be processed before it can be translated. Bacteria do not contain these processing enzymes. So in the bacterium, even if the human mRNA were translated, the protein it would code for would not be insulin. cDNA is the complementary copy of a functional mRNA molecule. So when this is the template, its mRNA transcript will be functional and when translated, human (pro-) insulin will be synthesized.

13.9 The restriction map is shown below in Figure A. The map is built by considering the large fragments first. For example, the 2500-bp B fragment is cut with A to produce 1900-bp and 600-bp fragments. The 2100-bp A fragment is cut with B to produce the same 1900-bp fragment, and a 200-bp fragment. Thus the 200-bp and 600-bp fragments must be on opposite sides of the 1900-bp fragment. The map is extended in a step-by-step fashion, by next considering other cuts which produced 200-bp fragments, 600-bp fragments, and so on.

13.10 The sequencing gel banding pattern is shown in the following figure:

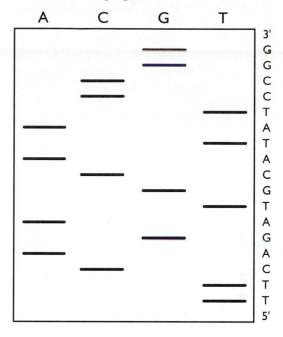

13.12 I-1 is heterozygous for the RFLP and the disease. Inspection of the blot figure (Figure 13B, p. 317) shows three different "haplotypes" (DNA types) are segregating in this family. Let us name them A, B, and C in decreasing order of size. Let us use *d* for the normal allele of the disease gene, and *D* for the allele causing the disease. I-1 is then *Dd* and AB, but we don't know whether he is *DA/dB* or *DB/dA*, so we cannot tell which of his offspring are recombinant and which are nonrecombinant. I-2 is *dB/dB*, so I-1 must be *DA/dB* in order to produce *dB/dB* progeny, assuming linkage. Given the genotypes of I-1 and I-2, the genotypes of II-2 and II-5 must both be *DA/dB* and II-3 and II-4 must both be *dB/dB*. Thus all four generation II progeny of I-1 and I-2 are nonrecombinant. II-2 (*DA/dB*) pairs with II-1 (*dC/dC*) and has six offspring.

■ *Figure A*

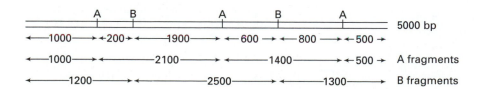

III-1 is dB/dC, III-2 is DA/dC, III-4 is DA/dC, III-5 is dB/dC, III-6 is dA/dC, and III-7 is DA/dC. Of these offspring, only III-6 is recombinant because it is not DA. The pairing of II-5 (DA/dB) and II-6 (dA/dA) produces six offspring, of which only III-12 (dA/dA) and III-13 (DB/dA) are recombinants. That is, III-12 has received the recombinant chromosome dA and III-13 has received the recombinant chromosomes DB. The pairing of III-2 (DA/dC) with III-3 (dB/dB) produces three nonrecombinant progeny: IV-1 is DA/dB, IV-2 is DA/dB, and IV-3 is dC/dB. The pairing of III-13 (DB/dA) with III-14 (dC/dC) produces four offspring, of which only IV-6 is recombinant (dB/dC). Overall there are 19 individuals whose recombinant or nonrecombinant status can be ascertained. Of these, four (III-6, III-12, III-13, and IV-6) are recombinant. 4/19 = 0.21, which is less than the 0.5 expected from independent assortment. But is it significantly less? To find out, we must do the χ^2 test. Independent assortment predicts 19/2 = 9.5 recombinants and the same number of non-recombinants. The value of χ^2 is thus 6.368, with one degree of freedom. The χ^2 table (Chapter 5, p. 102) indicates that this difference is significant, so we can conclude the two loci are linked.

Chapter 14

14.3 A constitutive phenotype can be the result of either a $lacI^-$ or a $lacO^c$ mutation.

14.5 $lacI^+ lacO^c lacZ^+ lacY^-/lacI^+ lacO^+ lacZ^- lacY^+$ (It cannot be ruled out that one of the repressor genes is $lacI^-$)

14.7 Answer is in Table 14.A.

14.9 The CAP, in a complex with cAMP, is required to facilitate RNA polymerase binding to the lac promoter. The RNA polymerase binding occurs only in the absence of glucose and only if the operator is not occupied by a repressor (i.e., if lactose is absent). A mutation in the CAP gene, then, would render the lac operon incapable of expression because RNA polymerase would not be able to recognize the promoter.

14.13 The cI gene codes for repressor protein that functions to keep the lytic functions of the phage repressed when lambda is in the lysogenic state. A cI mutant strain would be unable to establish lysogeny, and thus it would also follow the lytic pathway.

Chapter 15

15.1 The final product of the rRNA genes is an rRNA molecule. Hence a large number of genes are required to produce the large number of rRNA molecules required for ribosome biosynthesis. Ribosomal proteins, in contrast, are the end products of the translation of mRNAs, which can be "read" over and over to produce the large number

Table 14.A

	GENOTYPE	INDUCER ABSENT		INDUCER PRESENT	
		B-GALACTOSIDASE	PERMEASE	B-GALACTOSIDASE	PERMEASE
a.	$lacI^+$ $lacO^+$ $lacZ^+$ $lacY^+$	−	−	+	+
b.	$lacI^+$ $lacO^+$ $lacZ^-$ $lacY^+$	−	−	−	+
c.	$lacI^+$ $lacO^+$ $lacZ^+$ $lacY^-$	−	−	+	−
d.	$lacI^-$ $lacO^+$ $lacZ^+$ $lacY^+$	+	+	+	+
e.	$lacI^s$ $lacO^+$ $lacZ^+$ $lacY^+$	−	−	−	−
f.	$lacI^+$ $lacO^c$ $lacZ^+$ $lacY^+$	+	+	+	+
g.	$lacI^s$ $lacO^c$ $lacZ^+$ $lacY^+$	+	+	+	+
h.	$lacI^+$ $lacO^c$ $lacZ^+$ $lacY^-$	+	−	+	−
i.	$lacI^-$ $lacO^+$ $lacZ^+$ $lacY^+/$ $lacI^+ lacO^+ lacZ^- lacY^-$	−	−	+	+
j.	$lacI^-$ $lacO^+$ $lacZ^+$ $lacY^-/$ $lacI^+ lacO^+ lacZ^- lacY^+$	−	−	+	+
k.	$lacI^s$ $lacO^+$ $lacZ^+$ $lacY^-/$ $lacI^+ lacO^+ lacZ^- lacY^+$	−	−	−	−
l.	$lacI^+$ $lacO^c$ $lacZ^-$ $lacY^+/$ $lacI^+ lacO^+ lacZ^+ lacY^-$	−	+	+	+
m.	$lacI^s$ $lacO^+$ $lacZ^+$ $lacY^+/$ $lacI^+ lacO^c lacZ^+ lacY^+$	+	+	+	+

of ribosomal protein molecules required for ribosome biosynthesis.

15.3 The data show clearly that the synthesis of ovalbumin is dependent on the presence of the hormone estrogen. The data do not indicate at what level estrogen works. Theoretically, it could act to increase transcription of the ovalbumin gene, to stabilize the ovalbumin mRNA, to stabilize the ovalbumin protein, or to stimulate transport of the ovalbumin mRNA out of the nucleus. Experiments in which the levels of the ovalbumin mRNA were measured have shown that the production of ovalbumin is primarily regulated at the level of transcription.

15.6 four: A_3, A_2B, AB_2, B_3

15.9 Preexisting mRNAs (stored inthe oocyte) are recruited into polysomes as development begins following fertilization.

15.11

MUTANT	CLASS
a	segmentation gene (segment polarity)
b	maternal gene (anterior-posterior gradient)
c	segmentation gene (gap)
d	homeotic (eye to wing transdetermination)
e	segmentation gene (gap)

15.12 Preexisting mRNA that was made by the mother and deposited in the egg prior to fertilization is translated up until the gastrula stage. After gastrulation, new mRNA synthesis is necessary for production of proteins needed for subsequent embryo development.

15.13 The tissue taken from the blastula/gastrula has not yet been committed to its final differentiated state in terms of its genetic programming; that is, it is not yet determined. Thus when it is transplanted into the host, the determined tissues surrounding the transplant in the host communicate with the transplanted issue and cause it to be determined in the same way as they are; for example, tissue transplanted to a future head area will differentiate into head material, and so on. In contrast, tissues in the neurula stage are now determined as to their final tissue state once development is complete. Thus tissue transplanted from a neurula to an older embryo cannot be influenced by the determined, surrounding tissues and will develop into the tissue type for which it is determined, in this case an eye.

Chapter 16

16.1 e. None of these (i.e., all are classes of mutations).

16.3 a. The normal anticodon was 5′-CAG-3′,

while the mutant one is 5′-CAC-3′. The mutational event was a CG-to-GC transversion.

b. Presumably Leu

c. Val

d. Leu

16.5 Acridine is an intercalating agent, and so can be expected to induce frameshift mutations. 5BU, on the other hand, is incorporated in place of T, but is relatively likely to be read as C by DNA polymerase because of keto to enol shift. Thus 5BU induced mutations would be expected to be point mutations, usually TA-to-CG transitions. If these expectations are realized, *lacz-1* would probably contain a single amino acid difference from the normal β-galactosidase, although it could be a truncated normal protein due to a nonsense point mutation. *lacz-2*, on the other hand, should have a completely altered amino acid sequence after some point, and might also be truncated.

16.6 a. Six.

b. Three. The mutation results in replacement of the UGG codon in the mRNA by UAG, which is a chain termination codon.

16.7 a. UAG: CAG Gln; AAG Lys; GAG Glu; UUG Leu; UCG Ser; UGG Trp; UAU Tyr; UAC Tyr; UAA chain terminating.

b. UAA: CAA Gln; AAA Lys; GAA Glu; UUA Leu; UCA Ser; UGA chain terminating; UAU Tyr; UAC Tyr; UAG chain terminating.

c. UGA: CGA Arg; AGA Arg; GGA Gly; UUA Leu; UCA Ser; UAA chain terminating; UGU Cys; UGC Cys; UGG Trp

16.9

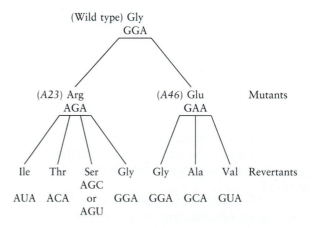

16.11 5BU in its normal form is a T analog; in its rare form it resembles C. The mutation is an AT-to-GC transition.

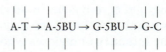

A-T → A-5BU → G-5BU → G-C

Chapter 17

17.1 a. pericentric inversion (DoE F inverted)

b. nonreciprocal translocation (BC moved to other arm)

c. tandem duplication (EF duplicated)

d. reverse tandem duplication (EF duplicated)

e. deletion (C deleted)

17.2 See text, p. 408

17.4 a.

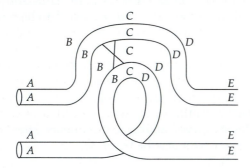

b. From part a the crossover between B and C is as follows:

c. Paracentric inversion, because the centromere is not included in the inverted DNA segment.

17.6 a. Mr. Lambert is heterozygous for a pericentric inversion of chromosome 6. One of the breakpoints is within the fourth light band up from the centromere, and the other is in the sixth dark band below the centromere. Mrs. Lambert's chromosomes are normal.

b. When Mr. Lambert's chromosome 6s paired, they formed an inversion loop which included the centromere. Crossing-over occurred within the loop, and gave rise to the partially duplicated, partially deficient 6, which the child received.

c. The child's abnormalities stem from having three copies of some, and only one copy of other, chromosome 6 regions. The top part of the short arm is duplicated, and there is a deficiency of the distal part of the long arm in this case.

d. The inversion appears to cover more than half of the length of chromosome 6, so crossing-over will occur in this region in the majority of meioses. In the minority of meioses where crossing-over has occurred outside the loop or where it has

occurred within the loop but the child receives an uncrossed-over chromatid, the child can be normal. There is significant risk for abnormality, so monitoring of fetal chromosomes should be done.

17.7 a. He has a paracentric inversion within the long arm of one of his number 12 chromosomes.

b. Crossing-over within the inversion loop should produce dicentric chromatids, which would form anaphase bridges. These chromatid bridges would be visible where they join the two chromatid masses at anaphase 1.

c. The inversion is large, so bridges will be formed in the majority of meioses. Cells that form a bridge do not complete meiosis or form sperm in mammals.

d. Nothing can be done to increase his sperm count.

17.8

$$a \longrightarrow c \longrightarrow e \longrightarrow d$$
$$\searrow$$
$$b$$

That is, the following sequence occurred:

a. *A B C D E F G H I*

⟶ Inversion

c. *A B F E D C G H I*

⟶ Inversion

e. *A B F E H G C D I*

d and b are derived from e by separate inversion events:

e. *A B F E H G C D I*

⟶ Inversion

d. *A B F C G H E D I*

and

e. *A B F E H G C D I*

⟶ Inversion

b. *H E F B A G C D I*

17.9

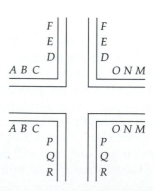

17.10 a. Mr. Denton has normal chromosomes. Mrs. Denton is heterozygous for a balanced reciprocal translocation between chromosomes 6 and 12. Most of the short arm of chromosome 6 has been reciprocally translocated onto the long arm of chromosome 12. The breakpoints appear to be in the thick dark band just above the centromere of 6 and in the third dark band below the centromere of 12.

b. The child received a normal 6 and a normal 12 from his father. In meiosis in Mrs. Denton, the 6s and 12s paired and formed a crosslike figure in prophase I, segregation of adjacent nonhomologous centromeres to the same pole occurred, and the child received a gamete containing a normal 6 and one of the translocation chromosomes. (See Figure 17.11.)

c. The child received a normal chromosome 6 and a normal chromosome 12 from Mr. Denton. The child is not phenotypically normal because of partial trisomy (it has three copies of part of the short arm of chromosome 6), and partial monosomy (it has only one copy of most of the long arm of chromosome 12).

d. Given that segregation of adjacent homologous centromeres to the same pole will be relatively rare in this case, there should be about a 50% chance that a given conception will be chromosomally unbalanced. Of the 50% balanced ones, 50% would be translocation heterozygotes.

e. Prenatal monitoring of fetal chromosomes could be done, followed by therapeutic abortion of chromosomally unbalanced fetuses.

17.13 a. 45

b. 47

c. 23

d. 69

e. 48

17.14 b. An individual with three instead of two chromosomes is said to be *trisomic*.

17.15 a. Mother. The color-blind Turner syndrome child must have received her X chromosome from her father because the father carries the only color-blind allele. Therefore, the mother must have produced an egg with no X chromosomes.

b. Father. Because the Turner syndrome child does not have color blindness, she must have received her X chromosome from the homozygous normal mother. Therefore, the father must have produced a sperm with no X or Y chromosome.

17.16 a. *AA aa*

b. If we label the four alleles *A1, A2, a1,* and *a2,* there are 6 possible gamete types: *A1 A2, A1 a1, A1 a2, A2 a1, A2 a2,* and *a1 a2.* i.e., 1/6 *AA*, 4/6 *Aa*, and 1/6 *aa*. The possible gamete pairings are as follows:

	1/6 AA	4/6 Aa	1/6 aa
1/6 AA	1/36 AAAA	4/36 AAAa	1/36 AAaa
4/6 Aa	4/36 AAAa	16/36 AAaa	4/36 Aaaa
1/6 aa	1/36 AAaa	4/36 Aaaa	1/36 aaaa

Phenotypically this gives 35/36 *A*: 1/36 *a*

17.18 (7N + 10N) × 2 = 34 (4N)

Chapter 18

18.5 Since a transposition results in the loss of the intron, it may be hypothesized that transposition in A occurs via an RNA intermediate. That is, intron removal occurs only at the RNA level.

The lack of intron removal during B transposition suggests that in B there is a DNA-DNA transposition mechanism, without any RNA intermediate.

18.6 In engineering the retrovirus in the way he plans, the investigator would probably be creating a new cancer virus, in which the cloned nerve growth factor gene would be the oncogene. It is, of course, an advantage that the engineered virus would not be able to reproduce itself, but we know that many "wild" cancer viruses are also defective and reproduce with the help of other viruses. If the engineered virus were to infect cells carrying other viruses (for example wild-type versions of itself) that could supply the *env* and *gag* functions, the new virus could be reproduced and spread. Presumably, infection of normal nerve cells *in vivo* by the engineered retrovirus would sometimes result in abnormally high levels of nerve growth factor, and thus perhaps in the production of nervous system cancers.

In Avian myeloblastosis virus (AMV) the *pol* and *env* genes are partially deleted. Thus like our investigator's virus, AMV needs a helper virus to reproduce. In AMV the *myb* oncogene has been inserted, which encodes a nuclear protein presumably involved in control of gene expression. In our new virus, the oncogene would be the cloned nerve growth factor gene.

Chapter 19

19.3 The two codes are different in codon designations. The mitochondrial code has more extensive wobble so that fewer tRNAs are needed to read all possible sense codons. As a consequence, fewer mitochondrial genes are necessary. The advantage is that fewer tRNAs are needed, and hence fewer tRNA genes need be present than for cytoplasmic tRNAs.

19.5 The features of extranuclear inheritance are differences in reciprocal-cross results (not related to sex), nonmappability to known nuclear chromosomes, Mendelian segregation not followed, and the indifference to nuclear substitution.

19.8 1/2 *petite*, 1/2 wild type (*grande*)

19.11 The first possibility is that the results are the consequence of a sex-linked lethal gene. The females would be homozygous for a dominant gene *L* that is lethal in males but not in females. In this case mating the F_1 females of an $L/L \times +/Y$ cross to $+/Y$ mates should give a sex ratio of 2 females: 1 male in the progeny flies.

The second possibility is that the trait is cytoplasmically transmitted via the egg and is lethal to males. In this case the same F_1 females should continue to have only female progeny when mated with $+/Y$ males.

19.15 a. If normal cytoplasm is [*N*] and male-sterile cytoplasm is [*Ms*], then the F_1 genotype is [*Ms*]*Rf/rf*, and the phenotype is male-fertile.

b. The cross is [*MS*]*Rf/rf* ♀ × [*N*]*rf/rf* ♂ giving 50% [*Ms*]*Rf/rf* and 50% [*Ms*]*rf/rf* progeny. Thus the phenotypes are 1/2 male-fertile and 1/2 male-sterile.

19.16 a. Carlos Mendoza will have inherited his mitochondrial DNA from his mother, and she will have inherited it from her mother. If Mrs. Mendoza and Mrs. Escobar have different mitochondrial RFLPs, then it can be determined which of them contributed mitochondria to Carlos.

b. None of the potential grandfathers need to be tested, since they will not have given any mitochondria to Carlos. In addition, there is no point in testing Mrs. Sanchez. She may have given mitochondria to Carlos' father, but the father did not pass them on to Carlos.

c. If Mrs. Mendoza and Mrs. Escobar do not differ in RFLP, the data will not be helpful. If they do differ, and if Carlos matches Mrs. Mendoza, the case should be dismissed. If Carlos matches Mrs. Escobar, then the Escobar and Sanchez couples are indeed the grandparents, and the Mendozas have claimed a stolen child.

19.18 There are various possibilities. You could isolate the minor peak and examine it using the electron microscope. If the molecules in this peak are circular, they are unlikely to be nuclear fragments. You could grow the yeast in the presence of an intercalating agent such as acridine, and see whether this treatment causes the minor peak species to disappear. If it does, the minor peak is organellar in origin. You could isolate the minor peak, label it (by nick translation, for example), and hybridize this labeled DNA to DNA within suitably prepared yeast cells. Then, using the electron microscope, you could determine whether the label is found over the nucleus or the mitochondria. Finally, you could isolate the minor peak DNA and study its homology to other yeast mitochondrial sequences.

19.19 The parental snails were *D/d* female and *d/−* male. The F_1 snail is *d/d*. (Given the F_1 genotype, the male can either be homozygous d or heterozygous, but the determination cannot be made from the data given.)

Chapter 20

20.2   252/1024

20.3   Each pure-breeding parent is homozygous for the genes (however many there are) controlling the size character, and hence each parent is homogeneous in type. A cross of two pure-breeding strains will generate an F_1 heterozygous for those loci controlling the size trait. Since the F_1 is genetically homogeneous (all heterozygotes), it shows no greater variability than the parents.

20.6 a.   8 cm

b.   1 (2 cm) :6 (4 cm) :15 (6 cm) :20 (8 cm) :15 (10 cm) :6 (12 cm) :1 (14 cm)

c.   Proportion of F_2 of *AA BB CC* is $(1/4)^3$; proportion of F_2 of *aa bb cc* is $(1/4)^3$; proportion of F_2 with heights equal to one or other is $(1/4)^3 + (1/4)^3 = 2/64$.

d.   The F_1 height of 8 cm can be produced only by maintaining heterozygosity at no less than one gene pair (e.g., *AA Bb cc*). Thus although 20 of 64 F_2 plants are expected to be 8 cm tall, none of these will breed true for this height.

20.9   It is a quantitative trait. Variation appears to be continuous over a range rather than falling into three discrete classes. Also, the pattern in which the F_1 mean falls between the parental means and in which the F_2 individuals show a continuous range of variation from one parental extreme to the other is typical of quantitative inheritance.

20.12   The cross is *aa bb cc* (3 lb) $\times$ *AA BB CC* (6 lb), which gives an F_1 that is *Aa Bb Cc* and which weighs 4.5 lb. The distribution of phenotypes in the F_2 is 1 (3 lb) :6 (3.5 lb) :15 (4 lb) :20 (4.5 lb) :15 (5 lb) :6 (5.5 lb) :1 (6 lb).

20.14 a.   The progeny are *Aa Bb Cc Dd*, which are 18 dm high.

b.   (1) The minimum number of capital-letter alleles is one and the maximum number is four, giving a height range of 12 to 18 dm. (2) The minimum number is one, and the maximum number is four, giving a height range of 12 to 18 dm. (3) The minimum number is four, and the maximum number is six, giving a height range of 18 to 22 dm. (4) The minimum number is zero, and the maximum number is seven, giving a height range of 10 to 24 dm.

20.17   Heritability is specific to a particular population and to a specific environment and cannot be used to draw conclusions about the basis of populational differences. Because the environments of the farms in Kansas and in Poland differ, and because the two wheat varieties differ in their genetic make-up, the heritability of yield calculated in Kansas cannot be applied to the wheat grown in Poland. Furthermore, the yield of TK138 would most likely be different in Poland, and might even be less than the yield of the Russian variety when grown in Poland.

20.18 Selection differential = 14.3 − 9.7 = 4.6
Selection response = 13 − 9.7 = 3.3
Narrow-sense heritability = selection response/selection differential = 3.3/4.6 = 0.72

20.19 Selection response = narrow-sense heritability × selection differential.
Selection response = 0.60 × 10 g = 6 g

Chapter 21

21.1 $p = f(B) = [(2 \times 52) + (40)]/200 = 0.72$
$q = 1 - p = 0.28$

21.4 a. $\sqrt{0.16} = 0.4 = 40\%$ = frequency of recessive alleles; $1 - 0.4 = 0.6 = 60\%$ = frequency of dominant alleles; $2pq = (2)(0.4)(0.6) = 0.48$ = probability of heterozygous diploids. Then $(0.48)/[(2 \times 0.16) + 0.48] = 0.48/0.80 = 60\%$ of recessive alleles are heterozygotes.

b. If $q^2 = 1\% = 0.01$, then $q = 0.1$, $p = 0.9$, and $2pq = 0.18$ heterozygous diploids. Therefore $(0.18)/[(2 \times 0.01) + 0.18] = 0.18/0.20 = 90\%$ of recessive alleles in heterozygotes.

21.5 $2pq/q^2 = 8$, and so $2p = 8q$; then $2(1 - q) = 8q$, and $2 = 10q$, or $q = 0.2$

21.8 a. Let p = the frequency of S and q = the frequency of s. Then

$$p = \frac{2(188)\ SS + 717\ Ss}{2(3146)} = \frac{1093}{6292} = 0.1737$$

$$q = \frac{717\ Ss + 2(2241)\ ss}{2(3146)} = \frac{5199}{6292} = 0.8263$$

b.

CLASS	OBSERVED	EXPECTED	d	d^2/e
SS	188	94.9	+93.1	91.235
Ss	717	903.1	−186.1	38.361
ss	2241	2147.9	+93.1	4.032
	3146	3145.9	0	133.628

There is only one degree of freedom because the three genotypic classes are completely specified by two gene frequencies, namely, p and q. Thus the number of degrees of freedom = number of genes − 1. The χ^2 value for this example is 133.628, which, for one degree of freedom, gives a P value less than 0.0001. Therefore the distribution of genotypes differs significantly from the Hardy-Weinberg equilibrium.

21.11 If the frequency of this gene is q, females would occur with the character at a frequency of q^2, and males with the character would occur at a frequency of q. The frequency of males is $q = 0.4$; thus we may predict that the frequency of females would be $(0.4)^2 = 0.16$ if this is a sex-linked gene. This result fits the observed data. Therefore the frequency of heterozygous females is $2pq = 2(0.6)(0.4) = 0.48$. For sex-linked genes no heterozygous males exist.

21.14

$$q = \frac{\mu}{\mu + \upsilon} = \frac{6 \times 10^{-7}}{(6 \times 10^{-7}) + (6 \times 10^{-8})}$$

$$= \frac{6 \times 10^{-7}}{(6 \times 10^{-7}) + (0.6 + 10^{-7})} = \frac{6}{6.6} = 0.91$$

$p = 1 - q = 1 - 0.91 = 0.09$

Thus the frequencies are 0.008 AA, 0.16 AA', and 0.828 $A'A'$

21.15 a. 100
   b. 96
   c. 36
   d. 7.8

21.17 $p'II = mpI + (1 - m)pII$
$pII = (0.20)(0.5) + (1 - 0.20)(0.70) = 0.66$

21.18 a. When selectively neutral, the genes distribute themselves according to the Hardy-Weinberg law, so 0.25 are AA, 0.05 are Aa, and 0.25 are aa.
   b. $q = 0.33$
   c. $q = 0.25$

21.20 a. $q = 0.63$
   b. $q = 0.64$
   c. $q = 0.66$

21.23 a. The data fit the idea that a single BamHI site varies. The probe is homologous to a region wholly within the 4.1-kb piece bounded on one end by the variable BamHI site, and on the other end by a constant site. When the variable site is present, the hybridized fragment is 4.1 kb. When the variable site is absent, the fragment extends to the next constant BamHI site, and is 6.7 kb long. People with only 4.1 or only 6.7 kb bands are homozygotes; people with both are heterozygotes.
   b. The "+" allele of the variable site is present in 2(6) + 38 = 50 chromosomes, and the "−" allele is present in 2(56) + 38 = 150 chromosomes. Thus $f(+)$ is 0.25 and $f(-)$ is 0.75.
   c. If the population is in Hardy-Weinberg equilibrium, we would expect $(0.25)^2$, or 0.0625 of the sample to show only the 4.1 kb band. This would be 6.25 individuals. We observed 6. We

expect $(0.75)^2$ or 0.5625 to be homozygous for the 6.7 kb band, which is 56.25 individuals. We saw 56. Finally, we would expect 2(0.25)(0.75) or 0.375 to be heterozygotes, or 37.5 individuals. We observed 38. The observed numbers are so close to the expected that a χ^2 test is unnecessary.

21.25 a. Mutation (i) leads to change in gene frequencies within a population if no other forces are acting; (ii) introduces new genetic variation; (iii) may lead to genetic differentiation among populations if population size is small.

b. Migration (i) increases population size; (ii) increases genetic variation within populations; (iii) equalizes gene frequencies among populations.

c. Genetic drift: (i) reduces genetic variation within populations and leads to genetic change over time; (ii) increases genetic differences among populations; (iii) increases homozygosity within populations.

d. Inbreeding: (i) increases homozygosity within populations; (ii) decreases genetic variation.

21.27 Comparison with the rates of nucleotide substitution in Table 21.6 indicates that sequence A has as high a rate as that typically observed in mammalian pseudogenes. In addition, the rates of synonymous and nonsynonymous substitutions are the same. These observations suggest that sequence A is either a pseudogene or is a sequence that provides no function, since high rates of substitution are observed when sequences are functionless.

Sequence B has a relatively low rate of nonsynonymous substitution but a relatively high rate of synonymous substitution. This is the pattern we expect when a sequence codes for a protein; thus sequence B probably encodes a protein.

Credits

by permission of Dr. Gerry Fink.

Figure 18.18: Based on data from T. Takeya and H. Hanafusa in *Cell* 32(1983):881–90.

Table 18.4: From R. Weiss et al., *RNA Tumor Viruses,* 2nd ed., 1985. Reprinted by permission.

Table 18.5: From "Cooperation Between Oncogenes" by T. Hunter, *Cell* 64: 249–70. Copyright © 1991 by Cell Press. Reprinted by permission of Cell Press and T. Hunter.

Figure 19.5, 19.6: From Henry Allan Gleason et al., *New Britton and Brown, Illustrated Flora of the Northeastern U.S. and Adjacent Canada.* Copyright © 1952 by The New York Botanical Garden. Reprinted by permission.

Figure 20.2: Adapted from C.M. Jackson, "The physique of male students at the University of Minnesota: A study in constitutional anatomy and physiology," in *The American Journal of Anatomy,* vol. 40, p. 66, September, November 1927; January 1928. Reprinted by permission of Allan Liss Inc.

Figure 20.5: From L.A. Snyder, D. Freifelder, and D.L. Hartl, *General Genetics,* 1985. Reprinted by permission of Jones and Bartlett Publishers, Inc.

Figure 20.6b: From A. H. Sturtevant and G. W. Beadle, *An Introduction to Genetics,* p. 265 (Philadelphia: W. B. Saunders Co., 1940). Reprinted by permission.

Figure 20.7: From Curtis and Barnes, *Biology,* 5th ed. Worth Publishers, New York, 1989. Reprinted by permission.

Figure 20.8: From D. S. Falconer, *Introduction to Quantitative Genetics,* 2nd ed., 1981. Reprinted by permission of Longman Group Ltd. London.

Figure 20.9: Adapted from *Biology: Life on Earth,* 3rd ed., by Gerald Audesirk and Teresa Audesirk. Copyright © 1993 by Macmillan Publishing Company. Reprinted by permission.

Figure 21.1: From E. B. Ford, *Ecological Genetics.* Reprinted by permission of Chapman and Hall Ltd.

Figure 21.6: From Philip W. Hedrick, *Genetics of Populations,* 1983. Reprinted by permission of Jones and Bartlett Publishers, Inc.

Photo Credits

p. 3: David Scharf/Peter Arnold, Inc.　**p. 4:** CNRI/SPL/Photo Researchers.　**p. 5:** (a) J. Forsdyke/Gene Cox/SPL/Photo Researchers; (b) J. B. Woolsey Associates; (c) Carolina Biological Supply; (d) Larry Lefever/Grant Heilman Photography; (e) Dr. Jeremy Burgess/SPL/Photo Researchers; (f) John Colwell/Grant Heilman Photography; (g) Courtesy of John Sulston, Medical

Research Council/Laboratory of Molecular Biology.　**p. 9:** Courtesy of Dr. William C. Earnshaw, The Johns Hopkins University. **p. 12:** Carolina Biological Supply Company. **p. 14:** J. R. Paulson and U. K. Laemmli, 1977. *Cell* 12: 817–28. © 1977 MIT. Photo courtesy of Dr. U. K. Laemmli.　**p. 15:** Courtesy of Elton Stubblefield. Produced from *Cytobios* 19:27 (1978), with permission of the Faculty Press, Cambridge, England.　**p. 17:** Courtesy of M. Westergaard and D. von Wettstein.　**p. 27:** Scott, Foresman. **p. 43:** Norman R. Lightfoot/Photo Researchers. **p. 44:** From the *Journal of Heredity* 23 (Sept. 1932), p. 345.　**p. 60:** (left) Digamber S. Borgaonkar, Ph.D. (right) Nadler/Children's Memorial, Chicago. **p. 61:** (left) Digamber S. Borgaonkar, Ph.D. (right) Nadler/Children's Memorial, Chicago. **p. 62:** Scott, Foresman.　**p. 65:** Scott, Foresman. **p. 66:** Courtesy of Dr. Carol Witkop, University of Minnesota.　**p. 79:** (a) Fritz Prenzel/Animals Animals; (b) International American Albino Association, Inc.; (c) Larry Lefever/Grant Heilman Photography.　**p. 80:** (a) Barbara J. Wright/ Animals Animals; (b) Larry Lefever/Grant Heilman Photography; (c) C. Prescott Allen/ Animals Animals; (d) Larry Lefever/Grant Heilman Photography.　**p. 82:** (a) Scott, Foresman; (b) Hans Reinhard/OKAPIA/Photo Researchers; (c) E. R. Degginger.　**p. 90:** Paul Conklin. **p. 129:** (top) David Scharf/Peter Arnold, Inc.; (bottom) Michael Gabridge/Custom Medical Stock. **p. 132:** Micrograph courtesy of David P. Allison, Biology Division, Oak Ridge National Laboratory. **p. 138:** Dr. Harold W. Fisher.　**p. 142:** Bruce Iverson.　**p. 163:** (left) Bill Longcore/SS/Photo Researchers; (right) Jackie Lewin, Royal Free Hospital/SPL/Photo Researchers.　**p. 174:** Dr. Edward J. Bottone/Mt. Sinai.　**p. 181:** Courtesy of Professor M. H. F. Willkins, Biophysics Dept., King's College, London.　**p. 183:** (a) Richard Pastor/Courtesy Food and Drug Administration. **p. 184:** Irving Geis.　**p. 188:** Dr. Gopal Kurti/Photo Researchers.　**p. 190:** Jack Griffith.　**p. 193:** Photo Researchers.　**p. 194:** © 1979 J. F. Gennaro/Photo Researchers.　**p. 195:** David M. Phillips/Visuals Unlimited.　**p. 198:** Jack Griffith.　**p. 199:** Barbara Hamakalo.　**p. 200:** J. R. Paulson and U. K. Laemmli, 1977. *Cell* 12: 817–28. © 1977 MIT. Photo courtesy of Dr. U. K. Laemmli.　**p. 222:** Scott, Foresman.　**p. 226:** Scott, Foresman. **p. 244:** Courtesy of O. L. Miller, Jr., Lewis and Clark Professor of Biology, University of Virginia. **p. 246:** Courtesy of Philip Leder, National Institutes of Health (From Tilghman, Curtis Tiemeier, Leder & Weisman, *Proceedings of the National Academy of Sciences,* vol. 75 (1979), p. 1309.　**p. 252:** Courtesy of

Index

Note: Page numbers in *italics* indicate material in figures and tables.